ANNALS OF
THE NEW YORK ACADEMY
OF SCIENCES

Volume 986

EDITORIAL STAFF

Director, Publishing and New Media
SARAH GREENE

Manging Editor
JUSTINE CULLINAN

Associate Editors
STEVEN E. BOHALL
TRUMBULL ROGERS

The New York Academy of Sciences
2 East 63rd Street
New York, New York 10021

THE NEW YORK ACADEMY OF SCIENCES
(Founded in 1817)

BOARD OF GOVERNORS, September 2002 – September 2003

TORSTEN N. WIESEL, *Chairman of the Board*
JOHN F. NIBLACK, *Vice Chairman*
JOHN T. MORGAN, *Treasurer*
ELLIS RUBINSTEIN, *Chief Executive Officer* [ex officio]

Honorary Life Governors
WILLIAM T. GOLDEN   JOSHUA LEDERBERG

Governors

ELEANOR BAUM	KAREN E. BURKE	PRAVEEN CHAUDHARI
R. BRIAN FERGUSON	GERALD D. FISCHBACH	RONALD L. GRAHAM
MARNIE IMHOFF	JACQUELINE LEO	BRUCE McEWEN
PAUL MARKS	RONAY MENSCHEL	SANDRA PANEM
PETER RINGROSE	LEE G. VANCE	DEBORAH WILEY

HELENE L. KAPLAN, *Counsel* [ex officio]

Na,K-ATPase AND RELATED CATION PUMPS

Structure, Function, and Regulatory Mechanisms

ANNALS OF THE NEW YORK ACADEMY OF SCIENCES
Volume 986

Na,K-ATPase AND RELATED CATION PUMPS

Structure, Function, and Regulatory Mechanisms

Edited by Peter Leth Jorgensen, Steven J.D. Karlish, and Arvid B. Maunsbach

The New York Academy of Sciences
New York, New York
2003

Copyright © 2003 by the New York Academy of Sciences. All rights reserved. Under the provisions of the United States Copyright Act of 1976, individual readers of the Annals are permitted to make fair use of the material in them for teaching or research. Permission is granted to quote from the Annals provided that the customary acknowledgment is made of the source. Material in the Annals may be republished only by permission of the Academy. Address inquiries to the Permissions Department (permissions@nyas.org) at the New York Academy of Sciences.

Copying fees: For each copy of an article made beyond the free copying permitted under Section 107 or 108 of the 1976 Copyright Act, a fee should be paid through the Copyright Clearance Center, Inc., 222 Rosewood Drive, Danvers, MA 01923 (www.copyright.com).

∞ The paper used in this publication meets the minimum requirements of the American National Standard for Information Sciences—Permanence of Paper for Printed Library Materials, ANSI Z39.48-1984.

Library of Congress Cataloging-in-Publication Data

International Conference on Na,K-ATPase and Related Cation Pumps (10th : 2002 ; Helsingør, Denmark)
   Na,K-ATPase and related cation pumps : structure, function, and regulatory mechanisms / edited by Peter Leth Jorgensen, Steven J.D. Karlish, and Arvid B. Maunsbach.
     p. ; cm. — (Annals of the New York Academy of Sciences ; v. 986)
   Includes index.
     ISBN 1-57331-401-3 (cloth : alk. paper) — ISBN 1-57331- 402-1 (paper : alk. paper)
   1. Sodium/potassium ATPase—Congresses. 2. Ion channels—Congresses.
   [DNLM: 1. Na (+) -K (+) - Exchanging ATPase—physiology—Congresses. 2. Cation Transport Proteins—pharmacology—Congresses. 3. Cation Transport Proteins—physiology—Congresses. 4. Na (+) -K (+) - Exchanging ATPase—pharmacology—Congresses. QU 136 I593n 2003] I. Jorgensen, Peter Leth. II. Karlish, Steven J.D. III. Maunsbach, Arvid Bernhard. IV. Title. V. Series.
   Q11.N5 vol. 986
   [QP609.S63]
   500 s—dc21
   [571.6]
                                                                            2003004815

GYAT/PCP
Printed in the United States of America
ISBN 1-57331-401-3 (cloth)
ISBN 1-57331-402-1 (paper)
ISSN 0077-8923

ANNALS OF THE NEW YORK ACADEMY OF SCIENCES

Volume 986
April 2003

Na,K-ATPase AND RELATED CATION PUMPS

Structure, Function, and Regulatory Mechanisms

Editors and Conference Organizers
P. L. JORGENSEN, S. J. KARLISH, AND A. B. MAUNSBACH

This volume is the result of a conference entitled the 10th International Conference on Na,K-ATPase and Related Cation Pumps held on August 8-14, 2002 in Elsinore, Denmark.

CONTENTS

Preface. *By* P. L. JORGENSEN, S. J. KARLISH, AND A. B. MAUNSBACH xix

Introduction: Reflections on Ten International Conferences on Na,K-ATPase and Related Cation Pumps. *By* J. C. SKOU xxiii

Part I. High-Resolution Protein Structure

Crystal Structures of Ca^{2+}-ATPase in Various Physiological States. *By* CHIKASHI TOYOSHIMA, HIROMI NOMURA, AND YUJI SUGITA 1

Renal Na,K-ATPase Structure from Cryo-electron Microscopy of Two-Dimensional Crystals. *By* HANS HEBERT, PASI PURHONEN, KAREN THOMSEN, HENRIK VORUM, AND ARVID B. MAUNSBACH 9

Projection Map of Covalently Phosphorylated Ca-ATPase from Tubular Crystals. *By* F. DELAVOIE, D. MCINTOSH, F. HENAO, G. PERANZI, P. CHAMPEIL, D. STOKES, AND J-J. LACAPÈRE 17

How Does Conformation Change Ouabain Binding from Rejection (E1) to Acceptance (E2)? *By* ROBERT LICKELY POST 20

Part II. Molecular Mechanism, Function, and Structure–Function Relationships

Transmission of E_1-E_2 Structural Changes in Response to Na^+ or K^+ Binding in Na,K-ATPase. *By* PETER L. JORGENSEN 22

Conformational Dynamics of Na^+/K^+- and H^+/K^+-ATPase Probed by Voltage Clamp Fluorometry. *By* SVEN GEIBEL, DIRK ZIMMERMANN, GIOVANNI ZIFARELLI, ANJA BECKER, JAN B. KOENDERINK, YI-KANG HU, JACK H. KAPLAN, THOMAS FRIEDRICH, AND ERNST BAMBERG 31

Investigating the Energy Transduction Mechanism of P-type ATPases with Fe^{2+}-Catalyzed Oxidative Cleavage. *By* STEVEN J. D. KARLISH 39

Importance of Transmembrane Segment M3 of Na^+,K^+-ATPase for Control of Conformational Changes and the Cytoplasmic Entry Pathway for Na^+. *By* BENTE VILSEN AND MADS TOUSTRUP-JENSEN 50

Insights into the Structural Basis for Modulation of $E_1 \leftrightarrow E_2$ Transitions by Cytoplasmic Domains of the Na,K-ATPase α Subunit. *By* LAURA SEGALL, LOIS K. LANE, AND RHODA BLOSTEIN 58

Characterization of Ca^{2+} ATPase Residues Involved in Substrate and Cation Binding. *By* GIUSEPPE INESI, HAILUN MA, SUMING HUA, AND CHIKASHI TOYOSHIMA 63

Mutagenesis of Residues Involved in Control of the Ca^{2+} Entry Pathway and Conformational Changes Associated with Ca^{2+} Binding in the SR Ca^{2+}-ATPase. *By* JENS PETER ANDERSEN, JOHANNES D. CLAUSEN, ANJA PERNILLE EINHOLM, AND BENTE VILSEN 72

Proteolytic Studies on the Transduction Mechanism of Sarcoplasmic Reticulum Ca^{2+}-ATPase: Common Features with Other P-Type ATPases. *By* JESPER V. MØLLER, GUILLAUME LENOIR, MARC LE MAIRE, BIRTE STÆHR JUUL, AND PHILIPPE CHAMPEIL 82

Involvement of the Cytoplasmic Loop L6-7 in the Entry Mechanism for Transport of Ca^{2+} through the Sarcoplasmic Reticulum Ca^{2+}-ATPase. *By* F. CORRE, C. JAXEL, J. FUENTES, T. MENGUY, P. FALSON, B. A. LEVINE, J. V. MØLLER, AND M. LE MAIRE 90

Site-Directed Mutagenesis of Amino Acids in the Cytoplasmic Loop 6/7 of Na,K-ATPase. *By* G. XU, R. A. FARLEY, D. J. KANE, AND L. D. FALLER 96

ATP Binding Residues of Sarcoplasmic Reticulum Ca^{2+}-ATPase. *By* D. B. MCINTOSH, J. D. CLAUSEN, D. G. WOOLLEY, D. H. MACLENNAN, B. VILSEN, AND J. P. ANDERSEN 101

Molecular Modeling of SCH28080 Binding to the Gastric H,K-ATPase and MgATP Interactions with SERCA- and Na,K-ATPases. *By* KEITH MUNSON, OLGA VAGIN, GEORGE SACHS, AND STEVE KARLISH .. 106

Inhibition Kinetics of the Gastric H,K-ATPase by K-Competitive Inhibitor SCH28080 as a Tool for Investigating the Luminal Ion Pathway. *By* OLGA VAGIN, KEITH MUNSON, SVETLANA DENEVICH, AND GEORGE SACHS 111

Ion Occlusion/Deocclusion Partial Reactions in Individual Palytoxin-Modified Na/K Pumps. *By* PABLO ARTIGAS AND DAVID C. GADSBY 116

Cation Stoichiometry and Cation Pathway in the Na,K-ATPase and
Nongastric H,K-ATPase. *By* JEAN-DANIEL HORISBERGER,
SAÏDA GUENNOUN, MURIEL BURNAY, AND KÄTHI GEERING 127

Toward an Understanding of Ion Transport through the Na,K-ATPase. *By*
HANS-JÜRGEN APELL 133

Na,K-Pump Reaction Kinetics at the Tip of a Patch Electrode: Derivation of
Reaction Kinetics for Electrogenic and Electrically Silent Reactions
during Ion Transport by the Na,K-ATPase. *By* R. DANIEL PELUFFO AND
JOSHUA R. BERLIN 141

Two-Electrode Voltage-Clamp Analysis of Na,K-ATPase Asparagine 776
Mutants. *By* JAN B. KOENDERINK, SVEN GEIBEL, EVA GRABSCH,
JAN JOEP H. H. M. DE PONT, ERNST BAMBERG, AND
THOMAS FRIEDRICH 150

Binding of 1 Rb^+ Accelerates Dephosphorylation of the Na^+,K^+-ATPase
without Leading to Rb^+ Occlusion. *By* SERGIO B. KAUFMAN,
RODOLFO M. GONZÁLEZ-LEBRERO, PATRICIO J. GARRAHAN,
AND ROLANDO C. ROSSI 155

Kinetic Investigations of the Mechanism of the Rate-Determining Step of the
Na^+,K^+-ATPase Pump Cycle. *By* RONALD J. CLARKE,
PAUL A. HUMPHREY, CHRISTIAN LÜPFERT, HANS-JÜRGEN APELL,
AND FLEMMING CORNELIUS 159

Homology Modeling of Na,K-ATPase: A Putative Third Sodium Binding Site
Suggests a Relay Mechanism Compatible with the Electrogenic Profile
of Na^+ Translocation. *By* K. O. HÅKANSSON AND P. L. JORGENSEN 163

Proton Pumps

Use of a Fluorescent Maleimide to Probe Structure–Function Relationships in
Stalk Segments 4 and 5 of the Yeast Plasma-Membrane H^+-ATPase. *By*
CAROLYN W. SLAYMAN, MANUEL MIRANDA, JUAN PABLO PARDO, AND
KENNETH E. ALLEN 168

The E_1/E_2-Preference of Gastric H,K-ATPase Mutants. *By*
JAN JOEP H. H. M. DE PONT, HERMAN G. P. SWARTS,
PETER H. G. M. WILLEMS, AND JAN B. KOENDERINK 175

Nongastric H,K-ATPase: Structure and Functional Properties. *By* NIKOLAI
MODYANOV, NIKOLAY PESTOV, GAIL ADAMS, GILLES CRAMBERT,
MANORANJANI TILLEKERATNE, HAO ZHAO, TATYANA KORNEENKO,
MIKHAIL SHAKHPARONOV, AND KÄTHI GEERING 183

Plant Cation Pumps

Mechanism of Proton Pumping by Plant Plasma Membrane H^+-ATPase: Role
of Residues in Transmembrane Segments 5 and 6. *By* M. G. PALMGREN,
M. J. BUCH-PEDERSEN, AND A. L. MØLLER 188

Function and Regulation of the Two Major Plant Plasma Membrane H^+-
ATPases. *By* MAGDALENA WOLOSZYNSKA, JUSTYNA KANCZEWSKA,
ARTEM DRABKIN, OLIVIER MAUDOUX, STÉPHANIE DAMBLY, AND
MARC BOUTRY 198

Heavy Metal Pumps

Functional Properties of the Human Copper-Transporting ATPase ATP7B (the Wilson's Disease Protein) and Regulation by Metallochaperone Atox1. *By* SVETLANA LUTSENKO, RUSLAN TSIVKOVSKII, AND JOEL M. WALKER 204

Heavy Metal Transport CPx-ATPases from the Thermophile *Archaeoglobus fulgidus*. *By* JOSÉ M. ARGÜELLO, ATIN K. MANDAL, AND SEBASTIAN MANA-CAPELLI 212

P-Type ATPase Superfamily: Evidence for Critical Roles for Kingdom Evolution. *By* HIDEYUKI OKAMURA, MASATSUGU DENAWA, RYOSUKE OHNIWA, AND KUNIO TAKEYASU 219

Extended Abstracts

The Na,K-ATPase S5-H5 Helix: Structural Link between Phosphorylation and Cation-Binding Sites. *By* ATIN K. MANDAL, LYUDMILA MIKHAILOVA, AND JOSÉ M. ARGÜELLO 224

Na,K-ATPase α-β Subunit Interactions in the Transmembrane Domain. *By* CIMING LI, GILLES CRAMBERT, UDO HASLER, AND KÄTHI GEERING ... 226

Negative Changes of the Membrane Capacitance due to Electrogenic Na Transport by the Na,K-ATPase. *By* V. S. SOKOLOV, A. A. LENZ, AND H.-J. APELL 229

Isolation of $(\alpha\beta)_4$-Tetraprotomer Having Half-of-Sites ATP Binding from Solubilized Dog Kidney Na^+/K^+-ATPase. *By* YUTARO HAYASHI, NOBUKO SHINJI, YOSHIKAZU TAHARA, EMI HAGIWARA, AND HITOSHI TAKENAKA 232

ATPase Activity and Oligomerization of Solubilized Na^+/K^+-ATPase Maintained by Synthetic Phosphatidylserine. *By* NOBUKO SHINJI, YOSHIKAZU TAHARA, EMI HAGIWARA, TAKAYUKI KOBAYASHI, KUNIHIRO MIMURA, HITOSHI TAKENAKA, AND YUTARO HAYASHI 235

Cation Requirement for Nucleotide Binding to Na,K-ATPase. *By* MIKAEL ESMANN AND NATALYA U. FEDOSOVA 238

Single Mutation of Lys or Arg Residue in ATP Binding Pocket in Rat Na/K-ATPase Alpha-1 Subunit Induces Different Affinity Change in High- and Low-Affinity ATP Binding. *By* TOSHIAKI IMAGAWA, SATOMI TERAMACHI, AND KAZUYA TANIGUCHI 240

Localization of Catalytic Active Sites in the Large Cytoplasmic Domain of Na^+/K^+-ATPase. *By* RITA KRUMSCHEID, KLÁRA SUÁNKOVÁ, RÜDIGER ETTRICH, JAN TEISINGER, EVEN AMLER, AND WILHELM SCHONER 242

Calorimetry of Na,K-ATPase. *By* M. STOLZ, E. LEWITZKI, E. SCHICK, M. MUTZ, AND E. GRELL 245

Expression of Na,K-ATPase in *P. pastoris*: Fe^{2+}-Catalyzed Cleavage of the Recombinant Enzyme. *By* DAVID STRUGATSKY, RIVKA GOLDSHLEGER, EITAN BIBI, AND STEVEN J. D. KARLISH 247

The Mechanism of Na-K Interaction on Na,K-ATPase. *By* CLAUDIA DONNET AND KATHLEEN J. SWEADNER 249

Salt Effects on the Kinetics of the Electrogenic Na^+ Transport in the Na,K-ATPase. *By* ARTEM G. AYUYAN, VALERIJ S. SOKOLOV, AND HANS-JÜRGEN APELL 252

Mutational Analysis of Ouabain Interaction with the M5–M6 Hairpin of Na,K-ATPase. *By* L. Y. QIU, J. B. KOENDERINK, H. G. P. SWARTS, P. H. G. M. WILLEMS, AND J. J. H. H. M. DE PONT 255

Role of the Isoform-Specific Region of the Na,K-ATPase Catalytic Subunit. *By* MARIE-JOSÉE DURAN, SANDRINE V. PIERRE, DEBORAH L. CARR, AND THOMAS A. PRESSLEY 258

Structure/Function Analysis of Na,K-ATPase α1 and α2 Central Isoform-Specific Regions Reveals Their Involvement in Regulation by Protein Kinase C. *By* S. V. PIERRE, M-J. DURAN, D. L. CARR, AND T. A. PRESSLEY 260

Characterization of the Electrostatic Component of the Nucleotide Binding to Na,K-ATPase. *By* N. U. FEDOSOVA, P. CHAMPEIL, AND M. ESMANN ... 263

Independent Access of Fluorescein Isothiocyanate and $Co(NH_3)_4ATP$ to Their Binding Sites on the Protomer of Na,K-ATPase. *By* J. D. CAVIERES AND J. HADDOW 265

Importance of Thr^{214} in the Conserved TGES Sequence of the Na^+,K^+-ATPase for Vanadate Binding and Hydrolysis of E_2P. *By* MADS TOUSTRUP-JENSEN AND BENTE VILSEN 267

Differential Inactivation of Na,K-ATPase by Erythrosin Isothiocyanate. *By* M. TAYLOR, D. OWEN, A. TARIQ, AND J. D. CAVIERES 270

Mutational Analysis of the Interactions of the Alpha and Beta Subunits of the Na,K-ATPase. *By* M. D. LAUGHERY, S. MCLOUD, AND J. H. KAPLAN .. 273

Mechanism of Phosphoryl Group Transfer. *By* L. D. FALLER, A. K. NAGY, D. J. KANE, AND R. A. FARLEY 275

Oligomeric Structure of P-Type ATPases Observed by Single Molecule Detection Technique. *By* SHUNJI KAYA, KAZUHIRO ABE, KAZUYA TANIGUCHI, MICHIO YAZAWA, TSUYOSHI KATOH, MAHITO KIKUMOTO, KAZUHIRO OIWA, AND YUTARO HAYASHI 278

K^+ Induced Simultaneous Liberation of Two Moles of P_i, One from One Mole of EP and the Other from EATP, of Oligomeric H/K-ATPase from Pig Stomach. *By* KAZUHIRO ABE, SHUNJI KAYA, TOSHIAKI IMAGAWA, AND KAZUYA TANIGUCHI 281

An Improved Method to Measure the Interactions of P-Type ATPases with the Lipidic Environment. *By* VALERIA LEVI, JUAN P. F. C. ROSSI, ANA M. VILLAMIL GIRALDO, PABLO R. CASTELLO, AND F. LUIS GONZÁLEZ FLECHA 283

Extracellularly Applied Br-TITU Inhibits the Na^+/K^+ Pump by Interacting with Tryptophan at the Entrance to the Cation Sites. *By* G. A. YUDOWSKI, M. BAR SHIMON, R. M. GONZÁLEZ-LEBRERO, R. C. ROSSI, P. J. GARRAHAN, S. J. D. KARLISH, AND L. BEAUGÉ 287

Interactions between Cations and Na,K-ATPase Membranes Studied with Solid-State NMR. *By* LOUISE ODGAARD JAKOBSEN, NIELS CHR. NIELSEN, AND MIKAEL ESMANN 290

Interaction between ATP and the Na/K-ATPase from Duck Supraorbital Salt Glands. *By* PROMOD R. PRATAP, NATALIE OLDEN-STAHL, OANA DEDIU, AND G. ULRICH NIENHAUS 293

Three-Dimensional Structure-Activity Relationship Modeling of Digoxin Inhibition and Docking to Na^+,K^+-ATPase. *By* W. JAMES BALL, JR., CAROL D. FARR, STEFAN PAULA, SUSAN M. KEENAN, ROBERT K. DELISLE, AND WILLIAM J. WELSH 296

A Parallel Study of Eosin-Fluorescence Change and Rb^+ Occlusion in the Na^+/K^+-ATPase. *By* M. R. MONTES, R. M. GONZÁLEZ-LEBRERO, P. J. GARRAHAN, AND R. C. ROSSI 298

The Sidedness of the Direct Route of Occlusion of K^+ in the Na^+/K^+-ATPase. *By* RODOLFO M. GONZÁLEZ-LEBRERO, SERGIO B. KAUFMAN, PATRICIO J. GARRAHAN, AND ROLANDO C. ROSSI 301

The Muscle-Specific βm Protein Is Functionally Different from Other Members of the X,K-ATPase β-Subunit Family. *By* NIKOLAY B. PESTOV, GILLES CRAMBERT, HAO ZHAO, TATYANA V. KORNEENKO, MIKHAIL I. SHAKHPARONOV, KÄTHI GEERING, AND NIKOLAI N. MODYANOV 304

Influence of Intramembrane Electric Charge on H,K-ATPase. *By* IRENA KLODOS 306

The Role of Lys^{791} and Asn^{792} in Gastric H,K-ATPase. *By* HERMAN G. P. SWARTS, PETER H. G. M. WILLEMS, JAN B. KOENDERINK, AND JAN JOEP H. H. M. DE PONT 308

Functional Consequences of Charge Reversals of Acidic Residues in M1 of the SR Ca-ATPase. *By* ANJA PERNILLE EINHOLM, BENTE VILSEN, AND JENS PETER ANDERSEN 310

Overexpression of SERCA1a Ca^{2+}-ATPase in Yeast. *By* PIERRE FALSON, GUILLAUME LENOIR, THIERRY MENGUY, FABIENNE CORRE, CÉDRIC MONTIGNY, PER A. PEDERSEN, DENYSE THINÈS, AND MARC LE MAIRE   312

$TNP-8N_3$-ADP Photoinactivation of the Phosphatase Activity of FITC-Modified SERCA. *By* G. BARRIENTOS, M. TAYLOR, C. HIDALGO, AND J. D. CAVIERES 315

Ca^{2+} Occlusion of Sarcoplasmic Reticulum Ca^{2+}-ATPase by CrATP. *By* BIRTE STÆHR JUUL AND JESPER V. MØLLER 318

A Model Accounting for the Simultaneous Transport of Calcium and Manganese in Sarcoplasmic Reticulum Membranes. *By* DÉBORA ALEJANDRA GONZÁLEZ, MARIANO ANÍBAL OSTUNI, JEAN-JACQUES LACAPÈRE, AND GUILLERMO LUIS ALONSO 320

Interaction of an Aromatic Dibromo-Isothiouronium Derivative with the Ca-ATPase of Sarcoplasmic Reticulum. *By* MERVYN C. BERMAN AND STEVEN J. KARLISH 323

Time-Resolved Partial Reactions of the SR Ca-ATPase Investigated with a Fluorescent Styryl Dye. *By* CHRISTINE PEINELT AND HANS-JÜRGEN APELL.. 325

Macrocyclic Carbon Suboxide Oligomers as Potent Inhibitors of the Na,K-ATPase. *By* ROBERT STIMAC, FRANZ KEREK, AND HANS-JÜRGEN APELL 327

The Inherent Energy in SR Ca^{2+}-ATPase Is Convertible into Chemical Work.
By MAKOTO USHIMARU AND YOSHIHIRO FUKUSHIMA............... 330

Purification of SERCA*1a* Ca^{2+}-ATPase Mutants Expressed in Yeast. *By*
GUILLAUME LENOIR, CÉDRIC MONTIGNY, MARC LE MAIRE, AND
PIERRE FALSON.. 333

Mutational Analysis of Lys^{252} and Its Interaction with Loop 6–7 in the SR
Ca^{2+}-ATPase. *By* JOHANNES D. CLAUSEN AND JENS PETER ANDERSEN. 335

Phospholamban Inhibits Ca^{2+} Pump Oligomerization and Intersubunit Free
Energy Exchange Leading to Activation of Cardiac Muscle SERCA2a.
By JAMES E. MAHANEY, R. WAYNE ALBERS, HOWARD KUTCHAI, AND
JEFFREY P. FROEHLICH... 338

ATP Regulation of Calcium Binding in Ca^{2+}-ATPase Molecules of the
Sarcoplasmic Reticulum. *By* JUN NAKAMURA, GENICHI TAJIMA, AND
CHIKARA SATO.. 341

The Nature of the Low-Frequency Normal Modes of the E1Ca Form of the
SERCA1 Ca^{2+}-ATPase. *By* N. REUTER, K. HINSEN, AND J-J. LACAPÈRE 344

Protonation of the *Neurospora crassa* Plasma Membrane H^+-ATPase as a
Function of pH Monitored by ATR-FTIR. *By* O. RADRESA,
V. RAUSSENS, J-M. RUYSSCHAERT, AND E. GOORMAGHTIGH.......... 347

Mutagenic Study of Residues in Transmembrane Helix 4, 5, and 6 of the Plant
Plasma Membrane P-Type H^+-ATPase. *By* M.J. BUCH-PEDERSEN,
A. L. MØLLER, AND M. G. PALMGREN................................ 349

Mutational Analysis of Charged Residues in the Putative KdpB-TM5 Domain
of the Kdp-ATPase of *Escherichia coli*. *By* MARC BRAMKAMP AND
KARLHEINZ ALTENDORF.. 351

Part III. Molecular Mechanisms of Regulation

Functional Roles of the α Isoforms of the Na,K-ATPase. *By* JERRY LINGREL,
AMY MOSELEY, IVA DOSTANIC, MARC COUGNON, SUIWEN HE, PAUL
JAMES, ALISON WOO, KYLE O'CONNOR, AND JONATHAN NEUMANN.. 354

Ion Pump–Interacting Proteins: Promising New Partners. *By* PHILIPP PAGEL,
ALESSANDRA ZATTI, TOHRU KIMURA, AMY DUFFIELD, VERONIQUE
CHAUVET, VANATHY RAJENDRAN, AND MICHAEL J. CAPLAN.......... 360

Amino Acids in the TM4-TM5 Loop of Na,K-ATPase Are Important for
Biosynthesis. *By* JESPER R. JØRGENSEN, JENS HOUGHTON-LARSEN,
METTE DORPH JACOBSEN, AND PER AMSTRUP PEDERSEN............. 369

Differential Degradation of the Na^+/K^+-ATPase Subunits in the Plasma
Membrane. *By* SHIGE H. YOSHIMURA AND KUNIO TAKEYASU.......... 378

Small Ion Transport Regulators: FXYD Proteins

FXYD Proteins as Regulators of the Na,K-ATPase in the Kidney. *By*
KATHLEEN J. SWEADNER, ELENA ARYSTARKHOVA, CLAUDIA DONNET,
AND RANDALL K. WETZEL... 382

FXYD Proteins: New Tissue- and Isoform-Specific Regulators of
  Na,K-ATPase. *By* KÄTHI GEERING, PASCAL BÉGUIN, HAIM GARTY,
  STEVEN KARLISH, MARIA FÜZESI, JEAN-DANIEL HORISBERGER, AND
  GILLES CRAMBERT 388

A Specific Functional Interaction between CHIF and Na,K-ATPase: Role of
  FXYD Proteins in the Cellular Regulation of the Pump. *By* HAIM GARTY,
  MOSHIT LINDZEN, MARIA FÜZESI, ROMAN AIZMAN, RIVKA
  GOLDSHLEGER, CAROL ASHER, AND STEVEN J. D. KARLISH 395

Immunocytochemical Localization of Na,K-ATPase Gamma Subunit and
  CHIF in Inner Medulla of Rat Kidney. *By* KAARINA PIHAKASKI-
  MAUNSBACH, HENRIK VORUM, ELSE-MERETE LØCKE, HAIM GARTY,
  STEVEN J. D. KARLISH, AND ARVID B. MAUNSBACH 401

Adaptation of Murine Inner Medullary Collecting Duct (IMCD3) Cell
  Cultures to Hypertonicity. *By* JUAN M. CAPASSO, CHRISTOPHER J.
  RIVARD, LAURA M. ENOMOTO, AND TOMAS BERL 410

Gamma Structural Variants Differentially Regulate Na,K-ATPase Properties.
  By ELENA ARYSTARKHOVA AND RANDALL K. WETZEL 416

Structure/Function Studies of the Gamma Subunit of the Na,K-ATPase. *By*
  RHODA BLOSTEIN, HELEN X. PU, ROSEMARIE SCANZANO, AND
  ATHINA ZOUZOULAS 420

Cell-Specific Expression of Three Members of the FXYD Family along the
  Renal Tubule. *By* NICOLETTE FARMAN, MICHEL FAY, AND
  FRANÇOISE CLUZEAUD 428

Dominant Isolated Renal Magnesium Loss Is Caused by Misrouting of the
  Na^+,K^+-ATPase γ-Subunit. *By* IWAN C. MEIJ, JAN B. KOENDERINK,
  JOKE C. DE JONG, JAN JOEP H. H. M. DE PONT, LEO A. H. MONNENS,
  LAMBERT P. W. J. VAN DEN HEUVEL, AND NINE V. A. M. KNOERS 437

FXYD7, the First Brain- and Isoform-Specific Regulator of Na,K-ATPase:
  Biosynthesis and Function of Its Posttranslational Modifications. *By*
  GILLES CRAMBERT, PASCAL BÉGUIN, MARC ULDRY, FLORIANNE
  MONNET-TSCHUDI, JEAN-DANIEL HORISBERGER, HAIM GARTY, AND
  KÄTHI GEERING 444

Phosphorylation of the Na^+,K^+-ATPase in Skeletal Muscle: Potential
  Mechanism for Changes in Pump Cell-Surface Abundance and Activity.
  By LUBNA AL-KHALILI, ANNA KROOK, AND ALEXANDER V. CHIBALIN . 449

Regulation of Active Calcium Transport

Physiological Functions of Plasma Membrane and Intracellular Ca^{2+} Pumps
  Revealed by Analysis of Null Mutants. *By* GARY E. SHULL,
  GBOLAHAN OKUNADE, LYNNE H. LIU, PETER KOZEL,
  MUTHU PERIASAMY, JOHN N. LORENZ, AND VIKRAM PRASAD 453

Characterization of PISP, a Novel Single-PDZ Protein That Binds to All
  Plasma Membrane Ca^{2+}-ATPase b-Splice Variants. *By* GEOFFREY M.
  GOELLNER, STEVEN J. DEMARCO, AND EMANUEL E. STREHLER 461

The Regulation of SERCA-Type Pumps by Phospholamban and Sarcolipin.
  By DAVID H. MACLENNAN, MICHIO ASAHI, AND A. RUSSELL TUPLING . 472

The Thermogenic Function of the Sarcoplasmic Reticulum Ca^{2+}-ATPase of Normal and Hyperthyroid Rabbit. *By* LEOPOLDO DE MEIS, ANA PAULA ARRUDA, WAGNER S. DA-SILVA, MARCELO REIS, AND DENISE P. CARVALHO 481

Receptor Function of the Ouabain Site

Na,K-ATPase as a Signal Transducer. *By* OLEG AIZMAN AND ANITA APERIA . 489

Molecular Mechanisms of Na/K-ATPase–Mediated Signal Transduction. *By* ZIJIAN XIE 497

Low Concentrations of Ouabain Activate Vascular Smooth Muscle Cell Proliferation. *By* JULIUS C. ALLEN, JOEL ABRAMOWITZ, AND ASLIHAN KOKSOY 504

Regulation of Ca^{2+} Signaling by Na^+ Pump Alpha-2 Subunit Expression. *By* VERA GOLOVINA, HONG SONG, PAUL JAMES, JERRY LINGREL, AND MORDECAI BLAUSTEIN 509

Visualization of Na,K-ATPase Interacting Proteins Using FRET Technique. *By* PER UHLÉN 514

Extended Abstracts

Na/K-ATPase Regulates Intracellular ROS Level in Cerebellum Neurons. *By* ALEXANDER BOLDYREV, ELENA BULYGINA, MARIA YUNEVA, AND WILHELM SCHONER 519

Intermolecular Interaction between Na^+/K^+-ATPase α Subunit and Glycogen Phosphorylase. *By* KUNIO TAKEYASU, TSUBASA KAWASE, AND SHIGE H. YOSHIMURA 522

Molecular Activity of Na^+,K^+-ATPase Relates to the Packing of Membrane Lipids. *By* PAUL L. ELSE, BEN J. WU, L. H. STORLIEN, AND A. J. HULBERT 525

Proteins Binding to $\alpha1\beta1$ Isozyme of Na,K-ATPase. *By* NATALIYA DOLGOVA, NATALIYA MAST, OLGA AKIMOVA, ALEXANDER RUBTSOV, AND OLGA LOPINA 527

Domains Involved in the Interactions between FXYD and Na,K-ATPase. *By* MOSHIT LINDZEN, ROMAN AIZMAN, YAEL LIFSHITZ, MARIA FÜZESI, STEVEN J. D. KARLISH, AND HAIM GARTY 530

Defining the Nature and Sites of Interaction between FXYD Proteins and Na,K-ATPase. *By* MARIA FÜZESI, RIVKA GOLDSHLEGER, HAIM GARTY, AND STEVEN J. D. KARLISH 532

NO Regulation of Na,K-ATPase: Nitric Oxide Regulation of the Na,K-ATPase in Physiological and Pathological States. *By* DORETTE Z. ELLIS AND KATHLEEN J. SWEADNER 534

Functional Expression of the $\alpha4$ Isoform of the Na,K-ATPase in Both Diploid and Haploid Germ Cells of Male Rats. *By* GUSTAVO BLANCO 536

Responses at the Translational Level to Heterologous Expression of the Na,K-ATPase. *By* LOTTE STEFFENSEN AND PER AMSTRUP PEDERSEN .. 539

Protein Kinase C Phosphorylation Directed at Novel C-Terminal Sites in Na,K-ATPase. *By* ANDERS KRÜGER, YASSER A. MAHMMOUD, AND FLEMMING CORNELIUS 541

Modification of the PKC Phosphorylation Site Ser-23 of the Rat α1 Subunit. *By* L. A. VASILETS, A. SPIELMANN, AND W. SCHWARZ 543

Expression of a Na,K-ATPase-EGFP Chimera in COS Cells: Can Internalization Explain PKA- or PKC-Mediated Inhibition of ^{86}Rb Uptake? *By* BO KRISTENSEN, SVEND BIRKELUND, AND PETER LETH JORGENSEN 546

PKA and PKC Phosphorylation of Gastric H,K-ATPase. *By* YASSER AHMED MAHMMOUD AND FLEMMING CORNELIUS 548

Seasonal Changes of Ca-ATPase Activity in Skeletal Muscle Sarcoplasmic Reticulum of the Ground Squirrel *Spermophilus undulatus*. *By* ALEXANDER S. KONDRASHEV-LUGOVSKII, ANNA N. MALYSHEVA, KENNETH B. STOREY, OLGA D. LOPINA, AND ALEXANDER M. RUBTSOV 550

Acidic-Lipid Responsive Regions of the Plasma Membrane Ca^{2+} Pump. *By* HUGO P. ADAMO, FELICITAS DE TEZANOS PINTO, LUIS M. BREDESTON, AND GERARDO R. CORRADI 552

Part IV. Cellular and Physiological Regulation

Short-Term Aldosterone Action on Na,K-ATPase Surface Expression: Role of Aldosterone-Induced SGK1? *By* FRANÇOIS VERREY, VANESSA SUMMA, DIRK HEITZMANN, DAVID MORDASINI, ALAIN VANDEWALLE, ERIC FÉRAILLE, AND MARIJA ZECEVIC 554

Renal Tubule Sodium Transporter Abundance Profiling in Rat Kidney: Response to Aldosterone and Variations in NaCl Intake. *By* MARK A. KNEPPER, GHEUN-HO KIM, AND SHYAMA MASILAMANI 562

Mechanism of Control of Na,K-ATPase in Principal Cells of the Mammalian Collecting Duct. *By* ERIC FÉRAILLE, DAVID MORDASINI, SANDRINE GONIN, GEORGES DESCHÊNES, MANLIO VINCIGUERRA, ALAIN DOUCET, ALAIN VANDEWALLE, VANESSA SUMMA, FRANÇOIS VERREY, AND PIERRE-YVES MARTIN 570

Themes in Ion Pump Regulation. *By* F. CORNELIUS AND Y. A. MAHMMOUD . . 579

Isoform-Specific Regulation of Na^+,K^+-ATPase Endocytosis and Recruitment to the Plasma Membrane. *By* VERA LUCAS TEIXEIRA, ADRIAN I. KATZ, CARLOS H. PEDEMONTE, AND ALEJANDRO M. BERTORELLO 587

The Sodium Pump Keeps Us Going. *By* TORBEN CLAUSEN 595

Extended Abstracts

Na,K-ATPase and the Significance of Sodium in the Mechanism of Potassium-Induced Relaxation of Rat-Isolated Mesenteric Arteries. *By* DIDIER X. P. BROCHET 603

Molecular Activity of Sodium Pumps in the Kidney of Mammals and Birds. *By* NIGEL TURNER, A. J. HULBERT, AND PAUL L. ELSE 606

Na$^+$,K$^+$-ATPase Subunit Isoforms of the Developing Central Nervous System of the Lizard *Gallotia galloti*. By M. F. ARTEAGA, J. AVILA, P. MARTÍN-VASALLO, AND C. M. TRUJILLO 608

Glutamate Receptors Regulate Na/K-ATPase in Cerebellum Neurons. By ELENA BULYGINA, OLGA GERASIMOVA, AND ALEXANDER BOLDYREV . 611

Na,K-ATPase Isoforms in Pregnant and Nonpregnant Rat Uterus. By RACHEL FLOYD, ALI MOBASHERI, PABLO MARTÍN-VASALLO, AND SUSAN WRAY 614

Using Na,K-ATPase Itself for Large-Scale Isolation and Purification of Endogenous Digitalis–Like Factors. By F. MANDEL, A. VASILIEV, AND I. KRIVOI 617

Inhibition of the Na,K-ATPase by the Antiarrhythmic Drug, Bretylium. By CRAIG GATTO, C. THEODORE BARKULIS, WILLIAM R. SCHNEIDER, JEREMY P. HOLDEN, KRISTA L. ARNETT, AND MARK A. MILANICK 620

Muscular K-Clearance Capacity *in Vivo* Must Be Evaluated on the Basis of K and Na,K-ATPase Concentrations. By HENNING BUNDGAARD AND KELD KJELDSEN 623

Modest K$^+$ Restriction Provokes Insulin Resistance of Cellular K$^+$ Uptake without Decrease in Plasma K$^+$. By LI E. YANG, PATRICK K. K. LEONG, JUAN P. GUZMAN, MICHAEL S. RHEE, AND ALICIA A. MCDONOUGH ... 625

Na$^+$,K$^+$-ATPase in the Marine Alga *Heterosigma akashiwo*. By YUKICHI HARA, YUKO MIKAMI, MARIKO SHONO, AND MASATO WADA 628

Positive Inotropic Effect Induced by Sequence-Specific Na$^+$,K$^+$-ATPase Antibody in Intact Cardiac Myocytes. By S. Q. WANG, H. CHENG, A. C. MYERS, B. J. CANNING, AND K. Y. XU 630

Inhibition of Purified Human Kidney Na$^+$,K$^+$-ATPase by Cyclosporine A: A Possible Mechanism for Drug Human Nephrotoxicity. By M. YOUNES-IBRAHIM, M. BARNESE, P. BURTH, AND M. V. CASTRO-FARIA 633

Regulation of Na,K-ATPase by cAMP-Dependent Protein Kinase Anchored on Membrane via A-Kinase Anchoring Protein Subtype, AKAP-150, in Rat Parotid Gland. By K. KURIHARA AND N. NAKANISHI 636

Porcine Kidney Extract Contains Factor(s) That Inhibit the Ouabain-Sensitive Isoform of Na,K-ATPase (α2) in Rat Skeletal Muscle: A Convenient Electrophysiological Assay. By I. KRIVOI, A. VASILIEV, V. KRAVTSOVA, M. DOBRETSOV, AND F. MANDEL 639

Rat Resistance Vessels Preferentially Contain the Ouabain-Insensitive α1 Isoform of Na,K-ATPase. By OTTO HANSEN 642

Na-Pump Kinetic Properties Are Differently Altered in the Brain Regions of the Cholecystokinin$_2$ Receptor–Deficient Mice. By KRISTIINA ROOTS, SULEV KÕKS, CZESLAVA KAIRANE, TIIT SALUM, ELLO KARELSON, EERO VASAR, AND MIHKEL ZILMER 644

Electrogenic Na$^+$/HCO$_3^-$ Cotransporter rkNBC1 Immunolocalized in Rat Eye. By HENRIK VORUM, CHRISTIAN AALKJÆR, HENRIK HAGER, SØREN NIELSEN, AND ARVID B. MAUNSBACH 646

Na,K-ATPase in the Regulation of Epithelial Cell Structure. *By* AYYAPPAN K. RAJASEKARAN, JEGAN GOPAL, AND SIGRID A. RAJASEKARAN 649

Expression of Na,K-ATPase β-Subunit in Transformed MDCK Cells Increases the Translation of the Na,K-ATPase α-Subunit. *By* SIGRID A. RAJASEKARAN, JEGAN GOPAL, AND AYYAPPAN K. RAJASEKARAN 652

Quality Control of Gastric Proton Pump in the Endoplasmic Reticulum by Ubiquitin/Proteasome System. *By* SHINJI ASANO, TOHRU KIMURA, HOKARA ISHIZUKA, MAGOTOSHI MORII, AND NORIAKI TAKEGUCHI 655

Mg-ATPase from Microsomal Fraction of Rabbit Gastric Mucosa Is Ecto-ATP-Diphosphohydrolase. *By* MARINA SMAGINA, NATALIYA DOLGOVA, ALEXANDER RUBTSOV, AND OLGA LOPINA 658

Part V. Pathophysiology and Pharmacology

Genetics of Hypertension: The Adducin Paradigm. *By* GIUSEPPE BIANCHI AND GRAZIA TRIPODI 660

Mechanisms of Pressure Natriuresis: How Blood Pressure Regulates Renal Sodium Transport. *By* ALICIA A. MCDONOUGH, PATRICK K. K. LEONG, AND LI E. YANG 669

Ouabain as a Mammalian Hormone. *By* WILHELM SCHONER, NATALI BAUER, JOCHEN MÜLLER-EHMSEN, ULRIKE KRÄMER, NJDE HAMBARCHIAN, ROBERT SCHWINGER, HANS MOELLER, HOLGER KOST, CHRISTINE WEITKAMP, THOMAS SCHWEITZER, ULRIKE KIRCH, HORST NEU, AND ERNST-GÜNTHER GRÜNBAUM 678

11-Hydroxylation in the Biosynthesis of Endogenous Ouabain: Multiple Implications. *By* JOHN M. HAMLYN, JAMES LAREDO, JUI R. SHAH, ZHUO REN LU, AND BRUCE P. HAMILTON 685

Antihypertensive Compounds That Modulate the Na-K Pump. *By* P. FERRARI, M. FERRANDI, L. TORIELLI, P. BARASSI, G. TRIPODI, E. MINOTTI, I. MOLINARI, P. MELLONI, AND G. BIANCHI 694

Myocardial Na,K-ATPase and Digoxin Therapy in Human Heart Failure. *By* KELD KJELDSEN AND HENNING BUNDGAARD 702

Expression and Cellular Localization of Na,K-ATPase Isoforms in Dog Prostate in Health and Disease. *By* A. MOBASHERI, I. EVANS, P. MARTÍN-VASALLO, AND C. S. FOSTER 708

Molecular Characterization of a Putative Sodium/Iodide Symporter in the South African Clawed Frog, *Xenopus laevis*. *By* D. L. CARR, F. LAHARRAGUE, B. KAHN, T. A. PRESSLEY, AND J. A. CARR 711

Index of Contributors 713

Assistance was received from:
- AARHUS UNIVERSITY RESEARCH FOUNDATION
- ASTRA ZENECA
- BYK GULDEN
- THE CARLSBERG FOUNDATION
- THE DANISH MEDICAL RESEARCH COUNCIL
- THE DANISH NATURAL SCIENCES RESEARCH COUNCIL
- THE DANISH WEIZMANN SOCIETY
- FIFTH FRAMEWORK PROGRAMME OF THE EUROPEAN COMMISSION FOR HIGH LEVEL SCIENTIFIC CONFERENCES
- THE LEO FOUNDATION
- THE LUNDBECK FOUNDATION
- THE NOVO-NORDIC FOUNDATION

The New York Academy of Sciences believes it has a responsibility to provide an open forum for discussion of scientific questions. The positions taken by the participants in the reported conferences are their own and not necessarily those of the Academy. The Academy has no intent to influence legislation by providing such forums.

Preface

PETER LETH JORGENSEN,[a] STEVEN J. D. KARLISH,[b]
AND ARVID B. MAUNSBACH[c]

[a]*Biomembrane Center, University of Copenhagen, Denmark,
August Krogh Institute, 2100 Copenhagen OE, Denmark*

[b]*Department of Biological Chemistry, Weizmann Institute of Science,
Rehovoth, 76100, Israel*

[c]*The Water and Salt Research Center, Department of Cell Biology,
Institute of Anatomy, University of Aarhus,
DK-8000 Aarhus C, Denmark*

The Tenth International Conference on Na,K-ATPase and Related Cation Pumps was held in Elsinore, Denmark on August 8–14, 2002. For the past thirty years the Na,K-ATPase conferences have brought together workers in this field to present new data and discuss recent developments. The field has undergone fundamental changes since 1957, when J. C. Skou described an ATPase from crab nerve activated by Na and K ions. However, in the past two years we have witnessed a dramatic development with the publication of the first crystal structure of a P-type cation pump, that of the sarcoplasmic reticulum Ca-ATPase, at 2.6 Å resolution, in an $E_1 \cdot Ca$ conformation.[1] In order to understand the mechanism of active cation transport by any P-type pump, we need to know how the protein structure changes in the different steps of the catalytic cycle.

A highlight of the meeting was the presentation of a new structure, at 3.1 Å resolution, of Ca-ATPase in an E_2 conformation, with bound Mg and the inhibitor, thapsigargin.[2] A comparison of the structures in the E_1 and E_2 conformations provides unique insights into the mechanism of active Ca transport. These structures also have extraordinary predictive and explanatory value for all P-type pumps, and facilitate planning of experiments to test details of the active transport mechanisms. The P-type pump family consists of five distinguishable groups (see the P-type ATPase database http://biobase.dk/~axe/Patbase.html), including Na,K-ATPase, gastric H,K-ATPase, Ca-ATPase of the sarcoplasmic and endoplasmic reticulum, plasma membrane Ca-ATPase, and plasma membrane H-ATPase of fungi and higher plants, as well as heavy metal pumps and phospholipid pumps.

At the conference the molecular structure and function of Na,K-ATPase, H,K-ATPases, H-ATPase, and Ca,ATPase were compared and contrasted in several symposia, and special sessions were held to discuss heavy metal pumps and plant proton pumps. Of course, despite the outstanding value of the SERCA Ca-ATPase structures for understanding the working of all pumps, these structures cannot provide complete insight into the working of Na,K-ATPase and the gastric and nongastric H,K-ATPase with additional beta subunits or regulatory domains and subunits (FXYD proteins) discussed below. Also, the basis for the different cation selectivities and detailed transport mechanisms of all pumps remains unknown, as do the mechanisms whereby the E_1/E_2 conformational transitions are coupled to active cation transport. Overall, we now view structure–function relationships of all P-type cation pumps as one subject in which we search for general principles of energy transduction, as well as unique structural, functional, and regulatory features of the individual classes of pumps. Further progress will require a combination of detailed functional analysis and additional structural information at low resolution, and new crystal structures of additional P-type pumps.

Mechanisms of physiological regulation of Na,K-ATPase, Ca-ATPase, and H,K-ATPase in normal cell function as well as pathophysiological processes involved in human disease and pharmacology have become a focus of great interest, particularly in view of the maturation of the classic structure–function studies of cation pumps. More than half of the conference time was dedicated to these areas in sessions entitled Molecular Mechanisms of Regulation, Regulation of the Na,K Pump in Kidney, Regulation of Na,K-ATPase by FXYD Proteins, Physiological Na,K-Pump Regulation: Muscle, the Na,K Pump and Hypertension, and Cation Pumps as Drug Targets. As could be expected for essential physiological processes that create the Na and K gradients across cell membranes, or Ca gradients across cell membranes and sarco- and endoplasmic reticulum, Na,K-ATPase and Ca-ATPase are regulated at several levels by hormones, in a developmental and tissue-specific fashion, by isoforms, and by additional regulatory proteins. Several novel regulatory phenomena and techniques for studying them were described. The use of transgenic animals to investigate the physiological functions of the isoforms of Na,K-ATPase and Ca-ATPases (SERCA and PMCA), or regulatory proteins (FXYD), is coming into wider use. Frequently, homozygous knockouts of the different isoforms are lethal, but analysis of the heterozygotes is proving to be very informative. Thus, for example, it has become clear that the α2 isoform of Na,K-ATPase is responsible for mediating the positive inotropic effects of cardiac glycosides, while the α1 isoform may mediate toxic effects.

Regulation of the Na,K-ATPase by the FXYD family of proteins was a major focus of the meeting. These one-trans-membrane segment proteins, named FXYD after the conserved signature sequence, include seven members (FXYD1-7) and are expressed in a tissue-specific fashion. FXYD2 is the well-known gamma subunit of renal Na,K-ATPase, but until recently its function was unknown. Other members such as phospholemman (FXYD1, PLM), CHIF (FXYD4), or FXYD7 were known to be expressed in muscle (PLM), colon and kidney (CHIF), and brain (FXYD7), respectively, but were thought to act as ion channels or regulators. A central message of the meeting is that all of these proteins are regulators of the Na,K-ATPase that modulates the pump kinetics that adapt the functional characteristics to the specific environment of different tissues. The important physiological role of these proteins is illustrated in the association between a mutation in the γ subunit and hereditary primary renal hypomagnesemia. It is of interest that a related regulatory mechanism of cardiac SERCA Ca-ATPases exists that involves one-trans-membrane proteins, phospholamban, and sarcolipin.

Several speakers discussed another novel regulatory mechanism, assigning a signal transducing role for the Na,K-ATPase, particularly in growth and hypertrophy of cardiac cells. The classic effects of cardiac glycosides in stimulating cardiac contractility result from changes in cellular Na and Ca concentrations, and it is now thought that the changes in ion concentration are mediated by the α2 isoform and occur in a localized space between the cardiac cell membrane and the sarcoplasmic reticulum. By contrast, long-term effects of ouabain to stimulate growth may involve induction of direct interactions of the Na,K-ATPase with multiple cellular signaling pathways and activation of transcription factors that control growth-related gene expression. The precise role of Na,K-ATPase in the complex network of effects involved in growth regulation will undoubtedly be worked out in the coming years. The significance of these contrasting roles of cardiac glycosides can be appreciated from the evidence for the involvement of endogenous ouabain-like compounds in the development of genetic hypertension. Endogenous ouabain (EO) appears to be an isomer of plant ouabain, but is synthesized in the adrenal glands, where it seems to act as a mammalian steroid hormone. In a session dedicated to hypertension, evidence was presented for a fundamental role of mutations in the cytoskeletal protein adducin in the genetic mechanism that influences renal Na and water retention and for endogenous ouabain in the adaptive mechanisms that correct blood volume at the expense of raised peripheral resistance and blood pressure. An important general message that emerged from these sessions on physiological regulation and pathophysiological mechanisms is that direct and specific interactions between Na,K-ATPase and other cellular proteins play a central role. Thus the identification of interacting proteins and a definition of the nature and specificity of such interactions will be a central challenge for the future.

Finally, novel approaches were presented for potential drug therapy for hypertension, on the basis of antagonists of endogenous oubain, and for new strategies, which bypass the clinical limitations on the use of the already successful proton pump inhibitors, for inhibition of gastric H,K-ATPase for treatment of ulcers and other acid-related gastric diseases.

An important organizational aspect of the conference was the invitation to a large number of young investigators and research students to present their work as short plenary lectures or as oral poster presentations. Funds from the Fifth Framework Programme of the European Commission for High Level Scientific Conferences allowed us to award grants to cover the participation of more than thirty young researchers from Europe. Basic support from the Research Foundation of the University of Aarhus allowed us to support many young scientists from other countries, for example, the United States, Argentina, and Japan. We hope that this strategy will continue to encourage young scientists to get involved and make contributions.

REFERENCES

1. TOYOSHIMA, C., M. NAKASAKO, H. NOMURA & H. OGAWA. 2000. Crystal structure of the calcium pump of sarcoplasmic reticulum at 2.6 Å resolution. Nature **405:** 647–655.
2. TOYOSHIMA, C. & H. NOMURA. 2002. Structural changes in the calcium pump accompanying the dissociation of calcium. Nature **418:** 605–611.

Introduction

Reflections on Ten International Conferences on Na,K-ATPase and Related Cation Pumps

JENS CHR. SKOU

Institute of Biophysics, University of Aarhus, 8000 Aarhus C, Denmark

When I started medical school in 1937, the cell membrane was impermeable to Na. Two years later, in 1939, Schmidt, Heppel, and Steinbach independently showed that the muscle membrane is permeable to Na. Subsequently Dean suggested that the cell membrane contains pumps that can pump out Na or pump in K, explaining the unequal distribution of Na and K across the cell membrane. To many, especially chemists, it was difficult to accept the idea of a membrane containing elements that could convert chemical energy into work; after all, the membrane was looked upon as a passive barrier. This new idea, however, was substantiated in the following decade. But what was the nature of the pump? In 1957 it was suggested that membrane-bound Na^+,K-activated ATPase was involved; the evidence was given by Robert Post and coworkers in 1960, who showed in experiments on red blood cells that the effect of Na and K on transport correlated with their effect on the enzyme activity. Since then extensive work has been done by scientists from many different countries in an effort to understand how this enzyme system can convert chemical energy from about 10 kg ATP per 24 hours into work, as in the transport of Na and K against electrochemical gradients.

In order to facilitate the exchange of information from this research, nine (and with the present meeting, ten) international conferences have been held on the subject. The first was held in 1973 in New York at the initiative of Amir Askari;[a] the second was convened in 1978 in Aarhus, Denmark. Since then it has been a tradition to have a conference every three years.

It is often said about meetings that the same people meet in different places, telling the same story. Looking at the list of speakers from these ten meetings, it is true that many of the names are the same; it is, however, not true that they have told the same story. The subject has been the same, but there has been considerable development and a steady increase in information. Many new speakers have presented new ways of attacking the problems. The knowledge that was available at the first meeting seems primitive compared to what we know today.

Address for correspondence: Professor Emeritus, Jens Chr. Skou, Institute of Biophysics, University of Aarhus, Universitetsparken 185, 8000C, Aarhus, Denmark. Voice: +45-89422950; fax: +45-86129599.

jcs@biophys.au.dk

[a]Properties and Functions of $(Na^+ + K^+)$-Activated Adenosinetriphosphatase. 1974. Annals of the New York Academy of Sciences. Amir Askari, Ed.: Vol. 242. New York, NY.

A wooden chair designed by Hans Wegner, the famous Danish furniture designer, was so elegant and simple in its design that American designers found its design unmatched; they named it "The Chair." Those of you who are old enough to have followed the televised presidential debates between Kennedy and Nixon in September 1960 may have seen it; it was the chair used by the candidates. You may also have seen it in a museum. When Wegner learned about the American designers' view on his chair, his only remark was, "Nothing is final."

"Nothing is final" could also be the motto for our work. Science is a steady revision of the accepted; we all know that after another 45 years today's views will seem primitive. In this process of revision, speculation plays an important role, and the basis for the speculation is imagination and knowledge.

In his book *Science and Human Values*, Jacob Bronowski says that the act of imagination includes finding likenesses between two things that are thought unlike. As an example he cites Newton's thinking of the similarity between a thrown apple and the moon turning around the earth—a most astonishing act of imagination, which turned out to be very productive. Placing known data in a new framework is another example of imaginative processes that have led to important breakthroughs in science—for example, when it was realized the sun, and not the earth, was the center of the solar system.

This is how modern science started: when the view on motion was changed from the doctrine of impetus to the doctrine of inertia. Considering that the new view on motion could not be tested experimentally, it took enormous intellectual and imaginative effort to arrive at the new view.

These are famous examples of how imagination has changed our picture of the world. Dean's suggestion of pumps in the membrane during a time when the membrane was looked upon as a passive barrier is an example from our own field. The act of imaginative process is important; by this we try to see connections and extend the borders of the system we are working inside.

What has this to do with the present meeting? Considering the importance of the imaginative process for the progress of science, and knowing that today's views are not tomorrow's, it is peculiar how little room is left for speculation in scientific journals. The discussion is the first section that the editor asks you to cut. This lack of speculative discussion can lead to a self-perpetuating process where you are trapped into a certain way of thinking, by which you try to explain your results, even if there are inconsistencies, instead of using your imagination to see whether it is possible to set up a new framework to explain your and others' results. As science history tells us, using the imagination is what leads to revision of the accepted and to development. It is therefore important that there be other possibilities, where you have opportunities to present your ideas and have them discussed. Meetings like the one recorded in this volume provide such a setting, where one can meet fellow scientists and engage in formal and informal discussions.

In science we build on each other' results; international cooperation is therefore important. From this point of view it is valuable to meet the persons whose names you know from their papers. Attending a meeting gives you a survey of the development in the field.

None of us likes to hear that our speculative suggestions are nonsense, so in order to share speculative suggestions that may lead to fruitful discussions and new views, it is very important that you know that the attitude of the listener/reader is positive and that your suggestions will be discussed seriously, respectfully, and in a friendly atmosphere. In my view, this has characterized our previous meetings. The background for this is that we meet regularly, get acquainted, and make friends.

It is society that pays for our research and has an interest in knowing what it gets in return. Even if society holds a positive attitude toward biological research, there is a growing fear for the ethical consequences of the application of new knowledge, and not least where the development of gene technology is concerned. Information and openness about our research and how it can be used is therefore important.

Another problem is the tendency to transfer money from basic research to applied research in order to obtain useful results quickly. It is important to say that this is a shortsighted policy. Applied research builds on the results of basic research, and without basic research the sources will soon dry up. It is therefore important that scientists feel the responsibility to inform the public and the decision makers about the importance of basic research—not for our own sake, but primarily for the sake of scientific development and its benefit to society. In this context international meetings are important, as they attract attention from the media, and thereby offer an opportunity to inform the public about our research. Regarding the usefulness of our research, you may answer as Faraday did when the English prime minister, Gladstone, asked what electricity could be used for, "Sir, one day you may tax it."

In my view these meetings have stimulated the progress of research in this field. And it is inspiring that not only have so many participated again and again, but that so many younger investigators are showing interest in the field. Over the years, the field has grown from meetings on the Na,K-ATPase to meetings on many different transport ATPases, and out of the same family, the P-type ATPases.

We are indebted to Peter Leth Jorgensen, Arvid Maunsbach, Steven Karlish, and their coworkers for arranging this conference.

I hope you will enjoy both the scientific and the social part of the meeting, and when you leave that you have formed new friendships with fellow scientists, which can be important for your further research in the field. I wish you a profitable and pleasant stay in Elsinore.

Crystal Structures of Ca^{2+}-ATPase in Various Physiological States

CHIKASHI TOYOSHIMA, HIROMI NOMURA, AND YUJI SUGITA

Institute of Molecular and Cellular Biosciences, The University of Tokyo, Bunkyo-ku, Tokyo 113-0032, Japan

> ABSTRACT: The structures of the Ca^{2+}-ATPase (SERCA1a) in different physiological states were determined by X-ray crystallography. Detailed comparison of the structures in the Ca^{2+}-bound form and unbound (but thapsigargin bound) form reveals that very large rearrangements of the transmembrane helices take place accompanying Ca^{2+} dissociation and binding and that they are mechanically linked with equally large movements of the cytoplasmic domains. The meaning of the rearrangement of the transmembrane helices becomes apparent by homology modeling of the Na^+K^+-ATPase.
>
> KEYWORDS: crystal structure; crystallography; Ca^{2+}-ATPase; SERCA1a; phospholipid; transmembrane helix; Ca^{2+}-binding sites; domain movements; homology modeling

The Ca^{2+}-ATPase from the rabbit skeletal muscle sarcoplasmic reticulum (SERCA1a) is structurally and functionally the best characterized member of the P-type ATPase superfamily. We have solved the crystal structures of this enzyme in the Ca^{2+}-bound ($E1Ca^{2+}$),[1] Ca^{2+}-unbound but thapsigargin (TG)-bound (E2(TG))[2], and Ca^{2+}-unbound but Mg^{2+}/F-bound (E2(Mg/F)) forms. The structure of the $E1Ca^{2+}$ form at 2.6 Å resolution (PDB entry code: 1EUL) was published in 2000,[1] and that of E2(TG) at 3.1 Å resolution (1IWO) in 2002.[2] The structure of the E2(Mg/F) complex[3] was solved before that of the E2(TG) form. However, the quality of the crystals of the Mg/F complex has been improving continuously: we now have data to 2.5 Å resolution of this form and are still refining the model. In working on the E2 forms, we realized that the 1EUL model contained some errors, and we are re-refining the model. At least one phospholipid molecule (likely to be phosphatidylcholine) can now be identified along the M7 helix.

Address for correspondence: Chikashi Toyoshima, Institute of Molecular and Cellular Biosciences, The University of Tokyo, Bunkyo-ku, Tokyo 113-0032, Japan. Voice: +81-3-5841-8492; fax: +81-3-5841-8491.
ct@iam.u-tokyo.ac.jp

Ca^{2+}-ATPase AS A MEMBRANE PROTEIN

Because P-type ATPases are membrane proteins, their relationships with the lipid bilayer may have critical importance. For example, the positions of the cation binding sites in the direction normal to the membrane may be a critical parameter in interpreting the results of physiological measurements and obviously depend on the orientation of the enzyme with respect to the membrane plane. In our original publication of the Ca^{2+}-bound form,[1] we specified the orientation as in Figure 1 of Ref. 1 but presented the model so that the M5 helix became upright (Fig. 2 of Ref. 1). This apparently caused much confusion. The orientation of the enzyme becomes even more important in comparing different forms and interpreting the meanings of the conformation changes.

Although the crystallization of SERCA1a required addition of exogenous phospholipid (PC), lipid bilayers were not visualized directly within the crystal lattice. This is because conventional crystallography discards low-resolution data that contain information on the lipid bilayer. Nevertheless, crystal structures give some information on the position of the bilayer. This is because the protein molecules are inserted in the membrane in alternating directions. This means that a symmetry axis (twofold rotation axis) runs through the center of the bilayer. In the E1Ca^{2+} crystals, there is only one such axis that determines the position of the M1 helix with respect to the membrane; that is, Cys70 comes to the center of the bilayer. Fortunately, one of the E2(TG) crystal forms belonged to a $P4_1$ space group and contained two monomers related by a noncrystallographic (i.e., local) twofold symmetry in an asymmetric unit. (That is the reason why 1IWO contains two molecules, although their conformations are assumed to be exactly the same.) Crystals of the E2(Mg/F) form grew with either $C2$ or $P2_1$ symmetry, providing other instances. Thus, we now have several points that will come to the center of the lipid bilayer. These positions indicate that, at least as a first approximation, the lipid bilayers run parallel to the crystallographic *ab* plane in all the crystals. The origin of the z-coordinate (along the axis normal to the membrane) in a $P4_1$ crystal is arbitrary but is adjusted to the center of the bilayer in 1IWO. The coordinates for the aligned models can be obtained from the authors' homepage (http://www.iam.u-tokyo.ac.jp/StrBiol/model).

In FIGURE 1, E1Ca^{2+} and E2(TG) are aligned and inserted into the bilayer of dioleoylphosphatidylcholine (DOPC), which supports full activity of Ca^{2+}-ATPase.[4] The coordinates of DOPC molecules are generated by a molecular dynamics calculation (Y. Sugita and C. Toyoshima, unpublished work). This figure alone shows many interesting features. For example, amphiphathic segments of the models (i.e., short loops connecting transmembrane helices and a short amphipathic part of M1 (M1′ in FIG. 1)) are located within the interface region of the lipid bilayer. This is an expected feature that supports the orientations of the models with respect to the bilayer.

Nevertheless, presumably the most important features are that the locations of the Ca^{2+}-binding sites are offset (4.4 and 7.3Å) to the cytoplasmic side and that site II

FIGURE 1. E1Ca^{2+} and E2(TG) forms of Ca^{2+}-ATPase in lipid bilayer. Cylinders represent α-helices, arrows β-strands. The lipid bilayer is generated by molecular dynamics simulation of DOPC molecules. Two spheres (*circled*) in the transmembrane region of E1Ca^{2+} represent bound Ca^{2+}. Transmembrane helices and P1 helix are numbered.

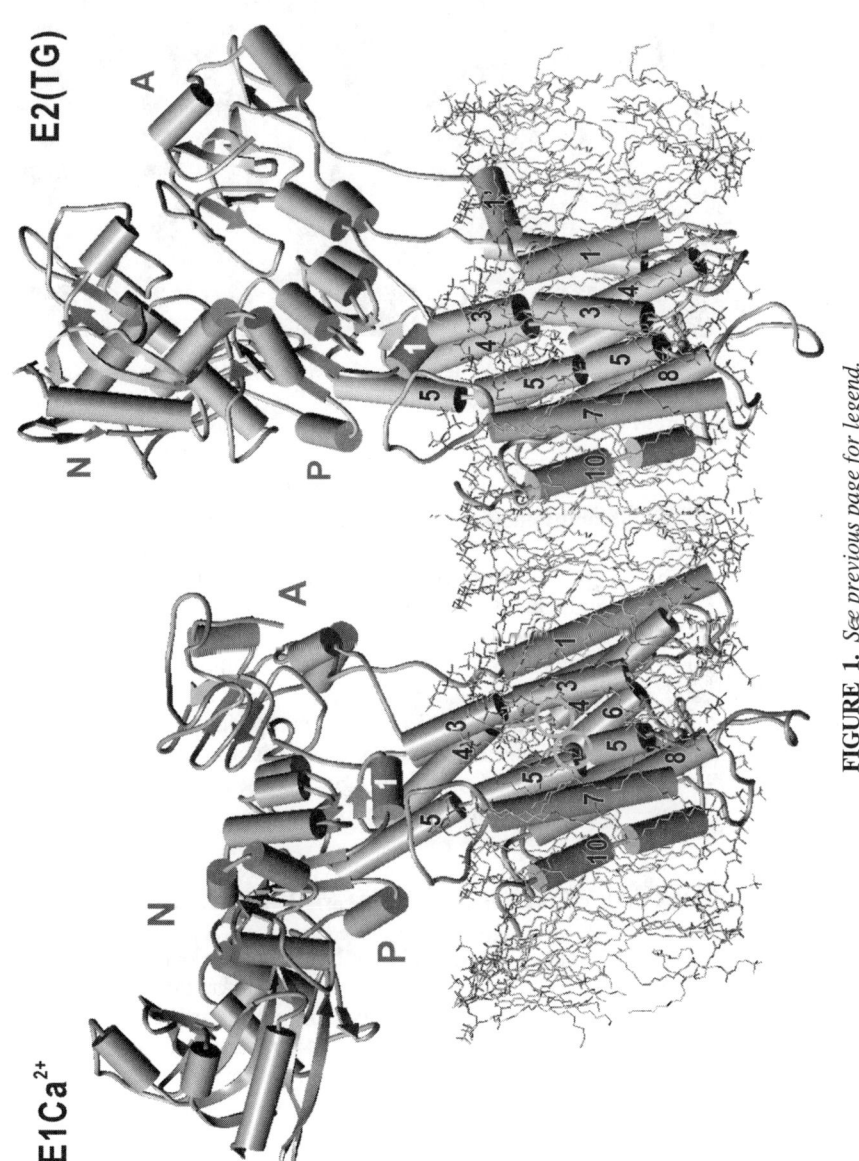

FIGURE 1. *See previous page for legend.*

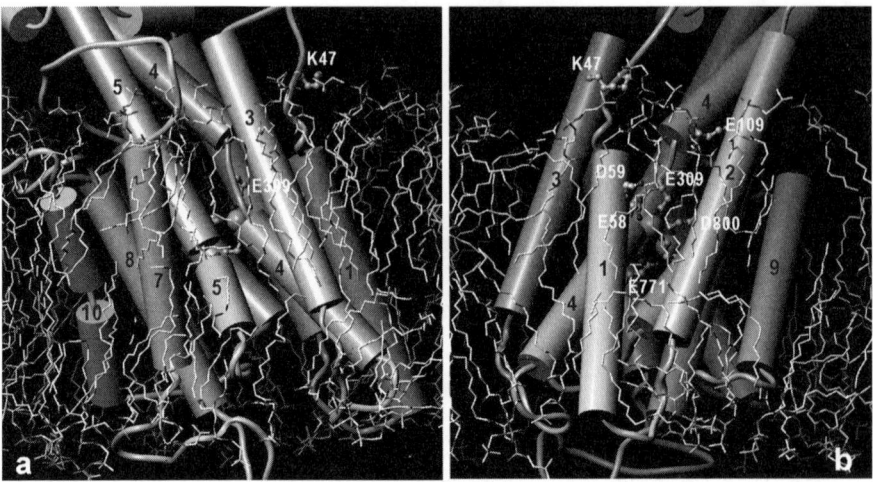

FIGURE 2. Orthogonal views of the Ca^{2+}-binding sites in $E1Ca^{2+}$. Viewed from the front (**a**) and right (**b**) along the membrane plane. Note that E309 is located within the hydrophobic core part of the lipid bilayer.

is closer to the cytoplasmic surface than site I. East and Lee[5] pointed out that there is a hole lined by Glu58 and Glu309 carboxyls leading to the Ca^{2+}-binding sites. They suggested that this hole might be the entry pathway. The question is whether this hole is accessible from bulk water or sealed by a lipid molecule (FIG. 2). The z-coordinate of the hole indicates that the hole is located within the hydrophobic core of the membrane. However, M1 and M2 helices might seclude the hole from lipids. We hope that the current re-refinement of the $E1Ca^{2+}$ model will provide the answer, yet crystal packing might destroy the proper lipid environment. In the E2(TG) form, Glu309 is clearly accessible from the cytoplasm through a different pathway.[2] Although there is an increasing number of results suggesting that Ca^{2+} enters via Glu309, to answer the question of how Ca^{2+} reaches the binding sites still needs careful studies, including full atom molecular dynamics simulation.

REARRANGEMENT OF TRANSMEMBRANE HELICES

Large-scale rearrangements of transmembrane helices take place during $E1Ca^{2+}$ → E2(TG) transition. Of the four transmembrane helices that form two Ca^{2+}-binding sites, M8 does not move, but the other three helices all move. Particularly important movements in $E1Ca^{2+}$ → E2(TG) are (1) a shift of M4 towards the lumenal (extracellular) side by one turn of an α-helix (FIG. 3, FIG. 4)), (2) bending of the upper part of M5 towards M4 (FIG. 3), and (3) a nearly 90° rotation of the unwound part of M6 (FIG. 4). Although it is clear that the Ca^{2+}-binding sites are destroyed, these movements are perplexingly large. Nevertheless, homology modeling the cation binding sites of the Na^+K^+-ATPase provides an explanation.[6]

It was surprising to see that very regular coordination geometry could be made immediately for the K^+-binding sites of the Na^+K^+-ATPase,[6] starting from the main

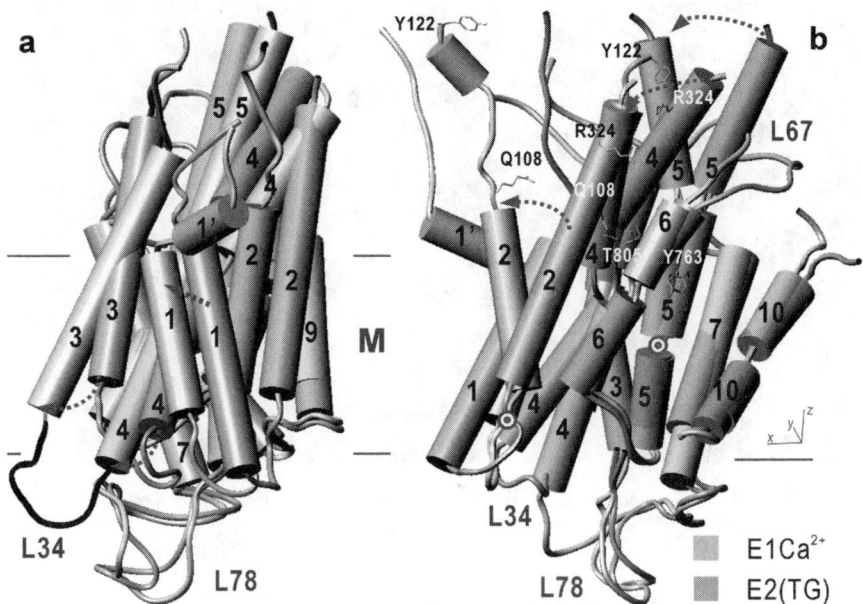

FIGURE 3. Rearrangements of the transmembrane helices between E1Ca^{2+} and E2(TG) forms. Rearrangement of transmembrane helices viewed from the right-hand side (**a**) and the back side (**b**) of that presented in FIG. 1. The models for E1Ca^{2+} and E2(TG) are superimposed, with the A domain located on the left-hand side. Viewing direction is chosen so that the M5 helix lies along the plane of the paper in the two states in **b**. M8 and M9 are removed in **b**, so that "domino" movements of the M5, M4, and M2 helices can be seen clearly. *Double circles* show pivot positions for M2 and M5. *Arrows* indicate the directions of movements in E1Ca^{2+} → E2(TG).

chain conformation derived from the E2(TG) model of the Ca^{2+}-ATPase.[2] The residues involved in the cation-binding sites appear essentially the same between Ca^{2+}- and Na$^+$K$^+$-ATPases (FIG. 5). The only critical difference is that Asn796 in Ca^{2+}-ATPase is substituted by Asp811 in Na$^+$K$^+$-ATPase. Asp811 is a key residue in K$^+$ binding, playing a similar role to Asp800 in Ca^{2+}-binding of the Ca^{2+}-ATPase and Asp815 in Na$^+$-binding of the Na$^+$K$^+$-ATPase, by contributing to both sites I and II. The position of Asp811 (Asn796 in Ca^{2+}-ATPase) is one turn below Asp815 (Asp800 in Ca^{2+}-ATPase) towards the extracellular side. To form a cation binding site there, M4 must move towards the extracellular side so that carbonyl oxygens can contribute to the binding. Hence, the movement of M4 ensures the release of one cation (Ca^{2+} or Na$^+$) and the binding of the other cation (H$^+$ or K$^+$) at a different level with respect to the membrane. This will certainly help in avoiding the competition between the binding cations.

This movement of M4 can be generated by the bending of the upper part of M5 (FIG. 3a). This bending allows different residues on different faces of the M5 helix to be used for cation binding (e.g., Ser782 is used instead of Glu786 in K$^+$ binding at site I) (FIG. 5). This will allow profound reorganization of the binding residues to

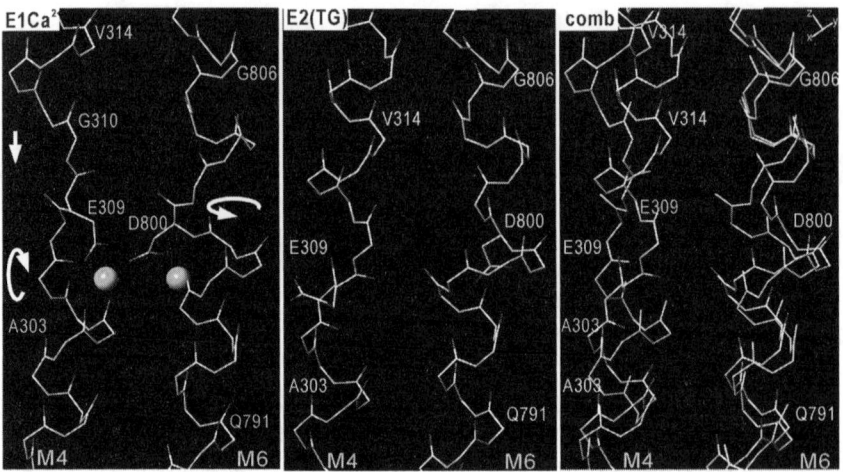

FIGURE 4. Movements of the transmembrane helices M4 and M6 between E1Ca^{2+} and E2(TG). Backbones of the M4 and M6 helices are shown together with the side chains of E309 (M4) and D800 (M6). They are superimposed in the rightmost panel ("comb"). Spheres represent bound Ca^{2+}. *Arrows* show the directions of the movements of M4 and the unwound part of M6 in E1Ca^{2+} → E2(TG). Note that whole M4 moves nearly one turn of α-helix in the vertical direction (along the helix), whereas the movement of M6 is rather localized around D800.

accommodate a larger cation (K$^+$) with a higher affinity, and also to ensure the release of Na$^+$.

These transmembrane helices are mechanically linked with each other. Also, M4 and M5 are integrated in the P domain as a part of a large β-sheet. The P domain is linked with M3 through the P1 helix (using Glu340) and thus indirectly with the A domain. A 110° rotation of the A domain is another key event during the E1-E2 transition[7] and most likely regulates the lumenal gate. In the E2(TG) form, the gate is closed at the very surface of the molecule. In the Mg/F complex, it is more opened but presumably not widely enough for bound Ca^{2+} to be released to the lumen. This is a reasonable feature, however, because Mg/F appears to be more like a "product" state analogue just after the hydrolysis of acylphosphate,[8] rather than a true E2P analogue.

MEANING OF THE OPEN AND CLOSED CONFIGURATIONS OF THE CYTOPLASMIC DOMAINS

Ca^{2+}-ATPase appears to undergo, in the absence of Ca^{2+} and Pi, large-scale thermal movements continuously involving both transmembrane and cytoplasmic domains, and the P domain is the coordinator of the movements. That must be the reason why the backward reaction (E1Ca^{2+} → E2H$^+$) is possible. If so, we would expect that the SR vesicles with loaded Ca^{2+} show significant leakage, and TG will be able to stop it. This has indeed been demonstrated (G. Inesi, private communica-

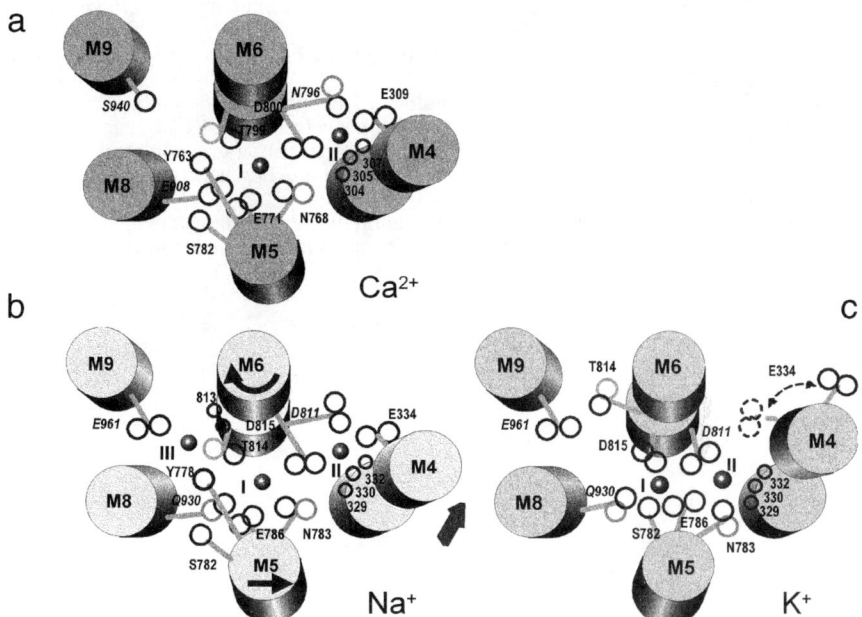

FIGURE 5. Models of the cation binding sites in Ca^{2+}- and Na^+K^+-ATPases. The transmembrane helices (cylinders) are viewed from the cytoplasmic side. Residue numbers are for rabbit SERCA1 (**a**) and human α1 (**b**, **c**). *Arrows* in **b** represent the directions of movements from the E1 (**b**) to E2 form (**c**). *Circles* appearing darker represent oxygen atoms (carbonyl oxygen atoms appear smaller). Residue numbers *in italic* are those different between Ca^{2+}-ATPase and Na^+K^+-ATPase α1.

tion). On the other hand, the closed configuration of the cytoplasmic domains will limit such movements and therefore the leakage. Another role of the closed configuration is to stop the reaction cycle, which is regulated essentially by Ca^{2+} alone. ATP can bind to the enzyme even when Ca^{2+} is absent, but without Ca^{2+} the reaction cycle cannot proceed. Although the N domain approaches the P domain, γ-phosphate of ATP cannot reach the phosphorylation residue Asp351. It requires an even deeper inclination of the N domain or the rotation of the A domain to release the N domain. This, in turn, requires the binding of Ca^{2+}. Whether this requires only the first Ca^{2+} or both Ca^{2+} to bind may be an interesting problem to explore.

REFERENCES

1. TOYOSHIMA, C., M. NAKASAKO, H. NOMURA & H. OGAWA. 2000. Crystal structure of the calcium pump of sarcoplasmic reticulum at 2.6 Å resolution. Nature **405:** 647–655.
2. TOYOSHIMA, C. & H. NOMURA. 2000. Structural changes in the calcium pump accompanying the dissociation of calcium. Nature **418:** 605–611.
3. MURPHY, A. & R.J. COLL. 1992. Fluoride binding to the calcium ATPase of sarcoplasmic reticulum converts its transport sites to a lower affinity, lumen-facing form. J. Biol. Chem. **267:** 16990–16994.

4. LEE, A.G. 1998. How lipids interact with an intrinsic membrane protein: the case of the calcium pump. Biochim. Biophys. Acta **1376:** 381–390.
5. LEE, A.G. & J.M. EAST. 2001. What the structure of a calcium pump tells us about its mechanism. Biochem. J. **356:** 665–683.
6. OGAWA, H. & C. TOYOSHIMA. 2002. Homology modeling of the cation binding sites of the Na^+K^+-ATPase. Proc. Natl. Acad. Sci. USA **99:** 15977–15982.
7. DANKO, S., K. YAMASAKI, T. DAIHO, *et al.* 2001. Organization of cytoplasmic domains of sarcoplasmic reticulum Ca^{2+}-ATPase in E_1P and E_1ATP states: a limited proteolysis study. FEBS Lett. **505:** 129–135.
8. WANG, W. *et al.* 2002. Structural characterization of the reaction pathway in phosphoserine phosphatase: crystallographic "snapshots" of intermediate states. J. Mol. Biol. **319:** 421–431.

Renal Na,K-ATPase Structure from Cryo-electron Microscopy of Two-Dimensional Crystals

HANS HEBERT,[a] PASI PURHONEN,[a,e] KAREN THOMSEN,[b,c] HENRIK VORUM,[d] AND ARVID B. MAUNSBACH[b,c]

[a]*Karolinska Institutet, Department of Biosciences, Center for Structural Biochemistry, Novum, S-141 57 Huddinge, Sweden*

[b]*The Water and Salt Research Center, and* [c]*Department of Cell Biology, Institute of Anatomy, and* [d]*Department of Biochemistry, University of Aarhus, DK-8000 Aarhus, Denmark*

ABSTRACT: The molecular structure of Na,K-ATPase was determined by electron crystallography from two-dimensional crystals induced in purified membranes isolated from the outer medulla of pig kidney. The P2 type unit cell contains two protomers in the E_2 conformation, each of them with a size of $65 \times 75 \times 150$ Å^3. The α, β, and γ subunits in the membrane crystals were demonstrated in the crystals with Western blotting and related to distinct domains in the density map. The α subunit corresponds to most of the density in the transmembrane region as well as to the large hydrophilic headpiece on the cytoplasmic side of the membrane. The headpiece is divided into three separated domains. One of these gives rise to an elongated projection onto the membrane plane, while the putative nucleotide binding and phosphorylation domains form compact densities in the rest of the cytoplasmic part of the structure. Density on the extracellular face corresponds to the protein part of the β subunit. Ten helices from the catalytic α subunit correspond to two groups of distinct densities in the transmembrane region. The structure of the lipid bilayer spanning part also suggests positions for the transmembrane helices from the β and γ subunits. The overall structure of the α subunit of Na,K-ATPase as determined here by cryo-electron microscopy is similar to the X-ray structure of Ca-ATPase. However, conformational changes between the E_1 and E_2 forms are suggested by different relative positions of cytoplasmic domains.

KEYWORDS: Na,K-ATPase 3-D structure; 2-D crystals; electron crystallography; subunit organization; membrane protein

---

Address for correspondence: Hans Hebert, Karolinska Institutet, Department of Biosciences, Center for Structural Biochemistry, Novum, S-141 57 Huddinge, Sweden. Voice: +46-8-6089219; fax: +46-8-6089290.
Hans.Hebert@csb.ki.se

[e]Permanent address: University of Tampere, IMT, FIN-33101 Tampere, Finland.

INTRODUCTION

Na,K-ATPase and other P-type ATPases convert the energy from ATP hydrolysis into ion translocation through a pumping mechanism that involves a phosphorylated intermediate.[1–3] The pump cycle of P-type ATPases goes through several distinct conformational states, of which the E_1 and E_2 forms are characterized by differences in ion affinity. All Na,K-ATPases have two subunits, the catalytic α subunit with a molecular mass of about 110 kDa and the glycosylated β subunit with a protein M_r of 31.5 kDa. In some tissues, including the kidney, the complete Na,K pump has an additional small γ subunit[4] with an M_r of about 7.3 kDa, which has been suggested to play a role in regulation of enzyme activity.[5,6] While the α subunit has ten transmembrane segments,[2] the β and γ subunits cross the phospholipd bilayer only once.[7] The Na,K-ATPase subunits are present in different isoforms in different tissues,[8] but in kidney the only isoforms of the two large subunits are $α_1$ and $β_1$.

By electron microscopy Na,K-ATPase was first identified in purified membranes isolated from the outer medulla of the kidney.[9,10] 2-D crystals induced in these membranes with vanadate, which is known to arrest the protein in the E_2 conformation,[11] have subsequently been used to characterize the 3-D shape of the enzyme by electron microscopy following negative staining[12,13] (for a review, see Ref. 14). Cryo-electron microscopy has more recently been applied both to flat 2-D crystals obtained from renal Na,K-ATPase[15–17] and tubular crystals prepared from duck salt glands,[18] but 3-D crystals of Na,K-ATPase suitable for X-ray crystallography have not yet been obtained.

Recently, the crystal structure of the related P-type ion pump Ca-ATPase of the skeletal muscle sarcoplasmatic reticulum was determined at 2.6 Å resolution.[19,20] From this atomic structure many conclusions concerning structure and function relationships of the calcium pump can be drawn, and some results may also be transferred onto other P-type ATPases. However, the Ca-ATPase lacks β and γ subunits, and significant differences from the Na,K-ATPase can be expected. Thus, an understanding of the specific properties of Na,K-ATPase must rely upon knowledge about the structure of this protein.

In the present work we report a structure determination of Na,K-ATPase from 2-D crystals induced in native kidney membranes and analyzed by cryo-electron microscopy and compare the results with the reported atomic structures of Ca-ATPase.

MATERIALS AND METHODS

Enzyme Purification

Na,K-ATPase was purified from the outer medulla of pig kidney by selective extraction of a microsomal fraction with sodium dodecyl sulphate in the presence of ATP followed by isopycnic zonal centrifugation.[21] The specific Na,K-ATPase activity of the enzyme peaks recovered from the zonal gradient and used for crystallization was 20–30 μmol P_i/min·mg protein. The purified membrane fragments were dialyzed against 1 mM NH_4VO_3, 5 mM $MgCl_2$, 5 mM $CaCl_2$, and 0.33 mM phospholipase A_2 in 10 mM imidazole-HCl, pH 7.3, at 4° C for 1 day.[15, 22]

Western Blot Analysis

In order to identify the individual subunits α, β, and γ of Na,K-ATPase, samples of purified membrane fragments, before and after crystallization, were analyzed by Western blotting. The samples were dissolved in sample buffer and processed using mouse monoclonal anti-α1 and anti-β1 antibodies and affinity-purified rabbit anti-γ (C-terminal) antibody, as previously described.[16]

Cryo-Electron Microscopy and Image Processing

Membrane crystals were applied to carbon-coated electron microscopy grids and rapidly plunged into liquid ethane in a Reichert-Jung KF80 cryo-fixation unit, transferred to a Philips CM120 electron microscope operated at 120 kV, and analyzed at a specimen temperature of −170° C. Low-dose recordings (< 15 e−/Å^2) were made on photographic film at a nominal magnification of 50,000. Crystalline areas were digitized using a Zeiss Scai scanner operated at a pixel size of 7 μm, corresponding to 1.4 Å at the specimen level. Image processing was performed with the MRC program suite.[23] Each crystalline area was corrected for lattice distortions and the contrast transfer function (CTF). The tilt parameters of individual projections were refined against 3-D data sets iteratively by minimizing the phase residual during the merging process. The procedure converged and showed phase residuals below 50° (random 90°). The 3-D map was visualized by surface rendering using Iris Explorer and by fishnet models using the program O. Alignment of the Na,K-ATPase map to the atomic model of Ca-ATPase[19] was partly made with Situs.[24]

RESULTS AND DISCUSSION

2-D Crystallization

The purified membranes treated with vanadate consisted of Na,K-ATPase molecules arranged on a crystalline lattice (FIG. 1a). From cryo-electron microscopy of crystals in ice (FIG. 1b), the size of the unit cell was determined to be a=146.5 Å, b=51.6 Å, and γ= 96.5°, similar to our previous studies at low resolution.[25] The two-sided plane group symmetry was p21, corresponding to the three-dimensional space group P2. The 2-D crystals typically consisted of 300–500 unit cells. Western blot analysis of specimens subjected to 2-D crystallization showed the presence of α-, β- and γ-subunits.

3-D Structure

The crystals from which the 3-D structure was determined originated from specimens at nominal tilt angles up to 60° and with isotropically distributed tilt axis directions. The tilt parameters for each of the crystals were refined during the analysis. On the basis of phase deviations between two 3-D reconstructions from two subsets of the complete data, information to 9.5 Å resolution was used for calculating the final 3-D map.

The 3-D structure of the αβγ Na,K-ATPase protomer has an extension perpendicular to the plane of the membrane of approximately 150 Å (FIG. 2). Approximately

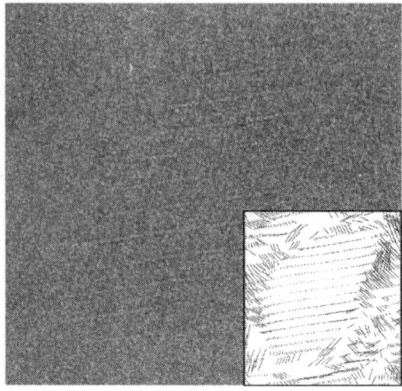

FIGURE 1. Two-dimensional crystals of renal Na,K-ATPase from negatively stained (*left*) and frozen hydrated specimens. The *inset* in the micrograph of the ice-embedded crystal shows the corresponding positions of the unit cells in the crystal and deviations from a perfect lattice. Scale bar: 500 Å.

30 Å and 90 Å of the structure protrude on the extra- and intracellular faces of the lipid bilayer, respectively. On the cytoplasmatic side of the membrane the structure is subdivided into separated domains: the nucleotide binding (N), the actuator (A), and the phosphorylation (P) domain.

Conformations of the Cytoplasmatic Domain

Automatic docking of the transmembrane domain from the Ca-ATPase model to the corresponding part of the Na,K-ATPase structure resulted in an orientation between the two proteins close to what was observed visually (FIG. 2). From the comparison it was clear that the N domain has different positions relative to the rest of the protein (FIG. 2). This is expected since the Ca-ATPase model was obtained from the E_1-form, while vanadate as used for 2-D crystallization of Na,K-ATPase is known to arrest P-type ATPases in the E_2 form. As observed from the cytoplasmatic side of the membrane, the transition from the E_1 to the E_2 form shows a movement of the N domain towards the A domain. The angular difference is about 40°. Thus, the structure is more closed in the E_2 form, and the contact surface between two symmetry-related protomers in the 2-D crystals gives rise to the characteristic groove along the short unit cell axis, as seen in projections perpendicular to the plane of the membrane (FIG. 3). A similar conformational change as suggested here has been observed also in Ca-ATPase[19,20,26] as well as in H-ATPase.[27]

The Transmembrane Region

At the present resolution level, it is not possible to trace the transmembrane helices of Na,K-ATPase independently. However, from comparisons with Ca^{2+}-ATPase it is clear that individual sections show a good correspondence (FIG. 2). In addition to the 10 predicted helices, Na,K-ATPase shows two additional densities in the transmembrane region as compared to Ca^{2+}-ATPase, indicating the presence of two ad-

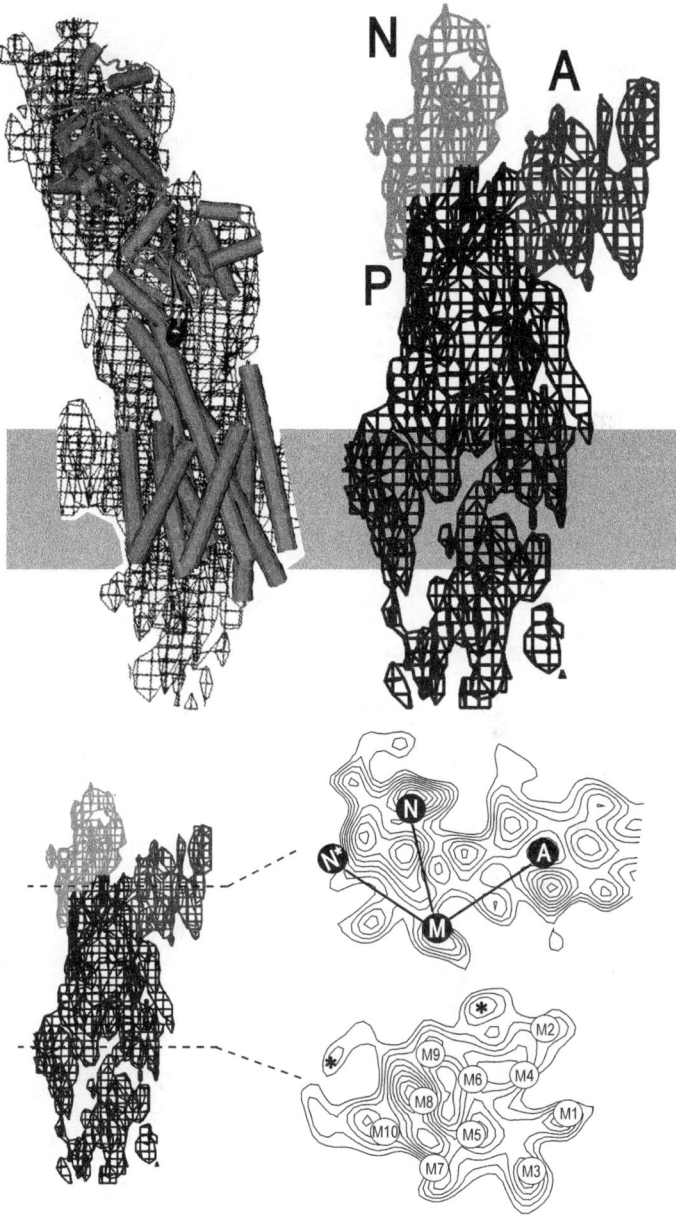

FIGURE 2. Three-dimensional structure of renal Na,K-ATPase at 9.5 Å resolution. The *upper part* shows two orientations of the protomer viewed along the plane of the membrane (*gray rectangle*) with the cytoplasmatic side up. The height of the molecule is approximately 150 Å. The nucleotide (N, *light gray*), actuator (A, *medium gray*), and phosphorylation (P) domains have been marked. The best fit of the atomic model of Ca-ATPase[19] is shown as an overlay in the view to the *left*. The *lower part* of the figure shows

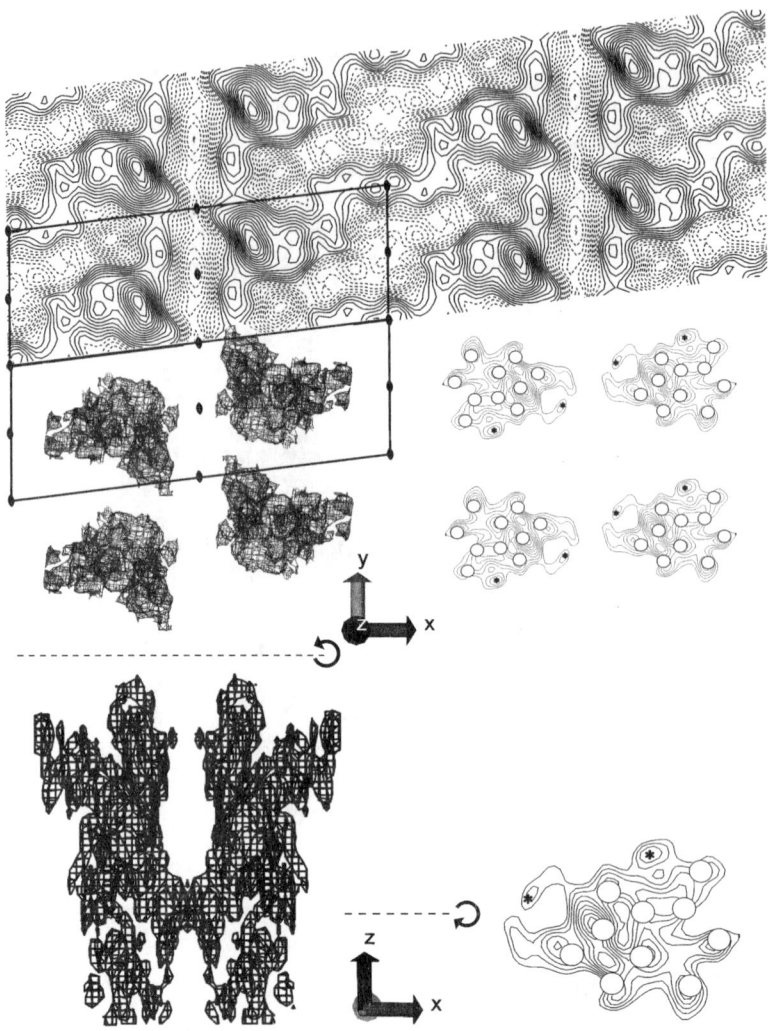

FIGURE 3. The packing of the αβγ protomers of Na,K-ATPase in the 2-D crystals. The *upper part* shows a contour plot of the projection map. In the *middle*, the 3-D structure (*left*) and a transmembrane section have been placed at the corresponding lattice positions. The *lower part* shows the 3-D model at a slightly lower contouring level, indicating the close contact between protomers in the transmembrane region at the level of the depicted section. The empty rings show the positions of the transmembrane helices in Ca-ATPase, as in FIG. 2.

---

two sections from the model at the indicated heights. The section cut on the cytoplasmatic side shows the positions of the M (membrane), A, and N domains in the present structure of Na,K-ATPase in the E_2 conformation (MAN) and the E_1 form of Ca-ATPase (MAN*). The positions of the helices of Ca-ATPase have been depicted in the transmembrane section. Putative positions for the two additional helices from the β and γ subunits present in the protomer have been marked with asterisks.

ditional helices. Since both the β- and γ-subunits are present in the crystals it is likely that these densities correspond to the single transmembrane span from each of these subunits. Several investigations have shown that the transmembrane helix of the β-subunit is close to M7 and M8 of the α-subunit.[28, 29] Thus, it is likely that the density to the left in a slice through the transmembrane region as depicted in FIGURE 2 arises from the β-subunit. If this assignment is correct the density facing the arc formed by M2, M4, M6, and M9 would correspond to the position of the γ-subunit.

Protein–Protein Interactions

From crystallogenesis studies of Na,K-ATPase we observed formation of dimers that subsequently stack into ribbons.[30] The closest contact between protomers of Na,K-ATPase in the 2-D crystals is about one of the twofold axes in the transmembrane region (FIG. 3). The putative β-subunit transmembrane α-helix is located at this position. Interestingly the β1-isoform present in the specimens used for 2-D crystallization has aligned glycine residues proposed to be involved in dimerization of membrane proteins.[29]

ACKNOWLEDGMENTS

This work was supported by grants from the Swedish Research Council, the Danish Medical Research Council, the Water and Salt Research Center established and supported by the Danish National Research Foundation (Danmarks Grundforskningsfond), the Aarhus University Research Foundation, the Karen Elise Jensen Foundation, and the Novo Nordisk Foundation. P.P. acknowledges a fellowship from the Graduate School of Electron Microscopy and Structural Characterization, Tampere, Finland.

REFERENCES

1. SKOU, J.C. 1957. The influence of some cations on an adenosine triphosphatase from peripheral nerves. Biochim. Biophys. Acta **23**: 394–401.
2. MØLLER, J.V., B. JUUL & M. LE MAIRE. 1996. Structural organization, ion transport, and energy transduction of P-type ATPases. Biochim. Biophys. Acta **1286**: 1–51.
3. JØRGENSEN, P.L. & P.A. PEDERSEN. 2000. Structure-function relationships of Na^+, K^+, ATP, or Mg^{2+} binding and energy transduction in Na,K-ATPase. Biochim. Biophys. Acta **45028**: 1–18.
4. FORBUSH III, B., J.H. KAPLAN & J.F. HOFFMAN. 1978. Characterization of a new photoaffinity derivative of ouabain: labeling of the large polypeptide and of a proteolipid component of the Na,K-ATPase. Biochemistry **17**: 3667–3676.
5. THERIEN, A.G., R. GOLDSHLEGER, S.J.D. KARLISH & R. BLOSTEIN. 1997. Tissue-specific distribution and modulatory role of the γ subunit of the Na,K-ATPase. J. Biol. Chem. **272**: 32628–32634.
6. ARYSTARKHOVA, E., R.K. WETZEL, N.K. ASINOVSKI & K.J. SWEADNER. 1999. The gamma subunit modulates Na^+ and K^+ affinity of the renal Na,K-ATPase. J. Biol. Chem. **274**: 33183–33185.
7. MERCER, R.W., D. BIEMESDERFER, D.P. BLISS, JR., *et al.* 1993. Molecular cloning and immunological characterization of the γ polypeptide, a small protein associated with the Na,K-ATPase. J. Cell Biol. **121**: 579–586.
8. LINGREL, J.B., J. ORLOWSKI, M.M.. SHULL & E.M. PRICE. 1990. Molecular genetics of Na,K-ATPase. Prog. Nucleic Acid Res. Mol. Biol. **38**: 37–89.

9. MAUNSBACH, A.B. & P.L. JØRGENSEN. 1974. Ultrastructure of highly purified preparations of (Na$^+$+K$^+$)-ATPase from the outer medulla of the rabbit kidney. Proc. VIII Internat. Congr. Electron Microscopy, Canberra, Vol II: 214–215.
10. DEGUCHI, N., P.L JØRGENSEN & A.B MAUNSBACH. 1977. Ultrastructure of the sodium pump. Comparison of thin sectioning, negative staining, and freeze-fracture of purified, membrane-bound (Na$^+$,K$^+$)-ATPase. J. Cell Biol. **75:** 619–634.
11. CANTLEY, L.C., L.G. CANTLEY & L. JOSEPHSON. 1978. A characterization of vanadate interactions with the (Na,K)-ATPase: mechanistic and regulatory implications. J. Biol. Chem. **253:** 7361–7368.
12. HEBERT, H., E. SKRIVER & A.B. MAUNSBACH. 1985. Three-dimensional structure of renal Na,K-ATPase determined by electron microscopy of membrane crystals. FEBS Lett. **187:** 182–186.
13. HEBERT, H., E. SKRIVER, M. SÖDERHOLM & A.B. MAUNSBACH. 1988. Three-dimensional structure of renal Na,K-ATPase determined from two-dimensional membrane crystals of the p1 form. J. Ultrastruct. Molec. Struct. Res. **100:** 86–93.
14. MAUNSBACH, A.B., E. SKRIVER & H. HEBERT. 1991. Two-dimensional crystals and three-dimensional structure of Na,K-ATPase analyzed by electron microscopy. *In* The Sodium Pump: Structure, Mechanism, and Regulation. J.H. Kaplan & P. De Weer, Eds.: 160–172. The Rockefeller University Press. New York.
15. MAUNSBACH, A.B., H. HEBERT & U. KAVÉUS. 1992. Cryo-electron microscope analysis of frozen-hydrated crystals of Na,K-ATPase. Acta Histochem. Cytochem. **25:** 279–285.
16. HEBERT, H., P. PURHONEN, H. VORUM, *et al.* 2001. Three-dimensional structure of renal Na,K-ATPase from cryo-electron microscopy of two-dimensional crystals. J. Mol. Biol. **314:** 479–494.
17. TAHARA, Y., A. OSHIMA, T. HIRAI, *et al.* 2000. The 11 Å resolution projection map of Na$^+$/K$^+$-ATPase calculated by application of single particle analysis to two-dimensional crystal images. J. Electron Microsc. **49:** 583–587.
18. RICE, W.J., H.S.YOUNG, D.W. MARTIN, *et al.* 2001. Structure of Na$^+$,K$^+$-ATPase at 11-Å resolution: Comparison with Ca^{2+}-ATPase in E$_1$ and E$_2$ states. Biophys. J. **80:** 2187–2197.
19. TOYOSHIMA, C., M. NAKASAKO, H. NOMURA & H. OGAWA. 2000. Crystal structure of the calcium pump of sarcoplasmic reticulum at 2.6 Å resolution. Nature **405:** 647–655.
20. TOYOSHIMA, C. & H. NOMURA. 2002. Structural changes in the calcium pump accompanying the dissociation of calcium. Nature **418:** 598–599.
21. JØRGENSEN, P.L. 1988. Purification of Na$^+$+K$^+$-ATPase: Enzyme sources, preparative problems, and preparation from mammalian kidney. Methods Enzymol. **156:** 29–43.
22. SKRIVER, E., A.B. MAUNSBACH & P.L JØRGENSEN. 1981. Formation of two-dimensional crystals in pure membrane-bound Na$^+$,K$^+$-ATPase. FEBS Lett. **131:** 219–222.
23. CROWTHER, R.A., R. HENDERSON & J.M. SMITH. 1996. MRC image processing programs. J. Struct. Biol. **116:** 9–16.
24. WRIGGERS, W., R.A. MILLIGAN & A. MCCAMMON. 1999. Situs: a package for docking crystal structures into low resolution maps from electron microscopy. J. Struct. Biol. **125:** 185–195.
25. HEBERT, H., P.L. JØRGENSEN, E. SKRIVER & A.B. MAUNSBACH. 1982. Crystallization patterns of membrane-bound (Na$^+$+K$^+$)-ATPase. Biochim. Biophys. Acta **689:** 571–574.
26. XU, C., W.J. RICE, H. WANZHONG & D.L. STOKES. 2002. A structural model for the catalytic cycle of Ca^{2+}-ATPase. J. Mol. Biol. **316:** 201–211.
27. KÜHLBRANDT, W., M. AUER & G.A. SCARBOROUGH. 1998. Structure of the P-type ATPases. Curr. Opin. Struct. Biol. **8:** 510–516.
28. FAMBROUGH, D.M., L. HUYNH & B. HWANG. 2000. The Na,K-ATPase α–β subunit assembly site. *In* Na/K-ATPase and Related ATPases. International Congress Series 1207. K. Taniguchi & S. Kaya, Eds.: 103–106. Elsevier. Amsterdam.
29. HASLER, U., G. CRAMBERT, J.-D. HORISBERGER & K. GEERING. 2001. Structural and functional features of the transmembrane domain of the Na,K-ATPase β subunit revealed by tryptophan scanning. J. Biol. Chem. **276:** 16356–16364.
30. SÖDERHOLM, M., H. HEBERT, E. SKRIVER & A.B. MAUNSBACH. 1988. Assembly of two-dimensional membrane crystals of Na,K-ATPase. J. Ultrastruct. Molec. Struct. Res. **99:** 234–243.

Projection Map of Covalently Phosphorylated Ca-ATPase from Tubular Crystals

F. DELAVOIE,[a] D. McINTOSH,[b] F. HENAO,[c] G. PERANZI,[a] P. CHAMPEIL,[d] D. STOKES,[e] AND J-J. LACAPÈRE[a]

[a]*INSERM U410, 75870 Paris, France*

[b]*Department of Chemical Pathology, Universtiy of Cape Town Medical School, Cape Town, South Africa*

[c]*Bioquímica y Biología Molecular, Universidad de Extremadura, 06080 Badajoz, Spain*

[d]*CNRS U2096, SBFM/DBJC, CEA-Saclay, 91191 Gif-sur-Yvette, France*

[e]*Skirball Institute, Cell Biology, New York University, New York, New York 10016, USA*

KEYWORDS: Ca-ATPase; electron microscopy; phosphorylated form; projection map

The covalent modification of Ca-ATPase with FITC (Lys-515) enabled the formation of a stable covalent phosphoenzyme.[1] Two-dimensional vesicular crystals of this phosphoenzyme have been produced from SR vesicles in the presence of decavanadate.[2] However, these crystals were too small to be analyzed for structural information. We devised a procedure to produce long tubular crystals from native SR vesicles[3] and report here the crystallographic analysis that yields a projection map for the phosphorylated ATPase.

In our crystallization conditions freeze–thaw cycles promote SR vesicle fusion and lead to formation of tubular crystals. Optimal growing time was generally two to four days. Electron micrographs of negatively stained samples showed crystalline arrays spreading over the entire length of the tube (FIG. 1A). These arrays appeared to be composed of double-stranded ribbons. Computed diffraction patterns recorded from spread-flattened tubular arrays (FIG. 1B) could be indexed according to two reciprocal lattices corresponding to the two sides of the tube. The unit cell parameters of phosphorylated ATPase crystals (a ≈ 112 Å, b ≈ 62 Å, c ≈ 77°) are very close to those for nonphosphorylated ATPase crystals. Analysis of the phase relationships indicated $p2$ symmetry for the phosphorylated ATPase crystals, as also observed for nonphosphorylated ATPase crystals formed in the presence of decavanadate. FIGURE 1C shows IQ plots of 22 average images, and IQ values of four are visible out to 20-Å resolution. The overall phase residual up to 20-Å resolution was ~21° for IQ 6.

Address for correspondence: J-J. Lacapère, INSERM U410, 16 rue Henri Huchard, 75870 Paris, France. Voice: +33-1-44-85-61-36; fax: +33-1-42-28-87-65.
lacapere@bichat.inserm.fr

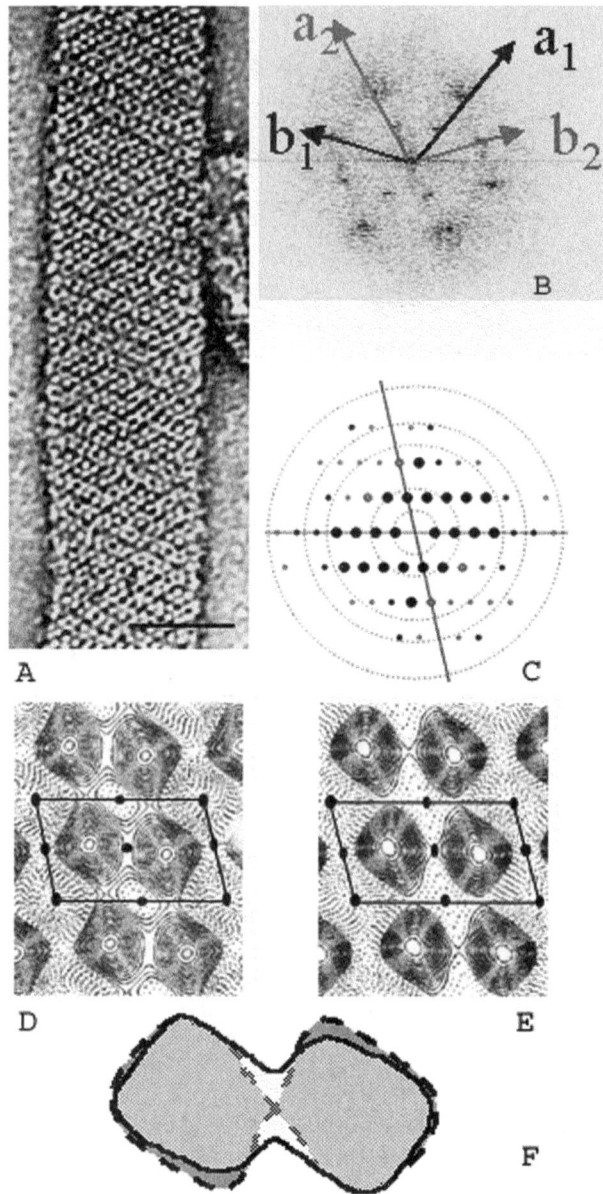

FIGURE 1. Tubular crystals of phosphorylated FITC-ATPase. (**A**) A negatively stained flattened tube. *Scale bar* represents 50 nm. (**B**) Computed diffraction pattern. (**C**) Fourier transform of the average. The size of the *circles* indicates the IQ value. The *rings* are at radii corresponding to 100-, 50-, 25-, and 20-Å resolution. (**D** and **E**) Projection maps at 20-Å resolution of phosphorylated and nonphosphorylated FITC-ATPase, respectively. Projection maps were obtained after enforcing the $p2$ symmetry. The *parallelogram* depicts the unit cell. (**F**) Superposition of the outer contours of phosphorylated and nonphosphorylated ATPase dimers.

The phase error of the Fourier components obtained after averaging indicated that all the symmetry-related reflections were reliable up to 20-Å resolution, and the projection density map was calculated at this resolution. This projection map for the phosphorylated ATPase (FIG. 1D) was compared with a projection at the same resolution of the nonphosphorylated ATPase (FIG. 1E) In both cases, the unit cell contains two protein monomers related to each other by a twofold axis of symmetry. FIGURE 1F exhibits a superposition of phosphorylated and nonphosphorylated dimers of Ca-ATPase. Intermolecular contacts between phosphorylated ATPase monomers seem to be stronger than between the nonphosphorylated monomers within the dimer. Phosphorylated monomers seem to be smaller than nonphosphorylated ones.

Many previously described protocols for producing tubular crystals point to the importance of the lipid-to-protein ratio. Destabilization and fusion of SR vesicles by freeze-thaw cycles are considered[3] to change this ratio locally and may permit easier fusion of phospholipid-rich areas. We found that decavanadate plays a significant role in the vesicle fusion, and the spatial redistribution of ATPase by decavanadate increases the segregation effect of the freeze–thaw process (data not shown). The fusion process permits formation of large sheets containing two distinct lipid-to-protein ratio areas. Sheet edges are protein rich and contain crystalline arrays, while the rest of the sheet is phospholipid rich. We hypothesize that sheets provide a reservoir of crystalline arrays of Ca-ATPase that leads to growth of long tubular crystals. The 20-Å projection maps of phosphorylated and nonphosphorylated Ca-ATPase show similar dimeric organization of protein in the unit cell. The outer contours of the two conformations are very similar, suggesting that only limited rearrangement of the cytosolic domains has occurred as a result of phosphorylation, at least in this conformation with empty calcium sites. The determination of three-dimensional structure of the phosphorylated form of the sarcoplasmic reticulum Ca-ATPase by cryoelectron microscopy is in progress.

REFERENCES

1. CHAMPEIL, P. *et al.* 2001. A remarkably stable phosphorylated form of Ca-ATPase prepared from Ca-loaded and FITC-labeled SR. J. Biol. Chem. **276:** 5795–5803.
2. HENAO, F. *et al.* 2001. Phosphorylated Ca-ATPase stable enough for structural studies. J. Biol. Chem. **276:** 24282–24285.
3. DELAVOIE, F. *et al.* 2001. Tubular crystals formation by fusion of crystallized SR Ca-ATPase proteoliposomes. Biol. Cell. **93:** 363–364.

How Does Conformation Change Ouabain Binding from Rejection (E1) to Acceptance (E2)?

ROBERT LICKELY POST

Department of Physiology, School of Medicine, University of Pennsylvania, Philadelphia, Pennsylvania 19104-6085, USA

KEYWORDS: Na,K-ATPase ouabain binding conformation E1 E2

The sodium pump, Na,K-ATPase, has two principal conformations, termed E1 and E2, as does the calcium pump of sarcoplasmic reticulum, SERCA. Ouabain binds to Na,K-ATPase, poorly to E1, and tightly to E2, and binding residues have been identified.[1] So how does the arrangement of ouabain-binding residues change between E1 and E2 to prevent or accept binding of ouabain?

The three-dimensional arrangement of residues is unknown for Na,K-ATPase but is known for both conformations of SERCA.[2,3] Although ouabain does not bind to SERCA, the primary structure of SERCA is homologous to that of Na,K-ATPase.[4] Thus residues that bind ouabain in Na,K-ATPase can be mapped to corresponding residues in SERCA. Changes in the arrangement of these residues between E1 and E2 in SERCA should be similar to corresponding changes in Na,K-ATPase.

Ouabain-binding residues of Na,K-ATPase were mapped to residues of SERCA as follows.[4] The residue before the equal sign is in Na,K-ATPase,[1,5] and the residue after the equal sign is in SERCA: C104 = S72 and Q111 = L75 in helix M1; D121 = I85 and N122 = T86 in loop L1-2; Y308 = G286 in loop L3-4; F786 = T778, L793 = E785, and T797 = P789 in loop L5-6; R880=A852 in helix M7; F982 = L968 in helix M10. The SERCA atomic coordinates (files 1EUL and 1IWO) were downloaded from the Protein Data Bank, and these structures were viewed in RasMol. The best view was that of the extracellular face showing only the residues previously listed. In order to view E1 and E2 in the same orientation, helices 8 and 9 were fiducial structures (not shown).

As the enzyme changes between E1 and E2 (FIG. 1), residues in L1-2, M1, and L3-4 revolve around M6, as indicated by the double-headed arrow. As E1 changes to E2, the motion is counterclockwise in this view corresponding to rotation of the

---

Address for correspondence: Robert Lickely Post, Department of Physiology, School of Medicine, University of Pennsylvania, 3700 Hamilton Walk, Philadelphia, PA 19104-6085. Voice: 215-898-3060; fax: 215-573-5851.

postrl@mail.med.upenn.edu

Ann. N.Y. Acad. Sci. 986: 20–21 (2003). © 2003 New York Academy of Sciences.

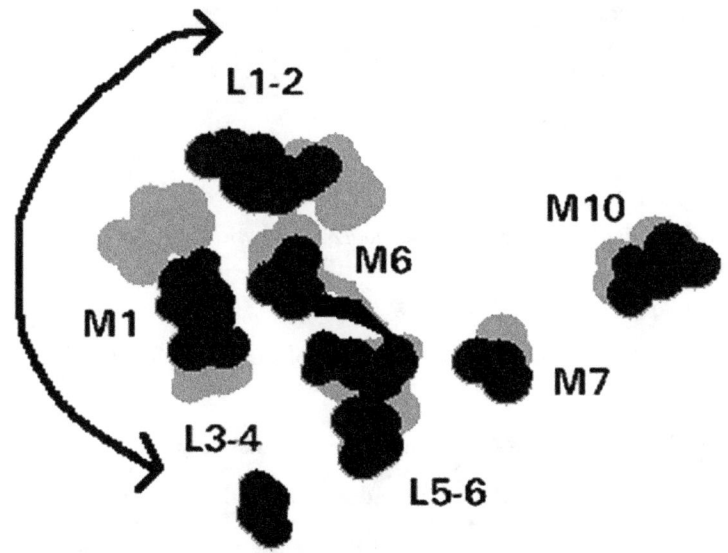

FIGURE 1. Change in arrangement of ouabain-binding residues as conformation changes between E1 (*gray*) and E2 (*overlapping black*). Residues are identified only by the helix or loop to which they are attached. Continuity of loop L5-6 with helix M6 is shown by a ribbon.

cation-binding residues on M6.[3] There is also reorientation of residues in L5-6 and M7.

In a side view of the E2 ouabain-binding center (not shown), residues in L5-6, M6, and M7 lie close to a plane almost parallel to the membrane, whereas residues in M1 and L1-2 project into extracellular space. This arrangement is consistent with binding of ouabain in a bent conformation.[6]

REFERENCES

1. PALASIS, M., T.A. KUNTZWEILER, J.M. ARGUELLO & J.B. LINGREL. 1996. Ouabain interactions with the H5-H6 hairpin of the Na,K-ATPase reveal a possible inhibition mechanism via the cation binding domain. J. Biol. Chem. **271:** 14176–14182.
2. TOYOSHIMA, C., M. NAKASAKO, H. NOMURA & H. OGAWA. 2000. Crystal structure of the calcium pump of sarcoplasmic reticulum at 2.6 Å resolution. Nature **405:** 647–655.
3. TOYOSHIMA, C. & H. NOMURA. 2002. Structural changes in the calcium pump accompanying the dissociation of calcium. Nature **418:** 605–611.
4. SWEADNER, K.J. & C. DONNET. 2001. Structural similarities of Na,K-ATPase and SERCA, the Ca^{2+}-ATPase of the sarcoplasmic reticulum. Biochem. J. **356:** 685–704.
5. CROYLE, M.L., A.L. WOO & J.B. LINGREL. 1997. Extensive random mutagenesis analysis of the Na^+/K^+-ATPase alpha subunit identifies known and previously unidentified amino acid residues that alter ouabain sensitivity—implications for ouabain binding. Eur. J. Biochem. **248:** 488–495.
6. MIDDLETON, D.A., S. RANKIN, M. ESMANN & A. WATTS. 2000. Structural insights into the binding of cardiac glycosides to the digitalis receptor revealed by solid-state NMR. Proc. Nat. Acad. Sci. USA **97:** 13602–13607.

Transmission of E_1-E_2 Structural Changes in Response to Na^+ or K^+ Binding in Na,K-ATPase

PETER L. JORGENSEN

Biomembrane Center, University of Copenhagen, Denmark, 2100 Copenhagen OE, Denmark

KEYWORDS: E_1-E_2 transition; Na^+- or K^+-ion binding; Na,K-ATPase

ABSTRACT: The extensive E_1-E_2 conformational changes in response to Na^+ or K^+ binding in the absence of other ligands must be driven by motion of the side chains contributing to cation coordination, but the differences in structure of Na^+ and K^+ sites have not been resolved. The recent high resolution structure model of the E_2 conformation of Ca-ATPase offers the first opportunity to examine and model the changes accompanying the adjustment of the cation sites from an E_1 form with specificity for Na^+ to an E_2 form with specificity for K^+. The model of the E_2 form provides a remarkable fit to the data of direct Tl^+ or K^+ binding after site-directed mutagenesis of residues Asp804 and Asp808 in M6, Glu 779, Gln776, and Ser775 in M5, and Glu327 in M4. Cytoplasmic domain movements during $E_1 \leftrightarrow E_2$ conformational transition can be monitored by proteolytic cleavage. Protection of the chymotrypsin-sensitive bond at Leu266 in L2/3 and rotation of the A domain is more complete in the E_2Mg-vanadate-ouabain complex than in the E_2[2K] form.

INTRODUCTION

The existence of two protein conformations of the α-subunit of Na,K-ATPase during the catalytic cycle was first detected as distinct patterns of proteolytic cleavage in Na^+ or K^+ medium. In the absence of other ligands than K^+ in the medium, trypsin cleaves at T1, Arg-438 in the N domain and subsequently at T2, Lys-30 in the N-terminus of the A domain.[1] In Na^+ medium, T1 is protected to trypsin, while T2 is cleaved rapidly, and T3 at Arg-262 in the L23 loop is cleaved more slowly by trypsin. Chymotrypsin cleaves the neighboring bond at Leu-266 (C3) in Na^+ medium but only slowly in K^+ medium.[2] Specific tryptic cleavage of the T2 site of the E_1[2Ca] form of Ca-ATPase is localized to the T2 site at Arg-198.[3] A systematic study of limited proteolysis in various states revealed that the patterns of cleavage in

---

Address for correspondence: Peter L. Jorgensen, Biomembrane Center, University of Copenhagen, Denmark, August Krogh Institute, Universitetsparken 13, 2100 Copenhagen OE, Denmark. Voice: +45-35321670; fax: +45-35321567.

pljorgensen@aki.ku.dk

Na⁺ medium, $E_1[3Na]$, resemble cleavage of the ATP bound form in K⁺ medium, ATP-E_1(2K). The pattern in the E_2 [2K] conformation is similar to that of the E_2 P[2Na] form, reflecting that these long-range conformational changes of the α-subunit couple the scalar processes of ATP binding, phosphorylation, and dephosphorylation to the vectorial extrusion of 3 Na⁺ ions and uptake of 2 K⁺ ions.[4] In the forward reaction, the $E_1P \rightarrow E_2P$ transition is coupled to Na⁺ extrusion, and the increase in binding energy for ATP of the E_2 [2K]→E_1 (2K) transition is the driving force for K⁺ uptake in cells.

The extensive long-range structural effects of K⁺ or Na⁺ binding in the absence of other ligands must be driven by motion of the side chains contributing to cation coordination, but the different structures of the Na⁺ and K⁺ sites have not been resolved. The exposure and protection of proteolytic cleavage sites in Na,K-ATPase can be interpreted in terms of movements of the cytoplasmic A, P, and N domains[5] in light of structural reconstructions of E_2 forms of Na,K-ATPase,[6,7] but these models did not visualize $E_1 \leftrightarrow E_2$ transitions of the intramembrane segments. The recent high-resolution structure model of the E_2 conformation of Ca-ATPase[8,9] offers the first opportunity to resolve the changes accompanying the adjustment of the cation sites from an E_1 form with specificity for Na⁺ to an E_2 form with specificity for high-affinity binding of K⁺ and its congeners Rb⁺ and Tl⁺. An important question is whether the cation-induced conformational changes of the cytoplasmic domains resemble those seen after phosphorylation with Mg^{2+} bound to the protein. Can binding of K⁺ ions alone elicit the full 110° horizontal rotation of the A domain to dock on to the P and N domains, or are Mg^{2+} binding and phosphorylation or vanadate binding required? Another important issue is the route of the structural changes. Are the Na⁺/K⁺-induced conformational changes transmitted directly through M4 and M5 to the P and N domains to influence ATP binding, phosphorylation, and dephosphorylation, or are they transmitted indirectly via the connections of the S4-helix1 to M3 and L2/3 to the A domain?

STRUCTURE–FUNCTION RELATIONSHIPS OF THE E_1[3Na] AND E_2 [2K] FORMS OF THE CATION SITES

In the model in FIGURE 1, two Na⁺ binding sites in Na,K-ATPase are assumed to be homologous to the two Ca^{2+} sites in the Ca-ATPase structure. The E_1 and E_1-P forms known to bind three Na⁺ ions, and models for localization of the third Na⁺ site are discussed in an accompanying article.[12] Na⁺ binding data after mutation support the role of residues Glu327, Asn776, Glu779, Asp804, and Asp808 in Na⁺ binding (TABLE 1).[11] In site I, one of the Na⁺ ions is coordinated to Asn776, Glu779, Thr807, and Gln923. In site II, the second Na⁺ ion coordinates to Glu327 and Asp804, as well as to the carbonyl groups of residues of the unwound loop, Val322, Ala323, and Val325,[9] while the side chain of Asp808 coordinates both Na⁺ ions.[9,10]

Site-directed mutagenesis and assay of direct cation binding of Rb⁺, or Tl⁺ and K⁺ displacement of ATP binding, provided evidence for involvement of four carboxylate groups in binding of K⁺ (TABLE 1).[11] In spite of the resemblance to the model for binding of Ca^{2+},[10] this interpretation of the mutagenesis data has been controversial, particularly with respect to the carboxylate residues of Glu327 in M4 and Glu779 in M5 (see review in Ref. 11). As seen from FIGURE 2, conservative muta-

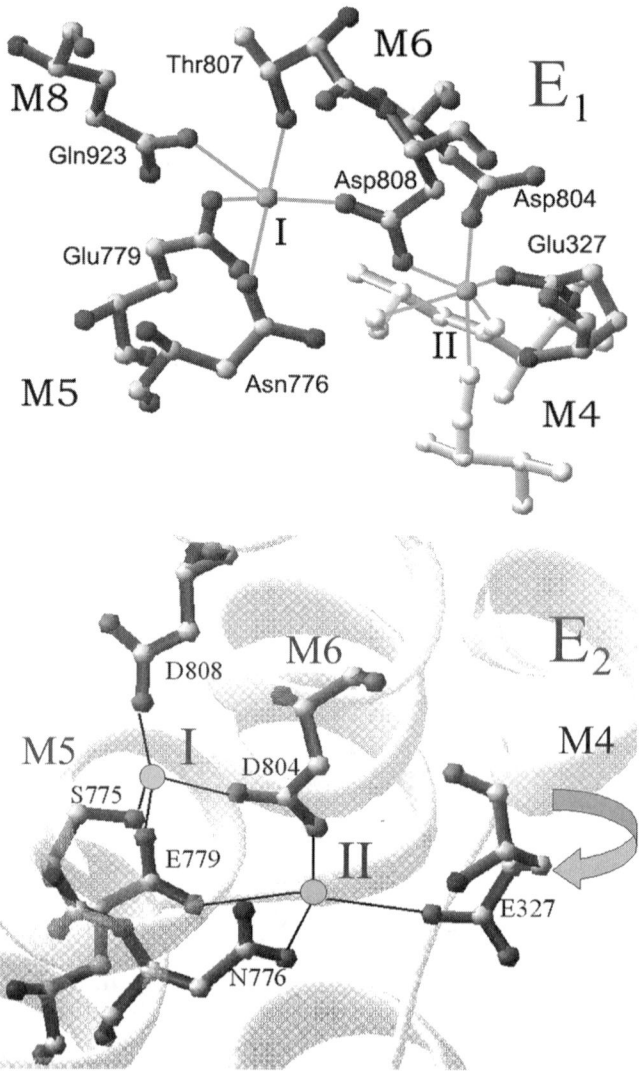

FIGURE 1. Representation of the Na^+-ion binding sites I and II. E_1: homology model of Na,K-pump side chains on the backbone of the high-resolution structure of SR Ca-ATPase (1EUL).[9] E_2: model of the K^+-ion binding sites I and II in a model based on the E_2 Mg-thapsigargin complex (1IWO).[8]

TABLE 1. Alterations in Gibbs free energy ($\Delta\Delta G_b$) of Tl$^+$- or Na$^+$-ion binding at 4°C of mutations of residues in M4, M5, or M6 of the α-subunit of Na,K-ATPase

Allele	Alterations in Gibbs free energy $\Delta\Delta G_b$ (kJ/mol)[a]		
	Tl$^+$	K$^+$ (K$^+$-ATP)	Na$^+$ (EP)
M4			
E327D	>+8	>+11	+3.0
E327Q	>+8	+7.9	+4.0
E327S		+9.3	+2.9
E327A		+5.6	+2.7
E327L		+3.0	+2.5
M5			
E779Q	+4.3	+4.9	+6.8
E779D	>+8	>+10	>+10
N776A	+6.7	+9.9	+6.8
N776Q	+1.9	+3.1	+8.0
S775A	+3.6	+7.7	+5.9
S775T	+6.3	+8.3	+4.1
M6			
D804A	>+8	>+12	
D804E	>+8	+4.8	>+12
D804N	>+8	>+12	+3.9
D808A	>+8	>+12	+3.2
D808E	>+8	>+12	
D808N	>+8	>+12	>+12

[a]$\Delta\Delta G_b$ values were calculated from binding data, as described in Ref. 11.

tions of either side chain abolished high-affinity binding of two Tl$^+$ ions per α-subunit, while the stoichiometry is reduced to one Tl$^+$ ion per α-subunit in the Glu779 → Gln mutation with $K_{1/2}$ = 27 μM as compared with 7.3 μM in wild type. To explain this, the substituted carboxamide group Gln779 was proposed to coordinate one Tl$^+$ ion but to compromise high-affinity binding of the second Tl$^+$ ion, while Glu779 is in a position to coordinate both tightly bound K$^+$ ions in the E_2[2K] form in FIGURE 1. Direct binding of Tl$^+$ or K$^+$ ions to a series of mutations of Glu327 also suggested that this residue is free to rotate.[11] It is interesting to see that there also is support for these predictions in the structure model.

The different structures of the two models in FIGURE 1 reflect that the $E_1 \leftrightarrow E_2$ transitions are large global changes in conformation involving nearly all intramembrane segments. During the $E_1 \rightarrow E_2$ transition, the transmembrane segments M4 and M5 move as a rigid body with the P domain, accompanied by movements of M3 and M6. A pivoting point for tilting and bending of M5 is Gly-770,[8] corresponding to Pro-778 in Na,K-ATPase. As neighbor to the pivoting point, the carboxylate group of Glu779 undergoes minimal movement towards the center of the cation-binding cavity by, at most, 2 Å, while Asn776 is displaced towards cation site II. Transmembrane segment M6 is close to the pivoting point and its lateral movement is small, but

M6 with Asp804 and Asp808 undergoes a large, almost 90-degree, clockwise rotation during the $E_1 \to E_2$ transition. After this rotation Asp808 contributes to coordination of the K^+ ion in site I, while Asp804 occupies a central position in E_2, as it is capable of coordinating two K^+ ions. In this position Asp804 in M6 is located in a helix tier close to the level of Glu779 in M5 and Glu327 in M4. M4 is farther away from the pivoting point and undergoes a large vertical shift with movement of Glu-327 by about 5 Å toward the extracellular face during transition to the E_2 form. The unwound loop is turned and the carboxylate group of Glu309 points away from site II in the E_2 form of Ca-ATPase.[8] It is probable that this represents a conformation for discharge of Ca^{2+} from the sites via an extracellular gate. In Na,K-ATPase, Glu327 is essential for high-affinity binding of $Tl^+(K^+)$,[11] and rotation of Glu327 in the unwound loop appears to be required for occlusion of the cations in Na,K-ATPase, as proposed in the model in FIGURE 1B and earlier.[11]

The driving force for this motion of the intramembrane segments is contributed by the adaptation of the side chains to high-affinity binding of the $Tl^+(K^+)$ ions. In the absence of phosphorylation, the intrinsic affinity of Na,K-ATPase for Na^+ is relatively low, and cholin-Cl or tris-Cl stabilizes the E_1 form as well as Na^+.[1,2] The Na,K pump has an exceptionally high intrinsic affinity for binding of K^+, as also seen from FIGURE 2 with $K_D = 7$ μM for direct binding of Tl^+ to wild type Na,K-ATPase at a capacity of two Tl^+ ions per α-subunit. This corresponds to a change in Gibbs free energy for binding of Tl^+ of $\Delta G_b = -29$ kJ/mol at 4°C for each of two ions.[11] Adjustment of Sites I and II from the E_1 [3Na] form to the E_2[2K] form with high-affinity binding of two K^+ or Tl^+ ions is accompanied by a change of the distances between coordinating groups to adapt from the diameter of Na^+ (1.9 Å) to the

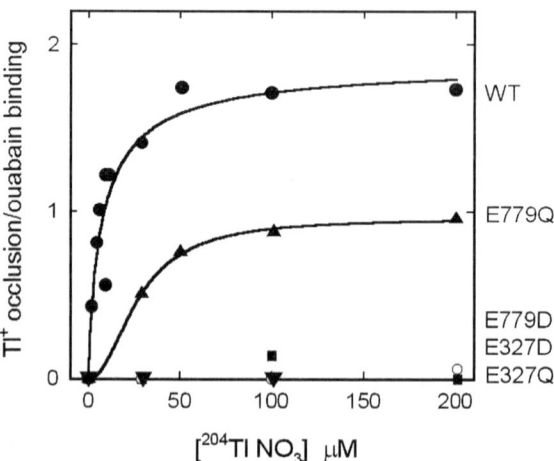

FIGURE 2. High-affinity Tl^+ binding in wild type (WT) (●) with maximum capacity of 1.9 ± 0.3 Tl^+ ion per ouabain-binding site, and $K_{1/2} = 7.3 \pm 3.3$ μM. Tl^+-ion binding is abolished after conservative mutations Glu327 → Gln (■), Glu327 → Asp (O). Tl^+-ion binding is also abolished in the Glu799 → Asp mutation (▼), while it is retained at 0.96 ± 0.03 Tl^+ ion per ouabain site, and $K_{1/2} = 27 \pm 1$ μM in the Glu779 → Gln (▲) mutation.

larger size of K^+ or Tl^+ (2.7–2.8 Å) and an increase of the number of coordinating groups per ion. High-affinity binding and occlusion of two K^+ ions thus elicits extensive tilting or twisting of transmembrane segments, and the force of this motion is sufficient for transmission of the structural change to the cytoplasmic domains of the protein.

MONITORING A AND P DOMAIN INTERACTIONS BY PROTEOLYTIC CLEAVAGE

To examine if cation-induced conformational changes in the cytoplasmic protrusion are similar to those elicited by other ligands, proteolytic cleavage can be used as a tool to monitor cytoplasmic domain movements. Exposure and protection of the chymotrypsin-sensitive bond at Leu266 in L2/3 monitors the rotation of the A domain (FIG. 3). Interaction between the tryptic cleavage site at Arg438 in the N domain and at Lys30 in the N-terminus provide evidence for interaction between the A and N domains in the $E_2[2K]$ conformation.[4] In the model in FIGURE 3, the segment containing the chymotrypsin cleavage site at Leu-266 is seen at the termination of a helix in the L2/3 loop connecting the A domain to M3. Dissociation of the A domain

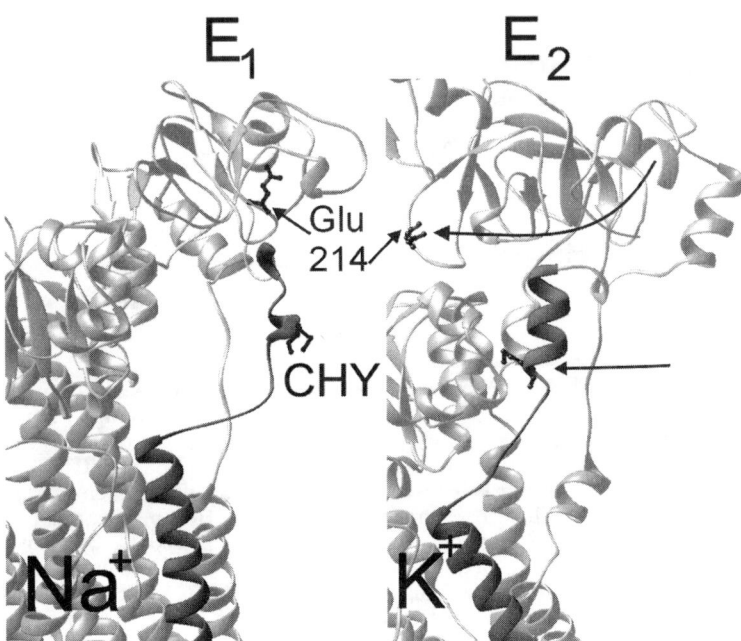

FIGURE 3. Model of the connections from the A domain via L2/3 to M3, P1-helix, and L6/7 in Na,K-ATPase, as modeled on the backbone of Ca-ATPase in the E_1[Ca] form[9] E_1, or in the E_2 Mg-thapsigargin form[8] E_2. CHYM denotes the position of the peptide bond (Leu266) that is cleaved by chymotrypsin in the E_1[3Na] form of the α-subunit of Na,K-ATPase.[2,4]

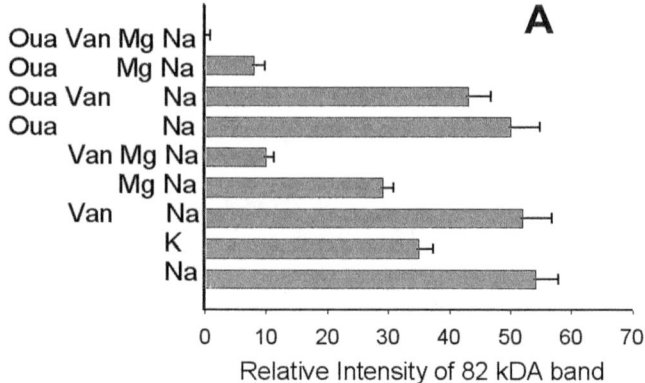
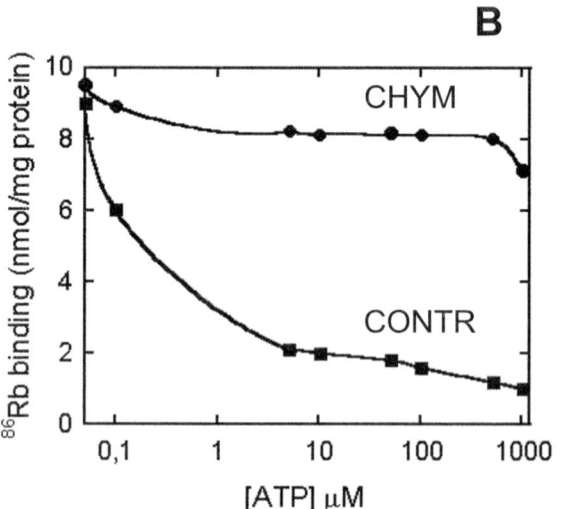

FIGURE 4. Intensity of the chymotryptic 82 kDa fragment of the α-subunit of Na,K-ATPase after cleavage in the presence of 10 mM Na$^+$ or 10 nM K$^+$, 3 mM Mg^{2+}, 1 mM VO$_3^-$, and 0.1 mM ouabain, as indicated. For chymotryptic cleavage, aliquots of yeast membranes were pelleted and resuspended in 15 mM Tris-HCl, pH 7.5; 1 mM Tris EDTA; 2 µg/mL leupeptin; and 2 µg/mL pepstatin and mixed with equal volumes of a solution containing the desired ligands at twice the final concentration in 15 mM Tris-HCl, pH 7.5, and 1 mM Tris EDTA at a final membrane protein concentration of 0.4 mg/mL. Five µL of chymotrypsin (0.1 mg/mL) was added, and the mixture was incubated at 37°C for 30 minutes. The digestion was stopped by precipitation by adding ice-cold 10% (w/v) TCA. Proteolytic fragments were analyzed by electrophoresis in SDS and Western blotting, as described before.[18]

from the P domain in E$_1$ [3Na] or E$_1$-P [3Na] forms of Na,K-ATPase exposes the bond at Leu 266 (Cl) in the L2/3 loop to chymotrypsin.[2,4] Similarly, sites for trypsin, V8, and proteinase-K are exposed in the E$_1$ [2Ca] form of Ca-ATPase,[3,13] suggesting that the conformational movements of the A domain are similar in the two enzymes.

The importance of the A domain movements and the L2/3 loop are underlined by the nature of its connections near the cytoplasmic end of M3 in a triangular hydrogen bonded arrangement between L2/3, helix-1 of S4 and L6/7 in both Ca- and Na,K-ATPase.[9,16] During the $E_1 \rightarrow E_2$ transition this segment is twisted around the N-terminus along with the large horizontal turn of the β–sheet structure of the A domain toward the P domain. In the E_2 form, the segment is closely opposed to both the P and the N domain, and the chymotryptic cleavage site is completely protected (FIG. 3). The exposure and protection of the chymotryptic cleavage site, therefore, monitors the rotation of the A domain during $E_1 \leftrightarrow E_2$ transition.

The data in FIGURE 4A show that in comparable conditions, the reduction in rate of cleavage is more moderate after exchanging Na^+ with K^+ in the medium (35%) than after addition of Mg^{2+} (46%) or incubation with Mg-vanadate (82%) or Mg-vanadate and ouabain (100%). Addition of Mg^{2+} to the Na^+ medium, thus, has almost the same effect as the exchange of Na^+ for K^+, while formation of the E_2 Mg-vanadate or E_2 Mg-vanadate-ouabain complexes is required to abolish cleavage of Leu-266 and to complete the horizontal turn of the A domain. As a corollary to this, the Fe^{2+}-catalyzed cleavage data suggest that the rate of cleavage of the 212-TGES segment is higher in the E_2P-Mg-ouabain form than in the E_2[2Rb] or E_2P conformations.[15] Ouabain binding to the E_2P-Mg form is known to cause a large reduction of the affinity for K^+ binding and also to quench fluorescence from FITC bound in the ATP site of the N domain.[4] High-affinity ouabain binding to the extracellular loops of the transmembrane segments therefore alters the adjustment of side chains in the K^+-ion sites, and the structural changes are transmitted via L2/3 and the A domain movement to the P and N domains.

The catalytic properties of Na,K-ATPase can be examined after complete and specific chymotryptic cleavage at Leu266. Cleavage almost prevents ATP displacement of occluded Rb^+, as shown in FIGURE 4B. Similarly, it prevents K^+ displacement of ATP binding, to allow high-affinity ATP binding in both NaCl and in KCl medium.[2] Chymotryptic cleavage at Leu266 also blocks Na^+ translocation and stabilizes the protein in the E_1-P[3Na] form,[4] thus preventing translocation of Na^+. Transmission of structural changes via L2/3 and the A domain movements are therefore crucial for energy transduction during active Na,K transport.

REFERENCES

1. JORGENSEN, P.L. 1975. Purification and characterization of (Na^+,K^+)-ATPase. V. Conformational changes in the enzyme transitions between the Na-form and the K-form studied with tryptic digestion as a tool. Biochim. Biophys. Acta **401:** 399–415.
2. JORGENSEN, P.L. & J. PETERSEN. 1985. Chymotryptic cleavage of alpha-subunit in E_1-forms of renal $(Na^+ + K^+)$-ATPase: effects on enzymatic properties, ligand binding and cation exchange. Biochim. Biophys. Acta **821:** 319–333.
3. ANDERSEN, J.P., et al. 1986. Localization of E_1-E_2 conformational transitions of sarcoplasmic reticulum Ca-ATPase by tryptic cleavage and hydrophobic labeling. J. Membr. Biol. **93:** 85–92.
4. JORGENSEN, P.L. & J.P. ANDERSEN. 1988. Structural basis for E_1-E_2 conformational transitions in Na,K-pump and Ca-pump proteins. J. Membr. Biol. **103:** 95–120.
5. JORGENSEN, P.L., J.R. JORGENSEN & P.A. PEDERSEN. 2001. Role of conserved TGDGVND-loop in Mg^{2+} binding, phosphorylation, and energy transfer in Na,K-ATPase. J. Bioenerg. Biomembr. **33:** 367–377.

6. RICE, W.J. et al. 2001. Structure of Na^+,K^+-ATPase at 11-Å resolution: comparison with $Ca2^+$-ATPase in E_1 and E_2 states. Biophys. J. **80:** 2187–2197.
7. HEBERT, H. et al. 2001. Three-dimensional structure of renal Na,K-ATPase from cryo-electron mcroscopy of two-dimansional crystals. J. Mol. Biol. **314:** 479–494.
8. TOYOSHIMA, C. & H. NOMURA. 2002. Structural changes in the calcium pump accompanying the dissociation of calcium. Nature **418:** 605–611.
9. TOYOSHIMA, C. et al. 2000. Crystal structure of the calcium pump of sarcoplasmic reticulum at 2.6 Å resolution. Nature **405:** 647–655.
10. ANDERSEN, J.P. 1995. Dissection of the functional domains of the sarcoplasmic reticulum $Ca(2^+)$-ATPase by site-directed mutagenesis. Biosci. Rep. **15:** 243–261.
11. JORGENSEN, P.L. & P.A. PEDERSEN. 2001. Structure-function relationships of Na(+), K(+), ATP, or Mg(2+) binding and energy transduction in Na,K-ATPase. Biochim. Biophys. Acta **1505:** 57–74.
12. HÅKANSSON, K. & P.L. JORGENSEN. 2003. Homology modeling of Na,K-ATPase: a putative third sodium binding site suggests a relay mechanism compatible with the electrogenic profile of Na^+ translocation. This volume.
13. MOLLER, J.V., B. JUUL & M. LE MAIRE. 1996. Structural organization, ion transport, and energy transduction of P-type ATPases. Biochim. Biophys. Acta **1286:** 1–51.
14. GOLDSHLEGER, R. & S.J. KARLISH. 1999. The energy transduction mechanism of Na,K-ATPase studied with iron-catalyzed oxidative cleavage. J. Biol. Chem. **274:** 16213–16221.
15. PATCHORNIK, G. et al. 2002. The ATP-Mg(2+) binding site and cytoplasmic domain interactions of Na(+),K(+)-ATPase investigated with Fe(2+)-catalyzed oxidative cleavage and molecular modeling. Biochemistry **41:** 11740–11749.
16. JORGENSEN, P.L., K.O. HÅKANSSON & S.J. KARLISH. 2003. Structure and mechanism of Na,K-ATPase: functional sites and their interactions. Annu. Rev. Physiol. In press.
17. GLYNN, I.M., Y. HARA & D.E. RICHARDS. 1984. The occlusion of sodium ions within the mammalian sodium-potassium pump: its role in sodium transport. J. Physiol. **351:** 531–547.
18. PEDERSEN, P.A., J.H. RASMUSSEN & P.L. JORGENSEN. 1996. Expression in high yield of pig alpha 1 beta 1 Na,K-ATPase and inactive mutants D369N and D807N in *Saccharomyces cerevisiae*. J. Biol. Chem. **271:** 2514–2322.

Conformational Dynamics of Na^+/K^+- and H^+/K^+-ATPase Probed by Voltage Clamp Fluorometry

SVEN GEIBEL,[a] DIRK ZIMMERMANN,[a] GIOVANNI ZIFARELLI,[a] ANJA BECKER,[a] JAN B. KOENDERINK,[c] YI-KANG HU,[b] JACK H. KAPLAN,[b] THOMAS FRIEDRICH,[a] AND ERNST BAMBERG[a]

[a]*Max-Planck-Institute of Biophysics, Kennedyallee 70, D-60596 Frankfurt/M., Germany*

[b]*Oregon Health Science University, 3181 S.W. Sam Jackson Park Road, Portland, Oregon 97201, USA*

[c]*Department of Biochemistry, Nijmegen Center for Molecular Life Sciences, University of Nijmegen, P.O. Box 9101, 6500 HB Nijmegen, the Netherlands*

ABSTRACT: We used the method of site-directed fluorescence labeling in combination with voltage-clamp fluorometry for time-resolved recording of localized conformational transitions of the Na^+/K^+-and H^+/K^+-ATPase. Therefore, single cysteine mutations were introduced into the extracellular TM5–TM6 loop of the sheep Na^+/K^+-ATPase α_1-subunit devoid of other extracellular cysteines. Upon expression in *Xenopus* oocytes and covalent attachment of tetramethylrhodamine-maleimide (TMRM) as a reporter fluorophore, Cys-mutant N790C showed large fluorescence changes of up to 5% in response to extracellular K^+ that were completely abolished by ouabain. When voltage jumps were applied under Na^+/Na^+-exchange conditions, we observed fluorescence changes that paralleled the transient currents originating from the $E_1P \leftrightarrow E_2P$ transition. These fluorescence changes were also completely inhibited by ouabain, as were the voltage jump–induced transient currents. Transient fluorescence changes could also be measured as a function of increasing K^+ concentrations, that is, under turnover conditions. As a result, the distribution between E_1 and E_2 states can be determined at any time and membrane potential. Very similar fluorescence signals were obtained for rat gastric H^+/K^+-ATPase upon expression in oocytes, when a single cysteine was introduced at a position homologous to N790 in Na^+/K^+-ATPase for attachment of the fluorophore. As to the high sequence similarity among P-type ATPases within the TM5 helix and the TM5-TM6 loop region, our results enable new means of kinetic investigation for these pumps under physiological conditions in living cells.

KEYWORDS: cysteine labeling; N790C; fluorescence; conformational transition; helix TM5

Address for correspondence: Ernst Bamberg, Max-Planck-Institute of Biophysics, Kennedyallee 70, D-60596 Frankfurt/M., Germany. Voice: +49-69-6303-300; fax: +49-69-6303-305.
bamberg@mpibp-frankfurt.mpg.de

INTRODUCTION

The transduction of primary energy from ATP hydrolysis to active ion transport is brought about by conformational changes.[1–3] Several approaches were undertaken to reveal Na^+/K^+-ATPase conformational changes;[4–9] however, assignment of the observed signals to specific enzymatic reaction steps was difficult, and only rearrangements of rather large domains of the Na^+/K^+-ATPase could be observed. Our aim was to follow site-specific conformational changes in real time and assign them to partial reactions of the Na^+/K^+-ATPase reaction cycle. Therefore, we used the method of site-directed fluorescence labeling in combination with voltage-clamp fluorometry, which was pioneered in the laboratories of Isacoff[10] and Bezanilla[11] to reveal conformational rearrangements corresponding to voltage sensor movements of the *Shaker* K^+-channel. This technique involves electrophysiological measurements on membrane proteins, in which single cysteines are introduced at putative reporter sites. A fluorescent dye with a reactive sulfhydryl-specific moiety is covalently linked to this cysteine. The fluorescence emission of the dye (in our case tetramethylrhodamine-maleimide, TMRM) responds to changes in the hydrophobicity of its environment, which are brought about by the conformational changes.

Modeling of the Na^+/K^+-ATPase primary sequence into the 3-D structure of the SERCA pump[12] shows that the TM5 helix extends from the site of ATP binding/phosphorylation, respectively, into the cation-binding region. Helix TM5 and the subsequent extracellular loop to transmembrane segment TM6 are highly conserved among all P-type ATPases[13] and may be important for the coupling of ATP hydrolysis to active ion transport. Therefore, we chose the extracellular TM5-TM6 loop for insertion of reporter cysteines into a sheep Na^+/K^+-ATPase construct devoid of other extracellular cysteines.

RESULTS/DISCUSSION

From the single cysteine mutants we expressed in *Xenopus* oocytes, most showed pump activity comparable to wild type. After labeling with TMRM (see FIG. 1B) only the N790C mutant responded to extracellular K^+ and voltage jumps with changes in fluorescence intensity. FIGURE 1C shows a parallel recording of pump current and fluorescence of TMRM-labeled mutant N790C. Addition of K^+ induces a stationary current that is accompanied by a fluorescence increase. Both stationary current and fluorescence increase could be inhibited by ouabain. This can be explained in terms of the Albers-Post scheme: In the absence of K^+ or in the presence of ouabain, the Na^+/K^+-ATPase mainly accumulates the E_2P conformation, while in the presence of K^+ the turnover is increased, a stationary current develops, and the E_2P conformation is depleted in favor of E_1 states. As this leads to an increase in fluorescence, the environment of the fluorophore is changed from a quenching, aqueous medium to a sheltered, hydrophobic environment during the E_2P–E_1 transition.

To allow for a more specific attribution of fluorescence signals to partial reactions of the Albers-Post cycle, voltage pulse experiments were carried out under K^+-free conditions, in which the protein is restricted to Na^+/Na^+ exchange (according to a reaction sequence $E_1P(Na^+) \leftrightarrow E_2P(Na^+) \leftrightarrow E_2P + Na^+$). The Na^+/K^+-ATPase then shuttles exclusively between E_1P and E_2P, because dephosphorylation in the absence

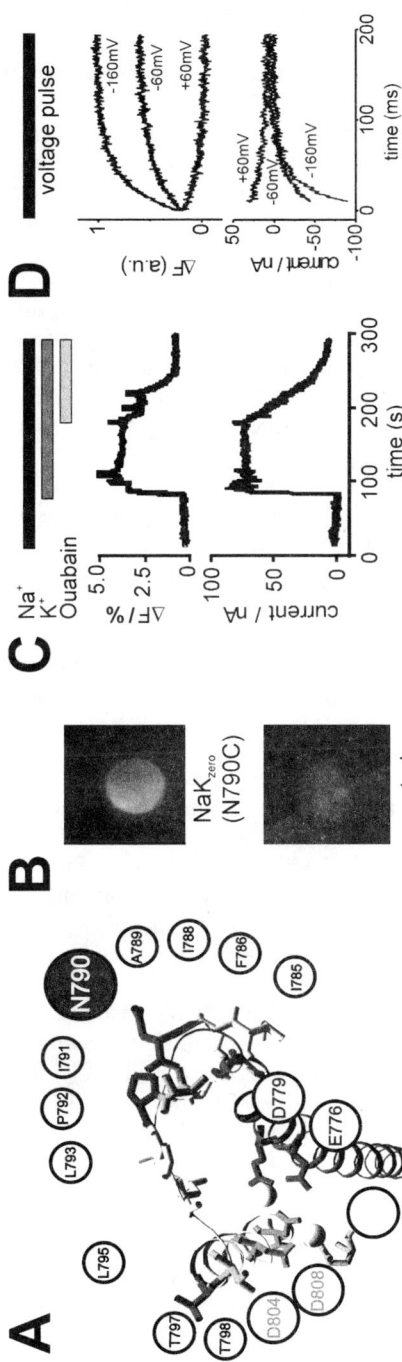

FIGURE 1. (**A**) Model of the TM5 helix and TM5-TM6 loop backbone of the Na+/K+-ATPase according to the 3-D structure of the SERCA pump with ball-and-stick representations of cation-coordinating residues and the residues of the extracellular TM5-TM6 loop. The 2 Ca^{2+} ions of the SERCA structure are represented by spheres. (**B**) Fluorescence images of *Xenopus* oocytes after labeling with TMRM (*upper*: oocyte expressing Na+/K+-ATPase mutant N790C; *lower*: uninjected control oocyte). (**C**) Parallel recording of pump current (*lower trace*) and fluorescence change (*upper trace*) from a single *Xenopus* oocyte expressing Na+/K+-ATPase mutant N790C in response to 10 mM K+ and 5 mM ouabain according to the perfusion protocol as stated. (**D**) *Upper panel*: Fluorescence changes in response to voltage pulses of TMRM-labeled Na+/K+-ATPase mutant N790C under extracellularly K+-free conditions to potential values as stated (from 0 mV holding potential). *Lower panel*: Transient currents in response to voltage pulses recorded in parallel to the above fluorescence traces.

of K^+ is very slow. The main electrogenic events, namely sodium binding/release steps, are intimately linked to the E_1P–E_2P conformational change.[14–17] Therefore, occupancy of these states changes in response to voltage pulses. After perturbation by a voltage jump, relaxation into a new equilibrium occurs, giving rise to transient currents. Positive potentials favor E_2P and release of Na^+ to the extracellular medium, whereas negative potentials enable the reuptake of extracellular Na^+ and subsequent conversion into E_1P.[17]

FIGURE 1D shows results from a parallel recording of transient currents and fluorescence changes of TMRM-labeled Na^+/K^+-ATPase mutant N790C in response to voltage jumps under Na^+/Na^+-exchange conditions ($[K^+]_o = 0$). Positive potentials favor the E_2P conformation driven by release of positive charges to the extracellular side, resulting in a positive transient current. At the same time the fluorescence signal decreases. This observation agrees with the findings of the stationary current/fluorescence data shown in FIGURE 1C. The kinetics of transient currents and fluorescence changes evoked by voltage pulses under these conditons are almost identical. We conclude that the fluorescence signals measure the E_1P–E_2P conformational change. This structural rearrangement is directly correlated to the physiological function of the protein, as extrusion of Na^+ ions is a key step during ion transport. The fluorescence label at position 790 demonstrates the direct involvement of the TM5–TM6 loop in this crucial conformational change. The same voltage jump experiment in the presence of the Na^+/K^+-ATPase–specific inhibitor ouabain did not lead to detectable transient charge movements, and also fluorescence changes were abolished (see FIG. 2A, panel g).

Fluorescence signals could also be measured in the presence of K^+. These conditions prevent the observation of transient currents because the E_2P state, which allows for electrogenic release and rebinding of Na^+, is depleted by K^+-stimulated dephosphorylation. Therefore fluorescence measurements allow one to determine the number of E_1 and E_2 states under turnover conditions of the enzyme (FIG. 2). The data in this FIGURE originated from a continuous recording from an oocyte expressing Na^+/K^+-ATPase mutant N790C, thus also allowing for comparison of the absolute fluorescence levels under the varying conditions of the experiment. In summary, the data provide three main observations: (1) the stationary fluorescence rises upon increase of $[K^+]$; (2) the relative fluorescence changes upon a given voltage jump become smaller with increasing $[K^+]$; and (3) the kinetics of the fluorescence changes are accelerated with increasing $[K^+]$.

The stationary fluorescence level at −80 mV, visible at the beginning of each set of fluorescence signals, rises with increasing K^+ concentration, indicating the onset of saturation between 10 mM and 30 mM extracellular $[K^+]$. This observation corresponds to the increase in stationary fluorescence from FIGURE 2C upon a solution exchange from 0 mM to 30 mM $[K^+]$. FIGURE 2A, panel a shows fluorescence responses to voltage pulses at extracellular $[K^+] = 0$ mM, the same conditions as for FIGURE 1D. Voltage jumps to positive potentials elicited only small fluorescence decreases, which did not change with voltage. By contrast, negative potentials induced strongly voltage-dependent fluorescence increases. With extracellular $[K^+] = 1$ mM (FIG. 2A, panel b), fluorescence signals started from an already increased stationary value. Positive voltage pulses induced fluorescence decreases, and jumps to negative potentials led to fluorescence increases with the onset of saturation becoming apparent at extremely positive or negative voltages. The kinetics of the fluorescence sig-

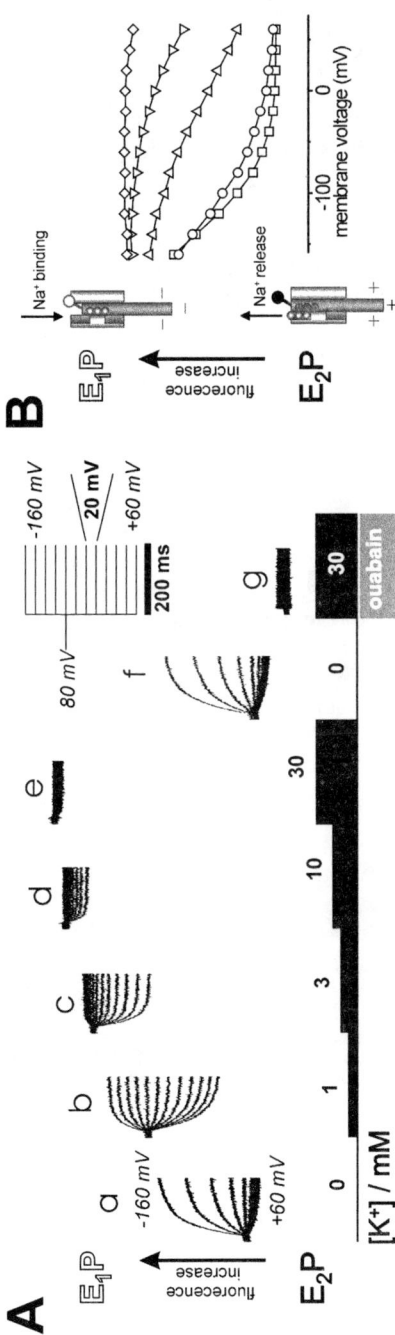

FIGURE 2. Voltage-dependent fluorescence signals of TMRM-labeled H$^+$/K$^+$-ATPase mutant S806C expressed in *Xenopus* oocytes. (**A**) Fluorescence changes in response to the stated voltage step protocol (*right panel*) at an external pH of 7.4 under extracellularly K$^+$-free conditions. (**B**) Complete inhibition of voltage-dependent fluorescence changes on the same oocyte after extracellular addition of 150 μM SCH28080 (voltage protocol as for **A**).

nals was accelerated compared to the data obtained in the absence of K$^+$. With further increasing extracellular [K$^+$] (FIG. 2A, panels c and d), negative voltage pulses only induced small and voltage-independent fluorescence increases, whereas positive potentials led to comparatively large fluorescence decreases. The absolute difference in voltage-dependent fluorescence (ΔF-V curve) continuously decreases with increasing [K$^+$] (see FIG. 2B), and the voltage dependence of the ΔF-V curve shows a progressively positive shift with increasing [K$^+$]. The evaluation of the [K$^+$] dependence of the stationary fluorescence (FIG. 2B) further shows that the slope of the ΔF-V curve declines with increasing [K$^+$]. This is indicative of a reduced voltage sensitivity under conditions in which the E$_2$P conformation is depleted in favor of E$_1$ states. At extracellular [K$^+$] = 30 mM, stationary fluorescence at –80 mV reaches a maximum, and voltage pulses in either direction can no longer induce any change in fluorescence (FIG. 2A, panel e), showing that the voltage dependence of the distribution between E$_1$ and E$_2$ states is completely lost when extracellular K$^+$ is present in saturating concentrations and E$_1$ states accumulate. To test for reversibility, the voltage protocol was applied in the absence of K$^+$ again (FIG. 2A, panel f). Fluorescence traces are equivalent to those obtained under starting conditions (compare FIG. 2A, panels a and f).

Addition of ouabain at [K$^+$] = 30 mM reduces fluorescence to a minimum (FIG. 2A, panel g), even slightly lower than obtained in the absence of potassium at depolarizing voltages, and voltage pulses do not induce fluorescence changes, showing that the pump is arrested in the E$_2$P conformation.

To test for applicability of our experimental approach for other P-type ATPases, we inserted a cysteine within the TM5–TM6 loop of the rat gastric H$^+$/K$^+$-ATPase at the site homologous to N790 of the Na$^+$/K$^+$-ATPase. The gastric H$^+$/K$^+$-ATPase does not carry out net electrogenic transport (presumably export of 2 protons versus

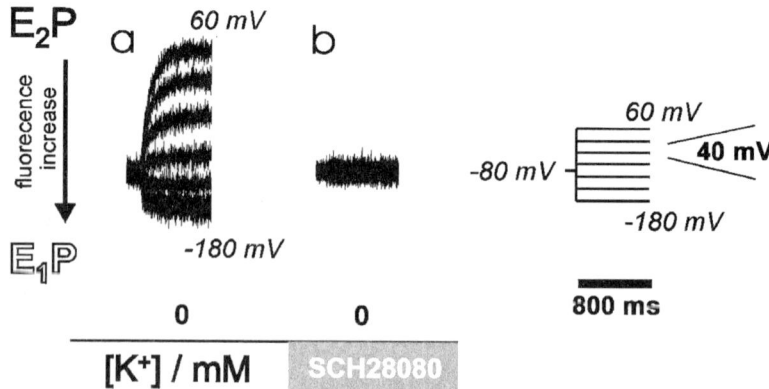

FIGURE 3. Voltage-dependent fluorescence signals of TMRM-labeled H$^+$/K$^+$-ATPase mutant S806C expressed in *Xenopus* oocytes. (**a**) Fluorescence changes in response to the stated voltage step protocol (*right panel*) at an external pH of 7.4 under extracellularly K$^+$-free conditions. (**b**) Complete inhibition of voltage-dependent fluorescence changes on the same oocyte after extracellular addition of 150 μM SCH28080 (voltage protocol as for **a**).

import of 2 K^+ ions), which makes it difficult to carry out kinetic investigations of this enzyme by electrophysiological methods. However, the H^+/K^+-ATPase reaction cycle contains at least two electrogenic steps, one of which had been correlated with H^+ transport in response to the formation of phosphoenzyme after ATP concentration jumps.[18,19] Therefore, under K^+-free conditions it should be possible to disturb the voltage-dependent equilibrium between E_1P and E_2P states by voltage jumps. Upon expression of H^+/K^+-ATPase mutant S806C in *Xenopus* oocytes and subsequent fluorescence labeling by TMRM, we observed transient fluorescence changes in response to voltage jumps under K^+-free conditions, which were very similar to the signals recorded for the Na^+/K^+-ATPase (FIG. 3a). These fluorescence changes could be completely abolished in the presence of the H^+/K^+-ATPase–specific inhibitor SCH28080 (FIG. 3b). This opens up new routes for kinetic investigation of this electroneutrally operating pump under *in vivo* conditions, especially under physiological pH gradients.

CONCLUSIONS

With the voltage clamp fluorometry method it is possible to perform time-resolved measurements of conformational changes of the fully active Na^+/K^+- and H^+/K^+-ATPase under physiological conditions. Furthermore, we are able to specifically locate this conformational change, which involves movement of a reporter fluorophore attached to amino acid position 790 at the extracellular end of helix TM5, to a single reaction step of the reaction cycle, the E_1P–E_2P conformational change. The fluorophore encounters a sheltered, hydrophobic environment in the E_1P state and a more exposed quenching, aqueous medium in the E_2P state. An obvious explanation for such a change in the fluorophore environment would be a movement of helix TM5 relative to the other helices accompanied by an opening of cation binding sites to the extracellular space.

Considering the high homology within the TM5 and TM5–TM6 loop regions among all P-type ATPases, this method offers the possibility for studying ion transport-related conformational transitions of other transporters of that type that have resisted a detailed kinetic investigation so far.

REFERENCES

1. JORGENSEN, P.L. 1975. Purification and characterization of (Na^+,K^+)-ATPase. V. Conformational changes in the enzyme Transitions between the Na-form and the K-form studied with tryptic digestion as a tool. Biochim. Biophys. Acta **401:** 399–415.
2. LUTSENKO, S., R. ANDERKO & J. H. KAPLAN. 1995. Membrane disposition of the M5-M6 hairpin of Na^+,K^+-ATPase alpha subunit is ligand dependent. Proc. Natl. Acad. Sci. USA **92:** 7936–40.
3. MIKHAILOVA, L., A.K. MANDAL & J.M. ARGUELLO. 2002. Catalytic phosphorylation of Na,K-ATPase drives the outward movement of its cation-binding H5-H6 hairpin. Biochemistry **41:** 8195–8202.
4. KARLISH, S.J. 1980. Characterization of conformational changes in (Na,K) ATPase labeled with fluorescein at the active site. J. Bioenerg. Biomembr. **12:** 111–136.
5. SMIRNOVA, I.N. & L.D. FALLER. 1993. Mechanism of K^+ interaction with fluorescein 5′-isothiocyanate-modified Na^+,K^+-ATPase. J. Biol. Chem. **268:** 16120–16123.

6. GLYNN, I.M. *et al.* 1987. Comparison of rates of cation release and of conformational change in dog kidney Na,K-ATPase. J. Physiol. **383:** 477–485.
7. KANE, D.J. *et al.* 1997. Stopped-flow kinetic investigations of conformational changes of pig kidney Na^+,K^+-ATPase. Biochemistry **36:** 13406–13420.
8. TANIGUCHI, K. & S. MARDH. 1993. Reversible changes in the fluorescence energy transfer accompanying formation of reaction intermediates in probe-labeled (Na^+,K^+)-ATPase. J. Biol. Chem. **268:** 15588–15594.
9. KLODOS, I. 1994. Partial reactions in Na^+/K^+- and H^+/K^+-ATPase studied with voltage-sensitive fluorescent dyes. *In* The Sodium Pump. E. Bamberg & W. Schoner, Eds.: 517–528. Steinkopff. Darmstadt.
10. MANNUZZU, L.M., M.M. MORONNE & E.Y. ISACOFF. 1996. Direct physical measure of conformational rearrangement underlying potassium channel gating. Science **271:** 213–216.
11. CHA, A. & F. BEZANILLA. 1997. Characterizing voltage-dependent conformational changes in the Shaker K^+ channel with fluorescence. Neuron **19:** 1127–1140.
12. TOYOSHIMA, C. *et al.* 2000. Crystal structure of the calcium pump of sarcoplasmic reticulum at 2.6 Å resolution. Nature **405:** 647–655.
13. SWEADNER, K.J. & C. DONNET. 2001. Structural similarities of Na,K-ATPase and SERCA, the Ca^{2+}-ATPase of the sarcoplasmic reticulum. Biochem. J. **356:** 685–704.
14. NAKAO, M. & D.C. GADSBY. 1986. Voltage dependence of Na translocation by the Na/K pump. Nature **323:** 628–630.
15. HILGEMANN, D.W. 1994. Channel-like function of the Na,K pump probed at microsecond resolution in giant membrane patches. Science **263:** 1429–132.
16. FRIEDRICH, T. & G. NAGEL. 1997. Comparison of Na^+/K^+-ATPase pump currents activated by ATP concentration or voltage jumps. Biophys. J. **73:** 186–194.
17. HOLMGREN, M. *et al.* 2000. Three distinct and sequential steps in the release of sodium ions by the Na^+/K^+-ATPase. Nature **403:** 898–901.
18. STENGELIN, M., K. FENDLER & E. BAMBERG. 1993. Kinetics of transient pump currents generated by the (H,K)-ATPase after an ATP concentration jump. J. Membr. Biol. **132:** 211–227.
19. VAN DER HIJDEN, H.T. *et al.* 1990. Demonstration of the electrogenicity of proton translocation during the phosphorylation step in gastric H^+K^+-ATPase. J. Membr. Biol. **114:** 245–256.

Investigating the Energy Transduction Mechanism of P-type ATPases with Fe^{2+}-Catalyzed Oxidative Cleavage

STEVEN J. D. KARLISH

Department of Biological Chemistry, Weizmann Institute of Science, Rehovoth, 76100, Israel

ABSTRACT: This paper discusses specific oxidative cleavage of renal Na^+,K^+-ATPase and gastric H^+,K^+-ATPase, catalyzed by bound Fe^{2+} or the complex ATP-Fe^{2+}, and its implication for the energy transduction mechanism of P-type ATPases. The cleavage technique provides information on the spatial organization of the proteins in different conformations and, since ATP-Fe^{2+} substitutes for ATP-Mg^{2+} in activating ATPase activity, on Mg^{2+}-ligating residues in different conformations. The experiments predict the existence of large movements of N, P, and A cytoplasmic domains accompanying E_1-E_2 and E_1P-E_2P conformational transitions—open in E_1 conformations and closed in E_2 conformations. These features fit well with the Ca^{2+}-ATPase crystal structures in E_1 or E_2 conformations and also provide evidence on ATP and Mg^{2+} binding sites that is not available from the structures. By combining information from cleavage experiments with molecular modeling, based on the Ca^{2+}-ATPase structure, features such as an N to P domain interaction in an E_1. ATP-Mg^{2+} conformation can be inferred. The organization of the N, P, and A domains and the ATP and Mg^{2+} binding sites in the different conformational states appears to be essentially similar for Na^+,K^+-ATPase, H^+,K^+-ATPase, and Ca^{2+}-ATPase. The oxidative cleavage technique may be a valuable tool to investigate long-range interactions that transduce the free energy of hydrolysis of ATP to active cation movements in P-type ATPases.

KEYWORDS: Fe^{2+}-cleavage; Na^+,K^+-ATPase; energy transduction mechanism; ATP-Mg^{2+} sites

INTRODUCTION

The Ca^{2+}-ATPase crystal structures, in an E_1Ca conformation[1] and the recently published E_2-thapsigargin bound state,[2] have focussed our attention on the major issues of the energy transduction mechanism of P-type ATPases, as well as providing criteria to judge and plan new experiments. These issues include long-range interactions between ATP and cation sites, by which E_1/E_2 transitions bring about cation movements, and by which cations occluded within trans-membrane segments catal-

---

Address for correspondence: Steven J.D. Karlish, Department of Biological Chemistry, Weizmann Institute of Science, Rehovoth, 76100, Israel. Voice: +972-8-934-2278; fax: +972-8-934-4118.

steven.karlish@weizmann.ac.il

Ann. N.Y. Acad. Sci. 986: 39–49 (2003). © 2003 New York Academy of Sciences.

yse phosphorylation and dephosphorylation. Of course, in the absence of molecular structure of pumps such Na^+,K^+-ATPase and H^+,K^+-ATPase, with additional beta subunits, or regulatory FXYD proteins in the case of Na^+,K^+-ATPase, additional techniques are needed to look at pump-specific features, such as cation occlusion sites and subunit interactions (reviewed in Ref. 3).

This article summarizes our work on Fe^{2+}- or ATP-Fe^{2+} catalyzed cleavage.[4–9] Bound Fe^{2+} or ATP-Fe^{2+} catalyze generation of hydroxyl radicals in the presence of ascorbate/hydrogen peroxide by the Fenton reaction, and the radicals specifically cleave the polypeptide chain. Because more than one bond can be cut, the different cleavage positions must all be in proximity to the bound Fe^{2+} and thus to each other, leading to conclusions on structural organization.

SELECTIVITY OF CLEAVAGE PATTERNS AND IDENTIFICATION OF FRAGMENTS

TABLE 1 catalogues the fragments of renal Na^+,K^+-ATPase, produced in different conditions, by their N-terminal sequences and domain locations (N, P, A, or M) as defined for Ca^{2+}-ATPase.[1] The data refer to pig kidney Na^+,K^+-ATPase but oxidative cleavages of gastric H^+,K^+-ATPase are essentially identical.[7] The exact N-terminus was determined by sequencing in a few cases, but usually this is impossible due to blocked N-termini. In the latter cases, the location is referred to as "near" an N-terminus, determined to within about 10 residues by comparison with standard proteolytic fragments. Recently we have determined the masses by MALDI-TOF mass spectrometry, after developing a procedure for extracting fragments from gels and exchanging SDS for deoxycholate, which is compatible with MALDI-TOF.[10] By comparing masses of intact α subunit (112,198±491 Da ($n=4$), theoretical 112,281 Da) and known Fe^{2+}-cleavage fragments (e.g., V^{712}-Y^{1017} 34,807±97 ($n=7$), theoretical 34,552 Da; G1-G213 23,275±62 ($n=7$), theoretical 23,449) with those of unknown fragments, the mass accuracy of blocked fragments is determined within 2–5 residues. This increased accuracy of the assignments is important since it allows one to make structural predictions with increased confidence.

We are also developing electrospray MS techniques with greatly improved accuracy. The chemistry of cleavages is thought to involve radical reactions in alternative "α-amidation" and "diamide" pathways.[11] In addition protein side-chains can be oxidized by OH radicals.[11] Thus, accurate ESI-MS mass information will probably reveal heterogeneity of the N-termini. Nevertheless, the high specificity of the cleavages implies that they occur at residues that interact directly with bound Fe^{2+} (see also Ref. 8 for recent evidence using a new probe fluorescein-DTPA).

CLEAVAGES MEDIATED BY BOUND Fe^{2+} IONS

Incubation of renal Na^+,K^+-ATPase in an $E_2(K)$ conformation with Fe^{2+}, ascorbate, and H_2O_2 leads to six fragments of the α subunit while only two of the fragments are seen in E_1Na and E_1 conformations (TABLE 1). β and γ subunits are not cleaved. Kinetic features and insensitivity to radical scavengers are indicative of a site-specific cleavage mechanism.[4] In fact, parallel behavior of four conformationally sensitive cleavages compared to two conformationally insensitive cleavages is indicative of two separate Fe^{2+} sites (1 and 2). Experimental support for a two Fe^{2+} site mechanism was obtained from observations that a bathophenanthroline-Cu^{2+} complex catalyzes oxidative cleavages exactly like Fe^{2+} at site 2, but not at site 1.[12]

The conformationally sensitive cleavages at Fe^{2+} site 1 are all located at or near highly conserved cytoplasmic sequences: ^{214}ESE in the A domain; and ^{712}VNDS, near ^{608}MVTGD, and ^{367}CSDK in the P domain (TABLE 1). On the basis of these data we predicted that in the E_2 conformation these sequences are in proximity, and that the cytoplasmic domains interact, while in E_1 the domains are separated and preclude cleavages. In view of the identical cleavage of Na^+,K^+-ATPase and H^+,K^+-ATPase and involvement of the most highly conserved cytoplasmic sequences, the mechanism should apply to all P-type pumps.

When the Ca^{2+}-ATPase crystal structure became available[1] and then molecular models in an E_2-vanadate conformation,[1,13] it became possible to test the predictions. In the E_1 conformation N, P, and A domains are well separated, whereas in the E_2 model the N, P, and A domains are gathered together, A docking onto both N and P domains. In E_2-vanadate models the relevant residues in site 1 are all close to each other, shown for the 1kju structure in FIGURE 1 (black backbone). In E_1 the domains

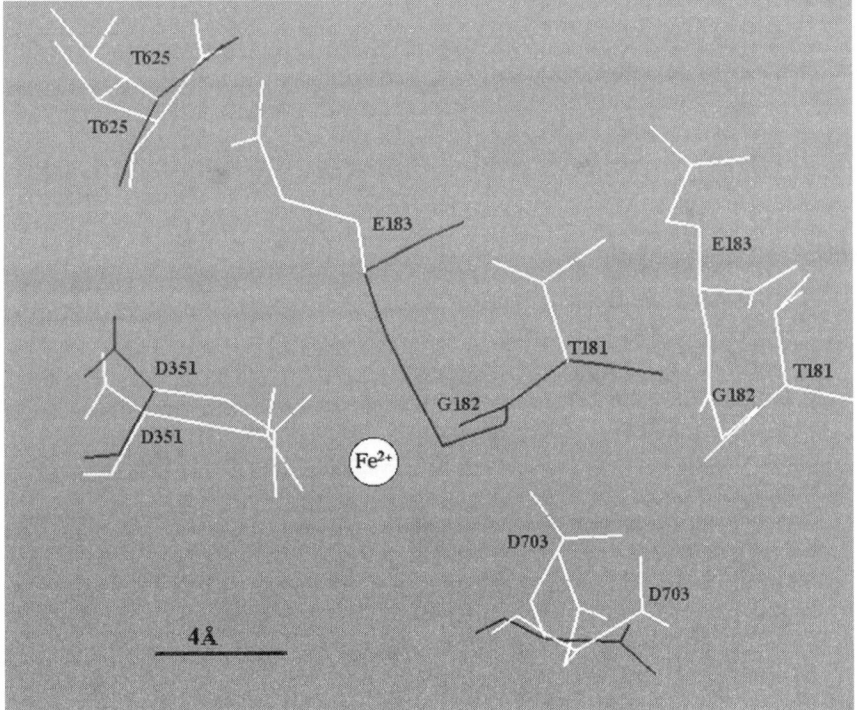

FIGURE 1. Proximity of conserved residues of P and A domains in a model of E_2-van (1kju) and the crystal structure in the E_2-TG conformation (1iwo). The backbone of the 1kju structure is in black, while that of the 1iwo structure is in white. The Fe^{2+} has been arbitrarily placed centrally between the P and A domains.

TABLE 1. Oxidative cleavage fragments of renal Na,K-ATPase observed in different conformational states[a]

Condition	Conformation	Fe^{2+} Site	N-termini of Fragments with Domain Locations	N-termini of Fragments with Domain Locations	Ref.
K, Rb, Cs, Tl or low μ	$E_2(X)_{occ}$ E_2	Fe^{2+} Site 1	^{214}ESE (A) *** nr^{608}MVTGD (P) **	nr^{367}CSDK (P) ^{712}VNDS (P) ***	4 5 7
		Site 2	nr^{81}EWVK (M1) **	nr^{283}HFIH (M3) ***	
Na or high μ	E_1Na E_1	Fe^{2+} Site 2	nr^{81}EWVK (M1) **	nr^{283}HFIH (M3) ***	4 5 7
Pi (low μ)	E_2·Pi	Fe^{2+} Site 1 Site 2	^{214}ESE (A) *** nr^{81}EWVK (M1) **	^{712}VNDS(P) *** nr^{283}HFIH (M3) ***	5 7
Pi/Mg/ouabain or Vanadate/Mg	E_2-P·Mg· ouabain E_2-V·Mg	Fe^{2+} Site 2	nr^{81}EWVK (M1) **	nr^{283}HFIH (M3) ***	5 7
ATP, Fe^{2+} (high μ)	E_1-ATP-Fe^{2+}	ATP-Fe^{2+}	^{712}VNDS (P) ***	nr ^{440}VAGDA (N) and 460–490 *	6 7 8
ATP, Fe^{2+} Na Steady state	E_1Na·ATPFe^{2+} E_1-P E_2-P	ATP-Fe^{2+} E_1P·Fe^{2+} E_2-P·Fe^{2+}	^{712}VNDS (P) *** nr ^{214}ESE (A) **	nr ^{440}VAGDA (N) *	6 7
ATP, Fe^{2+} Na Transient	E_2-P	E_2-P·Fe^{2+}	nr ^{214}ESE (A) ***	nr ^{712}VNDS (P) *	6
Rb (K) ATP, Fe^{2+}	E_2(Rb)·ATP-Fe^{2+}	ATP-Fe^{2+}	nr ^{214}ESE (A) **	n nr ^{440}VAGDA (N) and 460–490 *	8

[a]Numbering is for the pig α1 subunit of Na,K-ATPase. For essentially identical fragments of hog gastric H,K-ATPase, see Ref. 7. High μ = high ionic strength; low μ = low ionic strength.
Fragment intensity: *** = major; ** = intermediate; * = minor.

are well separated, as predicted (not shown). It is of interest that in the new E_2-thapsigargin structure, 1iwo,[2] the A domain does not come into the same degree of proximity, although the P domain residues are essentially superimposable (FIG. 1, white backbone). This could imply that bound vanadate/Mg^{2+} draws the A domain closer to the P domain, as a characteristic of E_2-vanadate or E_2-P forms. Similarly, bound Fe^{2+} may also draw the domains together in unphosphorylated E_2 conformations.

Phosphate selectively suppresses the cleavages near ^{367}CSDK and ^{608}MVTGD, while Pi/Mg^{2+}/ouabain or vanadate/Mg^{2+} also suppress cleavages at ^{214}ESE and ^{712}VNDS, leaving only Fe^{2+} site 2 cleavages intact (see TABLE 1).[5,7] These conditions stabilize E_2 conformations, but the cleavage patterns are typical of E_1. As an explanation of this paradox, Pi was proposed to bind to D369 and to a residue in ^{608}MVTGD, and Mg^{2+} to residues in ^{214}ESE and ^{712}VNDS, and sterically hinder access of the Fe^{2+}. This hypothesis is supported by the crystal structure of the haloacid dehalogenase (HAD) family of proteins, which also uses an active site carboxylate, and are homologous to P-type pumps in the P domain.[14,15] Pi indeed interacts with residues homologous to D351 and T624 of Ca^{2+}-ATPase, and Mg^{2+} with the residue homologous to D703 and also the Pi. Thus Fe^{2+} in site 1 may be equivalent to the Mg^{2+} site in the absence of Pi or ATP. The involvement of the TGES loop cannot be tested since HAD proteins lack this loop.

The N-termini of Fe^{2+} site 2 fragments, near ^{283}HFIH and ^{81}EWVK before the entrance of M3 and M1, respectively, were predicted to be in proximity in both E_1 and E_2 states.[4] The E_1Ca crystal structure shows proximity of M3 and M1 near the cytoplasmic surface, and a homology model, built with the trans-membrane sequence of Na^+,K^+-ATPase, confirms proximity at one position where the histidines in M3 (^{283}HFIH) meet E81 (^{81}EWVK) near M1.[13] Thus the M3 plus M1 arrangements of Na,K-, H, K-, and Ca^{2+}-ATPase are similar. In E_2 the stalk segments leading into M3 and especially M1 undergo substantial movements, but not M3 itself.[2] E258, which is homologous to H286 of Na^+,K^+-ATPase, still comes into proximity with M1, which may explain conformation-insensitive cleavages.

THE ATP-Fe^{2+} COMPLEX ACTS AS A SPECIFIC AFFINITY CLEAVAGE REAGENT

The ATP-Fe^{2+} complex can substitute for ATP-Mg^{2+} as a substrate for Na^+,K^+-ATPase. ATP-Fe^{2+} has been found to catalyze specific cleavages within the normal ATP-Mg^{2+} sites, permitting detection of Mg^{2+} binding sequences in the different states of the catalytic cycle.[6-8]

TABLE 1 (rows 6-9) catalogues fragments of Na^+,K^+-ATPase produced by specific ATP-Fe^2-mediated cleavages. Essentially similar results were observed for gastric H^+,K^+-ATPase.[7] In a high ionic strength medium, which stabilizes an E_1 conformation, a major fragment with N-terminus ^{712}VNDS (P domain) and two less prominent fragments near ^{440}VAGDA and between residues 460-490 (N domain) are produced. A similar result is observed with a nonhydrolyzable analogue AMPPNP-Fe^{2+}. In a Na-containing medium, E_1Na, the ^{712}VNDS and near ^{440}VAGDA fragments are observed, but the fragment between 460-490 is not observed. This small but distinct difference between the cleavage in E_1 and E_1Na conformation appears to reflect a Na-specific conformational change in the N domain. Overall ATP-Fe^{2+}-catalyzed cleavages in E_1 conformations lead to the prediction that the Fe^{2+} in bound ATP-Fe^{2+} interacts with residues near ^{712}VNDS (P domain) and also near ^{440}VAGDA (N domain), that is, P and N domains interact in the E_1-ATP-Fe^{2+} conformation.

In the presence of Na ions, ATP, and Fe^{2+}, the enzyme is phosphorylated.[17] In a steady-state condition, with both phosphorylated and unphosphorylated forms present, a band near ^{214}ESE is seen in addition to the ^{712}VNDS and near ^{440}VAGDA bands. The near ^{214}ESE band is not seen in the presence of ATP, Fe^{2+}, Na, and oligomycin, a condition in which E_1-P $\Rightarrow$ E_2-P is blocked, or with AMPPNP, while the ^{712}VNDS and near ^{440}VAGDA bands are produced. These observations indicated that the ^{214}ESE fragment is associated with cleavage of E_2-P. Mg^{2+} ions are tightly bound in E_2-P and dissociate upon E_2-P hydrolysis.[17] The same could be predicted for Fe^{2+} ions. By incubating the enzyme with ATP-Fe^{2+} and Na^+, then chelating free Fe^{2+} with Desferal to arrest phosphorylation, the E_2-P form can be isolated transiently. Addition of ascorbate/H_2O_2, even in the presence of the Fe^{2+} chelator, leads to cleavages of E_2-P prior to its hydrolysis. The major cleavage is near ^{214}ESE, while that at ^{712}VNDS is much less prominent. Addition of Rb together with the ascorbate/H_2O_2 prevents the cleavage, while addition of ouabain after ATP/Fe^{2+}, but before the Desferal, amplifies and stablizes the ^{214}ESE fragment. These characteristics are expected for tight binding of Fe^{2+} to E_2-P and release of bound Fe^{2+} upon E_2-P hydrolysis. In E_2-P there is a striking change in ligation of Fe^{2+} or Mg^{2+} ions, which bind primarily in the ^{212}TGESE sequence (A domain), while ^{712}VNDS (P domain) is less important. In addition, the near ^{440}VAGDA fragment is not observed, suggesting that the N and P domains do not interact, as in E_1.

ATP-Fe^{2+} binding to E_2(K) with a low affinity is the first step of the catalytic cycle.[18] In an E_2(Rb) conformation ATP-Fe^{2+} catalyzes cleavage near ^{214}ESE, near ^{440}VAGDA, and between 460-490.[8] Thus the prediction is that A and N domains interact and, with ATP-Fe^{2+} bound, residues near the sequences ^{214}ESE (A), ^{440}VAGDA, and 460-490 (N) are in proximity to each other and bound Fe^{2+}.

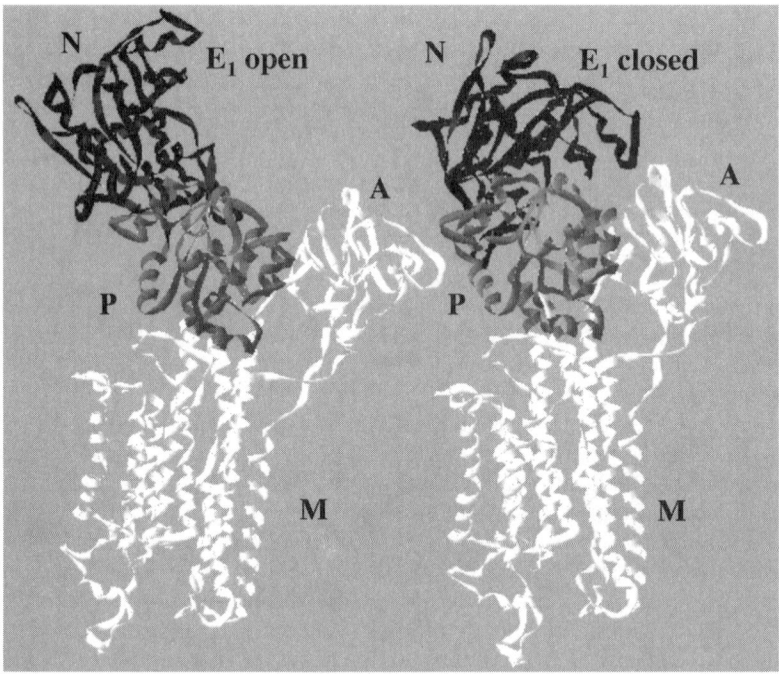

FIGURE 2. E_1 states of Ca^{2+}-ATPase with open (*left*) or closed (*right*) conformations of the N and P domains. The left-hand model is the standard 1eul structure. The right-hand model is the form obtained by docking ATP-Mg^{2+} and causing the N to P domain closure. The ATP- Mg^{2+} molecule is omitted.

MOLECULAR MODELING OF BOUND ATP-Fe^{2+} (See Ref. 19)

The N and P domains must close up when ATP is bound, because the purine ring of ATP is bound in the N domain and the gamma phosphate must approach D351.[1] Cleavages of E_1·ATP-Fe^{2+} provide a constraint to test this by molecular modeling. The procedure involved docking a molecule of ATP-Mg^{2+} in the crystal structure and rotating the N domain about 80° around bonds connecting the N and P domains.[8] FIGURE 2 illustrates the N to P domain movement by comparing the open and closed forms of E_1 (ATP-Mg^{2+} is omitted). The detailed view in FIGURE 3 depicts residues known to participate in binding ATP (N domain) and the γP (P domain) and putative Mg^{2+} binding residues. The purine ring of ATP is docked in the N domain and the γP is next to D351, although this is not shown. The Mg^{2+} is 3Å from D703. Mutation of the homologous D710 of Na^+,K^+-ATPase shows that it is crucial for Mg^{2+} binding,[20] and the homologous aspartate of HAD proteins is known to bind Mg^{2+}.[15] T353 is also close to the Mg^{2+} (see Ref. 21). E439 is located in the N domain at the end of a loop which lines the ATP binding pocket.[3] E439 comes within 3 Å of the Mg^{2+}. The homologue residue of Na^+,K^+-ATPase is D443 in VAGDA. Thus, the modeling strongly supports the prediction of N to P closure and points to D443 as a candidate for Mg^{2+} binding in Na^+,K^+-ATPase in addition to D710.

The E_2·thapsigargin structure[2] shows close approach of N and A domains, specifically E439 (N) to V185 (A), consistent with the prediction based on cleavage of

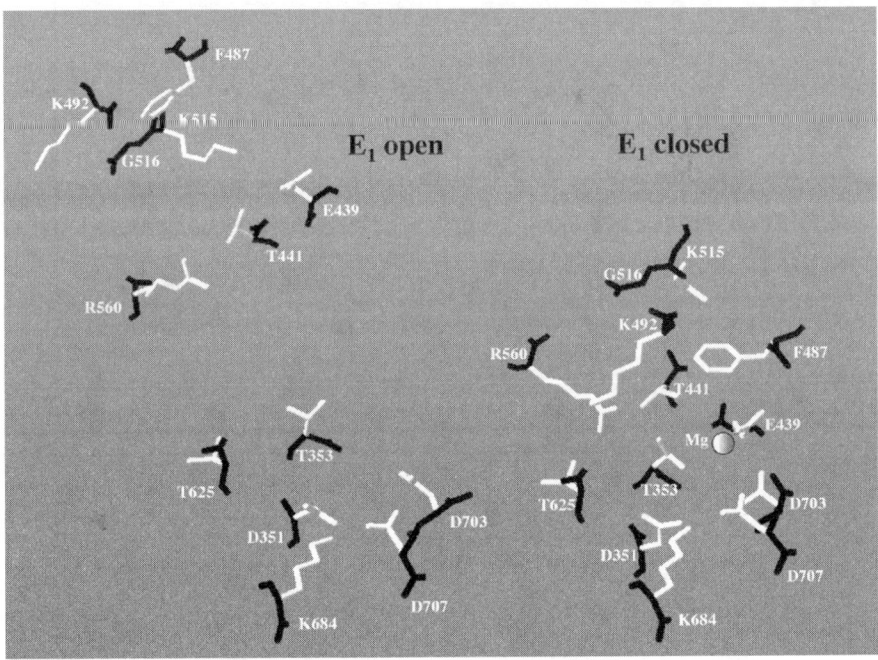

FIGURE 3. Residues predicted to be involved in binding of ATP-Mg^{2+} depicted in the open (*left*) or closed (*right*) forms of the E_1 conformation. Residue numbering is for SERCA1.

the $E_2(Rb) \cdot ATP-Fe^{2+}$ form (TABLE 1). Modeling of the E_2 model (1kju), with ATP-Mg^{2+} bound, suggests that the magnesium ion may chelate E216 and the D443 side chains of Na^+,K^+-ATPase.[18]

CONFORMATION-DEPENDENT DOMAIN INTERACTIONS AND ATP-Mg^{2+} BINDING SITES

FIGURE 4 depicts schematically domain interactions and the ATP-Mg^{2+} binding residues in the different conformations, as inferred from oxidative cleavages. In E_1

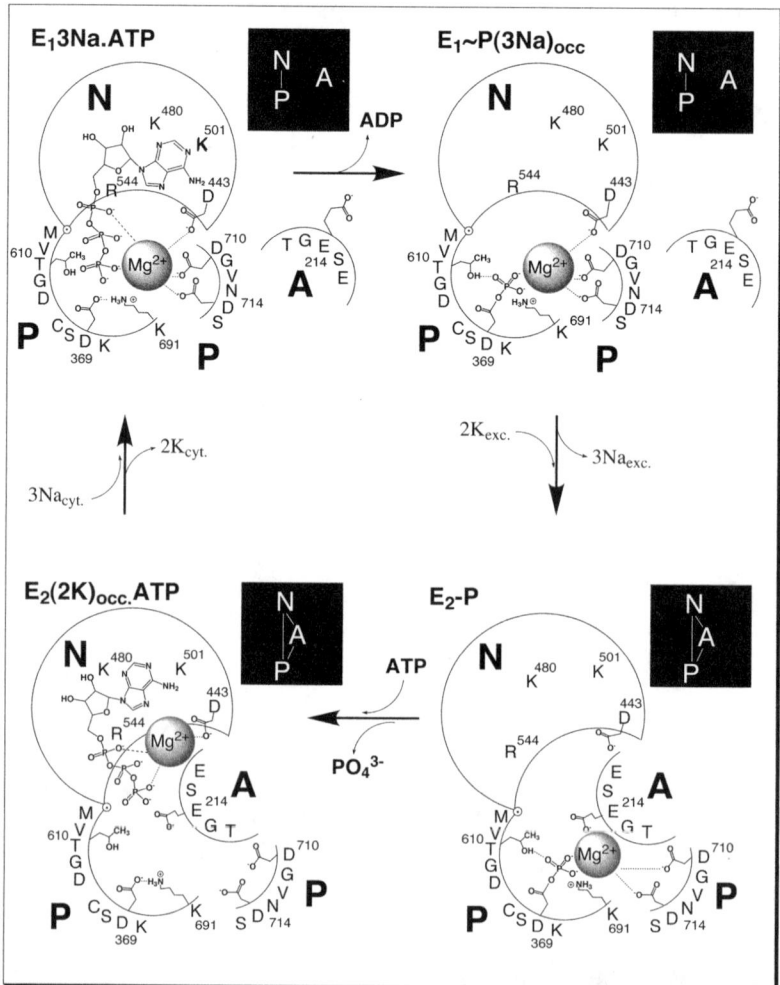

FIGURE 4. Schematic diagrams of N, P, A interactions and ATP-Mg^{2+} or Mg^{2+} binding sites in the different states of the catalytic cycle. The proposed role for D443 in Mg^{2+} binding is still hypothetical at his stage. (Reproduced with permission from Patchornik et al.[8])

and E_1-P, N and P interact, and A is displaced to one side. Mg^{2+} is bound to the gamma P, D710, and we propose to D443. In E_2-P, A interacts with both P and N and interferes with the close N-to-P interaction. The Mg^{2+} undergoes a major change in ligation towards the ^{212}TGES loop. The changed geometry of Mg^{2+} ligation, and a more hydrophobic environment in E_2-P^{22}, may be crucial for the change in chemical reactivity from ADP sensitivity of E_1-P to water sensitivity of E_2-P. This property is essential for any cation pump that maintains tight coupling of ATP hydrolysis and cation movements.[23] Upon E_2-P hydolysis, Mg^{2+} dissociates. In $E_2(K)$, ATP-Mg^{2+} then rebinds to the N and A domains. Low affinity ATP binding in $E_2(K)$ can be compared to the high affinity in E_1, in which ATP-Mg^{2+} binds to both N and P domains. In $E_2(K)$, the A domain hinders binding of ATP-Mg^{2+} to the P domain. In effect, the A domain competes with ATP for binding to the P domain, accounting at least in part for the low affinity.

ONE OR TWO ATP-Mg^{2+} SITES?

Suggestions that the Na^+,K^+-ATPase contain coexisting high- and low-affinity ATP sites (reviewed in Ref. 24) have been tested by comparing cleavages at high (500 μM) or low (<5 μM) ATP-Fe^{2+} concentrations. Cleavages are identical at both concentrations and thus do not support the hypotheses of two independent ATP-Mg^{2+} sites[8] (see also Ref. 25).

CONCLUSIONS AND PERSPECTIVES

Oxidative cleavages, particularly in combination with molecular modeling, provide novel insights into the domain movements and ATP-Mg^{2+} binding sites. The cleavage technique may be a valuable tool to investigate the mechanism whereby cytoplasmic domain interactions associated with E_1/E_2 conformational transitions bring about movements of trans-membrane segments and active cation transport. Predictions on roles of specific residues are now being tested by cleavage of mutant Na^+,K^+-ATPase expressed in *P. pastoris* (see Ref. 26), and recombinant Na^+,K^+-ATPase with newly engineered Fe^{2+} sites is being planned.

ACKNOWLEDGMENTS

This work was supported by a grant from the Israel Science Foundation 15/00.

REFERENCES

1. TOYOSHIMA, C., M. NAKASAKO, H. NOMURA & H. OGAWA. 2000. Crystal structure of the calcium pump of sarcoplasmic reticulum at 2.6 Å resolution. Nature **405:** 647–655.
2. TOYOSHIMA, C. & H. NOMURA. 2002. Structural changes in the calcium pump accompanying the dissociation of calcium. Nature **418:** 605–611.
3. JORGENSEN, P.L., K.O. HÅKANSSON & S.J.D. KARLISH. 2003. Structure and mechanism of Na^+,K^+-ATPase: Functional sites and their interactions. Annu. Rev. Physiology. In press.

4. GOLDSHLEGER, R. & S.J.D. KARLISH. 1997. Fe^{2+}-catalysed cleavage of the α subunit of Na/K-ATPase. Evidence for conformation-sensitive interactions between cytoplasmic domains. Proc. Natl. Acad. Sci. USA **94:** 9596–9601.
5. GOLDSHLEGER, R. & S.J.D. KARLISH. 1999. The energy transduction mechanism of Na^+,K^+-ATPase studied with Fe^{2+}-catalyzed oxidative cleavage. J. Biol. Chem. **274:** 16213–16221.
6. PATCHORNIK, G., R. GOLDSHLEGER & S.J.D. KARLISH. 2000. The complex ATP-Fe^{2+} serves as a specific affinity cleavage reagent in ATP-Mg^{2+} sites of Na^+,K^+-ATPase: altered ligation of Fe^{2+} (Mg^{2+}) ions accompanies the $E_1P \rightarrow E_2P$ conformational change. Proc. Natl. Acad. Sci. USA **97:** 11954–11959.
7. SHIN, J.M. et al. 2001. Selective Fe^{2+}-catalyzed oxidative cleavage of gastric H^+,K^+-ATPase. Implications for the energy transduction mechanism of P-type cation pumps. J. Biol. Chem. **276:** 48440–48450.
8. PATCHORNIK, G. et al. 2002. The ATP-Mg^{2+} binding site and cytoplasmic domain interactions of Na^+,K^+-ATPase investigated with Fe^{2+}-catalyzed oxidative cleavage and molecular modeling. Biochemistry **41:** 1740–11749.
9. GOLDSHLEGER, R. et al. 2001. Structural organization and energy transduction mechanism of Na^+,K^+-ATPase studied with transtion metal-catalyszed oxidative cleavage. J. Bioenerg. Biomembr. **35:** 387–399.
10. GOLDSHLEGER, R. et al. 2002. Application of a "top-down" approach for the analysis of selective Fe^{+2}-catalyzed oxidative cleavages of Na,K-ATPase from pig kidney by mass spectrometry. Proceedings of the 50th ASMS Conference on Mass Spectrometry and Allied Topics. Orlando, Florida. June 2–6, 2002.
11. BERLETT, B.S. & E.R. STADTMAN. 1997. Protein oxidation in aging, disease, and oxidative stress. J. Biol. Chem. **272:** 20313–20316.
12. TAL, D.M. et al. 2001. Proximity of transmembrane segments M3 and M1 of the alpha subunit of Na^+,K^+-ATPase revealed by specific oxidative cleavage mediated by a complex of Cu^{2+} ions and 4,7-diphenyl-1,10-phenanthroline. Biochemistry **40:** 12505–12514.
13. XU, C., W.J. RICE, W. HE & D.L. Stokes. 2002. A structural model for the catalytic cycle of Ca^{2+}-ATPase. J. Mol. Biol. **61:** 884–894.
14. ARAVIND, L., M.Y. GALPERIN & E.V. KOONIN. 1998. The catalytic domain of the P-type ATPase has the haloacid dehalogenase fold. Trends Biol. Sci. **23:** 127–129.
15. WANG, W. et al. 2001. Crystal structure of phosphoserine phosphatase from Methanococcus jannaschii, a hyperthermophile, at 1.8 resolution. Structure **9:** 65–71.
16. RENDI, R. & M.L UHR. 1964. Sodium, potassium requiring adenosinetriphosphatase activity. I. Purification and properties. Biochim. Biophys. Acta **89:** 520–531.
17. FUKUSHIMA, Y. & R.L. POST. 1978. Binding of divalent cation to phosphoenzyme of sodium- and potassium-transport adenosine triphosphatase. J. Biol. Chem. **253:** 6853–6862.
18. POST, R.L., C. HEGYVARY & S. KUME. 1972. Activation by adenosine triphosphate in the phosphorylation kinetics of sodium and potassium ion transport adenosine triphosphatase. J. Biol. Chem. **247:** 6530–6540.
19. MUNSON, K. et al. 2003. Molecular modeling of SCH28080 binding to the gastric H,K-ATPase and MgATP interaction with the SERCA- and Na,K-ATPases. Ann. N. Y. Acad. Sci. This volume.
20. PEDERSEN, P.A., J.R. JORGENSEN & P.L. JORGENSEN. 2000. Importance of conserved alpha-subunit segment 709GDGVND for Mg^{2+} binding, phosphorylation, and energy transduction in Na^+,K^+-ATPase. J. Biol. Chem. **275:** 37588–37595.
21. CLAUSEN, J.D. et al. 2001. Importance of Thr-353 of the conserved phosphorylation loop of the sarcoplasmic reticulum Ca^{2+}-ATPase in MgATP binding and catalytic activity. J. Biol. Chem. **276:** 35741–35750.
22. DE MEIS, L. 1985. Role of water in processes of energy transduction: Ca^{2+}-transport ATPase and inorganic pyrophosphatase. Biochem. Soc. Symp. **50:** 97–125.
23. JENCKS, W.P. 1989. How does a calcium pump pump calcium? J. Biol. Chem. **264:** 18855–18858.
24. TANIGUCHI, K. et al. 2001. The oligomeric nature of Na/K-transport ATPase. J. Biochem. (Tokyo) **129:** 335–342.

25. MARTIN, D.W. & J.R. SACHS. 2000. Ligands presumed to label high affinity and low affinity ATP binding sites do not interact in an (alpha beta)$_2$ diprotomer in duck nasal gland Na$^+$,K$^+$-ATPase, nor do the sites coexist in native enzyme. J. Biol. Chem. **275:** 24512–24517.
26. STRUGATSKY, D. 2003. Expression of Na,K-ATPase in *P. pastoris*: Fe^{2+}-catalyzed cleavage of the recombinant enzyme. Ann. N. Y. Acad. Sci. This volume.

Importance of Transmembrane Segment M3 of Na$^+$,K$^+$-ATPase for Control of Conformational Changes and the Cytoplasmic Entry Pathway for Na$^+$

BENTE VILSEN AND MADS TOUSTRUP-JENSEN

Department of Physiology, University of Aarhus, DK-8000 Aarhus C, Denmark

ABSTRACT: A series of mutations were introduced into the sequence Glu^{282}-Ile-Glu-His-Phe-Ile-His^{288} of the NH$_2$-terminal part of M3 of the rat kidney Na$^+$,K$^+$-ATPase, and the resulting mutant pumps were analyzed functionally. Several of the mutations affected the conformational transitions between E_1 and E_2 forms of dephospho- and phosphoenzyme. Mutations to Glu^{282} and Phe^{286} affected the E_1-E_2 and E_1P-E_2P equilibria in parallel, indicating a role for these two residues in both conformational changes. Mutation to His^{285} preferentially affected the E_1P-E_2P equilibrium, and mutation to Ile^{283} affected only the E_1-E_2 equilibrium of the dephosphoenzyme, demonstrating that the conformational changes of M3 in the phospho- and dephospho-forms are not identical. Several of the mutants showed a reduced apparent affinity for Na$^+$, pointing to an important role of the region in optimizing the Na$^+$ binding properties of the enzyme. It is possible that this part of M3 is closely associated with an entry pathway through which the Na$^+$ ions pass from the cytoplasmic surface to reach the cation-binding pocket. Some of the mutants also displayed an increased Na$^+$-ATPase activity, and a good correlation was observed between the turnover rate for Na$^+$-ATPase activity and the rate of dephosphorylation in the absence of K$^+$, indicating an increased ability of Na$^+$ to activate dephosphorylation of E_2P by binding in place of K$^+$ at the extracellularly facing sites. Thus, M3 also seems to be a part of the signaling pathway between the external cation sites and the catalytic site.

KEYWORDS: Na$^+$,K$^+$-ATPase; mutagenesis; M3; cation interaction; cytoplasmic entry pathway; conformational changes; Na$^+$-ATPase activity

Previous mutational studies of P-type ATPases have identified highly conserved residues with oxygen-containing side chains in the transmembrane segments M4, M5, and M6 as crucial to cation binding and occlusion.[1] Indeed, the high-resolution crystal structure of the Ca^{2+}-ATPase with bound Ca^{2+} revealed that these residues donate

---

Address for correspondence: Dr. Bente Vilsen, Department of Physiology, University of Aarhus, Ole Worms Allé 160, DK-8000 Aarhus C, Denmark. Voice: +45-8942-2832; fax: +45-8612-9065.

bv@fi.au.dk

Ca^{2+} ligands in the binding pocket.[2] Although considerable information has been obtained concerning the cation binding residues, much remains to be learned about the conformational changes that couple the binding and release of the cations to the phosphorylation and dephosphorylation reactions occurring in the catalytic site more than 40 Å away. Furthermore, the pathways for migration of the cations to and from the membrane-embedded ligands have not been identified, and these pathways most likely contribute to define the cation-binding properties, including the ability to select particular ions.

The NH$_2$-terminal part of transmembrane helix M3 is located about midway between the catalytic site and the cation-binding residues in M4, M5, and M6. In the Ca^{2+}-ATPase the NH$_2$-terminal part of M3 seems to occupy a strategic position that allows it to interact with domain P as well as the cytoplasmic loop ("L6-7") connecting M6 and M7.[2] In addition to a few conserved residues, M3 contains several polar and charged residues that are not conserved between the Ca^{2+}-ATPase and the Na$^+$,K$^+$-ATPase, and which, therefore, might contribute to defining the cation selectivity or other cation-specific aspects of these enzymes. A particularly interesting sequence in the NH$_2$-terminal part of M3 of the Na$^+$,K$^+$-ATPase rat α_1-isoform is Glu^{282}-Ile-Glu-His-Phe-Ile-His^{288} (FIG. 1). It is remarkable that Glu^{282} of the Na$^+$,K$^+$-ATPase is substituted by a side chain of opposite charge, Lys^{252}, in the Ca^{2+}-ATPase. In the E_1Ca$_2$ crystal structure of the Ca^{2+}-ATPase, Lys^{252} is within hydrogen bonding/van der Waals interaction distance of loop L6-7,[2] whereas in the recently published E_2 crystal structure, Lys^{252} has moved away from L6-7.[3] The

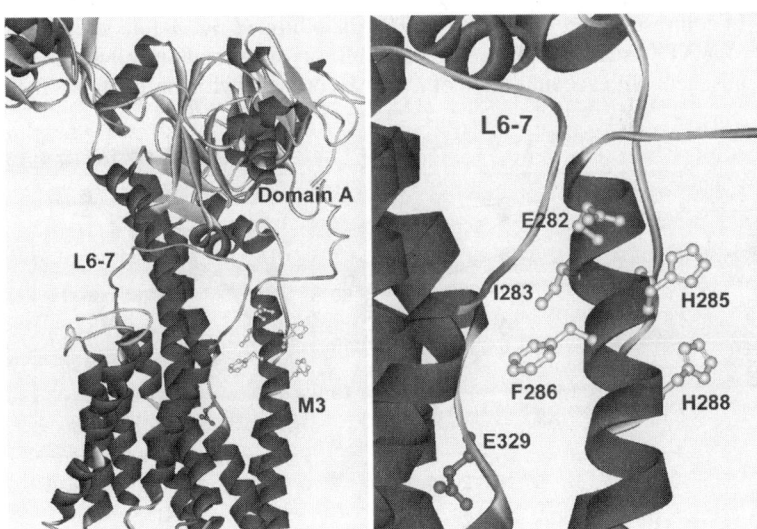

FIGURE 1. Structural model showing the location of L6-7 and the M3 residues described here (except Glu^{284}, which is on the backside). Glu^{329} is indicated as well to show the cation-binding pocket. The model was made on the basis of the Ca^{2+}-ATPase crystal structure in the E_1Ca$_2$ form (accession code 1EUL).[2] The SWISS-MODEL Comparative Protein Modeling facility[7] was used to replace L6-7 and M3 residues of the Ca^{2+}-ATPase with the corresponding residues of the Na$^+$,K$^+$-ATPase.

positions equivalent to His^{285} and His^{288} in the Na^+,K^+-ATPase are occupied by the negatively charged glutamic acid residues Glu^{255} and Glu^{258} in the Ca^{2+}-ATPase. Phe^{286} of the Na^+,K^+-ATPase is, on the other hand, conserved in the Ca^{2+}-ATPase, where it interacts with the specific inhibitor, thapsigargin, which prevents Ca^{2+} binding.[3]

In the present paper, we review studies of point mutants with alterations to residues in the M3 sequence just mentioned. The methods used have previously been described.[4,5] The mutants include Glu^{282}-Ala, Glu^{282}-Lys, Glu^{282}-Asp, Ile^{283}-Ala, Glu^{284}-Ala, His^{285}-Glu, His^{285}-Lys, Phe^{286}-Ala, and His^{288}-Ala. The mutational analysis revealed that the NH_2-terminal part of M3 is critical to at least three important functions of the Na^+,K^+-ATPase: (i) the conformational transitions between E_1 and E_2 forms of dephospho- and phosphoenzyme; (ii) Na^+ binding at the cytoplasmically facing sites; and (iii) the communication between the extracellularly facing cation binding sites and the catalytic site.

E_1-E_2 AND E_1P-E_2P CONFORMATIONAL EQUILIBRIA

In titrations of the ATP and vanadate dependencies of Na^+,K^+-ATPase activity, the mutants Glu^{282}-Lys, Glu^{282}-Asp, and Ile^{283}-Ala, displayed a 6–8-fold increase of the apparent affinity for ATP and a 6-fold decrease of the apparent affinity for vanadate, relative to the wild type. These findings suggest that the distribution of E_2 and E_1 is changed in favor of E_1, since it is well known that in the wild-type enzyme ATP binds with high affinity to E_1 and with low affinity to E_2, and vanadate binds solely to E_2. In contrast, Glu^{282}-Ala and Phe^{286}-Ala exhibited a 2-3–fold decrease of the affinity for ATP and a 3-fold increase in the affinity for vanadate, relative to the wild type, consistent with accumulation of E_2. In accordance with these effects, measure-

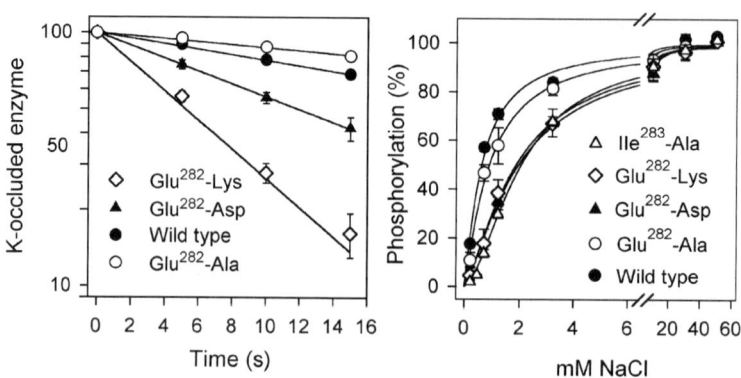

FIGURE 2. Time course of K^+ deocclusion (*left*) and Na^+-dependence of steady-state phosphorylation from ATP in presence of oligomycin (*right*). Methods were as previously described.[4,5] K^+ deocclusion was measured by phosphorylation of enzyme equilibrated with 100 μM KCl. The time course[5] was replotted semi-logarithmically. The relative $K_{0.5}$ values for Na^+ activation of phosphorylation corresponding to those in the *right panel* are given in TABLE 1 for all the mutants.

TABLE 1. Relative affinities for Na$^+$ in activation of phosphorylation from ATP and classification of mutants according to shift of conformational equilibrium

Mutation	$K_{0.5}$ (Na$^+$) for Phosphorylation	Shift of Conformational Equilibrium
Wild type	1	—
Glu^{282}-Ala	1.44	E_2/E_2P
Glu^{282}-Lys	2.98	E_1/E_1P
Glu^{282}-Asp	3.14	E_1/E_1P
Ile^{283}-Ala	3.72	E_1
Glu^{284}-Ala	0.95	Wild type
His^{285}-Glu	0.86	E_1P
His^{285}-Lys	0.94	E_1P
Phe^{286}-Ala	3.65	E_2/E_2P
His^{288}-Ala	2.07	Wild type

ments of K$^+$ deocclusion (see examples in FIG. 2, left panel) showed that the rate constant characterizing the K$^+$-deoccluding $E_2(K_2) \rightarrow E_1$ conversion was increased in Glu^{282}-Lys, Glu^{282}-Asp, and Ile^{283}-Ala and decreased in Glu^{282}-Ala and Phe^{286}-Ala, relative to wild type. The remaining mutants, His^{285}-Glu, His^{285}-Lys, Glu^{284}-Ala, and His^{288}-Ala, showed wild type–like behavior in these assays.

Studies of the phosphoenzyme showed a significant change in the steady-state distribution of ADP-sensitive E_1P and ADP-insensitive E_2P in favor of E_1P for Glu^{282}-Lys, Glu^{282}-Asp, His^{285}-Lys, and His^{285}-Glu, whereas for mutant Phe^{286}-Ala a slightly higher level of the ADP-insensitive E_2P intermediate, relative to wild type, was noticed. Dephosphorylation experiments initiated from E_1P showed that for Glu^{282}-Lys, His^{285}-Glu, and His^{285}-Lys, the rate constant characterizing the $E_1P \rightarrow E_2P$ conversion was reduced relative to wild type, explaining the change of the steady-state distribution of the phosphoenzyme intermediates in favor of E_1P indicated by the increased ADP sensitivity. For Glu^{282}-Ala and Phe^{286}-Ala, the rate constant corresponding to the $E_1P \rightarrow E_2P$ conversion was enhanced relative to wild type. It may be concluded that Glu^{282}-Lys and Glu^{282}-Asp displace the conformational equilibria of dephosphoenzyme and phosphoenzyme in parallel in favor of E_1 and E_1P, respectively, and mutants His^{285}-Glu and His^{285}-Lys displace the conformational equilibrium of the phosphoenzyme in favor of E_1P, whereas mutant Ile^{283}-Ala displaces the conformational equilibrium of the dephosphoenzyme in favor of E_1 without disturbing the conformational equilibrium of the phosphoenzyme. Mutants Glu^{282}-Ala and Phe^{286}-Ala also displace the conformational equilibria of dephosphoenzyme and phosphoenzyme in parallel, but in the opposite direction, that is, in favor of the E_2 and E_2P forms, respectively. Taken as a whole, the data indicate that the NH$_2$-terminal part of M3 plays a central role in control of the E_1-E_2 and E_1P-E_2P conformational equilibria. The various effects on the conformational equilibria described are indicated in TABLE 1 by the classification of the mutants as E_1-, E_2-, E_1P-, and E_2P-type.

Na$^+$ BINDING AT THE CYTOPLASMICALLY FACING SITES

In titrations of the Na$^+$ concentration–dependence of phosphorylation by ATP presented in TABLE 1 and FIGURE 2 (right panel), the mutants Glu^{282}-Ala, Glu^{282}-Lys, Glu^{282}-Asp, Ile^{283}-Ala, Phe^{286}-Ala, and His^{288}-Ala displayed a significantly reduced apparent affinity for Na$^+$ relative to the wild-type enzyme. Because only the E_1 form, and not E_2, binds Na$^+$ with high affinity, the reduced apparent affinity for Na$^+$ in Glu^{282}-Ala may at least in part be attributed to the displacement of the E_1-E_2 conformational equilibrium in favor of E_2. For the mutants Glu^{282}-Lys, Glu^{282}-Asp, and Ile^{283}-Ala with displacement of the E_1-E_2 equilibrium in favor of E_1, and for His^{288}-Ala with wild type-like E_1-E_2 equilibrium, there must, however, be another reason for the 2–4-fold reduction of the apparent Na$^+$ affinity. Because the shift of the conformational equilibrium in Glu^{282}-Lys, Glu^{282}-Asp, and Ile^{283}-Ala in favor of E_1 would *per se* tend to *increase* the apparent affinity for cytoplasmic Na$^+$, it may be concluded that the presence of lysine or aspartate at position 282, or alanine at position 283, not only disturbs the E_1-E_2 equilibrium, leading to a higher concentration of the E_1 form, but also affects the intrinsic Na$^+$ binding properties of the E_1 form. The latter effect may actually be more pronounced than revealed by the 3–4-fold decrease in apparent Na$^+$ affinity, because it is masked by the displacement of the conformational equilibrium in favor of E_1. In Phe^{286}-Ala the conformational equilibrium was shifted to the same extent in favor of E_2 as in Glu^{282}-Ala, but the reduction of Na$^+$ affinity was much more pronounced than in Glu^{282}-Ala (4-fold vs. 1.4-fold). This may suggest a role of Phe^{286} in the Na$^+$-binding properties of the E_1 form as well. In conclusion, the studies point to an important role of the NH$_2$-terminal part of M3 in optimizing the Na$^+$-binding properties of the enzyme, even though the M3 residues are located far away from the cation-binding sites in the membrane.

COMMUNICATION BETWEEN EXTRACELLULAR CATION SITES AND THE CATALYTIC SITE

FIGURE 3 (left panel) exemplifies a remarkable feature seen for some of the M3 mutants, also related to Na$^+$ binding, but this time at the extracellular sites. In the wild-type enzyme, the Na$^+$-activated ATPase activity observed in the absence of K$^+$ is rather low, corresponding to 400–500 min^{-1}, because K$^+$ binding at the extracellularly facing sites of the E_2P form is required for activation of dephosphorylation. However, for His^{285}-Glu and Glu^{282}-Ala, this so-called "Na$^+$-ATPase activity" reached extraordinary levels, 6- and 10-fold higher, respectively, than that of the wild type. In Glu^{282}-Ala, this high Na$^+$-ATPase activity constituted about 90% of the maximal Na$^+$,K$^+$-ATPase activity. For comparison, the Na$^+$-ATPase activity constituted about 6% of the maximal Na$^+$,K$^+$-ATPase activity in the wild type. The Na$^+$-ATPase activity displayed by Glu^{282}-Lys and Glu^{282}-Asp was only moderately increased corresponding to 2–3-fold relative to wild type. Previously, a few other mutants with alterations to transmembrane residues in M4, M5, and M6 located in or close to the cation-binding pocket were shown to display increased Na$^+$-ATPase activity. The maximum Na$^+$-ATPase turnover rate of 4534 min^{-1} displayed by Glu^{282}-Ala is, however, higher than described for any other Na$^+$,K$^+$-ATPase mutant.

To elucidate the underlying mechanism, a series of dephosphorylation experiments was carried out with dephosphorylation initiated either from E_1P or E_2P. FIGURE 3 (right panel) shows examples in which the dephosphorylation of E_1P was studied in the presence of 200 mM Na$^+$ for Glu^{282}-Ala and Glu^{282}-Lys, exhibiting high and moderate Na$^+$-ATPase activity, respectively. The accumulation of E_1P was achieved by performing the phosphorylation from ATP in the presence of 600 mM NaCl. The dephosphorylation is rather slow in the wild type in absence of K$^+$. However, the dephosphorylation rate was increased 2.5- and 5.4-fold for Glu^{282}-Lys and Glu^{282}-Ala, respectively, explaining the increased Na$^+$-ATPase activity. Comparison of the wild type, Glu^{282}-Ala, and Glu^{282}-Lys, indicates a good correlation between the turnover rate for Na$^+$-ATPase activity (FIG. 3, left panel) and the rate of dephosphorylation in the absence of K$^+$ (FIG. 3, right panel). To test whether the high dephosphorylation rate observed in the absence of K$^+$ was due to spontaneous dephosphorylation of E_1P bypassing E_2P, dephosphorylation experiments were, furthermore, carried out with enzyme phosphorylated at a lower Na$^+$ concentration of 20 mM to predominantly accumulate the E_2P phosphoenzyme intermediate.[5] In Glu^{282}-Ala, the dephosphorylation of E_2P was 5.2-fold increased relative to wild type. These results indicate that the high Na$^+$-ATPase activity is due to an increased ability of Na$^+$ to activate dephosphorylation of E_2P by binding in place of K$^+$ at the extracellularly facing sites. Thus, M3 seems to be part of a signaling pathway between the external cation sites and the catalytic site.

CORRELATION WITH STRUCTURE

Conformational Changes

Our data demonstrate the importance of the NH$_2$-terminal part of M3 for the conformational transitions between E_1 and E_2 forms of the dephospho- as well as phosphoenzyme. Of relevance for the understanding of the observed mutational effects on conformational changes is the fact that M3 is linked through the peptide backbone

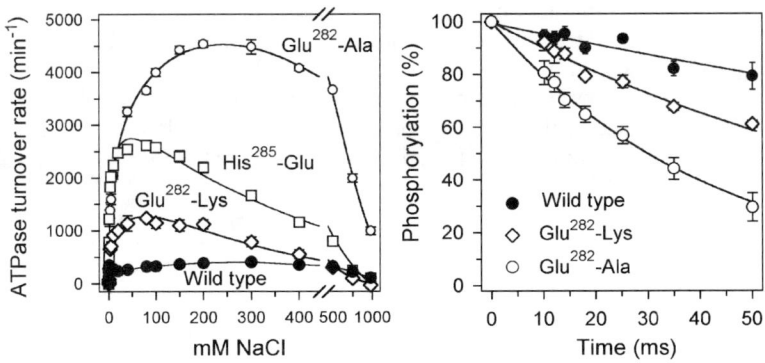

FIGURE 3. Na$^+$-dependence of ATPase activity in the absence of K$^+$ (*left*) and time course of dephosphorylation of E_1P phosphoenzyme (*right*). Methods were as previously described.[5]

to the minor cytoplasmic domain ("domain A"; see FIG. 1) and through non-covalent bonds at the NH_2-terminal end to domain P.[2] Proximity relations determined by specific affinity cleavage have shown that the conserved $T^{214}GES$ sequence in domain A moves toward the catalytic site in domain P during the $E_1 \rightarrow E_2$ and $E_1P \rightarrow E_2P$ transitions,[6] and previous mutagenesis analysis has implicated Gly^{263}, located in the cytoplasmic extension of M3, in this movement.[4] Due to the links between M3 and domains A and P, the positions of domains A and P must be influenced by the orientation of M3 and vice versa. Hence, changes in the ability of domain A to move toward domain P, resulting from changes in the orientation and relations of M3, might contribute to the observed effects of the M3 mutations on the conformational transitions. Comparison of the crystal structures of the Ca^{2+}-ATPase in E_1Ca_2 and E_2 forms shows that M3 is shifted downward together with M4 up to about 5 Å in connection with the movements of domains A and P.[3] In addition M3 inclines and becomes strongly curved in the E_2 state. These movements may account for the finding that the NH_2-terminal part of M3 is important for the conformational transition between the E_1 and E_2 forms of the dephosphoenzyme. Since our data show that mutations to Glu^{282} and Phe^{286} affect the E_1-E_2 and E_1P-E_2P equilibria in parallel, these two residues may be involved in both conformational changes. On the other hand, mutation to His^{285} preferentially affected the E_1P-E_2P equilibrium, and mutation Ile^{283}-Ala affected only the E_1-E_2 equilibrium of the dephosphoenzyme (cf. TABLE 1 for the classification of the mutants as E_1-, E_2-, E_1P-, and E_2P-type), demonstrating that the conformational changes of the phospho- and dephospho-forms are not identical.

In contradiction to our results, it has recently been concluded, based on metal-catalyzed cleavage experiments, that M3 is static during the E_1-E_2 transition, because the metal-catalyzed cleavage thought to occur near His^{285} was unaffected by the conformational state.[6]

In analogy with the equivalent Lys^{252} residue of the Ca^{2+}-ATPase, it is possible that the side chain of Glu^{282} interacts with L6-7 (either with the backbone or with side chains such as Lys^{838}) in the E_1 form, and that this interaction is disrupted in the E_2 form. Thus a change of the interaction induced by replacement of Glu^{282} may give rise to the observed displacement of the conformational equilibrium in the Glu^{282} mutants. The interaction with L6-7 would be interrupted by replacement with the small hydrophobic residue alanine, thus explaining the shift toward the E_2 form in Glu^{282}-Ala (TABLE 1). In the Glu^{282}-Asp and Glu^{282}-Lys mutants the bonding to L6-7 may have become even stronger than in the wild type as a consequence of formation of contacts with new hydrogen bond acceptors/donors in L6-7, thus explaining the shift toward the E_1 form in these two mutants (TABLE 1).

A Possible Entry Pathway for Na^+

The explanation of the reduced affinity for Na^+ at the cytoplasmically facing sites seen for Glu^{282}-Asp, Glu^{282}-Lys, Ile^{283}-Ala, His^{288}-Ala, and Phe^{286}-Ala, some of which are E_1-type mutants (TABLE 1), must take into consideration that the NH_2-terminal part of M3 is located at the cytoplasmic surface of the membrane, 20–25 Å from the residues generally believed to make up the cation-binding pocket. One explanation of the reduced Na^+ affinity could be a long-range effect on the cation-binding residues in M6 exerted through a change in the interaction between M3 and

L6-7. It may also be speculated that M3 is closely associated with an entry pathway, through which the Na^+ ions have to pass from the cytoplasmic surface to reach the cation-binding pocket. This would be analogous to the recent proposal of a migration pathway for Ca^{2+} located between M1 and M3 in the Ca^{2+}-ATPase on the basis of the existence of a water-accessible channel at this location in the E_2 crystal structure.[3] The interaction between Glu^{282} and L6-7 could be important for control of the channel inlet. Furthermore, a change in the orientation of M3 induced by the mutations might change the proportions of the channel and, thus, affect the migration of Na^+. Interestingly, the Ca^{2+}-ATPase residue Leu^{253}, equivalent to Ile^{283} of the Na^+,K^+-ATPase, forms an integral part of the hydrophobic wall of the channel seen in the E_2 crystal structure, and Ile^{283}-Ala is the E_1-type mutant showing the most conspicuous reduction of Na^+ affinity in the present study (TABLE 1).

ACKNOWLEDGMENTS

This work was supported by grants from the Danish Medical Research Council, the Lundbeck Foundation, the Novo Nordisk Foundation, and the Research Foundation of Aarhus University.

REFERENCES

1. VILSEN, B., D. RAMLOV & J. P. ANDERSEN. 1997. Functional consequences of mutations in the transmembrane core region for cation translocation and energy transduction in the Na^+,K^+-ATPase and the SR Ca^{2+}-ATPase. Ann. N. Y. Acad. Sci. **834:** 297–309.
2. TOYOSHIMA, C., M. NAKASAKO, H. NOMURA & H. OGAWA. 2000. Crystal structure of the calcium pump of sarcoplasmic reticulum at 2.6 Å resolution. Nature **405:** 647–655.
3. TOYOSHIMA, C. & H. NOMURA. 2002. Structural changes in the calcium pump accompanying the dissociation of calcium. Nature **418:** 605–611.
4. TOUSTRUP-JENSEN, M., M. HAUGE & B. VILSEN. 2001. Mutational effects on conformational changes of the dephospho- and phospho-forms of the Na^+,K^+-ATPase. Biochemistry **40:** 5521–5532.
5. TOUSTRUP-JENSEN, M. & B. VILSEN. 2002. Importance of Glu^{282} in transmembrane segment M3 of the Na^+,K^+-ATPase for control of cation interaction and conformational changes. J. Biol. Chem. **277:** 38607–38617.
6. PATCHORNIK, G., R. GOLDSHLEGER & S.J.D. KARLISH. 2000. The complex ATP-Fe^{2+} serves as a specific affinity cleavage reagent in ATP-Mg^{2+} sites of Na,K-ATPase: Altered ligation of Fe^{2+} (Mg^{2+}) ions accompanies the $E_1P \rightarrow E_2P$ conformational change. Proc. Natl. Acad. Sci. USA **97:** 11954–11959.
7. GUEX, N & M. C. PEITSCH. 1997. SWISS.MODEL and the Swiss-PdbViewer: an environment for comparative protein modeling. Electrophoresis **18:** 2714–2723.

Insights into the Structural Basis for Modulation of $E_1 \leftrightarrow E_2$ Transitions by Cytoplasmic Domains of the Na,K-ATPase α Subunit

LAURA SEGALL,[a] LOIS K. LANE,[b] AND RHODA BLOSTEIN[a]

[a]*Department of Biochemistry, McGill University, Montreal, Quebec, Canada H3G 1Y6*

[b]*Department of of Pharmacology and Cell Biophysics, University of Cincinnati College of Medicine, Cincinnati, Ohio 45267-0575, USA*

KEYWORDS: Na,K-ATPase; N-terminus; deletion; isoform; chimera

Earlier studies have shown that a 32-residue deletion of the N-terminus of the α1 catalytic subunit of the rat Na,K-ATPase (α1M32), corresponding to a trypsin-sensitive site first described by Jorgensen,[1,2] together with mutation E233K in the M2-M3 cytoplasmic loop, results in a remarkably synergistic shift in E_1/E_2 poise towards E_1.[3-5] These findings were interpreted to indicate that interactions between these two cytoplasmic regions and the M4-M5 catalytic loop are critical for conformational coupling. Interestingly, the α2 enzyme is an E_1-shifted conformational isoform of α1, kinetically similar to α1M32.[3,6] This paper summarizes the results of experiments relevant to the role of cytoplasmic domains of α1 and α2 in conformational coupling.

RESULTS AND DISCUSSION

Role of the N-terminus in Conformational Transitions

The N-terminus of the rat Na,K-ATPase is a highly charged, flexible structure containing three putative helices predicted to span at least residues 27-33 (H1), 42-50 (H2), and 61-68 (H3) (FIG. 1). N-terminal deletion mutants of α1 were constructed corresponding to either disruption (α1M32) or loss (α1M40) of H1 and disruption (α1M46 and α1M49) or loss (α1M56) of H2 (FIG. 1). Effects of these mutations on $E_1 \leftrightarrow E_2$ shifts were determined by analyzing the reaction under steady-state conditions as well as assays of partial reactions $E_2(K) \rightarrow E_1 + K$ and $E_1 P \cdot Na \rightarrow E_2 P + Na$.

Address for correspondence: Dr. Rhoda Blostein, Montreal General Hospital Research Institute, 1650 Cedar Avenue - L11-132, Montreal, Que. Canada H3G1A4.
Rhoda.Blostein@mcgill.ca

TABLE 1. Summary of kinetic behavior of Na,K-ATPase N-terminal mutants

Enzyme	[a]Vanadate IC_{50}	Steady-state catalysis Na,K-ATPase/ Na-ATPase[b]	$K'_{ATP(L)}$	Catalytic turnover	Partial reaction rates $E_2(K) \to E_1$	$E_1P \to E_2P$
α1	1	60 ± 5	1	1	1	1
α1M32	>500	289 ± 16*	0.4*	0.5*	3.4*	0.2*
α1M40	>1000	423 ± 26	0.2 *	0.2 *	6.4*	0.2*
α1M46	1	78 ± 4	1	0.7 *	0.2*	1
α1M49	1	60 ± 1	1	0.7 *	0.4*	1
α1M56	1	65 ± 5	0.4*	0.7 *	1	1

[a]Normalized.
[b]Measured at 20 mM NaCl, and 1 μM ATP ± 1 mM KCl.
*$P<0.05$.

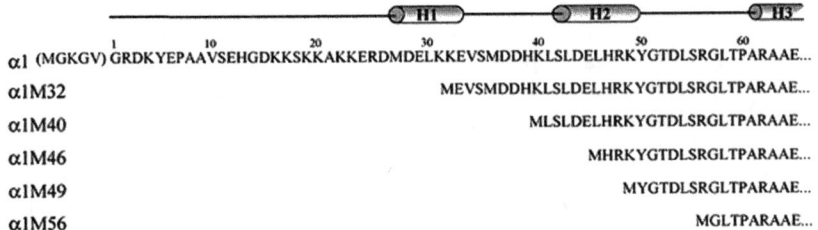

FIGURE 1. Schematic representation of the amino terminus of the rat α1 Na,K-ATPase and deletion mutants thereof. (Reprinted from Ref. 11, with permission from the American Society of Biochemistry and Molecular Biology.)

Analysis of the Overall Catalytic Cycle

As shown in TABLE 1, α1M32 and α1M40 are both less sensitive to vanadate inhibition compared to WT α1, suggesting a progressive E_1 shift in the E_1/E_2 equilibrium. Interestingly, this shift is reversed by deleting ≥46 residues. Evidence for similar shifts is observed when $E_2(K) \to E_1+K$ is rate limiting. Thus, at micromolar ATP, ≤1mM K^+ inhibits Na-ATPase of the WT α1 but stimulates α1M32 and α1M40. Reversal to α1-like inhibition is seen with further deletions (≥46 residues).

These effects are mirrored by changes in low-affinity ATP binding ($K'_{ATP(L)}$). Thus, α1M32 and α1M40 have lower $K'_{ATP(L)}$ values, whereas α1M46 and α1M49 have $K'_{ATP(L)}$ values similar to that of α1, consistent with the aforementioned reversal of the K^+ effect on Na-ATPase. Furthermore, both α1M32 and α1M40 have progressively decreased catalytic turnovers (measured as V_{max}/EP_{max}); turnovers of α1M46, α1M49, and α1M56 revert to near that of α1. Whether the lower $K'_{ATP(L)}$ of α1M56 relative to that of α1 reflects an additional role of this domain on access of ATP to its binding site has not been pursued.

Partial Reactions Relevant to Conformational Transitions

We first measured the release of K^+ from $E_2(K)$ via $E_2(K) \rightarrow E_1 \cdot K \rightarrow E_1 + K$ using an indirect approach.[7,8] Thus, the rate constants for α1M32 and α1M40 are significantly higher than that of α1, suggesting a right-shift in $E_2(K) \leftrightarrow E_1 + K$ (TABLE 1). Conversely, K^+ deocclusion rates of α1M46, α1M49, and α1M56 are similar or even slower than that of α1.

To measure the rate of $E_1P \rightarrow E_2P$, the enzyme was first phosphorylated by [γ-^{32}P]-ATP at high NaCl to stabilize E_1P, and then dephosphorylated by simultaneously decreasing the NaCl concentration and adding KCl and unlabeled ATP as described elsewhere.[9,10] Compared to α1, the $E_1P \rightarrow E_2P$ rate is 5-fold slower for α1M32 and a1M40, consistent with a shift toward E_1P, which is reversed by deleting ≥46 residues.

The Distinct Role of the N-terminus

The above analysis combined with structure predictions identifies a self-regulatory domain within the N-terminus that modulates conformational transitions via novel intramolecular interactions. A hypothetical model is proposed in FIGURE 2. In the case of WT α1 (FIG. 2A), H2 can interact with H1 through helix–helix interactions and/or salt-bridge formation in E_2, allowing the M2-M3 loop and the catalytic loop to come together (shaded ovals). H2 can also interact with a domain of the M2-M3 loop in E_1 to keep it apart from the catalytic loop. When H1 is disrupted or lost (α1M32 or α1M40; FIG. 2B), H2 preferentially interacts with the M2-M3 loop, placing a constraint on the latter, thereby weakening M2-M3/M4-M5 interaction and consequently stabilizing the E_1 conformation. The result is a shift in poise in favor of E_1. Lastly, disruption (α1M46 and α1M49) or loss of H2 (α1M56) (FIG. 2C)

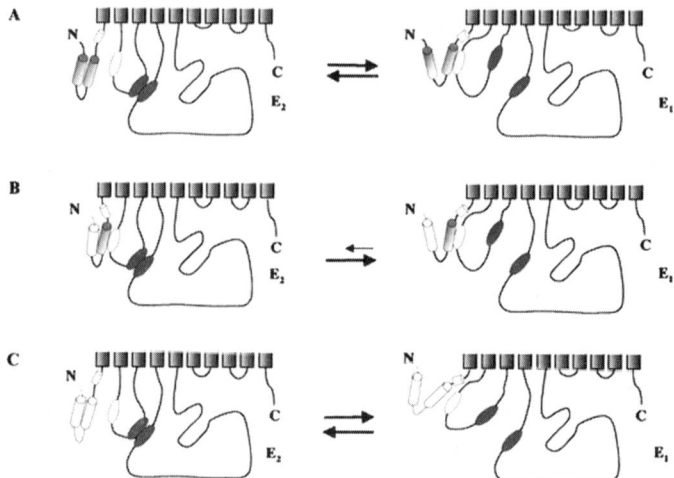

FIGURE 2. Model depicting N-terminal regulation of conformational equilibrium. See text for details. (Reprinted from Ref. 11, with permission from the American Society of Biochemistry and Molecular Biology.)

alleviates the constraint on M2-M3, allowing it once again to freely interact with the catalytic M4-M5 loop, thus resembling WT $\alpha 1$. It is noteworthy that the $\alpha 2$ and $\alpha 3$ isoforms of Na,K-ATPase have similar helical structures in their N-termini. Furthermore, analogous disruptions of the first helix of $\alpha 2$ and $\alpha 3$ shift the poise for each enzyme toward E_1 forms, suggesting that the N-terminus of Na,K-ATPase encompasses an auto-regulatory domain present in all three isoforms.

Structural Basis for α1/α2 Isoform-Specific Kinetic Differences

To investigate the structural basis for the E_1 shift of $\alpha 2$, cytoplasmic mutants of $\alpha 2$ analogous to those previously analyzed for $\alpha 1$ were first analyzed to determine whether the cytoplasmic interactions of $\alpha 2$ were similar to those of $\alpha 1$. The results show that as for $\alpha 1$, the mutations of $\alpha 2$ ($\alpha 2$M30 and $\alpha 2$E231K) further shift the E_1/E_2 balance of $\alpha 2$ toward E_1. Therefore, $\alpha 1/\alpha 2$ chimeras were constructed in which domains of divergent primary sequence were interchanged. Thus, the N-terminus (residues 1-85) and a portion of the M4-M5 loop near the ATP binding site (residues 429-575) of $\alpha 1$ were substituted with the analogous regions of $\alpha 2$. In two additional chimeras, the entire A domain (residues 1-346) was interchanged. Preliminary results indicate that the E_1 shift of $\alpha 2$ results from an increase in $E_2 \rightarrow E_1$, but not $E_2P \rightarrow E_1P$, which is mainly due to diversity in the A domain beyond the N-terminus, with a lesser contribution of the isoform-distinct M4-M5 loop.

ACKNOWLEDGMENTS

This work was supported by grants from CIHR (No. MT-3876) and QHSF (to R.B.), the NIH (No. HL-49204 to L.K.L.) and a scholarship from the HSF of Canada (to L.S.).

REFERENCES

1. JORGENSEN, P.L. 1975. Purification and characterization of Na,K-ATPase. V. Conformational changes in the enzyme transitions between the Na-form and the K-form studied with tryptic digestion as a tool. Biochim. Biophys. Acta **401:** 399–415.
2. JORGENSEN, P.L. 1977. Purification and characterization of Na,K-ATPase. VI. Differential tryptic modification of catalytic functions of the purified enzyme in presence of NaCl and KCl. Biochim. Biophys. Acta **466:** 97–108.
3. DALY, S.E., L.K. LANE & R. BLOSTEIN. 1994. Functional consequences of amino-terminal diversity of the catalytic subunit of the Na,K-ATPase. J. Biol. Chem. **269:** 23944–23948.
4. DALY, S.E., R. BLOSTEIN & L.K. LANE. 1997. Functional consequences of a posttransfection mutation in the H2-H3 cytoplasmic loop of the alpha subunit of Na,K-ATPase. J. Biol. Chem. **272:** 6341–6347.
5. BOXENBAUM, N. et al. 1998. Changes in steady-state conformational equilibrium resulting from cytoplasmic mutations of the Na,K-ATPase alpha-subunit. J. Biol. Chem. **273:** 23086–23092.
6. SEGALL, L., S.E. DALY & R. BLOSTEIN. 2001. Mechanistic basis for kinetic differences between the rat $\alpha 1$, $\alpha 2$, and $\alpha 3$ isoforms of the Na,K-ATPase. J. Biol. Chem. **276:** 31535–31541.
7. DALY, S.E., L.K. LANE & R. BLOSTEIN. 1996. Structure/function analysis of the amino-terminal region of the $\alpha 1$ and $\alpha 2$ subunits of Na,K-ATPase. J. Biol. Chem. **271:** 23683–23689.

8. THERIEN, A.G. & R. BLOSTEIN. 1999. K^+/Na^+ antagonism at cytoplasmic sites of Na,K-ATPase: a tissue-specific mechanism of sodium pump regulation. Am. J. Physiol. **277:** C891–898.
9. KLODOS, I., R.L. POST & B. FORBUSH 3RD. 1994. Kinetic heterogeneity of phosphoenzyme of Na,K-ATPase modeled by unmixed lipid phases: competence of the phosphointermediate. J. Biol. Chem. **269:** 1734–1743.
10. VILSEN, B. 1997. Leucine 332 at the boundary between the fourth transmembrane segment and the cytoplasmic domain of Na,K-ATPase plays a pivotal role in the ion translocating conformational changes. Biochemistry **36:** 13312–13324.
11. SEGALL, L., Z.Z. JAVAID, S.L. CARL, *et al.* 2003. Structural basis for $\alpha 1$ versus $\alpha 2$ isoform distinct behavior. J. Biol. Chem. **278**. In press.

Characterization of Ca^{2+} ATPase Residues Involved in Substrate and Cation Binding

GIUSEPPE INESI,[a] HAILUN MA,[a] SUMING HUA,[a] AND CHIKASHI TOYOSHIMA[b]

[a]*Department of Biochemistry, University of Maryland, Baltimore, Maryland 21201, USA*

[b]*Institute of Molecular and Cellular Biosciences, University of Tokyo, Tokyo, Japan*

ABSTRACT: The role of amino acid residues involved in substrate and cation binding was investigated in complementary experiments on Fe^{2+}-catalyzed oxidation and cleavage, limited digestion with proteinase K, and mutational analysis. Cleavage at Ser346 was produced by Fe^{2+} in the presence of substrate (ATP or AMP-PNP) and Ca^{2+}, and was attributed to Fe^{2+} bound to a Mg^{2+} site near Ser346 and neighboring Glu696. Ca^{2+}- and ATP-dependent oxidation of the Thr441 side chain was also observed and attributed to Fe^{2+} substituting for Mg^{2+} in the Mg^{2+}-ATP complex bound to the N domain. Mutation of Arg560 or Glu439 within the N domain interfered with nucleotide-dependent ATPase resistance to digestion with proteinase K. Furthermore, mutation of Lys352, Lys684, Thr353, Asp703, or Asp707 within the P domain produced similar interference, consistent with a role of these residues in substrate stabilization at the catalytic site. In a third group of experiments, equilibrium isotherms were obtained with Asn796Ala and Glu309Gln mutants, demonstrating non-cooperative binding of one Ca^{2+} per ATPase, as opposed to cooperative binding of two Ca^{2+} by WT enzyme. No high-affinity binding by Asp800Asn, Glu771Gln, and Thr799Ala mutants was detected. It was also demonstrated that the conformational transitions involved in enzyme activation and interconversion of Ca^{2+} binding and phosphorylation energy, are triggered by Ca^{2+} binding to site II and stabilization of Glu309 (M4) and N796 (M6).

KEYWORDS: nucleotide binding; proteinase K; iron cleavage

INTRODUCTION

The sarcoplasmic reticulum (SR) Ca^{2+} ATPase is a rather favorable system for studies of cation transport ATPases. Its advantages include high yield of native SR vesicles for parallel measurements of ATP utilization and Ca^{2+} transport, abundance and density of ATPases molecules within the membrane for biochemical and struc-

Address for correspondence: Dr. Giuseppe Inesi, Department of Biochemistry, University of Maryland, Baltimore, MD 21201. Voice: 410-706-3220; fax: 410-706 8297.
ginesi@umaryland.edu

tural characterization, and favorable equilibrium and kinetic constants for definition of the partial reactions constiting the catalytic cycle. We report here experiments on localization of Mg^{2+} and Mg^{2+}-ATP binding within the ATPase protein, identification of amino acid residues involved in stabilization of nucleotide at the substrate site, and characterization of the triggering mechanism for enzyme activation and interconversion of Ca^{2+} binding and phosphorylation energy.

METHODS

Native SR vesicles were obtained with the microsomal fraction of rabbit leg muscle homogenate, as described by Eletr and Inesi.[1] Recombinant ATPase was obtained by exogenous gene expression in COS-1 cells infected with adenovirus vectors carrying chicken WT or mutant SERCA1 cDNA. Adenovirus vector construction, site-directed mutations, COS-1 cell cultures and infections, and preparation of microsomal fractions were previously described.[2] The conditions for Fe^{2+}-catalyzed oxidation, Ca^{2+} binding, and ATP generation are described by Hua et al.[3] and Inesi et al.[4] Limited digestion of ATPase with proteinase K, and protection by nucleotides and Ca^{2+}, were performed in media containing 50 mM MOPS, pH 7.0, 50 mM NaCl, 2 mM EGTA, and 5 mM MgCl2. The protein concentration was 1.2 mg microsomal protein, and 0.04 mg proteinase K per mL. The concentration of recombinant ATPase was adjusted to the same level in all experiments, by compensation with empty microsomes, and based on ATPase detection by Western blotting. The reaction was quenched with trichloroacetic acid (TCA). The quenched protein was subjected to electrophoretic analysis, followed by staining with Coomassie Blue or Western blotting.

RESULTS

Studies of Fe^2-Catalyzed Oxidation and Cleavage and Location of Mg^{2+} Binding

The expedient of using Fe^{2+} was first employed by Patchornik et al.[5] to determine the location of Mg^{2+} binding in the Na^+,K^+ ATPase. As a variation to the conditions described by Patchornik et al.[5], we used only ascorbic acid to assist the oxidation reaction by Fe^{2+}, rather than both ascorbic acid and hydrogen peroxide. Thereby we obtained a slower and more localized cleavage, yielding a 70–75 kDa band on SDS electrophoretic gels (FIG. 1), with an amino terminus sequence attributed to ^{346}SVICS. This cleavage is strictly Ca^{2+}- and ATP-dependent, as demonstrated by withholding either ligand (FIG. 1), or by site-directed mutations that interfere with Ca^{2+} (Glu771) or ATP (Arg560) binding. Such requirements suggest that the presence of Fe^{2+} (or Mg^{2+}) near Ser346 is catalytically relevant, consistent with previous reports of ATPase inactivation following Ser346 mutation.[6] Ser346 resides within the P domain near the L67 loop in close proximity of Glu696, surrounded by several oxygen atoms in a structural arrangement that is critical for the occurrence of catalytic phosphorylation.[7,8] Its location, however, is relatively distant (16 Å) from the phosphorylation site (i.e., Asp351), precluding direct contact with the Mg^{2+}-ATP sub-

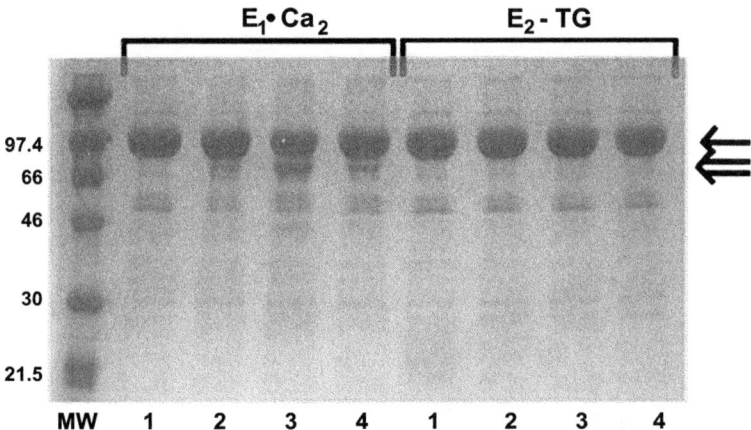

FIGURE 1. Patterns of ATPase protein cleavage following incubation with Fe^{2+} and ascorbate. Incubation with Fe^{2+} and ascorbate and solubilization and electrophoresis are described in METHODS. A set of samples was incubated for 50 minutes in the presence of Ca^{2+} (E_1-Ca_2), and another set in the absence of Ca^{2+} (E_2-TG). *Lane* 1: control; 2: no ATP or Mg^{2+}; 3: 0.2 mM ATP; 4: 0.2 mM ATP and 5 mM $MgCl_2$. *Single arrow* corresponds to the ATPase band, the *double arrow* to the cleavage product. (Original data from Hua et al.[3])

strate complex. It is possible that the observed cleavage is due to Fe^{2+} presence in a Mg^{2+} site that is required for catalytic activation, but is not in direct contact with ATP.

Peptide cleavage such as shown in FIGURE 1 is produced by Fe^{2+}-catalyzed oxidative attack resulting in cleavage of the polypeptide backbone, as in the experiment shown in FIGURE 1. On the other hand, we considered that in addition to this general oxidation and cleavage reaction, specific oxidation of threonine or serine side chains may occur, followed by very slow or no cleavage. In this case, the oxidation product can be reduced with NaB^3H4, resulting in radioactive labeling. Extensive proteolysis, followed by chromatographic separation and sequencing of the radioactive peptide, then reveals the labeled amino acid. In fact, we observed Fe^{2+}-catalyzed oxidation of Thr441, with a requirement for Ca^{2+} and ATP. Thr441 resides within the N domain, and its oxidation is likely to be catalyzed by Fe^{2+} substituting for Mg^{2+} in the Mg^{2+}-ATP substrate complex. The related structural arrangement can be visualized by modeling the Mg^{2+}-ATP complex conformation derived from the high-resolution structure of other enzymes (1KAX, 1A82, 1F2U, 1MJH, in the PDB), according to the ADP location detected by crystallographic studies of the SR ATPase in the E2-TG state.[9] In the model shown in FIGURE 2 the γ-phosphate of the Mg^{2+}-ATP complex in a folded configuration was fitted to that of ADP detected in the E2-TG crystal, resulting in very close approximation of Mg^{2+} to the side chain oxygen of Thr441. The prevalent interaction of the Mg^{2+}-ATP complex with the N domain suggests that such a folded conformation may be present on the the first encounter of the substrate with the open conformation of the ATPase headpiece[10] in the presence of Ca^{2+}.

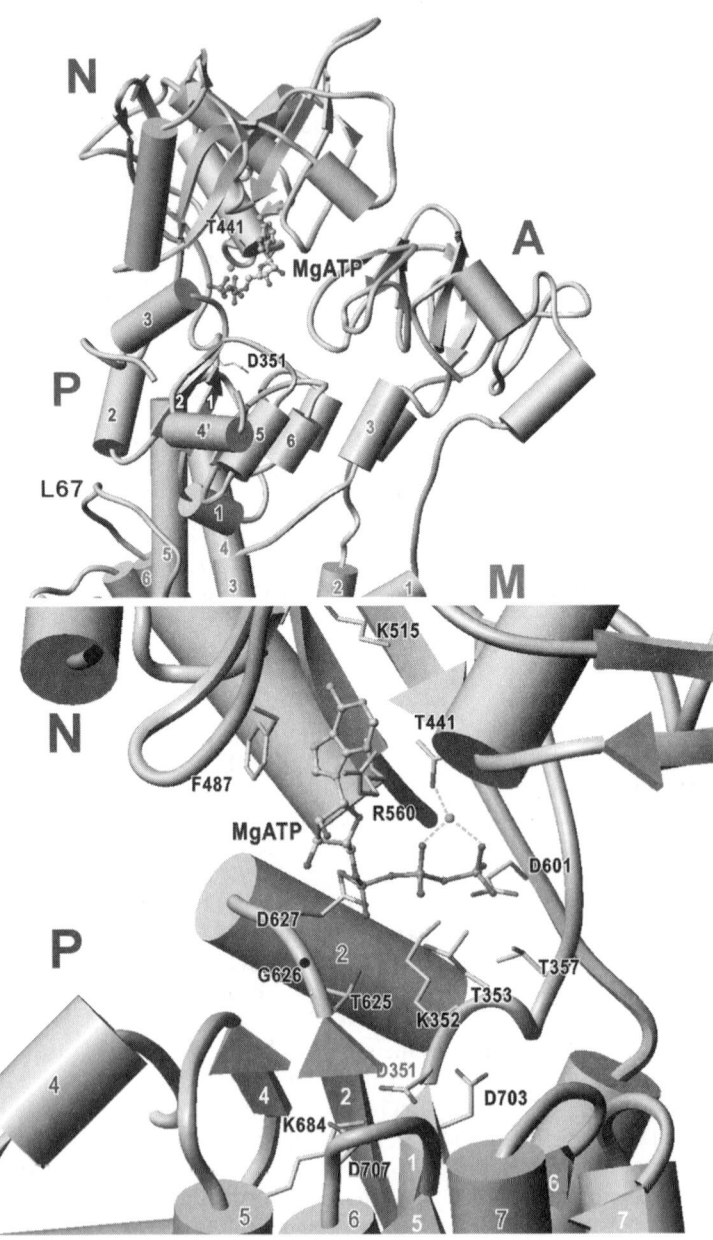

FIGURE 2. Modeling of the Mg-ATP complex within the ATPase structure. *Top:* Mg^{2+}-ATP in Ca^{2+}-ATPase as deduced from the crystal structure of Ca^{2+}-ATPase in the E_2-TG state (2) and oxidation with Fe^{2+}-ATP. Mg^{2+}-ATP is in ball-and-stick. Helices appear as *cylinders* and the strands as *arrows*. The side chain for the phosphorylation residue, Asp351, in the P domain is also shown. P4 helix in the P domain is removed. When 2 Ca^{2+} bind, the N domain

Residues Involved in ATP Stabilization at the Phosphorylation Site

In the model shown in FIGURE 2, the ATP terminal phosphate does not reach Asp351 with sufficient approximation to allow formation of the phosphorylated enzyme intermediate. Therefore, progression of the catalytic cycle requires that the bound ATP acquires an extended configuration, thereby permitting γ-phosphate approximation to Asp351. We then performed mutational analysis in an attempt to reveal the identity of amino acid residues that may be involved in nucleotide binding and consequent stabilization of the ATPase headpiece in a suitable conformation for phosphoryl transfer from ATP to Asp351.

It was shown by Danko et al.[11] that nucleotide binding protects the ATPase from limited digestion with proteinase K, presumably by the closure of the N and P domains. This protection requires the presence of both nucleotide (i.e., ATP or AMP-PCP) and Ca^{2+}, and is likely to correspond to the activated enzyme state. We then tested whether single mutations of residues within the catalytic site interfere with nucleotide protection of the ATPase from digestion with proteinase K.[12] We found that single mutation of Arg560 or Glu439 in the N domain interferes with nucleotide-dependent ATPase resistance to digestion by proteinase K. Interference is also observed following mutation of Lys684, Lys352, Thr353, Asp703, or Asp707 in the P domain. Furthermore, nucleotide-dependent resistance to proteinase K is not observed following mutations that interfere with Ca^{2+} binding. These experiments demonstrate that amino acid residues within the N domain as well as the P domain (FIG. 2) are involved in stabilization of the bound nucleotide substrate and extension of the ATP γ-phosphate to reach Asp351.

Residues Triggering Interconversion of Ca^{2+} Binding and Phosphorylation Potentials

Ca^{2+} binding is required for enzyme activation and occurs on two neighboring sites formed by amino acid residues contributed by the M4 (E309 and A305), M5 (E771 and N768), M6 (N796, T799, and D800), and M8 (E908) transmembrane helices (10, 13; FIG. 3). We have now obtained equilibrium binding isotherms demonstrating that binding of two Ca^{2+} to WT recombinant ATPase occurs with a cooperative mechanism, as previously shown in native SR vesicles.[14] Furthermore, while single mutation of E771 or D800 in site I interferes completely with Ca^{2+} binding, single mutation of E309 or N796 in site II yields non-cooperative binding of only one Ca^{2+} per ATPase, within the μM concentration range (FIG. 3A). This demonstrates that Ca^{2+} binding to site I is required for cooperative Ca^{2+} binding to site II, while high-affinity binding to site I can still occur if site II is inactivated by the E309 or N796 mutations. In either case, no Ca^{2+}-dependent phosphoenzyme formation by ATP in the forward direction of the catalytic cycle is observed following mutation of either site I or site II,[13] while Ca^{2+}-independent formation of phospho-

---

will approach closer to the P domain, whereas A domain will be detached to allow approach of the N domain. *Bottom:* a close-up of the structure around Mg^{2+}-ATP at the N-P interface. The direction of the movement of the N domain is approximately normal to the plane of the paper. Side chains are shown for some of the critical residues. *Dotted lines* connect the Mg^{2+} (central sphere) and oxygen atoms within a 2.5-Å distance.

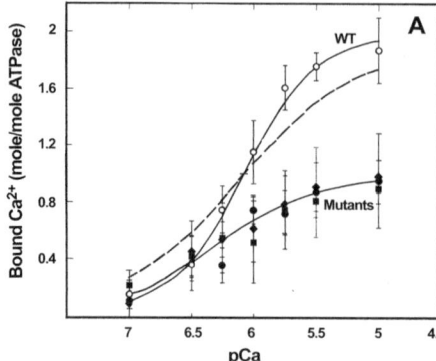
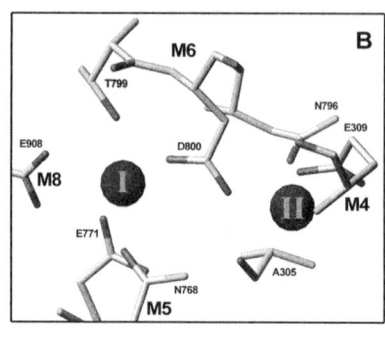

FIGURE 3. Ca^{2+} binding by WT, E309Q, T796A, and E309Q/T796A mutants (**A**) and molecular graphics representation of the two Ca^{2+} binding sites in WT ATPase (**B**). (**A**) Recombinant ATPase protein was obtained from COS-1 cells infected with adenovirus vectors, and Ca^{2+} binding was measured in the absence of ATP as described by Zhang et al.[2] WT (*open circles*); mutants: E309Q (*solid squares*), N796A (*solid diamonds*), and E309Q/N796A (*solid circle*). The experimental points (Ca^{2+} bound/E) obtained with WT ATPase required fitting with a cooperative two-site equation, while those for the E309Q, T796A, and E309Q/T796A mutants could be fitted simply with an independent binding equation. The *dashed line* shows the poor fitting of the WT data using an independent binding equation. No significant binding was observed with the E771Q, T799A, or D800N mutants. (**B**) Representation of the residues involved in Ca^{2+} binding was obtained directly from the crystallographic structure of the SR ATPase with Ca^{2+} bound[10] using an SGI system with Turbo-FRODO software. M4, M5, M6, and M8 refer to the transmembrane segments originating the binding residues. The section is along the plane of the membrane, viewed from the cytosolic side. (Original data of Inesi et al.[4])

enzyme intermediate by utilization of P_i is obtained in the reverse ATPase direction. In reverse single-cycle experiments, we now find that no ATP is obtained by addition of ADP and mM Ca^{2+} to phosphoenzyme formed by utilization of P_i, when either site I or site II mutants are used (TABLE 1). This demonstrates that Ca^{2+} binding to site II and the consequent stabilization of E309 (M4) and N796 (M6) are required for enzyme activation and interconversion of Ca^{2+} binding and phosphorylation energy in either direction of the ATPase cycle.

DISCUSSION

A sequence of partial reactions making up the ATPase and Ca^{2+} transport cycle, with forward and reverse kinetic constants, is given in FIGURE 4. The sequence states clearly that occupancy of site I and site II by Ca^{2+} is sequential, both in the forward and the reverse direction of the cycle. Utilization of ATP occurs only after binding of the second Ca^{2+} (single arrow on the cycle). On the other hand, interconversion of Ca^{2+} binding and phosphorylation energy occurs by an isomeric transition of the phosphoenzyme with both Ca^{2+} bound (double arrow on the cycle). The equilibrium constant for this reaction is nearly 1, indicating conservation of potential energy and

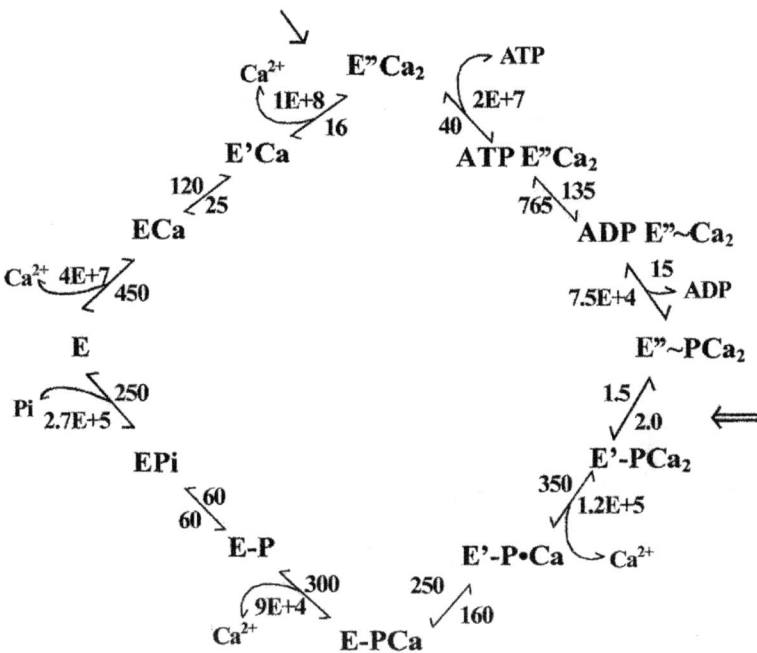

FIGURE 4. The catalytic and transport cycle of the SR ATPase. This sequence of partial reactions, with their microscopic constants, is required to fit the observed ATPase behavior, in equilibrium or kinetic experiments.[15] Most partial reactions, and their constants, were determined experimentally. The units are in $\sec^{-1}$ for first-order reactions, and $\sec^{-1} M^{-1}$ for second-order reactions. The overall Keq is 4.9^{e+5}, and is the result of ATP terminal phosphate hydrolytic cleavage under standard conditions. E' and E" indicate the conformations of the enzyme with one or two calcium ions bound with high affinity. E-P and E~P indicate phosphoenzyme with low or high phosphorylation energy.

interconversion through a conformational transition. The experimental evidence (TABLE 1) indicates that the conformational transition producing high phosphorylation potential is triggered by Ca^{2+} binding to site II, and stabilization of E309 (M4) and N796 (M5). An important feature of the Ca^{2+}-induced conformational transition in the forward direction of the cycle is the open conformation of the cytosolic ATPase headpiece[10] followed by ATP binding. Our experiments suggests that the first encounter of the nucleotide substrate with the N domain occurs in the form of Mg^{2+}-ATP complex in a folded configuration, with the adenosine moiety residing within a pocket delimited by Lys492 and Lys515, and stabilized by Arg560. In this complex, Mg^{2+} is stabilized by the phosphate oxygen atoms as well by the Thr441 side chain (FIG. 2). Rotation of the A domain and closure of the N and P domain would then yield the compact conformation of the headpiece,[11] favoring an extended configuration of the γ and β phosphates of ATP, and approximation of the γ-phosphate to Asp351. Several amino acid residues, including Lys352, Lys684, Thr353, Asp703, and Asp707, are involved, directly or indirectly, in substrate stabi-

TABLE 1. Effects of mutations of site I or site II residues

Sample	Ca^{2+} Site Occupancy	ATPase Activity	EP Formation (with ATP)	EP Formation (from P_i inhibition)	ATP Synthesis
WT	I and II	100	100	100	100
E309A	I	0	0	100	0
N796A	I	0	0	100	0
E309A/ N796A	I	0	0	100	0
E771Q	none	0	0	100	0
T799A	none	0	0	100	0
D800N	none	0	0	100	0

NOTE: Ca^{2+} binding, ATPase activity, phosphoenzyme formation by ATP in the presence of Ca^{2+}, phosphoenzyme formation with P_i in the absence of Ca^{2+}, and ATP synthesis by addition of ADP and Ca^{2+} to phosphoenzyme formed with P_i, were measured as described previously.[2,16] The values refer to nmol/mg protein/min for ATPase activity, nmol/mg for EP levels, and to percentage of EP (P_i) used for ATP synthesis. The functional values obtained with mutants were corrected based on ATPase expression as defined by western blots, with reference to microsomes of cells expressing WT ATPase. As the ATPase content of the microsomal preparation is 0.9–1.0 nmol/mg protein, the EP obtained with ATP or P_i reflects steady-state or equilibrium levels under these conditions, respectively.

lization within the P domain, favoring its catalytic utilization. Furthermore, the experiments on Fe^{2+}-catalyzed cleavage suggest that an independent Mg^{2+} is bound in the P domain, in close proximity of Ser346 and Glu696, near the L67 loop.

REFERENCES

1. ELETR, S. & G. INESI. 1972. Phospholipid orientation in sarcoplasmic reticulum membranes: spin label and proton NMR studies. Biochim. Biophys. Acta **282:** 174–179.
2. ZHANG, Z., D. LEWIS, C. STROCK, et al. 2000. Detailed characterization of the cooperative mechanism of Ca^{2+} binding and catalytic activation in the Ca^{2+} transport (SERCA) ATPase. Biochemistry **39:** 8758–8767.
3. HUA, S., G. INESI, H. NOMURA & C. TOYOSHIMA. 2002. Fe^{2+} catalyzed oxidation and cleavage of sarcoplasmic reticulum ATPase reveals Mg^{2+} and Mg^{2+}-ATP sites. Biochemistry **41:** 11405–11410.
4. INESI, G., Z. ZHANG & D. LEWIS. 2002. Cooperative setting for long range linkage of Ca^{2+} binding and ATP synthesis in the Ca^{2+} ATPase (SERCA). Biophys. J. **83:** 2327–2332.
5. PATCHORNIK, G., R. GOLDSHLEGER & J.D. KARLISH. 2000. The complex ATP-Fe^{2+} serves as a specific affinity cleavage reagent in ATP-Mg^{2+} sites of Na, K-ATPase: altered ligation of Fe^{2+} (Mg^{2+}) ions accompanies the $E_1P \rightarrow E_2P$ conformational change. Proc. Natl. Acad. Sci. USA **97:** 11954–11959.
6. ZHANG, Z., C. SUMBILLA, D. LEWIS, et al. 1995. Mutational analysis of the peptide segment linking phosphyorylation and Ca^{2+}-binding domains in the sarcoplasmic reticulum Ca^{2+}-ATPase. J. Biol. Chem. **270:** 1–8.
7. SORENSEN, T. & J.P. ANDERSEN. 2000. Importance of stalk segment S5 for intramolecular communication in the sarcoplasmic reticulum Ca^{2+}-ATPase. J. Biol. Chem.. **275:** 28954–28961.

8. ZHANG, Z., D. LEWIS, C. SUMBILLA, et al. 2001. The role of the M6-M7 loop (L67) in stabilization of the phosphorylation and Ca^{2+} binding domains of the sarcoplasmic reticulum Ca^{2+}-ATPase (SERCA). J. Biol. Chem. **276:** 15232–15239.
9. TOYOSHIMA, C. & H. NOMURA. 2002. Structural changes in the calcium pump accompanying the dissociation of calcium. Nature **418:** 605–611.
10. TOYOSHIMA, C., M. NAKASAKO, H. NOMURA & H. OGAWA. 2000. Crystal structure of the calcium pump of sarcoplasmic reticulum at 2.6 Å resolution. Nature **405:** 647–655.
11. DANKO, S., K. YAMASAKI, T. DAIHO, et al. 2001. Organization of cytoplasmic domains of sarcoplasmic reticulum Ca^{2+}-ATPase in E_1P and E_1ATP states: a limited proteolysis study. FEBS Lett. **505:** 129–135.
12. MA, H., S. HUA, G. INESI & C. TOYOSHIMA. 2002. In preparation.
13. CLARKE, D.M., T.W. LOO, G. INESI & D.H. MACLENNAN. 1989. Location of high affinity Ca^{2+}-binding sites within the predicted transmembrane domain of the sarcoplasmic reticulum Ca^{2+}-ATPase. Nature **339:** 476–478.
14. INESI, G., M. KURZMACK, C. COAN & D. LEWIS. 1980. Cooperative calcium binding and ATPase activation in sarcoplasmic reticulum vesicles. J. Biol. Chem. **255:** 3025–3031.
15. INESI, G., M. KURZMACK & D. LEWIS. 1988. Kinetic and equilibrium characterization of an energy-transducing enzyme and its partial reactions. Meth. Enzymol. **157:** 154–190.
16. DE MEIS, L. & G. INESI. 1982. ATP synthesis by sarcoplasmic reticulum ATPase following Ca^{2+}, pH, temperature, and water activity jumps. J. Biol. Chem. **257:** 1289–1294.

Mutagenesis of Residues Involved in Control of the Ca^{2+} Entry Pathway and Conformational Changes Associated with Ca^{2+} Binding in the SR Ca^{2+}-ATPase

JENS PETER ANDERSEN, JOHANNES D. CLAUSEN, ANJA PERNILLE EINHOLM, AND BENTE VILSEN

Department of Physiology, University of Aarhus, DK-8000 Aarhus C, Denmark

ABSTRACT: Rapid kinetic measurements were used to study the rate of Ca^{2+} dissociation from the high-affinity Ca^{2+} sites of the dephosphoenzyme (i.e., from the E_1Ca_2 form toward the cytoplasmic side) as well as the rate of Ca^{2+} binding with associated conformational changes ($E_2 \rightarrow E_1Ca_2$ transition) in the wild type and mutants of the sarcoplasmic reticulum Ca^{2+}-ATPase expressed in mammalian cells. Cluster mutations as well as single mutations in transmembrane segment M3 resulted in conspicuous effects on the rate of Ca^{2+} migration. Furthermore, mutation of Asp^{59} in transmembrane segment M1 to arginine exerted a profound effect on Ca^{2+} interaction. The data demonstrate an important role for M3 residues in control of the Ca^{2+} entry pathway and provide functional evidence in support of a close relationship between this pathway and the water-accessible channel leading between transmembrane segments M1 and M3 in the thapsigargin stabilized E_2 structure. In addition, rapid kinetic measurements demonstrated that the hydrogen bond network involving Asp^{813} of loop L6-7 and Lys^{758} of M5 is important for the $E_2 \rightarrow E_1Ca_2$ transition.

KEYWORDS: sarcoplasmic reticulum; Ca^{2+}-ATPase; M3; mutagenesis; Ca^{2+} dissociation; Ca^{2+} entry; rapid kinetics

The precision with which site-directed mutagenesis can reveal structure–function relationships is well illustrated by the remarkable agreement between the crystal structures of the Ca^{2+}-ATPase[1,2] and predictions based on early mutagenesis work. Thus, mutagenesis analysis assigned the Ca^{2+} binding residues to the two sites exactly as found in the E_1Ca_2 structure: Glu^{771} and Thr^{799} to site 1, Glu^{309} and Asn^{796} to site 2, and Asp^{800} to both sites.[3] Glu^{309} was predicted to be a gating residue at the inlet of the binding pocket,[4] as now indicated by its position in the E_2 structure.[5] There are numerous other examples: Mutation to Phe^{487} reduced nucleotide affinity,[6] and

---

Address for correspondence: Jens Peter Andersen, Department of Physiology, Ole Worms Allé 160, DK-8000 Aarhus C, Denmark. Voice: +45-8942-2814; fax: +45-8612-9065.
jpa@fi.au.dk

Ann. N.Y. Acad. Sci. 986: 72–81 (2003). © 2003 New York Academy of Sciences.

this residue ligates TNP-AMP in the E_1Ca_2 structure. The highly conserved Gly^{233} and $T^{181}GES$ of domain A were found essential to the E_1PCa_2 to E_2P transition,[7,8] and the crystal structures document the involvement of these residues in the conformational rearrangement of domain A that brings it into contact with domain P. Mutational analysis predicted Lys^{684} to be part of the catalytic site,[9,10] although in the primary structure Lys^{684} was located far away from the phosphorylated Asp^{351}, and in both crystal structures the side chain amino group of Lys^{684} is now seen to be located within 3 Å from Asp^{351}. Sensitivity to the specific inhibitor thapsigargin was lost upon replacement of M3 residues[11–13] that in the E_2 crystal structure are seen to interact with thapsigargin. These and many other correlations provide a solid foundation to build upon in future mutagenesis studies.

Due to the wealth of structural information now available from the crystal structures,[1,2] mutagenesis has entered a new era, in which the task is to explain the functional relevance of the structural features seen, rather than to predict structure.

CYTOPLASMIC ENTRY PATHWAY FOR Ca^{2+}

Recently, the focus has been on the nature and location of pathways for the initial migration of the Ca^{2+} ions between the cytoplasm and the intramembranous high-affinity binding sites. Here, we review evidence for an important role of transmembrane segment M3 in control of the gateway to the Ca^{2+} sites. To study the kinetics

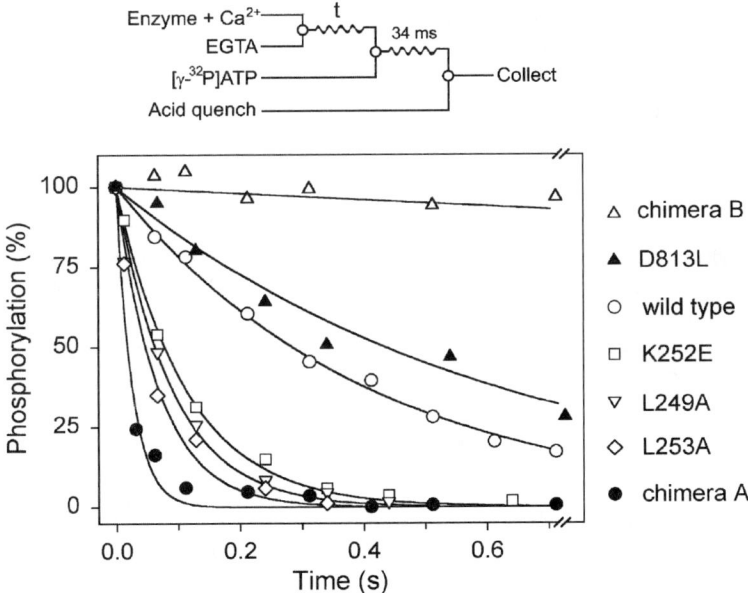

FIGURE 1. Rapid kinetic phosphorylation measurements of Ca^{2+} dissociation at 25°C. Chimera A: $Lys^{252}LeuAspGlu \rightarrow GluIleGluHis$. Chimera B: $Gly^{257}GluGlnLeu \rightarrow IleHisLeuIle$.

of Ca^{2+} interaction in wild-type and mutant Ca^{2+}-ATPases expressed in mammalian cells, we have applied rapid kinetic measurements of phosphorylation from [γ-^{32}P]ATP, using mixing protocols that take advantage of the dependence of the phosphorylation reaction on the occupancy of the Ca^{2+} sites.[10,13] With these methods, it is possible to obtain information on the rate of Ca^{2+} dissociation from the high-affinity Ca^{2+} sites of the dephosphoenzyme (i.e., from the E_1Ca_2 form toward the cytoplasmic side) (FIG. 1), as well as on the rate of Ca^{2+} binding with associated conformational changes ($E_2 \rightarrow E_1Ca_2$ transition). FIGURE 1 shows measurements of Ca^{2+} dissociation. These data were obtained by mixing the Ca^{2+}-saturated enzyme with an excess of EGTA to allow Ca^{2+} to dissociate for the indicated time intervals, followed by testing the ability to phosphorylate by a 34-ms incubation with [γ-^{32}P]ATP prior to acid quenching. Under these conditions, the ability to phosphorylate disappears at a rate corresponding to that of the first Ca^{2+} dissociation step in the sequential mechanism, because activation of phosphorylation requires the binding of both Ca^{2+} ions.[14]

Ca^{2+} Dissociation in Chimeras A and B

The first evidence for the importance of transmembrane segment M3 in connection with Ca^{2+} interaction came from studies of a chimera in which the complete M3 of the Ca^{2+}-ATPase was replaced with the corresponding M3 segment of the Na^+,K^+-ATPase, actually with the primary aim of defining the thapsigargin binding region. This chimera displayed a more than 100-fold reduction of the apparent Ca^{2+} affinity relative to wild type.[11] In continuation of this work, we examined chimeras where only short sequences of the Ca^{2+}-ATPase were replaced with the corresponding sequences of the Na^+,K^+-ATPase.[13] Two of these chimeras showed particularly interesting kinetics of Ca^{2+} interaction. Chimera A has the segment Lys^{252}-Leu-Asp-Glu^{255} at the cytoplasmic end of M3 replaced by the corresponding Na^+,K^+-ATPase residues, Glu-Ile-Glu-His. Chimera B has Gly^{257}-Glu-Gln-Leu^{260} replaced by the corresponding Na^+,K^+-ATPase residues, Ile-His-Leu-Ile (actually reducing thapsigargin sensitivity 100-fold). As seen in FIGURE 1, the effects on Ca^{2+} dissociation were conspicuous, chimera A showing a 17-fold enhancement of the Ca^{2+} dissociation rate and chimera B a 23-fold reduction, relative to wild type. To obtain maximum time resolution, these studies were conducted at pH 6, where Ca^{2+} dissociation is rather slow in the wild type. It was also possible to determine the rate of Ca^{2+} dissociation pH 7, based on a single time point obtained 34 ms after mixing Ca^{2+}-saturated enzyme simultaneously with [γ-^{32}P]ATP and EGTA. Under these conditions, a 6-fold increase of the Ca^{2+} dissociation rate was found for chimera A and a 4-fold decrease for chimera B.[13]

Ca^{2+} Dissociation in M3 Point Mutants

The results just described encouraged us to study a series of point mutants with alterations to M3 residues. It is seen in FIGURE 1 that at pH 6, mutants $Leu^{249} \rightarrow$ Ala, $Lys^{252} \rightarrow$ Glu, and $Leu^{253} \rightarrow$ Ala displayed a 4- to 7-fold increase of the rate of Ca^{2+} dissociation relative to wild type. At pH 7, a 3- to 5-fold increase was seen for these mutants. Thus, it was not only the cluster mutation in chimera A that increased Ca^{2+} dissociation, but single mutations near the cytoplasmic end of M3 were effective as well. Other mutations to Lys^{252}, including alanine and methionine substitutions, had less influence (up to 2-fold increase of the Ca^{2+} dissociation rate; see extended abstract by Clausen and Andersen elsewhere in this book), and mutations to Asp^{254} and Glu^{255} (even $Glu^{255} \rightarrow His$) showed no effect at all at pH 7.0. It follows that the enhanced Ca^{2+} dissociation seen for chimera A is largely due to the replacement of Lys^{252} with glutamate, with a possible contribution from the replacement of Leu^{253} with isoleucine (depending on whether isoleucine present in chimera A substitutes better for leucine than alanine).

On-rate for Ca^{2+} in M3 Chimeras and Point Mutants

A general finding for chimeras A and B, as well as the point mutants, was that the apparent Ca^{2+} affinity observed by titration of the Ca^{2+} dependence of phosphorylation at steady state was not changed as expected on the basis of the observed Ca^{2+} dissociation rates. In fact, at 25°C chimera A, as well as the point mutants $Leu^{249} \rightarrow$ Ala, $Lys^{252} \rightarrow$ Glu, and $Leu^{253} \rightarrow$ Ala, displayed wild type–like apparent Ca^{2+} affinity, although a reduced Ca^{2+} affinity was expected from the increase of Ca^{2+} dissociation rate, and chimera B, surprisingly, displayed reduced apparent affinity for Ca^{2+}, although an increased Ca^{2+} affinity was expected on the basis of the reduced Ca^{2+} dissociation rate. This suggests that the on-rate for Ca^{2+} was affected in the same direction as the off-rate (increased for chimera A and the point mutants and reduced for chimera B). Direct evidence supporting an increased on-rate for Ca^{2+} was obtained for chimera A[13] and point mutants $Leu^{249} \rightarrow$ Ala and $Lys^{252} \rightarrow$ Glu in rapid quench phosphorylation experiments, where $[\gamma-^{32}P]ATP$ and Ca^{2+} were added simultaneously to enzyme deprived of Ca^{2+} (FIG. 2).

Correlation with Structure

The E_1Ca_2 crystal structure[1] does not reveal the entry pathway to the Ca^{2+} sites (FIG. 3A), presumably because the pathway is closed following the binding of Ca^{2+} ("occlusion"). However, in the E_2 structure,[2] a water-accessible pocket, or channel, that is a candidate for a Ca^{2+} entry pathway, has opened up between M1 (which is

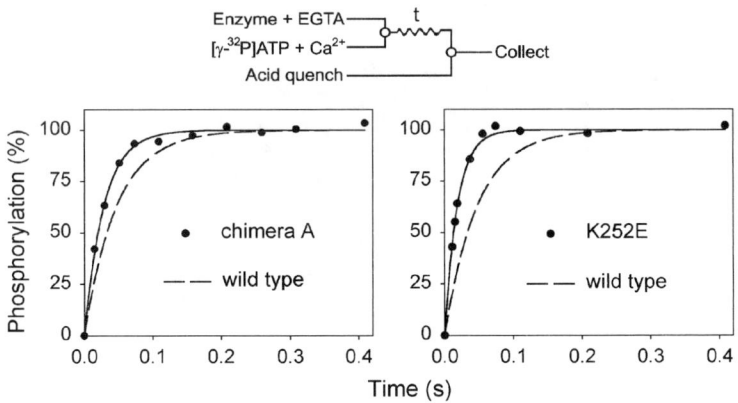

FIGURE 2. Rapid kinetic phosphorylation measurements of Ca^{2+} binding at 25°C.

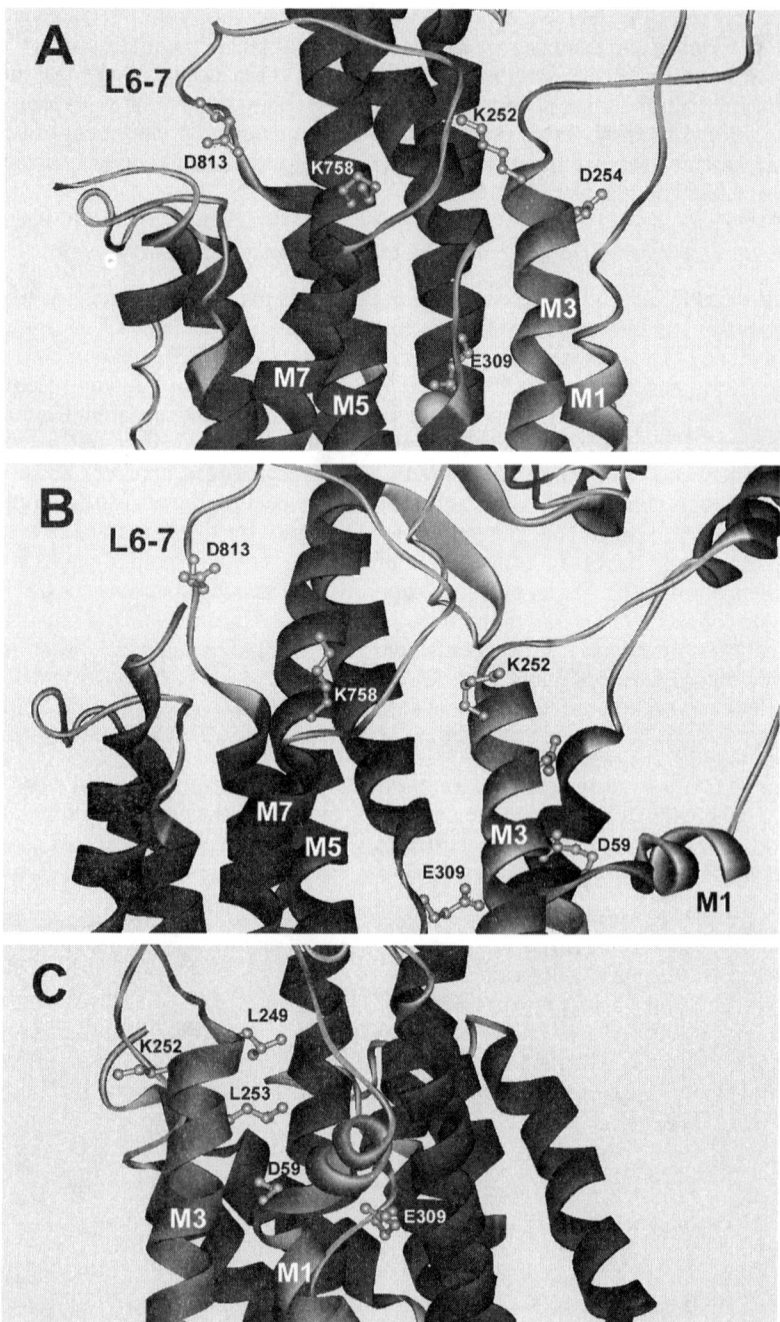

FIGURE 3. *See following page for legend.*

bent in E_2) and M3 (FIG. 3B and FIG. 3C) and leads to the Ca^{2+} binding M4 residue, Glu^{309}. The two hydrophobic side chains of Leu^{249} and Leu^{253} form an integral part of the hydrophobic wall lining this channel (FIG. 3C). The prominent effects on Ca^{2+} interaction of substitution of Leu^{249} and Leu^{253} with the smaller, but still hydrophobic, alanine may result from a destabilization of the wall of the channel, changing channel proportions or allowing the Ca^{2+} ions to slip away from the E_1Ca_2 form by an alternative route.

In the E_1Ca_2 structure, Lys^{252} seems to form hydrogen bonds to the two backbone carbonyls of Pro^{824} and Glu^{826} in the C-terminal part of the cytoplasmic loop connecting M6 and M7 ("L6-7") (FIG. 3A), whereas in the E_2 structure, the side chain of Lys^{252} has moved away from L6-7 (FIG. 3B). This movement involves the whole M3 helix, which inclines and moves downwards (toward the lumen). Simultaneously, M1 moves upwards and bends at the suggested Ca^{2+} inlet (FIG. 3C), possibly as a consequence of steric collision with M3.[2] The finding that the alanine substitution of Lys^{252} exerted less functional perturbation than the glutamate substitution of Lys^{252} shows that the prominent effects of the glutamate substitution can only be partially accounted for by breakage of the hydrogen bonds to the L6-7 backbone. In addition, the negative charge of the glutamate must play an important role, possibly by electrostatically repelling the backbone carbonyls of L6-7, thereby bringing about a movement of M3 resembling to some extent that occurring when the channel opens in the wild-type enzyme during transition from E_1Ca_2 to E_2.

A Possible Role in Cation Selectivity

The very low rate of Ca^{2+} migration seen for chimera B is interesting in relation to the problem of understanding the specific ion selectivity of P type ATPases. The residues replaced with corresponding Na^+,K^+-ATPase residues in chimera B are rather close to the channel inlet between M1 and M3 seen in the E_2 structure, and the effect on Ca^{2+} migration could be due to the replacement of Glu^{258} with the positively charged histidine which would tend to repel the Ca^{2+} ion. Elsewhere in this book, Toustrup-Jensen and Vilsen note that residues in the same region of the Na^+,K^+-ATPase as discussed here for Ca^{2+}-ATPase, including Glu^{282}, Ile^{283}, and His^{288} at the positions equivalent to Lys^{252}, Leu^{253}, and Glu^{258} of the Ca^{2+}-ATPase are important for the Na^+ binding properties of the E_1 form. This could be a sign that these residues are related to the entry pathway for Na^+, thus suggesting a common location of the cation inlet in Ca^{2+}-ATPase and Na^+,K^+-ATPase. The region around the opening between M1 and M3 may play a role in determining the specific cation selectivity, thus explaining the multiple amino acid differences between the Ca^{2+}-ATPase and the Na^+,K^+-ATPase in this region.

FIGURE 3. Structural organization of Ca^{2+}-ATPase residues discussed. (**A**) E_1Ca_2 structure (Protein Data Bank accession code 1EUL: note the interactions of Lys^{252} and Lys^{758} with the backbone of L6-7. (**B**) E_2 structure (Protein Data Bank accession code 1IWO): same view as in (**A**); the suggested Ca^{2+} entry pathway is from the right, between M1 and M3 close to Asp^{59}. Note that the opening of this pathway involves a change of orientation of Lys^{252}, away from L6-7. (**C**) E_2 structure (Protein Data Bank accession code 1IWO): view right into the water-accessible pocket between M1 and M3 suggested as Ca^{2+} entry pathway. Note Leu^{249} and Leu^{253} lining the pocket.

Asp^{59} of M1

As described in the extended abstract by Einholm *et al.* elsewhere in this book, we have also studied a series of mutations to acidic residues in M1. Mutation of Asp^{59} to arginine induced a 10-fold increase of the Ca^{2+} dissociation rate (observed using the same method and conditions as for FIG. 1). Thus, the effect of the $Asp^{59} \rightarrow$ Arg mutation on Ca^{2+} interaction is even more pronounced than seen for the M3 point mutants. Asp^{59} is located right at the channel inlet between M1 and M3 seen in the E_2 structure (FIGS. 3B and 3C).

Conclusions Regarding the Cytoplasmic Entry Pathway

Our data demonstrate an important role for M3 residues in control of the Ca^{2+} entry pathway and provide functional evidence in support of a close relationship between this pathway and the water-accessible channel leading between transmembrane segments M1 and M3 in the thapsigargin-stabilized E_2 structure.[2] A crucial question that needs to be answered is, however, whether we should actually expect to see an open entry pathway for Ca^{2+} in this structure. Kinetic experiments have shown that Ca^{2+} binding is rather slow (rate constant 1.5 s^{-1} at 20°C), when Ca^{2+} is added at pH 6 where the E_2 form prevails,[15] and thapsigargin, moreover, prevents Ca^{2+} binding.[16] Therefore, the channel could be an exit pathway for countertransported protons rather than an entry pathway for Ca^{2+}, and it may have to undergo further rearrangement to be able to receive Ca^{2+}. On the other hand, such a rearrangement could involve only small movements, mainly of Glu^{309} at the bottom of the channel, functioning as a gating residue involved in the binding of Ca^{2+} as well as the release of countertransported protons.[4,17] In this scenario, the extensive conformational rearrangement related to the $E_2 \rightarrow E_1Ca_2$ transition, as seen by comparison of the two crystal structures, would occur after binding of the first Ca^{2+} ion. This hypothesis needs further confirmation.

CONFORMATIONAL CHANGES ASSOCIATED WITH Ca^{2+} BINDING

As an alternative to a water-accessible channel for Ca^{2+} migration to the membrane sites, one could imagine that Ca^{2+} ions located at specific pre-binding sites are moved to their final destination by conformational changes induced by Ca^{2+} binding. Three aspartic acid residues in the N-terminal part of L6-7 have been suggested to make up such a pre-binding site, with L6-7 acting as a switch that by undergoing conformational changes guides Ca^{2+} to intramenbranous binding.[18] On the basis of the observation of a low steady-state phosphorylation level after triple mutation of these aspartates, it has, however, been argued that their roles in connection with Ca^{2+} binding is of a more indirect nature, involving a disturbance of catalytic activation.[19]

We have found that mutation of only one of the aspartates, Asp^{813}, to leucine, is sufficient to reduce both the apparent Ca^{2+} affinity and the steady-state phosphorylation level. Measurement of the rate of Ca^{2+} dissociation showed a slight (1.5-fold) reduction, relative to wild type (FIG. 1). FIGURE 4 shows transient kinetics of phosphorylation of enzyme preincubated with a saturating Ca^{2+} concentration to ensure that Ca^{2+} was already bound at the moment when [γ-^{32}P]ATP was added. In contrast to the wild type, $Asp^{813} \rightarrow$ Leu showed a large phosphorylation overshoot. Phospho-

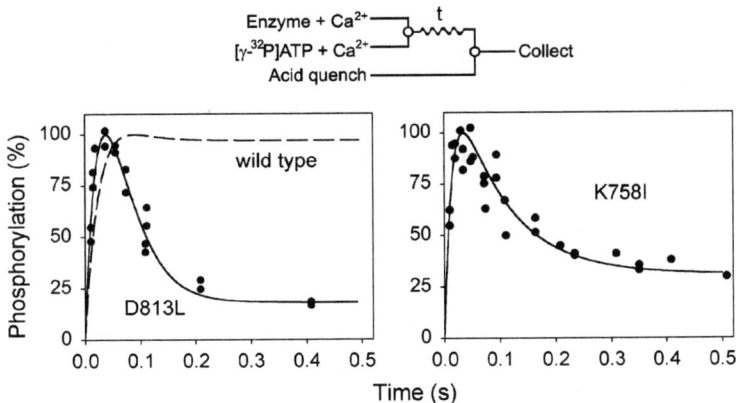

FIGURE 4. Transient kinetics of phosphorylation at 25°C of enzyme preincubated with 100 mM Ca^{2+}.

rylation was actually rapid and reached a high initial value, but then declined to a low steady-state value. This overshoot indicates that dephosphoenzyme accumulates at steady state, likely because the $E_2 \to E_1Ca_2$ transition, which has to take place after dephosphorylation, is slow in the mutant. Because the phosphorylation was rapid, there seems to be no defect in catalytic activation ($E_1Ca_2 \to E_1PCa_2$). The slow $E_2 \to E_1Ca_2$ transition could in principle result either from disruption of the suggested pre-binding site for Ca^{2+} in L6-7,[18] or from interference with conformational change(s). For comparison, we have included data for a previously characterized mutant, $Lys^{758} \to Ile$,[20,21] which also showed a reduced apparent affinity for Ca^{2+} and a large phosphorylation overshoot as a result of a slow $E_2 \to E_1Ca_2$ transition (FIG. 4). Because of the positive charge, the lysine is unlikely to be directly involved in Ca^{2+} binding. It is interesting that both Asp^{813} and Lys^{758} participate in a network of hydrogen bonds involving L6-7, which changes in relation to the $E_2 \to E_1Ca_2$ transition (compare FIG. 3A with FIG. 3B). In the E_1Ca_2 structure,[1] Asp^{813} hydrogen bonds to Asn^{755} in M5 and Ser^{917} in L8-9, whereas in the E_2 structure[2] these hydrogen bonds are broken and replaced by others (to Arg^{819} in L6-7). Lys^{758} hydrogen bonds to the backbone carbonyl of Leu^{828} in L6-7 in the E_1Ca_2 structure, whereas in E_2 it interacts with other L6-7 residues. Hence, Asp^{813}, as well as Lys^{758}, are active participants in the conformational changes occurring in L6-7 in relation to the $E_2 \to E_1Ca_2$ transition.

ACKNOWLEDGMENTS

This study was supported by grants from the Danish Medical Research Council, the Novo Nordisk Foundation, the Lundbeck Foundation, and the Research Foundation of Aarhus University, Denmark.

REFERENCES

1. TOYOSHIMA, C., M. NAKASAKO, H. NOMURA & H. OGAWA. 2000. Crystal structure of the calcium pump of sarcoplasmic reticulum at 2.6 A resolution. Nature **405**: 647–655.
2. TOYOSHIMA, C. & H. NOMURA. 2002. Structural changes in the calcium pump accompanying the dissociation of calcium. Nature **418**: 605–611.
3. ANDERSEN, J. P. & B. VILSEN. 1994. Amino acids Asn^{796} and Thr^{799} of the Ca^{2+}-ATPase of sarcoplasmic reticulum bind Ca^{2+} at different sites. J. Biol. Chem. **269**: 15931–15936.
4. VILSEN, B. & J. P. ANDERSEN. 1998. Mutation to the glutamate in the fourth membrane segment of Na^+,K^+-ATPase and Ca^{2+}-ATPase affects cation binding from both sides of the membrane and destabilizes the occluded enzyme forms. Biochemistry **37**: 10961–10971.
5. GREEN, N. M. & D. H. MACLENNAN. 2002. Calcium calisthenics. Nature **418**: 598–599.
6. MCINTOSH, D. B., D. G. WOOLLEY, B. VILSEN & J. P. ANDERSEN. 1996. Mutagenesis of segment $^{487}FSRDRK^{492}$ of sarcoplasmic reticulum Ca^{2+}-ATPase produces pumps defective in ATP binding. J. Biol. Chem. **271**: 25778–25789.
7. ANDERSEN, J. P., B. VILSEN, E. LEBERER & D. H. MACLENNAN. 1989. Functional consequences of mutations in the b-strand sector of the Ca^{2+}-ATPase of sarcoplasmic reticulum. J. Biol. Chem. **264**: 21018–21023.
8. CLARKE, D. M., T. W. LOO & D. H. MACLENNAN. 1990. Functional consequences of mutations of conserved amino acids in the b-strand domain of the Ca^{2+}-ATPase of sarcoplasmic reticulum. J. Biol. Chem. **265**: 14088–14092.
9. VILSEN, B., J. P. ANDERSEN & D. H. MACLENNAN. 1991. Functional consequences of alterations to amino acid residues located in the hinge domain of the Ca^{2+}-ATPase of sarcoplasmic reticulum. J. Biol. Chem. **266**: 16157–16164.
10. SØRENSEN, T. L.-M., Y. DUPONT, B. VILSEN & J. P. ANDERSEN. 2000. Fast kinetic analysis of conformational changes in mutants of the Ca^{2+}-ATPase of sarcoplasmic reticulum. J. Biol. Chem. **275**: 5400–5408.
11. NØRREGAARD, A., B. VILSEN & J. P. ANDERSEN. 1994. Transmembrane segment M3 is essential to thapsigargin sensitivity of the sarcoplasmic reticulum Ca^{2+}-ATPase. J. Biol. Chem. **269**: 26598–26601.
12. YU, M., L. ZHANG, A. K. RISHI, et al. 1998. Specific substitutions at amino acid 256 of the sarcoplasmic/endoplasmic reticulum Ca^{2+} transport ATPase mediate resistance to thapsigargin in thapsigargin-resistant hamster cells. J. Biol. Chem. **273**: 3542–3546.
13. ANDERSEN, J. P., T. L.-M. SØRENSEN, K. POVLSEN & B. VILSEN. 2001. Importance of transmembrane segment M3 of the sarcoplasmic reticulum Ca^{2+}-ATPase for control of the gateway to the Ca^{2+}-sites. J. Biol. Chem. **276**: 23312–23321.
14. PETITHORY, J. R. & W. P. JENCKS 1988. Sequential dissociation of Ca^{2+} from the calcium adenosinetriphosphatase of sarcoplasmic reticulum and the calcium requirement for its phosphorylation by ATP. Biochemistry **27**: 5553–5564.
15. FORGE, V., E. MINTZ & F. GUILLAIN. 1993. Ca^{2+} binding to sarcoplasmic reticulum ATPase revisited. II. Equilibrium and kinetic evidence for a two-route mechanism. J. Biol. Chem. **268**: 10961–10968.
16. SAGARA, Y. & G. INESI 1991. Inhibition of the sarcoplasmic reticulum Ca^{2+} transport ATPase by thapsigargin at subnanomolar concentrations. J. Biol. Chem. **266**: 13503–13506.
17. ANDERSEN, J. P. & B. VILSEN 1992. Functional consequences of alterations to Glu^{309}, Glu^{771}, and Asp^{800} in the Ca^{2+}-ATPase of sarcoplasmic reticulum. J. Biol. Chem. **267**: 19383–19387.
18. MENGUY, T., F. CORRE, B. JUUL, et al. 2002. Involvement of the cytoplasmic loop L6-7 in the entry mechanism for transport of Ca^{2+} through the sarcoplasmic reticulum Ca^{2+}-ATPase. J. Biol. Chem. **277**: 13016–13028.
19. ZHANG, Z., D. LEWIS, C. SUMBILLA & G. INESI. 2001. The role of the M6-M7 loop (L67) in stabilization of the phosphorylation and Ca^{2+} binding domains of the sarcoplasmic reticulum Ca^{2+}-ATPase (SERCA). J. Biol. Chem. **276**: 15232–15239.

20. SØRENSEN, T., B. Vilsen & J. P. Andersen. 1997. Mutation Lys758→Ile of the sarcoplasmic reticulum Ca^{2+}-ATPase enhances dephosphorylation of E2P and inhibits the E_2 to E_1Ca_2 transition. J. Biol. Chem. **272:** 30244–30253.
21. SØRENSEN, T. L-M., Y. DUPONT, B. VILSEN & J. P. ANDERSEN. 2000. Fast kinetic measurements of mutational effects on conformational changes in SR Ca^{2+}-ATPase. *In* Na/K ATPase and Related ATPases. K. Taniguchi & S. Kaya, Eds.: 405–408. Excerpta Medica–Elsevier Science. Amsterdam.

Proteolytic Studies on the Transduction Mechanism of Sarcoplasmic Reticulum Ca^{2+}-ATPase

Common Features with Other P-Type ATPases

JESPER V. MØLLER,[a] GUILLAUME LENOIR,[b] MARC LE MAIRE,[b] BIRTE STÆHR JUUL,[a] AND PHILIPPE CHAMPEIL[b]

[a]*Department of Biophysics, University of Aarhus, DK-8000 Aarhus C, Denmark*

[b]*Section de Biophysique des Fonctions Membranaires, Dépatement de Biologie Joliot Curie, CEA, et CNRS, Unité de Recherche Associée 2096, F-91191, Orsay Cedex, France*

> ABSTRACT: After proteinase K–induced excision of five amino acid residues in the semiconserved polypeptide chain linking the end of the A domain with the S3/M3 transmembrane segment we find that Ca^{2+} transport is blocked while partial reactions like Ca^{2+} binding, ATP phosphorylation, and Ca^{2+}-occlusion are left intact. However, formation of the so-called E2P state (either from the phosphorylated species formed in the presence of ATP and Ca^{2+} or from the Ca^{2+}-depleted unphosphorylated species) is blocked. We conclude that the proteinase K–treated ATPase, while maintaining many of the partial reactions, is incapable of energy transduction because of the absence of an E2P state with Ca^{2+} binding sites exposed to the intravesicular space. Sequence comparisons and mutagenesis data point to an important role in energy transduction of P-type ATPases of a conserved motif located at the end of the A domain.
>
> KEYWORDS: Ca^{2+}-ATPase; transduction; proteinase K; Na^+,K^+-ATPase; proton ATPases

INTRODUCTION

Studies on the relationship between structure and function of P-type ATPases have benefitted from the use of both molecular biology techniques as well as more classical protein–chemical techniques. Proteolysis was introduced by Jørgensen[1,2] as a means to investigate functional properties of Na^+,K^+-ATPase. Three well defined N-terminal- and conformation-sensitive tryptic splits (T1-T3) were described.[1] When chymotrypsin was made to react with the ATPase in its E1 conformation, a single cleavage site (C3), localized at Leu-266 in the same region as the T3 tryptic site, was found, and the corresponding complementary peptides were formed in

---

Address for correspondence: Jesper V. Møller, Department of Biophysics, University of Aarhus, Ole Worms Allé 185, DK-8000 Aarhus C, Denmark. Voice: +45-8942-2938; fax: +45-8612-9599.

jvm@biophys.au.dk

large amounts, permitting a study of the intermediary reactions of the cleaved ATPase.[2] This cleavage led to a drastic decrease in enzyme activity and transport, despite the fact that the ability to form an ADP-sensitive phosphorylated intermediate from ATP was left intact, as was the ability to occlude K^+. It was concluded that the cleavage led to a defect in the E1P to E2P transition, resulting in a block of cation transport and enzyme activity.[2]

For the SERCA1a isoform of sarcoplasmic reticulum (hereafter referred to as Ca^{2+}-ATPase) proteolytic splits by V8 protease at Glu-232 (Ref. 3) and by proteinase K at Glu-243 (Ref. 4) (both located in the N-terminal part) have also been found to result in enzymatic inactivation, despite the fact that the ability of the cleaved ATPase to be phosphorylated from ATP in the presence of Ca^{2+} was intact. Similar findings have been reported by trypsinolysis in the same region, C-terminal to the well-recognized A2 cleavage site at Arg-198 (Ref. 5). All of these cleavage sites, leading to a block of ATP hydrolysis in Ca^{2+}-ATPase, are localized at the end of the A domain or beginning of the polypeptide chain that links the A domain with the S3-M3 helical segment. This link, which in the Ca^{2+} bound ATPase crystal form is in an exposed position,[6] therefore seems to be of crucial importance for energy transduction. This view is supported by site-directed mutagenesis of selected amino acid residues in yeast proton ATPase[7] and of Gly-233 in Ca^{2+}-ATPase.[8] In the following we review recent data[9] that we have obtained by proteolysis of Ca^{2+}-ATPase with proteinase K (PK), which bears on the role in energy transduction of this N-terminal link between the A domain and the membraneous sector of Ca^{2+}-ATPase.

DEFINED N-TERMINAL CLEAVAGE OF Ca^{2+}-ATPase BY PROTEINASE K

A frequent problem encountered in the use of proteolytic enzymes for functional studies is the formation of a large variety of cleavage products, arising from the fact that several parts of the polypeptide chain are likely to undergo proteolysis concurrently. However, with PK we have succeeded in preparing in large yield a well-defined proteolytic cleavage product of Ca^{2+}-ATPase.[9] This was done by adding to the proteolysis medium both Ca^{2+} and ATP analogues or ADP as protective ligands. As can be seen from FIGURE 1 this treatment results in the formation of two major cleavage products of Ca^{2+}-ATPase, a C-terminal fragment (denoted as p83C) and an N-terminal fragment (denoted as p28N) that were identified by their reactivities with C-terminal and N-terminal sequence-specific Ca^{2+}-ATPase antibodies, respectively. Edman degradation of p83C identified the N-terminus as Asp-244, and p83C thus represents the 244-994 fragment of Ca^{2+}-ATPase. The C-terminus of p28N was determined after further treatment of the N-terminal peptide with trypsin to cleave the peptide at Arg-198; the identity of the released p4 fragment was found to be A199-Q237 by mass spectrometry, identifying p28N as peptide 1-237. The ends of p28N and p83C are thus not contiguous, and we conclude that the intervening gap, comprising the ^{238}MAATE^{243} sequence in intact ATPase, has been excised from the ATPase during the extensive treatment with PK.

The fact that five amino acid residues have been excised does not appear to affect gross structural features. Thus the p28N/p83C complex remains associated after solubilization with $C_{12}E_8$ and elutes at the same position as intact ATPase by size

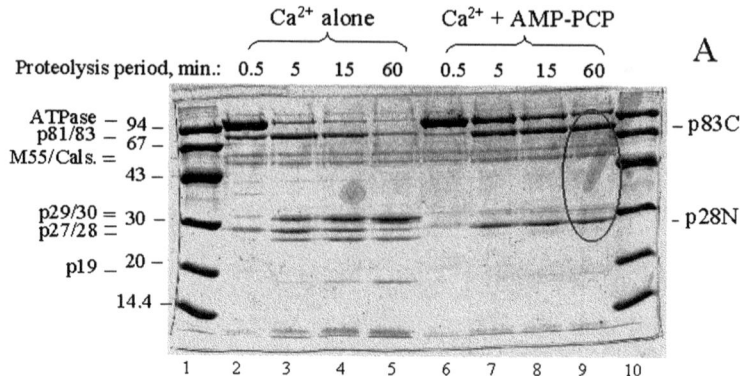

FIGURE 1. Treatment of Ca^{2+}-ATPase with proteinase K in the presence of Ca^{2+} results in the formation of p83C and p28N fragments (lanes 2–5), which can be stabilized against further degradation by addition of AMP-PCP (lanes 6–9). (Reaction conditions: sarcoplasmic reticulum vesicles, 1 mg protein/mL; proteinase K: 0.03 mg/mL; Ca^{2+}: 0.3 mM; AMPPCP: 0.5 mM pH 6.5; temp: 20°C). (From Møller et al.[9] Reproduced by permission.)

exclusion chromatography on a HPLC column. After thorough proteolytic treatment in the presence of 0.3 mM Ca^{2+} and 0.5 mM AMPPCP, about 50–70% of the original ATPase can be recovered as the p28N/p83C complex, the remainder being made up of further cleaved and inactive peptide together with a small (5–10%) fraction of uncleaved ATPase, which resists proteolytic cleavage after prolonged treatment with PK even in the absence of protecting ligands. It should be emphasized that the simultaneous binding of Ca^{2+} and nucleotide is necessary to obtain a high yield of p28N and p83C. We conclude that under these particular conditions the region linking the A domain with the S3/M3 segment remains sufficiently exposed to be subject to proteolytic attack, while at the same time the remainder of the Ca^{2+}-ATPase polypeptide is efficiently protected from proteolytic cleavage.

FUNCTIONAL PROPERTIES OF THE PROTEINASE K–CLEAVED Ca^{2+}-ATPase

Detailed analysis of the functional properties of the PK-treated preparations indicates that ATP hydrolysis by the p83C/p28N complex is virtually blocked (being reduced to a level below 1% of that of the intact enzyme), leading to cessation of active transport.[9] However, phosphorylation by ATP remains intact and full phosphorylation is obtained at an even lower concentration of Ca^{2+} than in intact ATPase. Direct binding measurements indicate that there is, in fact, little difference in the affinity of the intact enzyme and the p83C/p28N complex for binding of Ca^{2+} (half-maximal binding was observed at pCa 5.66 ± 0.05 and 5.47 ± 0.05, respectively, at pH 7.0 and 5 mM Mg^{2+}). The cleaved ATPase also responds to binding of Ca^{2+} by changes in intrinsic (tryptophan) and extrinsic (FITC) fluorescence in very nearly the same manner as the intact enzyme, both in kinetic and steady-state experiments.[9] The phosphorylated enzyme is rapidly dephosphorylated by addition of (ADP + EGTA), but after addition of EGTA alone, release of Ca^{2+} from the cleaved ATPase is delayed, consistent with intramembranous occlusion of Ca^{2+}, similar to what is found with intact ATPase. Occlusion of Ca^{2+} by the p83C/p28N complex could also be detected after reaction with CrATP, followed by solubilization with $C_{12}E_8$ and size exclusion chromatography.

On the other hand, substantial differences in behavior were observed in experiments designed to explore other partial reactions of the cleaved ATPase. Thus, the formation of E2P under conditions optimal for its formation by the intact ATPase (absence of Ca^{2+} and presence of 40% dimethylsulfoxide; see FIG. 2A) was severely curtailed (an appreciable part of what remains in the experiment with the PK-treated preparation in FIGURE 2B is accounted for by the aforementioned presence of intact ATPase in our PK-treated preparation). After phosphorylation with ATP in the pres-

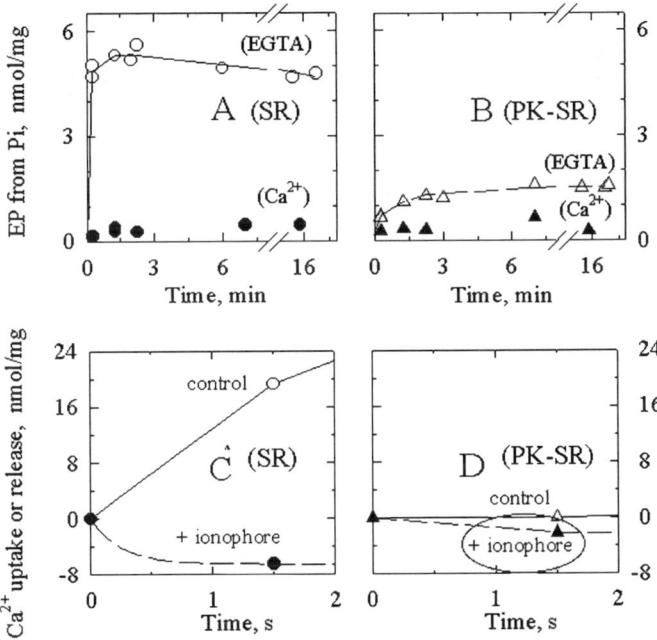

FIGURE 2. Deficient formation of E2P by the p83C/p28N complex (PK–SR), compared to that of intact ATPase (SR), either after reaction with inorganic phosphate in absence of Ca^{2+} (**A, B**) or from E1P formed by reaction with ATP (**C, D**). Formation of E2P from Pi (**A, B**) was measured by Millipore filtration in a standard assay. (Reaction conditions: 40% dimethylsulfoxide (v/v), 0.2 mM ^{32}Pi, 5 mM Mg^{2+}, and pH 7.0 buffer with either 1 mM EGTA (*open symbols*) or 0.1 mM Ca^{2+} [for control]). Formation of E2P from E1P (**C, D**) Vesicles, deposited on Millipore filters in the presence of 20% (v/v) dimethylsulfoxide, were perfused at pH 7.0 with 50 μM $^{45}Ca^{2+}$, and then in addition with 50 μM MgATP and 5 mM Mg^{2+} for various periods, in the presence (i, *closed symbols*) or absence (ii, *open symbols*) of ionophore (A23187), to reveal (i) changes in the binding of Ca^{2+} by the phosphorylated ATPase as a result of conversion to the E2P state, or (ii) changes in the medium concentration caused by uptake into the vesicles. (The figure is based on data presented in Møller[9] and reproduced by permission.)

ence of Ca^{2+} and dimethylsulfoxide, conversion of E1P to E2P, as monitored by the release of bound Ca^{2+}, was also strongly reduced and occurred at a much slower rate for the cleaved (FIG. 2D) as compared to the intact ATPase (FIG. 2C). In experiments with FITC-labeled ATPase we obtained spectrofluorometric evidence for the inability of the cleaved enzyme to bind orthovanadate tightly, and also found that it reacted with thapsigargin in a different way than in intact ATPase.[9]

THE STRUCTURAL BASIS FOR ENERGY TRANSDUCTION

Our data obtained by limited proteolytic cleavage of Ca^{2+}-ATPase suggest that ATP hydrolysis and Ca^{2+} transport are blocked because the cleaved ATPase is unable to form the E2P species, at least at a reasonable rate, either from the E1P or the E2 state. This is the case, although conformational changes associated with other processes like Ca^{2+} binding and ATP phosphorylation seem to proceed in a normal fashion. According to the scheme for Ca^{2+} transport shown in FIGURE 3, this means that the translocation sites for Ca^{2+} after PK treatment do not open toward the lumen, neither in the forward nor in the backward direction (i.e., after reaction with Pi). As noted in the INTRODUCTION, a phenotype with related properties was previously described for Na^+,K^+-ATPase after cleavage with chymotrypsin at Leu-266 (Ref. 2).

Recent studies on the mechanism of energy transduction in P-type ATPases have focused on the role played by the A domain. By oxidative Fe^{2+}-cleavage, evidence

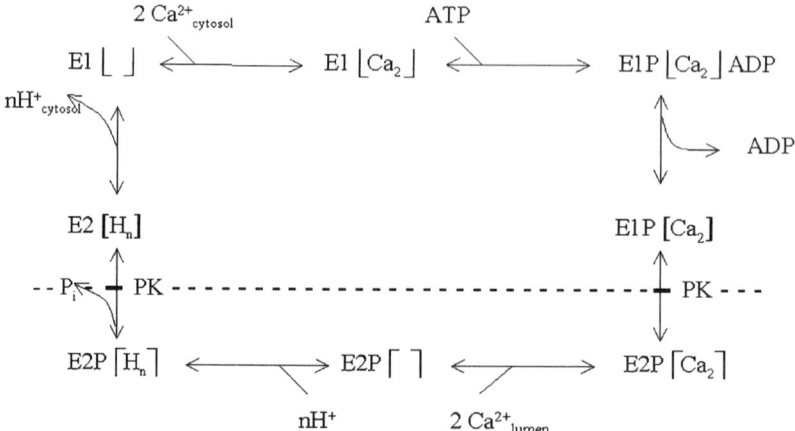

FIGURE 3. Scheme summarizing essential features of the Ca^{2+}-ATPase transport cycle and the effect of proteinase K on the partial reactions. The *upper part* shows partial reactions with the Ca^{2+} binding sites exposed toward the outside (cytosol), whereas the *lower part* depicts partial reactions associated with Ca^{2+} binding sites exposed toward the luminal side. The electrogenicity of the Ca^{2+} transport process is assumed to be partially compensated by exchange with protons in the opposite direction. The forms E1P[Ca_2] and E2[H_n] denote intermediates with occluded Ca^{2+} and protons. Symbols ⌊ ⌋ and ⌈ ⌉ refer to sites accessible from the cytosol or lumen, respectively.

of reorientation of cytosolic domains in the E2 and E2P states (as compared to E1-like states) has been obtained for Na^+,K^+-ATPase[10] and H^+,K^+-ATPase[11] that brings the conserved TGES motif of the A domain in juxtaposition to the phosphorylation site. Direct evidence for such a transition is now available in the high-resolution structure for Ca^{2+}-ATPase in E2 (thapsigargin) crystals.[12] This interaction results in a compact structure because of rotation of the A domain in a plane parallel to the membrane and movement of the N domain from a tilted to an upright position. As a result the previously exposed link between the A domain and the S3/M3 segment moves from a peripheral to a more central position in the structure at which some of the amino acid residues at the PK cleavage site (mainly Gln-244, Thr-242 and Met-239) seem to be stabilized by hydrogen bonds with Lys-712 in the P6 helix, located at the end of the P domain.[12]

It is therefore possible that these interactions, which will be lost after PK treatment, are involved in the processes leading to energy transduction. However, previously published studies of ATPase mutants in the MAATE[243] sequence, by which up to two of the aforementioned hydrogen bonds should be removed, did not reveal much effect of these modifications on the ATPase transport properties.[8] Another possibility would be that the link between the A domain and the S3/M3 segment is critical for the transduction event because it transmits conformational changes originating from the cytosolic domains further on toward the membraneous sector, resulting in the deeper insertion of the S3/M3 segment as seen in the new E2 structure,[12] an effect that, of course, would be disrupted by the PK cleavage.

However, as an alternative possibility excision of the MAATE[243] sequence by the proteinase K treatment also may have more indirect structural consequences that could lead to interference with the function of the A domain. Evidence that this is actually the case was obtained by comparing the effect of trypsin proteolysis on intact ATPase and the p83C/p28N complex. In the former case we confirmed that cleavage at Arg-198, resulting in the formation of the N-terminal A2 fragment, was protected in the E2 state,[13] in agreement with the more central location of the A domain in the new E2 (thapsigargin) structure.[12] However, this protection by EGTA addition was not observed in the PK-cleaved ATPase where the p28N peptide underwent more rapid degradation as the result of cleavage at Arg-198 and at other unidentified sites. An enhanced general sensitivity to proteolytic attack by trypsin was also found in the presence of Ca^{2+}, and during cleavage with V8 protease.

With the notable exception of plasma membrane Ca^{2+}-ATPase the amino acid sequence of the link between the A domain and S3/M3 segment has conserved features among P2-type ATPases (FIG. 4). Mutagenesis studies on various P-type ATPases have pointed to the presence of a critical region slightly N-terminal to the PK cleavage site,[7,8] located at the C-terminal end of the A domain. In this border region there is a conserved Gly and a conserved basic (lysine or arginine) residue, flanked by two aliphatic residues; mutations of the central residues in this motif have generally been associated with reduced enzyme activity and by reduced interaction with vanadate or reaction with inorganic phosphate in Ca^{2+}-ATPase[8] and proton ATPase.[7] In addition V8 protease cleavage at Glu-231 (Ref. 3) seems to lead to Ca^{2+}-ATPase with similar phenotypic characteristics as the PK cleaved enzyme. It could be that the conserved motif is of primary importance by participating in the conformational changes associated with energy transduction which are transmitted to the membranous domain from the A domain and the phosphorylation site (perhaps mainly via

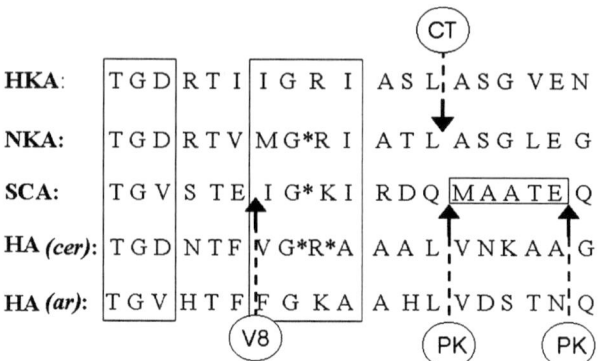

FIGURE 4. Alignment for various P-type ATPases of the end of the A domain and the N-terminal part of the polypeptide chain linking the A domain with S3. The T G X turn motif is followed by the motif here suggested to be functionally important in energy transduction, as also supported by site-directed mutagenesis studies (residues marked by an asterisk). The figure also shows the location of V8, chymotryptic (CT), and proteinase K proteolytic sites (in the latter case also indicating the location of the MAATE[243] excised sequence of amino acid residues). HKA, gastric H^+,K^+-ATPase (*pig*); NKA, Na^+,K^+-ATPase (α1-*sheep*); SCA, sarcoplasmic reticulum Ca^{2+}-ATPase (*rabbit la*); HA(*cer*), proton ATPase (*Saccharomyces cerevisiae*); HA(*ar*), proton ATPase (*Arabidopsis thalianae*).

the highly conserved S4/M4 segment) to disrupt the high-affinity sites for Ca^{2+} and to open the internal gate to release Ca^{2+} to the intravesicular space.

Finally, we note that a distinctive feature of the reaction scheme in FIGURE 3 is that it places emphasis on the uniqueness of the E2P conformation in having a lumenal open state as compared to the nonphosphorylated state, where the binding sites are either in an occluded and protonated state (E2) or ready to open toward the cytosolic medium (E1). This can be seen as evidence that after proteolytic cleavage we are witnessing the performance of an enzyme that is incapable of opening the internal gate to release bound Ca^{2+} toward the lumen, but which otherwise retains most of its normal features. Exploration of the structural properties of this state in Ca^{2+}-ATPase and other P-type ATPases remains one of the important tasks ahead.

REFERENCES

1. JØRGENSEN, P.L. 1992. Na,K-ATPase, structure and transport mechanism. *In* Molecular Aspects of Transport Proteins. J.J.H.H.M. de Pont, Ed.: 1–28. Elsevier Science Publishers BV. Amsterdam.
2. JØRGENSEN, P.L. & J. PETERSEN. 1985. Chymotryptic cleavage of alpha-subunit in E1-forms of renal ($Na^+ + K^+$)-ATPase: effects on enzymatic properties, ligand binding and cation exchange. Biochim. Biophys. Acta **821**: 319–333.
3. LE MAIRE, M. *et al.* 1990. Ca^{2+}-induced conformational changes and location of Ca^{2+} transport sites in sarcoplasmic reticulum Ca^{2+}-ATPase as detected by the use of proteolytic enzyme (V8). J. Biol. Chem. **265**: 1111–1128.
4. JUUL, B. *et al.* 1995. Do transmembrane segments in proteolyzed sarcoplasmic reticulum Ca^{2+}-ATPase retain their functional Ca^{2+} binding properties after removal of cytoplasmic fragments by proteinase K? J. Biol. Chem. **270**: 20123–20134.

5. IMAMURA, Y. & M. KAWAKITA. 1986 Limited tryptic digestion of Ca^{2+}, Mg^{2+}-adenosine triphosphatase of the sarcoplasmic reticulum: enzymatic properties of A1b + B complex. J. Biochem. (Tokyo) **100:** 133–141.
6. TOYOSHIMA, C. et al. 2000. Crystal structure of the calcium pump of sarcoplasmic reticulum at 2.6 A resolution. Nature **405:** 647–655.
7. MORSOMME, P., C.W. SLAYMAN & A. GOFFEAU. 2000. Mutagenic study of the structure, function and biogenesis of the yeast plasma membrane H^+-ATPase. Biochim. Biophys. Acta **1469:** 133–157.
8. ANDERSEN, J.P. et al. 1995. Dissection of the functional domains of the sarcoplasmic reticulum Ca^{2+}-ATPase by site-directed mutagenesis. Biosci. Rep. **15:** 243–261.
9. MØLLER, J.V. et al. 2002. Calcium transport by sarcoplasmic reticulum Ca^{2+}-ATPase. Role of the A domain and its C-terminal link with the transmembrane region. J. Biol. Chem. **277:** 38647–38659.
10. PATCHORNIK, G., R. GOLDSHLEGER & S.J.D. KARLISH. 2000. The complex ATP-Fe^{2+} serves as a specific affinity cleavage reagent in ATP-Mg^{2+} sites of Na,K-ATPase: altered ligation of Fe^{2+} (Mg^{2+}) ions accompanies the $E_1P \rightarrow E_2P$ conformational change Proc. Natl. Acad. Sci. USA **97:** 11954–11959.
11. SHIN, J.M. et al. 2001. Selective Fe^{2+}catalyzed oxidative cleavage of gastric H^+,K^+-ATPase. J. Biol. Chem. **276:** 48440–48450.
12. TOYOSHIMA, C. & H. NOMURA. 2002. Structural changes in the calcium pump accompanying the dissociation of calcium. Nature **418:** 605–611.
13. DANKO, S. et al. 2001. Organization of cytoplasmic domains of sarcoplasmic reticulum Ca^{2+}-ATPase in E_1P and E_1ATP states: a limited proteolysis study. FEBS Lett. **505:** 129–135.

Involvement of the Cytoplasmic Loop L6-7 in the Entry Mechanism for Transport of Ca^{2+} through the Sarcoplasmic Reticulum Ca^{2+}-ATPase

F. CORRE,[a] C. JAXEL,[a] J. FUENTES,[a,b] T. MENGUY,[a] P. FALSON,[a]
B. A. LEVINE,[c] J. V. MØLLER,[d] AND M. LE MAIRE[a]

[a]*Section de Biophysique, Departement de Biologie Joliot Curie,
CEA et CNRS URA 2096 and LRA17V Université de Paris XI,
CE Saclay, 91191 Gif sur Yvette, France*

[b]*Dpto. Bioquímica y Biología Molecular y Genética, E.U. Enfermería y T.O.,
Universidad de Extremadura, Cáceres, Spain*

[c]*School of Biosciences, University of Birmingham, B15 2TT United Kingdom*

[d]*Department of Biophysics, University of Aarhus, DK-8000 Aarhus C, Denmark*

ABSTRACT: We have found that despite a markedly low calcium affinity the D813A/D818A mutant is capable, after complexation with Cr.ATP, of occluding Ca^{2+} to the same extent (1–2 Ca^{2+} per ATPase monomer) as wild-type ATPase. The inherent ability of the synthetic L6-7 loop peptide to bind Ca^{2+} was demonstrated with murexide and mass spectrometry. NMR analysis indicated the formation of specific 1:1 cation complexes of the peptide with calcium and lanthanum with coordination by all three aspartate residues D813/D815/D818 that resulted in an altered conformation of the peptide chain. Overall our observations suggest that, in addition to mediating contact between the intramembranous Ca^{2+} binding sites and the cytosolic phosphorylation site as previously suggested, the L6-7 loop, in a preceding step, participates in the formation of an entrance port important for lodging Ca^{2+} at a high-affinity binding site inside the membrane.

KEYWORDS: Ca^{2+} ATPase; loop NMR structure

INTRODUCTION

We previously obtained evidence for involvement of the L6-7 loop of sarcoplasmic reticulum (SR) Ca^{2+}-ATPase in the mechanism of calcium binding. This was initially indicated by changes in the Ca^{2+}-dependent rates of electrophoretic migration of a C-terminal proteolytic fragment (p20C) of Ca^{2+}-ATPase.[1] The data suggested that three conserved aspartic residues (Asp-813, -815, -818) at the N-terminal

Address for correspondence: M. le Maire, Bât. 528, DBJC, CE Saclay, 91191 Gif sur Yvette, France. Voice: +33-169086243; fax: +33-169088139.
lemairem@dsvidf.cea.fr

part of the L6-7 loop (p808-818) interacted with Ca^{2+}.[1] Cluster mutation Ca^{2+}-ATPase constructs, D813A-D818A (later called ADA) and D813A-D815A-D818A, displayed a marked reduction in the apparent affinity with which Ca^{2+} controls ATPase phosphorylation and turnover.[2,3] We have rationalized our findings in terms of involvement of the conserved aspartic residues of the L6-7 loop in the interaction with calcium ions during the initial steps of the Ca^{2+} binding process.[4] The crystal structures of SERCA 1a,[5,6] however, show the acidic residues of the L6-7 loop in a conformation unsuitable for multiple coordination by Ca^{2+}. An alternative possibility proposed that the decrease in Ca^{2+} binding after mutation of the acidic residues results from allosteric effects on intramembranous Ca^{2+}-binding.[2] In this view the essential function of the L6-7 loop is to mediate long-range interactions between the intramembranous Ca^{2+} binding sites and the phosphorylation site. To address questions on the function of the L6-7 loop in more detail, we report here data obtained on the contribution of the acidic residues of the loop in Ca^{2+} binding and occlusion.

RESULTS AND DISCUSSION

Cr.ATP-induced Calcium Occlusion by the Native SR or Yeast-expressed Wild-type or ADA-mutated Ca^{2+}-ATPase

To measure Cr.ATP-induced calcium occlusion by Ca^{2+}-ATPase, we combined pretreatment with Cr.ATP in the presence of $^{45}Ca^{2+}$ with subsequent solubilization by $C_{12}E_8$ and molecular sieve chromatography. We measured the elution of protein-bound $^{45}Ca^{2+}$ and compared it with the Ca^{2+}-ATPase concentration estimated in the same fractions by Western blot analysis. We found that both the ADA and WT of expressed Ca^{2+}-ATPase occluded Ca^{2+} to the same extent as Ca^{2+}-ATPase prepared from native SR (corresponding to 1-2 Ca^{2+} per ATPase monomer). This level of occlusion is close to what can be expected by the HPLC procedure and indicates that the ability of the ADA mutant to occlude Ca^{2+} is intact.

Solubilization by $C_{12}E_8$ of Various Cr.ATP-reacted Ca^{2+}-ATPase Species as a Function of Calcium

In similar experiments we found that, as previously reported,[7] the E309Q mutant does not occlude Ca^{2+}, but several attempts to measure Cr.ATP-induced Ca^{2+} occlusion by the E771Q (previously also reported to display no calcium occlusion[7]) and the D813A-D815A-D818A mutants failed, because we were unable to recover any Ca^{2+}-ATPase after elution from the HPLC column. This must mean that the mutated Ca^{2+}-ATPase (i) either forms large non-soluble aggregates in the yeast membrane; or (ii) that $C_{12}E_8$ solubilization converts the Ca^{2+}-ATPase to a labile state that readily undergoes rapid denaturation and aggregation. To examine these possibilities we tested the effect of the solubilization conditions of the various ATPase forms after pretreatment with Cr.ATP (FIG. 1): Using SR vesicles as a control we see that efficient solubilization (>80%) is obtained both at 1 mM and 0.12 mM Ca^{2+} as well as in 2 mM EGTA (which rapidly causes inactivation after $C_{12}E_8$ solubilization). But after addition of yeast control membranes and $C_{12}E_8$ to the same preparation of SR vesicles, solubilization is reduced to about 50% in the presence of 0.1 and 1 mM

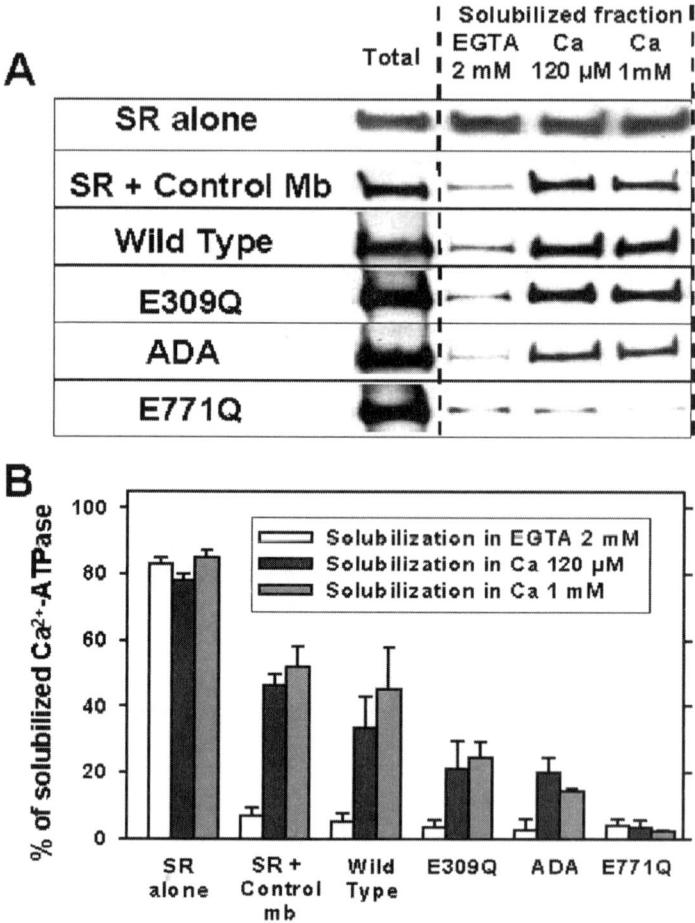

FIGURE 1. Ca^{2+}-dependent $C_{12}E_8$ solubilization of Cr.ATP-reacted wild-type and mutant Ca^{2+}-ATPases. The experiments were performed at several levels of calcium concentration (EGTA 2 mM, Ca^{2+} 120 µM and 1 mM) during the occlusion reaction and subsequent solubilization.[4] In (**A**), recoveries of Ca^{2+}-ATPase in the total fraction are compared with those of the solubilized fractions. In (**B**), histograms corresponding to the relative Ca^{2+}-ATPase content of the solubilized fractions (in %) versus the total content in the preparation from at least three experiments are shown for each set of solubilization condition. Ca^{2+}-ATPase from SR vesicles, solubilized by $C_{12}E_8$ in the absence and presence of yeast membranes with empty vector, served as controls.

Ca^{2+}, and to almost nothing in the presence of 2 mM EGTA. The same pattern is observed with the E309Q and ADA mutants, but here only 20–2% remains solubilized in the presence of Ca^{2+}. This indicates that Ca^{2+}-ATPase, fully solubilized at the moment of $C_{12}E_8$ addition, subsequently is prone to undergo aggregation by interaction with one or more component(s) in the yeast system. The nature of this interaction is unclarified, but may either be non-specific or depend on the presence of chaperones like *cer1p,* which have been reported to interact with unfolded proteins under conditions of yeast stress. Finally, the last line of FIGURE 1A shows that it was not possible under any conditions to solubilize or to retain in a soluble state the Cr.ATP-treated E771Q mutated Ca^{2+}-ATPase, irrespective of the Ca^{2+} concentration, in agreement with an increased susceptibility of this mutant to proteolytic digestion.[3]

Cation Binding Characteristics of the Synthetic L6-7 Peptide Analyzed by Murexide, Mass Spectrometry, and NMR

By use of murexide we found that the synthetic L6-7 peptide (p808-827) and the longer M7-L (p808-847) peptide bind appreciable amounts of Ca^{2+} unlike the L6-7 peptide with alanine substitution of the acidic residues or a version of the M7-L peptide (p818-847) lacking the conserved aspartic residues. High-resolution ESI-FTICR mass spectrometric studies of p808-827 identified specific cation complexes of the peptide with both Ca^{2+} and trivalent lanthanide and yielded a dissociation constant (half-maximal binding) for Ca^{2+} of approximately 0.4–1 mM. This Ca^{2+} affinity is lower than that for the whole C-terminal fragment p20C (p808-994), which, as estimated from the change of electrophoretic migration rate as a function of Ca^{2+} concentration, has a dissociation constant of about 100 µM. The relatively low affinity of the isolated L6-7 peptide for Ca^{2+} may reflect the accessibility of the free peptide to a larger range of conformations, only some of which bind Ca^{2+}. However, even an affinity as low as 1 mM does not rule out an important role of the loop as an entrance port for cytosolic Ca^{2+}, being involved in high-affinity intramembranous Ca^{2+} binding by a "back-up" mechanism.[4] Our NMR studies showed that the central heptapeptide segment of the isolated L6-7 loop (D813-P819) has a defined structure in solution and that this conformation alters upon alanine substitution of the aspartate residues. Titration of the native 808-827 loop peptide with the lanthanum (La^{3+}) cation as Ca^{2+} analogue led to upfield shifts of all three aspartate sidechain resonances and indicated binding of one lanthanum ion per peptide with an affinity greater than 10^{-3} M. The simultaneous coordination of the cation by the three carboxylates entails reorganization of the peptide conformation so as to enable proximity and a correct aspartate sidechain orientation as indicated by the structure (FIG. 2, center), which was derived from paramagnetic lanthanide and NOE data.[4] Competition experiments with increasing concentrations of Ca^{2+} indicated progressive displacement of the bound lanthanide, suggesting isomorphous coordination by the calcium ion. In FIGURE 2 the NMR-derived cation-bound structure is compared with the structure of the loop in the Ca^{2+}-ATPase structures of Toyoshima *et al.*:[5,6] it provides an illustration of the proposed transfer of aspartate bound Ca^{2+} from the outside to the inside by a flip-flop motion of the loop.

CONCLUDING COMMENTS

The present study[4] shows that the central part of the isolated peptide corresponding to the L6-7 loop in solution adopts an independently folded structure, similar to that of the published X-ray structures[5,6] of the Ca^{2+}-ATPase, but in which the aspartate groups D813, D815, D818 can readily form a Ca^{2+} binding site by reorientation of the central peptide bonds (cf. FIG. 2). The fact that mutation of these aspartate res-

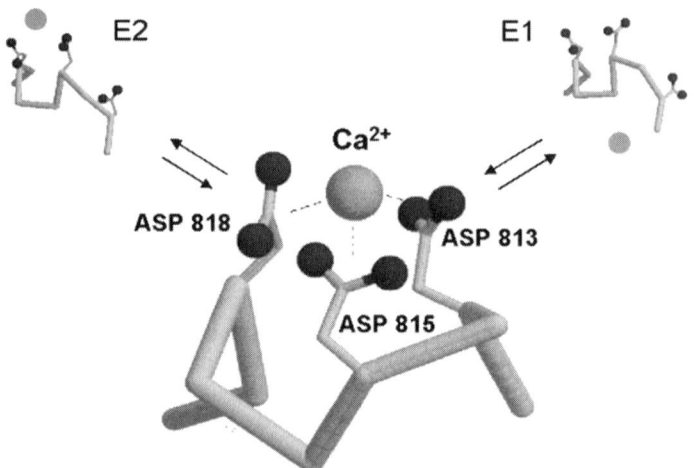

FIGURE 2. Proposed mechanism for the interaction of Ca^{2+} with the L6-7 loop, associated with the E2 to E1 conversion. The figure compares the NMR-derived average conformation derived for residues 813-819 of the L6-7 peptide in stick representation (central foreground structure), when bound to lanthanum, with the loop configuration observed in the crystal structures (background structures). Note the sidechain orientation of Asp-813, Asp-815 and Asp-818 in the NMR-based structure that coordinates cation (lanthanum or calcium) binding. Transient Ca^{2+} binding by the three conserved aspartate residues may act as a switch to internalize Ca^{2+}. This transition is depicted with the loop structure in the crystallographic E2 conformation on the left (PDB coordinate file 1IWO[6]), with one external calcium and the E1 conformation on the right with intramembraneously bound Ca^{2+} shown in the background (PDB coordinate file 1eul[5]).

idues in Ca^{2+}-ATPase to alanine does not affect Ca^{2+} occlusion while decreasing the turnover rate of the ATPase is consistent with their suggested participation in Ca^{2+} binding in an early part of the Ca^{2+} transport cycle. We propose that the loop, in addition to its probable role in mediating Ca^{2+}-dependent phosphorylation from ATP,[2,5,6] may act as a switch by which one of the two Ca^{2+} ions from the cytosol are guided toward intramembranous binding.

REFERENCES

1. FALSON, P., T. MENGUY, F. CORRE, et al. 1997. The cytoplasmic loop between putative transmembrane segments 6 and 7 in sarcoplasmic reticulum Ca^{2+}-ATPase binds Ca^{2+} and is functionally important. J. Biol. Chem. **272:** 17258–17262.
2. ZHANG, Z., D. LEWIS, C. SUMBILLA, et al. 2001. The role of the m6-m7 loop (L67) in stabilization of the phosphorylation and Ca^{2+} binding domains of the sarcoplasmic reticulum Ca^{2+} ATPase J. Biol. Chem. **276:** 15232–15239.
3. MENGUY, T., F. CORRE, L. BOUNEAU, et al. 1998. The cytoplasmic loop located between transmembrane segments 6 and 7 controls activation by Ca^{2+} of sarcoplasmic reticulum Ca^{2+}-ATPase. J. Biol. Chem. **273:** 20134–20143.
4. MENGUY, T., F. CORRE, B. JUUL, et al. 2002. Involvement of the cytoplasmic loop L6-7 in the entry mechanism for transport of Ca^{2+} through the sarcoplasmic reticulum Ca^{2+}-ATPase. J. Biol. Chem. **277:** 13016–13028.

5. TOYOSHIMA, C., M., NAKASAKO, H. NOMURA, & H. OGAWA. 2000. Crystal structure of the calcium pump of sarcoplasmic reticulum at 2.6 resolution. Nature **405:** 647–655.
6. TOYOSHIMA, C. & H. NOMURA. 2002. Structural changes in the calcium pump accompanying the dissociation of calcium. Nature **418:** 605–611.
7. VILSEN, B. & J.P. ANDERSEN. 1992. CrATP-induced Ca^{2+} occlusion in mutants of the $Ca(^{2+})$-ATPase of sarcoplasmic reticulum. J. Biol. Chem. **267:** 25739–25743.

Site-Directed Mutagenesis of Amino Acids in the Cytoplasmic Loop 6/7 of Na,K-ATPase

G. XU,[a] R. A. FARLEY,[a] D. J. KANE,[a,b] AND L. D. FALLER[b]

[a]*University of Southern California School of Medicine, Los Angeles, California 90033, USA*

[b]*UCLA School of Medicine, Los Angeles, California 90033, USA*

ABSTRACT: The loop between transmembrane helices 6 and 7 (L6/7) of P-type ATPases has been suggested to be important for the functional linkage of ion binding and enzyme phosphorylation or to be a site of initial cation binding. To investigate the role of L6/7 in Na,K-ATPase, alanine substitutions were made for charged and conserved residues in L6/7 of the human α1 subunit and the proteins were expressed in yeast for analysis. All mutants except the triple mutant E825A/E828A/D830A bound ouabain. Although the equilibrium dissociation constant for ouabain binding by most mutants was similar to the wild-type value, the K_d of R837A for ouabain binding was ~15-fold higher than the wild-type K_d. ^{18}O exchange measurements indicated that the apparent affinity of this mutant for Pi was reduced about 3-fold. The concentration dependence of KCl inhibition of ouabain binding or of NaCl inhibition of ouabain binding revealed 2–4-fold changes in the apparent affinity for cations in the E825A, E828A, and R837A mutants. The E825A and E828A mutants lost the ability to bind ouabain after extraction with 0.1% SDS or after brief heating, indicating that these mutations affected the stability of the enzyme. The ATPase activity of the other mutants was measured after extraction of crude yeast membranes with 0.1% SDS. For all mutants except R834A, R837A, and R848A, the activity was at least 50% of wild-type activity.

KEYWORDS: Na,K-ATPase; site-directed mutagenesis; sodium; potassium; ouabain

Amino acids in the cytoplasmic loop connecting transmembrane segments 6 and 7 of P-type ATPases have been suggested to be important as either initial recognition sites for the transported cations[1–3] or as components of the mechanism that links ion binding to enzyme phosphorylation.[4] There is considerable amino acid conservation among loop 6/7 sequences found in Ca-ATPase, Na,K-ATPase, and H,K-ATPase, including several acidic residues that have been implicated in Ca^{2+} binding by Ca-ATPase. In this report, alanine substitutions were made for charged and conserved

Address for correspondence: Robert A. Farley, Ph.D., University of Southern California, Keck School of Medicine, 1333 San Pablo Street, MMR 250, Los Angeles, CA 90033. Voice: 323-442-1240; fax: 323-442-2283.
 rfarley@hsc.usc.edu

amino acids in the human Na,K-ATPase α1 subunit and the interaction of the protein with various ions and ouabain was measured in order to determine whether the role of loop 6/7 in sodium pump function could be identified.

RESULTS

Mutant and wild-type Na,K-ATPase pumps were expressed in the yeast *Saccharomyces cerevisiae*.[5] Binding of ouabain by yeast membranes in a Mg^{2+}- and Pi-dependent reaction indicated that all of the mutants, with the exception of the triple mutant E825A/E828A/D830A, were present in the yeast membranes. A list of the mutations that were made is shown in TABLE 1, which also summarizes the ouabain K_d and B_{max} values and the ouabain-inhibited ATPase activity obtained for each mutant. From these data it can be seen that all mutants bound ouabain with a K_d similar to that of the wild-type control, except for the R837A mutant, which had a K_d value almost 15 times higher than the wild-type enzyme. Since the mutation was made in a loop of the protein that is exposed in the cytoplasm, it is unlikely that R837 participates directly in the ouabain binding site. Rather, the high K_d value is likely to be the result of an effect of the mutation on phosphorylation of the protein by Pi. This interpretation was supported by measurements of ^{18}O exchange between inorganic phosphate and water catalyzed by the protein.[6] When the Mg^{2+} and Pi dependence of ^{18}O exchange of the α1β1 control was compared with results for the mutants E825A, D830A, R834A, P836A, and R837A, the apparent phosphate dissociation constant ($K'_p/[1 + K_{H2O}]$) for R837A wassignificantly higher than the control value (5.3 ± 1.1 mM vs. 1.2 ± 0.1 mM). Thus, mutation of an amino acid in the cytoplasmic loop 6/7 affects the equilibrium constant between non-covalently and covalently bound Pi (K_{H2O} = [E-P]/[E·Pi]) and/or the dissociation constant of Pi from metalloenzyme (K'_p).

TABLE 1 also shows that most of the mutants retained at least 50% of wild-type Na,K-ATPase activity. The exceptions were R834A, R837A, and R848A. R834 corresponds to K819, R837 corresponds to R822, and R848 corresponds to W832 in chicken SERCA1. Like the R834A and the R837A mutants of Na,K-ATPase, both the K819A and the R822A mutants of SERCA1 were found to retain less than 50% of Ca-ATPase activity.[1,2,4] The SERCA1 mutants also had a marked reduction in phosphoenzyme formation from ATP or Pi, and these results obtained with SERCA1 were interpreted as evidence for an important role for K819 and R822 in determining the functional integrity of the phosphorylation domain of Ca-ATPase. The double mutation D813A/D818A and the triple mutation D813A/D815A/D818A in SERCA1, corresponding to amino acids E825, E828, and D830 in Na,K-ATPase, resulted in the loss of about 90% of Ca-ATPase activity. The E825A and E828A mutants of Na,K-ATPase lost all ouabain binding capacity after extraction with 0.1% SDS, and, therefore, it was not possible to measure any ouabain-sensitive Na,K-ATPase activity in the SDS-extracted membranes. Similarly, the triple mutant E825A/E828A/D830A did not bind ouabain and could not be examined further. The D830A mutant, however, retained nearly wild-type level activity. The loss of ouabain binding by the E825A and E828A mutants may have been due to a global destabilization of the mutants such that their sensitivity to denaturation by perturbations like exposure to detergents was increased relative to the wild-type en-

TABLE 1. Ouabain binding and Na,K-ATPase activity of loop 6/7 mutants

Allele	Turnover Number (% of control)	Ouabain K_d (nM)	Ouabain B_{max} (pmol/mg)
α1β1	100	17 ± 1	21 ± 2
E825A	ND	13 ± 3	5 ± 3
E828A	ND	19 ± 2	8 ± 4
D830A	70 ± 7	9 ± 1	9 ± 1
I831A	72 ± 14	13 ± 3	12 ± 1
M832A	ND	11 ± 3	3 ± 1
K833A	102 ± 4	27	31
R834A	25 ± 5	14 ± 1	13 ± 5
P836A	57 ± 3	22 ± 6	15 ± 3
R837A	11 ± 1	240 ± 35	11 ± 6
P839A	76 ± 2	21	23
K840A	102 ± 1	25	27
D842A	92 ± 7	26	15
K843A	164 ± 9	44	18
E847A	94 ± 1	20	26
R848A	34 ± 7	23 ± 10	18 ± 7

NOTE: Ouabain binding was measured as previously described[9] using yeast membranes extracted with 0.1% SDS, except for membranes containing E825A and E828A, which were not extracted. K_d and B_{max} values were obtained from nonlinear regression fits of a self-competition model[10] to the data. Values shown are mean ± SD from two or three different membrane preparations, except where no SD is shown and values were from a single measurement. Na,K-ATPase activity measurements were made using a NADH-coupled assay.[6] Turnover numbers (ATPase V_{max}/ouabain B_{max}) were calculated for one membrane preparation from each sample and are shown as percent of wild type. V_{max} was obtained from two to four measurements of ATP hydrolysis at ATP concentrations up to 3 mM; B_{max} was measured once for each allele.
ND = not determined.

zyme. Support for this interpretation was obtained by measuring the amount of ouabain bound by the membranes containing wild-type or mutant pumps before and after heating the samples to 50°C for 90 seconds. Wild-type and the D830A, M832A, and R837A mutants lost less than 25% of ouabain binding capacity after heating, whereas the E825A and the E828A mutants lost 80–90% of their capacity for ouabain binding after heating. Residues of Ca-ATPase corresponding to E825 and E828 in Na,K-ATPase participate in an extensive network of hydrogen bonds both to other amino acids within the loop 6/7 and also to amino acids in the cytoplasmic loops linking transmembrane segments 4 and 5, and linking transmembrane segments 8 and 9.[7] Hence, destabilization of the protein structure after substitution of alanine for E825 and E828 was not unexpected.

Loop 6/7 has been suggested to be part of an initial recognition site for transported cations in Ca-ATPase[1,2] and Na,K-ATPase,[3] and substitution of the acidic D813 and D818 residues in Ca-ATPase with alanine resulted in nearly 1000-fold increases in K_{Ca} for Ca-dependent phosphoenzyme formation from ATP. In order to test whether loop 6/7 mutations in Na,K-ATPase had similarly large effects on the inter-

TABLE 2. Effects of alanine substitutions in loop 6/7 on ion interactions with Na,K-ATPase

	Apparent K_m (Na)[a] (mM)	IC$_{50}$ (Na)[b] (mM)	Apparent K_m (K)[b] (mM)
α1β1	1.0 ± 0.1	3.3 ± 0.6	0.3 ± 0.1
E825A	3.1 ± 0.4	11.4 ± 0.8	0.7 ± 0.1
E828A	1.1 ± 0.2	18.3 ± 2.5	0.9 ± 0.2
R834A	1.6 ± 0.1	ND	0.3 ± 0.1
R837A	2.7 ± 0.2	1.0 ± 0.2	0.8 ± 0.1
R848A	1.4 ± 0.2	2.7 ± 0.5	0.3 ± 0.1

[a]Apparent K_m (Na) was determined as previously described[5] from the Na-dependence of ouabain binding in the presence of 3 mM MgCl$_2$ and 3 mM ATP. Data are mean ± SD (n = 3–5).
[b]The IC$_{50}$ (Na) and the apparent K_m (K) values were obtained from Na$^+$ or K$^+$ antagonism of equilibrium ouabain binding[5] measured in the presence of 4 mM MgCl$_2$ and 4 mM Pi. Data are mean ± SD, n = 2 for IC50 (Na); n = 4–6 for apparent K_m (K).
ND = not determined.

action of Na$^+$ and K$^+$ with the enzyme, the ability of Na$^+$ to promote ouabain binding to phosphoenzyme formed from ATP, and the ability of either Na$^+$ or K$^+$ to antagonize Mg^{2+}- and Pi-dependent ouabain binding, were measured. Most of the mutants were characterized by apparent K_m or IC$_{50}$ values that were similar to those of wild type. The apparent K_m and IC$_{50}$ values for mutants that behaved differently from wild type are compared to the wild-type values in TABLE 2. Several points should be noted from these data. First, the same amino acids whose substitution with alanine resulted in greater instability or a greater than 50% reduction in Na,K-ATPase activity also seem to have altered interactions with cations. Next, the magnitude of the effects on Na$^+$ or K$^+$ interaction with the mutants is small, with only a 3- to 5-fold increase in apparent K_m or IC$_{50}$. Finally, the mutants are not consistently altered in their interaction with the cations. For example, the E828A mutant had the largest increase in IC$_{50}$ (Na), yet showed an apparent K_m (Na) similar to wild type, and whereas R837A was characterized by a 2.7-fold increase in apparent K_m (Na), the IC$_{50}$ (Na) was only about one-third of wild type. Also, the apparent K_m (Na) was reduced by neutralization of both positively charged (R837) and negatively charged (E825) amino acid side chains.

CONCLUSION

The small changes in apparent K_m and IC$_{50}$ values for Na$^+$ and K$^+$ shown in TABLE 2 seem inconsistent with a role for loop 6/7 of Na,K-ATPase in the initial recognition of the transported cations. Unlike the Ca-ATPase, in which large changes in the K_{Ca} for Ca-dependent phosphoenzyme formation from ATP were observed, the changes in K_m and IC$_{50}$ for Na,K-ATPase are only 3–5-fold, and are also smaller than the 100-fold increases in K0.5 (K) found by Nielsen et al.[8] after mutagenesis of intramembrane carboxyl groups known to interact directly with occluded cations. Instead, the observed differences between wild-type and loop 6/7 mutants seem likely to be indirect effects of the mutations on interactions between the protein and the cations. Similarly, the effects of the R837A mutation on enzyme phosphorylation and/or Pi binding are probably indirect. The results obtained in the present study with Na,K-ATPase are consistent with the suggestion of Zhang et al.[4] that loop 6/7 is important in the transmission of the activation signal initiated by cation binding to the phosphorylation domain of the protein, and may, therefore, be a structural component of the mechanism that connects the ion binding sites to the phosphorylation site on P-type ATPases.

ACKNOWLEDGMENTS

This work was supported by NIH Grants GM28673 and DK52802.

REFERENCES

1. FALSON, P. et al. 1997. The cytoplasmic loop between putative transmembrane segments 6 and 7 in sarcoplasmic reticulum Ca^{2+}-ATPase binds Ca^{2+} and is functionally important. J. Biol. Chem. **272:** 17258–17262.
2. MENGUY, T. et al. 1998. The cytoplasmic loop located between transmembrane segments 6 and 7 controls activation by Ca^{2+} of sarcoplasmic reticulum Ca^{2+}-ATPase. J. Biol. Chem. **273:** 20134–20143.
3. SHAINSKAYA, A. et al. 2000. Entrance port for Na+ and K+ ions on Na,K-ATPase in the cytoplasmic loop between transmembrane segments M6 and M7 of the α subunit. Proximity of the cytoplasmic segment of the β subunit. J. Biol. Chem. **275:** 2019–2028.
4. ZHANG, Z. et al. 2001. The role of the M6-M7 loop (L67) in stabilization of the phosphorylation and Ca^{2+} binding domains of the sarcoplasmic reticulum Ca^{2+}-ATPase (SERCA). J. Biol. Chem. **276:** 15232–15239.
5. MULLER-EHMSEN, J. et al. 2001. Ouabain and substrate affinities of human Na^+-K^+-ATPase α1β1, α2β1, and α3β1 when expressed separately in yeast cells. Am. J. Physiol. **281:** C1355–C1364.
6. FARLEY, R.A. et al. 2001. ^{18}O Exchange evidence that mutations of arginine in a signature sequence for P-type pumps affect inorganic phosphate binding. Biochemistry **40:** 6361–6370.
7. TOYOSHIMA, C. et al. 2000. Crystal structure of the calcium pump of sarcoplasmic reticulum at 2.6A resolution. Nature **405:** 647–655.
8. NIELSEN J. M. et al. 1998. Importance of intramembrane carboxylic acids for occlusion of K+ ions at equilibrium in renal Na,K-ATPase. Biochemistry **37:** 1961–1968.
9. EAKLE, K.A. et al. 1992. High-affinity ouabain binding by yeast cells expressing Na,K-ATPase subunits and the gastric H,K-ATPase subunit. Proc. Natl. Acad. Sci. USA **89:** 2834–2838.
10. JOHNSON, C.L. et al. 1995. Comparison of the effects of potassium on ouabain binding to native and site-directed mutants of Na,K-ATPase. Arch. Biochem. Biophys. **317:** 133–141.
11. LOWRY, O.H. et al. 1951. Protein measurement with the Folin phenol reagent. J. Biol. Chem. **193:** 265–275.

ATP Binding Residues of Sarcoplasmic Reticulum Ca^{2+}-ATPase

D. B. McINTOSH,[a] J. D. CLAUSEN,[b] D. G. WOOLLEY,[a] D. H. MacLENNAN,[c] B. VILSEN,[b] AND J. P. ANDERSEN[b]

[a]*Division of Chemical Pathology, Department of Clinical Laboratory Sciences, Faculty of Health Sciences, University of Cape Town, Cape Town 7925, South Africa*

[b]*Department of Physiology, University of Aarhus, DK-8000 Aarhus C, Denmark*

[c]*Banting and Best Department of Medical Research, University of Toronto, Toronto, Ontario M5G1L6, Canada*

> ABSTRACT: ATP-binding residues in the N and P domains of sarcoplasmic reticulum Ca-ATPase have been investigated using mutagenesis in combination with a binding assay based on the photolabeling of Lys^{492} with $[\gamma-^{32}P]$ 2′,3′-O-(2,4,6 trinitrophenyl)-8-azido-ATP and competition with nucleotide. In the N domain, mutations to several residues in conserved motifs, ^{438}GEATE, ^{487}FSRDRK, ^{515}KGAPE, and ^{560}RCLALA produce nucleotide-binding defects. Key residues include Thr^{441}, Glu^{442}, Phe^{487}, Arg^{489}, Lys^{492}, Lys^{515}, Arg^{560}, and Leu^{562}. In the absence of Mg^{2+}, Arg^{489}, Lys^{492}, and Arg^{560} are most important, whereas in its presence Thr^{441} and Glu^{442} also play a crucial role. In the P domain, Asp^{351} is striking for its strong electrostatic repulsion of the γ-phosphate, especially in the presence of Mg^{2+}. Lys^{352} is a key residue, and Asp^{627} and Lys^{684} must come close to the nucleotide. Thr^{353}, Asn^{359}, Asp^{601}, and Asp^{703} interact only in the presence of Mg^{2+}. Asn^{706} and Asp^{707} are unimportant for nucleotide binding. The results identify several ATP binding residues in the N and P domains and suggest that Mg^{2+} changes the nucleotide/protein interaction in both. Models of bound ATP and MgATP are presented.
>
> KEYWORDS: Ca^{2+}-ATPase; P-type ion pumps; ATP binding; mutagenesis; TNP-ATP; photolabeling

The crystal structure of sarcoplasmic reticulum Ca^{2+}-ATPase with bound Ca^{2+} shows that the protein consists of four distinct domains: a membrane and stalk entity containing two Ca^{2+} ions, a phosphorylation (P) portion positioned on the stalk, a nucleotide binding (N) region, and a smaller actuator (A) domain loosely associated with the rest of the protein.[1] A nucleotide-binding site is located in the N domain through the binding of TNPAMP, and its position is substantiated by many studies

---

Address for correspondence: Dr. D.B. McIntosh, Faculty of Health Sciences, University of Cape Town, Cape Town 7925, South Africa. Voice: +27-21-4066187; fax: +27-21-4488150.
davidmci@chempath.uct.ac.za

implicating Lys^{492} and Lys^{515} in ligating ATP, as well as more recent site-directed mutagenesis showing the importance of Phe^{487}, Arg^{489}, and Lys^{492}.[2] Placing the adenine of ATP here in the N domain puts it some distance from Asp^{351}, the residue phosphorylated in the P domain during the catalytic cycle, and the N domain must roll, and close over, the phosphorylation site to bring the γ-phosphate next to the aspartate. Bound ATP must bridge the two domains, with binding residues located in both structures. While the location of TNP-AMP in the binding pocket in the N domain gave an indication of binding residues, the physical separation of the N and P domains and the fact that the TNP nucleotide is not ATP cast doubt on which residues in the N domain actually interact with ATP, which in the P domain interact, the orientation of the nucleotide, and the influence of Mg^{2+}.

Here we report on ATP and MgATP binding to several new mutants in both the N and P domains and reappraise pertinent previous results. Details of the mutagenesis and expression in COS-1 cells and of the nucleotide binding assay, which utilizes the specific photolabeling of Lys^{492} by [γ–^{32}P]TNP-8N$_3$-ATP and inhibition by ATP, are provided in previous work.[2,3]

FIGURE 1 summarizes the ATP and MgATP binding data for the following mutants: N domain: Thr^{441}→Ala, Glu^{442}→Ala, Phe^{487}→Leu, Arg^{489}→Leu, Lys^{492}→Tyr, Lys^{515}→Ala, Arg^{560}→Leu, and Leu^{562}→Phe; Hinge: Asn^{359}→Ala and Asp^{601}→Asn; P domain: Asp^{351}→Ala, Lys^{352}→Gln, Thr^{353}→Ala, Asp^{627}→Ala, Lys^{684}→Met, Asp^{703}→Ala, Asn^{706}→Ala, and Asp^{707}→Asn. Mutation Lys^{515}→Ala blocked photolabeling and the binding by this assay could not be assessed. In the figure the locations of the amino acid residues in the E1Ca$_2$ atomic structure are shown, and the observed changes in binding affinity relative to wild type are indicated diagrammatically.

NUCLEOTIDE-BINDING RESIDUES IN THE ABSENCE OF Mg^{2+}

Examining the results obtained in the absence of Mg^{2+} first (FIG. 1A), basic residues Arg^{489}, Lys^{492}, and Arg^{560} are the most critical for ATP binding. Phe^{487} is also important whereas, rather surprisingly, Leu^{562} is not. The leucine was substituted with phenylalanine, and it is possible that the aromatic ring interacts favorably with that of the adenine. Thr^{441} and Glu^{442} seem less important (only 4–5-fold change of affinity in the mutants). Thus, there is an axis of positively charged residues on the one side of the binding pocket that could interact with the negatively charged phosphates of the nucleotide.

Looking at the Hinge segments, Asn^{359} and Asp^{601} seem to play no role in ATP binding.

In the P domain, Lys^{352} is very important, while interaction of ATP, probably in the form of electrostatic repulsion, can be seen with Asp^{351} and Asp^{627}, as mutations to these residues increased the binding affinity. Surprisingly, the binding affinity was increased with mutation of Lys^{684}, which may be an indirect effect through increased freedom of Asp^{351}. Conserved residues Thr^{353}, Asp^{703}, Asn^{706}, and Asp^{707} seem to play no role in binding nucleotide.

A model of bound ATP is shown in the figure. The starting point is the strong interaction with the basic residues in the N domain, which is likely to be through the phosphoryl groups. The basic cluster may include Arg^{678} in the P domain, as this

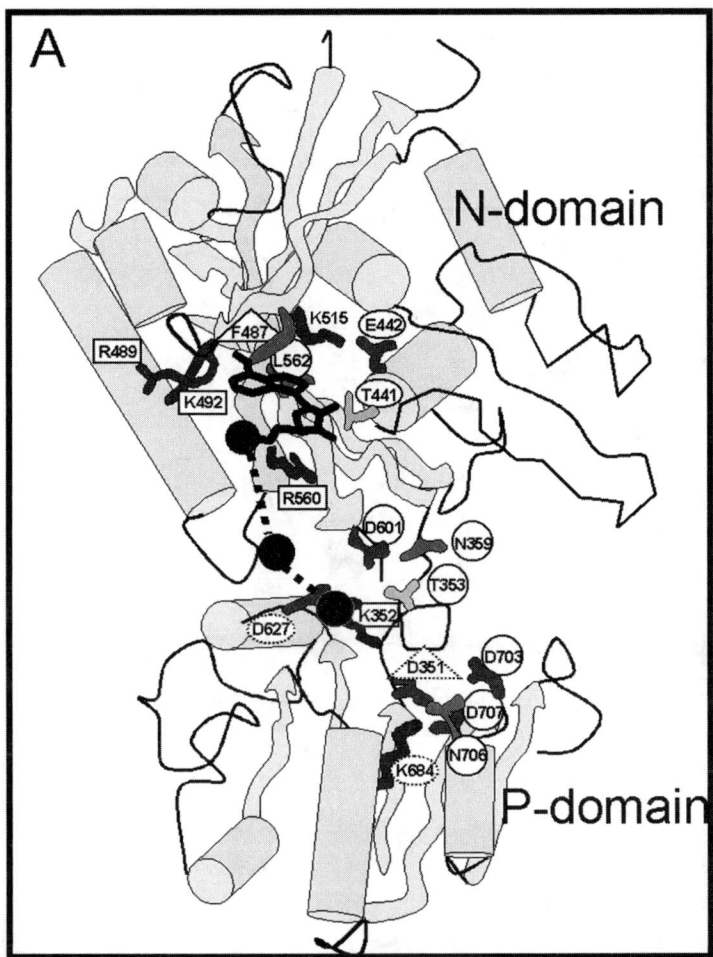

FIGURE 1. Schematic diagram of the N domain and truncated P domain in the E_1Ca_2 structure, displaying the amino acid residues mutated and approximate binding changes relative to wild type. (**A**) absence of Mg^{2+}. (**B**) presence of Mg^{2+}. Labels of residues are enclosed according to the following key: *circles*, no change in affinity relative to wild type; *ellipse*, a change of 3–10-fold; *triangle*, a change of 10–100-fold; *rectangle*: a change greater than 100-fold. Solid lines = an increase in K_d; dotted lines = a decrease in K_d.

residue is close to Lys^{492}, at least in the absence of nucleotide.[4] The orientation of the phosphates on this side of the binding pocket suggests that the ribose hydroxyls may interact with Thr^{441}. Placing the phosphates on this basic side is also suggested from the lack of interaction with residues Thr^{353}, Asp^{601}, Asn^{359}, and Asp^{703}. It also fits with the apparent electrostatic repulsion with Asp^{627}.

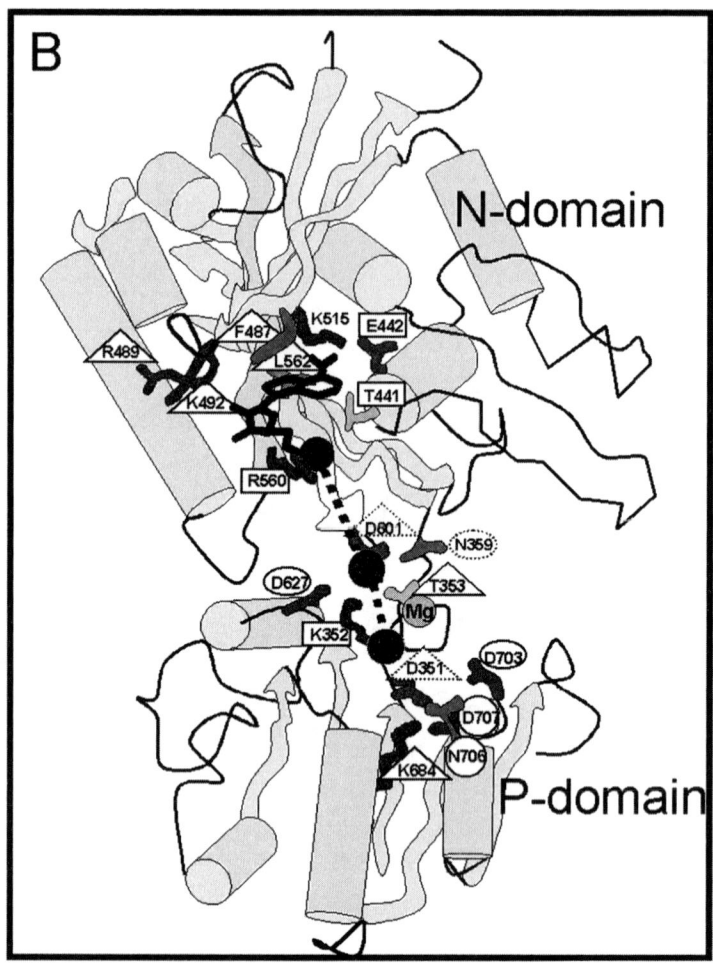

FIGURE 1. *Continued.*

NUCLEOTIDE-BINDING RESIDUES IN THE PRESENCE OF Mg^{2+}

The situation in the presence of Mg^{2+} is quite different. In the N domain, the emphasis shifts to Thr^{441} and Glu^{442}. Mutation to either of these residues affects the ATP binding affinity more than 100-fold in the presence of Mg^{2+}. Arg^{560} remains critical, but the importance of Phe^{487}, Arg^{489}, Lys^{492} is diminished.

In the Hinge segments, mutations to Asn^{359} and Asp^{601} increase the affinity for MgATP, so these residues possibly interact with the nucleotide.

In the P domain, Lys^{352} remains a key residue, Thr^{353} and Lys^{684} assume importance, and Asp^{627} and Asp^{703} come into play. Asn^{706} and Asp^{707} still do not participate in nucleotide binding.

In the model of bound MgATP, the phosphates are now orientated toward the Thr^{441}, Asp^{601}, Asn^{359}, Thr^{353}, and Asp^{703} axis. We place the Mg^{2+} ion close to Thr^{353},5 and Thr^{441} may approach Mg^{2+} in the closed state of the binding pocket.

ACKNOWLEDGMENTS

We thank Lene Jacobsen and Karin Kracht (Aarhus), and Irene Mardarowicz (Cape Town) for expert technical assistance. The research was supported by grants from the Danish Medical Research Council, the Novo Nordisk Foundation, Denmark, the Research Foundation of Aarhus University (to J.P.A.), and the National Research Foundation of South Africa and University of Cape Town (to D.B.M.). We are grateful to Dr. Chikashi Toyoshima for information on unpublished work and valuable discussions.

REFERENCES

1. TOYOSHIMA, C., M. NAKASAKO, H. NOMURA & H. OGAWA. 2000. Crystal structure of the calcium pump of sarcoplasmic reticulum at 2.6 Å resolution. Nature **405:** 647–655.
2. MCINTOSH, D.B., D.G. WOOLLEY, B. VILSEN & J.P. ANDERSEN. 1996. Mutagenesis of segment ^{487}Phe-Ser-Arg-Asp-Arg-Lys^{492} of sarcoplasmic reticulum Ca^{2+}-ATPase produces pumps defective in ATP binding. J. Biol. Chem. **271:** 25778–25789.
3. MCINTOSH, D.B., D.G. WOOLLEY, D.H. MACLENNAN, et al. 1999. Interaction of nucleotides with Asp^{351} and the conserved phosphorylation loop of sarcoplasmic reticulum Ca^{2+}-ATPase. J. Biol. Chem. **274:** 25227–25236.
4. MCINTOSH, D.B. 1992. Glutaraldehyde cross-links Lys-492 and Arg-678 at the active site of sarcoplasmic reticulum Ca^{2+}-ATPase. J. Biol. Chem. **267:** 22328–22335.
5. CLAUSEN J.D., D.B. MCINTOSH, D.G. WOOLLEY & J.P. ANDERSEN. 2001. Importance of Thr-353 of the conserved phosphorylation loop of the sarcoplasmic reticulum Ca^{2+}-ATPase in MgATP binding and catalytic activity. J. Biol. Chem. **276:** 35741–35750.

Molecular Modeling of SCH28080 Binding to the Gastric H,K-ATPase and MgATP Interactions with SERCA- and Na,K-ATPases

KEITH MUNSON,[a,b] OLGA VAGIN,[a,b] GEORGE SACHS,[a,b] AND STEVE KARLISH[c]

[a]*University of California at Los Angeles, Los Angeles, California, USA*

[b]*Veterans Administration Greater LA Healthcare System, Los Angeles, California 90073, USA*

[c]*Weizmann Institute of Science, Rehovot, Israel*

ABSTRACT: We have used homology molecular modeling based on the srCaATPase E_2 conformation, pdb1kju, to predict side chains involved in docking the K^+ competitive inhibitor, SCH28080, to the H,K-ATPase. A model for SCH28080 binding between residues L809 and A335 in the same space utilized by omeprazole is proposed. We also describe modeling MgATP binding to the E_1 structure of the srATPase, pdb1eul, as a paradigm for the Na,K- and H,K-ATPases. The resulting model, E_1·MgATP, visualizes a conformation not yet available by crystallization and successfully predicts a range of published results, including backbone cleavages near V440 (N domain) and V712 (P domain) mediated by FeATP in the Na,K-ATPase. A separate model for MgATP docked to E_2 (pdb1kju) shows that access of the γ phosphate to D351 is blocked by the A domain. The E_2· MgATP model explains FeATP-mediated cleavages of the Na,K-ATPase near V440 and E214 (A domain) and homologous results in the H,K-ATPase.

KEYWORDS: H,K-ATPase; SCH28080; molecular modeling

Research over three decades has shown that the P_2 type ATPases share distinct conformational states. A common molecular mechanism is supported further by homologies and similarities found throughout their aligned sequences, implying the feasibility of homology modeling if the molecular structure of at least one member of the family were known.

The first molecular structures associated with known conformational states have been elucidated for the SERCA-ATPase. X-ray diffraction data yielded a structure (pdb1eul) for the E_1·2Ca state[1] at 2.6 Å resolution which defined five distinct domains: N, for nucleotide binding; A or "actuator"; P or phosphorylation domain; M for the membrane domain comprising the 10 transmembrane (TM) helices; and E the

Address for correspondence: Dr. Keith Munson, VAGLAHS, Building 113, Room 324, 11301 Wilshire Boulevard, Los Angeles, CA 90073. Voice: 310-478-3711 x42056; fax: 310-312-9478.
kmunson@ucla.edu

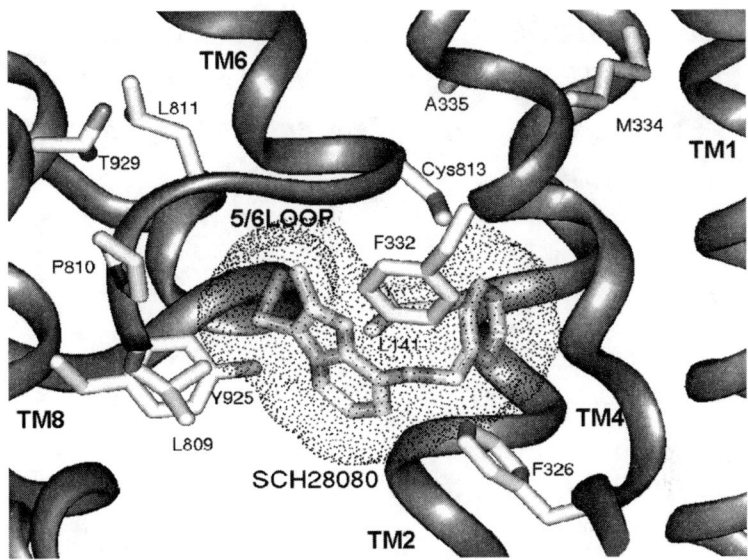

FIGURE 1. View of the proposed SCH28080 binding site with the TM3 and TM5 helices deleted for clarity. SCH28080 (stick and dot surface rendered) is docked between A335 and F332 (TM4) and C813 (TM6) with one end toward TM2 in the vicinity of L141 and the other end enclosed by the TM5/TM6 loop and Y925 (TM8).

extracytoplasmic domain. A structure at 6 Å resolution (pdb1kju) reported for E_2 was derived from cryoelectron microscopy.[2] Neither of these accounted for ATP phosphorylation owing to the large distance separating the phosphorylated aspartic acid 351 and the N domain. A theoretical model for $E_1 \cdot MgATP$ has been published, however, with N containing the docked nucleotide and then tilted about an apparent hinge at residues N359 and R604 to bring the γ phosphate into proximity of D351.[2] No crystallization of the Na,K- or H,K-ATPases has yet been possible, however, and we therefore built homology models to provide a framework for understanding experimental data and designing further investigation of these pumps.

This approach was first applied to model specific inhibitor interaction with the membrane domain of the gastric H,K-ATPase, the biological target for the treatment of ulcer disease. A sequence alignment of the transmembrane segments[3] was made between the rabbit SERCA-ATPase (P04191) and the H,K-ATPase (P27112) and the H,K side chains placed on the Ca-ATPase backbone (pdb1kju). Energy minimization led to a chimeric structure in which the membrane domain was H,K- and the cytosolic N, P, and A domains were still Ca-ATPase. Model side chains corresponding to L141 in TM2 and A335 in TM4 of the H,K-ATPase could be substituted with cysteine and cross-linked to C813 with little change in the backbone position. Cross-linking is observed experimentally in mutants L141C and A335C, but not in double mutants L141C/C813A and A335C/C813A (manuscript in preparation), illustrating how the model and experimental data are complementary.

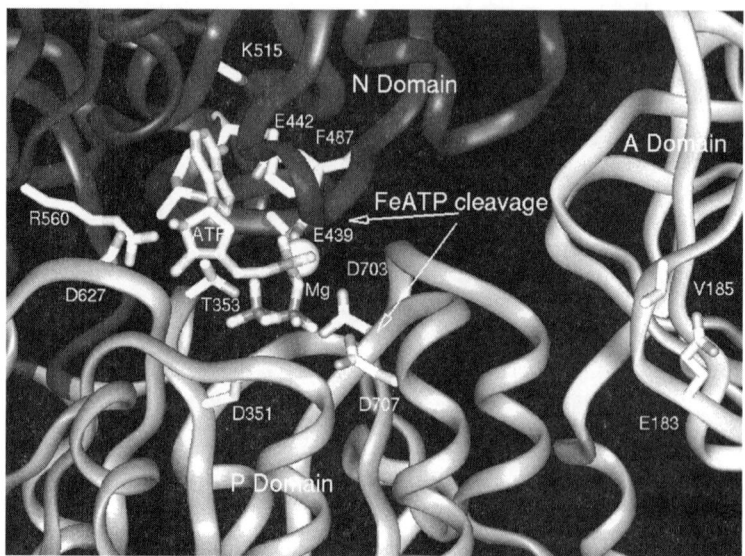

FIGURE 2. Model for MgATP binding to form E_1·MgATP of the SERCA-ATPase. Backbone positions (*arrows*) cleaved in the presence of FeATP in the Na,K-ATPase are near D443 and D710 (E439 and D703 in the model). Side chains close enough to form ligands of Mg^{2+} are T353, D703, and E439. D351 attacks the γ phosphate from the side opposite Mg^{2+}.

Specific H,K-ATPase inhibitors, the substituted pyridiyl methylsulfinyl benzimidazoles; for example, omeprazole, and the 1,2α imidazopyridines (e.g., SCH28080) each bind from the luminal side of the enzyme. Omeprazole inhibition is proportional to disulfide bond formation with cys813 (TM8), while the imidazopyridines are reversible, K^+-competitive inhibitors. A photoaffinity derivative of SCH28080 with an azido group in the *para* position of the phenyl ring labels in the region enclosing TM1 and TM2,[4] thus restricting the position of one end of the bound inhibitor. Inhibitor binding was further defined by kinetic analyses of site-specific mutants where mutations at L809, P810, L811, C813, Y925, T929, F332, A335, or M334[3,5] decreased SCH28080 affinity (K_i increased). Significantly, K_i was elevated 10^2- and 10^5-fold, respectively, by increasing side chain volume in L809F in the TM5/TM6 loop[6] and A335C in TM4.[7] We propose that antagonistic SCH28080 and omeprazole binding[8] occurs in the space between C813 and A335. The chimeric model, energy minimized with SCH28080 docked in this space (FIG. 1), shows that the side chains identified by mutagenesis surround the site. The model suggests further targets for mutation at L141 (giving increased volume) and possibly F326 where the Na,K-ATPase, which is insensitive to SCH28080, substitutes the much larger tryptophan.

We also modeled MgATP binding to E_1 and E_2 forms of the SERCA-ATPase as a paradigm for the Na,K- and H,K-ATPases. The separated N domain of E_1 (pdb1eul), containing docked MgATP in a position close to that shown for TNP-AMP,[1] was raised and tilted ~80° to allow proximity of the γ phosphate to D351 and then reattached and energy minimized (FIG. 2). The model,[9] developed in parallel to that of Xu et al.,[2] explains a variety of experimental results.[10–13] In particular, peptide cleavages near D443 (N domain) and D710 (P domain) mediated by FeATP in the Na,K-ATPase[10] are close to the corresponding SERCA-ATPase residues E439 and D703 that appear to contact Mg^{2+}.

MgATP docking in the N domain of the SERCA-ATPase E_2 conformation (pdb.1kju) and energy minimization yielded a model for $E_2 \cdot$ MgATP which explains changes in the pattern of FeATP-mediated cleavages in the Na,K-ATPase (and the H,K-ATPase[14]) where the split near V712 (CA-ATPase V705) seen in $E_1 \cdot$ FeATP disappears and a new cleavage near E214 arises.[9] These changes result when the A domain moves over P, placing the region near E214 (E183 in the model) next to the KVGE439 sequence in N and covering the segment enclosing D703 and D707 (D710 and D714 in the Na,K-ATPase). When MgATP binds to E_2 of the Na,K-ATPase, the magnesium ion (or iron of FeATP) may chelate E216 and D443 side chains. This would explain the observed cleavages near V440 and E214.

More details of inhibitor and MgATP binding to E_2 should be provided by homology modeling with the new high-resolution srCa-ATPase structure, pdb1iwo.[15]

REFERENCES

1. TOYOSHIMA, C., M. NAKASAKO, H. NOMURA & H. OGAWA. 2000. Crystal structure of the calcium pump of sarcoplasmic reticulum at 2.6 Å resolution. Nature **405:** 647–655.
2. XU, C., W.J. RICE, W. HE & D.L. STOKES. 2002. A structural model for the catalytic cycle of Ca(2+)-ATPase. J. Mol. Biol. **316:** 201–211.
3. VAGIN, O., K. MUNSON, N. LAMBRECHT, et al. 2001. Mutational analysis of the K(+)-competitive inhibitor site of gastric H,K-ATPase. Biochemistry **40:** 7480–7490.
4. MUNSON, K.B., C. GUTIERREZ, V.N. BALAJI, et al. 1991. Identification of an extracytoplasmic region of H(+),K(+)-ATPase labeled by a K(+)-competitive photoaffinity inhibitor. J. Biol. Chem. **266:** 18976–18988.
5. MUNSON, K.B., N. LAMBRECHT & G. SACHS. 2000. Effects of mutations in M4 of the gastric H(+),K(+)-ATPase on inhibition kinetics of SCH28080. Biochemistry **39:** 2997–3004.
6. VAGIN, O. et al. 2002. SCH28080, a K(+)-competitive inhibitor of the gastric H,K-ATPase, binds near the M5-6 luminal loop, preventing K(+) access to the ion binding domain. Biochemistry **41:** 12755–12762.
7. MUNSON, K.B., R. Garcia & G. Sachs. Manuscript in preparation.
8. HERSEY, S.J., L. STEINER, J. MENDLEIN, et al. 1988. SCH28080 prevents omeprazole inhibition of the gastric H+/K+-ATPase. Biochim. Biophys. Acta **956:** 49–57.
9. PATCHORNIK, G., K. MUNSON, R. GOLDSHLEGER, et al. 2002. The ATP-Mg(2+) binding site and cytoplasmic domain interactions of Na(+),K(+)-ATPase investigated with Fe(2+)-catalyzed oxidative cleavage and molecular modeling. Biochemistry **41:** 11740–11749.
10. PATCHORNIK G., R. GOLDSHLEGER & S.J. KARLISH. 2000. The complex ATP-Fe(2+) serves as a specific affinity cleavage reagent in ATP-Mg(2+) sites of Na,K-ATPase: altered ligation of Fe(2+) (Mg(2+)) ions accompanies the E(1) → E(2) conformational change. Proc. Natl. Acad. Sci. USA **97:** 11954–11959.
11. CLAUSEN, J.D., D.B. MCINTOSH, D.G. WOOLLEY & J.P. ANDERSEN. 2001. Importance of Thr-353 of the conserved phosphorylation loop of the sarcoplasmic reticulum Ca2+-ATPase in MgATP binding and catalytic activity. J. Biol. Chem. **276:** 35741–35750.
12. MCINTOSH, D.B., D.G. WOOLLEY, B. VILSEN & J.P. ANDERSEN. 1996. Mutagenesis of segment 487Phe-Ser-Arg-Asp-Arg-Lys492 of sarcoplasmic reticulum Ca2+-ATPase produces pumps defective in ATP binding. J. Biol. Chem. **271:** 25778–25789.

13. TRAN, C.M., E.E. HUSTON & R.A. FARLEY. 1994. Photochemical labeling and inhibition of Na,K-ATPase by 2-Azido-ATP. Identification of an amino acid located within the ATP binding site. J. Biol. Chem. **269:** 6558–6565.
14. SHIN, J.M., R. GOLDSHLEGER, K.B. MUNSON, *et al*. 2001. Selective Fe2+-catalyzed oxidative cleavage of gastric H+,K+-ATPase: implications for the energy transduction mechanism of P-type cation pumps. J. Biol. Chem. **276:** 48440–48450.
15. TOYOSHIMA, C. & H. NOMURA. 2002. Structural changes in the calcium pump accompanying the dissociation of calcium. Nature **418:** 605–611.

Inhibition Kinetics of the Gastric H,K-ATPase by K-Competitive Inhibitor SCH28080 as a Tool for Investigating the Luminal Ion Pathway

OLGA VAGIN, KEITH MUNSON, SVETLANA DENEVICH, AND GEORGE SACHS

University of California at Los Angeles and VA Greater Los Angeles Healthcare System, Los Angeles, California 90073, USA

ABSTRACT: The gastric H,K-ATPase and the Na,K-ATPase both are stimulated by luminal K^+, but differ in sensitivity to K^+-competitive inhibitors (ouabain and SCH28080), which implies a difference in structure near the luminal ion pathways in these two pumps. Knowledge of the amino acids in the H,K-ATPase that affect the mode of inhibition by SCH28080 and inhibitor affinity should provide insight into the regions of the membrane domain influencing the inhibitor selectivity and the luminal route to the ion transport site. Mutational scans in M4, 5, 6, and 8 have shown that amino acid residues affecting ion affinity (E343, K791, E795, E820, D824, E936) with either no or a lesser effect on the inhibitor affinity are located in the middle of the membrane domain. The residues significantly reducing inhibitor affinity, but not ion affinity (L809, P810, L811, T813, I816, Y925, T929), are located in the exoplasmic 5-6 loop and the luminal ends of M6 and M8. This suggests that the binding domain for SCH28080 contains the surface between L809 in the 5-6 loop and C813 at the luminal end of M6, ~two helical turns out from the ion binding region, where it blocks an ion access pathway. The mutations that change inhibitor kinetics are on the opposing faces of M6 and M8 and apparently modify the normal ion pathway or, perhaps, create an alternate ion pathway.

KEYWORDS: H,K-ATPase; SCH28080; inhibition kinetics

The gastric H,K-ATPase, as does the homologous Na,K-ATPase, binds luminal K^+ in the E_2 P conformation. Very little is known about the structure of a luminal ion entry pathway and the regions in the membrane domain that are responsible for the conformational changes of this pathway in the E_2-E_1 transition.

Inhibition kinetics of the gastric H,K-ATPase by SCH28080 is one of the useful tools for investigating the luminal ion entry path. SCH28080 is a K^+-competitive inhibitor of the gastric H,K-ATPase. The difference in size and shape of the inhibitor

---

Address for correspondence: Dr. Olga Vagin, VAGLAHS, Building 113, Room 324, 11301 Wilshire Boulevard, Los Angeles, CA 90073. Voice: 310-478-3711 X42055; fax: 310-312-9478.
olgav@ucla.edu

and transported cation as well as experimental data suggest that the ion transport site and inhibitor site are separate. It is likely the inhibitor blocks ion access, and the residues binding SCH28080 would potentially lie in or near the luminal ion entry path distal to the ion site. The structure of this path would be different both in the E_1 and in the E_2K conformations of the H,K-ATPase to which SCH28080 does not bind, and also will be different in a homologous Na,K-ATPase and nongastric H,K-ATPase since SCH28080 is highly selective for the gastric H,K-ATPase. Therefore, using SCH28080 as a probe can define a luminally accessible ion entry vestibule within the H,K-ATPase membrane domain and should provide insight into the regions changing the conformation of the vestibule due to K-binding and E_1-E_2 transition and explain the selectivity of the H,K-ATPase for SCH28080. Accordingly, the aim of the present study was to define the amino acid residues in the membrane domain affecting the mode of inhibition of the gastric H,K-ATPase by SCH28080 and inhibitor affinity using site-directed mutagenesis.

Wild-type and mutant H,K-ATPases were expressed in HEK 293 cells. Microsomal membranes were isolated and H,K-ATPase activity was measured as NH_4^+ stimulated activity in the absence of the inhibitor and at two or three fixed inhibitor concentrations, and V_{max} and $K_{m,app}$ values were determined. The type of inhibition—competitive, noncompetitive, mixed—was determined using inverse plots. $K_{m,app}/V_{max}$ were plotted against the inhibitor concentration, and the inhibitor affinity, K_i, was determined at the intercept on the X-axis. The ratios $K_{m,app}/V_{max}$ of expressed mutant enzymes were used to estimate the effect of mutations on the ability of the enzyme to interact with ion.

Since side chains from M4, 5, 6, and perhaps M8 form the ion transport domain, SCH28080 must block the ion pathway between these transmembrane helices. Scanning site-directed mutagenesis was performed on residues of the inner surface of membrane helices, M5, 6, and 8, as predicted from the model of the Ca-ATPase in E2 conformation (1kju.pdb). TABLE 1 and FIGURE 1 summarize the data discussed in this paper.

Most of the mutants only slightly affected ion and inhibitor affinities and retained purely competitive type of inhibition. Mutations of the residues aligned to the ion-binding residues of the Ca-ATPase reduced ion affinity with lesser effect on the inhibitor affinity. While mutations affecting ion binding are concentrated more toward the center of the membrane domain, the mutations significantly reducing SCH28080 affinity are located closer to the luminal face of the enzyme. In contrast to the mutations of ion-binding residues, these mutations resulted in loss of inhibitor affinity with either no effect or a lesser effect on ion affinity. The 100- and 4-fold reduction of the inhibitor affinity in the L809F and L809V mutants, without a significant change in ion affinity, suggests a direct involvement of the space adjacent to this leucine in SCH28080 binding. Substitution of neighboring residues in 5-6 loop region—proline 810 and leucine 811—also significantly reduced SCH28080 affinity. The mutation of cysteine 813 to the threonine present in SCH28080-insensitive Na,K-ATPase resulted in a 9-fold loss of the inhibitor affinity, and even conservative mutations to a serine and the smaller alanine resulted in a 3-fold decrease of the affinity. Cysteine 813 also forms a disulfide with covalent proton pump inhibitors such as omeprazole, demonstrating the presence of a space adjacent to cysteine 813 with a sufficient volume to bind SCH28080. The mutation of I816L near the luminal end of M6 resulted in 5-fold reduction of SCH28080 affinity. We also found two muta-

TABLE 1. The mutations in the M5-6 loop, M6, and M8 significantly reducing SCH28080 affinity

Location	Mutant	K_i[SCH] ± SE (nM)	Type of Inhibition	V_{max} ± SE (μmol/mg/h)	$K_{m,app}$[NH$_4^+$] ± SE (μM)	H,K-ATPase (% of total protein)	$V_{max}/K_{m,app}$ [NH$_4^+$] ±SE
5-6 loop	L809F	6150±210	noncomp.	39.0±1.3	1.9±0.30	3.0	20.5±3.0
	L809V	288±52	noncomp.	85.0±2.3	2.2±0.17	8.4	38.3±4.0
	P810A	281.16	comp.	87.0±2.1	2.2±0.20	12.0	39.3±4.5
	P810G	563±17	comp.	108.5±2.5	2.4±0.24	11.0	44.7±5.4
	L811F	625±63	mixed	113.0±1.9	2.0±0.12	2.5	56.5±4.3
M6	C813A	182±18	comp.	150.0±3.5	5.5±0.47	11.4	27.3±3.0
	C813S	169±30	comp.	62.9±	4.9±0.96	9.0	12.8±3.3
	C813T	586±69	comp.	40.0±4.9	6.6±1.80	0.4	6.1±2.4
	I816L	309±24	noncomp.	30.5±1.9	1.4±0.36	6.0	21.7±7.0
M8	Y925F	372±62	noncomp.	30.6±0.9	1.6±0.16	4.6	19.1±2.5
	T929L	512±69	comp.	26.6±1.0	1.2±0.20	3.0	22.6±4.7
	WT	64±11	comp.	132.0±3.2	2.4±0.05	8.0	55.0±1.3

tions in M8 that reduced the inhibitor affinity. According to recently published results,[3] the chimera containing a fragment 905-930 of the gastric H,K-ATPase placed into the Na,K-ATPase acquired some SCH28080 sensitivity. We mutated the residues that are not conserved between H,K-ATPase and Na,K-ATPase in this fragment to the corresponding Na,K residues and found 6- and 8-fold loss of SCH28080 affinity in Y925F and T929L mutants.

Considering the dimensions of SCH28080 (11 × 16 Å) and the distances between the corresponding residues in Ca-ATPase structure, we suggest that the SCH28080 binding region in the space adjacent to leucine 809 and cysteine 813 could include a surface contributed by the residues from M8 on one side and a portion of the M4 on the other with the phenyl ring of the inhibitor pointing toward M2. This representation is compatible with previously published results concerning mutations lowering SCH28080 affinity, M334I in M4,[1] and (I816A) in M6,[2] and is also compatible with published data on a possible contribution of Q905-V930 region to SCH28080 binding.[3] The postulated region of binding of SCH28080 also allows interaction of the inhibitor with the M1/M2 region as previously suggested by photo-affinity labeling of the M1/loop/M2 region of the H,K ATPase.[4] Further, the binding site illustrated here predicts an overlap between SCH28080 binding site and the omeprazole binding site explaining the reduction of omeprazole inhibition by the presence of SCH28080.[5]

Several mutations resulted in a noncompetitive or mixed SCH28080 inhibition kinetics. Such data require a change from mutually exclusive binding of cation and inhibitor to simultaneous binding of both cation and inhibitor. These mutations did not affect the ion affinity, and several of them also were without effect on inhibitor affinity. The simplest explanation is a change in accessibility of the cation to the binding domain in the presence of bound inhibitor in these mutants. These mutations

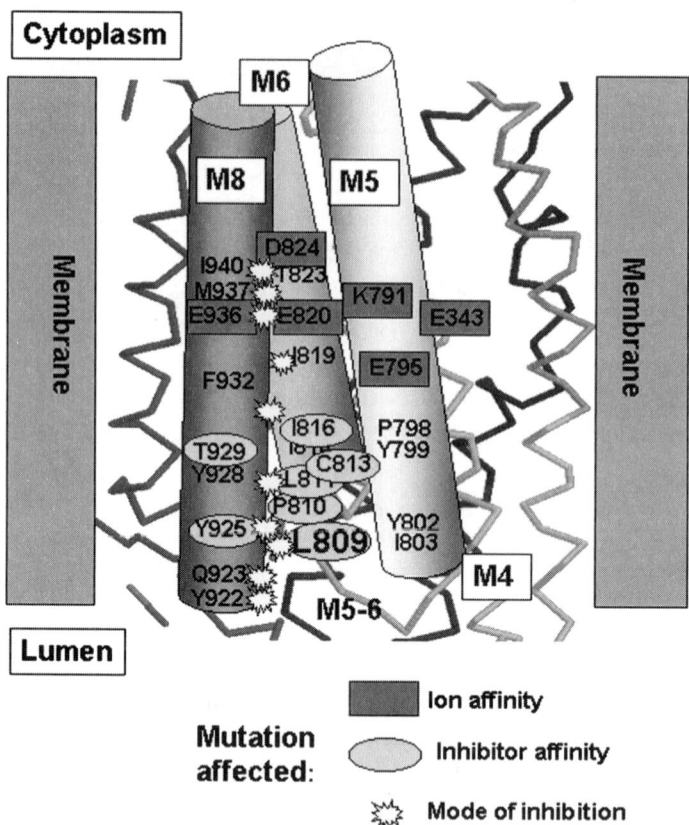

FIGURE 1. An illustration of the M5, M6, and M8 membrane and the M5-6 loop region of the gastric H,K-ATPase based on the Ca-ATPase in the E_2 conformation (pdb.1kju). The residues whose mutations resulted in a significant decrease in SCH28080 affinity (5–100-fold) are marked by ovals, and noncompetitive or mixed inhibition kinetics by the stars. Ion-binding residues defined by mutagenesis are marked by rectangles.

are placed on the opposing faces of the M6 and M8 helices, as illustrated by stars in FIGURE 1. Each of the mutations could alter the spacing between M6 and M8 that then could allow ion access to the transport sites in the middle of the membrane despite the presence of the inhibitor on the enzyme bound in the luminal vestibule.

In summary, the illustration in FIGURE 1 suggests that SCH28080 binds near the L809 in the 5-6 loop and C813 at the luminal end of M6, about two helical turns distal to the ion transport site, where it blocks the luminal ion access pathway. The mutations that change inhibitor kinetics are on the opposing faces of M6 and M8 and apparently modify the normal ion pathway or, perhaps, create an alternate ion pathway.

REFERENCES

1. MUNSON, K.B., N. LAMBRECHT & G. SACHS. 2000. Effects of mutations in M4 of the gastric H+,K+-ATPase on inhibition kinetics of SCH28080. Biochemistry **39:** 2997–3004.
2. ASANO, S., S. MATSUDA, S. HOSHINA, *et al.* 1999. Chimeric gastric H+,K+-ATPase inhibitable with both ouabain and SCH 28080. J. Biol. Chem. **274:** 6848–6854.
3. FARLEY, R.A., S. SCHREIBER, S.G. WANG & G. SCHEINER-BOBIS. 2001. A hybrid between Na+,K+-ATPase and H+,K+-ATPase is sensitive to palytoxin, ouabain, and SCH 28080. J. Biol. Chem. **276:** 2608–2615.
4. MUNSON, K.B., C. GUTIERREZ, V.N. BALAJI, *et al.* 1991. Identification of an extracytoplasmic region of H+,K(+)-ATPase labeled by a K(+)-competitive photoaffinity inhibitor. J. Biol. Chem. **266:** 18976–18988.
5. HERSEY, S.J., L. STEINER, J. MENDLEIN, *et al.* 1988. SCH28080 prevents omeprazole inhibition of the gastric H+/K+-ATPase. Biochim. Biophys. Acta **956:** 49–57.

Ion Occlusion/Deocclusion Partial Reactions in Individual Palytoxin-Modified Na/K Pumps

PABLO ARTIGAS AND DAVID C. GADSBY

Laboratory of Cardiac/Membrane Physiology, The Rockefeller University, New York, New York 10021, USA

ABSTRACT: In P-type ion-motive ATPases, transported ions approach their binding sites from one membrane surface, become buried deep within "occluded" conformations in which the sites are inaccessible from either membrane side, and are then deoccluded and released to the opposite membrane surface. This describes an alternating-gate transport mechanism, in which the pump acts like an ion channel with two gates that open and close alternately. The occluded states ensure that one gate closes before the other can open, thus preventing the large electrodiffusive ion fluxes that would otherwise quickly undo the pump's electrochemical work. High-resolution crystal structures of two conformations of the SERCA (sarcoplasmic and endoplasmic reticulum Ca^{2+}) P-type ATPase, together with mutagenesis results and analyses of structural models based on homology, have begun to provide a picture of the ion coordination sites in related P-type ATPases, including the Na/K pump. However, in no P-type ATPase are the structures and mechanisms of the gates known. The marine toxin, palytoxin (PTX), is known to bind to the Na/K pump and elicit a nonselective cation leak pathway, possibly by disrupting the strict coupling between the pump's inner and outer gates, allowing them to both be open. We recently found that ion flow through PTX-modified Na/K pump-channels appears to be modulated by two gates that can be regulated by the pump's physiological ligands in a manner suggesting that gating reflects underlying ion occlusion/deocclusion partial reactions. We review that work here and provide evidence that the pore of the PTX-induced pump-channel has a diameter > 6 Å.

KEYWORDS: Na,K-ATPase; Na/K pump; ion channel; intracellular gate; extracellular gate; gating; conformational change; ion occlusion; deocclusion; palytoxin; ventricular myocyte; HEK293 cell; outside-out patch; macroscopic current; microscopic current; single-channel current

INTRODUCTION

According to the alternating-access model of ion-motive pump function,[1,2] the Na,K-ATPase may be represented as a transmembrane ion channel that incorporates (at least) two gates that open alternately to regulate access to the ion-binding sites

Address for correspondence: David C. Gadsby, Laboratory of Cardiac/Membrane Physiology, The Rockefeller University, 1230 York Avenue, New York, NY 10021-6399. Voice: 212-327-8680; fax: 212-327-7589.
gadsby@mail.rockefeller.edu

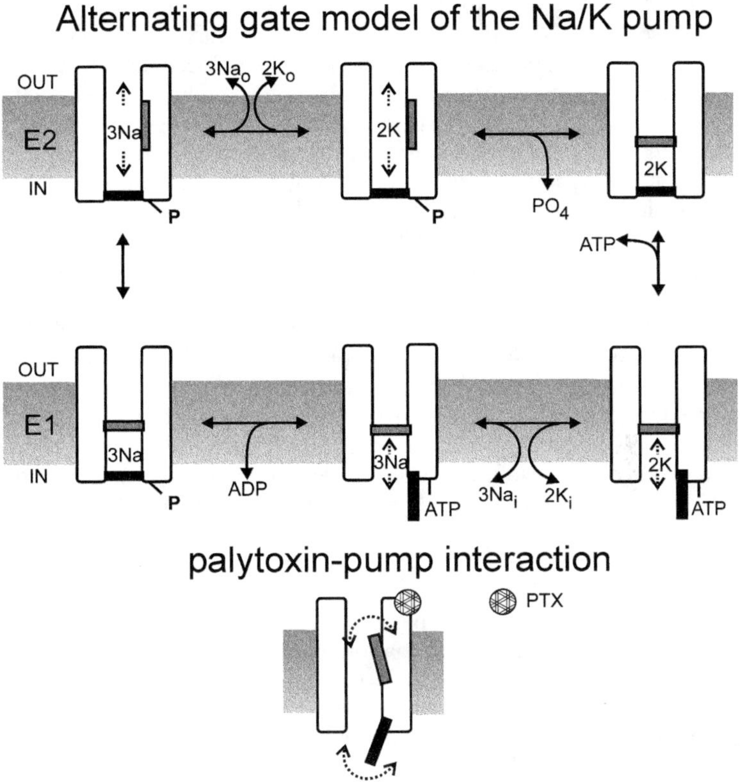

Alternating gate model of the Na/K pump

FIGURE 1. (Top) Alternating-gate model of the Post–Albers[4,5] cycle of the Na/K pump, cartooned as an ion channel with an extracellular-side (labeled OUT) gate (*gray*) and a cytoplasmic-side (IN) gate (*black*) that open alternately, but never simultaneously, during normal 3Na$^+$/2K$^+$ transport. Occluded states (both gates shut) follow binding of 3 internal Na$^+$ and phosphorylation (*lower left*), and binding of 2 external K$^+$. (Reproduced with permission from Ref. 3.) **(Bottom)** Cartoon of channel-like Na/K pump with bound palytoxin (PTX; represented as a ball), allowing the pump's two gates to be open at the same time.

along the ion-translocation pathway (FIG. 1, top). Because ion flow through an open channel can occur orders of magnitude faster than usual rates of pumped ion transport, the probability of the pump's two gates being open simultaneously must be extremely low (<10^{-6})[3] to avoid dissipation of the Na$^+$ and K$^+$ electrochemical gradients. The sequences of conformational changes cartooned in FIGURE 1 that, in each complete cycle,[4,5] effect extrusion of 3 Na$^+$ from, and uptake of 2 K$^+$ into, the cell at the expense of hydrolysis of 1 ATP fulfill this constraint.

The Na/K pump alternates between two principal conformations: E1, with cytoplasmic access (FIG. 1, lower row), and E2, with extracellular access (FIG. 1, upper row), to the ion-binding sites. Following binding of the third intracellular Na$^+$ to E1, the pump becomes phosphorylated ("P" in FIG. 1) from ATP (acting with $K_{0.5}$ < 1 µM) and both gates close, occluding the 3 Na$^+$ ions within the pump. The phosphorylated pump relaxes to the E2 conformation, in which the external gate opens to release the 3 Na$^+$. After Na$^+$ dissociation, binding of 2 K$^+$ prompts closing of the external gate, leading to dephosphorylation[6] and occlusion of the K$^+$ inside the E2 conformation[7] (FIG. 1, top right). Subsequent binding of ATP (acting with $K_{0.5} \approx 100$ μM) changes the conformation back to E1, opening the intracellular gate and deoccluding the K$^+$ to the cytoplasm.[8] The high- and low-affinity effects of ATP can be further distinguished because nonhydrolyzable ATP analogues, and even ADP, can mimic the low-affinity ATP action.[9–11]

A channel-like structure in the Na/K pump, which transported ions must transit on their way to or from their binding sites, has indeed been proposed on the basis of functional studies,[12–15] and also for the closely related SERCA- and H,K-ATPases.[15,16] However, no such channel was obvious in recent high-resolution crystal structures of the E1Ca-TNPAMP[17] or E2-thapsigargin[18] states of the SERCA pump, although an access channel-like cavity leading from the extracellular surface was apparent in a lower resolution (8 Å) structure of the E2-decavanadate-thapsigargin complex of the SERCA pump.[19]

No comparably high-resolution structures are available for Na,K-ATPases, but a toxin called palytoxin (PTX), isolated from marine coelenterates,[20] binds specifically to the Na/K pump, opening an ion channel[21,22] that likely shares at least part of the pathway normally traversed by the pumped ions.[23,24] Following the alternating-gate model (FIG. 1), a simple hypothesis is that PTX disrupts the usual strict coupling between the pump's two gates, allowing them to sometimes be open at the same time (FIG. 1, bottom). We recently found that the Na/K pump's principal physiological ligands influence the gating of PTX-induced channels, in a manner consistent with the alternating-gate model.[3] Here, we review that work and briefly describe new findings about the influence of phosphorylation on the PTX-induced channel and about its ionic selectivity.

MATERIALS AND METHODS

Currents were recorded at 22–25°C in outside-out membrane patches using an Axopatch 200B amplifier and pClamp7 software (Axon Instruments, Inc.). The patches were excised from guinea-pig ventricular myocytes (isolated as described[25]) or from HEK293 cells (from ATCC) seeded on polylysine-coated coverslips 1–3 days before recordings. Thin-walled pipettes coated with Sylgard and used without fire-polishing had resistances of 2–8 MΩ for macroscopic and 15–30 MΩ for microscopic current recording. Extracellular solution contained (in mM) 140 sulfamic acid, 10 HEPES, 10 HCl, 1 MgCl$_2$, 1 CaCl$_2$, 0.5–5 BaCl$_2$ (to block K$^+$-channels), and 160 Na$^+$, K$^+$, Rb$^+$, Cs$^+$, or TMA$^+$ (tetramethylammonium); the Tl$^+$ solution contained 155 mM TlNO$_3$ instead of the sulfamic acid and HCl. Pipette (cytoplasmic surface) solution had (in mM) 150 Na$^+$, 130 L-glutamic acid, 10 HEPES, 10 EGTA, 10 TEACl, and 1 MgCl$_2$, with or without nucleotide as specified. MgATP, Na$_2$ADP, or Li$_4$AMPPNP [AMPPNP: adenosine 5′-(β,γ-imido)triphosphate] were added from 200 mM stock solutions (pH 7.4 with NMDG), and equimolar MgCl$_2$ was added with ADP and AMPPNP. All solutions had pH 7.4 and osmolality of 285–305 mOsm/kg. An aliquot of 100 μM PTX (*Palythoa toxica*; Calbiochem) stock solution (in H$_2$O) was defrosted and diluted into external solutions just before each experiment; all such PTX-containing solutions also included 0.002% BSA to minimize PTX binding to nonglass surfaces.

RESULTS AND DISCUSSION

Permeation

To assay PTX interactions with Na/K pumps and characterize the resulting ion channels, we measured membrane conductance. FIGURE 2A illustrates the large inward (downward) current shift (at -20 mV) caused by saturating [PTX] (100 nM) in an outside-out HEK293-cell patch, with almost symmetrical [Na], and 5 mM pipette MgATP. That inward current reflects Na^+-ion flow through many PTX-induced channels. The very small vertical deflections about the baseline current caused by 100-ms changes in membrane potential (-80 to $+60$ mV) shortly after the first exposure to Na^+ solution (FIG. 2A; arrow labeled "Na"), but before the first addition of PTX (FIG. 2A; horizontal bar), indicate that membrane conductance was initially small. The much larger current deflections shortly after adding PTX to the external Na^+ solution reflect the substantial conductance caused by PTX-induced opening of ion channels in all the Na/K pumps in the patch. Subtraction of the current level at a given membrane potential before PTX from that after PTX addition yielded the PTX-induced currents (I_{PTX}) plotted against potential in the current-voltage curve labeled "Na" (●) in FIGURE 2B. The PTX-induced current reversed sign near 0 mV, as expected for current carried by the principal cation, Na^+, at roughly equal intracellular and extracellular concentrations.

Before PTX, only small current changes accompanied switches between test monovalent cations (each at ~160 mM), and voltage-induced current deflections were also small (FIG. 2A). However, after PTX exposure, the voltage-induced currents became much bigger and some solution switches caused large holding current shifts (FIG. 2A), indicating permeability differences among the test cations in the PTX-induced channels. For example, the large inward current shift on switching from Na^+ to Tl^+ and the large outward shift on switching to TMA^+, after PTX addition (FIG. 2A), suggest that PTX-induced channels are more permeable to Tl^+, but less permeable to TMA^+, than they are to Na^+. To quantify these effects, we determined I_{PTX}-voltage relationships (FIG. 2B), by subtraction, for each external cation. From the shifts (ΔV_{rev}) in I_{PTX} reversal potential (V_{rev}) for each cation, X^+, relative to V_{rev} in Na^+, permeability ratios P_X/P_{Na} were calculated from $\Delta V_{rev} = V_{revX} - V_{revNa} = (RT/F) \cdot \ln\{P_X[X]/P_{Na}[Na]\}$. The sequence was $P_{Tl} \gg P_K \approx P_{Rb} > P_{Cs} \approx P_{Na} \gg P_{TMA}$, corroborating previous reports of poor selectivity among K^+, Rb^+, Cs^+, and Na^+ ions.[26,27] However, the further large negative shift of V_{rev} on reducing external [TMA] 10-fold, by isosmotic substitution with sucrose (labeled "suc" in FIG. 2), confirms that TMA^+ ions do permeate through PTX-induced channels. The narrowest region must therefore be wide enough to pass TMA^+, that is, must have a diameter ≥ 6 Å (geometric mean of approximate dimensions of smallest box enclosing the ion).[28] That such a wide pore selects poorly between alkali metal ions (FIG. 2) is not surprising, but the much higher Tl^+ permeability ($P_{Tl}/P_{Na} \approx 3$) was unexpected. Interestingly, Tl^+ also activates Na,K-ATPase[29] with higher apparent affinity than K^+, Rb^+, or Cs^+.

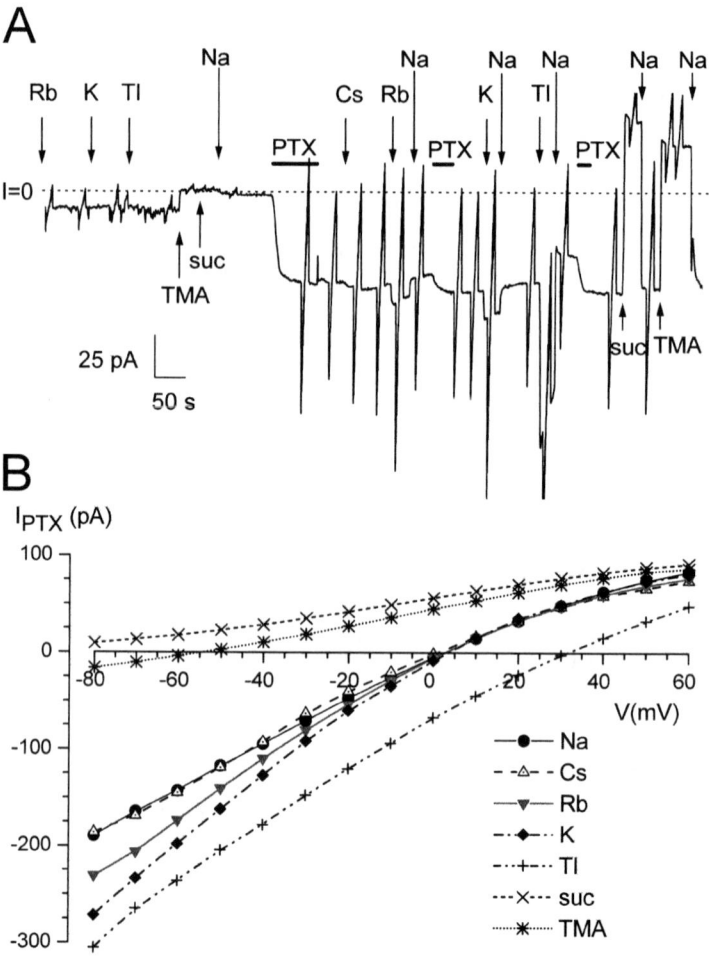

FIGURE 2. PTX-induced conductance is poorly selective among small cations. (**A**) Macroscopic current, at −20 mV, in HEK293-cell outside-out patch, with 150 mM Na and 5 mM ATP in the intracellular (pipette) solution. The principal (~160 mM) external cation was exchanged, as indicated, before and after addition of 100 nM PTX (Cs$^+$ was tested earlier), and conductance was assayed with 100-ms steps (vertical deflections) to voltages (V) of −80 mV ≤ V ≤ +60 mV. (**B**) PTX-induced current-voltage (I_{PTX}–V) relationships ($I_{PTX} = I$ after PTX − I before PTX) obtained with indicated extracellular solutions yielded permeability ratios for each cation relative to Na$^+$.

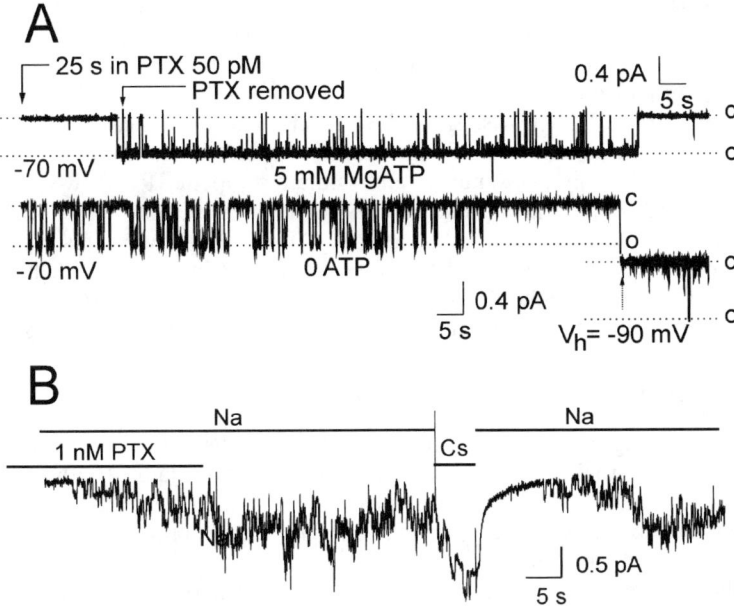

FIGURE 3. Modulation of gating of PTX-bound channels by Na/K pump ligands. (**A, top trace**) Current, at −70 mV, in outside-out myocyte patch, with ~160 mM Na^+ solutions and 5 mM internal MgATP, exposed to 50 pM PTX until one channel opened. (**A, bottom trace**) Current from outside-out myocyte patch at −70 mV, with Na^+ solutions, but no ATP; gating after PTX removal showed shorter open bursts and more frequent long closed periods than with ATP. (Reproduced with permission from Ref. 31.) (**B**) Record from outside-out myocyte patch at −40 mV, with Na^+ solutions, but no ATP, exposed to 1 nM PTX until several (≥4) channels opened. Replacement of all external Na^+ by Cs^+ elicited a rapid drop in channel P_o, accompanied by a baseline current shift due to Na^+ vs. Cs^+ differences in seal current; reapplication of Na^+ reopened the channels. (Reproduced with permission from Ref. 3.)

Gating of PTX-Bound Na/K Pump-Channels: Interior Gate

In a key experiment that helped pinpoint the Na/K pump as the receptor for PTX, Wu and collaborators[21] found that ATP activated channels in inside-out patches with saturating [PTX] in the pipette. However, whether this reflected ATP-mediated increases in open probability (P_o; fraction of time spent open) of pump-channels with PTX already bound, or in apparent affinity for PTX that then quickly bound and opened pump-channels, was unclear. To resolve this, we used outside-out patches and applied low [PTX] until a channel opened, and then washed away unbound PTX to prevent further channel activation (FIG. 3A). The upper trace in FIGURE 3A shows typical gating of a PTX-activated channel in an outside-out patch, with 5 mM intracellular MgATP and symmetrical [Na^+] solutions; activity persisted long after PTX washout. In the presence of ATP, the PTX-induced channel displayed long open bursts and high open probability (with ATP, average $P_o \approx 0.95$).[3]

In outside-out patches without pipette ATP (FIG. 3A, lower trace), PTX activated a channel of the same conductance that also remained active after PTX washout. However, with no ATP, channel open probability was low (without ATP, average $P_o \approx 0.2$),[3] in part due to briefer openings. Because no channel activity was seen before exposure to PTX (FIGS. 2A and 3A) and because PTX cannot rebind after washout, we can conclude that ATP increases P_o of PTX-bound channels. In inside-out patches with many Na/K pumps and saturating pipette [PTX], we found that ATP, ADP, or AMPPNP all similarly increased macroscopic current (~6-fold) over similar concentration ranges, half-maximally at 10–100 μM.[3] This nucleotide-dependent increase in P_o therefore likely reflects low-affinity nucleotide binding to PTX-modified pumps, at the site promoting opening of the pump's cytoplasmic gate, and K^+ deocclusion and release into the cell, during normal Na/K pumping (FIG. 1).

Gating of PTX-Bound Na/K Pump-Channels: Exterior Gate

The alternating-gate cartoon (FIG. 1) implies that the extracellular-side gate is controlled by external K^+ and the pump's phosphorylation status. Binding of extracellular K^+ promotes closing of the external gate,[30] and hence occlusion of the K^+ and pump dephosphorylation,[6,7] whereas phosphorylation promotes opening of the external gate with concomitant K^+ release.[30] Closure of PTX-bound Na/K pump-channels by external K^+ is illustrated in FIGURE 3B. The outside-out patch, in Na^+ solutions, but with no pipette ATP, was exposed to 1 nM PTX until several channels (at least four) opened, whereupon the toxin was removed. Gating of PTX-induced channels continued after PTX washout, but, on replacing external Na^+ with the K^+ congener Cs^+, P_o immediately dropped until, near the end of the 5-s exposure to Cs^+, only a single opening was seen. The Cs^+ also caused a temporary baseline shift due to Na^+ vs. Cs^+ differences in pipette seal resistance that reversed on restoring Na^+. There was no channel activity when Na^+ was first readmitted, but within a few seconds openings reappeared and channel activity soon reached a level comparable to that before the brief exposure to Cs^+. Because there was no PTX in the extracellular solution, this channel closing by Cs^+ and reopening by Na^+ must reflect changes in the gating of PTX-bound channels.

In the particular instance depicted in FIGURE 3B, there seemed little or no PTX unbinding in Cs^+ because the current levels in Na^+ were roughly the same before and after application of Cs^+. However, as illustrated in FIGURE 4, K^+ and its congeners not only close the external gate, but also greatly speed PTX unbinding.[3] The records in FIGURE 4A and 4B are from outside-out patches, with intracellular (pipette) solution containing Na^+ and either 2 mM AMPPNP (FIG. 4A) or 1 μM ATP (FIG. 4B), and in both cases saturating [PTX] (≥100 nM) was first applied in Na^+ external solution (shown only in FIG. 4A). When PTX was then removed in the presence of Na^+, the unbinding of PTX (monitored as decay of the activated current) proceeded extremely slowly (FIG. 4A). However, the PTX-induced current decayed rapidly during brief exposures to K^+ (following instantaneous, transient current increases reflecting the slightly higher permeability of the channels to K^+ compared to Na^+; FIG. 2). This current decay was not due to mere K^+-induced closing of PTX-bound channels (similar to the action of Cs^+ in FIG. 3B) because switching back to Na^+ after each brief exposure to K^+ did not fully restore the current, so eventually the current in Na^+ solution returned to its level before PTX application (FIG. 4A). This suggests

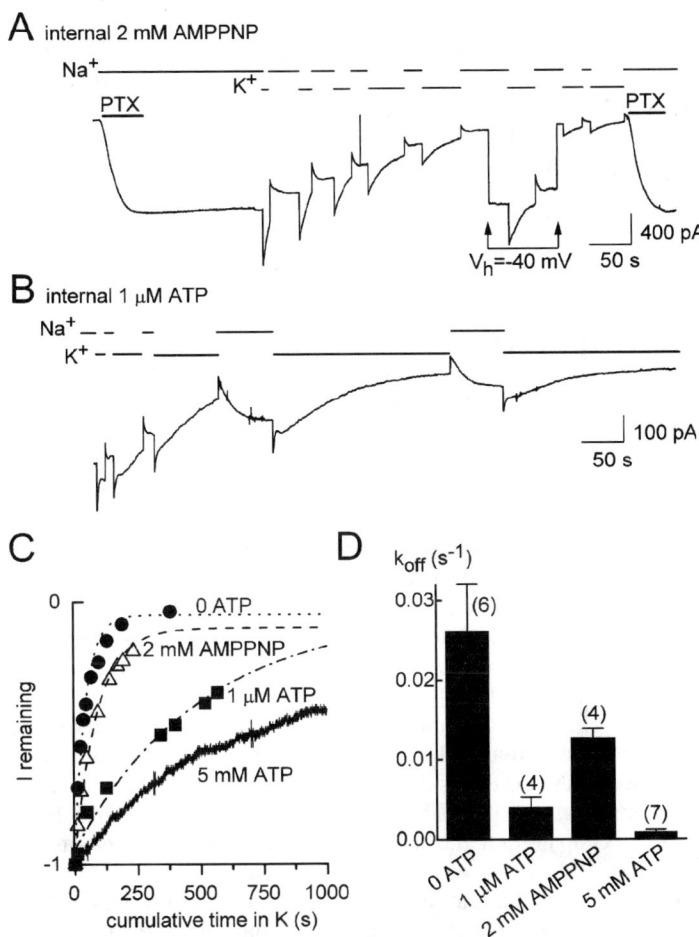

FIGURE 4. K^+-mediated acceleration of PTX unbinding is countered by pump phosphorylation. **(A)** Current from HEK293-cell outside-out patch at −20 mV (except where indicated), with Na^+ solutions and 2 mM pipette MgAMPPNP, exposed to 100 nM PTX. K^+ sped current decline, and restoring Na^+ partly restored inward current as PTX-bound channels reopened; the decline of current in Na^+ reflects PTX unbinding in K^+ (part C, *open triangles*). **(B)** Current as in part A, but with 1 μM MgATP, which slowed both channel closing and PTX unbinding in K^+ (part C, *solid squares*). **(C)** Residual current in Na^+ vs. cumulative time in K^+, without nucleotide (0 ATP, *solid circles*; from Ref. 3), with 2 mM MgAMPPNP, or with 1 μM MgATP; discontinuous lines show exponential fits. The thick continuous line shows current decay in K^+ with 5 mM pipette MgATP, directly reflecting PTX unbinding as no reopening accompanies switches back to Na^+. **(D)** Average apparent PTX dissociation rates, $k_{off} = 1/\tau$, from exponential fits to time courses as in part C.

that PTX had unbound during the exposures to K^+, an interpretation supported by the finding that reapplication of PTX reactivated the same amplitude current as observed initially (FIG. 4A, right). This relatively rapid decline of PTX-induced current during exposure to K^+, with AMPPNP in the pipette, is comparable to that observed without any pipette nucleotide.[3] In contrast, the current decay under otherwise similar conditions, but with 1 µM ATP in the pipette, appeared somewhat slower (FIG. 4B).

To separate PTX unbinding from K^+-induced channel closure, we further analyzed the results of switching back and forth between Na^+ and K^+ solutions after PTX washout. Like the temporary closing of the few PTX-bound channels by Cs^+ and their reopening by Na^+ in FIGURE 3B, in FIGURE 4A and 4B each exposure to K^+ closed up to thousands of channels, and each reapplication of Na^+ reopened those PTX-bound channels that had been merely closed by the K^+. Therefore, the reduced current amplitude in Na^+ after each application of K^+ represents loss of that fraction of pump-channels from which PTX unbound during the preceding exposure to K^+. Accordingly, plotting residual current amplitude in Na^+ against cumulative time in K^+ (FIG. 4C) yields the PTX unbinding rate (k_{off}) in K^+, that is, reciprocal time constant from exponential fit (FIG. 4C, broken lines). Without nucleotide (FIG. 4C, solid circles), PTX unbinds rapidly in K^+ solution,[3] comparable to the result with 2 mM AMPPNP (FIG. 4C, open triangles). In contrast, with 5 mM ATP in the pipette, channel closing in K^+ was extremely slow (FIG. 4C, continuous recording) and no reopening was observed on switching back to Na^+ (not illustrated), implying that, with saturating ATP, current decay in K^+ directly mirrors PTX unbinding. However, with only 1 µM ATP (FIG. 4C, solid squares), a concentration capable of supporting maximal Na/K pump phosphorylation in the absence of K^+,[6] PTX unbinding also appears slowed, that is, intermediate between the rates observed with 0 and 5 mM ATP.

FIGURE 4D summarizes average k_{off} measurements for PTX, which suggest that ATP-dependent pump phosphorylation slows closing of PTX-induced channels and hence PTX unbinding, thereby antagonizing effects of extracellular K^+. The slower PTX unbinding at 5 mM than at 1 µM ATP presumably reflects saturation of the low-affinity site at 5 mM ATP, resulting in more rapid rephosphorylation of PTX-modified pumps than at 1 µM. The weaker influence of 2 mM AMPPNP is likely attributable to occupancy of the low-affinity nucleotide-binding site, but with no phosphorylation.[3]

In the context of the alternating-gate model (FIG. 1), these findings suggest that extracellular K^+ ions promote closing of the external gate of PTX-induced channels, just as they promote K^+-ion occlusion during the pump's normal ion-transport cycle. In keeping with this interpretation, phosphorylation appears to antagonize the K^+ effects on PTX-induced channels, consistent with the known effects of phosphorylation to promote external gate opening to release occluded Na^+ during normal pumping or to release occluded K^+ when the pump is reverse ("back-door") phosphorylated from P_i plus Mg^{2+}.[30]

Overall, these studies imply that PTX neither irreversibly nor grossly modifies Na/K pump structure because the effects are readily reversible and repeatable, and because the P_o of PTX-induced channels can be regulated by the pump's physiological ligands in a manner consistent with expectations based on the alternating-gate mechanism of ion transport. In addition to thus supporting the alternating-gate model, our findings indicate the importance of further characterizing the PTX-opened pore. In particular, our demonstration that TMA$^+$ can permeate the PTX-induced channel implies that the minimum diameter of the ion-translocation pathway in the unmodified Na/K-ATPase may be >6 Å.

ACKNOWLEDGMENTS

We thank Claudia Basso and Miguel Holmgren for help with preliminary experiments and much useful discussion, and Paola Vergani for help with the illustrations. Work was supported by NIH Grant No. HL-36783.

REFERENCES

1. PATLAK, C.S. 1957. Contributions to the theory of active transport. II. The gate type non-carrier mechanism and generalizations concerning tracer flow, efficiency, and measurement of energy expenditure. Bull. Math. Biophys. **19:** 209–235.
2. LÄUGER, P. 1979. A channel mechanism for electrogenic ion pumps. Biochim. Biophys. Acta **552:** 143–161.
3. ARTIGAS, P. & D.C. GADSBY. 2003. Na$^+$/K$^+$-pump ligands modulate gating of palytoxin-induced ion channels. Proc. Natl. Acad. Sci. USA **100:** 501–505.
4. POST, R.L., A.K. SEN & A.S. ROSENTHAL. 1965. A phosphorylated intermediate in adenosine triphosphate–dependent sodium and potassium transport across kidney membranes. J. Biol. Chem. **240:** 1437–1445.
5. ALBERS, R.W. 1967. Biochemical aspects of active transport. Annu. Rev. Biochem. **36:** 727–756.
6. POST, R.L., C. HEGYVARY & S. KUME. 1972. Activation by adenosine triphosphate in the phosphorylation kinetics of sodium and potassium ion transport adenosine triphosphatase. J. Biol. Chem. **247:** 6530–6540.
7. BEAUGÉ, L.A. & I.M. GLYNN. 1979. Occlusion of K ions in the unphosphorylated sodium pump. Nature **280:** 510–512.
8. FORBUSH, B., III. 1987. Rapid release of ^{42}K and ^{86}Rb from an occluded state of the Na,K-pump in the presence of ATP or ADP. J. Biol. Chem. **262:** 11104–11115.
9. GLYNN, I.M. & D.E. RICHARDS. 1982. Occlusion of rubidium ions by the sodium-potassium pump: its implications for the mechanism of potassium transport. J. Physiol. **330:** 17–43.
10. KAPLAN, J.H. & L.J. KENNEY. 1982. ADP supports ouabain-sensitive K-K exchange in human red blood cells. Ann. N.Y. Acad. Sci. **402:** 292.
11. SIMONS, T.J.B. 1975. The interaction of ATP-analogues possessing a blocked gamma-phosphate group with the sodium pump in human red cells. J. Physiol. **244:** 731–739.
12. GADSBY, D.C., P. DE WEER & R.F. RAKOWSKI. 1993. Extracellular access to the Na,K pump: pathway similar to ion channel. Science **260:** 100–103.
13. HEYSE, S., I. WUDDEL, H.J. APELL & W. STURMER. 1994. Partial reactions of the Na,K-ATPase: determination of rate constants. J. Gen. Physiol. **104:** 197–240.
14. HOLMGREN, M., J. WAGG, F. BEZANILLA, *et al.* 2000. Three distinct and sequential steps in the release of sodium ions by the Na$^+$/K$^+$-ATPase. Nature **403:** 898–901.
15. APELL, H.J. & A. DILLER. 2002. Do H(+) ions obscure electrogenic Na(+) and K(+) binding in the E(1) state of the Na,K-ATPase? FEBS Lett. **532:** 198–202.
16. PEINELT, C. & H.J. APELL. 2002. Kinetics of the Ca(2+), H(+), and Mg(2+) interaction with the ion-binding sites of the SR Ca-ATPase. Biophys. J. **82:** 170–181.
17. TOYOSHIMA, C., M. NAKASAKO, H. NOMURA & H. OGAWA. 2000. Crystal structure of the calcium pump of sarcoplasmic reticulum at 2.6 Å resolution. Nature **405:** 647–655.
18. TOYOSHIMA, C. & H. NOMURA. 2002. Structural changes in the calcium pump accompanying the dissociation of calcium. Nature **418:** 605–611.
19. ZHANG, P., C. TOYOSHIMA, K. YONEKURA, *et al.* 1998. Structure of the calcium pump from sarcoplasmic reticulum at 8-Å resolution. Nature **392:** 935–939.

20. MOORE, R.E. & P.J. SCHEUER. 1971. Palytoxin: a new marine toxin from coelenterate. Science **172:** 495–498.
21. KIM, S.Y., K.A. MARX & C.H. WU. 1995. Involvement of Na,K-ATPase in the induction of ion channels by palytoxin. Naunyn-Schmiedeberg's Arch. Pharmacol. **351:** 542–554.
22. REDONDO, J., B. FIEDLER & G. SCHEINER-BOBIS. 1996. Palytoxin-induced Na^+ influx into yeast cells expressing the mammalian sodium pump is due to the formation of a channel within the enzyme. Mol. Pharmacol. **49:** 49–57.
23. GUENNOUN, S. & J-D. HORISBERGER. 2000. Structure of the 5th transmembrane segment of the Na,K-ATPase alpha subunit: a cysteine-scanning mutagenesis study. FEBS Lett. **482:** 144–148.
24. GUENNOUN, S. & J.-D. HORISBERGER. 2002. Cysteine-scanning mutagenesis study of the sixth transmembrane segment of the Na,K-ATPase alpha subunit. FEBS Lett. **513:** 277–281.
25. GADSBY, D.C. & M. NAKAO. 1989. Steady-state current-voltage relationship of the Na/K pump in guinea-pig ventricular myocytes. J. Gen. Physiol. **94:** 511–537.
26. TOSTESON, M.T., J.A. HALPERIN, Y. KISHI & D.C. TOSTESON. 1991. Palytoxin induces an increase in the cation conductance of red cells. J. Gen. Physiol. **98:** 969–985.
27. MURAMATSU, I., M. NISHIO, S. KIGOSHI & D. UEMURA. 1988. Single ionic channels induced by palytoxin in guinea pig ventricular myocytes. Br. J. Pharmacol. **93:** 811–816.
28. DWYER, T.M., D.J. ADAMS & B. HILLE. 1980. The permeability of the endplate channel to organic cations in frog muscle. J. Gen. Physiol. **75:** 469–492.
29. SCHUURMANS-STECKHOVEN, F. & S.L. BONTING. 1981. Transport adenosine triphosphatases: properties and functions. Physiol. Rev. **61:** 1–76.
30. FORBUSH, B., III. 1987. Rapid release of ^{42}K or ^{86}Rb from two distinct transport sites on the Na,K-pump in the presence of Pi or vanadate. J. Biol. Chem. **262:** 11116–11127.
31. ARTIGAS, P. & D.C. GADSBY. 2002. Ion channel–like properties of the Na^+/K^+ pump. Ann. N.Y. Acad. Sci. **976:** 1–10.

Cation Stoichiometry and Cation Pathway in the Na,K-ATPase and Nongastric H,K-ATPase

JEAN-DANIEL HORISBERGER, SAÏDA GUENNOUN, MURIEL BURNAY, AND KÄTHI GEERING

Institute of Pharmacology and Toxicology, University of Lausanne, Switzerland

ABSTRACT: The mechanism of cation translocation by the Na,K-ATPase was investigated by cysteine scanning mutagenesis and measurements of accessibility through exposure to cysteine reagents. In the native protein, accessible residues were found only at the most extracellular residues of the 5th and 6th transmembrane segments (TMS) and the short loop between them. However, after modification by palytoxin a number of residues became accessible along the whole length of the 5th TMS and in the outer half of the 6th TMS, showing the contribution of each of these segments to the "channel" formed by the palytoxin-transformed Na,K-pump. Assuming that this structure is similar in the native and the palytoxin-transformed pump, our data allow us to determine the residues lining the cation pathway from the extracellular solution to their binding sites. A critical position in the 5th TMS contains a lysine conserved in all known nonelectrogenic H,K-ATPases, and a serine in all known electrogenic Na,K-ATPase sequences. Wild-type or mutant Na,K- or H,K-ATPase α subunits were coinjected with the *Bufo* β2 subunit in *Xenopus* oocytes and Rb[86] uptake and electrophysiological measurements were performed. An electrogenic activity was recorded for the H,K-ATPase mutants in which the positively charged lysine had been replaced by neutral or negatively charged residues, while nonelectrogenic transport was observed with the $S_{782}R$ mutant of the Na,K-ATPase. The presence or the absence of a positively charged residue at the S_{782} position appears to be critical for the stoichiometry of cation exchange.

KEYWORDS: Na,K-ATPase; H,K-ATPase; electrogenic activity; cysteine scanning mutagenesis; palytoxin

INTRODUCTION

Even though several members of the P-ATPase family such as the calcium ATPase of the sarcoplasmic and endoplasmic reticulum (SERCA), the Na,K- and H,K-ATPases, and the yeast H-ATPases have been very extensively studied and a high-resolution structure has even been available for SERCA for the past 2 years,[1] the mechanism of cation translocation and the principle of energy transduction from ATP hydrolysis to uphill ion transport is not yet understood.

Address for correspondence: J.-D. Horisberger, Institut de Pharmacologie, Bugnon 27, CH-1005 Lausanne, Switzerland. Voice: +41 2 692 5362; fax: +41 2 692 5355.
Jean-Daniel.Horisberger@ipharm.unil.ch

In order to explore the structure of the cation pathway across one of these P-ATPases, we have applied the substituted cysteine accessibility method to the Na,K-ATPase using the native enzyme or the enzyme transformed into a cation channel by exposure to palytoxin. We have also taken advantage of the high degree of similarity between the H,K-ATPases, which exchange a symmetrical number of cations and thus have a nonelectrogenic function, and the Na,K-ATPases, which exchange 3 Na$^+$ for 2 K$^+$ and are thus electrogenic, to identify a critical residue located in the middle of the 5th transmembrane segment (TMS) and the role of a positively charged residue at this position in the stoichiometry of cation transport.

METHODS

Cysteine-scanning mutagenesis experiments were performed with the *Bufo marinus* Na,K-ATPase α1 subunit coexpressed with the β1 subunit of the same species and the experiments testing the electrogenicity were performed with wild-type or mutant *Bufo marinus* Na,K-ATPase α1 or a bladder H,K-ATPase (the "nongastric" H,K-ATPase of *Bufo*) expressed with the β2 isoform of the *Bufo* β subunit. These isoforms were chosen because of the relative resistance to ouabain, which allows them to be studied after inhibition of the endogenous Na,K-pump by low concentrations of ouabain (0.2 µM). Both isoforms can be inhibited by high concentrations of ouabain (2 mM).

Xenopus oocytes were coinjected with α and β subunits (or β subunit alone as control), incubated for 2–3 days to allow for expression of the protein, and then studied either by electrophysiology or in ^{86}rubidium uptake experiments as described in detail before.[2,3] Palytoxin was used at concentrations of 2 to 4 nM in order to obtain a large increase in the oocyte membrane conductance in oocytes expressing functional Na,K-ATPases. All the experiments using palytoxin were done in the presence of 10 µM ouabain in order to inhibit completely the endogenous Na,K-ATPase of the oocyte.

RESULTS AND DISCUSSION

Cysteine Scanning Studies

Each amino acid position from K774 to Y824 was mutated to a cysteine, with the exception of C809, which is already a cysteine (note that the residue numbering is that of the *Bufo marinus* Na,K-ATPase α1 isoform). Of all these mutants, only three, D811C, T814C, and D815C, had apparently no functional expression as judged from the capacity to generate a K$^+$-activated outward current. D815 is completely conserved in all animal Na,K- and H,K-ATPases, while D811 is conserved in the Na,K- and nongastric H,K-ATPases and is substituted by a glutamic acid in the gastric H,K-ATPases. The side chains of the homologous of these three residues in SERCA (N796, T799, and D800) are associated with the cation binding sites according to Toyoshima's SERCA structure.[1]

Palytoxin, at a concentration of 2 or 4 nM, induced a large increase of the membrane conductance in oocytes expressing all the mutants, but not in oocytes injected with the cRNA of the β subunit alone, indicating that all the mutants were synthesized and expressed at the plasma membrane.

The accessibility of the substituted cysteines was first tested in a number of mutants by measuring the K^+-activated current before and after a 2-min exposure to a 100-μM MTSEA solution in the absence of K^+. A partial but significant inhibition was observed for the most extracellularly located position of the 5th TMS A796, and the first of the two conserved proline P799 in the extracellular loop. A stronger inhibition was observed for the two following positions L800 and P801 as well as for the first three positions of the 6th TMS: G803, T804, and V805.

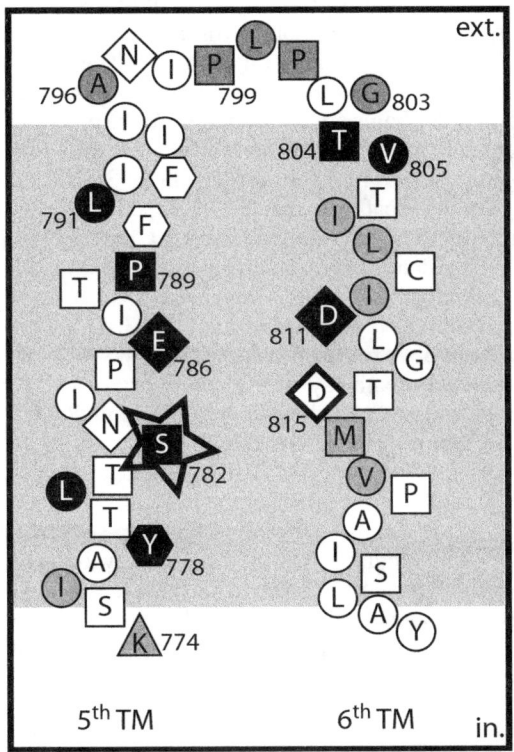

FIGURE 1. Scheme of the structure of the 5th and 6th transmembrane segments and the short extracellular loop between these two segments. The sequence and the numbering are that of the *Bufo marinus* α1 Na,K-ATPase. The membrane width is indicated by the gray background band. Amino acid positions that become readily accessible after treatment with palytoxin are indicated with white letters on black, and positions with a weaker effect of MTSEA are shown in gray. The positions of the extracellular loop where cysteine substitutions make the native Na,K-pump sensitive to MTSEA are also indicated in gray. The chemical characteristics of the residues are indicated as follows: *circles* for lipophilic residues; *squares* for more hydrophilic residues; *hexagons* for aromatic residues; *triangles* for positively charged residues; and *diamonds* for negatively charged residues. The extracellular side of the membrane is on the top. The *star* indicates the S782, homologous to K800 in the *Bufo* bladder H,K-ATPase; the position of a positively charged residue at this position seems to determine a electroneutral cation exchange mode of transport.

The effect of MTSEA exposure after treatment with palytoxin has been reported in detail elsewhere.[4,5] Briefly, we tested the effect of a 1-min exposure to a 100-μM MTSEA solution on the palytoxin-induced conductance. The results of these measurements are summarized in FIGURE 1. Six positions in the 5th TMS and 3 positions in the 6th TMS were clearly accessible, as shown by a MTSEA-induced inhibition of 20% or larger. A few other positions showed a significant but lower degree of inhibition, as shown in FIGURE 1.

Among the readily accessible positions of the 5th TMS, E786 was expected to be part of the cation binding site, as its homologous residue in SERCA (E771) is clearly involved in calcium binding site I,[1] and this residue has been shown earlier to be involved in cation binding in the Na,K-ATPase by site-directed mutagenesis experiments.[6,7] More surprisingly three lipophilic residues I776, L780, and L791 are also accessible to MTSEA after treatment with palytoxin. The homologous residues in SERCA (I761, I765, and F776) are not located on the same side of the 5th TMS α helix, but rather their side chains are pointing toward the exterior of the protein, that is, toward the membrane lipids. We do not have any explanation for these observations, but we can propose two hypothesis. First, these 3 residues might be reached by MTSEA through another pathway than the hypothetical "central" cation pathway located between the 4th, 5th, and 6th TMS. Second, the 5th TMS might undergo a rotation during the E1-E2 conformation change that would expose another side of this α helix to the cation pathway.

In the 6th TMS, the three outermost positions, G803, T804, and V805, are readily accessible, which might be expected because of their position close to the extracellular "mouth" of the cation pathway. The accessibility of D811 was also expected from the well-known role of this residue in cation binding,[6,8] and from the position of its homologous N796 in SERCA structure in direct contact with calcium site II. In contrast, D815 was not accessible to MTSEA, despite the fact that its homologous D800 in SERCA is directly participating in the structure of the two calcium binding sites.[1]

In summary, our results indicate that the cation pathway opened by palytoxin in the Na,K-ATPase comprises several amino acids that are expected to form the cation binding sites. These data support the idea that palytoxin does not completely disturb the Na,K-ATPase structure, but rather modifies the "physiological" pathway for cations across the pumps.

Mutagenesis Studies of the S782/K800 Position

If one does not include the residues located at the end of the transmembrane helices, the 10 TMS of the Na,K-ATPase α subunit contains only seven negatively charged residues, five of which (one in TMS 4, TMS 5, and TMS 8, and 2 in TMS 6) are also present in the H,K-ATPases. These residues are highly conserved and most important for cation binding and transport. Not a single positively charged residue is found in any of the TMS of the Na,K-ATPase α subunit sequences, while a lysine is present in all known gastric or nongastric H,K-ATPase sequences (K800 in the *Bufo marinus* bladder H,K-ATPase α subunit), close to the center of the 5th TMS among a number of conserved residues. At the position corresponding to K800, a serine is present in all known Na,K-ATPase sequences. The nature of the residue at this position (S782) has been shown to have a strong influence on the apparent affinity for extracellular K^+.[7,9,10]

Serine 782 was substituted with alanine, arginine, lysine, or a glutamic acid. Oocytes expressing these mutants did not exhibit a significant ^{86}Rb uptake when studied with a 5-mM extracellular K^+ concentration. This result was not unexpected because of the low K^+ affinity already reported in similar mutants.[7,9,10] We thus used the ouabain-sensitive uptake in a 40-mM K^+ solution as transport assay. Oocytes expressing the S782K and S782E mutants did not show a significant ^{86}Rb uptake either in this condition, but the S782A and S782R mutants expressed a small transport activity amounting to about 30% of the wild-type Na,K-ATPase, and only these 2 mutants were studied further, and the electrogenic activity was measured in the same conditions (i.e., the current inhibited by 2 mM ouabain in a 40-mM extracellular solution). A significant ouabain-sensitive outward current (about 41% of the value recorded with the wild type) was recorded with the S782A mutant, but not with the S782R mutant. Thus, among those expressed, the mutant with a neutral residue at position 782 was able to produce an electrogenic activity, while the mutant with a positively charged residue at that position was not electrogenic.

In the nongastric type *Bufo* bladder H,K-ATPase, the corresponding lysine was substituted with alanine, serine, arginine, and glutamic acid. In this case, all mutants expressed a significant ^{86}Rb uptake activity when studied in a 5-mM K^+ solution. Only a small inward current was recorded with the oocytes expressing the K800R or K800S mutants, similar to what was observed with the wild-type *Bufo* bladder H,K-ATPase.[11] In contrast, a significant outward current was activated by addition of K^+ in oocytes expressing the K800A (see FIGURE 2 for an example) or the K800E mutant.

The electrogenic activity of the same mutants was also studied in the absence of extracellular Na^+, a condition in which the apparent affinity for extracellular K^+ is higher.[3] Under those circumstances, the behavior of the K800S mutant was different:

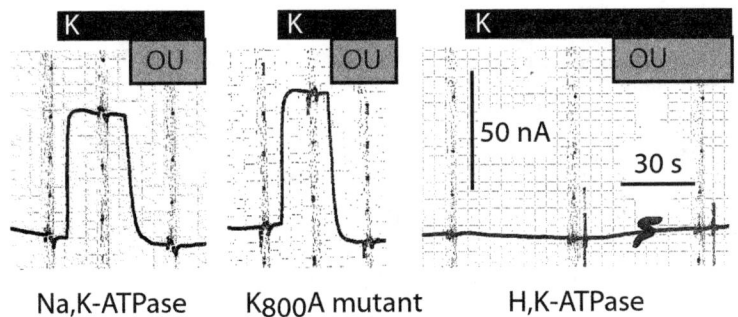

FIGURE 2. Representative current traces obtained under voltage clamp conditions (–50 mV) with oocytes expressing the Na,K-ATPase (*left*), the nongastric H,K-ATPase (*right*), or the K800A mutant of the H,K-ATPase (*middle trace*). Addition of 5 mM K^+ resulted in an outward (positive) current in the oocytes expressing the Na,K and the K800A mutant, but only in a small inward current with those expressing the H,K-ATPase. The currents induced by K^+ were inhibited by 2 mM ouabain (OU).

at low K^+ concentrations (0.1 and 0.5 mM), a significant outward current was recorded, while at higher K^+ concentrations, the current reverted to negative values.

In conclusion, we showed that when K800 was substituted with another positively charged amino acid, the mutant expressed a K^+ transport activity which was not electrogenic, and thus transported a similar number of cations in and out of the cell during each transport cycle, in a way similar to the wild-type gastric and nongastric H,K-ATPases and also to the S782R mutant of the Na,K-ATPase. In contrast, when K800 was substituted with a neutral or a negatively charged amino acid (K800A, K800E), an electrogenic cation pump was operating in a way very similar to the wild-type Na,K-ATPase or the Na,K-ATPase mutant in which the charge of the residue's side chain was not modified (S782A). The only apparent exception was the K800S mutant, which showed a more complex behavior. While this mutant was able to produce a robust but entirely nonelectrogenic activity in Na^+-rich extracellular solutions, it carried out an electrogenic transport at low K^+ concentrations in a Na^+-free solution, but reverted to nonelectrogenic transport or even an electrogenicity of reverse polarity at higher K^+ concentrations. Thus, the K800S mutant does not seem to have a fixed stoichiometry and may also provide a "leak," that is, a flow of cations not tightly coupled to the transport cycle.

REFERENCES

1. TOYOSHIMA, C., M. NAKASAKO, H. NOMURA & H. OGAWA. 2000. Crystal structure of the calcium pump of sarcoplasmic reticulum at 2.6 A resolution. Nature **405:** 647–655.
2. JAISSER, F., J.-D. HORISBERGER, K. GEERING & B.C. ROSSIER. 1993. Mechanisms of urinary K^+ and H^+ excretion: primary structure and functional expression of a novel H,K-ATPase. J. Cell Biol. **123:** 1421–1429.
3. JAISSER, F., P. JAUNIN, K. GEERING, et al. 1994. Modulation of the Na,K-pump function by the β-subunit isoforms. J. Gen. Physiol. **103:** 605–623.
4. GUENNOUN, S. & J.-D. HORISBERGER. 2000. Structure of the 5th transmembrane segment of the Na,K-ATPase a subunit: a cysteine-scanning mutagenesis study. FEBS Lett. **482:** 144–148.
5. GUENNOUN, S. & J.-D. HORISBERGER. 2002. Cysteine-scanning mutagenesis study of the sixth transmembrane segment of the Na,K-ATPase a subunit. FEBS Lett. **513:** 277–281.
6. NIELSEN, J.M., P.A. PEDERSEN, S.J.D. KARLISH & P.L. JORGENSEN. 1998. Importance of intramembrane carboxylic acids for occlusion of K^+ ions at equilibrium in renal Na,K-ATPase. Biochemistry **37:** 1961–1968.
7. JORGENSEN, P.L. & P.A. PEDERSEN. 2001. Structure-function relationships of Na+, K+, ATP, or Mg2+ binding and energy transduction in Na,K-ATPase. Biochim. Biophys. Acta Bio-Energ. **1505:** 57–74.
8. JEWELL-MOTZ, E.A. & J.B. LINGREL. 1993. Site-directed mutagenesis of the Na,K-ATPase: consequences of substitutions of negatively-charged amino acids localized in the transmembrane domains. Biochemistry **32:** 13523–13530.
9. PELUFFO, R.D., J.M. ARGÜELLO & J.R. BERLIN. 2000. The role of Na,K-ATPase alpha subunit Serine 775 and Glutamate 779 in determining the extracellular K^+ and membrane potential-dependent properties of the Na,K-pump. J. Gen. Physiol. **116:** 47–59.
10. PEDERSEN, P.A., J.M. NIELSEN, J.H. RASMUSSEN & P.L. JORGENSEN. 1998. Contribution to Tl^+, K^+, and Na^+ binding of asn(776), ser(775), thr(774), thr(772), and tyr(771) in cytoplasmic part of fifth transmembrane segment in alpha-subunit of renal Na,K-ATPase. Biochemistry **37:** 17818–17827.
11. BURNAY, M., K. GEERING & J.-D. HORISBERGER. 2001. The *Bufo marinus* bladder H,K-ATPase carries out electroneutral ion transport. Am. J. Physiol. **281:** F869–F874.

Toward an Understanding of Ion Transport through the Na,K-ATPase

HANS-JÜRGEN APELL

Department of Biology, University of Konstanz, Fach M635, 78457 Konstanz, Germany

ABSTRACT: In the Na,K-ATPase the charge-translocating reaction steps were found to be binding of the third Na^+ ion to the cytoplasmic side and the release of all three Na^+ ions to the extracellular side as well as binding of the two K^+ ions on the extracellular side. The conformation transition $E_1 \to E_2$ was only of minor electrogenicity; all other reaction steps produced no significant charge movements. In the SR Ca-ATPase and the gastric H,K-ATPase, all ion-binding and -release steps were identified to move charge through the membrane. The high-resolution structure of the SR Ca-ATPase in state E_1 revealed the position of the ion-binding sites in the transmembrane part of the protein. If the same arrangement is assumed for the Na pump, the missing expected charge movements in state E_1 may to be assumed to be apparent effects. With the proposal that binding of 2 Na^+ or 2 K^+ is compensated correspondingly by H^+ ions, agreement between structural and functional aspects is obtained. Investigations of the pH-dependence of ion-binding steps indicate competition between the ions and electrogenic H^+ binding in support of this concept.

KEYWORDS: binding sites; ion transport; pH effects; electrogenicity; styryl dyes; fluorescence; competitive inhibition of ion binding

INTRODUCTION

The function of the Na,K-ATPase in the membrane of cells maintains the electrochemical potential gradient of Na^+ and K^+ ions.[1-3] Ion transport is facilitated by coupling the energy-providing enzymatic process with a ping-pong mechanism of ion translocation.[4] This process is described by the so-called post-Albers cycle[5,6] (FIG. 1A). If, for example, the Na^+ translocating pathway of the cycle is examined, we see that four reaction steps could contribute to the charge translocation: (1) ion binding; (2) ion occlusion; (3) conformation transition; and (4) ion release to the opposite side (FIG. 1B). To quantify the "dielectric" distance over which the ion is moved, dielectric coefficients were introduced that describe the fraction of the membrane dielectric over which the charge is shifted perpendicular to the plane of the membrane.[3] If the dielectric coefficient is non-zero, the accompanying reaction is termed "electrogenic." In FIGURE 1B the dielectric coefficient for cytoplasmic binding of a Na^+ ion would be α' (and $\alpha' + \beta' + \beta'' + \alpha'' = 1$ for the transfer across the whole membrane). In the last decade numerous studies were performed with differ-

---

Address for correspondence: Hans-Jürgen Apell, Department of Biology, University of Konstanz, Fach M635, 78457 Konstanz, Germany. Voice: +49-7531-882253; fax: +49-7531-883183.
h-j.apell@uni-konstanz.de

Ann. N.Y. Acad. Sci. 986: 133–140 (2003). © 2003 New York Academy of Sciences.

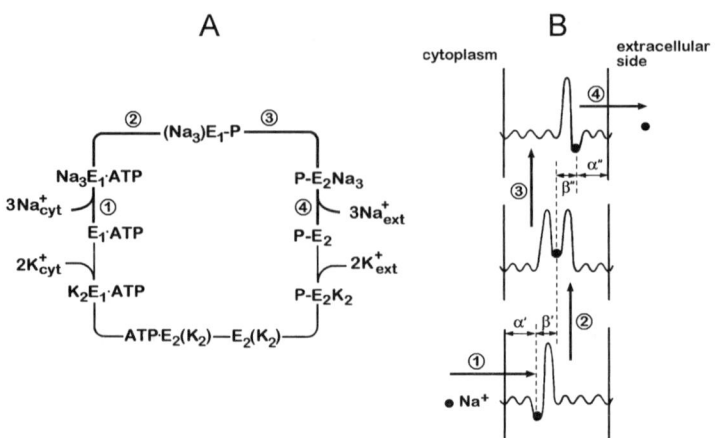

FIGURE 1. (**A**) Post-Albers cycle of the Na,K-ATPase. The Na$^+$-translocating pathway is characterized by four partial reactions: (1) Na$^+$ binding; (2) ion occlusion and enzyme phosphorylation; (3) conformation transition $E_1 \to E_2$; and (4) Na$^+$ deocclusion and release. (**B**) Schematic representation of the energy profile of the Na$^+$ pathway for the various states of Na$^+$ translocation. The Greek letters indicate the corresponding dielectric coefficients (see text).

ent electrophysiological and fluorescence spectroscopical techniques to determine the dielectric coefficients of all electrogenic partial reactions.[7–10]

On the basis of these results, a structure–function concept was constructed in which on the extracellular side, the first Na$^+$ ion is released through a narrow access channel, the step with the highest dielectric coefficient ($\alpha'' \approx 0.7$);[7,8] then a conformational relaxation occurs[11] before the other two Na$^+$ ions reach the aqueous phase with lower dielectric coefficients (0.1–0.2), probably caused by reduction of the dielectric coefficient of the transmembrane part of the pump protein, for example, by intrusion of water molecules.[7] On the cytoplasmic side only a single reaction step was detected to be electrogenic, binding of the third Na$^+$ ion.[12,13]

When similar studies of the electrogenicity were performed with the SR Ca-ATPase[14,15] and the gastric H,K-ATPase (unpublished data), it was found that in these P-type ATPases all ion-binding and -release reactions were electrogenic. This difference from the Na,K-ATPase is significant, since all three ATPases are believed to have closely related structures.[16] An important input into the considerations of structure–function relations was produced by the 3-D structure of the E_1 conformation of the SR Ca-ATPase at atomic resolution,[17] which revealed that the two Ca^{2+} ions were bound to extremely well coordinating ion sites in the middle of the transmembrane domains.

When we studied "backdoor phosphorylation" of the Na,K-ATPase in the absence of K$^+$ ions,[18] we found that the population of the state of E_2 that could be phosphorylated by P_i needed to have occluded two H$^+$ ions, so that the reaction sequence has to be:

$$E_1 + 2H^+_{cyt} \to H_2E_1 \to E_2(H_2) \to P\text{-}E_2(H_2) \to P\text{-}E_2 + 2H^+_{ext}$$

Because of this finding it was important to investigate the interaction of H^+ ions with the ion-binding sites to scrutinize whether instantaneous H^+ binding to empty sites of the Na,K-ATPase in state E_1 conceals otherwise electrogenic Na^+ and K^+ movements.

COMPETITION OF H^+ WITH Na^+ AND K^+ IONS AT THE CYTOPLASMIC SITES

Na^+ binding in state E_1 can be measured by equilibrium titration experiments with high accuracy using the fluorescent styryl dye RH421.[13] This dye, and others of this family, are hydrophobic substances with one polar end, so they insert into lipid membranes in an aligned manner.[19] Due to the electrochromic mechanism of their chromophore they detect changes of local electric fields in the membrane dielectric and, therefore, report charge movements in membrane preparations which are packed with ion pumps, such as Na,K-ATPase[13,18] and SR Ca-ATPase.[14,15] From Na^+-titration experiments with the Na,K-ATPase in its E_1 conformation the half-saturating concentration, $K_{1/2}$, was determined by fitting a Hill function to the Na^+-dependent fluorescence changes. These experiments were performed in buffers with a pH set between 6 and 8.5. In FIGURE 2 the $K_{1/2}$ values obtained from such experiments are shown as function of pH. They show clearly that binding of Na^+ is affected by H^+ concentration. $K_{1/2}$ increases between pH 8.5 and 6 by a factor of 5. $K_{1/2}$ corresponds approximately to the K_M value of the second of three Na^+ ions bound.[13] The fluorescence change observed in these experiments was generated by the binding of the third Na^+ ion to the "Na^+-specific site,"[12,13] and it was found that the maximum fluores-

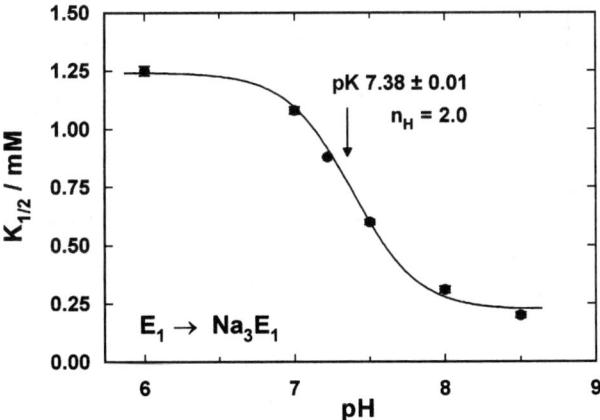

FIGURE 2. Effect of buffer pH on cytoplasmic Na^+ binding as detected by RH421 fluorescence changes. Buffers contained 25 mM histidine and 0.5 mM EDTA, and pH was adjusted by Tris or HCl. The half-saturating Na^+ concentration, $K_{1/2}$, increased with the H^+ concentration, which can be explained by competition between Na^+ and H^+ at the same site(s). The drawn line through the data represents a fit with a Hill function with a half-saturating concentration of 41.7 nM (or pK 7.38).

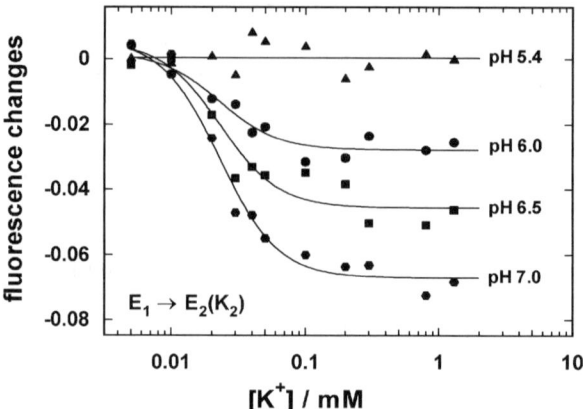

FIGURE 3. Cytoplasmic K^+ binding to the Na,K-ATPase detected by fluorescence changes of RH421 at various pH in a buffer containing 25 mM histidine, and 0.5 mM EDTA, when pH was adjusted by HCl. Data were fitted with a Hill function (*solid lines*). The Hill coefficient of 2.0 ± 0.1 was constant over a pH range of 5.4–7.

cence change was not significantly affected by the pH applied (not shown). This indicates that no competition between Na^+ and H^+ occurred at the third binding site. Therefore, it may be proposed that H^+ is able to bind to side groups of amino acids which are part of (or close to) the two binding sites which are not Na^+ specific. If so, this should be reflected also in pH effects on K^+ binding, which shows barely significant electrogenic effects when investigated at physiological pH.[12,20] Results of pH-dependent cytoplasmic K^+ binding are shown in FIGURE 3. Although the fluorescence changes were much smaller than in the case of Na^+ binding, it is clear that electrogenic K^+ binding could be seen when the buffer pH was increased from 5 to 7. The Hill fits drawn through the data had a Hill coefficient, n_H, of 2, which indicated interactions with more than one H^+ in this process. At pH 5.4 no apparent or net charge movement could be detected. Experiments with buffer pH higher than 7 could be obtained only by addition of Tris, but this addition reduced the fluorescence changes for so far unknown reasons and these data were therefore not included in the analysis.

CYTOPLASMIC AND EXTRACELLULAR H^+ BINDING

pH-titration experiments were performed in the absence of other monovalent cations in E_1, and in P-E_2 conformations of the Na,K-ATPase. The latter were obtained in the presence of Mg^{2+} ions either by addition of 500 μM P_i or by addition of 10 mM NaCl + 100 μM ATP. The pH-induced fluorescence changes are shown in FIGURE 4. Although the fluorescence changes were small, when in the E_1 conformation, H^+ ions were added in the pH range 7.2–5.5, these changes ($\Delta F = -13\%$) were nevertheless significant. In control experiments with completely blocked enzyme $\Delta F < 5\%$ was found for the same pH jump. When the reaction sequence $E_1 \rightarrow H_2E_1 \rightarrow E_2(H_2)$ was studied during backdoor phosphorylation,[18] a pK value of 8.6 was estimated from the pH-dependent population of state $E_2(H_2)$, so that below pH 7.2 more than 95% of the ion sites in E_1 would have bound protons. Therefore it must be expected that further addition of H^+ would produce only minor fluorescence changes.

When the enzyme was phosphorylated, the fluorescence increased by between 30% (pH 5.5) and 40% (pH 7.2), indicating the release of at least one H^+ ion from inside the transmembrane region of the protein. This may be explained as a consequence of a dramatic decrease of the pK of one (or more) amino acid side groups to which the protons bind in the E_1 conformation when no Na^+ or K^+ ions are present. Reducing the pH in state P-E_2 by addition of aliquots of HCl resulted in a decrease of the RH421 fluorescence, which represents H^+ binding to a site within the protein dielectric (FIG. 4). The obvious differences in the fluorescence levels in the two protein conformations may be explained by the above-mentioned large shift of the pK values of the side chains of acidic amino acids to which H^+ are able to bind in, or close to, the ion-binding sites. Thus, whereas in E_1 the pK (about 8.6) is so high that in the physiological pH range the "sites" are mostly occupied by H^+ ions in the absence of other cations, the pK in P-E_2 is proposed to drop so significantly that the same sites are then largely unprotonated. A rough estimate from the titration experiment in FIGURE 4 suggests a pK < 5.5 for P-E_2.

CONSEQUENCES FOR THE STRUCTURE–FUNCTION RELATIONSHIP

On the basis of these results, together with the generally accepted constraint that all P-type ATPases have closely related structures,[16] a concept for the position and characteristics of the ion-binding sites may be developed that is able to describe the

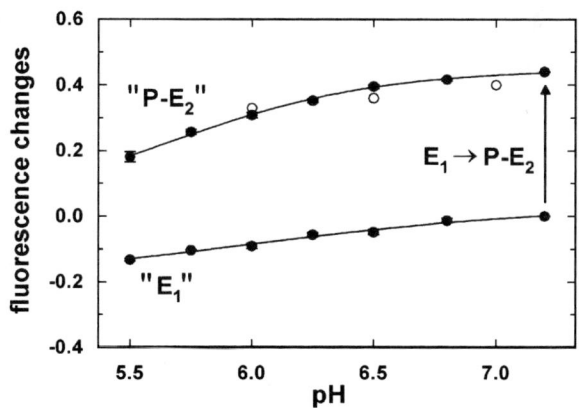

FIGURE 4. pH titration of the ion sites in the two principal conformations, E_1 and P-E_2, in the absence of other monovalent cations. The *vertical arrow* indicates the RH421 fluorescence change induced by phosphorylation of the Na,K-ATPase with P_i in the absence of other monovalent cations, or by addition of 10 mM NaCl + 100 µM ATP. Initial buffer composition was 25 mM histidine, 0.5 mM EDTA, and 10 mM $MgCl_2$, at pH 7.2.

electrogenicity as well as the detected interaction, or competition, of the cations binding to these sites. To maintain the homology of ion sites in the P-type ATPases, it is assumed that in the E_1 conformation two sites are in the middle of the membrane dielectric, where they were recently identified for the SR Ca-ATPase.[17] In the case of the Na,K-ATPase these sites are able to bind various monovalent cations[21] including H^+, as shown above. Binding of a third (Na^+) ion, which is a special feature of the Na,K-ATPase, occurs only after two Na^+ ions have already bound.[21] This third site is positioned about 25% of the dielectric thickness into the membrane from the cytoplasmic side[12] and is virtually exclusively selective for Na^+. Such a topographical arrangement, however, appears to contradict the observation that the cytoplasmic binding or release of the two K^+ or of the first two Na^+ ions is not electrogenic.[12,20] To resolve this problem, the reported H^+ binding to the cytoplasmic sites may be brought into play.

When under physiological conditions the Na,K-ATPase reaches state K_2E_1 in the pump's cycle, the subsequent release of K^+ ions may be accompanied (and supported) by an uptake of one H^+ per K^+ ion that binds to an acidic side group of an amino acid at the binding site: $K_2E_1 + 2H^+ \rightarrow H_2E_1 + 2K^+$. Such a reaction would be apparently electroneutral. Only at high pH approaching the pK of the H^+ binding groups, would a significant electrogenic contribution of K^+ binding and release become detectable (cf. FIG. 4).

A corresponding reaction is to be expected for binding of the first two Na^+ ions to H_2E_1. Moreover, any small effect of incompletely matched exchange of Na^+ for H^+ on the RH421 fluorescence at physiological pH would be concealed because binding of the third Na^+ is electrogenic and generates a large and pH-independent contribution. Nevertheless, the competition between Na^+ and H^+ ions for the same sites is evident in the pH dependence of the half-saturating Na^+ concentration for binding of the first two Na^+ ions (FIG. 2). The fact that, despite the almost complete occupancy of the sites by H^+ under physiological conditions, Na^+ binding is so fast that it could not be resolved so far may be explained by the fact that H^+ ions are extremely small and can exchange between a carboxylate and a water molecule without steric hindrance for (or by) cation that is shedding off its hydration shell to enter the binding site. In addition, a multiply coordinated (alkali) cation and a "free" H_3O^+ ion are an energetically much more favorable combination than a free cation and a protonated carboxylate.

CONCLUSIONS

The placement of two ion-binding sites of the Na,K-ATPase inside of the membrane dielectric, as suggested by the structure of the SR Ca-ATPase and the apparent electroneutrality of cytoplasmic K^+ release and binding of the first two Na^+ ions under physiological conditions, can be explained by a transient binding of two H^+ ions to carboxylate groups in or close to the ion binding sites. This leads to a modification of the post-Albers cycle (cf. FIG. 1A) for the E_1 conformation in the following way:

$$\cdots \rightarrow K_2E_1ATP \rightarrow H_2E_1ATP \rightarrow Na_2E_1ATP \rightarrow Na_3E_1ATP \rightarrow \cdots$$

After binding of two Na^+ ions in exchange for two H^+ ions the highly specific third site becomes available and the third Na^+ ion binds electrogenically. This nonsingle file mechanism allows the last ion bound to be released first after the conformation transition to the P-E$_2$ states. In P-E$_2$ the ion sites are modified such that they may have shifted slightly within the dielectric, but certainly such that the affinity for Na$^+$ and H$^+$ ions has decreased by orders of magnitude, whereas that for K$^+$ remains almost constant.[10] Due to the low affinity of the binding sites for H$^+$ in P-E$_2$ under physiological conditions the sites remain unoccupied, and consequently release of Na$^+$ and binding of K$^+$ are found to be electrogenic.

ACKNOWLEDGMENTS

This work is based on collaborations with Milena Roudna and Anna Diller. This work was supported by the Deutsche Forschungsgemeinschaft (Ap 45/4) and INTAS 2001-0224.

REFERENCES

1. GLITSCH, H.G. 2001. Electrophysiology of the sodium-potassium-ATPase in cardiac cells. Physiol. Rev. **81:** 1791–1826.
2. JORGENSEN, P.L. & P.A. PEDERSEN. 2001. Structure-function relationships of Na$^+$, K$^+$, ATP, or Mg^{2+} binding and energy transduction in Na,K-ATPase. Biochim. Biophys. Acta **1505:** 57–74.
3. LÄUGER, P. 1991. Electrogenic Ion Pumps: 1–313. Sinauer. Sunderland, MA.
4. APELL, H.-J. 1997. Kinetic and energetic aspects of Na$^+$/K$^+$-transport cycle steps. Ann. N. Y. Acad. Sci. **834:** 221–230.
5. ALBERS, R.W. 1967. Biochemical aspects of active transport. Annu. Rev. Biochem. **36:** 727–756.
6. POST, R.L., C. HEGYVARY & S. KUME. 1972. Activation by adenosine triphosphate in the phosphorylation kinetics of sodium and potassium ion transport adenosine triphosphatase. J. Biol. Chem. **247:** 6530–6540.
7. WUDDEL, I. & H.-J. APELL. 1995. Electrogenicity of the sodium transport pathway in the Na,K-ATPase probed by charge-pulse experiments. Biophys. J. **69:** 909–921.
8. HOLMGREN, M. *et al.* 2000. Three distinct and sequential steps in the release of sodium ions by the Na$^+$/K$^+$-ATPase. Nature **403:** 898–901.
9. DE WEER, P., D.C. GADSBY & R.F. RAKOWSKI. 2000. The Na/K-ATPase: a current-generating enzyme. *In* The Na/K Pump and Related ATPases. K. Taniguchi & S. Kaya, Eds.: 27–34. Elsevier. Amsterdam.
10. APELL, H.J. & S.J. KARLISH. 2001. Functional properties of Na,K-ATPase, and their structural implications, as detected with biophysical techniques. J. Membr. Biol. **180:** 1–9.
11. HILGEMANN, D.W. 1994. Channel-like function of the Na,K pump probed at microsecond resolution in giant membrane patches. Science **263:** 1429–1432.
12. DOMASZEWICZ, W. & H.-J. APELL. 1999. Binding of the third Na$^+$ ion to the cytoplasmic side of the Na,K-ATPase is electrogenic. FEBS Lett. **458:** 241–246.
13. SCHNEEBERGER, A. & H.-J. APELL. 1999. Ion selectivity of the cytoplasmic binding sites of the Na,K-ATPase: I. Sodium binding is associated with a conformational rearrangement. J. Membr. Biol. **168:** 221–228.
14. BUTSCHER, C., M. ROUDNA & H.-J. APELL. 1999. Electrogenic partial reactions of the SR-Ca-ATPase investigated by a fluorescence method. J. Membr. Biol. **168:** 169–181.
15. PEINELT, C. & H.-J. APELL. 2002. Kinetics of the Ca^{2+}, H$^+$ and Mg^{2+} interaction with the ion-binding sites of the SR-Ca-ATPase. Biophys. J. **82:** 170–181.
16. SWEADNER, K.J. & C. DONNET. 2001. Structural similarities of Na,K-ATPase and SERCA, the Ca^{2+}-ATPase of the sarcoplasmic reticulum. Biochem. J. **356:** 685–704.

17. TOYOSHIMA, C. et al. 2000. Crystal structure of the calcium pump of sarcoplasmatic reticulum at 2.6 Å resolution. Nature **405:** 647–655.
18. APELL, H.-J. et al. 1996. Kinetics of the phosphorylation of Na,K-ATPase by inorganic phosphate detected by a fluorescence method. Biochemistry **35:** 10922–10930.
19. PEDERSEN, M. et al. 2001. Detection of charge movements in ion pumps by a family of styryl dyes. J. Membr. Biol. **185:** 221–236.
20. PINTSCHOVIUS, J., K. FENDLER & E. BAMBERG. 1999. Charge translocation by the Na^+/K^+-ATPase investigated on solid supported membranes: cytoplasmic cation binding and release. Biophys. J. **76:** 827–836.
21. SCHNEEBERGER, A. & H.-J. APELL. 2001. Ion selectivity of the cytoplasmic binding sites of the Na,K-ATPase: II. Competition of various cations. J. Membr. Biol. **179:** 263–273.

Na,K-Pump Reaction Kinetics at the Tip of a Patch Electrode

Derivation of Reaction Kinetics for Electrogenic and Electrically Silent Reactions during Ion Transport by the Na,K-ATPase

R. DANIEL PELUFFO AND JOSHUA R. BERLIN

Department of Pharmacology/Physiology, UMDNJ-New Jersey Medical School, Newark, New Jersey 07103, USA

ABSTRACT: Patch-clamp electrophysiological techniques allow manipulations of electrochemical driving forces for ion transport by the Na,K-ATPase. For this reason, this technique has been used to study steady-state ion transport properties of the Na,K-ATPase. High temporal resolution during these manipulations also permits rapid reactions, such as extracellular ion-binding reactions, to be measured as charge movements when the enzyme is engaged in electroneutral ion exchange modes. Just as useful, but less widely recognized, is the ease with which electrophysiological techniques can be used to critically study reaction steps that do not directly involve ion binding. Three studies are briefly presented to show how pre-steady-state and/or steady-state electrophysiological techniques can be used to study ion-binding reactions in a novel fashion and the kinetics of electrically silent reaction steps of this enzyme. The reaction kinetics derived from each of these studies can be used to attain detailed mechanistic information about ion transport by the Na,K-ATPase.

KEYWORDS: patch-clamp electrophysiological techniques; ion transport; electrophysiological characterization

INTRODUCTION

A wide array of experimental techniques has been employed to study Na,K-ATPase function. Among these techniques, electrophysiological methods are well recognized for their application to investigate membrane potential (V_M)-dependent ion-binding reactions of this enzyme. Much of the focus of these studies has been directed at the kinetics of extracellular Na^+ and K^+ binding reactions that are most critically sensitive to the electric field across the membrane. As we demonstrate below, electrophysiological methods can also be applied in novel ways to study the mecha-

---

Address for correspondence: Dr. R. Daniel Peluffo, Department of Pharmacology and Physiology, UMDNJ-New Jersey Medical School, 185 S. Orange Avenue, Newark, NJ 07103. Voice: 973-972-1490; fax: 973-972-7950.

peluffrd@umdnj.edu

nism of ion-binding reactions and as a versatile approach to investigating mechanistic and kinetic questions beyond those pertaining specifically to ion binding.

MECHANISM OF EXTRACELLULAR K$^+$ BINDING

The existence of ion wells for binding of extracellular Na$^+$ and K$^+$ is now well accepted. Experimental data show that Na$^+$ and K$^+$ binding reactions move charge in the membrane electric field,[1–5] and the ion concentration and V_M dependence of these charge-moving reactions demonstrate that ion binding is the V_M-dependent step. For this reason, the question has now been raised as to what type of enzyme structural features might underpin the mechanistic nature of these charge-moving ion-binding steps. Two alternative mechanisms have been proposed to explain V_M-dependent ion-binding reactions. One hypothesis postulates that ion binding occurs in a high-field access channel in the E_2-P conformation of the Na,K-ATPase, while the alternative proposal is that intrinsic protein charges move during rapid transitional occlusion/deocclusion steps as part of extracellular ion-binding/dissociation reactions.[6] Distinguishing between these alternative mechanisms is not experimentally practical by manipulating concentrations of transported ions and membrane potential.

To test the validity of these alternative mechanisms, we took advantage of previously reported observations that quaternary organic amines, such as tetraethylammonium ions (TEA), inhibit steady-state Na,K-pump activity, specifically ouabain-sensitive ^{24}Na$^+$ fluxes, in a manner that is competitive with extracellular K$^+$ activation of the Na,K-ATPase.[7] Experiments examining ^{86}Rb$^+$ occlusion by the Na,K-ATPase then showed that TEA and other structurally related quaternary amines are not occluded by the Na,K-ATPase.[8] Instead, the amines appeared to block Rb$^+$ occlusion reactions. Given these data, one could then make the prediction that inhibition of Na,K-pump current by quaternary organic amines will be V_M-dependent only if extracellular K$^+$ (K^+_o) binding occurs in a high-field access channel, because the amine could enter the access channel to produce competitive inhibition. On the other hand, inhibition would be V_M-independent if rapid rearrangements of intrinsic charges occur during occlusion because the amines would block the occlusion steps.

To test this prediction, we first determined the effect of quaternary amines on steady-state Na,K-pump currents measured in voltage-clamped cardiac myocytes. FIGURE 1A shows the effect of 25 mM TEA on 0.2 mM K^+_o-activated Na,K-pump current (I_{pump}) measured in cardiac myocytes superfused in extracellular Na$^+$ (Na^+_o)-free solutions. The cells were subjected to whole-cell voltage clamp techniques using patch electrodes filled with a 115 mM Na$^+$-containing salt solution, as previously described.[9] TEA decreased Na,K-pump current density in these cells at all potentials tested, but the negative slope of the current–voltage relationship in Na^+_o-free solutions was observed in the presence (open circles) and absence of the quaternary amine. Additional experiments[10] showed that maximal Na,K-pump current density was similar in the presence and absence of TEA, although the concentration of K^+_o needed to half-maximally activate Na,K-pump current was increased at all V_M tested. Together, these results suggested that current inhibition was competitive with K^+_o activation. Analyzing the data with a pseudo-3-state model for current activation by K^+_o and inhibition by TEA allowed us to conclude that TEA

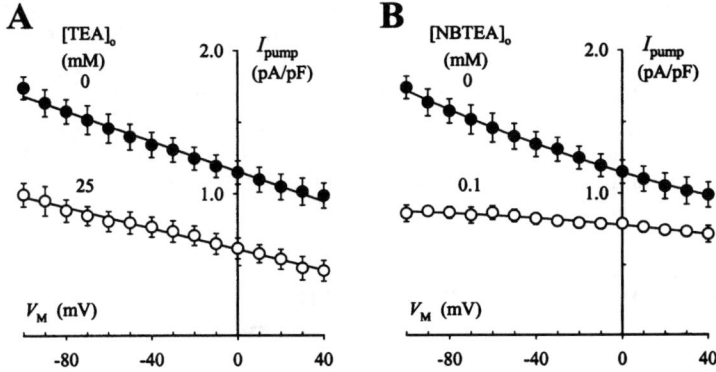

FIGURE 1. Effect of organic quaternary amines on steady-state Na,K-pump current. Solid curves were fit to the data using arbitrary polynomial functions. (**A**) Effect of 25 mM TEA ($n = 5$). (**B**) Effect of 0.1 mM NBTEA ($n = 5$).

inhibition was independent of membrane potential. The V_M dependence of Na,K pump current in FIGURE 1A then represents the V_M dependence of Na,K-ATPase activation by K^+_o. These conclusions are similar to those of Eckstein-Ludwig et al.,[11] who studied TEA inhibition of steady-state Na,K-pump current in *Xenopus* oocytes.

The potency of quaternary amines to block $^{86}Rb^+$ release from the Na,K-ATPase has been reported to be greatly enhanced by using TEA-related compounds with a benzylic moiety substituting for one of the alkyl groups.[8] For this reason, we investigated the effects of two quaternary amines, benzyltriethylammonium ion (BTEA) and *para*-nitrobenzyltriethylammonium ion (NBTEA) on Na,K-pump current.

FIGURE 1B shows that 0.1 mM NBTEA effectively inhibited 0.2 mM K^+_o-activated Na,K-pump current measured under experimental conditions similar to those used above. Clearly, NBTEA inhibited the Na,K-ATPase at much lower concentrations than TEA. Further experiments[10] showed by BTEA had a similar effect on Na,K-pump current and that this amine increased the K^+_o concentration needed for half-maximal activation of Na,K-pump current, more so at negative V_M, without affecting maximal current density. Thus, these quaternary amines also appeared to inhibit the Na,K-ATPase in a manner that was competitive with K^+_o.

The membrane potential dependence of the Na,K-pump current became very shallow in the presence of NBTEA over the range of V_M studied (FIG. 1B). Similar results were observed with BTEA block of pump current, suggesting that the apparent V_M dependence of Na,K-pump current was altered by these compounds. Analyzing the data (not shown) with a pseudo-3-state model allowed us to conclude that V_M-dependent activation of Na,K-pump current by K^+_o was not altered by BTEA, since the fraction of the membrane electric field dissipated by K^+_o-dependent reactions (λ_K), calculated to be 0.37 ± 0.02, was similar to λ_K measured in the absence of BTEA. Interestingly, the same calculation showed that BTEA inhibition of current also dissipated a fraction of the membrane electric field, 0.40 ± 0.05, that was similar to λ_K. In other words, BTEA inhibited Na,K-pump current in a V_M-

dependent manner at a site situated at the same electrical distance in the membrane as the site for K^+_o activation. Similar results were observed with NBTEA. Given the competitive character of quaternary amine inhibition with respect to K^+_o, one simple interpretation of these results is that BTEA and NBTEA inhibit the Na,K-ATPase at the site for voltage-dependent K^+_o binding. In regards to our prediction, these results are consistent with BTEA and NBTEA blocking the Na,K-ATPase in a high-field access channel and with any rapid occlusion/deocclusion reactions moving little charge through the membrane dielectric.

KINETICS OF ELECTRICALLY SILENT REACTION STEPS OF THE Na,K-ATPase

Transient charge movements can be used as a tool to study electrically silent reactions that are kinetically distinguishable from associated charge-moving reaction steps. An example of such a reaction is ADP release from the Na,K-ATPase phosphoenzyme, a step closely related to electrogenic Na^+_o release. To investigate these reactions, cardiac myocytes were whole-cell voltage-clamped with patch electrodes containing a 115 mM Na^+, 15 mM MgATP electrode solution (pH 7.3), and the cells were superfused with a 145 mM Na^+, K^+-free extracellular solution at 23°C (pH 7.4), conditions that confine the enzyme to states involved in V_M-dependent electroneutral Na^+-Na^+ exchange.[1,3–5] Upon application of a voltage-jump protocol in the absence and the presence of 1 mM ouabain,[4] transient currents (i.e., transient charge movements) were obtained. FIGURE 2A shows superimposed traces of ouabain-sensitive transient charge movements in response to voltage pulses from –40 mV to various potentials between –140 and +60 mV in the presence of 145 mM Na^+_o and no added ADP. FIGURE 2B shows currents that result from the same maneuvers in the presence of 2 mM intracellular ADP. Currents decayed in an exponential fashion to a zero steady-state level, consistent with a V_M-dependent electroneutral Na^+-Na^+ exchange process. It is clear that ADP slowed the kinetics of current relaxation at depolarizing potentials as compared to the control zero-ADP condition. In fact, fitting exponential functions to the decaying portion of the current traces revealed that the apparent rate constant for current relaxation (k_{tot}) reaches a minimum at approximately 0 mV with plateau values of ~140 s^{-1} in the absence and ~70 s^{-1} in the presence of 2 mM ADP.

The observation that ADP decreased the rate of current decay at depolarizing potentials (that is, where k_{tot} becomes V_M-independent) is consistent with the reaction scheme depicted in FIGURE 2C. The basic features of the scheme are (1) the entire pool of Na,K-ATPase is distributed among phosphorylated states; (2) ADP release from and rebinding to the phosphoenzyme are electroneutral reactions; (3) ADP dissociates before the electrogenic release of Na^+, that is, Na^+ is still bound (and probably occluded) when ADP is released; and (4) Na^+_o is in rapid equilibrium with Na^+ bound in an ion well.

Values of forward and reverse rate constants for the entire reaction scheme in FIGURE 2C, together with electrical coefficients, were calculated by measuring the apparent rate constant for current relaxation (k_{tot}) over a 240-mV range of V_M at various concentrations of intracellular ADP and Na^+_o.[12] The set of differential equations describing this scheme was also solved to derive an explicit expression for

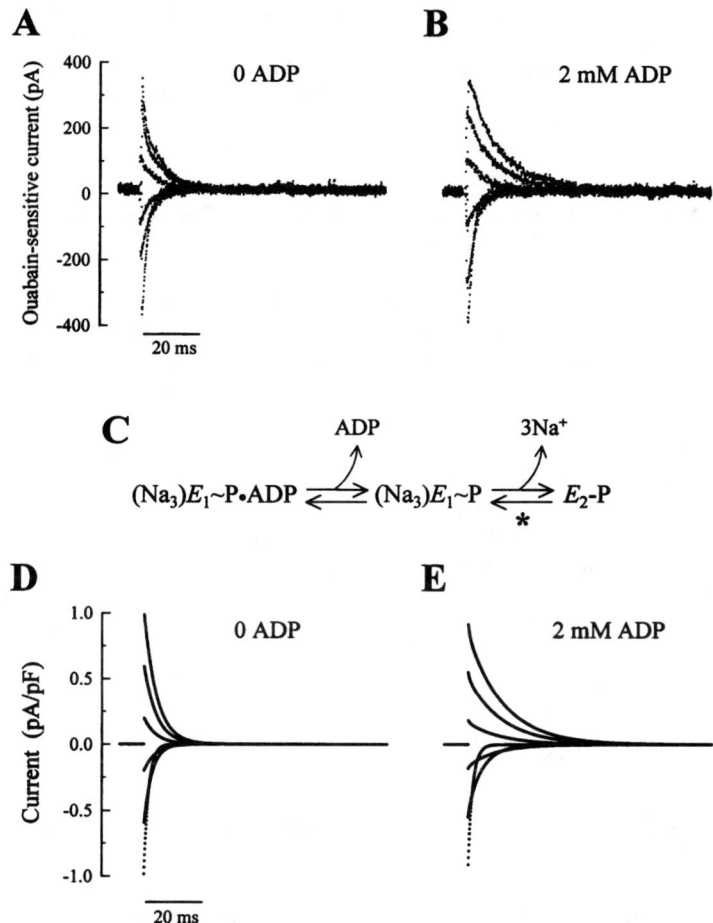

FIGURE 2. Effect of intracellular ADP on the kinetics of Na^+-dependent transient charge movements. (**A**) Superimposed ouabain-sensitive transient currents elicited by voltage jumps from –40 mV to –140, –100, –60, –20, 20, and 60 mV in the absence of added ADP. (**B**) Superimposed transient currents recorded with 2 mM ADP added to the electrode solution. (**C**) Three-state reaction scheme: *parentheses* indicate Na^+ occluded in the enzyme and the *asterisk* indicates a V_M-dependent reaction in the scheme. (**D**) Currents simulated for the 0 mM ADP condition. Superimposed traces are for the same voltage jumps shown in panel **A**. (**E**) Simulated currents for 2 mM ADP.

time- and V_M-dependent current, $I_M = f(t; V_M)$, in which k_{tot} is one of the exponential coefficients. To test how well this scheme describes the data, calculated values of rate constants and electrical coefficients were fed into $I_M = f(t; V_M)$, and simulations were run using experimental ranges of time and V_M. The simulated currents are shown in FIGURE 2D (0 ADP) and 2E (2 mM ADP), and clearly, they resemble current traces obtained experimentally (FIG. 2A and 2B). Thus, ADP is released from the E_1~P phosphoenzyme in an electroneutral reaction prior to the electrogenic release of Na^+. The calculated rate constants indicate that ADP release and rebinding are fast reactions, while Na^+ reocclusion is likely the rate-limiting step following V_M-dependent Na^+_o rebinding to E_2-P.

COMPARISON OF FUNCTIONAL Na,K-ATPase EXPRESSION IN DIFFERENT CELL TYPES

Increasingly, molecular methods are making it possible to understand the regulation of Na,K-ATPase expression and trafficking. The end point of these investigations, nonetheless, must be how functional expression is regulated in the cell. One difficulty in these investigations is then to compare functional expression between different cell types. Biochemical methods allow for such a comparison, but usually only after disruption of the cell membrane, which always leaves open the possibility that some membrane-bound component important in regulating Na,K-ATPase function is lost with preparatory steps. For this reason, we tested whether steady-state Na,K-pump current measurements could be combined with measurements of transient charge movements to determine both the cell surface expression and functional properties of the enzyme in the same cell preparation (or even the same cell). For this purpose, we compared Na,K-ATPase in rat versus guinea pig ventricular myocytes.

Single ventricular myocytes were whole-cell voltage-clamped with patch electrodes containing a salt solution with 115 mM Na^+, as previously described,[9] and superfused in a 145 mM Na^+-containing solution with 15 mM K^+ at 35°C to maximally activate Na,K-ATPase activity. The V_M dependence of K^+_o-activated, cardiac glycoside (CG)-inhibitable Na,K-pump currents measured with rat (filled circles) and guinea pig ventricular myocytes (open circles) are shown in FIGURE 3A. As in previous publications,[9,13] steady-state Na,K-pump current in Na^+-containing superfusion solutions displayed a positive slope at negative V_M that reached a plateau at positive V_M in the presence of saturating K^+_o. Scaling the current–voltage relationships to maximal current (FIG. 3B) showed that the V_M dependence of Na,K-pump current was quite similar in both cell types. In addition, the K^+_o dependence of current activation (FIG. 3C) was the same, since the K^+_o concentration for half-maximal activation ($K_{0.5}$) for both rat and guinea pig myocytes was 2.3 ± 0.2 ($n = 4$ for each), similar to the apparent $K_{0.5}$ for activation of Na,K-pump current published in previous reports.[9,13] Most notable, however, was the difference in maximal Na,K-pump current density, which was more than 2 times greater in rat myocytes, 4.67 ± 0.13 pA/pF, than in guinea pig cardiac cells, 1.93 ± 0.06 pA/pF. A review of published literature seems to support this finding.[9,13]

The difference in Na,K-pump current densities in rat and guinea pig myocytes could be due to a difference in the number of functional Na,K-ATPase molecules expressed in the cell membrane and/or the reaction kinetics of the enzymes in the membrane. To examine the former possibility, we measured transient charge movements under conditions promoting Na^+-Na^+ exchange at 25°C, as previously described.[4] FIGURE 3D and 3E show CG-sensitive charge movements produced by depolarizations from –40 mV to +40 mV in rat and guinea pig myocytes, respectively. What is

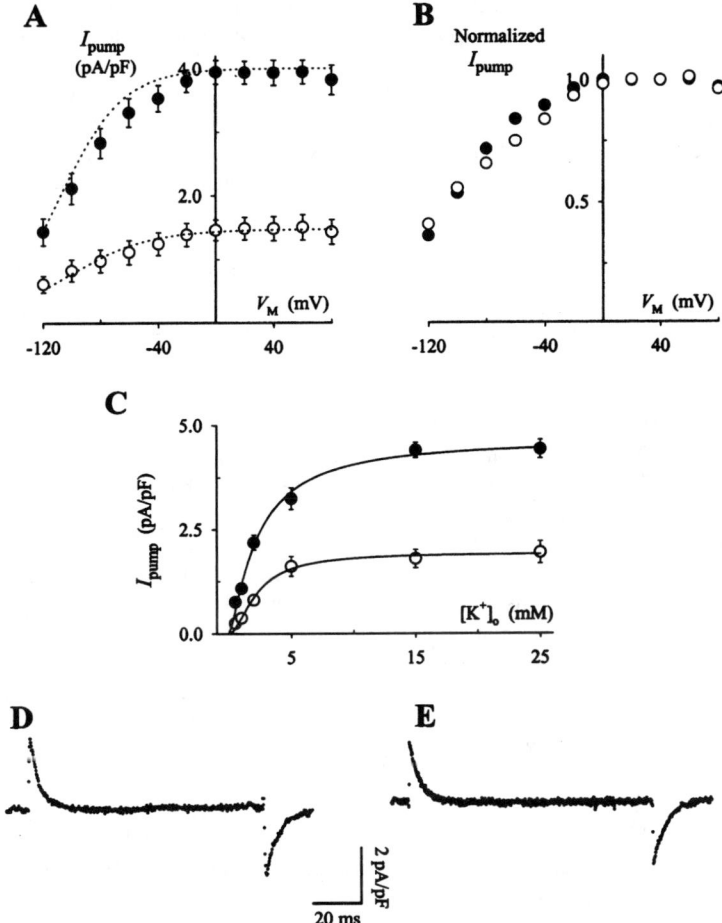

FIGURE 3. Comparison of Na,K-ATPase current properties in rat (*solid circles*) and guinea pig (*open circles*) myocytes. (**A**) Voltage dependence of Na,K-pump current at various V_M in rat ($n = 8$) and guinea pig ($n = 6$) cells. The *dashed curves* were calculated using a pseudo-2-state model for the Na,K-ATPase. (**B**) Currents were normalized to values at +40 mV. (**C**) K^+_o dependence of Na,K-pump current at 0 mV. The curves are functions derived from fitting the data with a Hill equation. Calculated Hill coefficients are 1.30 and 1.75 for rat and guinea pig, respectively. (**D**) Ouabain (1 mM) -sensitive transient current measured in a rat myocyte. (**E**) Strophanthidin (0.25 mM) -sensitive transient current measured in a guinea pig myocyte.

most obvious about these current tracings is their relative similarity in the amount of charge moved. By examining the V_M dependence of CG-sensitive charge movements and fitting the resulting data (not shown) with Boltzmann functions, we were able to calculate the maximal amount of mobile charge (ΔQ_{max}) to be 24.7 ± 1.2 ($n = 3$) and 22.4 ± 1.5 fC/pF ($n = 3$) for rat and guinea pig cells, respectively. The data, therefore, show that the functional Na,K-ATPase expression density is similar in rat and guinea pig myocytes and the differences in current density shown in FIGURE 3A must be due to differences in enzyme turnover rate.

The slope (k) of these Boltzmann functions, 26 mV per e-fold change in charge, suggested that V_M-dependent reactions during Na^+-Na^+ exchange move essentially one charge through the membrane dielectric (i.e., 1 charge moves per enzyme). This value allowed us to approximate the density of functional Na,K-ATPase expression as follows:

Na,K-ATPase density (μm^{-2}) = ΔQ_{max} * 0.01 pF/μm^2 * N/F * 1 enzyme/charge

where N is Avogadro's number and F is the Faraday constant. With this calculation, we determined that Na,K-pump density was 1500 ± 67 μm^{-2} for rat and 1404 ± 94 μm^{-2} for guinea pig myocytes.

The electrophysiological data shown in FIGURE 3 also allowed us to gain more information about the underlying differences in reaction kinetics of enzymes in these cells. By examining the kinetics of transient charge movements and the temperature dependence of those kinetics (Q_{10} = 3.1), we determined that the pseudo-first order forward and backward rate constants describing transient charge movements at 0 mV and 35°C were 747 ± 12 and 7 ± 1 s^{-1}, respectively, for rat (n = 3), and 415 ± 13 and 12 ± 1 s^{-1}, respectively, for guinea pig myocytes (n = 3). Thus, the forward rate of Na^+-Na^+ exchange reactions is slightly faster in rat cells.

Could this difference explain the greater than 2-fold higher steady-state Na,K-pump current in rat myocytes? Using a pseudo-2-state model containing one V_M-dependent reaction[14] to describe the data in FIGURE 3A (dashed curves), we could show that the observed differences in forward and backward rate constants for Na^+-Na^+ exchange reactions were not sufficient to account for the higher current levels in rat myocytes. Instead, the model calculations require the rate of another V_M-*independent* reaction, rate-limiting for steady-state Na,K-ATPase function, to be three-fold faster in rat myocytes to correctly fit the experimental data. Given the conditions of our experiments, this rate-limiting reaction is most likely K^+ deocclusion and release to the cell interior.[15]

CONCLUSION

Three applications of electrophysiological characterization of the Na,K-ATPase have been briefly presented here. The purpose of this presentation is to stress the point that similar electrophysiological approaches are useful in a wide range of applications beyond the well-recognized use of these techniques in studying ion-binding reactions to the enzyme. These techniques will also continue to find further applications when combined with other techniques, such as fluorescence microscopy, to study the function of the Na,K-ATPase.

ACKNOWLEDGMENT

This work was supported by a Grant-in-Aid from the American Heart Association (R.D.P.), and by the National Institutes of Health (J.R.B.).

REFERENCES

1. NAKAO, M. & D.C. GADSBY. 1986. Voltage dependence of Na translocation by the Na/K pump. Nature **323:** 628–630.
2. STÜRMER, W., R, BÜHLER, H.-J. APELL & P. LÄUGER. 1991. Charge translocation by the Na,K-pump: II. Ion binding and release at the extracellular side. J. Membr. Biol. **121:** 163–176.
3. HILGEMANN, D.W. 1994. Channel-like function of the Na,K pump probed at microsecond resolution in giant membrane patches. Science **263:** 1429–1432.
4. PELUFFO, R.D. & J.R. BERLIN. 1997. Electrogenic K^+ transport by the Na^+-K^+ pump in rat cardiac ventricular myocytes. J. Physiol. (London) **501:** 33–40.
5. HOLMGREN, M., J. WAGG, F. BEZANILLA, *et al.* 2000. Three distinct and sequential steps in the release of sodium ions by the Na^+/K^+-ATPase. Nature **403:** 898–901.
6. HILGEMANN, D.W. 1994. Flexibility and constraint in the interpretation of Na^+/K^+ pump electrogenicity: What is an access channel? *In* The Sodium Pump. E. Bamberg & W. Schoner, Eds.: 507–517. Steinkopff. Darmstadt, Germany.
7. SACHS, J.R. & M.E. CONRAD. 1968. Effect of tetraethylammonium on the active cation transport system of the red blood cell. Am. J. Physiol. **215:** 795–798.
8. FORBUSH, B., III. 1988. The interaction of amines with the occluded state of the Na,K-pump. J. Biol. Chem. **263:** 7979–7988.
9. ISHIZUKA, N., A.J. FIELDING & J.R. BERLIN. 1996. Na pump current can be separated into ouabain-sensitive and -insensitive components in single rat ventricular myocytes. Jap. J. Physiol. **46:** 215–223.
10. BERLIN, J.R. & R.D. PELUFFO. 1998. Inhibition of Na,K pump current in cardiac myocytes by organic quaternary amines. Biophys. J. **74:** A338.
11. ECKSTEIN-LUDWIG, U., J. RETTINGER, L.A. VASILETS & W. SCHWARZ. 1998. Voltage-dependent inhibition of the Na^+,K^+ pump by tetraethylammonium. Biochim. Biophys. Acta **1372:** 289–300.
12. PELUFFO, R.D. 1998. Effect of ADP on extracellular Na^+-dependent transient charge movements by the Na pump from rat cardiomyocytes. Biophys. J. **74:** A192.
13. NAKAO, M. AND D.C. GADSBY. 1989. [Na] and [K] dependence of the Na/K pump current-voltage relationship in guinea pig ventricular myocytes. J. Gen. Physiol. **94:** 539–565.
14. SAGAR, A. & R.F. RAKOWSKI. 1994. Access channel model for the voltage dependence of the forward-running Na^+/K^+ pump. J. Gen. Physiol. **103:** 869–894.
15. GLYNN, I.M. 1985. The Na^+, K^+-transporting adenosine triphosphatase. *In* The Enzymes of Biological Membranes, 2nd ed. A.N. Martonosi, Ed.: 35–114. Plenum. New York.

Two-Electrode Voltage-Clamp Analysis of Na,K-ATPase Asparagine 776 Mutants

JAN B. KOENDERINK,[a,b] SVEN GEIBEL,[a] EVA GRABSCH,[a] JAN JOEP H. H. M. DE PONT,[b] ERNST BAMBERG,[a] AND THOMAS FRIEDRICH[a]

[a]*Department of Biophysical Chemistry, Max-Planck-Institute of Biophysics, Frankfurt am Main, Germany*

[b]*Department of Biochemistry, Nijmegen Center for Molecular Life Sciences, Nijmegen, the Netherlands*

ABSTRACT: Steady-state and pre-steady-state currents of Asn[776] mutants of Na,K-ATPase are presented. The stationary current generated by N776Q strongly depends on the membrane potential, but has a negative slope, opposite to that of the wild-type enzyme. The apparent rate constant of the reaction sequence $E_1P(Na^+) \leftrightarrow E_2P + Na^+$ of this mutant is rather independent of the membrane potential and is at resting and depolarizing membrane potential higher than that of the wild-type enzyme. Thus, the voltage-dependent increase of the rate coefficient of the wild type that is associated with extracellular Na^+ rebinding is almost absent in the N776Q mutant. These findings indicate that dislocating the carboxamide group of Asn[776] decreases the affinity of sodium at its extracellular binding site.

KEYWORDS: *Xenopus laevis*; electrophysiology; transient currents; Asn776; mutagenesis

INTRODUCTION

Each reaction cycle, Na,K-ATPase transports three sodium ions out and two potassium ions into the cell. Although several putative cation-coordinating amino acids are identified in Na,K-ATPase, their influence on electrogenic cation transport is ambiguous. Because Asn[776] is probably involved in cation binding,[1,2,8] we wanted to analyze its function in electrophysiological experiments.

Fendler *et al.*[3] were the first to record pre-steady-state transient currents of Na,K-ATPase upon photolysis of caged ATP. Transient currents in response to voltage steps were first shown by Nakao and Gadsby.[4] Kinetic analysis of transient currents yields information about the rate of partial reactions of the catalytic cycle. The use of *Xenopus laevis* oocytes for investigation of Na,K-ATPase transient currents[5] opened up new avenues for analysis of Na,K-ATPase mutants. Here we present steady-state and pre-steady-state currents of Asn[776] mutants of Na,K-ATPase.

Address for correspondence: Jan B. Koenderink, Department of Biochemistry, Nijmegan Center for Molecular Life Sciences, P.O. Box 9101, 6500 HB Nijmegan, the Netherlands. Voice: +31-24-361-35-17; fax: +31-24-361-64-13.
J.Koenderink@ncmls.kun.nl

RESULTS AND DISCUSSION

Asn^{776} was replaced by Gln, Asp, and Ala. Ten ng of (mutated) rat Na,K-ATPase α-subunit and 2 ng of the β-subunit mRNAs were injected into *X. laevis* oocytes. After 3 days the oocytes were used for two-electrode voltage-clamp experiments.

Xenopus oocytes express endogenous Na,K-ATPase (~25 nA) that can be inhibited by 10 µM ouabain. The expressed rat Na,K-ATPase is insensitive for ouabain.[6] The specific rat Na,K-ATPase current is defined as the difference between currents measured in the presence of 10 µM and 10 mM ouabain. Before measurement the oocytes were loaded with Na^+ to increase the internal Na^+ concentration.[5] Extracellular addition of 5 mM K^+ generated a maximum current for wild-type Na,K-ATPase at a holding potential of −20 mV (183 ± 26 nA, n = 18). The mutants N776Q, N776D, and N776A showed currents [188 ± 35 nA (n = 8), 108 ± 33 nA (n = 7), and 232 ± 64 nA (n = 7), respectively] that were not substantially different from that of the wild type. This indicates that Asn^{776} is not essential for enzymatic activity.

The current–voltage (*IV*) relationship of the wild-type rat Na,K-ATPase was investigated at different K^+ concentrations. The current without K^+ was subtracted from that with K^+ at membrane potentials between −160 mV and +40 mV, and the maximal current was set at 100%. Currents of the wild-type enzyme increased with increasing K^+ concentrations and membrane potential (FIG. 1). At positive membrane potentials, the currents decreased, forming a bell-shaped curve that is shifted to higher membrane potentials with increasing [K^+]. Negative voltages promote K^+ transport into the cell. The outward transport of Na^+ works more efficiently at positive voltages. Thus, the membrane potential at which the current is maximal depends on the cation concentrations.

The N776A mutant behaved similarly to the wild-type enzyme. The current generated by N776D is virtually voltage-independent. The currents generated by the N776Q mutant strongly depended on the membrane potential, and the slope of the *IV* curve was negative, in contrast to the wild type. This indicates that the competition of Na^+ at negative membrane potentials is absent. Pedersen *et al.*[7] found that the carboxamide group of Asn^{776} was equally important for Tl^+ (K^+) or Na^+, whereas a shift in the position of the carboxamide of Asn^{776} (N776Q) caused a large depression of Na^+ binding without affecting Tl^+ binding. These results agree with our findings.

The kinetics of pre-steady-state currents upon voltage jumps were analyzed for the mutants to investigate reaction kinetics of Na^+-dependent reaction steps.[4] With high Na^+ concentrations on both sides of the membrane, Na,K-ATPase is restricted to the Na^+–Na^+ exchange mode [$E_1P(Na^+) \leftrightarrow E_2P + Na^+$]. FIGURE 2A shows ouabain-sensitive (10-mM) transient currents of the rat Na,K-ATPase. These currents were fit by a monoexponential function starting 5 ms after onset of the voltage pulse to exclude artefacts arising from capacitive charging of the membrane. In FIGURE 2B the resulting apparent relaxation rate constants are plotted against the membrane potential. This distribution is only weakly voltage-dependent at positive voltages and increases strongly at negative voltages. This is interpreted in terms of a voltage-dependent rebinding of Na^+ [$E_1P(Na^+) \leftarrow E_2P + Na^+$], whereas the forward reaction [$E_1P(Na^+) \rightarrow E_2P + Na^+$] is weakly or not voltage dependent.[4,8]

Transient currents of mutant N776A decayed with apparent rate constants greater than ~250 s^{-1} and are difficult to resolve in voltage-clamp experiments on oocytes.

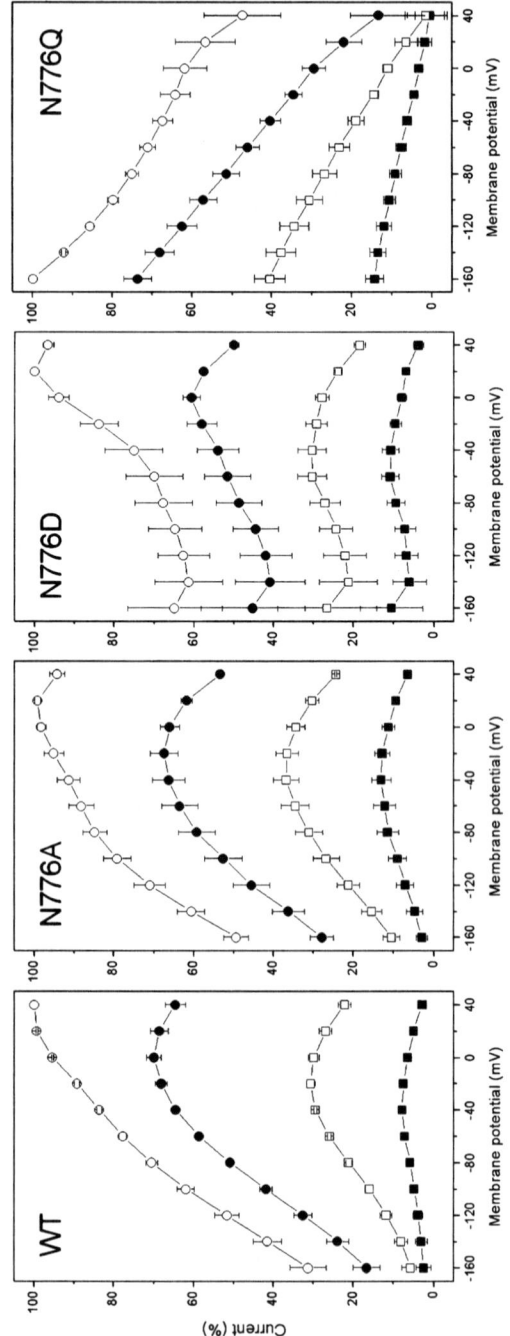

FIGURE 1. Voltage dependence of extracellular K^+ activation of Na,K-ATPase current. Steady-state current–voltage relationships of mutant and wild-type pumps were obtained by subtracting current measured in K^+-free solution from that measured in K^+-containing solution at each potential. Values are the mean ± S.E. ($n = 3$–4), at the following KCl concentrations: 0.15 mM, ■; 0.5 mM, □; 1.5 mM, ●; 5 mM, ○.

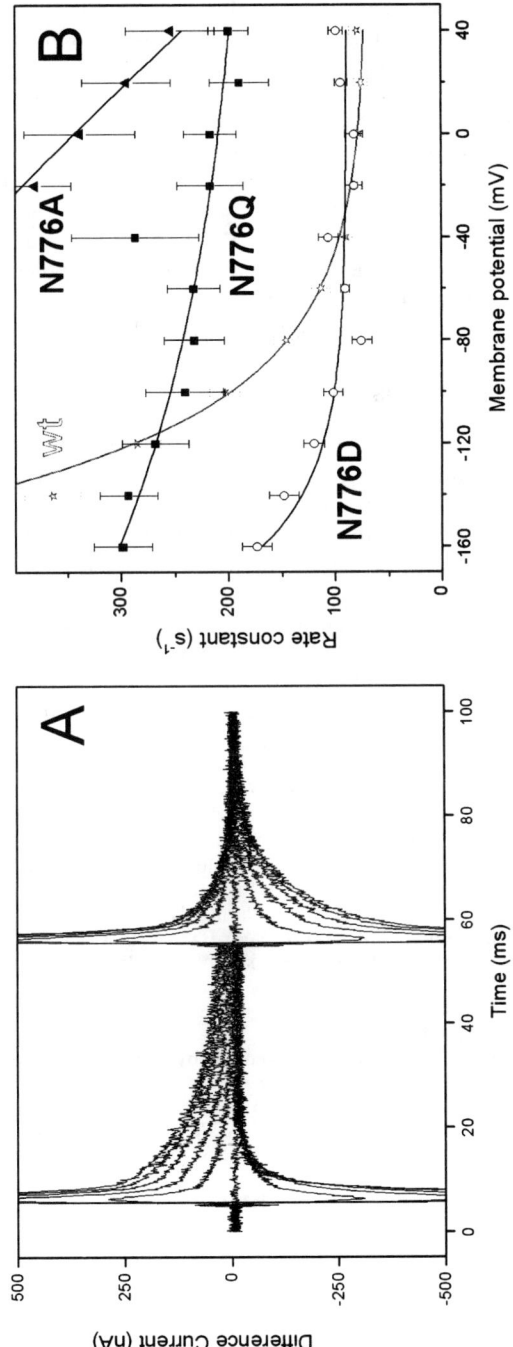

FIGURE 2. (**A**) Recordings of voltage–current traces of the wild-type Na,K-ATPase (difference between 10 μM and 10 mM ouabain). (**B**), Voltage dependence of the relaxation rate constant of the pre-steady-state currents of N776A, N776D, N776Q, and wild-type Na,K-ATPase.

However, together with comparably high stationary currents of this mutant, this can be interpreted as a fast $E_1P(Na^+) \leftrightarrow E_2P + Na^+$ relaxation.

The N776D mutant exhibited a lower and only weakly voltage-dependent rate constant at negative voltages. As the voltage-dependent increase in apparent rate constants at negative potentials, which is associated with Na^+ rebinding $[E_1P(Na^+) \leftarrow E_2P + Na^+]$, is less pronounced, this suggests a decreased apparent affinity for Na^+. This deviation from the wild-type is in agreement with the small, voltage-independent stationary current.

The apparent rate constant of the N776Q mutant was nearly completely voltage independent and at resting and depolarizing membrane potential higher than that of the wild-type enzyme. Because the voltage-dependent increase of the rate constant was almost absent in the N776Q mutant, this finding strengthens our initial hypothesis that dislocating the $CONH_2$ group of Asn^{776} decreases the Na^+ affinity at the extracellular sodium-binding site.

It is remarkable that complete removal of the carboxamide group in the N776A mutant had no measurable effect on the voltage sensitivity of the Na,K-ATPase, although accelerating the $E_1P(Na^+) \leftrightarrow E_2P + Na^+$ relaxation drastically. Replacement of the carboxamide group by an acid group without changing the length of the side chain (N776D) makes the current rather potential-independent at negative membrane potentials. The largest effect was seen in mutant N776Q, in which the carboxamide group is kept, but in which the side chain is elongated by one CH_2-group. It might be that by this mutation the carbonyl residue comes further in the cation-binding pocket, and so destroys (part of) cation binding. Future experiments using a steady-state voltage–current relationship and pre-steady-state charge movements may shed more light on the structure–function relationship of Na,K-ATPase.

REFERENCES

1. TOYOSHIMA, C., M. NAKASAKO, H. NOMURA & H. OGAWA. 2000. Crystal structure of the calcium pump of sarcoplasmic reticulum at 2.6 Å resolution. Nature **405:** 647–655.
2. TOYOSHIMA, C. & H. NOMURA. 2002. Structural changes in the calcium pump accompanying the dissociation of calcium. Nature **418:** 605–611.
3. FENDLER, K., E. GRELL, M. HAUBS & E. BAMBERG. 1985. Pump currents generated by the purified Na^+K^+-ATPase from kidney on black lipid membranes. EMBO J. **4:** 3079–3085.
4. NAKAO, M. & D.C. GADSBY. 1986. Voltage dependence of Na translocation by the Na/K pump. Nature **323:** 628–630.
5. RAKOWSKI, R.F. 1993. Charge movement by the Na/K pump in *Xenopus* oocytes. J. Gen. Physiol. **101:** 117–144.
6. KOENDERINK, J.B., H.P.H. HERMSEN, H.G.P. SWARTS, et al. 2000. High-affinity ouabain binding by a chimeric gastric H^+,K^+-ATPase containing transmembrane hairpins M3-M4 and M5-M6 of the alpha 1-subunit of rat Na^+,K^+-ATPase. Proc. Natl. Acad. Sci. USA **97:** 11209–11214.
7. PEDERSEN, P.A., J.M. NIELSEN, J.H. RASMUSSEN & P.L. JORGENSEN. 1998. Contribution to Tl^+, K^+, and Na^+ binding of Asn776, Ser775, Thr774, Thr772, and Tyr771 in cytoplasmic part of fifth transmembrane segment in alpha-subunit of renal Na,K-ATPase. Biochemistry **37:** 17818–17827.
8. DE WEER, P. 1990. The Na/K pump: a current generating enzyme. *In* Regulation of Potassium Transport across Biological Membranes. L. Reuss, G. Szabo, and J.M. Russell, Eds.: 5–22. University of Texas Press. Austin.

Binding of 1 Rb$^+$ Accelerates Dephosphorylation of the Na$^+$,K$^+$-ATPase without Leading to Rb$^+$ Occlusion

SERGIO B. KAUFMAN, RODOLFO M. GONZÁLEZ-LEBRERO, PATRICIO J. GARRAHAN, AND ROLANDO C. ROSSI

Instituto de Química y Fisicoquímica Biológicas and Departamento de Química Biológica, Facultad de Farmacia y Bioquímica, Universidad de Buenos Aires, Junín 956, C1113AAD Buenos Aires, Argentina

ABSTRACT: In steady-state conditions and for concentrations of the K$^+$-congener Rb$^+$ less than 2.5 mM, Rb$^+$-dependent ATPase activity is significantly higher than the steady-state rate of breakdown of Rb$^+$-occluded states, a discrepancy that disappears at sufficiently high [Rb$^+$]. Direct experimental evidence is provided that supports the explanation that the binding of a single Rb$^+$ to the phosphoenzyme conformer E_2P accelerates dephosphorylation without leading to the occlusion of the cation.

KEYWORDS: dephosphorylation; occlusion; kinetics

INTRODUCTION

We have reported[1] that at Rb$^+$ concentrations equal to or higher than 2.5 mM, the measured ATPase activity, v, was very similar to that calculated as the product between the concentration of the intermediate containing occluded Rb$^+$, E_{occ}, and the value of the apparent rate constant of Rb$^+$ deocclusion, k_{deocc}. This result was expected since, under the conditions of the experiments, all the flow of the ATPase reaction passed through the formation and breakdown of occluded intermediates, and the deocclusion of Rb$^+$ is an irreversible reaction. However, when Rb$^+$ was present in low concentrations (less than 2.5 mM), v was higher than the value calculated as $[E_{occ}]$ times k_{deocc}. The difference between the measured and the calculated ATPase activities could in principle be explained by the turning over of the Na$^+$-ATPase cycle, which of course does not lead to occlusion of Rb$^+$, but even if we subtracted from the measured activity that obtained in the absence of Rb$^+$, v_{Na} (which is the maximal possible value for the Na$^+$-ATPase activity), that difference remained, that is, $v-v_{Na}$ was still much higher than $[E_{occ}]$ times k_{deocc}.

Address for correspondence: Sergio B. Kaufman, Instituto de Química y Fisicoquímica Biológicas and Departamento de Química Biológica, Facultad de Farmacia y Bioquímica, Universidad de Buenos Aires, Junín 956, C1113AAD Buenos Aires, Argentina. Voice: +54-11-4962-5506; fax: +54-11-4962-5457.
sbkauf@qb.ffyb.uba.ar

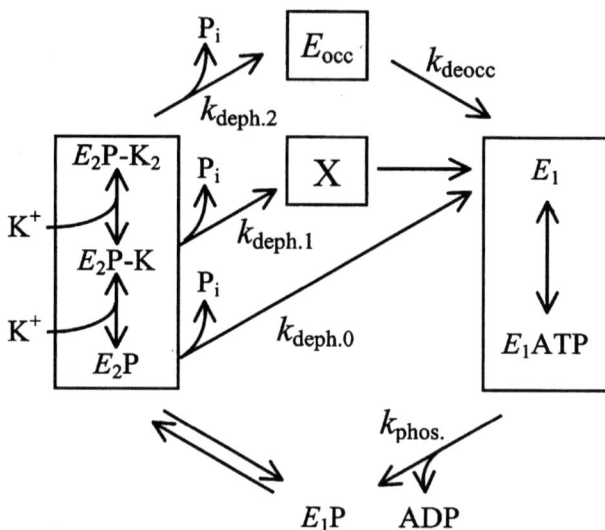

FIGURE 1. A simplified version of the Albers–Post scheme for the Na$^+$,K$^+$-ATPase, including the binding of a unique K$^+$ to E_2P. E_{occ} stands for the sum of species that are known to occlude K$^+$.

This led us to postulate a version of the Albers–Post scheme for the Na$^+$,K$^+$-ATPase that broadens the possibilities for a quantitative analysis of our results by displaying in more detail the steps of binding of K$^+$ to the phosphoenzyme (FIG. 1). The scheme shows explicitly the binding of a single K$^+$ to E_2P, with the formation of E_2P-K, which dephosphorylates to a hypothetical intermediate X. This is rarely included in kinetic schemes (but see Ref. 1) although, unless there was a strong positive interaction between the binding of the first and the second K$^+$, the state E_2P-K should be present in significant amounts at low [K$^+$], and its kinetic properties would eventually be reflected on the rates of both dephosphorylation and K$^+$ occlusion. In the scheme, all three states E_2P, E_2P-K, and E_2P-K$_2$ participate in the dephosphorylation reaction, but only E_2P-K$_2$ produces occluded K$^+$, with a stoichiometry of 2 K$^+$ per enzyme unit. According to this, the rate of dephosphorylation and of K$^+$ occlusion should be, respectively:

$$v_{(Dephosphorylation)} = [E_2P] \, k_{deph.0} + [E_2P{-}K] \, k_{deph.1} + [E_2P{-}K_2] \, k_{deph.2}$$

$$v_{(Occlusion)} = 2 \, [E_2P{-}K_2] \, k_{deph.2}$$

Calculations based on the scheme in FIGURE 1 predict that, assuming rapid equilibrium for the binding and release of K$^+$ to the phosphoenzyme, addition of K$^+$ in very low concentrations to a medium where the enzyme is performing steady-state Na$^+$-ATPase activity will either decrease, leave unchanged, or increase the initial rate of dephosphorylation, depending on whether the value of $k_{deph.1}$ is lower than, equal to, or higher than that of $k_{deph.0}$. Considering K$^+$ occlusion under identical conditions, the initial rate should not increase at very low concentrations of K$^+$, but only at concentrations of the cation at which the intermediate E_2P-K$_2$ starts to be formed in significant amounts.

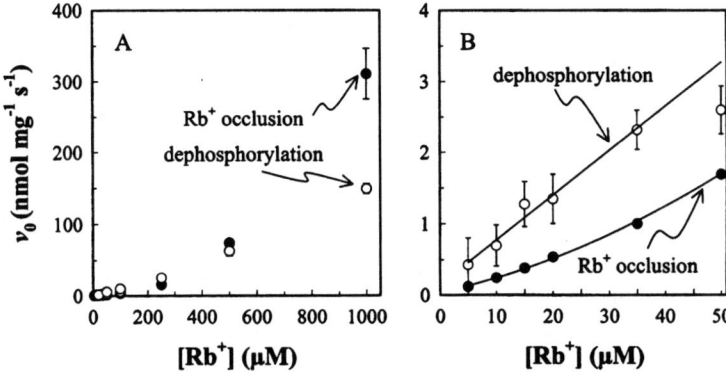

FIGURE 2. The initial rates of dephosphorylation and of Rb^+ occlusion as a function of the concentration of Rb^+.

MATERIALS AND METHODS

Na^+,K^+-ATPase

Purified preparations from pig kidney, following Jensen el al.,[2] were kindly provided by the Department of Biophysics, University of Aarhus, Denmark. All reactions took place at 25°C.

Time Courses of Dephosphorylation and Occlusion

Reactions were started by adding Rb^+ in different concentrations (see FIG. 2) to an enzyme suspension performing steady-state Na^+-ATPase activity in media with 150 mM NaCl, 10 μM ATP, 0.7 mM $MgCl_2$, 0.2 mM EDTA, and Imidazole-HCl 25 mM, pH = 7.4 at 25°C. We used $[\gamma^{32}P]ATP$ or $[^{86}Rb]Rb^+$ to measure the time courses of dephosphorylation or occlusion, respectively. Phosphorylated enzyme and occluded Rb^+ were measured as described by Schwarzbaum et al.[3] or by Rossi et al.,[4] respectively.

RESULTS AND DISCUSSION

In order to calculate the initial rates of dephosphorylation and of Rb^+ occlusion, we performed experiments measuring the time courses of breakdown of EP and of the formation of occluded Rb^+ in parallel experiments, using media of the same composition and temperature. Time courses were measured after addition of Rb^+ in different concentrations to a medium where the enzyme was performing Na^+-ATPase activity in steady state, as just described under MATERIALS AND METHODS.

From the absolute value of the initial slope of the time courses we calculated the initial rates (v_0) of dephosphorylation and of Rb^+ occlusion. The results of these initial rates are plotted as a function of $[Rb^+]$ in FIGURE 2A and 2B. It can be seen that both the rate of dephosphorylation and of Rb^+ occlusion increased along curves that tended to bend upwards as [Rb$^+$] increased (FIG. 2A), probably signifying the initial part of sigmoid curves. Also, for [Rb$^+$] equal to 250 µM or lower, the values for the initial rate of dephosphorylation were higher than those for Rb$^+$ occlusion, a picture that was reversed for Rb$^+$ concentrations equal or above 500 µM. At 1000 µM Rb$^+$, the initial rate of occlusion was almost twice as high as that for dephosphorylation, which is expected if a stoichiometry of 2 Rb$^+$ occluded per dephosphorylated unit held at nonlimiting Rb$^+$ concentrations. In FIGURE 2B, we plotted part of the results up to 50 µM Rb$^+$, which shows that the values of v_0 for dephosphorylation tended to increase with addition of Rb$^+$, even at the smallest concentrations tested. This indicates that dephosphorylation of E_2P-Rb must be faster than that of E_2P, and therefore that $k_{deph.1}$ must have a higher value than $k_{deph.0}$. A precise value for $k_{deph.1}$ is difficult to assess, but from calculations based on simulations fitting an important amount of data, we found that this value could lie within a range between 20 and 50 s^{-1}.[1]

Regarding the initial rate of Rb$^+$ occlusion, the values are clearly below those of dephosphorylation, and the slope for [Rb$^+$] tending to zero of the curve is significantly lower than that observed for dephosphorylation. If the intermediate X in the scheme in Figure 1 contained Rb$^+$, with a stoichiometry of 1 Rb$^+$ per enzyme unit, then at very low concentrations of Rb$^+$, the values of the initial rate of formation of occluded Rb$^+$ should equal those of the initial rate of dephosphorylation. As this is not the case, the obtained results indicate that E_2P-Rb does not lead to Rb$^+$ occlusion.

The significance of the effect of binding of a single Rb$^+$ (K$^+$) to the phosphoenzyme must still be evaluated, but it should be taken into account when considering a possible transport of a single K$^+$ by the Na$^+$,K$^+$-ATPase at low concentrations of the cation. It seems worthwhile to mention that, under the conditions of our experiments, the binding of a single Rb$^+$ to the phosphoenzyme could occur by replacement of one Na$^+$, probably bound to external transport sites of the enzyme.

ACKNOWLEDGMENT

This work was supported by grants from Fundación Antorchas, Agencia Nacional de Promoción Científica y Tecnológica, Consejo Nacional de Investigaciones Científicas y Técnicas, and Universidad de Buenos Aires, Argentina.

REFERENCES

1. KAUFMAN, S.B. *et al.* 1999. Are the states that occlude rubidium obligatory intermediates of the Na$^+$/K$^+$-ATPase reaction? J. Biol. Chem. **274:** 20779–20790.
2. JENSEN J., J.G. NØRBY & P. OTTOLENGHI. 1984. Binding of sodium and potassium to the sodium pump of pig kidney evaluated from nucleotide-binding behaviour. J. Physiol. (London) **346:** 219–241.
3. SCHWARZBAUM, P.J., S.B. KAUFMAN, R.C. ROSSI & P.J. GARRAHAN. 1995. An unexpected effect of ATP on the ratio between activity and phosphoenzyme level of Na$^+$/K$^+$-ATPase in steady state. Biochim. Biophys. Acta **1233:** 33–40.
4. ROSSI, R.C. *et al.* 1999. An attachment for non-destructive, fast quenching of samples in rapid mixing experiments. Anal. Biochem. **270:** 276–285.

Kinetic Investigations of the Mechanism of the Rate-Determining Step of the Na$^+$,K$^+$-ATPase Pump Cycle

RONALD J. CLARKE,[a] PAUL A. HUMPHREY,[a] CHRISTIAN LÜPFERT,[b] HANS-JÜRGEN APELL,[c] AND FLEMMING CORNELIUS[d]

[a] *School of Chemistry, University of Sydney, Sydney, NSW 2006, Australia*

[b] *Department of Biophysical Chemistry, Max-Planck-Institut für Biophysik, D-60596 Frankfurt am Main, Germany*

[c] *Department of Biology, University of Konstanz, D-78435 Konstanz, Germany*

[d] *Department of Biophysics, University of Aarhus, Ole Worms Allé 185, DK-8000 Aarhus C, Denmark*

ABSTRACT: The kinetics of the $E_2 \to E_1$ conformational change of unphosphorylated Na$^+$,K$^+$-ATPase from rabbit kidney were investigated via the stopped-flow technique using the fluorescent label RH421 (pH 7.4, 24°C). The enzyme was preequilibrated in a solution containing 25 mM histidine and 0.1 mM EDTA to initially stabilize the E_2 conformation. On mixing enzyme with NaCl alone, tris-ATP alone, or NaCl and tris-ATP simultaneously, a fluorescence decrease was observed. The reciprocal relaxation time, $1/\tau$, of the fluorescence transient was found to increase with increasing NaCl concentration and reached a saturating value in the presence of 1 mM tris-ATP of 54 ($\pm$3) s^{-1}. The experimental behavior could be described by a binding of Na$^+$ to the enzyme in the E_2 state with a dissociation constant of 31 ($\pm$7) mM, which induces a subsequent rate-limiting conformational change to the E_1 state. Similar behavior, but with a decreased saturating value of $1/\tau$, was found when NaCl was replaced by choline chloride. Experiments performed with enzyme from shark rectal gland showed similar effects, but with a lower amplitude of the fluorescence change and a higher saturating value of $1/\tau$ for both the NaCl and choline chloride titrations. The results suggest that Na$^+$ ions or salt in general play a regulatory role, similar to ATP, in enhancing the rate of the rate-limiting $E_2 \to E_1$ conformational transition by interaction with the E_2 state.

KEYWORDS: stopped-flow; fluorescence; voltage-sensitive dye; rate constant; regulation

The kinetics of the $E_2 \to E_1$ conformational change of unphosphorylated Na$^+$,K$^+$-ATPase from rabbit kidney and shark rectal gland were investigated[1] via the stopped-flow technique using the fluorescent label RH421 (pH 7.4, 24°C). This

Address for correspondence: Ronald J. Clarke, School of Chemistry, University of Sydney, Sydney, NSW 2006, Australia. Voice: +61 2 9351 4406; fax: +61 2 9351 3329.
r.clarke@chem.usyd.edu.au

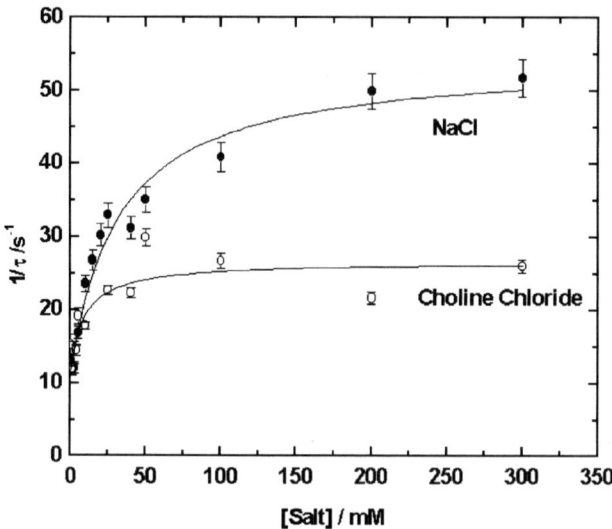

FIGURE 1. Effect of varying salt concentration, that is, NaCl (*filled circles*) and choline chloride (*open circles*), on the reciprocal relaxation time ($1/\tau$) of RH421 fluorescence transients of rabbit kidney Na^+,K^+-ATPase induced by mixing simultaneously with salt plus 2 mM of tris-ATP; excitation wavelength = 577 nm, emission wavelength ≥ 665 nm. The *solid lines* represent nonlinear least-square fits of equation 1 to the data.

method has the advantage over other techniques that the probe RH421 does not interfere with the ATP binding site, so that measurements can be made with high time resolution in very close to physiological conditions. The enzyme was preequilibrated in a solution containing 25 mM histidine and 0.1 mM EDTA to stabilize initially the E_2 conformation. On mixing rabbit kidney enzyme with NaCl alone, tris-ATP alone, or NaCl and tris-ATP simultaneously, a fluorescence decrease was observed. The reciprocal relaxation time, $1/\tau$, of the fluorescent transient was found to increase with increasing NaCl concentration and reached a saturating value in the presence of 1 mM tris-ATP of 54 ± 3 s^{-1} in the case of rabbit kidney enzyme (see FIG. 1). Similar behavior, but with a decreased saturating value of $1/\tau$ was found when NaCl was replaced by choline chloride. Analogous experiments performed with enzyme from shark rectal gland showed similar effects, but with a significantly lower amplitude of the fluorescence change and a higher saturating value of $1/\tau$ for both the NaCl and choline chloride titrations.

The observed increase in $1/\tau$ with increasing NaCl or choline chloride implies that Na^+, choline$^+$, or salt bind to the enzyme in the E_2 state and stimulate the $E_2 \rightarrow E_1$ conformational change. Similar behavior has previously been observed with ATP,[2] which is thought to bind to a low-affinity regulatory site and stimulate the $E_2 \rightarrow E_1$ transition. The effects of both Na^+ and ATP on the conformational transition can be summarized in the generalized reaction scheme shown in FIGURE 2. Under conditions of saturating ATP concentrations, that is, the experimental condition for

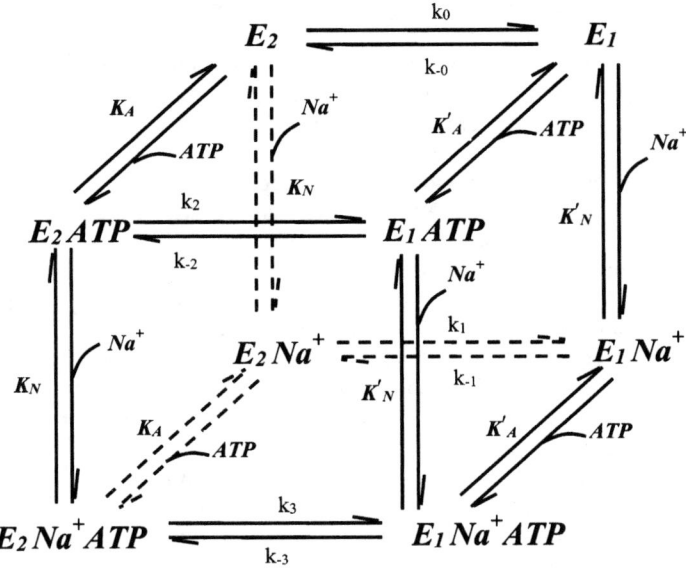

FIGURE 2. Generalized reaction scheme for the conformational change, Na⁺ binding and ATP binding of unphosphorylated Na⁺,K⁺-ATPase.

the results shown in FIGURE 1, this reaction scheme predicts the following dependence of $1/\tau$ on the Na⁺ concentration:

$$\frac{1}{\tau} \approx \frac{k_2 + k_{-2} + (k_3 + k_{-3})[\text{Na}^+]/K}{1 + ([\text{Na}^+]/K)} \quad (1)$$

where K is the dissociation constant of the enzyme for Na⁺ ions, k_2 and k_{-2} are the forward and backward rate constants of the transition $E_2 \cdot \text{ATP} \leftrightarrow E_1 \cdot \text{ATP}$, and k_3 and k_{-3} are the forward and backward rate constants of the transition $E_2 \cdot \text{ATP} \cdot \text{Na}^+ \leftrightarrow E_1 \cdot \text{ATP} \cdot \text{Na}^+$. Fitting of the NaCl titration data shown in FIGURE 1 to this equation yields the values $k_2 + k_{-2} \approx 11 \pm 1 \text{ s}^{-1}$, $k_3 + k_{-3} \approx 54 \pm 3 \text{ s}^{-1}$, and $K \approx 31 \pm 7$ mM. Similarly, replacing [Na⁺] by [choline⁺] in the preceding equation and fitting the choline chloride titration data yields $k_2 + k_{-2} \approx 13 \pm 2 \text{ s}^{-1}$, $k_3 + k_{-3} \approx 27 \pm 2 \text{ s}^{-1}$, and $K \approx 9 \pm 5$ mM.

The conclusion that Na⁺ can bind to the E_2 state would at first sight seem to be contrary to the generally accepted view that Na⁺ ions bind to transport sites of the enzyme in the E_1 state of the enzyme. Forbush[3] and Hasenauer et al.[4] also found, however, from rapid filtration studies that Na⁺ ions increased the rate of radioactive K⁺ or Rb⁺ release by enhancing the rate of the $E_2 \rightarrow E_1$ transition. From their measurements they concluded that other cation sites can be occupied at the same time as the K⁺ transport sites. We propose, therefore, that Na⁺ ions or salt in general play a regulatory role, similar to low-affinity ATP binding, in enhancing the rate of the rate-limiting $E_2 \to E_1$ conformational transition by interaction with the E_2 state.

Since the experiments reported here were carried out on open membrane fragments with both sides of the enzyme accessible to Na^+ ions, it is not possible from the experiments described here alone to decide whether the Na^+ ions are binding to the enzyme from the cytoplasmic or the extracellular face. Experiments of van der Hijden and de Pont[5] on rabbit kidney enzyme reconstituted into lipid vesicles, however, showed that extracellular Na^+ ions increase the steady-state phosphorylation level of the enzyme, most likely by inducing a transition of the enzyme to the E_1 conformation, from which it can be phosphorylated. Based on these results, it therefore appears very likely that the Na^+-induced stimulation of the $E_2 \to E_1$ transition found here is in fact due to Na^+ binding to regulatory extracellular sites.

Whether or not such extracellular Na^+ binding plays an important regulatory role under physiological conditions depends on the extracellular Na^+ concentration. According to electron microprobe results of Thurau,[6] the Na^+ concentration in the extracellular fluid of kidney tubule cells is approximately 160 mM. Comparison with the results shown in FIGURE 1 shows that at this concentration the Na^+ effect on the reciprocal relaxation time has virtually reached saturation. Variations in the extracellular Na^+ concentration over a fairly wide range therefore would be expected to have a relatively minor influence on the rate of the $E_2 \to E_1$ transition. Thus, it seems unlikely that Na^+-induced stimulation of the $E_2 \to E_1$ transition would play an important regulatory role *in vivo*. The effect could be seen rather as an evolutionary optimization of enzyme operation under the normal physiological conditions. The normally high extracellular Na^+ or salt concentration can be thought of as facilitating the spontaneous relaxation of the enzyme back into the E_1 state following its dephosphorylation, so that it is ready once again to become phosphorylated by ATP and transport Na^+ ions.

REFERENCES

1. HUMPHREY, P.A. *et al.* 2002. Mechanism of the rate-determining step of the Na^+,K^+-ATPase pump cycle. Biochemistry **41**: 9496–9507.
2. CLARKE, R.J. *et al.* 1998. Kinetics of Na^+-dependent conformational changes of rabbit kidney Na^+,K^+-ATPase. Biophys. J. **75**: 1340–1353.
3. FORBUSH, B., III. 1987. Rapid release of ^{42}K and ^{86}Rb from an occluded state of the Na,K-pump in the presence of ATP or ADP. J. Biol. Chem. **262**: 11104–11115.
4. HASENAUER, J. *et al.* 1993. Allosteric regulation of the access channels to the Rb^+ occlusion sites of $(Na^+ + K^+)$-ATPase. J. Biol. Chem. **268**: 3289–3297.
5. VAN DER HIJDEN, H.T.W.M. *et al.* 1989. Cation sidedness in the phosphorylation step of Na^+/K^+-ATPase. Biochim. Biophys. Acta **983**: 142–152.
6. THURAU, K. 1979. Nephrology—A look into the future. Kidney Int. **15**: 1–6.

Homology Modeling of Na,K-ATPase

A Putative Third Sodium Binding Site Suggests a Relay Mechanism Compatible with the Electrogenic Profile of Na^+ Translocation

K. O. HÅKANSSON AND P. L. JORGENSEN

Biomembrane Center, August Krogh Institute, University of Copenhagen, 2100 Copenhagen OE, Denmark

ABSTRACT: Identification of the third Na^+ binding site would be crucial in interpretation of the electrophysiological behavior of Na,K-ATPase. To address this question a three-dimensional homology model of Na,K-ATPase was built from the known crystallographic structure of Ca-ATPase (1EUL). Phe760, which is conserved in virtually all Ca-ATPases, is replaced by Ser768 in Na,K-ATPase, resulting in a small cavity between M4, M5, and M6. A partially hydrated Na^+ ion can be bound at this third site on the cytoplasmic side of cation binding sites 1 and 2. This leads to the proposal that the conductance of the "third Na^{+}" ion across ~70% of the membrane dielectric may be achieved by adding up the passage of one Na^+ ion from the described cytoplasmic cavity to cation site 1 and the further conductance of the previously bound Na^+ ion from cation site 1 to the extracellular phase. This relay mechanism may therefore be compatible with the electrogenic profile of Na^+ translocation.

KEYWORDS: Na,K-ATPase; Na^+ binding; electrogenic transport; homology modeling

INTRODUCTION

The members of the P-type ATPase family differ not only in their cation specificities, but also in the stoichiometry and mechanistic details of ion transfer. Na,K-ATPase exports three Na^+ ions from the cytoplasm for each hydrolyzed ATP molecule, while the transport cycle of Ca-ATPase involves displacement of only two Ca^{2+} ions from cytoplasm to the luminal side of the membrane. Two of the Na^+ binding sites in Na,K-ATPase can be assumed to be homologous with the two Ca^{2+} binding sites in Ca-ATPase, but the position of the third Na^+ binding site has not been elucidated. In the cation transport cycle of Na,K-ATPase, the uptake and release of the three Na^+ ions are associated with varying degrees of charge transfer. Binding of the third Na^+ ion from the cytoplasmic side and release of the first Na^+ ion on the extra-

Address for correspondence: K.O. Håkansson, Biomembrane Center, August Krogh Institute, University of Copenhagen, Universitetsparken 13, 2100 Copenhagen OE, Denmark. Voice: +45-3532-1677; fax: +45-3532-1567.
kohakansson@aki.ku.dk

Ann. N.Y. Acad. Sci. 986: 163–167 (2003). © 2003 New York Academy of Sciences.

cellular side are more electrogenic than the other steps in the transport cycle. Identification of the third Na$^+$ binding site would be crucial in interpretation of the electrophysiological behavior of Na,K-ATPase. We addressed this question by building a three-dimensional homology model of Na,K-ATPase from the known crystallographic structure of Ca-ATPase (1 EUL),[1] followed by energy minimization with CNS.[2]

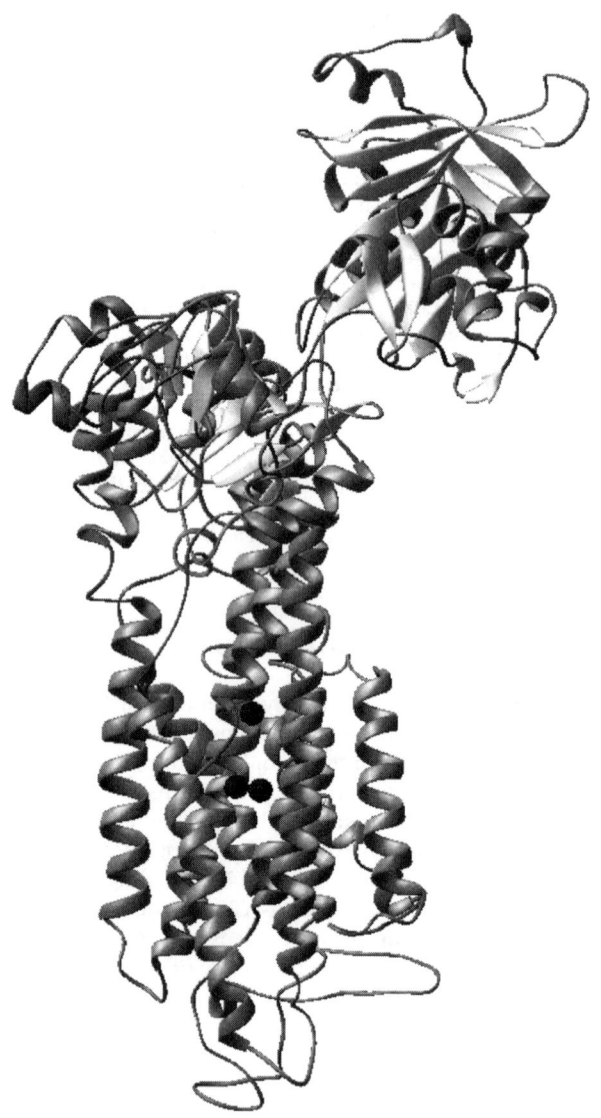

FIGURE 1. The position of the third putative Na binding site is shown above the other two sites in the overall structure of the Na,K-ATPase model. All Na$^+$ ions are shown as spheres.

PROPOSED SODIUM BINDING SITE

There is an interesting difference between the two proteins in the transmembrane region where metal transport is thought to take place. Phe760, which is conserved in virtually all Ca-ATPases is replaced by Ser768 in Na,K-ATPase, resulting in a small cavity between M4, M5, and M6.

A Na^+ ion can be docked into this site, 9–10 Å to the cytoplasmic side of the other two Na^+ ions, as shown in FIGURES 1 and 2. The modeled Na^+ ion is coordinated by the side-chain and main-chain oxygens of Ser768 and three water molecules, and makes van der Waals contacts with Thr772. The main- and side-chain oxygens of the serine residue and the three water molecules can be built with ideal octahedral symmetry and metal to ligand distances of 2.3 Å. The distance to the sixth ligand, Thr772, is short (2.9 Å to Cγ2 and 3.6 Å to the Oγ1) for a van der Waals distance, but a direct coordination to the Oγ1 was abandoned, since energy minimization returned the original conformation. In contrast to the first two sites, which are nega-

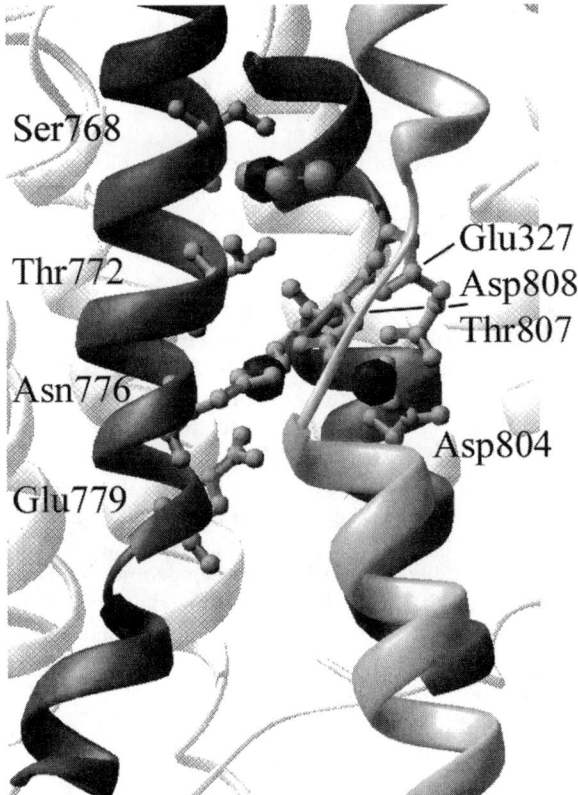

FIGURE 2. Closer view showing the coordinating side chains. Hydration water in the third site is shown as spheres.

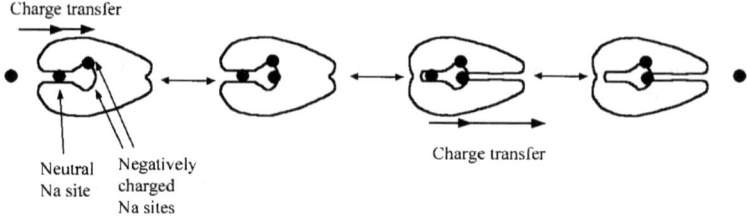

FIGURE 3. Mechanistic model for Na$^+$ translocation where the two highly electrogenic steps proceed through a relay mechanism, and the total charge transfer reflects the sum of the simultaneous movements of two Na$^+$ ions.

tively charged, the third site is neutral, which is in agreement with kinetic and electric studies.[3,4]

ELECTROPHYSIOLOGICAL IMPLICATIONS

Based on the structural observations in the homology model and the available electrophysiological observations, we propose a novel mechanism for Na$^+$ transfer through the Na-K-ATPase, as illustrated in FIGURE 2. In this scheme, a partially hydrated Na$^+$ ion is bound at the third site on the cytoplasmic side of cation binding sites 1 and 2. This Na$^+$ ion is replaced and displaced to site 1 by the last Na$^+$ ion to enter, in a relay or bucket-brigade mechanism. The resulting charge translocation corresponds to the sum of the paths traveled by both ions.

During the conformational changes that take place during and after phosphate transfer from ATP to protein, the third site is lost and the Na$^+$ ion is dehydrated and translocated toward cation site 1. This will result in mutual repulsion and allow conduction again to occur through a relay mechanism where the Na$^+$ ion in cation site 1 is translocated to the extracellular side of the membrane and replaced with the Na$^+$ ion from the third, putative Na$^+$ binding site. In the three-dimensional structure model of the α-subunit of Na,K-ATPase, the total ion conduction associated with the release of the first Na$^+$ ion is the sum of the movement of the third Na$^+$ ion from the third site to cation site 1 plus the conduction of a Na$^+$ ion from cation site 1 to the extracellular phase. Since the electric field through a medium of mixed chemical composition is not uniform, the relationship between geometric distance through the channel and current will not be linear. However, since the distance traveled by these two ions together will correspond to 70–80% of the hydrophobic bilayer thickness, this step is likely to be associated with a substantial current. The demonstration of a cavity suitable for partial dehydration and binding of a Na$^+$ ion therefore leads to the proposal that the conductance of the "third Na$^+$" ion across ~70% of the membrane dielectric[5,6] can be achieved by adding up the passage of one Na$^+$ ion from the described cytoplasmic cavity to cation site 1 and the further conductance of the previously bound Na$^+$ ion from cation site 1 to the extracellular phase. The release of any of the remaining two ions will result in a much smaller electrogenic signal. In contrast, earlier mechanistic models of Na,K-ATPase ion transport have explained the electrogenic profile through fluctuations between a narrow channel (high current) and a widely open aqueous cavity (low current).[6] Such movements seem highly unlikely in view of the high-resolution structure of the E_1 form and the available models of the E_2 form of Ca-ATPase.[1,7]

ACKNOWLEDGMENT

We thank the Carlsberg Foundation and the Danish Research Council for financial support.

REFERENCES

1. TOYOSHIMA, C. et al. 2000. Crystal structure of the calcium pump of sarcoplasmic reticulum at 2.6 A resolution. Nature **405:** 647–655.
2. BRÜNGER, A.T. et al. 1998. Crystallography and NMR system: a new software suite for macromolecular structure determination. Acta Crystallogr. **D54:** 905–921.
3. HEYSE, S. et al. 1994. Partial reactions of the Na,K-ATPase: determination of rate constants. J. Gen. Physiol. **104:** 197–240.
4. OR, E., R. GOLDSHLEGER & S.J.D. KARLISH. 1996. An effect of voltage on binding of Na^+ at the cytoplasmic surface of the Na(+)-K^+ pump. J. Biol. Chem. **271:** 2470–2477.
5. HOLMGREN, M. et al. 2000. Three distinct and sequential steps in the release of sodium ions by the Na^+/K^+-ATPase. Nature **403:** 898–901.
6. APELL, H.J. & S.J. KARLISH. 2001. Functional properties of Na,K-ATPase, and their structural implications, as detected with biophysical techniques. J. Membr. Biol. **180:** 1–9.
7. XU C. et al. 2002. A structural model for the catalytic cycle of Ca(2+)-ATPase J. Mol. Biol. **316:** 201–211.

Use of a Fluorescent Maleimide to Probe Structure–Function Relationships in Stalk Segments 4 and 5 of the Yeast Plasma-Membrane H^+-ATPase

CAROLYN W. SLAYMAN, MANUEL MIRANDA, JUAN PABLO PARDO,[a] AND KENNETH E. ALLEN

Departments of Genetics, and Cellular & Molecular Physiology, Yale University School of Medicine, New Haven, Connecticut 06510, USA

ABSTRACT: In the yeast plasma-membrane H^+-ATPase and other P-type ATPases, conformational changes are transmitted between cytoplasmic and membrane-embedded domains via a stalk region composed of cytoplasmic extensions of membrane segments 2, 3, 4, and 5. The present study has used a fluorescent maleimide (Alexa-488) to probe Cys residues introduced into stalk segments 4 and 5 of the yeast enzyme. In the case of S5, Cys substitutions along one face led to a constitutive, 5- to 10-fold activation of the ATPase in the absence of glucose. Based on homology with SERCA Ca^{2+}-ATPase, this face is likely to be buried in the interior of the protein, close to the P domain. Three Cys residues on the opposite face of S5 (A668C, S672C, and D676C) were accessible to Alexa-488 under all conditions tested. In addition, three other Cys residues at or near the boundary between the two faces reacted with Alexa-488 only (V665C, L678C) or preferentially (Y689C) in plasma membranes from glucose-metabolizing cells; this result provides the first direct evidence for a change in conformation of S5 during glucose activation. For stalk segment 4, site-directed mutagenesis gave no sign of a role in glucose-dependent regulation. Rather, substitutions at 13 consecutive positions along S4 caused kinetic changes consistent with a shift in equilibrium from E_2 to E_1. Four Cys residues along this stretch of S4 (Q357C, K362C, S364C, and S368C) reacted with Alexa-488, indicating that they are exposed to the aqueous medium as predicted in the SERCA-based structural model.

KEYWORDS: H^+-ATPase; yeast; stalk domain; maleimides; vanadate resistance; glucose regulation

Address for correspondence: Carolyn W. Slayman, Departments of Genetics, and Cellular & Molecular Physiology, Yale University School of Medicine, 333 Cedar Street, New Haven, CT 06510. Voice: 203-737-1770; fax: 203-737-1771.
carolyn.slayman@yale.edu
[a]Current address: Universidad Nacional Autonoma de Mexico, Departamento de Bioquimica, Facultad de Medicina, Mexico, D.F., 04510 Mexico.

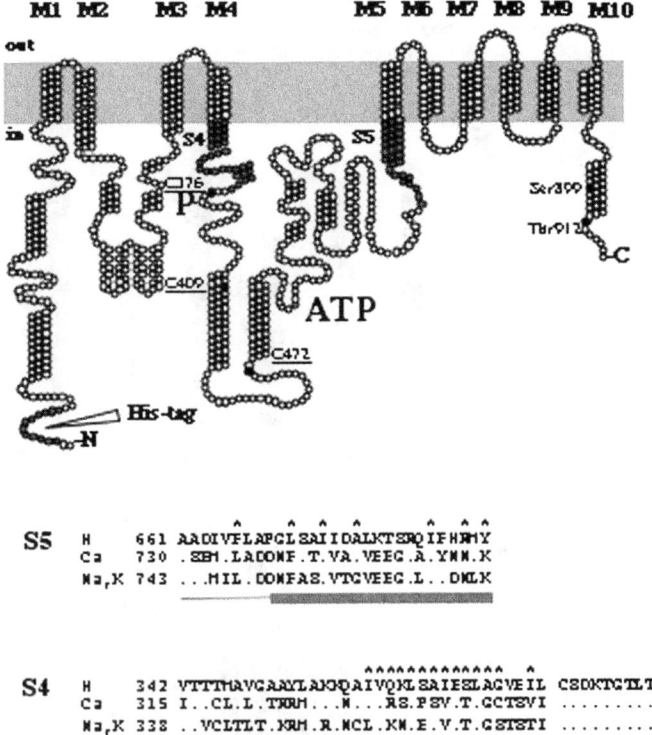

FIGURE 1. Topological diagram of the yeast H^+-ATPase (**top**) and alignment of stalk segments 4 and 5 of the yeast H^+-ATPase, SERCA Ca^{2+}-ATPase, and Na^+,K^+-ATPase (**bottom**). In the topological diagram, stalk segments 4 and 5 and an N-terminal 10 histidine tag are highlighted in *gray*; three Cys residues present in the modified 3C version of the ATPase (C376, C409, and C472) are represented in *black*, as are two C-terminal residues (S899 and T912), thought to be involved in glucose regulation. In the alignment, residues identical to the yeast Pma1 sequence are indicated by a *period*, and residues in Pma1 at which substitutions led to vanadate resistance or altered glucose regulation are labeled by an *asterisk*. *Gray solid rectangles* represent α-helical regions, as seen in the structure of Ca^{2+}-ATPase.[11]

INTRODUCTION

The yeast plasma-membrane H^+-ATPase, which is encoded by the *PMA1* gene,[1] pumps protons out of the cell and supports the H^+-coupled uptake of amino acids, sugars, and inorganic ions. It is an abundant constituent of the surface membrane, accounting for ca. 10% of membrane protein and hydrolyzing 20% or more of total cellular ATP.[2] Recent work in our laboratory has focused on stalk segments 4 and 5 of the H^+-ATPase, which seem likely to play a central role in transmitting conformational changes between the cytoplasmic and membrane-embedded domains (FIG. 1). Both S4 and S5 display significant sequence conservation with other P-type ATPases, including the Na^+,K^+- and Ca^{2+}-ATPases of animal cells (FIG. 2).

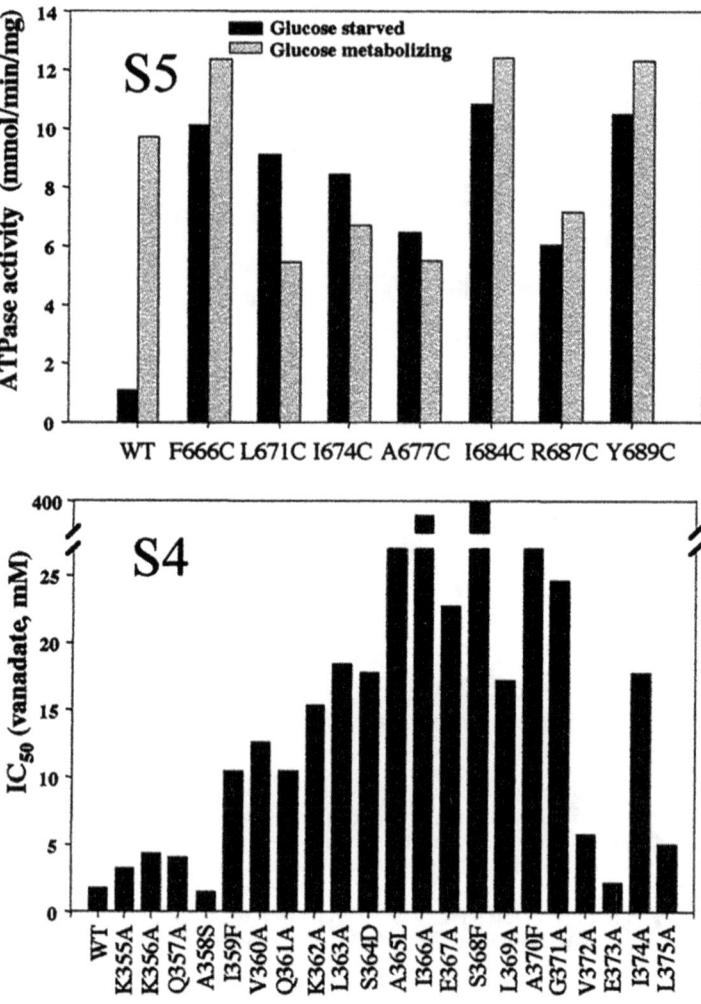

FIGURE 2. Effect of mutations along stalk segments 4 and 5 of the yeast H^+-ATPase. (**Top panel**) Mutations in S5 were integrated into the chromosomal copy of the *PMA1* gene and expressed at the plasma membrane. Purified plasma membranes from glucose-starved (GS) and glucose-metabolizing cells (GM) were then assayed for ATP hydrolysis. (**Bottom panel**) ATPases with mutations in S4 were expressed in secretory vesicles[13] and assayed for ATP hydrolysis in the presence of increasing concentrations of vanadate; the resulting IC_{50} values were plotted as a function of the linear sequence along S4. The IC_{50} value for S368F was obtained from a previous study.[14]

In the present study, the fluorescent maleimide Alexa-488 has been used to probe the accessibility and reactivity of cysteines introduced into both stalk segments. Because the Alexa dye is membrane-impermeant,[3] the results serve to define residues in both segments that are exposed to the aqueous medium. They also give direct evidence for a change in conformation of stalk segment 5 during activation of the H⁺-ATPase by glucose.

METHODS

Two strains of *Saccharomyces cerevisiae* were used in this study: SY4, to express mutant forms of the H⁺-ATPase in secretory vesicles, and NY13, to express them at the plasma membrane.[4] For labeling studies, Cys substitutions were introduced into a version of the *PMA*1 gene that contained only three of the nine native cysteines (C376, C409, and C472) in order to reduce background labeling by Alexa-488; a 10-His tag was included at the N terminus to allow purification of the ATPase by Ni-NTA chromatography (Miranda *et al.*, submitted for publication). Cells were grown and membrane fractions were isolated as described previously.[4] To determine the ability of introduced Cys residues to react with Alexa-488 (Alexa-Fluor-488 C_5 maleimide sodium salt; Molecular Probes), secretory vesicles were suspended at a protein concentration of 1 mg/mL in 0.8 M sorbitol, 1 mM EDTA, 10 mM TEA, pH 7.2, and plasma membranes, at 0.5 mg/mL in 50 mM HEPES, pH 7.0. In both cases, Alexa-488 was added from a fresh stock solution to a final concentration of 1 mM; the membrane suspension was incubated at 30°C for the desired time; and the reaction was stopped by dilution into ATPase assay mixture (50 mM MES/tris, pH 6.25, 5 mM KN_3, 5 mM Na_2ATP, 10 mM $MgCl_2$, 5 mM phosphoenolpyruvate, 50 μg/mL pyruvate kinase, 1 mM β-mercaptoethanol) or purification mixture (5 mM HEPES, pH 7.5, 10% glycerol, 20 mM β-mercaptoethanol). ATPase activity was assayed as described previously.[5] Alternatively, after purification of the ATPase by Ni-NTA chromatography,[6] labeling of the 100-kDa polypeptide was analyzed by SDS-polyacrylamide gel electrophoresis and fluorography.

RESULTS

Role of Stalk Segment 5 in the Glucose-Dependent Regulation of H⁺-Atpase Activity

The yeast plasma-membrane H⁺-ATPase has been known for many years to be regulated by glucose. In carbon-starved cells, ATPase activity falls by a factor of 5 to 10, and when glucose is added back to the medium, the activity rebounds rapidly to the starting level.[7] Several lines of evidence have implicated kinase-mediated phosphorylation of the C terminus in this response to glucose: (1) the speed of the response, which is complete within several minutes; (2) the fact that glucose activation is prevented by mutations of two or more Ser/Thr residues near the C-terminus;[8] and (3) the discovery, in thermolysin digests of purified ATPase, of two as yet unidentified phosphopeptides that decrease in amount during carbon starvation and increase upon readdition of glucose.[9]

If, as seems likely, the dephosphoryated C terminus exerts an autoinhibitory effect on H⁺-ATPase activity,[10] one would like to know the molecular mechanism by which the inhibition takes place. A partial answer came recently from a scanning mutagenesis study of stalk segment 5,[4] in which Cys substitutions along one face of

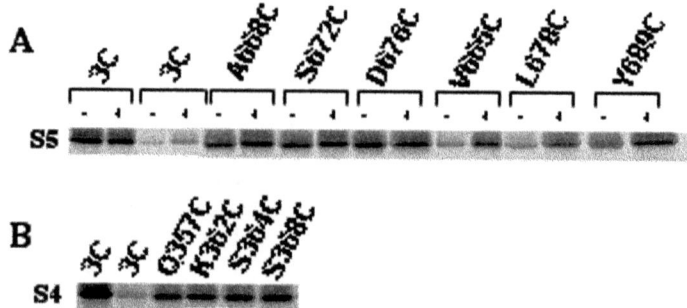

FIGURE 3. Fluorescent labeling of the H$^+$-ATPase by Alexa-488. (**A**) Plasma membranes were purified from glucose-starved and glucose-metabolizing cells and labeled by Alexa-488 for 8 min (3C control, D676C, and L678C) or 20 min (V665C, S672C, Y689C). (**B**) Secretory vesicles containing 3C, Q357C, K362C, S364C, or S368C-ATPases were labeled by Alexa-488 for 20 min.

the S5 α-helix led to a constitutive, 5- to 10-fold activation of the ATPase in the absence of glucose (FIG. 3A). When the H$^+$-ATPase was modeled using the three-dimensional structure of SERCA1 Ca^{2+}-ATPase[11] as a template (see below), the "regulatory" face of S5 was seen to be buried in the interior of the protein, close to the mechanistically essential P domain.

To ask whether this part of the stalk region does in fact undergo a conformational change in the presence of glucose, a fluorescent maleimide (Alexa-488) has been used to probe the accessibility of Cys residues introduced along S5. Cys mutations were first transferred into a specially engineered form of the ATPase gene in which six of the nine native cysteines had been replaced by alanine; as hoped, the resulting "3C" ATPase was almost fully active and exhibited little or no background labeling with Alexa-488. Plasma membranes from glucose-starved (GS) or glucose-metabolizing (GM) cells were incubated with 1 mM Alexa dye, and the 100 kDa ATPase polypeptide was purified and analyzed for fluorescence.

Under these conditions, six cysteines (A673C, K679C, T680C, Q683C, H686C, M688C) failed to react with Alexa-488 during incubations of up to 20 min (not shown). Three cysteines reacted equally well in GS and GM membranes: D676C, which was fully labeled in 8 min, and A668C and S672C, which required 20 min (FIG. 4A). In a helical-wheel diagram (not shown), these three Cys residues are located opposite to the "regulatory" face of S5, and given the hydrophilic nature of Alexa-488, must be exposed to the aqueous medium. Indeed, the SERCA-based structural model of the yeast H$^+$-ATPase shows all three to be situated at the surface of the protein (FIG. 5A). Finally, and of greatest interest, three additional Cys residues were labeled only (V665C, L678C) or much more rapidly (Y689C) in plasma membranes from glucose-metabolizing cells (FIG. 4A). In the structural model, V665C and L678C are located immediately below the protein surface, while Y689C lies at the base of a deep cavity. Thus, the results point to a glucose-induced conformational change in S5 that enhances the accessibility of these three cysteines to the hydrophilic probe.

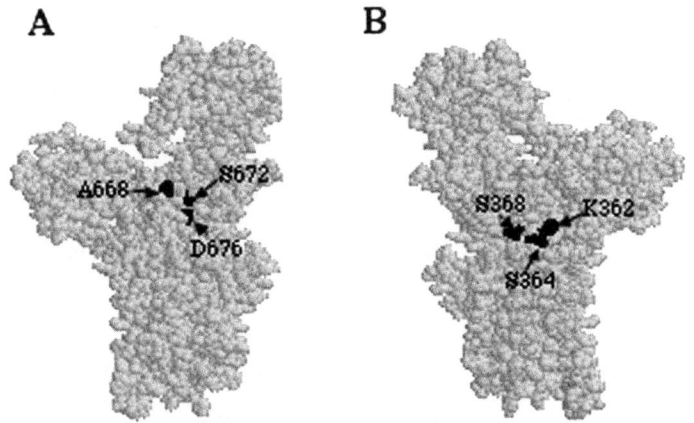

FIGURE 4. Structural model for the yeast H^+-ATPase, based on homology with SERCA Ca^{2+}-ATPase.[11] Residues labeled by Alexa-488 in stalk segment 5 (**A**) and stalk segment 4 (**B**) are highlighted in *black*.

Role of Stalk Segment 4 in the E_1–E_2 Conformational Change

By contrast with the results on S5, scanning mutagenesis has given no indication of a role for stalk segment 4 (S4) in glucose regulation. Instead, amino acid substitutions at 13 consecutive sites along S4 led to a 5- to 200-fold increase in the IC_{50} for inorganic orthovanadate (FIG. 3B), consistent with the idea that the mutant forms of the H^+-ATPase have undergone a shift in equilibrium from the vanadate-sensitive E_2 state toward the vanadate-resistant E_2 state.[12]

When Cys residues were introduced at eight positions along S4 and probed with Alexa-488, four of them (Q357C, K362C, S364C, and S368C) proved to be reactive under all conditions tested (FIG. 4B), indicating that they are exposed to the aqueous environment. As in the case of S5, these four residues are located at the surface of the H^+-ATPase in the SERCA-based structural model (FIG. 4B). Work is now in progress to ask whether the accessibility of these or other S4 cysteines can be modified by physiological ligands such as ATP, ADP, or inorganic vanadate.

CONCLUSIONS

The results of this study, along with previously published results on the yeast plasma-membrane H^+-ATPase[15] and SERCA Ca^{2+}-ATPase,[16,17] reveal that stalk segments 4 and 5 are indeed conformationally active regions, playing an important part in the reaction cycle and the regulation of ATPase activity. The fact that a fluorescent maleimide such as Alexa-488 can be introduced specifically into these regions provides a useful tool for future studies to look directly at the mechanistic role of S4 and S5.

ACKNOWLEDGMENTS

This work was supported by research grant GM15761 from the National Institute of General Medical Sciences. We are grateful to Tony Ambesi, Valery Petrov, Brett Mason, and Silvia Lecchi for helpful discussions.

REFERENCES

1. SERRANO, R., M.C. KIELLAND-BRANDT & G.R. FINK. 1986. Yeast plasma membrane ATPase is essential for growth and has homology with (Na^+-K^+), K^+ and Ca^{2+}-ATPases. Nature **319:** 689–693.
2. GRADMANN, D., U.P. HANSEN, W.S. LONG & C.L SLAYMAN. 1978. Current-voltage relationships for the plasma membrane and its principal electrogenic pump in *Neurospora crassa*: I. Steady-state conditions. J. Membr. Biol. **39:** 333–367.
3. CHIU, I., D.M. DAVIS & J.L. STROMINGER. 1999. Trafficking of spontaneously endocytosed MHC proteins. Proc. Natl. Acad. Sci. USA **96:** 13944–13949.
4. MIRANDA, M., K.E. ALLEN, J.P. PARDO & C.W. SLAYMAN. 2001. Stalk segment 5 of the yeast plasma membrane H^+-ATPase: mutational evidence for a role in glucose regulation. J. Biol. Chem. **276:** 22485–22490.
5. AMBESI, A., K.E. ALLEN & C.W. SLAYMAN. 1997. Isolation of transport-competent secretory vesicles from *Saccharomyces cerevisiae*. Anal. Biochem. **251:** 127–129.
6. JANKNECHT, R. *et al.* 1991. Rapid and efficient purification of native histidine-tagged protein expressed by recombinant vaccinia virus. Proc. Natl. Acad. Sci. USA **88:** 8972–8976.
7. SERRANO, R. 1983. *In vivo* glucose activation of the yeast plasma membrane ATPase. FEBS Lett. **156:** 11–14.
8. PORTILLO, F., P. ERASO & R. SERRANO. 1991. Analysis of the regulatory domain of yeast plasma membrane H^+-ATPase by directed mutagenesis and intragenic suppression. FEBS Lett. **287:** 71–74.
9. CHANG, A. & C.W. SLAYMAN. 1992. Maturation of the yeast plasma membrane H^+-ATPase involves phosphorylation during intracellular transport. J. Cell Biol. **115:** 289–295.
10. PORTILLO, F. 2000. Regulation of plasma membrane H^+-ATPase in fungi and plants. Biochim. Biophys. Acta **1469:** 31–42.
11. TOYOSHIMA, C., M. NAKASAKO, H. NOMURA & H. OGAWA. 2000. Crystal structure of the calcium pump of sarcoplasmic reticulum at 2.6 A resolution. Nature **405:** 647–655.
12. AMBESI, A., M. MIRANDA, K.E. ALLEN & C.W. SLAYMAN. 2000. Stalk segment 4 of the yeast plasma membrane H^+-ATPase. Mutational evidence for a role in the E1-E2 conformational change. J. Biol. Chem. **275:** 20545–20550.
13. NAKAMOTO, R.K., R. RAO & C.W. SLAYMAN. 1991. Expression of the yeast plasma membrane H^+-ATPase in secretory vesicles. J. Biol. Chem. **266:** 7940–7949.
14. HARRIS, S.L., D.S. PERLIN, D. SETO-YOUNG & J.E. HABER. 1991. Evidence for coupling between membrane and cytoplasmic domains of the yeast plasma membrane H^+-ATPase. J. Biol. Chem. **266:** 24439–24445.
15. SOTEROPOULOS, P., A. VALIAKHMETOV, R. KASHIWASAKI & D.S. PERLIN. 2001. Helical stalk segments S4 and S5 of the plasma membrane H^+-ATPase from *Saccharomyces cerevisiae* are optimized to impact catalytic site environment. J. Biol. Chem. **276:** 16265–16270.
16. SORENSEN, T. L.-M. & J.P. ANDERSEN. 2000. Importance of stalk segment S5 for intramolecular communication in the sarcoplasmic reticulum Ca^{2+}-ATPase. J. Biol. Chem. **275:** 28954–28961.
17. STROCK, C. *et al.* 1998. Direct demonstration of Ca^{2+} binding defects in sarco-endoplasmic reticulum Ca^{2+} ATPase mutants overexpressed in COS-1 cells transfected with adenovirus vectors. J. Biol. Chem. **273:** 15104–15109.

The E_1/E_2-Preference of Gastric H,K-ATPase Mutants

JAN JOEP H. H. M. DE PONT, HERMAN G. P. SWARTS,
PETER H. G. M. WILLEMS, AND JAN B. KOENDERINK

Department of Biochemistry, Nijmegen Center for Molecular Life Sciences, University of Nijmegen, Nijmegen, the Netherlands

ABSTRACT: Gastric H,K-ATPase has, in the absence of ATP and added ions, a preference for the E_2 conformation. Mutations in the cation-binding pocket often result in a preference for the E_1-conformation. This can be paralleled by the occurrence of K^+-*independent* ATPase activity. These two phenomena could be separated by combined mutagenesis of several residues in and around the cation-binding pocket. Models of the three-dimensional structure of H,K-ATPase visualize the relationship between the E_1/E_2 preference and the structure.

KEYWORDS: gastric H,K-ATPase; SCH 28080; vanadate; mutagenesis; Sf9 cells; baculovirus; conformational equilibrium.

INTRODUCTION

Since the establishment in 1985 of the primary structure of P-type ATPases,[1,2] the presence of negatively charged amino acid residues within transmembrane segments has received much attention. A role for these residues in cation binding and transport seemed likely, and was supported by mutagenesis studies of these residues[3] in SR Ca^{2+}-ATPase. The establishment of the crystal structure of this ion pump in the $E_1(Ca)_2$ conformation[4] provided clear evidence for this idea.

After we succeeded in functional expression of gastric H,K-ATPase, using the baculovirus expression system,[5] we first mutagenized the negatively charged amino acids present in and around transmembrane domains five and six of the catalytic subunit.[6] The most interesting mutant we obtained was the one in which Glu^{820}, present in transmembrane domain six, was converted into a Gln residue. This mutant lacked K^+-stimulated ATPase activity, but was still phosphorylated by ATP. Moreover, when the enzyme was preincubated with 10 mM KCl, the phosphorylation level was still nearly maximal (FIG. 1). This suggests that Glu^{820} is either directly involved in K^+-binding or in signal transmission between the K^+-filled binding site and the phosphorylation site. The lack of K^+-sensitivity of this mutant was confirmed in dephosphorylation studies.

Address for correspondence: Jan Joep H.H.M. de Pont, Department of Biochemistry, Nijmegen Center for Molecular Life Sciences, University of Nijmegen, Nijmegen, the Netherlands. Voice: +31-24-3614260; fax: +31-24-3616413.
 J.dePont@ncmls.kun.nl

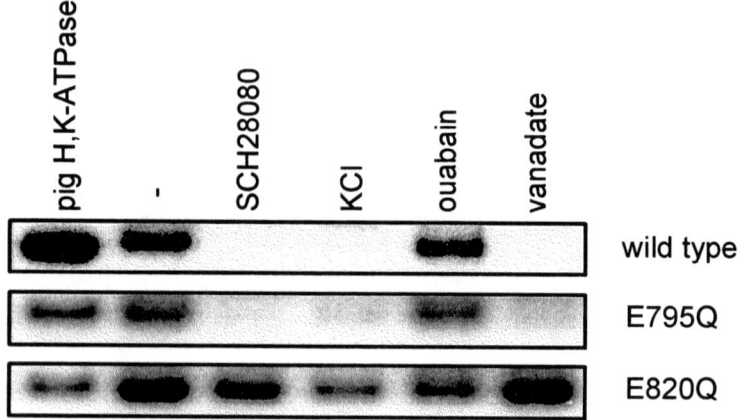

FIGURE 1. Effect of SCH 28080, KCl, ouabain, and vanadate on ATP-phosphorylation of wild-type gastric H,K-ATPase and the mutants E795Q and E820Q. (Adapted from Swarts et al.[6])

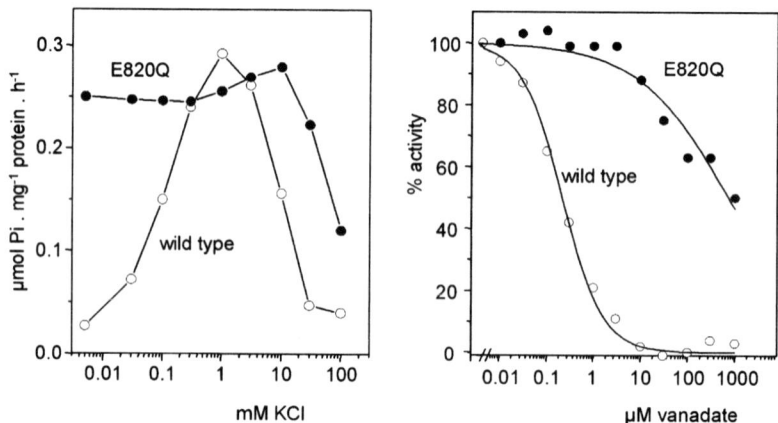

FIGURE 2. K^+-dependence of the ATPase activity (**left**) and effect of vanadate on it (**right**). Results are given for the wild-type enzyme and the E820Q mutant. (Adapted from Swarts et al.[7] and Hermsen et al.[9])

We also found that the ATP-phosphorylation level of this mutant hardly decreased by either the specific H,K-ATPase inhibitor SCH 28080 or vanadate (FIG. 1). At that time, we could not give a satisfactory explanation for these findings. The situation became even more complex when we discovered that the *K^+-independent* ATPase activity of the E820Q mutant, which we previously assumed to be an endogenous property of the Sf9 cell membranes, could be largely inhibited by SCH 28080 (FIG. 2).[7] The sensitivity of this *K^+-independent* (or constitutive) ATPase activity of

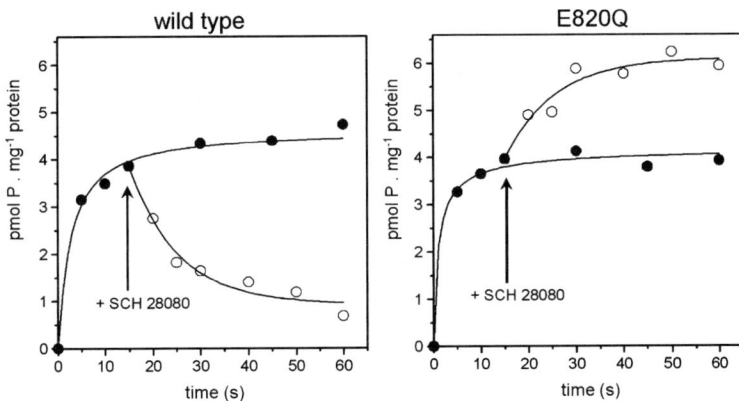

FIGURE 3. Effect of addition of SCH 28080 15 s after start of the ATP-phosphorylation reaction with wild-type ATPase and the E820Q mutant. (From Swarts et al.[8])

the E820Q mutant for SCH 28080 was high and comparable to that of the K^+-sensitive ATPase activity of the wild-type enzyme. Thus by mutation of a single residue, an enzyme was produced with an SCH 28080-sensitive, but *K^+-independent,* ATPase activity.

The *K^+-independent* activity of the E820Q mutant could be explained by the enhanced spontaneous dephosphorylation rate of the phosphorylated intermediate. The latter process, however, needed a temperature of 21°C, whereas K^+-stimulated dephosphorylation of the wild-type enzyme already occurred at 0°C.

The difference in sensitivity of the E820Q mutant for SCH 28080 of the ATPase activity (high) and the ATP-phosphorylation level (low) prompted further study.[8] FIGURE 3 shows that when the phosphorylation experiment was carried out at 21°C, the presence of SCH 28080 increased rather than decreased the phosphorylation level of the E820Q mutant. This finding can be explained by assuming a shift to the right in the $E_2 \leftrightarrow E_1$ equilibrium (see FIG. 4). According to this assumption, the mutant does not react with SCH 28080 during preincubation, and ATP-phosphorylation normally takes place. The increase in the phosphorylation level is due to formation of a complex between E_2-P and SCH 28080 that is more stable than E_2-P alone. Dephosphorylation rate measurements confirmed this idea.[8] These experiments were the first indication that upon mutagenesis of this residue, the $E_2 \leftrightarrow E_1$ equilibrium shifted to the right. This idea was supported by studies with vanadate (FIG. 2; see also FIG. 4). The IC_{50} value for vanadate inhibition of the K^+-insensitive ATPase activity of the E820Q mutant is more than 100 times higher than for inhibition of the K^+-sensitive ATPase activity of the wild-type enzyme, confirming the E_1 preference of the E820Q mutant.

In addition, we showed that the *K^+-independent* ATPase activity of several other single (E820A, E795L) and double (E795A/E820A, E795L/E820Q, E795Q/E820A, E795Q/E820D, E795Q/E820Q) mutants has a low affinity for vanadate.[9] The phosphorylation level of all these mutants was not decreased by SCH 28080, but rather was enhanced. This suggests that these mutants all have a high preference for the E_1

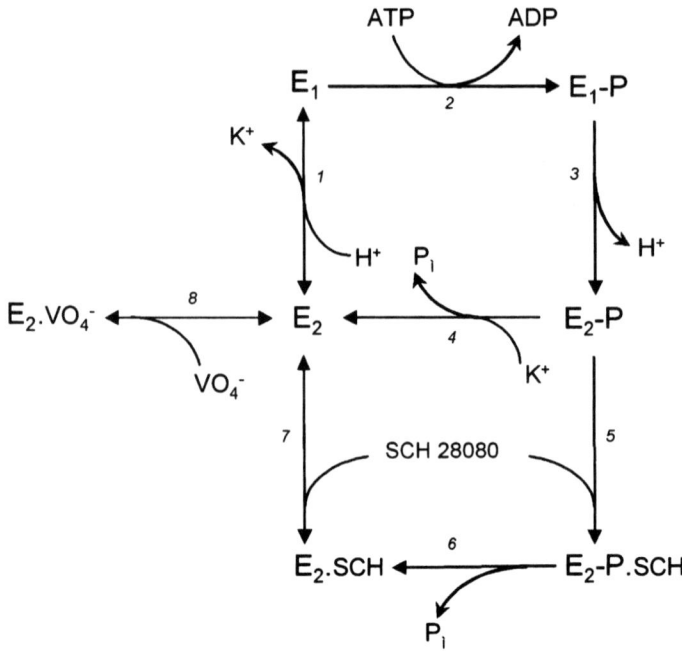

FIGURE 4. Simplified Albers–Post reaction scheme for gastric H,K-ATPase, indicating the reaction sites for SCH 28080 and vanadate.

conformation. Mutants with K^+-stimulated ATPase activity, such as E820D and E795Q, had similar vanadate affinities as the wild-type enzyme, and their phosphorylation level was decreased by SCH 28080.

The most interesting of these double mutants was E795Q/E820D. This mutant had K^+-independent ATPase activity, a low vanadate sensitivity of the ATPase reaction, and no lowering effect of SCH 28080 on the phosphorylation level, whereas the individual single mutants had wild-type behavior. The residues analogous to Glu^{795} and Glu^{820} are in the E_2 structure of Ca^{2+}-ATPase[10] located at opposite sides of the cation-binding pocket. Thus, two minor changes at both sides of the cation-binding pocket had a similar effect on the ATPase activity and the E_1/E_2-preference as the E820Q mutation.

COUPLING BETWEEN K^+-INDEPENDENT ATPase ACTIVITY AND E_1-PREFERENCE

The question now arose whether a preference for the E_1-conformation was always coupled to *K^+-independent* ATPase activity. We could approach this question when we discovered[11] that several mutants of Glu^{343}, located in transmembrane domain four, formed a phosphorylated intermediate but did not dephosphorylate. Moreover, the phosphorylation level decreased by preincubation with SCH 28080, suggesting that these mutants had a preference for the E_2 form. For mutants that could be phosphorylated, but without ATPase activity (E343D, E343A and E343L), we measured vanadate sensitivity of the phosphorylation level.[12] The ATP-phosphorylation levels of these mutants were lowered by vanadate, with a similar sensitivity as the wild-type enzyme confirming that these mutants prefered E_2.

We next constructed double and triple mutants in which, in addition to E343Q or E343D, the glutamates on positions 795 and/or 820 were replaced by glutamines.[11] These double and triple mutants could be phosphorylated by ATP, but showed neither (K^+-activated) ATPase activity nor a K^+-stimulated or a spontaneously enhanced dephosphorylation rate. Measurement of the effects of SCH 28080 and vanadate on the phosphorylation level of these mutants revealed that the kind of residue on position 795 had no effect on the direction of the equilibrium. All three amino acids on position 343 (Glu, Gln, or Asp) kept the E_2 preference as long as the original Glu was present on position 820. However, as soon as a Gln was introduced on this position, the equilibrium shifted toward the E_1 form. This was independent of the presence of K^+-independent ATPase activity, which was always blocked with an Asp on position 343. Thus in the double mutant E343D/E820Q, the first mutation determined the (lack of) ATPase activity, and the second one the direction of the E_1/E_2 equilibrium.

RESIDUES Lys^{791} AND Asn^{792}

We recently studied two other residues in M5 by site-directed mutagenesis. Asn^{792} is conserved within the P-type ATPase family, and its homologue in SR Ca^{2+}-ATPase is involved in Ca^{2+} binding.[4] The positively charged Lys^{791} is specific for H,K-ATPases, as a Ser is present on this position in Na,K-ATPases and SR Ca^{2+}-ATPase. Although the K^+-stimulated ATPase activities of several single Asn^{792} and Lys^{791} mutants were very low (for Lys^{791} mutants, see also Refs. 12 and 13), the enzyme preparations could be phosphorylated by ATP (see also Swarts et al., in this volume).

The phosphorylation level of mutant K791A decreased by low SCH 28080 and vanadate concentrations, suggesting that the mutant had a similar E_2 preference as the wild-type enzyme. The phosphorylation level of mutant N792A was only inhibited for 50% by 100-μM SCH 28080 and very little by 100-μM vanadate. So the equilibrium position of this mutant was shifted in the direction of the E_1 form. Combination of the N792A with the E343D mutation did not change the E_1 preference of the former mutation, while the hydrolysis of the phosphorylated intermediate was blocked, as it is in all E343D mutants. In the N792A/E820Q double mutant SCH 28080 increased the phosphorylation level considerably, while vanadate barely reduced the phosphorylation level, suggesting a clear E_1 preference. This double mutant also showed K^+-independent ATPase activity. The phosphorylation level of the triple mutant E343D/N792A/E820Q increased by SCH 28080 and decreased only with very high vanadate concentrations. Thus the effect of the E820Q mutant on the E_1/E_2 equilibrium was dominant over that of the two other mutations. However, like the E343D/E820Q mutant, the triple mutant had no (K^+-independent) ATPase activity.

Combination of the mutation K791A with either E343D or E820Q or both always resulted in an enzyme with a strong preference for the E_2 conformation. Thus the

K791A mutation was the only one that overruled the E_1 preference of the E820Q mutation. In other words, the equilibrium shift of the E820Q mutation only occurred when a positively charged Lys residue was present on position 791.

DISCUSSION

In the crystal structure of the E_1-form of SR Ca^{2+}-ATPase a, number of side chains present in transmembrane domains 4, 5, 6, and 8 have been shown to be involved in Ca^{2+} binding.[4] In this structure a central role that corresponds to Asp^{824} in H,K-ATPase is attributed to Asp^{800}. This residue appears to be involved in binding of both Ca^{2+} ions.

Toyoshima and Tomura[10] showed in Ca^{2+}-ATPase that the residues corresponding to Glu^{820} and Asp^{824} rotated about 90°, going from the E_1 to the E_2 form, so that in the E_2 structure the residue corresponding to Glu^{820} took over the central position of the residue corresponding to Asp^{824}. They suggested from homology modeling that the residue corresponding to Glu^{820} would coordinate to both K^+ ions in Na,K-ATPase.[10] It is therefore likely that there is a similar role for Glu^{820} in K^+ binding in gastric H,K-ATPase.

This postulate agrees very well with most previous findings on mutants of this enzyme. Glu^{820} could be replaced by an Asp without many changes in the enzymatic properties of this ATPase.[7,14] However, any other replacement of this residue abolished K^+-stimulated activity.[14] Since binding of both K^+ ions is apparently needed for K^+-stimulated dephosphorylation, this explains the latter effect. However, these replacements yielded K^+-*independent* ATPase activity.[7] We therefore postulated that the filled K^+-binding pocket was apparently mimicked by these mutations. In addition, we found that the K^+-*independent* ATPase activity of these mutants was only inhibited by very high vanadate concentrations,[9] suggesting that these mutants favored the E_1 form. The latter preference also explained the increasing effect of SCH 28080 on the phosphorylation level of mutant E820Q[8].

Several other residues postulated to be involved in the K^+-binding sites have been mutated in gastric H,K-ATPase. Mutation of Glu^{795} into a Gln has no effect on the ATPase activity.[12,13,15] However, mutation into Asp or Asn residues markedly reduced the ATPase activity and the apparent K^+ affinity, suggesting that only the side chain geometry, and not the charge, of this residue is important.[13,15,16] Our recent studies show that Asn^{792} plays a similar role. Mutation of this residue resulted in a 10-fold reduction in K^+ affinity in the dephosphorylation reaction. Asp^{824} (corresponding to Asn^{796} in Ca^{2+}-ATPase that is involved in the binding of both Ca^{2+} ions[15]) was not investigated in depth by us, since the mutated enzyme D824N barely phosphorylated. Vagin et al.[13] found no change in the apparent NH_4^+ affinity of the D824E mutant, whereas they found an increase in the apparent NH_4^+ affinity for the mutants D824A and D824N, and that both had a very low enzyme activity. Thus a specific role for Asp^{824} in K^+ binding has not yet been established. The role of the residue in M8 (Glu^{936}) is uncertain, because none or only minor changes in apparent K^+ affinity of the ATPase reaction were found with mutants on this position.[12,13] It might be that the presence of a β-subunit in H,K-ATPase that is likely to be present in this area of the enzyme, makes the situation different from that of Ca^{2+}-ATPase. Interestingly, the apparent K^+ affinities in the ATPase reaction of the mutants T823A and T823L located in M6, were unchanged or even increased,[12,13] which is in line with the assumed location outside the cation-binding pocket in the E_2 form.

According to the E_2-model of SR Ca^{2+}-ATPase, the residue equivalent to Glu^{343} points away from the ion-binding site. It is striking that with the exception of E343Q, which has a 10-fold lower K^+ affinity, all other mutants (E343D, E343L, E343V, E343I, and E343A) show neither K^+-dependent nor -independent ATPase activity,[11,17,18] nor a shift to the E_1 form.[11] If the same structure holds for the gastric H,K-ATPase as for the E_2 form of Ca^{2+}-ATPase, this might suggest that this residue is necessary for coupling between the cation-binding site and the phosphorylation domain.

The residue analogous to Lys^{791} in SR Ca^{2+}-ATPase (Ser^{767}) is in the E_2 model positioned close to the residues corresponding to both Glu^{820} and Glu^{795}, and was thought to be involved in hydrogen binding with the latter two residues.[10] The presence of a Lys residue with its positive charge is specific for H,K-ATPases. Since Lys^{791} is, like Glu^{343}, located between the cation-binding site and the phosphorylation domain, it might have a similar role in transmission of the signal to the phosphorylation domain as Glu^{343}. We found that, just like the E343D mutant, the K791A blocked the *K^+-independent* ATPase activity caused by the E820Q mutation.

However, there is an important difference between the results with mutation of these two residues. Whereas mutation of Glu^{343} had no effect on the E_1-promoting property of the E820Q mutation, mutation of Lys^{791} preserved the E_2 conformation, whether or not E820Q was present. Suppose that the E820Q mutation indeed converts the cation-binding pocket in a structure that resembles the K^+-filled pocket, it would give two signals to the intracellular domains. One signal, showing a preference for the E_1 conformation, could be blocked by the K791A, but not by the E343D mutation. The signal to the phosphorylation site, resulting in an enhanced dephosphorylation rate, could be blocked by both mutations. Further biophysical and biochemical experiments are necessary to test the preceding interpretation of these fascinating findings.

REFERENCES

1. SHULL, G.E., A. SCHWARTZ & J.B. LINGREL. 1985. Amino-acid sequence of the catalytic subunit of the (Na^+,K^+)ATPase deduced from a complementary DNA. Nature **316**: 691–695.
2. MACLENNAN, D.H., C.J. BRANDL, B. KORCZAK & N.M. GREEN. 1985. Amino-acid sequence of a $Ca^{2+}+Mg^{2+}$-dependent ATPase from rabbit muscle sarcoplasmic reticulum, deduced from its complementary DNA sequence. Nature **316**: 696–700.
3. CLARKE, D.M., T.W. LOO, G. INESI & D.H. MACLENNAN. 1989. Location of high affinity Ca^{2+}-binding sites within the predicted transmembrane domain of the sarcoplasmic reticulum Ca^{2+}-ATPase. Nature **339**: 476–478.
4. TOYOSHIMA, C., M. NAKASAKO, H. NOMURA & H. OGAWA. 2000. Crystal structure of the calcium pump of sarcoplasmic reticulum at 2.6 Å resolution. Nature **405**: 647–651.
5. KLAASSEN, C.H.W. et al. 1993. Functional expression of gastric H,K-ATPase using the baculovirus expression system. FEBS Lett. **329**: 277–282.
6. SWARTS, H.G.P. et al. 1996. Role of negatively charged residues in the fifth and sixth transmembrane domains of the catalytic subunit of gastric H^+,K^+-ATPase. J. Biol. Chem. **271**: 29764–29772.
7. SWARTS, H.G.P. et al. 1998. Constitutive activation of gastric H^+,K^+-ATPase by a single mutation. EMBO J. **17**: 3029–3035.

8. SWARTS, H.G.P. et al. 1999. Conformation-dependent inhibition of gastric H^+,K^+-ATPase by SCH 28080 demonstrated by mutagenesis of glutamic acid 820. Mol. Pharmacol. **55:** 541–547.
9. HERMSEN, H.P.H. et al. 2001. Mimicking of K^+-activation by double mutation of glutamate-795 and glutamate-820 of gastric H^+,K^+-ATPase. Biochemistry **40:** 6527–6533.
10. TOYOSHIMA, C. & H. NOMURA. 2002. Structural changes in the calcium pump accompanying the dissociation of calcium. Nature **418:** 605–611.
11. SWARTS, H.G.P. et al. 2001. K^+-independent gastric H^+,K^+-ATPase: dissociation of K^+-independent dephosphorylation and preference for the E_1-conformation by combined mutagenesis of transmembrane glutamate residues. J. Biol. Chem. **276:** 36909–36916.
12. RULLI, S.J., N.M. LOUNEVA, E.V. SKRIPNIKOVA & E.C. RABON. 2001. Site-directed mutagenesis of cation coordinating residues in the gastric H,K-ATPase. Arch. Biochem. Biophys. **387:** 27–34.
13. VAGIN, O. et al. 2001. Mutational analysis of the K^+-competitive inhibitor site of gastric H,K-ATPase. Biochemistry **40:** 7480–7490.
14. HERMSEN, H.P.H., H.G.P. SWARTS, J.B. KOENDERINK & J.J.H.H.M. DE PONT. 1998. The negative charge of glutamic acid 820 in the gastric H^+,K^+-ATPase α-subunit is essential for K^+ activation of the enzyme activity. Biochem. J. **331:** 465–472.
15. HERMSEN, H.P.H., J.B. KOENDERINK, H.G.P. SWARTS & J.J.H.H.M. DE PONT. 2000. The carbonyl group of glutamic acid-795 is essential for gastric H^+,K^+-ATPase activity. Biochemistry **39:** 1330–1337.
16. PELUFFO, R.D., J.M. ARGUELLO, J.B. LINGREL & J.R. BERLIN. 2000. Electrogenic sodium-sodium exchange carried out by Na,K-ATPase containing the amino acid substitution Glu779Ala. J. Gen. Physiol. **116:** 61–73.
17. ASANO, S. et al. 2000. Mutational analysis of the putative K^+-binding site on the fourth transmembrane segment of the gastric H^+,K^+-ATPase. J. Biochem. (Tokyo) **127:** 993–1000.
18. ASANO, S. et al. 1996. Functional expression of gastric H^+,K^+-ATPase and site-directed mutagenesis of the putative cation binding site and the catalytic center. J. Biol. Chem. **271:** 2740–2745.

Nongastric H,K-ATPase: Structure and Functional Properties

NIKOLAI MODYANOV,[a] NIKOLAY PESTOV,[a,b] GAIL ADAMS,[a]
GILLES CRAMBERT,[c] MANORANJANI TILLEKERATNE,[a]
HAO ZHAO,[a] TATYANA KORNEENKO,[b] MIKHAIL SHAKHPARONOV,[b]
AND KÄTHI GEERING [c]

[a]*Department of Pharmacology, Medical College of Ohio,
Toledo, Ohio 43614, USA*

[b]*Shemyakin and Ovchinnikov Institute of Bioorganic Chemistry,
Moscow,117871, Russia*

[c]*Institute of Pharmacology, University of Lausanne,
Lausanne, CH-1005, Switzerland*

> ABSTRACT: Nongastric H,K-ATPases whose catalytic subunits (AL1) encoded by human ATP1AL1 and homologous animal genes comprise the third distinct group within the X,K-ATPase family. No unique nongastric β has been identified. Precise *in situ* colocalization and strong association of AL1 with β1 of Na,K-ATPase was detected in apical membranes of rodent prostate epithelium. In this tissue, β1NK serves as an authentic subunit of both the Na,K- and nongastric H,K-pumps. Upon expression in *Xenopus* oocytes the human AL1 can assemble with β1NK, and more efficiently with gastric βHK, into functional H,K-pumps. Both AL1/β complexes exhibit a similar K-affinity, and their K-transport depends on intra- and extracellular Na. These data provide new evidence that nongastric H,K-ATPase can perform Na/K-exchange, and indicate that β does not significantly affect this ion-pump function. Analysis of human nongastric H,K-ATPase expressed in Sf-21 insect cells revealed that AL1/βHK exhibits substantial enzymatic activities in K-free medium and K stimulates, but Na has inhibitory effect on ATP hydrolysis. Thus, although the nongastric H,K-ATPase can function as Na/K exchanger, its reaction mechanism is different from that of the Na,K-ATPase. Human nongastric H,K-ATPase is highly sensitive to bufalin, digoxin, and digitoxin, but almost resistant to digoxigenin and ouabagenin.
>
> KEYWORDS: nongastric H,K-ATPase; subunit composition; catalytic functions; ion-transport properties

INTRODUCTION

Nongastric H,K-ATPases are presumed to be involved in the maintenance of electrolyte homeostasis under pathophysiological conditions such as, for example,

---

Address for correspondence: Nikolai Modyanov, Department of Pharmacology, Medical College of Ohio, Toledo, OH 43614. Voice: 419-383-4182; fax 419-383-2871.
nmodyanov@mco.edu

chronic hypokalemia.[1,2] However, the exact role of these ATPases under normal conditions *in vivo* remains unclear, mainly because of the still insufficient characterization of their structural and functional properties. In previous studies (for review, see Ref. 3) we determined complete structures of the human ATP1AL1gene[4] and corresponding cDNAs;[5] and for the first time characterized the encoded human protein (AL1) as a catalytic subunit of the ouabain-sensitive nongastric H,K-ATPase using expression in *Xenopus* oocytes.[6] In this brief review, we outline the current status of our studies.

IDENTIFICATION OF THE β-SUBUNIT

Similar to Na,K- and gastric H,K-ATPases, nongastric H,K-ATPases require participation of the β-subunit for formation of the active enzyme.[7] However, it remained for many years unclear whether one of the known β-subunits or a hitherto unidentified member of the X,K-ATPase β-subunit family is a real subunit of nongastric H,K-ATPases, because none of these enzymes has been isolated and analyzed directly, and available experimental data were rather controversial, as discussed in detail previously.[2,3,8,9]

To find an appropriate source for the structural characterization of the nongastric H,K-ATPase on the protein level, we recently performed detailed Western blot analysis of the expression of different X,K-ATPases in rodent prostate complexes, where high levels of the AL1 mRNA were revealed by RT-PCR.[10,11] It was shown that the AL1 protein is present at the highest level in the coagulating gland (anterior prostate), where its abundance was found to be ~2.5-fold higher than in the distal colon.[11] Among other X,K-ATPases, only Na,K-ATPase α1 and β1 were detected in significant amounts in coagulating gland membranes.

Immunohistochemical experiments revealed that in the epithelium of rat and mouse coagulating glands, the AL1 has strictly apical polarization, while α1NK is localized only in basolateral membranes.[11] Labeling with β1-specific antibodies showed a bright apical labeling and a weak labeling of basolateral membranes (FIG. 1). (The same results have been obtained with several different β1 antibodies, including monoclonal antibodies kindly provided by Drs. M. Caplan, and

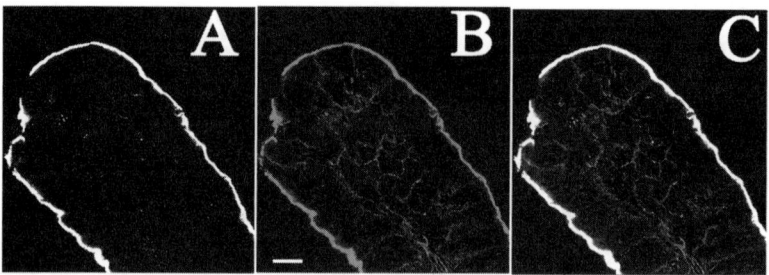

FIGURE 1. (A) Confocal images of rat coagulating gland infolding labeled with nongastric H,K-ATPase AL1 antibodies, and (B) β1NK monoclonal antibody IEC 1/48 (kindly provided by Dr. A. Quaroni). (C) Merged images of A and B; bar, 10 μM.

A. Quaroni.) Double labeling with antibodies against AL1 and with β1, demonstrates precise colocalization of the AL1 with the β1NK in the apical membranes *in situ*. No labeling was detected in the epithelium upon staining with antibodies against any other X,K-ATPase α or β isoforms. Existence of the stable AL1/β1 complex in coagulating glands has also been demonstrated by coimmunoprecipitation of the AL1 with the β1-specific antibodies, but not with antibodies against β2, β3, and β*m* (data not shown).

These findings clearly demonstrated that β1 is not only a subunit of the Na,K-ATPase, but also serves as an authentic subunit of nongastric H,K-ATPase, a real counterpart of the AL1 in apical membranes of rodent coagulating gland.

ION TRANSPORT AND ENZYMATIC FUNCTIONS

We use the *Xenopus* oocyte expression system to analyze ion-translocating pump activity of the human nongastric H,K-ATPase and the baculovirus expression system in Sf-21 insect cells to measure its catalytic functions. Analysis of the capability of the AL1 to form a stable and functionally active ATPase complex upon coexpression in *Xenopus* oocytes together with each of the known X,K-ATPase β subunits revealed that gastric βHK and β2-like *Bufo* bladder β are able to associate with AL1 much more efficiently than its real counterpart β1NK.[8,9] Formation of the active ATPase complex of AL1 with βHK, but not with β1NK or β3NK, was observed in baculovirus expression system.[12] To explain this phenomenon, one can suggest that heavy glycosylation (7–8 sites in βHK and *Bufo* bladder β vs. 3 sites in β1NK) is an essential structural feature required for more efficient formation of the stable and functional AL1/β complexes in heterologous expression systems.

To further investigate the capability of the human nongastric H,K-ATPase to function as a Na/K exchanger, and to examine the influence of different β subunits on the functional properties of the nongastric H,K-ATPase, we compared ion-transport functions of the AL1 coexpressed in *Xenopus* oocytes with β1NK, βHK, and *Bufo* bladder β.[9] It was shown that despite its moderate association efficiency, β1NK can produce functional pumps at the cell surface. Most significantly, the $K_{1/2}$ values for K^+-activation of the $^{86}Rb^+$ uptake mediated by the AL1/βHK and AL1-β1NK were similar, indicating that β does not significantly influence the K^+ affinity of nongastric H,K-ATPase. The Na^+-transport activities of the AL1/βHK and AL1-β1NK were compared in oocytes, in which intracellular Na^+ was elevated and controlled by the coexpressed amiloride-sensitive epithelial Na^+ channel. Both complexes were able to produce a similar and significant decrease in intracellular Na^+ concentration that was comparable to that produced by Na,K-ATPase. Another important finding was that ^{86}Rb influx mediated by the AL1/βHK into oocytes loaded with different Na^+ concentrations was stimulated by intracellular Na^+ with the $K_{1/2}$ of ~9 mM Na^+. These data demonstrate for the first time that Na-efflux is indeed mediated by nongastric H,K-ATPase, since it is directly coupled to K-uptake.[9]

These results provide further evidence that the human nongastric H,K-ATPase can perform Na/K exchange and that, in contrast to gastric H,K-ATPase or Na,K-ATPase,[7] functions of the recombinant human nongastric H,K-ATPase are not significantly affected by the association with the real partner—β1NK or its effective surrogate—βHK.[9] This indicates that the kinetic and transport properties of the recombinant human nongastric H,K-ATPase determined in different expression systems with different β subunits, are likely to be physiologically relevant.

To gain insight into the enzymatic functions of the human nongastric H,K-ATPase, we have developed a procedure for the large-scale preparation of the recombinant enzyme using the baculovirus expression system.[3,12] In this study, the ATP-hydrolyzing activity of human nongastric H,K-ATPase represented by the AL1/βHK, has been analyzed with respect to K^+, Na^+, pH, and ATP dependencies and inhibitor sensitivity for the first time.[12] One of the most interesting findings was that the AL1/βHK exhibits a substantial, specific ATPase activity in nominally K^+-free medium. The addition of K^+ stimulates the ATP hydrolysis up to 4-fold. The ATPase activity was moderately sensitive to ouabain and to SCH 28080 with Ki ~65 μM and ~95 μM, respectively. The shift of Ki values to the low millimolar range at higher K^+ indicates a competition between K^+ and both inhibitors on the extracytoplasmic surface of the enzyme.[12]

Since the transport modes of the nongastric ATPases also include Na/K exchange,[2,9] it was reasonable to expect that Na^+ may serve as an activator of ATP hydrolysis by the AL1/βHK complex. However, the highest level of ATPase activity was measured in nominally Na^+-free medium, and the inhibition by Na^+ was detected at all K^+ concentrations, but was the strongest in K^+-free medium (Ki ~24 mM). These results clearly indicate that there is an antagonism between K^+ stimulation and Na^+ inhibition in this reaction.[12] Thus, although the nongastric H,K-ATPase, similar to the Na,K-ATPase, is able to perform Na/K exchange, their reaction mechanisms with respect to effects of Na^+ on the enzyme kinetics are clearly different.

We examined the sensitivity of the AL1/βHK toward different cardiac glycosides. Bufalin and proscillaridin A, bufadienolides with 6-member lactone rings, have been identified as the most powerful inhibitors with a Ki of ~1 μM. It was also shown that digoxin and digitoxin, which are more lipophilic than ouabain, exhibit relatively strong inhibitory effects on ATPase reaction (Ki ~10 μM) and on phosphatase activity (Ki ~5 μM). However, in contrast to Na,K-ATPase, the AL1/βHK is practically resistant to highly hydrophobic genins such as digoxigenin and ouabagenin. These findings suggest that the specific structural features of the lactone ring and carbohydrate moieties of cardenolides play an essential role in the efficiency of interactions between inhibitors and nongastric H,K-ATPase.

ACKNOWLEDGMENT

These studies were supported by grants from National Institutes of Health (HL-36573, GM-54997), from the Swiss National Fund for Scientific Research (31-64793.01, 31-53721.98), and from the Russian Foundation for Basic Research (00-04-48153, 98-04-48408).

REFERENCES

1. SILVER, R.B. & M. SOLEIMANI. 1999. H-K-ATPases: regulation and role in pathophysiological states. Am. J. Physiol. **276:** F799–F811.
2. JAISSER, F. & A.T. BEGGAH. 1999. The nongastric H-K-ATPases: molecular and functional properties. Am. J. Physiol. **276:** F812–F824.

3. MODYANOV, N. et al. 2000. Structural and functional properties of human ouabain-sensitive H,K-ATPase. *In* The Na,K-ATPase and Related ATPases. K. Taniguchi and S. Kaya, Eds.: 139–146. Elsevier. Amsterdam.
4. SVERDLOV, V., M. KOSTINA & N. MODYANOV. 1996. Genomic organization of the human ATP1AL1 gene encoding a ouabain-sensitive H,K-ATPase. Genomics **32:** 317–327.
5. GRISHIN, A.V. et al. 1994. Cloning and characterization of the entire cDNA encoded by ATP1AL1—A member of the human Na,K/H,K-ATPase gene family. FEBS Lett. **349:** 144–150.
6. MODYANOV, N.N. et al. 1995. The human ATP1AL1 gene encodes a ouabain-sensitive H,K-ATPase. Am. J. Physiol. **269:** C992–C997.
7. GEERING, K. 2001. The functional role of β subunits in oligomeric P-type ATPases. J. Bioenerg. Biomembr. **33:** 425–438.
8. GEERING, K. et al. 2000. Role of membrane domains M9 and M10 in the assembly process and association efficiency of human, non-gastric H,K-ATPase α subunits (ATP1AL1) with known β subunits. Biochemistry **39:** 12688–12698.
9. CRAMBERT, G. et al. 2002. Human nongastric H,K-ATPase: transport properties of ATP1al1 assembled with different β subunits. Am. J. Physiol **283:** C305–C314
10. PESTOV, N.B. et al. 1998. Ouabain-sensitive H,K-ATPase: tissue-specific expression of the mammalian genes encoding the catalytic α-subunit. FEBS Lett. **40:** 320–324.
11. PESTOV, N.B. et al. 2002. Nongastric H,K-ATPase in rodent prostate: lobe-specific expression and apical localization. Am. J. Physiol. **282:** C907–C916
12. ADAMS, G. et al. 2001. Catalytic function of nongastric H,K-ATPase expressed in Sf-21 insect cells. Biochemistry **40:** 5765–5776.

Mechanism of Proton Pumping by Plant Plasma Membrane H$^+$-ATPase

Role of Residues in Transmembrane Segments 5 and 6

M. G. PALMGREN,[a,b] M. J. BUCH-PEDERSEN,[a] AND A. L. MØLLER[a]

[a]*Department of Plant Biology, The Royal Veterinary and Agricultural University, Thorvaldsensvej 40, DK-1871 Frederiksberg C, Denmark*

[b]*CNRS, Universite Pierre et Marie Curie Paris VI, Case courrier 154, 4, place Jussieu, F-75252 Paris, France*

ABSTRACT: The mechanism of proton pumping by P-type plasma membrane H$^+$-ATPases is not well clarified. Site-directed mutagenesis studies suggest that Asp684, situated in transmembrane segment M6, is involved in coordination of proton(s) in plant plasma membrane H$^+$-ATPase. This hypothesis is supported by atomic models of H$^+$-ATPases built on the basis of the crystal structure of the related SERCA1a Ca^{2+}-ATPase. However, more biochemical, genetic, and structural studies are required before we will be able to understand the nature of the proton binding site(s) in P-type H$^+$-ATPases and the mechanism of action of these pumps.

KEYWORDS: proton pumping; P-type ATPase; plasma membrane H$^+$-ATPase; mechanism; *Arabidopsis thaliana*; PMA1; AHA2

INTRODUCTION

Plasma membrane H$^+$-ATPases in plant and fungal cells have an analogous physiological function to Na$^+$,K$^+$-ATPases in animal cells.[1–2] Thus, both proteins generate electrochemical ion gradients across the plasma membrane that in turn drive secondary active transport of solutes into and out of cells. However, instead of employing Na$^+$, in plant and fungal cells this gradient is made up of H$^+$. Both types of pumps are P-type ATPases,[3] predicted to have ten transmembrane segments and with most of the protein, including the nucleotide binding and phosphorylation sites, exposed to the cytoplasmic side of the membrane.

Substantial progress has been made to map the transport pathways in representative P-type ATPases such as Ca^{2+}- and Na$^+$,K$^+$-ATPases.[4–7] These studies have established the presence of cation binding sites formed by several coordinating groups

Address for correspondence: Prof. Michael Gjedde Palmgren, Université Pierre et Marie Curie, CNRS/Paris VI, Tour 53, 4e etagé, case 154, 4, place Jussieu, F-75252 Paris, France. Voice: +33-1-4427- 5913; fax: +33-1-4427-6151.

palmgren@biobase.dk

situated in transmembrane helices M4, M5, M6, and M8. However, the molecular determinants of cation specificity ion P-type proton pumps have not been identified so far and, as a consequence, the mechanism of H^+-transport by P-type ATPases is not well characterized.

PROTON PUMPS CAN HAVE A VERY DIFFERENT STRUCTURAL ORGANIZATION

Proton pumps are found in virtually all cells, and many cells have different proton pumps in different intracellular membranes. The 3D high-resolution structures of a number of proton pumps have demonstrated completely different architectures. Two examples can be used to illustrate this observation. First, bacteriorhodopsin, a light-powered proton pump with a molecular weight of only 27 kDa, has a single subunit equipped with a prosthetic group.[8–10] Second, F_0F_1 ATPases, which can operate as ATP-driven proton pumps, are large complexes with molecular weights up to 650 kDa composed of eight subunits in prokaryotes and 16–18 in higher eukaryotes, none of which have prosthetic groups.[11–14] These findings indicate that different strategies for proton transport operate in these pumps, even though the basic principles might be the same.[10,15]

As with other integral membrane proteins involved in transport, residues of P-type H^+-pumps involved in ion coordination are likely to be situated in membrane-spanning segments. In related P-type ATPases, residues in M4, M5, M6, and M8 have been shown be contribute to ion coordination.[4–6] In H^+-ATPases, residues in transmembrane segments that can become protonated, such as Asp, Glu, Arg, and Lys, have attracted special attention.[16,17] Charged residues in AHA2, believed to be situated within the plane of suggested membrane spanning segments, are Glu74 (M1), Asp95 (M2), Arg655 (M5), and Asp684 (M6). All of these residues are conserved among plant plasma membrane H^+-ATPases.

MECHANISM OF H^+-TRANSPORT BY PLASMA MEMBRANE H^+-ATPases

At least two types of mechanism can be envisioned for any proton pump. One mechanism could be the funneling of protons along a so-called proton wire. According to this model, protons are transported in several discrete steps by sequential transfer from one protonable group to another along a pathway. One characteristic of this type of proton pumping is that, in a single pump catalytic cycle, the proton entering the protein complex is not necessarily the same as the one leaving it. Such a mechanism is evident in the simple proton channel gramicidin, in which the proton wire consists of water molecules.[18] In proton pumps, unidirectional transport along proton wires has been proposed to occur by protein conformational changes, which induce local changes in the pK_as of protonable groups.[19–21]

Alternatively, an alternating access binding-site mechanism, involving alternating access of the site to each side of the membrane, could be hypothesized. The binding site could be a cavity in which the cation is coordinated by several groups, each one contributing to formation of ion specificity. As a result of conformational changes, this site is exposed to either side of the membrane due to the formation of "half-channels," in this way allowing for directional transport.[10,13,22,23]

All available evidence for well-characterized P-type pumps, such as Na^+,K^+- and Ca^{2+}-ATPases, suggest that they operate by an alternating access mechanism.[26] Thus, each cation to be transported binds to a single ion binding site in the membraneous part of the pump molecules. When the pump alternates between the E_1 and the E_2 conformations, Na^+ or Ca^{2+} binding sites are either exposed to the cytoplasm as high-affinity sites or exposed to the extracytoplasmic side as low-affinity sites.[24] Alternating site mechanisms have also been proposed for other proton pumps, such as bacteriorhodopsin[10] and F_0F_1 ATPase.[22]

ATOMIC MODELS OF PLASMA MEMBRANE H^+-ATPases

The SERCA1 Ca^{2+}-ATPase is so far the only P-type ATPase that has been crystalized to generate high-resolution structures.[6,7] This has prompted comparative protein structure modeling of related P-type ATPases.[25–27] Whereas detection of significant sequence similarity to a protein of known three-dimensional structure implies a structure prediction by homology, finding a true alignment is the crucial step for success in homology modeling of related protein structures.[28–30] Thus, just a single residue gap or insertion can alter dramatically the relative positions of side chains in the predicted structure.

Sequence alignments between 159 diverse P-type ATPases allowed for the identification of eight (a–h) conserved core segments comprising a total number of 265 residues.[31] These sequence motifs comprise only one transmembrane segment, namely M4 (FIG. 1). While sequence identity between proteins within these motifs is high, sequence identity outside these regions can be very low, making it difficult to align sequences.

Compared to the well-characterized Ca^{2+}-ATPases, H^+-ATPases are relatively homologous with respect to the A and P domains. The N domains, involved in nucleotide binding, show significant homology and share conserved motifs, although large deletions are observed in H^+-ATPases. Among transmembrane segments, M1, M2, and M4 are relatively easy to align. M5 and M6 can be aligned with somewhat different results, depending on the algorithm used (FIG. 2). M7 to M10 have low sequence identity, and sequence alignments become less reliable.

Sequence alignment of plasma membrane H^+-ATPases with the SERCA1 Ca^{2+}-ATPase has been the basis for comparative structure modeling of Arabidopsis AHA2,[25] *S. cerevisiae* Pma1p,[25] and *Neurospora crassa* PMA1.[26] The latter model was, in addition, fitted to an electronmicroscopic 8-Å map of the *Neurospora* H^+-ATPase.[32] It remains to be shown whether this low-resolution structure represents the *Neurospora* H^+-ATPase in the E_1 conformation. The fact that the *Neurospora* enzyme is strongly stimulated by a C-terminal peptide[26] might indeed suggest that it is in the low-affinity E_2 conformation.

The N-terminal domain, transmembrane segments M7 to M10 and the C-terminal regulatory domain were not predicted in the models of Arabidopsis AHA2 and *S. cerevisiae* Pma1p, since these regions do not align well the Ca^{2+}-pump. The 8-Å crystal structure of the *Neurospora* H^+-ATPase was used as a basis for modeling of regions with no identity to the Ca^{2+}-ATPase.

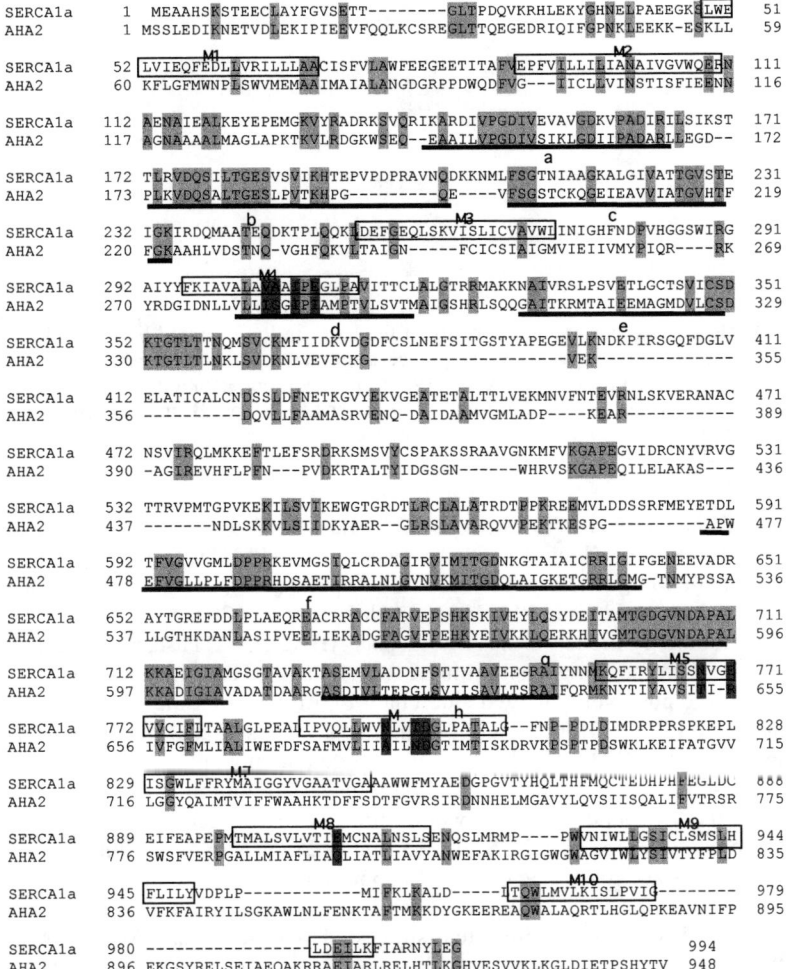

FIGURE 1. Sequence comparison between the sarcoplasmic Ca^{2+}-ATPase SERCA1a and the Arabidopsis thaliana H^+-ATPase AHA2. Conserved residues are indicated by a *light-gray underlay*. Residues involved in calcium coordination in SERCA1a are marked with a *dark-gray underlay* and letters colored *white*. The ten transmembrane segments (M1–M10) from SERCA1a and their corresponding sequence in AHA2 are highlighted in *rectangles*. Eight segments conserved in all P-type ATPases (numbered a–h) are presented by *black lines*. The alignment was generated by ClustalW (http://www.ebi.ac.uk/clustalw/) using default parameters.

A COMPARISON OF HYPOTHETICAL PROTON BINDING SITE IN DIFFERENT H^+-ATPase MODELS

In the structure of the SERCA1 Ca^{2+}-ATPase in the E_1 conformation, two Ca^{2+} binding sites are evident.[6] Site I is formed by M5, M6, and M8, with the Ca^{2+} ion

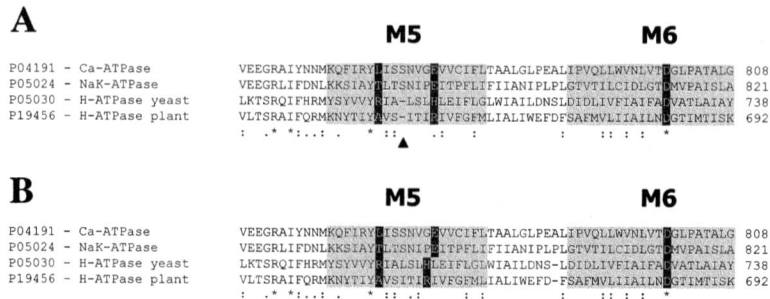

FIGURE 2. Two possible alignments of the hair-pin loop formed by M5 and M6 in a number of related P-type ATPases. Sequence alignments with *Arabidopsis thaliana* H⁺-ATPase AHA2 were generated by ClustalW (http://www.ebi.ac.uk/clustalw/) by applying different gap penalties. A gap is introduced in either (**A**) M5 or (**B**) in the loop connecting M5 and M6. The position of the *arrow* indicates gaps in the alignments.

being coordinated by Asn768(M5), Glu771 (M5), Thr799 (M6), Asp800 (M6), and Glu908 (M8). Site II, which is located between M4 and M6, is formed by side-chain oxygens of Glu309 (M4), Asn796 (M6), and Asp800 (M6), and main-chain carbonyl oxygens of three residues in M4 (Val304, Ala305, and Ile307).

In the homology model of the Arabidopsis AHA2 H⁺-ATPase,[25] a single proton binding site corresponding to site II in SERCA1 Ca^{2+}-ATPase is suggested. Thus, in the model, the positively charged side chain of Arg655 fills the cavity that corresponds to site I in the SERCA1 Ca^{2+}-pump. It is suggested that the carboxyl group of Asp684 contributes to coordination of a single ion, most likely H$_3$O$^+$, positioned between M4 and M6. At this position, the transported ion might be positioned in a saddle on the unwound M4. This would allow for hydrogen-bonding with main-chain carbonyl oxygens of Ile282, Gly283, and Ile285 in M4. According to this model, Arg655 does not contribute directly to coordination of H$^+$/H$_3$O$^+$.

In the model of *N. crassa* PMA1 H⁺-ATPase, a single proton binding site is likewise suggested. This ion binding site is suggested to be defined by Asp730 (M6) and Glu805 (M8). The former of these residues corresponds to Asp684 in the AHA2 H⁺-ATPase. In addition, the putative proton binding site contains the basic residues Arg695 and His701, corresponding to AHA2, Ala649, and Arg655, respectively. Additional polar groups in the vicinity of the proton binding site are the side chains of Tyr694 and Ser699, and the main-chain carbonyls of Ile331, Ile332, and Val334 in the unwound M4, the latter three residues corresponding to AHA2 residues Ile282, Gly283, and Ile285, respectively.

Thus, both homology models of H⁺-ATPases suggest a proton binding site defined by a conserved aspartate in M6 (Asp684$_{AHA2}$/Asp730$_{PMA1}$) and three carbonyl oxygens in the unwound M4 (Ile282$_{AHA2}$/Ile331$_{PMA1}$, Gly283$_{AHA2}$/Ile332$_{PMA1}$, and Ile285$_{AHA2}$/Val334$_{PMA1}$).

On the other hand, the two models disagree with respect to the role of additional charged residues in the vicinity of the proton binding site. Thus, in the *Neurospora* PMA1 model, three charged residues are close to the predicted proton binding site (Glu805$_{PMA1}$, Arg695$_{PMA1}$, and His701$_{PMA1}$). None of these residues are conserved in plant proton pumps, except for the basic side chain at position $Arg655_{AHA2}/His701_{PMA1}$.

When the two models are examined more closely, the primary sequences of transmembrane segments M5 of AHA2 and PMA1, respectively, are aligned differently with those of SERCA1. Thus, a single residue gap is introduced in the alignment of AHA2 and SERCA1 M5's, whereas such a gap is absent in the alignment of PMA1 and SERCA1 M5's.

If the gap in M5 is omitted in the AHA2 alignment, this would reposition most of the side chains in the model of this transmembrane segment. Thus, $Arg655_{AHA2}/His701_{PMA1}$ would not be protruding into the cavity corresponding to site I, but rather would be colliding with the nearby M7 (FIG. 3). In the SERCA1 Ca^{2+} pump, M5 and M7 are tilted relative to each other, but come very close at one point, separated only by a glycine residue in each segment (FIG. 4). Although most residues in the Ca^{2+} pump move relative to each other when the pump alternates between E_1 and E_2, interestingly, this contact point is preserved in the two conformations (FIG. 4). The glycine in M5 ($Gly770_{SERCA1}$) would correspond to $Arg655_{AHA2}/His701_{PMA1}$ in the alternative alignment. Introducing the large side chains of these residues at this position imposes severe structural constraints on the atomic model. Keeping the gap

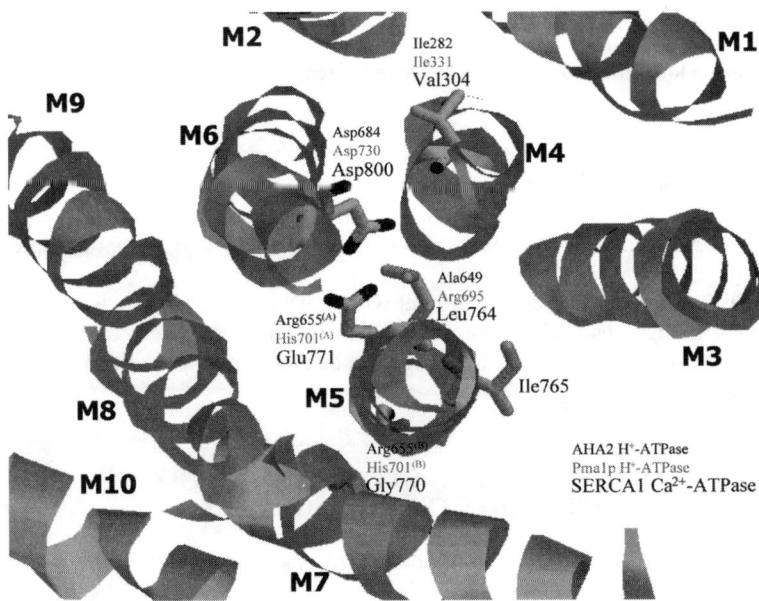

FIGURE 3. A top view of the transmembrane region of SERCA1a Ca^{2+}-ATPase. The structure is deposited in the protein database (PDB) with the accession number 1EUL. Residues in the ribbon model that are discussed in the text are indicated by *balls and sticks*. **Top rows**: Corresponding residues in Arabidopsis AHA2 H^+-ATPase; **middle rows**: corresponding residues in *Saccharomyces cerevisiae* Pma1p H^+-ATPase; **lower rows**: residues in SERCA1a Ca^{2+}-ATPase. (**A**) According to alignment in FIGURE 2A. (**B**) According to alignment in FIGURE 2B.

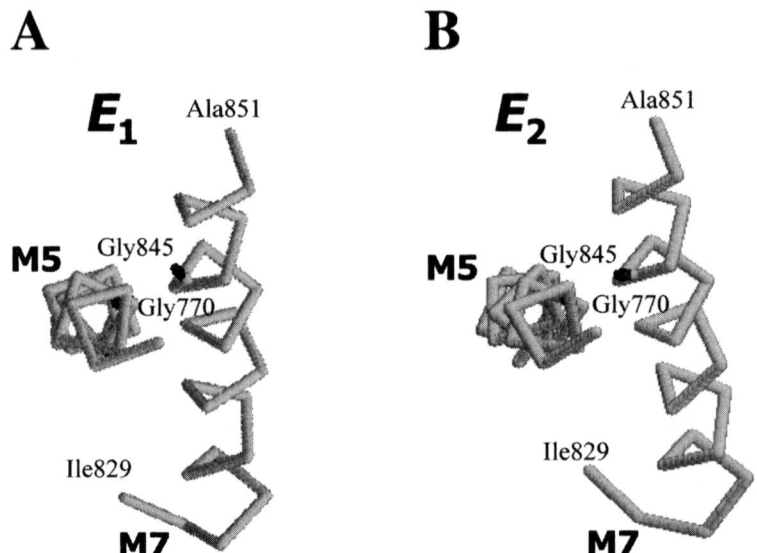

FIGURE 4. Top view of M5 and M7 of SERCA1a Ca^{2+}-ATPase in the conformations E_1 and E_2. Protein database (PDB) accession numbers are 1EUL and 1IWO, respectively. Two glycine residues discussed in the text are indicated by *balls and sticks*.

in M5, but positioning it further upstream than indicated in FIGURE 2, would move the side chains of $Ala649_{AHA2}$/$Arg695_{PMA1}$ out of the predicted proton binding site toward the region between M3 and M7, a position corresponding to that occupied by $Ile765_{SERCA1}$.

Obviously, it is very difficult to predict which of these alignments is likely to be correct. Thus, high-resolution structures supported by site-directed mutagenesis studies are required before any conclusions can be drawn regarding the nature of the proton binding site(s) in H^+-ATPases.

TESTING OF STRUCTURAL MODELS BY SITE-DIRECTED MUTAGENESIS

To explore the functional role of Asp684 (M6) in a plant H^+-ATPase, a conservative D684N substitution was constructed, expressed in yeast, and purified to near homogeneity.[16] Proton pumping by the reconstituted mutant enzyme was completely abolished, whereas ATP was still hydrolyzed. During catalysis, the D684N enzyme accumulated a phosphorylated intermediate whose stability was sensitive to the addition of ADP. It was concluded that the mutant enzyme is locked in the E_1 conformation and is unable to proceed through the E_1P–E_2P transition. It thus appears that Asp684 in the plasma membrane H^+-ATPase is required for coupling between ATP hydrolysis, enzyme conformational changes, and H^+-transport.

Unfortunately, substitutions of Asp730, the corresponding residue in *S. cerevisiae* plasma membrane H^+-ATPase Pma1p, led to misfolded proteins that have proven recalcitrant to biochemical analysis.[17,35] In fact, homologous expression of Pma1p mutants in *S. cerevisae* often led to problems with protein biogenesis, a situation that can be associated with dominant negative effects. Still, substititution of Asp730$_{Pma1p}$ is possible provided that the positively charged Arg695$_{Pma1p}$ (M5) is substituted simultaneously. Thus, the double mutants R695D/D730R and R695A/D730A are both functional H^+-ATPases.[35] This would suggest that neither charges nor oxygen atoms at these two positions are essential for proton coordination by the Pma1p H^+-ATPase.[35]

All residues in M4 and M5 of *S. cerevisiae* Pma1p have been investigated systematically by alanine-scanning mutagenesis.[33,34] Although various phenotypes were observed, the substitutions in M4 had minor impacts on proton pumping by Pma1p.[33] In M5 of Pma1p, single substitutions of the positively charged residues Arg695 and His701 result in problems with protein biogenesis.[35,36]

A number of residues situated in other transmembrane segments have been mutagenized in *S. cerevisiae* Pma1p,[37] but their role, if any, in proton transport is not clear. In M1, the conserved residue Glu129 in *S. cerevisiae* Pma1p, corresponding to Glu74 in AHA2 H^+-ATPase, has been extensively mutagenized without any significant effect on ATP hydrolysis or ATP-dependent proton pumping.[17,38]

ACKNOWLEDGMENTS

Work in the laboratory of the authors was supported by the European Union's Biotechnology Program and the Human Frontier Science Program Organization,

REFERENCES

1. SERRANO, R. 1989. Structure and function of plasma membrane ATPase. Annu. Rev. Plant Physiol. Plant. Mol. Biol. **40:** 61–94.
2. PALMGREN, M.G. 1998. Proton gradients and plant growth: role of the plasma membrane H^+-ATPase. Adv. Bot. Res. **28:** 1–70.
3. MOLLER, J.V., B. JUUL & M. LE MAIRE. 1996. Structural organization, ion transport, and energy transduction of P-type ATPases. Biochim. Biophys. Acta **1286:** 1–51.
4. JORGENSEN, P.L., J.M. NIELSEN, J.H. RASMUSSEN & P.A. PEDERSEN. 1998. Structure-function relationships of E_1-E_2 transitions and cation binding in Na,K-pump protein. Biochim. Biophys. Acta **1365:** 65–70.
5. RICE, W. J. & D.H. MACLENNAN. 1996. Scanning mutagenesis reveals a similar pattern of mutation sensitivity in transmembrane sequences M4, M5, and M6, but not in M8, of the Ca^{2+}-ATPase of sarcoplasmic reticulum (SERCA1a). J. Biol. Chem. **271:** 31412–31419.
6. TOYOSHIMA, C., M. NAKASAKO, H. NOMURA & H. OGAWA. 2000. Crystal structure of the calcium pump of sarcoplasmic reticulum at 2.6 Å resolution. Nature **405:** 647–655.
7. TOYOSHIMA, C. & H. NOMURA. 2002. Structural changes in the calcium pump accompanying the dissociation of calcium. Nature **418:** 605–611.
8. HAUPTS, U., J. TITTOR & D. OESTERHELT. 1999. Closing in on bacteriorhodopsin: progress in understanding the molecule. Annu. Rev. Biophys. Biomol. Struct. **28:** 367–399.
9. HENDERSON, R. *et al.* 1990. Model for the structure of bacteriorhodopsin based on high-resolution electron cryo-microscopy. J. Mol. Biol. **213:** 899–929.

10. LANYI, J.K. 1997. Mechanism of ion transport across membranes. Bacteriorhodopsin as a prototype for proton pumps. J. Biol. Chem. **272:** 31209–31212.
11. ABRAHAMS, J.P., A.G. LESLIE, R. LUTTER & J.E. WALKER. 1994. Structure at 2.8 A resolution of F1-ATPase from bovine heart mitochondria. Nature **370:** 621–628.
12. CAPALDI, R.A. & R. AGGELER. 2002. Mechanism of the F_1F_0-type ATP synthase, a biological rotary motor. Trends Biochem. Sci. **27:** 154–160.
13. FILLINGAME, R.H., W. JIANG & O.Y. DMITRIEV. 2000. Coupling H^+ transport to rotary catalysis in F-type ATP synthases: structure and organization of the transmembrane rotary motor. J. Exp. Biol. **203:** 9–17.
14. STOCK, D., A.G. LESLIE & J.E. WALKER. 1999. Molecular architecture of the rotary motor in ATP synthase. Science **286:** 1700–1705.
15. KARLIN, A. 1997. Transport bicycles. Proc. Natl. Acad. Sci. USA **94:** 5508–5509.
16. BUCH-PEDERSEN, M.J., K. VENEMA, R. SERRANO & M.G. PALMGREN. 2000. Abolishment of proton pumping and accumulation in the E_1P conformational state of a plant plasma membrane H^+-ATPase by substitution of a conserved aspartyl residue in transmembrane segment 6. J. Biol. Chem. **275:** 39167–39173.
17. PETROV, V.V. et al. 2000. Functional role of charged residues in the transmembrane segments of the yeast plasma membrane H^+-ATPase. J. Biol. Chem. **275:** 15709–15716.
18. POMES, R. & B. ROUX. 2002. Molecular mechanism of H conduction in the single-file water chain of the gramicidin channel. Biophys. J. **82:** 2304–2316.
19. DIOUMAEV, A.K. et al. 1998. Existence of a proton transfer chain in bacteriorhodopsin: participation of Glu-194 in the release of protons to the extracellular surface. Biochemistry **37:** 2496–2506.
20. LUECKE, H. 2000. Atomic resolution structures of bacteriorhodopsin photocycle intermediates: the role of discrete water molecules in the function of this light-driven ion pump. Biochim. Biophys. Acta **1460:** 133–156.
21. ZSCHERP, C. et al. 1999. In situ determination of transient pK_a changes of internal amino acids of bacteriorhodopsin by using time-resolved attenuated total reflection Fourier-transform infrared spectroscopy. Proc. Natl. Acad. Sci. USA **96:** 5498–5503.
22. VIK, S.B. & B.J. ANTONIO. 1994. A mechanism of proton translocation by F_1F_0 ATP synthases suggested by double mutants of the a subunit. J. Biol. Chem. **269:** 30364–30369.
23. VIK, S.B., A.R. PATTERSON & B.J. ANTONIO. 1998. Insertion scanning mutagenesis of subunit a of the F_1F_0 ATP synthase near His245 and implications on gating of the proton channel. J. Biol. Chem. **273:** 16229–16234.
24. JORGENSEN, P.L. & J.P. ANDERSEN. 1988. Structural basis for E_1-E_2 conformational transitions in Na,K-pump and Ca-pump proteins. J. Membr. Biol. **103:** 95–120.
25. BUKRINSKY, J.T., M.J. BUCH-PEDERSEN, S. LARSEN & M.G. PALMGREN. 2001. A putative proton binding site of plasma membrane H^+-ATPase identified through homology modelling. FEBS Lett. **494:** 6–10.
26. KUHLBRANDT, W., J. ZEELEN & J. DIETRICH. 2002. Structure, mechanism, and regulation of the neurospora plasma membrane H^+-ATPase. Science **297:** 1692–1696.
27. MENSE, M., V. RAJENDRAN, R. BLOSTEIN & M.J. CAPLAN. 2002. Extracellular domains, transmembrane segments, and intracellular domains interact to determine the cation selectivity of Na,K- and gastric H,K-ATPase. Biochemistry **41:** 9803–9812.
28. ALEXANDROV, N.N. & R. LUETHY. 1998. Alignment algorithm for homology modeling and threading. Protein Sci. **7:** 254–258.
29. AL-LAZIKANI, B., J. JUNG, Z. XIANG & B. HONIG. 2001. Protein structure prediction. Curr. Opin. Chem. Biol. **5:** 51–56.
30. MARTÍ-RENOM, M.A. et al. 2000. Comparative protein structure modeling of genes and genomes. Annu. Rev. Biophys. Biomol. Struct. **29:** 291–325.
31. AXELSEN, K.B. & M.G. PALMGREN. 2000. Evolution of substrate specificities in the P-type ATPase superfamily. J. Mol. Evol. **46:** 84–101.
32. AUER, M., G.A. SCARBOROUGH & W. KÜHLBRANDT. 1998. Three-dimensional map of the plasma membrane H^+-ATPase in the open conformation. Nature **392:** 840–843.
33. Ambesi, A., R.L. Pan & C.W. Slayman. 1996. Alanine-scanning mutagenesis along membrane segment 4 of the yeast plasma membrane H^+-ATPase. Effects on structure and function. J. Biol. Chem. **271:** 22999–23005.

34. DUTRA, M.B., A. AMBESI & C.W. SLAYMAN. 1998. Structure-function relationships in membrane segment 5 of the yeast Pma1 H^+-ATPase. J. Biol. Chem. **273:** 17411–17417.
35. GUPTA, S.S., N.D. ALLEN, K.E. DEWITT & C.W. SLAYMAN. 1998. Evidence for a salt bridge between transmembrane segments 5 and 6 of the yeast plasma-membrane H^+-ATPase. J. Biol. Chem. **273:** 34328–34334.
36. WACH, A., P. SUPPLY, J.P. DUFOUR & A. GOFFEAU. 1996. Amino acid replacements at seven different histidines in the yeast plasma membrane H^+-ATPase reveal critical positions at His285 and His701. Biochemistry **35:** 883–890.
37. MORSOMME, P., C.W. SLAYMAN & A. GOFFEAU. 2000. Mutagenic study of the structure, function and biogenesis of the yeast plasma membrane H^+-ATPase. Biochim. Biophys. Acta **1469:** 133–157.
38. SETO-YOUNG, D. *et al.* 1996. Genetic probing of the first and second transmembrane helices of the plasma membrane H^+-ATPase from *Saccharomyces cerevisiae*. J. Biol. Chem. **271:** 581–587.

Function and Regulation of the Two Major Plant Plasma Membrane H^+-ATPases

MAGDALENA WOLOSZYNSKA,[a] JUSTYNA KANCZEWSKA,[a]
ARTEM DRABKIN, OLIVIER MAUDOUX, STÉPHANIE DAMBLY,
AND MARC BOUTRY

*Unité de Biochimie physiologique, Institut des Sciences de la Vie,
Université catholique de Louvain, Croix du Sud, 2-20
B-1348 Louvain-la-Neuve, Belgium*

ABSTRACT: Plant plasma membrane H^+-ATPases are encoded by a family of about ten genes organized into five subfamilies. Subfamilies I and II contain the most widely and highly expressed genes. In *Nicotiana plumbaginifolia*, they are represented, respectively, by *pma2* (plasma membrane H^+-ATPase) and *pma4*. When expressed in the yeast *Saccharomyces cerevisiae*, the two isoforms show different kinetics and are differently regulated by phosphorylation of the penultimate threonine residue and binding of regulatory 14-3-3 proteins. To determine if these differences also occurred in plant tissues, we developed an experimental approach allowing the characterization of a single isoform in the plant. When PMA2 bearing a 6-His tag was expressed under a strong transcription promoter in *Nicotiana tabacum* BY2 cells, solubilized from microsomal membranes and purified, the penultimate threonine was found to be phosphorylated, thus validating the model.

KEYWORDS: plant; plasma membrane; H^+-ATPase; regulation; 14-3-3 protein; phosphorylation

INTRODUCTION

In plants, as in fungi, the plasma membrane H^+-ATPase is involved in establishing the transmembrane potential and pH gradient that activate various secondary transporters. This enzyme is thus functionally and structurally homologous to animal Na^+,K^+-ATPase.[1,2]

Plant plasma membrane H^+-ATPases are encoded by a gene family of about ten members organized into five subfamilies. The expression of each gene is differentially regulated according to cell type, developmental stage, and the environment.[3] In *Nicotiana plumbaginifolia*, the two most widely expressed genes are *pma2* (plasma membrane H^+-ATPase) and *pma4*, belonging, respectively, to subfamilies I and II.[3] Since different isoforms are expressed in a single cell type, this prevents

---

Address for correspondence: Marc Boutry, Unité de Biochimie physiologique, Institut des Sciences de la Vie, Université catholique de Louvain, Croix du Sud, 2-20, B-1348 Louvain-la-Neuve, Belgium. Voice: +32-10-473621; fax: +32-10-473872.
boutry@fysa.ucl.ac.be
[a]These two authors made an equal contribution.

their individual purification and the comparison of their biochemical and regulatory properties. To overcome this problem, we expressed *pma2* and *pma4* in the yeast, *Saccharomyces cerevisiae*, depleted of its own two H^+-ATPase genes, and found that each of the plant H^+-ATPase isoforms allowed yeast growth, thus demonstrating their functional expression in the heterologous host and allowing comparison of their biochemical properties.[4] However, the data obtained with the yeast expression system have to be validated in the plant, and we therefore developed an expression system adapted to plant cells that allowed us to characterize a single isoform.

MATERIALS AND METHODS

1. Yeast Strains, Growth Conditions, and Plasma Membrane Preparation

The yeast strains expressing His-tagged H^+-ATPases and mutants used were PMA2, PMA2-E14D[5] and PMA4, PMA4-A129P.[6] Yeast cells were grown as described by Dambly and Boutry,[6] and plasma membranes were prepared as described by Morsomme *et al.*[7]

2. Plasmid Construction and Transformation and Growth of Plant Cells

The sequence coding for PMA2 tagged with six histidines at the N-terminus[5] was placed under the control of the CaMV 35S promoter and introduced, via *Agrobacterium tumefaciens*, into cultured tobacco BY-2 cells, transformed cells being selected on a solid Murashige and Skoog medium[8] containing 70 μg/mL kanamycin. The cells were subcultured at intervals of 7 days and kept at 26°C in the dark on a rotary shaker (90 rpm).

3. Purification of the Histidine-tagged H^+-ATPase Expressed in Plant Cells

The 7-day-old cells were collected and suspended in 250 mM sorbitol, 60 mM tris-HCl, pH 8.0, 10 mM EDTA, 5 mM EGTA, 20% ethylene-glycol, 30 mM β-glycero-phosphate, 10 mM NaF, 5 mM Na molybdate, 5 mM DTT, 0.6% PVP, 7 M urea. The cells were homogenized with glass beads and centrifuged at 13,000 × g for 15 min, then the supernatant was centrifuged at 110,000 × g for 40 min and the pelleted microsomal fraction suspended in 250 mM sorbitol, 60 mM tris-HCl, pH 8.0, 20% ethylen-glycol, 30 mM β-glycero-phosphate, 10 mM NaF, 7 M urea. The H^+-ATPase was solubilized and purified by nickel chromatography, essentially as described by Maudoux *et al.*[5]

RESULTS

The regulation of plasma membrane H^+-ATPase activity involves phosphorylation, followed by the binding of 14-3-3 proteins.[1,2] These regulatory proteins are present in all organisms and, acting as a dimer, bind to phosphorylated targets. We previously showed that PMA2 expressed in yeast was partially phosphorylated on the penultimate amino acid, a threonine residue, and that subsequent binding of regulatory 14-3-3 proteins occurred, resulting in conversion from a low active to a high

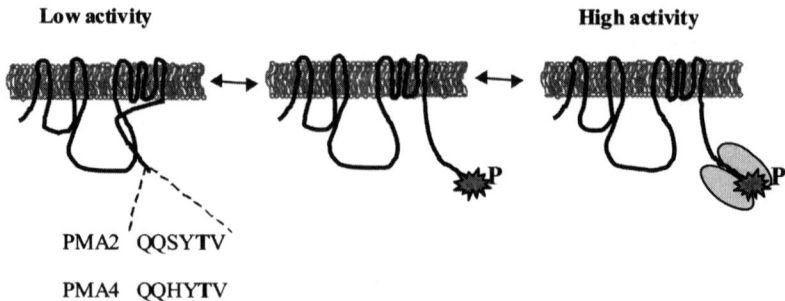

FIGURE 1. Model of plant plasma membrane H⁺-ATPase regulation. The low-activity state is maintained by interaction of the C-terminal autoinhibitory domain with the enzyme. H⁺-ATPase is activated when the autoinhibitory domain is displaced as the result of phosphorylation at the penultimate threonine residue, followed by the binding of 14-3-3 proteins. The last six residues of PMA2 or PMA4, which participate in the 14-3-3 binding motif, are indicated. The penultimate threonine residue is shown in *bold*.

active H⁺-ATPase[5] (FIG. 1). In contrast, PMA4 expressed in yeast was poorly phosphorylated and little binding of 14-3-3 proteins was observed[6] (FIG. 2). This difference was traced to a single residue two positions upstream from the phosphorylated residue;[6] in PMA2, this is a serine residue, while in PMA4, it is a histidine, and conversion of the PMA2 serine to histidine or of the PMA4 histidine to serine results in the opposite phenotype.[6] These data suggest that PMA2 and PMA4 are regulated differently.

Although the expression of PMA2 and PMA4 allowed yeast to grow in the absence of yeast H⁺-ATPase genes, the cells did not grow as well as wild-type yeast. This led to the selection of spontaneous mutants that were able to grow faster, especially when the pH of the external medium was decreased to pH 4.[7,9] Forty-two PMA2 mutants were characterized as single-point mutations, half of which were concentrated in the first part of the C-terminal region, thus defining the inhibitory domain, while the remainder were scattered in different domains of the H⁺-ATPase. Since this second group of mutants also had an improved ATPase activity,[7] one hypothesis would be that they induce a structural modification that makes the C-terminus more accessible to kinase and 14-3-3 proteins. To test this hypothesis, we used Western blotting to measure levels of 14-3-3 proteins in the plasma membrane fractions from two yeast strains expressing H⁺-ATPases containing mutations outside the C-terminal region, a PMA2 mutant (Asp14Glu; PMA2-E14D) and a PMA4 mutant (Ala129Pro; PMA4-A129P). We found higher levels of 14-3-3 proteins in both mutants compared to yeast strains expressing the wild-type proteins (FIG. 2). The higher level seen with the PMA2-E14D mutant compared to the PMA4-A129P mutant probably results from the combined effect of a more accessible C-terminal region (due to the mutation) and to the fact that, as explained previously, PMA2 has a better context for phosphorylation than PMA4.

Although heterologous expression of plant proteins in yeast is a powerful tool, it raises the question of whether the results obtained also apply to plant cells. To address this question we investigated the regulation of an individual H⁺-ATPase iso-

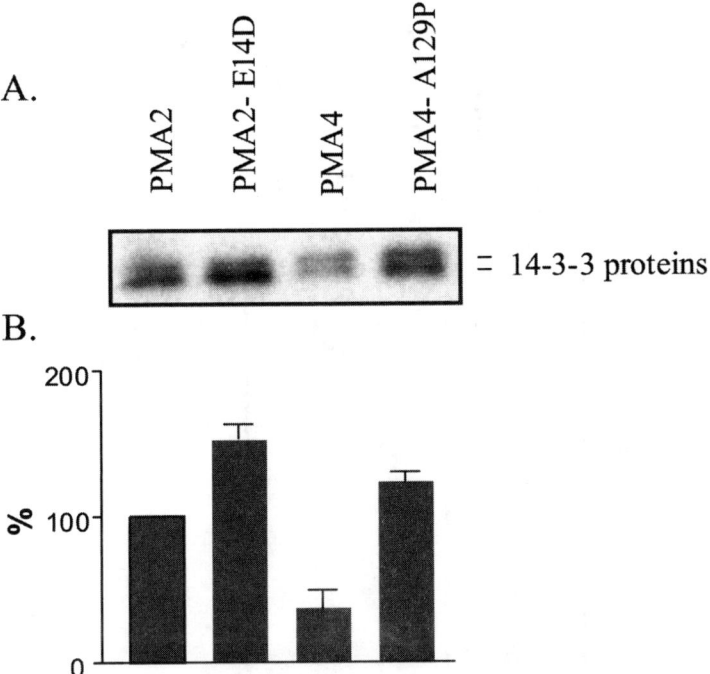

FIGURE 2. Binding of 14-3-3 proteins to PMA2, PMA2-E14D, PMA4, and PMA4-A129P. (**A**) Plasma membranes (2 µg) purified from yeast strains expressing the indicated PMA isoforms were analyzed by Western blotting using antibodies against 14-3-3 proteins. (**B**) Quantification of 14-3-3 binding using a PhosphoImager (BioRad). 100% corresponds to the signal obtained for PMA2.

form of *N. plumbaginifolia* in plant cells. *N. tabaccum* BY-2 cells were transformed with a plasmid containing the gene coding for PMA2 equipped with 6-His tag at the N-terminus, and PMA2 was selectively purified by nickel chromatography (FIG. 3) under denaturing conditions (7 M urea). After Coomassie Blue staining, a major band with a molecular weight of approximately 100 kDa was detected in the purified fraction, and Western blot analysis (FIG. 3) confirmed that this band was PMA2. To check whether PMA2 expressed in BY-2 cells was phosphorylated at the threonine residue, as in yeast, the membrane was probed with antiphosphothreonine antibodies (FIG. 3). The positive result indicated that PMA2 can be phosphorylated *in vivo* in plant cells and could be purified in the phosphorylated form. We now have a system that makes it possible to study how a single plant H^+-ATPase isoform expressed in plant cells, is regulated by phosphorylation in response to various environmental stresses.

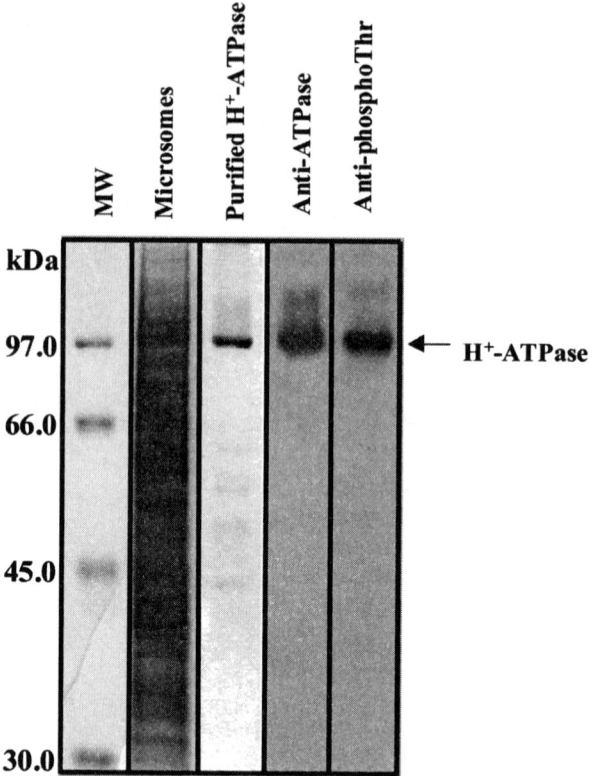

FIGURE 3. Purification and immunodetection of the *N. plumbaginifolia* PMA2 isoform of H^+-ATPase expressed in *N. tabaccum* BY-2 cells. The 6-His tagged PMA2 was purified from the microsomal fraction, analyzed by SDS-PAGE, and stained with Coomassie Blue (*lane 3*). The same sample was also transferred to a membrane and probed with antibodies against ATPase (*lane 4*) or phosphothreonine (Zymed Laboratories) (*lane 5*). The *arrow* indicates PMA2 H^+-ATPase. MW corresponds to molecular-weight markers.

ACKNOWLEDGMENTS

The work performed in this laboratory was supported by grants from the Interuniversity Poles of Attraction Program (Belgian State, Scientific, Technical and Cultural Services) the European Community, the Belgian Fund for Scientific Research, and the Human Frontier Science Program.

REFERENCES

1. PALMGREN, M.G. 2001. Plant plasma membrane H^+-ATPases: powerhouses for nutrient uptake. Annu. Rev. Plant Physiol. Plant Mol. Biol. **52:** 817–845.
2. MORSOMME, P. & M. BOUTRY. 2000. The plant plasma membrane H^+-ATPase: structure, function and regulation. Biochim. Biophys. Acta **1465:** 1–16.

3. ARANGO, M. *et al.* 2003. The plasma membrane proton pump-ATPase: the significance of gene subfamilies. Planta **216**: 355–365.
4. LUO, H., P. MORSOMME & M. BOUTRY. 1999. The two major types of plant plasma membrane H^+-ATPases show different enzymatic properties and confer differential pH sensitivity of yeast growth. Plant Physiol. **119**: 627–634.
5. MAUDOUX, O. *et al.* 2000. A plant plasma membrane H^+-ATPase expressed in yeast is activated by phosphorylation at its penultimate residue and binding of 14-3-3 regulatory proteins in the absence of fusicoccin. J. Biol. Chem. **275**: 17762–17770.
6. DAMBLY, S. & M. BOUTRY. 2001. The two major plant plasma membrane H+-ATPases display different regulatory properties. J. Biol. Chem. **276**: 7017–7022.
7. MORSOMME, P. *et al.* 1998. Single point mutations distributed in 10 soluble and membrane regions of the *Nicotiana plumbaginifolia* plasma membrane PMA2 H^+-ATPase activate the enzyme and modify the structure of the C-terminal region. J. Biol. Chem. **273**: 34837–34842.
8. MURASHIGE, T. & F. SKOOG. 1962. A revised medium for rapid growth and bioassay with tobacco tissue cultures. Physiol. Plant. **15**: 473–497.
9. MORSOMME, P. *et al.* 1996. Single point mutations in various domains of a plant plasma membrane H^+-ATPase expressed in *Saccharomyces cerevisiae* increase H^+-pumping and permit yeast growth at low pH. EMBO J. **15**: 5513–5526.

Functional Properties of the Human Copper-Transporting ATPase ATP7B (the Wilson's Disease Protein) and Regulation by Metallochaperone Atox1

SVETLANA LUTSENKO, RUSLAN TSIVKOVSKII, AND JOEL M. WALKER

Department of Biochemistry and Molecular Biology, Oregon Health & Science University, Portland, Oregon 97239, USA

ABSTRACT: Wilson's disease protein (WNDP) is a copper-transporting P_1-type ATPase which plays a key role in normal distribution of copper in a number of tissues, particularly in the liver and the brain. Copper has numerous effects on WNDP, altering its structure, activity, and intracellular localization. To better understand the function of this copper-transporting ATPase and its regulation by copper, we have recently developed the functional expression systems for WNDP and for Atox1, a cytosolic protein that serves as an intracellular donor of copper for WNDP. Here we summarize the results of our experiments on characterization of the enzymatic properties of WNDP and the effects of Atox1 on the WNDP activity.

KEYWORDS: ATP7B; copper; Wilson's disease; metallochaperone; Atox1; regulation

INTRODUCTION

Wilson's disease protein (WNDP) is a copper-transporting P-type ATPase with an essential role in human physiology. Mutations in this protein lead to copper accumulation in a number of tissues and to severe hepatic and neurological abnormalities.[1,2] Despite its important role, the function and regulation of WNDP remain poorly understood. Cell-biological experiments revealed that WNDP is located in the *trans*-Golgi network (TGN), where it delivers copper to such secreted copper-dependent enzymes as ceruloplasmin (FIG. 1). This primary location for WNDP is altered when concentration of copper in a cell is elevated.[3–5] In the presence of high copper, WNDP traffics to the vesicles located in close proximity to the apical plasma membrane (FIG. 1). Although the exact nature of this vesicular compartment is still un-

---

Address for correspondence: Svetlana Lutsenko, Department of Biochemistry and Molecular Biology, Oregon Health & Science University, Portland, OR 97239. Voice: 503-494-6953; fax: 503-494-8393.

lutsenko@ohsu.edu

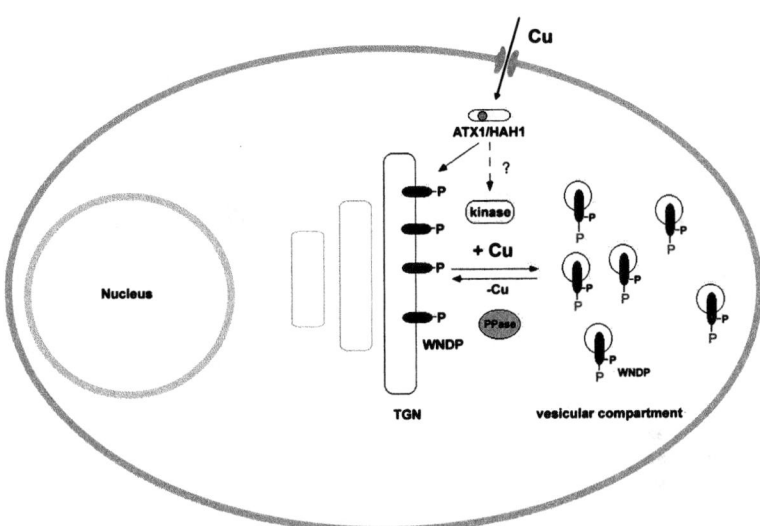

FIGURE 1. Copper regulates the intracellular distribution of WNDP. WNDP is predominantly located in the *trans*-Golgi network (TGN), where it delivers copper to the copper-dependent enzymes, and where it has a basal level of a kinase-mediated phosphorylation. When concentration of copper increases, more copper is delivered to WNDP by metallochaperone Atox1. Under these conditions, WNDP relocates to the vesicular compartment, where it is hyperphosphorylated. In this compartment, the major function of WNDP is copper detoxification. *Black and gray letters* "P" represent phosphorylation of different sites in WNDP.

known, the transport of copper into the vesicles appears to be the first step leading to removal of excess copper from the cell. Thus, in a cell WNDP has a dual role: (1) the delivery of a cofactor to the copper-dependent enzymes, and (2) metal detoxification. To accomplish these functions, WNDP responds to modulations in copper concentration by relocating from one compartment to the other and, possibly, by changing its activity.

In the past several years, our group has been interested in how changes in copper concentration regulate the intracellular localization and activity of WNDP. Our studies suggest that the copper-dependent phosphorylation of WNDP by a kinase could be involved in determining the intracellular distribution of WNDP.[6] Specifically, we found that in response to a change in copper concentration, WNDP becomes hyperphosphorylated (most likely on Ser or Thr residue), while a subsequent decrease in copper concentration leads to protein dephosphorylation. These changes in protein phosphorylation levels coincide with the intracellular location of WNDP.[6] Low phosphorylation is observed in TGN, while hyperphosphorylation occurs in the vesicular compartment, suggesting that phosphorylation could be a mechanism that controls intracellular distribution of the copper-transporting ATPase.[6] These findings raised many new questions. What is the molecular basis of copper signaling? Which kinase(s) and phosphatase(s) are involved in this process? What is the functional activity of WNDP in distinct cell compartments? and How is this activity regulated by copper? This paper summarizes the results of recent experiments investigating how changes in copper concentration are detected by WNDP and how WNDP responds to these changes.

METALLOCHAPERONES ARE ESSENTIAL COMPONENTS OF THE COPPER-TRANSPORTING MACHINERY

One of the interesting aspects of copper homeostasis in eucaryotic cells is the lack of free copper ions in the cytosol.[7] It is believed that copper transported into cells quickly becomes bound to the soluble cytosolic proteins named metallochaperones. These proteins deliver the metal to specific protein targets in several cell compartments.[8,9] Recent genetic and biochemical studies revealed that the human copper chaperone Atox1 (also known as HAH1) is important for delivery of copper to the secretory pathway,[10] suggesting that it works together with the copper-transporting ATPases (FIG. 1). Atox1 is an 8-kDa cytosolic protein that contains a single metal-binding motif MxCxxC. In contrast, WNDP is a 165-kDa membrane protein, which has seven metal-binding sites: six sites in the N-terminal domain, each containing the GMxCxxC motif, and one CPC site in the membrane portion of WNDP (FIG. 2A). The high-resolution structure of Atox1 was recently determined and was shown to be very similar to the structure of the N-terminal metal-binding subdomains of the copper-transporting ATPases.[11] Atox1 was also shown to interact with the N-terminal domain of WNDP[12] suggesting that Atox1 may transfer copper to the copper-ATPase via ligand exchange mechanism. We hypothesized that Atox1 could work as an intracellular copper sensor that regulated the WNDP copper occupancy and possibly the enzymatic activity of WNDP. To test this hypothesis, we carried out a series of experiments using heterologously expressed Atox1 and WNDP.[13]

FUNCTIONAL EXPRESSION AND ENZYMATIC PROPERTIES OF WNDP

For functional analysis of WNDP, our group has recently developed an expression system using baculovirus-mediated infection of insect cells.[14] This expression system allowed us to produce substantial amounts of protein and to characterize the major enzymatic properties of WNDP. Using this system, we found that upon addition of γ-[^{32}P]ATP WNDP formed a transient phosphorylated intermediate, which was sensitive to basic pH and to treatment with hydroxylamine.[14] WNDP can be dephosphorylated by the addition of ADP, indicating that the phosphorylation reaction is reversible. In addition, WNDP can be phosphorylated from inorganic phosphate (Pi) in the presence of Mg^{2+} (FIG. 2B). The activity of WNDP depends on the presence of the transported ion: addition of the specific copper chelator bathocuproine disulphonate (BCS) inhibits catalytic phosphorylation, and this effect can be reversed by addition of free copper.[14] All these properties are typical for the members of the P-type ATPase family.[15] Like all other P-type ATPases, WNDP also has a signature motif DKTG (FIG. 2A). The invariant aspartyl residue in this motif is predicted to be a target of catalytic phosphorylation in all P-type ATPases. In agreement

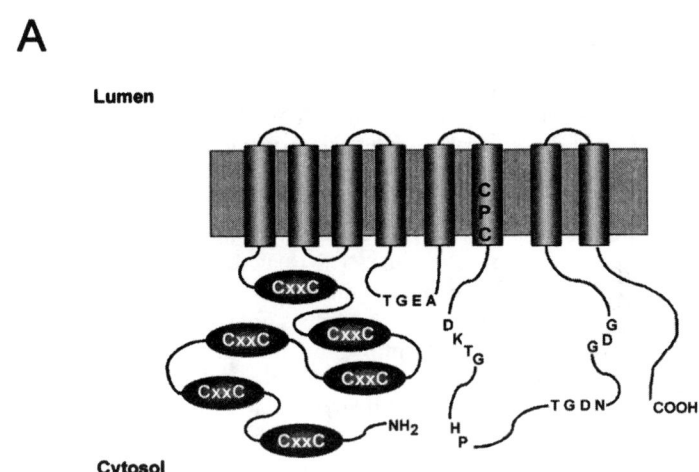

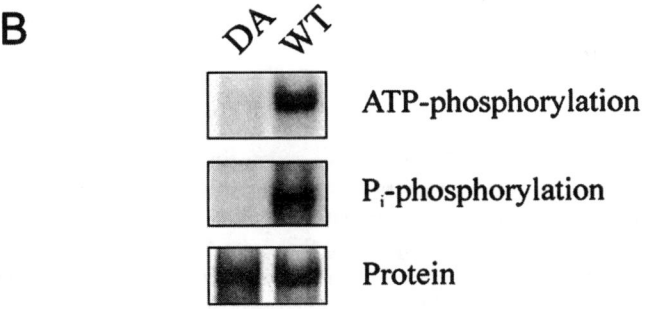

FIGURE 2. The characteristic structural features and catalytic phosphorylation of WNDP. (**A**) The transmembrane organization of WNDP. The ovals with the CxxC sequence represent six N-terminal metal-binding sites containing the conserved copper-binding motif GMxCxxC. "C" is a Cys residue involved in copper coordination; DKTG, TGEA, TGDN, and GDG are the sequence motifs characteristic for all P-type ATPases; CPC and HP sequences are conserved in the P_1-type ATPases. (**B**) Formation of phosphorylated intermediate following incubation with either γ-[^{32}P]-ATP,Mg (ATP-phosphorylation) or with Mg[^{32}Pi] (Pi-phosphorylation). WT is a wild-type WNDP; DA is the WNDP mutant in which aspartate in the DKTG motif is replaced with alanine. Both the DA mutant and the wild-type WNDP are expressed at a comparable level, as indicated by Coomassie-R250 staining (Protein).

with this prediction, the $D > A$ substitution abolishes catalytic phosphorylation of WNDP either from ATP or from Mg^{2+},Pi (FIG. 2B).

ATOX1 REGULATES THE COPPER OCCUPANCY AND CATALYTIC ACTIVITY OF WNDP

As was mentioned earlier, in a cell all copper is thought to exist in a bound form, most likely as a complex with several metallochaperones. To test whether metallochaperone Atox1 serves as a copper donor for WNDP, we produced a highly pure Atox1 using the intein-based protein-expression system (New England Biolabs) and demonstrated that Atox1 could be loaded with copper in the presence of glutathione. This procedure generates the Cu^{+1}–Atox1 complex, in which copper is bound to protein with stoicheometry close to 1:1.[13] Further experiments revealed that the Cu^{+1}–Atox1 complex restored the ability of the BCS-treated WNDP to form a catalytic intermediate as efficiently as free copper, indicating that copper was transferred from the metallochaperone to WNDP. Apo–Atox1 does not stimulate the catalytic activity of the BCS-inactivated WNDP, confirming that transfer of copper and not just the metallochaperone is essential for the WNDP function.[13]

Additional experiments determined that stimulation of the catalytic activity of WNDP by Cu^{+1}–Atox1 is likely to be a result of copper transfer from the chaperone to the N-terminal domain of WNDP (N-WNDP). To demonstrate the ability of Atox1 to transfer copper *in vitro* we developed the simple but efficient protocol shown in FIGURE 3. In this procedure, the increasing amounts of purified Cu^{+1}–Atox1 complex are incubated with the maltose-binding protein fusion of N-WNDP (N-WNDP-MBP) bound to amylose resin. This leads to a rapid and saturable copper transfer reaction. Then, Atox1 is washed off the resin and the N-WNDP is eluted with the maltose-containing buffer. The amount of copper bound to either partner before and after the reaction can be measured spectrophoptometrically. Using this procedure, we found that Atox1 could fill all six metal-binding sites in N-WNDP with copper.[13]

It is currently unknown whether Atox1 exists in a cell in the predominantly copper-bound form or in the apo form. It seems likely that both forms are present and that their ratio depends on how much copper is delivered into the cell by the copper-uptake system(s). We hypothesized that apo–Atox1, when present in excess over WNDP, could remove copper from WNDP and downregulate its activity. Further experiments confirmed this hypothesis. By using the procedure shown in FIGURE 3 in reverse, that is, by incubating apo–Atox1 with the copper-bound N-WNDP-MBP, we demonstrated that apo–Atox1 was able to strip copper from N-WNDP. In contrast to the forward copper transfer reaction, the reverse reaction is incomplete and one copper remains tightly bound to N-WNDP, even when a large excess of apo–Atox1 is used.[13]

Incubation of the full-length copper-bound WNDP with apo–Atox1 leads to a decrease in the WNDP activity, suggesting that Atox1 can regulate the WNDP function by controlling the amount of copper bound to this copper-transporting ATPase. At the same time, the activity of WNDP remains substantial (about 50%), even when apo–Atox1 is used in quantities sufficient to remove most of the copper from the N-terminal domain of WNDP.[13] This result indicates that occupation of all copper-

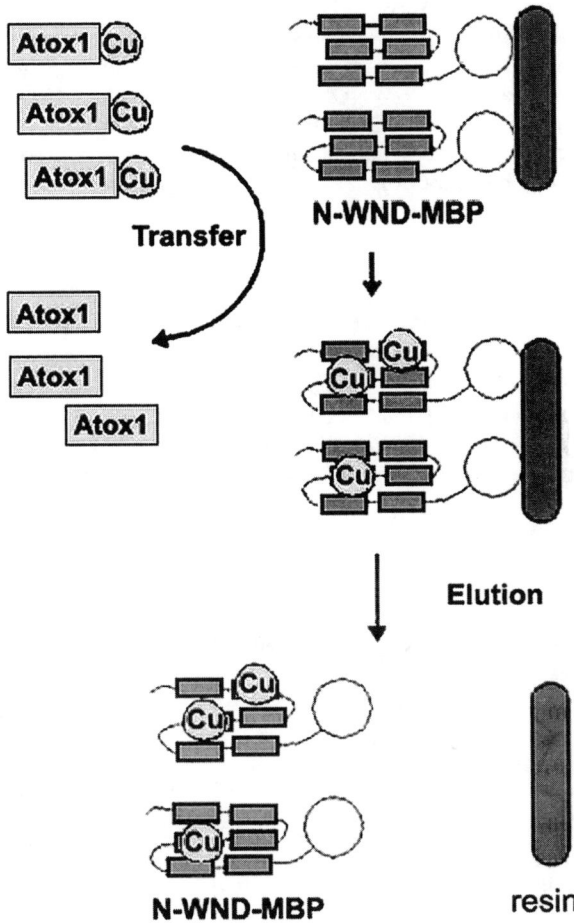

FIGURE 3. Schematic representation of the copper-transfer protocol. N-WNDP-MBP is the fusion of the maltose-binding protein and the N-terminal copper-binding domain of WNDP, and the six *gray block*s represents six-metal-binding sites. Atox1 is the metallochaperone; "Cu" in a *circle* indicates copper ion.

binding sites is not necessary to support the catalytic activity of WNDP, and that WNDP can function fairly efficiently in a wide range of copper concentrations.

Altogether, the current results can be described by the following working model. When copper concentration in a cell is low, only a limited amount of copper–Atox1 complexes is available. Under these conditions the copper transfer reaction proceeds with low efficiency and metal occupancy of the multiple copper-binding sites in WNDP is partial. Partial loading of WNDP with copper is sufficient to sustain basal catalytic and transport activity and maintain necessary delivery of copper into TGN. A rise in copper uptake increases the amount of copper–Atox1 complexes and stimulates delivery of copper to WNDP. The change in copper occupancy of N-WNDP is associated with conformational changes,[16,17] which in turn may serve as a triggering signal for the WNDP phosphorylation by the kinase and trafficking. Also, increased metal occupancy of N-WNDP and the copper-induced structural changes can stimulate the activity of WNDP in the vesicular compartment, helping efficient copper detoxification. Decrease in the intracellular copper would drive all these steps in reverse. Our current efforts are focused on careful testing of various aspects of this model.

ACKNOWLEDGMENTS

This work was supported by National Institute of Health Grant DK55719 to S.L. Another of the authores (R.T.) is a recipient of a postdoctoral fellowship from the American Heart Association (#0120573Z). J.M.W. was supported by NIH Training Grant HL07781. The authors thank Ms.Tina Purnat for help with the preparation of the manuscript.

REFERENCES

1. SCHEINBERG, I.H. & I. STERNLIEB. 1984. Wilson's Disease, Vol. 23, W. B. Saunders. Philadelphia.
2. LOUDIANOS, G. & J.D. GITLIN. 2000. Wilson's disease. Semin. Liver Dis. **20:** 353–364.
3. PAYNE, A.S., E.J. KELLY & J.D. GITLIN. 1998. Functional expression of the Wilson disease protein reveals mislocalization and impaired copper-dependent trafficking of the common H1069Q mutation. Proc. Natl. Acad. Sci. USA **95:** 10854–10859.
4. SCHAEFER, M., R.G. HOPKINS, M.L. FAILLA & J.D. GITLIN. 1999. Hepatocyte-specific localization and copper-dependent trafficking of the Wilson's disease protein in the liver. Am. J. Physiol. **276:** G639–G646.
5. ROELOFSEN, H. *et al.* 2000. Copper-induced apical trafficking of ATP7B in polarized hepatoma cells provides a mechanism for biliary copper excretion. Gastroenterology **119:** 782–793.
6. VANDERWERF, S.M., M.J. COOPER, I.V. STETSENKO & S. LUTSENKO. 2001. Copper specifically regulates intracellular phosphorylation of the Wilson's disease protein, a human copper-transporting ATPase. J. Biol. Chem. **276:** 36289–36294.
7. RAE, T.D. *et al.* 1999. Undetectable intracellular free copper: the requirement of a copper chaperone for superoxide dismutase. Science **284:** 805–808.
8. O'HALLORAN, T.V. & V.C. CULOTTA. 2000. Metallochaperones: an intracellular shuttle service for metal ions. J. Biol. Chem. **275:** 25057–25060.
9. HARRISON, M.D., C.E. JONES, M. SOLIOZ & C.T. DAMERON. 2000. Intracellular copper routing: the role of copper chaperones. Trends Biochem. Sci. **25:** 29–32.
10. HAMZA, I., M. SCHAEFER, L.W.J. KLOMP & J.D. GITLIN. 1999. Interaction of the copper chaperone HAH1 with the Wilson disease protein is essential for copper homeostasis. Proc. Natl. Acad. Sci. USA **96:** 13363–13368.
11. WERNIMONT, A.K. *et al.* 2000. Structural basis for copper transfer by the metallochaperone for the Menkes/Wilson disease proteins. Nat. Struct. Biol. **7:** 766–771.
12. LARIN, D. *et al.* 1999. Characterization of the interaction between the Wilson and Menkes disease proteins and the cytoplasmic copper chaperone, HAH1P. J. Biol. Chem. **274:** 28497–28504.
13. WALKER, J.M., R. TSIVKOVSKII & S. LUTSENKO. 2002. Metallochaperone atox1 transfers copper to the NH2-terminal domain of the Wilson's disease protein and regulates its catalytic activity. J. Biol. Chem. **277:** 27953–27959.

14. TSIVKOVSKII, R., J.F. EISSES, J.H. KAPLAN & S. LUTSENKO. 2002. Functional properties of the copper-transporting ATPase ATP7B (the Wilson's disease protein) expressed in insect cells. J. Biol. Chem. **277:** 976–983.
15. MOLLER, J.V., B. JUUL & M. LE MAIRE. 1996. Structural organization, ion transport, and energy transduction of P-type ATPases. Biochim. Biophys. Acta **1286:** 1–51.
16. DIDONATO, M. *et al.* 2000. Copper-induced conformational changes in the N-terminal domain of the Wilson disease copper-transporting ATPase. Biochemistry **39:** 1890–1896.
17. TSIVKOVSKII, R., B.C. MACARTHUR & S. LUTSENKO. 2001. The Lys1010-Lys1325 fragment of the Wilson's disease protein binds nucleotides and interacts with the N-terminal domain of this protein in a copper-dependent manner. J. Biol. Chem. **276:** 2234–2242.

Heavy Metal Transport CPx-ATPases from the Thermophile *Archaeoglobus fulgidus*

JOSÉ M. ARGÜELLO, ATIN K. MANDAL, AND SEBASTIAN MANA-CAPELLI

Department of Chemistry and Biochemistry, Worcester Polytechnic Institute, Worcester, Massachusetts 01609, USA

ABSTRACT: PIB-type ATPases transport diverse heavy metals (Cu^+, Ag^+, Cu^{2+}, Zn^{2+}, Cd^{2+}, Pb^{2+}, Co^{2+}) across membranes. Toward understanding their mechanisms of metal selectivity, we are studying thermophilic archaeal PIB-type ATPases. Like other PIB ATPases, these are characterized by the presence of a cation binding CPX sequence in their 6th transmembrane segment and by cytoplasmic N-terminus metal binding domains (N-MBDs). CopA and CopB from the thermophile *Archaeoglobus fulgidus* were cloned and expressed in *E. coli*. The resulting proteins were purified in a soluble active form. Typical yields were in the order of 3–5 mg of pure protein per liter of bacterial culture. Both enzymes showed maximum activity at 75–85°C. CopA was activated by $Ag^+ > Cu^+$ while CopB was activated by $Cu^{2+} > Ag^+ > Cu^+$. The differences in enzyme selectivity can be explained by different consensus sequences in the transmembrane cation binding domain (CopA: CPC, CopB: CPH). Mutagenesis studies show that the cysteines in the transmembrane CPC site of CopA are necessary for enzyme function, while those in the N-MBD (CXXC), although not essential, are required for maximum enzyme activity. Different from CopA, CopB has a His-rich N-MBD. Removal of this domain reduced enzyme activity without affecting enzyme selectivity. These studies show that these enzymes are an excellent system for structural functional studies directed to explain the mechanisms of metal selectivity by PIB ATPases.

KEYWORDS: heavy metals; CPx-ATPases; CopA; CopB; copper; thermophilic ATPase; *Archaeoglobus fulgidus*

INTRODUCTION

More than ninety genes encoding PIB ATPases have been identified in archaea, bacteria, and eukaryotes.[1] CPx-ATPases (so-called for the sequence, CP[C/H/S], present in their putative transmembrane metal binding site) confer metal tolerance to microorganisms, and are essential elements in the absorption, distribution, and bioaccumulation of metal micronutrients by higher organisms.[2] Their relevance is evident when the two human Cu^+-ATPases are considered. Mutations in these ATPases are responsible for Menkes' and Wilson's diseases.[3] The abundance of CPx-ATPases in plants (eight genes in *Arabidopsis thaliana*) also suggests fundamental and com-

---

Address for correspondence: Dr. José Argüello, Department of Chemistry and Biochemistry, Worcester Polytechnic Institute, Worcester, MA 01609. Voice: 508-831-5326; fax: 508-831-5933. arguello@wpi.edu

plex roles in micronutrient metal metabolism.[1] The formation of the phosphorylated intermediate and basic transport properties have been established for some of CPx-ATPases.[2,4–9] All tested enzymes drive the export of ions out of cells. Although one report showed that CopA, a Cu^+-ATPase from *E. hirae*, appears to pump Cu^+ into cells under conditions of limiting Cu^+,[7] studies of homologous enzymes from *E. coli*[6] and *A. fulgidus*[5] indicated that CopA drives Cu^+ efflux.

A particular characteristic of CPx-ATPase is its capacity to transport diverse metals. It has been proposed that the CPx sequences in H6 (CPC, CPH, SPC or CPS) participate in metal binding and transport.[1–3] The mutation of CPC→CPA in the *C. elegans* Cu-ATPase yielded a protein unable to rescue a Cu-ATPase deficient yeast mutant (Δccc2),[9] while *E. hirae* CopB carrying replacement CPH→SPH cannot rescue a CopB knockout mutant.[8] However, the relationship between ion specificity and the various signature sequences remains to be established. For instance, *in vivo* functional complementation (conferring metal resistance to bacteria) and *in vitro* functional assays, have shown that Ag^+ can also activate Cu^+-ATPases[4–6] while Zn^{2+}-ATPases can use Cd^{2+} and Pb^{2+} as substrates.[10,11] Moreover CoaT from *Synechocystis* PCC 6803 appears to transport Co^{2+}.[12] It is also puzzling that enzymes that transport monovalent (Cu^+, Ag^+), as well as those specific for divalent metals (Zn^{2+}, Cd^{2+}, Pb^{2+}) carry a CPC sequence in H6.[4,5,10,11] Alternatively, enzymes carrying CPC and others having the CPH sequence in H6 have been reported as specific for Cu^+/Ag^+.[4,5]

Most CPx-ATPases have metal binding domains in their N-terminus (N-MBDs). Eukaryotic enzymes have 2-6 repeats (each ≈ 60 aa) carrying the metal binding consensus GMTCXXC, while simpler, single CXXC N-MBDs are present in most prokaryotic proteins. However, N-MBDs do not appear to participate in determining the metal specificity. Several lines of evidence suggest that the N-MBDs have a regulatory role,[13,14] but a direct link between domain interactions and change in a rate-limiting step has not been established. A different type of N-MBD found in archaea and bacteria is characterized by a high histidine content.[2,7] There is no experimental evidence of metal binding to these "His-rich" N-MBDs and their function is unknown. Finally, a significant number of putative CPx-ATPases have no evident N-MBD and a few have a C-terminal MBD.

Our goal is to understand the mechanisms of metal selectivity in PIB ATPases. Toward this end, we initiated the study of representative enzymes with particular metal selectivity and structural features, CopA and CopB from *A. fulgidus*.

METHODS

Cloning, Expression, and Site-Directed Mutagenesis

The cDNA of genes AF0473 (CopA) and AF0153 (CopB) in the *A. fulgidus* genome were generated by polymerase chain reaction, using genomic DNA (ATCC, Manassas, VA) as template. The resulting cDNAs were subcloned into pBADTOPO/His vector (Invitrogen, Carlsbad, CA) that introduces a carboxyl terminal hexahistidine tag. An additional construct encoding a truncated CopB (lacking the first 51 amino acids) was also made. The Quick Change™ kit (Stratagene, La Jolla, CA) was used for site-directed mutagenesis. *E. coli* Top10 cells (Invitrogen, Carlsbad,

TABLE 1. Subgroups of CPx-ATPases based on their structure and ion specificity

Type	Sequences	Length (approx.)	No. of TM	H6 Cons.	N-MBD	No. of CxxC	Transported Ion
IB-1	16	1500	8	CPC	GMTCxxC	up to 6	Cu^+, Ag^+
IB-2	42	750	8	CPC	CxxC	up to 3	Cu^+, Ag^+
IB-3	15	700	8	CPC	CxxC	1	$Zn^{2+}, Cd^{2+}, Pb^{2+}$
IB-4	7	750	8	CPH	His-rich	0	Cu^{2+}, Cu^+, Ag^+
IB-5	9	700	6/8?	SPC	none	0	Co^{2+}
IB-6	3	750	8	CPS	CxxC/none	0–1	?

CA) were transformed with these constructs and expression induced with 0.002% L-arabinose.

Protein Purification

Protein purification was carried out as described.[5] In brief: *E. coli* membranes were treated with 0.75% dodecyl-β-D-maltoside (DDM) (Calbiochem, San Diego, CA), and subsequently solubilized membrane proteins were subject to Ni-affinity chromatography to isolate the expressed proteins.

Functional Assays

ATPase activity, phosphorylation, and dephosphorylation determinations were performed as previously described.[5]

RESULTS

To choose adequate models for our studies, we analyzed the available PIB ATPase sequences (<http://biobase.dk/~axe/Patbase.html>).[1] We mainly considered the two structural characteristics that differentiate CPx-ATPases: (*a*) the signature sequence (CPC, CPH, SPC or CPS) in H6; and (*b*) the ion metal binding sites in the cytoplasmic N-terminal region (N-MBD). As a result six protein groups were identified: Group IB-1 includes eukaryote proteins carrying 2-6 N-MBDs. All tested proteins in IB-1 seem to transport Cu^+.[2,3,9,13] Enzymes of group IB-2 are smaller bacterial proteins that appear to transport Cu^+/Ag^+.[5,6] Similar to enzymes of group IB-2, proteins in group IB-3 have CPC in H6 and CXXC in their N-MBD. Significantly, enzymes from group IB-3 transport Zn^{2+}, Cd^{2+}, and Pb^{2+}.[10,11] A His-rich N-MBD (≈30 aa containing >10 His in no apparent pattern) instead of CXXC distinguished the IB-4 group. Proteins in this group also carry a His in the consensus sequence in H6 (CPH). It has been reported that the group IB-4 enzyme CopB from *E. hirae* is activated by μM Cu(I).[4] However, as we report below, *A. fulgidus* CopB is mainly activated by Cu(II). The proteins in group IB-5 are unique, since they have the sequence SPC in H6. The only member of this group that has been studied is CoaT from *Synechocystis* PCC 6803.[12] This appears to confer Co^{2+} tolerance. Group IB-6 includes three proteins with the CPS sequence in H6. None of them has been characterized.

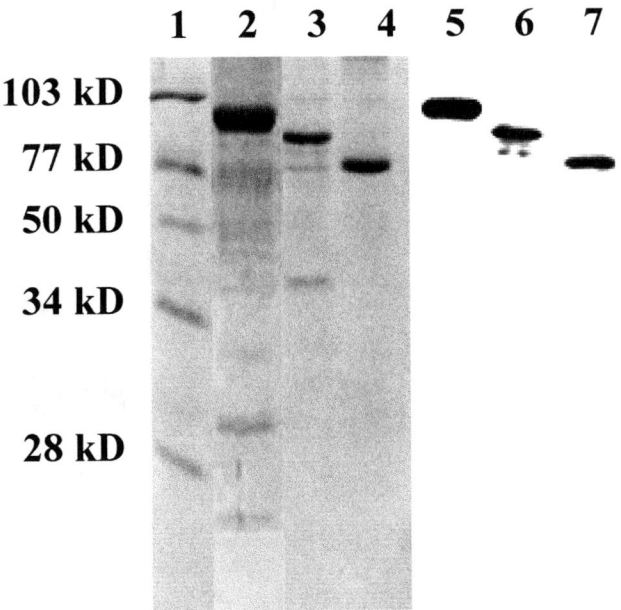

FIGURE 1. Expression and purification of CopA, CopB, and truncated CopB from *A. fulgidus*. Lanes 2 and 5: CopA; lanes 3 and 6: CopB; lanes 4 and 7: truncated CopB. 5 µg of protein were loaded in each lane. Lanes 1–3, and 4: 10% SDS-PAGE stained with Coomassie Brilliant Blue. Lanes 5, 6, and 7: blot immunostained with anti-V5 epitope antibody.

Considering these groups, we initiated the characterization of a representative protein from group IB-2 (CopA) and one from IB-4 (CopB). We chose to work thermophilic PIB-ATPases from *Archaeoglobus fulgidus* because there is no information on thermophilic P-type ATPases and because these might present advantages in structural studies. However, little information is available on the expression of thermophilic membrane proteins in mezophilic organisms. The uncertainty is relevant, taking into account the different lipid composition of thermophilic and mezophilic organisms. FIGURE 1 shows that both wild-type enzymes CopA and CopB, as well as the truncated form of CopB, can be produced in *E. coli* and obtained in a pure form after solubilization and affinity chromatography. Yields on the order of 3–5 mg of purified protein per liter of bacterial culture were routinely obtained. The isolated proteins were active at high temperature (CopA: max. activity at 75°C, CopB: max. activity at 85°C), with energies of activation of 100–105 kJ/mol.

As expected from its homology with other members of group IB-2, CopA was activated by Ag^+ and Cu^+ (FIG. 2A) and was not activated by Cu^{2+}, Zn^{2+}, Pb^{2+}, or Cd^{2+} (not shown). The V_{max} in the presence of Ag^+ was approximately four times larger than that observed in the presence of Cu^+ (in both cases 100 µM metal). This was correlated with a faster rate of dephosphorylation of the Ag^+-bound phosphoenzyme compared to the Cu^+-bound phosphoenzyme. Dephosphorylation was initiated by adding 1 mM ATP to phosphorylation mix after 9-sec incubation; $E2(Ag^+)P$ and

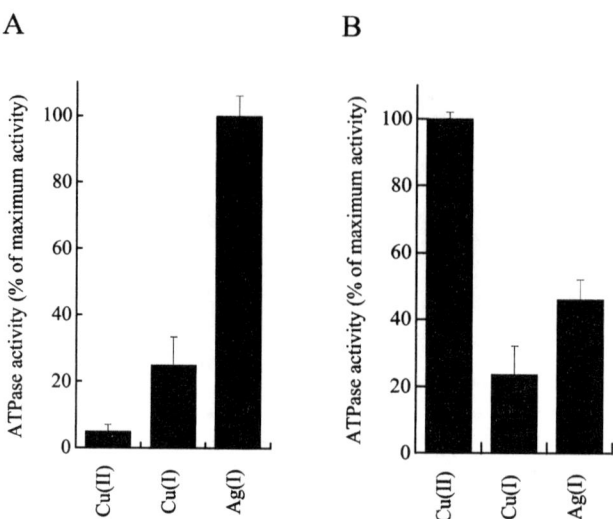

FIGURE 2. (A) Activation of CopA ATPase by metals (100 μM). (B) Activation of CopB ATPase by metals (1 μM). In both cases 2.5 mM dithiotreitol was included in Cu(I) assay mixture.

E2(Cu$^+$)P levels as a percentage of maximum phosphorylation were 41 ± 6% and 60 ± 5% respectively. Thus, it is apparent that the Ag$^+$-bound form of CopA is less stable that the physiological Cu$^+$-bound form of the enzyme. These results suggest that although CPx-ATPases can function with alternative substrates (Ag$^+$/Cu$^+$), the interaction of these ions with the binding site is different, perhaps using distinct coordinating atoms. In addition, the formation of ion bound phosphoenzyme forms indicate that CopA is likely to drive metal efflux. This is in agreement with studies of *E. coli* CopA, indicating that this enzyme transports Cu$^+$ outwardly.[7]

It has been described that Cu$^+$ and Ag$^+$ activate CopB from *E. hirae*.[4] We observed that *A. fulgidus* CopB was only partially activated by Cu$^+$ and Ag$^+$ and largely activated by Cu^{2+} (FIG. 2B). This is not surprising considering that CopB has a CPH metal binding site in the transmembrane region and a His-rich metal binding domain. Weak Lewis acids (like Cu$^+$) prefer to interact with weak Lewis bases (–SH groups), while stronger Lewis acid (like Cu^{2+}) tend to bind sites formed by strong Lewis bases (imidazolium group). Consequently, it is reasonable that CopB has a preference for Cu^{2+} instead of Cu$^+$. The different selectivity observed for CopA [Cu(**I**)-ATPase] and CopB [Cu(**II**)-ATPase] solves the apparent redundant copper ATPase activities present in a single organism (i.e., *A. fulgidus*, *E. hirae*, *Aquifex aeolicus*, *Methanobacterium thermoautotrophicum*, etc.).

Having established the ion specificity of CopA and CopB, we evaluated the functional relevance of Cys in the transmembrane cation binding site. FIGURE 3 shows that replacement of either Cys in the CPC sequence of CopA yields inactive enzymes. We also analyzed the effect of removing the cytoplasmic metal binding domains. CopA is singular since in addition of a C^{27}AMC^{30} N-MBD it has a putative

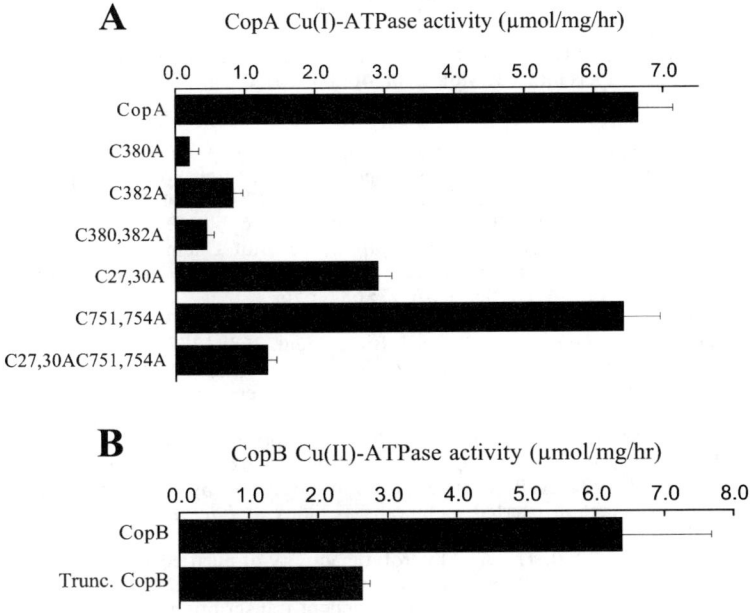

FIGURE 3. (**A**) Cu(I)-ATPase activity of CopA proteins carrying the indicated amino acid replacements. Activity was measured in the presence of 100 µM Cu$^+$, 2.5 mM dithiotreitol. (**B**) Cu(II)-ATPase activity of CopB and truncated CopB (lacking the His-rich N-MBD). Activity was measured in the presence of 1 µM Cu^{2+}.

C-teminus MBD (C^{751}HHC^{754}). Mutation of Cys in the N-MBD lead to significant reduction in the enzyme ATPase activity, while replacement of Cys at the putative C-MBD has no apparent effect on activity (FIG. 3). Removal of both cytoplasmic MBDs has no larger consequences than removing the N-MBD. These results are in agreement with previous studies of N-MBD of eukaryotic CPx-ATPase, and suggest that CopA preparations can be used to study the structural–functional role of this domain.

The N-terminal 51–amino acid fragment of CopB contains 17 His. To evaluate the functional role of this fragment we constructed a truncated CopB lacking this fragment. FIGURE 3 shows that the resulting protein was functional but with a reduced turnover rate. The truncated protein showed ion activation characteristics similar to wild-type protein (not shown). Consequently, this fragment, while not involved in enzyme selectivity, might also function as a regulatory domain, sensing the metal concentration and activating the enzyme upon ion binding.

In summary: we have developed a bacterial expression system where milligram amounts of thermophilic PIB-type ATPases can be obtained. We have shown the ion specificity of both PIB ATPases from *A. fulgidus*. Our studies indicated that while the transmembrane Cys in the signature sequence CPC is essential for enzyme function, the N-MBD of CopA appears to have a regulatory role. Similarly, the role of the His-rich MBD in CopB appears to be regulatory.

REFERENCES

1. AXELSEN, K.B. & M.G. PALMGREN. 1998. Evolution of substrate specificities in the P-type ATPase superfamily. J. Mol. Evol. **46:** 84–101.
2. RENSING, C. *et al.* 1999. Families of soft-metal-ion-transporting ATPases. J. Bacteriol. **181:** 5891–5897.
3. VULPE, C.D & S. PACKMAN. 1995. Cellular copper transport. Annu. Rev. Nutr. **10:** 293–322.
4. SOLIOZ, M. & A. ODERMATT. 1995. Copper and silver transport by CopB-ATPase in membrane vesicles of *Enterococcus hirae*. J. Biol. Chem. **270:** 9217–9221.
5. MANDAL, A.K. *et al.* 2002. Characterization of a thermophilic P-type Ag+/Cu+-ATPase from the extremophile *Archaeoglobus fulgidus*. J. Biol. Chem. **277:** 7201–7208.
6. RENSING, C. *et al.* 2000. CopA: An *Escherichia coli* Cu(I)-translocating P-type ATPase. Proc. Natl. Acad. Sci. USA **97:** 652–656.
7. ODERMATT, A. *et al.* 1993. Primary structure of two P-type ATPases involved in copper homeostasis in *Enterococcus hirae*. J. Biol. Chem. **268:** 12775–12779.
8. BISSIG, K.D. *et al.* 2001. Structure-function analysis of purified *Enterococcus hirae* CopB copper ATPase: effect of Menkes/Wilson disease mutation homologues. Biochem. J. **357:** 217–223.
9. YOSHIMIZU, T. *et al.* 1998. Essential Cys-Pro-Cys motif of *Caenorhabditis elegans* copper transport ATPase. Biosci. Biotechnol. Biochem. **62:** 1258–1260.
10. TSAI, K.J. *et al.* 1992. ATP-dependent cadmium transport by the cadA cadmium resistance determinant in everted membrane vesicles of *Bacillus subtilis*. J. Bacteriol. **174:** 116–121.
11. SHARMA, R. *et al.* 2000. The ATP hydrolytic activity of purified ZntA, a Pb(II)/Cd(II)/Zn(II)-translocating ATPase from *Escherichia coli*. J. Biol. Chem. **275:** 3873–3878.
12. RUTHERFORD, J.C. *et al.* 1999. Cobalt-dependent transcriptional switching by a dual-effector MerR-like protein regulates a cobalt-exporting variant CPx-type ATPase. J. Biol. Chem. **274:** 25827–25832.
13. VOSKOBOINIK, I. *et al.* 1999. Functional analysis of the N-terminal CXXC metal-binding motifs in the human Menkes copper-transporting P-type ATPase expressed in cultured mammalian cells. J. Biol. Chem. **274:** 22008–22012.
14. TSIVKOVSKII, R. *et al.* 2001. The Lys(1010)-Lys(1325) fragment of the Wilson's disease protein binds nucleotides and interacts with the N-terminal domain of this protein in a copper-dependent manner. J. Biol. Chem. **276:** 2234–2242.

P-Type ATPase Superfamily

Evidence for Critical Roles for Kingdom Evolution

HIDEYUKI OKAMURA, MASATSUGU DENAWA, RYOSUKE OHNIWA, AND KUNIO TAKEYASU

Graduate School of Biostudies, Kyoto University, Kitashirakawa-oiwake-cho, Sakyo-ku, Kyoto, 606-8502, Japan

ABSTRACT: The P-type ATPase has become a protein superfamily. On the basis of sequence similarities, the phylogenetic analyses, and substrate specificities, this superfamily can be classified into 5 families and 11 subfamilies. A comparative phylogenetic analysis demonstrates the relationship between the molecular evolution of these subfamilies and the establishment of the kingdoms of living things.

KEYWORDS: P-type ATPase superfamily; evolution of kingdom; subunit assembly; ouabain binding

INTRODUCTION

The existence of the P-type ATPase isoform was biochemically demonstrated first by Sweadner,[1] and then molecularly by Shull and Lingrel.[2] Takeyasu and Fambrough showed the isoform conservation throughout the evolution.[3] Since the whole genome sequences of *Haemophilus influenzae* Rd were determined in 1995,[4] a total of 5 Eukaryotes, 16 Archaea, and 68 Proteobacteria species have been added to the complete genome lists. Phylogenetic analyses of the primary sequences among them have classified the P-type superfamily into several types.[5–7] The distribution of the subtypes can now explain the physiological prerequisites for "life," "bacteria," "archae," "eukaryotes," "fungi," and "animal."

RESULTS AND DISCUSSION

Distribution and Significance of the P-type ATPase Subtypes in Life

Type 1B (heavy metal transporter) and type 2A (intracellular Ca-ATPase) are the most fundamental for life: These two subtypes are found in every kingdom (TABLE 1). They appeared very early in the evolution of life to establish an ancestral form of the cell (FIG. 1), and are expected to play critical roles for ion homeostasis in cell

---

Address for correspondence: Dr. Kunio Takeyasu, Graduate School of Biostudies, Kyoto University, Kitashirakawaoiwake-cho, Sakyo-ku, Kyoto 606-8502, Japan. Voice: +81-75-753-6852; fax: +82-75-753-6852.

takeyasu@lif.kyoto-u.ac.jp

Ann. N.Y. Acad. Sci. 986: 219–223 (2003). © 2003 New York Academy of Sciences.

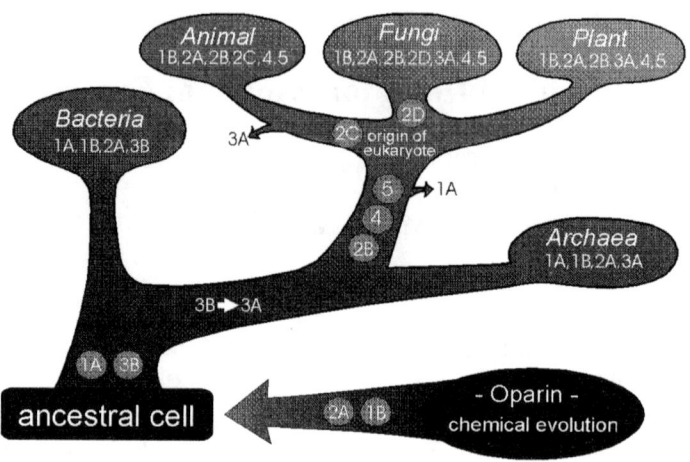

FIGURE 1. Evolution of the kingdoms and its correlation of the P-type ATPase superfamily.

physiology and gene regulation. The type 1B genes are separated into two large clades: Clade I includes bacerial Zn^{2+}-, Cd^{2+}-, Pb^{2+}-transporters and plant heavy metal transporters; and Clade II includes the prokaryotic and eukaryotic Cu^{2+} transporters. In bacteria, additional subtypes are required, that is, type 1A (K-ATPase) and type 3B (Mg-ATPase). Type 1A is unique to bacteria and archae, and type 3B is later converted to type 3A in the lineage to plants and fungi, and eventually lost in the lineage to animals (FIG. 1).

Type 2B (plasma membrane Ca-ATPase), type 4 (lipid translocator), and type 5 (unknown function) are the sure signs for eukaryotes: An ancestral form of the eukaryotic cells had gained these three subtypes possibly to develop the intracellular membrane systems, and, at the same time, lost type 1A. Type 5 ATPases are divided into two major subfamilies, 5A and 5B, that originated at least before the duplication between animal and fungi. Both subfamilies include the isoforms of *S. cerevisiae, S. pombe, C. elegans, D. melanogaster*, and *H. sapiens*, but type 5B has an insertion of long amino acid residues between the first and second transmembrane regions. This structural difference may lead to functional differences.

The lost and gained with respect to animals: The ancestor of animals found type 2C (Na/K-ATPase) and got rid of type 3A (proton pump). After the final set of the enzyme families and subfamilies were established independently in each kingdom or phylum of organisms, the isoform diversification occurred in parallel.

Type 2C Evolution in the Animal Kingdom

Characteristics of type 2C: Type 2C α-subunit tree includes two characteristic clades (FIG. 2a). The major one, Clade II, includes mammalian Na/K- and H/K-ATPases and their orthologue in invertebrates. The mammalian Na/K- and H/K-ATPases have a partner subunit, the β–subunit, and are inhibited by ouabain and

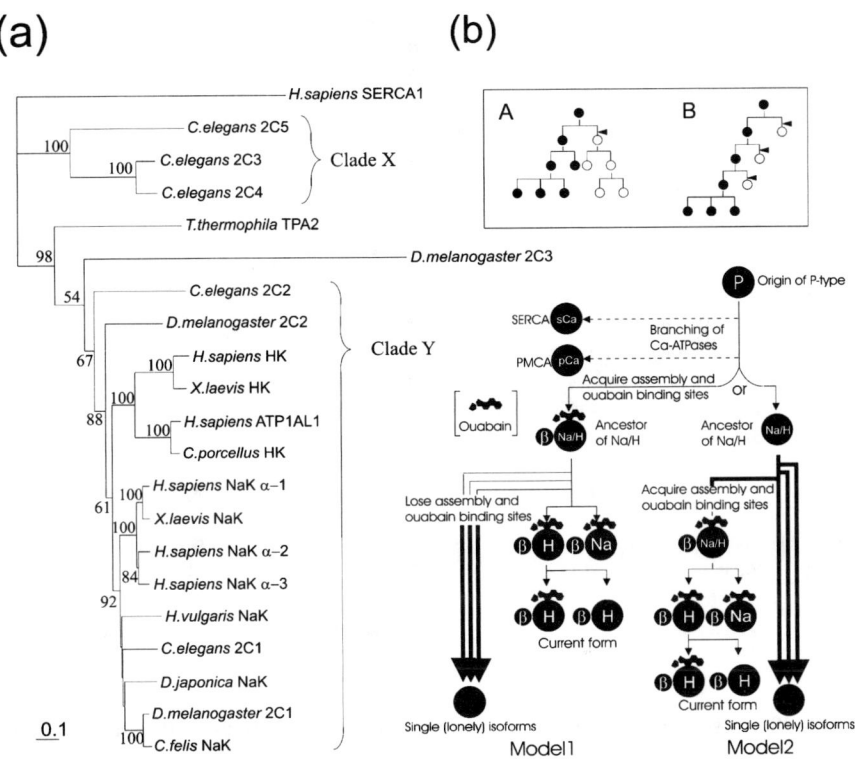

FIGURE 2. (a) Rooted phylogenetic trees of the type 2C P-type ATPase α-subunits. The out-group is the human SERCA1. *C. elegans* three isoforms in Clade I (Ce2C3, Ce2C4 and Ce2C5), *Tetrahymena* TPA2 and *Drosophila* Dm2C3 lack a set of amino-acyl residues required for ouabain binding. Moreover, *C. elegans* isoforms and TPA2 lack the α,β-subunit assembly domain. On the other hand, the isoforms in Clade II have both motives. The tree is inferred by the neighbor-joining method based on the alignment of the conserved regions, and the numbers at the nodes represent the bootstrap value in 1000 replications. **(b)** Models of the Na,K- (H,K-) ATPase α-subunit evolution in terms of the sites for ouabain-binding and subunit assembly. The models deal with the possibilities of either acquiring or losing these sites. With the assumption that the ancestral form had already accommodated with the sites, our analysis supports the idea that these sites might have lost in multiple steps (B), but not in a single step (A) (Model 1). The loss of assembly site would result in a relaxation of functional constraint that leads to the increase in the evolutionary rate (*thick arrows*). On the other hand, with the assumption that the ancestor obtained the sites during its evolution, the results in **(a)** suggest a single-step-acquisition model (D; Model 2). The resultant subunit assembly posed functional constraint on the pump molecule and led to a decrease in the evolutionary rate (*thin arrows*). Model 2 is more plausible than Model 1 because of its simplicity.

TABLE 1. The P-type ATPase superfamily and the occurrence in kingdoms

		1A	1B	2A	2B	2C	2D	3A	3B	4	5A	5B
Bacteria	Bacillus halodurans	●	●									
	Bacillus subtilis	●	●									
	Campylobacter jejuni	●	●									
	Chlamydia pneumoniae		●									
	Deinococcus radiodurans	●	●									
	Escherichia coli	●	●							●		
	Haemophilus influenziae		●									
	Heliobacter pylori		●									
	Pseudomonas aeruginosa	●	●	●						●		
	Ureaplasma urealyticum	●	●									
Archaea	Aeropyrum permix	●										
	Archaeoglobus fulgidis		●									
	Halobacterium sp. NCR-1											
	Methanobacterium thermoautotrophicum	●	●									
	Methanococcus jannaschii								●			
	Pyrococcus abyssi	●										
	Thermoplasma acidophilum	●	●						●			
Eukaryota	Fungi			●	●	●		●	●	●	●	●
	Plant			●	●	●			●		●	●
	Invertebrate			●	●	●	●			●	●	●
	Vertebrate			●	●	●	●			●	●	●

SCH28080, respectively. The domains required for the α,β-subunit assembly[8] and for the inhibitor binding have been determined.[9,10] On the other hand, some homologues in invertebrates lack such motifs (FIG. 2a). These isoforms might exist by themselves without the β-subunit. The *Tetrahymena thermophila* TPA2 also lacks them, but *D. melanogaster* Dm2C3 lacks only the ouabain-binding domain. Because these isoforms are close to the root than Clade II and because Clade II included both vertebrate and invertebrate, these isoforms have arisen before the separation of vertebrates and invertebrates and also before the separation of the Na/K-ATPase and the H/K-ATPase.

Origins of subunit assembly and inhibitory mechanisms: In type 2C β-subunits, some *D. melanogaster* isoforms (β-4, -5 and -6) lack one or two Cys-Cys bonds that are required for proper expression and assembly of the subunits on the plasma membrane in the case of the mammalian β-subunits[11] These *D. melanogaster* β-subunits might exist by themselves, too. The mutation rates for the "un-assembling α-/β-subunits" are much faster than those for the subunits that would assemble.[12] These findings support the idea that a relaxation of functional constraints would increase the rate of evolution and provide clues for identifying the origins of inhibitor sensitivity, subunit assembly, and separation of the Na/K- and H/K-ATPases (FIG. 2b). First, an ancestral member of the Na/K- and H/K-ATPase family lacked the inhibitor-binding and the subunit-assembly domains, and it existed as a single subunit P-type ATPase in protists and in lower invertebrate organisms. Second, in its evolution, this ancestral form acquired the abilities to bind ouabain and to assemble with a β-subunit, becoming an immediate ancestor of the currently well-known Na/K- and H/K-ATPase family. Third, from this family, the Na/K- and the H/K-ATPases diverged and became established in vertebrate organisms. Fourth, after the establishment of the H/K-ATPase subfamily, some of the members lost their ouabain-binding ability. In addition to these evolutionary steps, gene duplication of both α- and β-subunit genes within various animal phyla has independently led to the occurrence of α- and β-subunit isoform diversity in these phyla.

REFERENCES

1. SWEADNER, K.J. & R.C. GILKESON. 1985. Two isozymes of the Na,K-ATPase have distinct antigenic determinants. J. Biol. Chem. **260:** 9016–9022.
2. SHULL, G.E., J. GREEB & J.B. LINGREL. 1986. Molecular cloning of three distinct forms of the Na^+,K^+-ATPase alpha-subunit from rat brain. Biochemistry **25:** 8125–8132.
3. TAKEYASU, K., V. LEMAS & D.M. FAMBROUGH. 1990. Stability of Na^+/K^+-ATPase alpha-subunit isoforms in evolution. Am. J. Physiol. **259:** C619–630.
4. FLEISCHMANN, R.D., et al. 1995. Whole-genome random sequencing and assembly of Haemophilus influenzae Rd. Science **269:** 496–512.
5. AXELSEN, K.B. & M.G. PALMGREN. 1998. Evolution of substrate specificities in the P-type ATPase superfamily. J. Mol. Evol. **46:** 84–101.
6. MOLLER, J.V., B. JUUL & M. LE MAIRE. 1996. Structural organization, ion transport, and energy transduction of P-type ATPases. Biochim. Biophys. Acta **1286:** 1–51.
7. WANG, S. & K. TAKEYASU. 1997. Primary structure and evolution of the ATP-binding domains of the P-type ATPases in Tetrahymena thermophila. Am. J. Physiol. **272:** C715–728.
8. LEMAS, M.V., et al. 1994. 26 amino acids of an extracellular domain of the Na,K-ATPase alpha-subunit are sufficient for assembly with the Na,K-ATPase beta-subunit. J. Biol. Chem. **269:** 8255–8259.
9. PRICE, E.M. & J.B. LINGREL. 1988. Structure-function relationships in the Na,K-ATPase alpha subunit: site-directed mutagenesis of glutamine-111 to arginine and asparagine-122 to aspartic acid generates a ouabain-resistant enzyme. Biochemistry **27:** 8400–8408.
10. SACHS, G., et al. 1995. The pharmacology of the gastric acid pump: the H^+,K^+ ATPase. Annu. Rev. Pharmacol. Toxicol. **35:** 277–305.
11. NOGUCHI, S., Y. MUTOH & M. KAWAMURA. 1994. The functional roles of disulfide bonds in the beta-subunit of (Na,K)ATPase as studied by site-directed mutagenesis. FEBS Lett. **341:** 233–238.
12. TAKEYASU, K. et al. 2001. P-type ATPase diversity and evolution: the origins of ouabain sensitivity and subunit assembly. Cell. Mol. Biol. **47:** 325–333.

The Na,K-ATPase S5-H5 Helix

Structural Link between Phosphorylation and Cation-Binding Sites

ATIN K. MANDAL, LYUDMILA MIKHAILOVA, AND JOSÉ M. ARGÜELLO

Department of Chemistry and Biochemistry, Worcester Polytechnic Institute, Worcester, Massachusetts 01609, USA

KEYWORDS: conformational transitions; cysteine mutagenesis, $HgCl_2$; MTSET; maleimide-biotin; stalk region

The Na,K-ATPase undergoes significant conformational changes (E1↔E2) during its cycle. These are critical for the energy transduction between phosphorylation and cation-binding sites. However, these sites are separated by ≈50Å, raising the question of how the necessary long-distance interaction is mediated. Analogy modeling of Na,K-ATPase based on the Ca-ATPase structure[1] suggests that the phosphorylation and cation-binding sites are communicated by the long helix extending from Ala749 till Phe786 (sheep α1). This helix starts only ≈5Å of the phosphorylation site, constitutes the fifth transmembrane segment (H5), and provides residue side chains to the cation-binding sites.

We hypothesized that enzyme phosphorylation drives the movement of the H5 helix leading to a local conformational change at the cation-binding site modifying cation affinity. To test this hypothesis, using as a background a protein devoid of Cys in the transmembrane region (All-TM-Cys),[2] Cys was introduced in the extracellular ends of H5 and H6 and their joining loop, from T781 till L805. Mutated proteins were expressed in COS cells. Cys-substituted enzymes that retained functionality are listed in TABLE 1. These replacements did not lead to major alterations in protein expression, ATPase activity, cation and ATP dependence. Cells expressing mutated proteins were probed with extracellularly applied Hg^{2+}, MTSET or biotin-maleimide, while placing the enzyme in one of two conformations: E1(K-medium) and E2P(Na) (Na-medium).[3]

We observed that the enzyme modification was dependent on residue accessibility and, most importantly, on the enzyme phosphorylation. Modification of exposed Cys in mutants I787C, A789C, G796C, T797C, and V798C with Hg^{+2} and G796C, T797C, V798C with MTSET led to significant enzyme inactivation (>50%). Biotin-

Address for correspondence: Dr. José Argüello, Department of Chemistry and Biochemistry, Worcester Polytechnic Institute, Worcester, MA 01609. Voice: 508-831-5326; fax: 508-831-5933.
arguello@wpi.edu

TABLE 1. Summary of relative accessibility of residues in the H5-H6 loop

Reagent	Hg^{2+}		MTSET		Biotin Maleimide	
Conformation	E1	E2P	E1	E2P	E1	E2P
Residue						
All-TM-Cys	–	–	–	–	–	–
T781C	–	–	–	–	ND	ND
I785C	–	–	–	+[a]	+	+
I787C	+	++[a]	–	+[a]	+	++[a]
A789C	+	+	+	+	+	++[a]
I791C	+	+	–	–	–	+[a]
G796C	+	+	+	+	+	+
T797C	+	+	+	++[a]	–	++[a]
V798C	+	+	+	+	+	++[a]
T799C	–	–	–	–	ND	ND
C802C	–	–	–	–	+/–	+/–
L805C	–	+[a]	–	–	ND	ND

[a]Higher apparent reactivity when the enzyme is in the phosphorylated conformation.
ND = not determined.

maleimide reacted to various extents with all introduced Cys. However, preincubation with membrane-impermeable MTSET prevented labeling of the substituted proteins with biotin-maleimide, indicating the reaction of the maleimide with the extracellular facing Cys. The reactivity of Cys located at the end of H5 (I785C, I787C) and H6 (T797C, V798C) was particularly affected by the enzyme conformation (E1 or E2P(Na)). In all cases the residues appeared more accessible when the enzyme was placed in a phosphorylated conformation.

Our results suggest that the catalytic phosphorylation of D369, close to one end of the H5 helix, by displacing H5 influences the reactivity of Cys residues located ≈50 Å away, on the extracellular end of H5. Consequently, it can be assumed that cation coordinating amino acids in H5 are similarly displaced, thus, changing the cation-binding site geometry, its exposure to the intra- and extracellular compartments, and the affinity for the transported cation.

ACKNOWLEDGMENT

This work was supported by Grant-in-Aid 9750102N from the American Heart Association.

REFERENCES

1. TOYOSHIMA, C. *et al.* 2000. Crystal structure of the calcium pump of sarcoplasmic reticulum at 2.6 Å resolution. Nature **405:** 647–655.
2. SHI, H.G. *et al.* 2000. Functional role of cysteine residues in the Na,K-ATPase α subunit. Biochim. Biophys. Acta **1464:** 177–187.
3. ZICHITTELLA, A.E. *et al.* 2000. Reactivity of cysteines in the transmembrane region of the Na,K-ATPase α subunit probed with Hg^{2+}. J. Membr. Biol. **177:** 187–197.

Na,K-ATPase α-β Subunit Interactions in the Transmembrane Domain

CIMING LI, GILLES CRAMBERT, UDO HASLER, AND KÄTHI GEERING

Institute of Pharmacology and Toxicology of the University, CH-1005 Lausanne, Switzerland

KEYWORDS: *Xenopus* oocytes; Na,K-ATPase α-β interaction; cross-linking

Tyr^{40} and Tyr^{44} located on one face of the β's TM helix influence the transport kinetics of the Na,K-pump.[1] Moreover, cross-linking experiments on 19 kDa tryptic fragments of Na,K-ATPase suggest a close proximity of the β's TM domain and the α's TM8.[2] We have attempted to reveal which of the two cysteine residues in the α's TM8 cross-links with Cys^{46} in the β1's TM domain, and which α TM helix interacts with the tyrosine-containing face of the β's TM.

METHODS

Wild-type or mutant α and β subunits were expressed in *Xenopus* oocytes and metabolically labeled for 24 hours. Oxidative cross-linking was performed on microsomes in the presence of 10 mM $CuSO_4$ for 30 minutes at 37°C. Proteins were separated by SDS-PAGE and visualized by fluorography, and β subunits were quantified by laser densitometry.

RESULTS AND DISCUSSION

Oxidative cross-linking, performed on microsomes of *Xenopus* oocytes expressing wild-type α1 and β1 subunits, eliminated α and reduced β by about 50%, indicating that the two subunits had been cross-linked into high molecular mass species (FIG. 1, lane 1 and 2). Substitution of Cys^{46} in the β1's TM domain preserved cross-linking of the α, but nearly abolished cross-linking of the β (lanes 3 and 4). Substitution of cysteine residues in TM6-10 did not reduce the cross-linking efficiency of the associated β (lanes 5–10). Thus, our cross-linking experiments do not reveal

---

Address for correspondence: Käthi Geering, Institute of Pharmacology and Toxicology, Rue du Bugnon 27, CH-1005 Lausanne, Switzerland. Voice: +041-21-692-54-10; fax: +041-21-692-53-55.
 kaethi.geering@ipharm.unil.ch

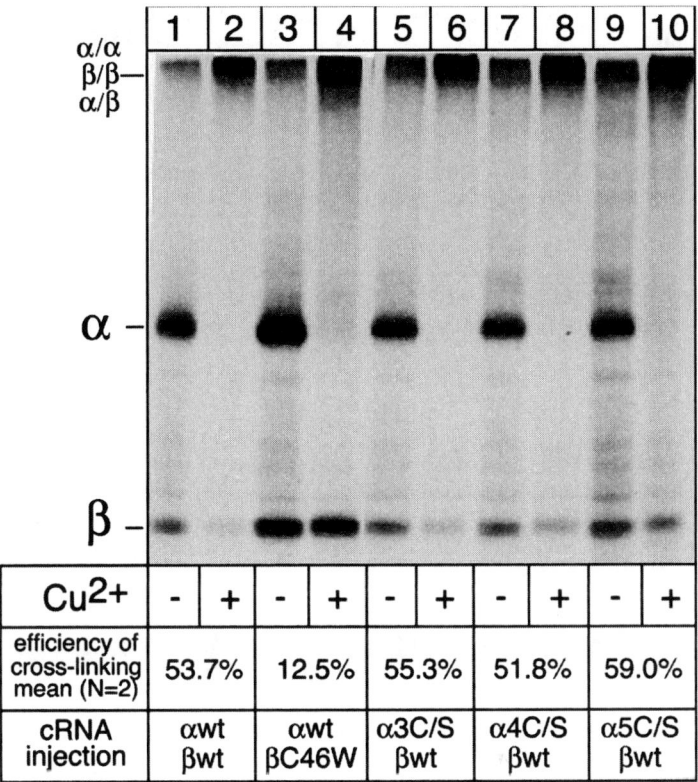

FIGURE 1. Mutation of cysteine residues in TM6–TM10 of the α subunit does not abolish cross-linking of the β subunit. *Xenopus* oocytes were injected with wild type (wt) or mutant α and β cRNA and metabolically labeled. Microsomes were prepared and subjected or not to oxidative cross-linking as described in METHODS. α3C/S = αC918S/C937S/C971S (TM8, 9); α4CS = αC918S/C937S/C971S/C990S (TM8, 9, 10); α5C/S = αC809S/C918S/C937S/C971S/C990S (TM6, 7, 8, 9, 10).

close positioning of the β's TM domain and the C-terminal α TM domains. Our results suggest that the cysteine-containing face of the β's TM helix is cross-linked with a cysteine residue in an N-terminal α TM domain or with one of the cytosolic loops. We can also not exclude the possibility of β-β cross-linking.[2]

Substitution of Tyr[40] and Tyr[44] with cysteine in a β subunit lacking Cys[46] (Y40C/Y44C/C46W) permits cross-linking with the α subunit, though to a lesser extent than with wild-type β subunit (data not shown). Cysteine substitution of Gln[858] and Gln[863] in the α's TM7, which potentially can interact with tyrosines, did not improve β1 cross-linking, indicating that the tyrosine-containing face of the β's TM helix is not closely positioned to glutamines in the α's TM7.

In conclusion, more experiments, using an α subunit lacking all cysteine residues, at least in the membrane domain, are needed to predict the actual positioning of the β subunit's TM domain in the context of the organization of the 10 TM domains of the α subunit.

REFERENCES

1. HASLER, U. *et al.* 2001. Structural and functional features of the transmembrane domain of the Na,K-ATPase β subunit revealed by tryptophan scanning. J. Biol. Chem. **276:** 16356–16364.
2. IVANOV, A., H. ZHAO & N. N. MODYANOV. 2000. Packing of the transmembrane helices of Na,K-ATPase: direct contact between β-subunit and H8 segment of α-subunit revealed by oxidative cross-linking. Biochemistry **39:** 9778–9785.

Negative Changes of the Membrane Capacitance due to Electrogenic Na Transport by the Na,K-ATPase

V. S. SOKOLOV,[a] A. A. LENZ,[a] AND H.-J. APELL[b]

[a]*A. N. Frumkin Institute of Electrochemistry RAS, 117071, Moscow, Russia*
[b]*Department of Biology, University of Konstanz, 78457 Konstanz, Germany*

KEYWORDS: Na,K-ATPase; electrogenicity; capacitance; Na^+ movements; ion binding

Electrogenic Na^+ transport was investigated in membrane fragments containing Na,K-ATPase adsorbed to bilayer lipid membranes (BLM) triggered by fast ATP release from caged ATP.[1] The influence of voltage on the transport after ATP release was determined as small increments of capacitance and conductance by applying an alternating voltage to the membrane.[2] An electrogenic Na^+ transport through a cytoplasmic access channel of the Na,K-ATPase was detected that disappeared after enzyme phosphorylation, and thus produced a negative capacitance and conductance increments at Na^+ concentrations below 5 mM.[3] This effect was studied now in more detail by measuring the frequency dependence of the capacitance and conductance increments at different Na^+ concentrations (FIG. 1). Fitting these data by the sum of Lorentzians allowed the discrimination of the separate steps in Na^+ transport and the determination of their parameters.[4]

$$\Delta C = C_0 \frac{\omega_0^2}{\omega^2 + \omega_0^2} + C_1 \frac{\omega_1^2}{\omega^2 + \omega_1^2} - C_2 \frac{\omega_2^2}{\omega^2 + \omega_2^2} + C_{\lim} \quad (1)$$

$$\Delta G = C_0 \omega_0 \frac{\omega^2}{\omega^2 + \omega_0^2} + C_1 \omega_1 \frac{\omega^2}{\omega^2 + \omega_1^2} - C_2 \omega_2 \frac{\omega^2}{\omega^2 + \omega_2^2} \quad (2)$$

At high Na^+ concentration the frequency dependence could be fitted by the sum of two Lorentzians with the amplitudes C_0 and C_1, and a constant term, $C_{\lim}$. The Lorentzian with the lowest corner frequency ω_0 (about 30 s^{-1}) corresponds to the

Address for correspondence: Dr. V.S. Sokolov, A.N. Frumkin Institute of Electrochemistry RAS, Leninsky pr., 31, 117071, Moscow, Russia. Voice and fax: +7-095-952-5582.
sokolov@netra.elchem.ac.ru

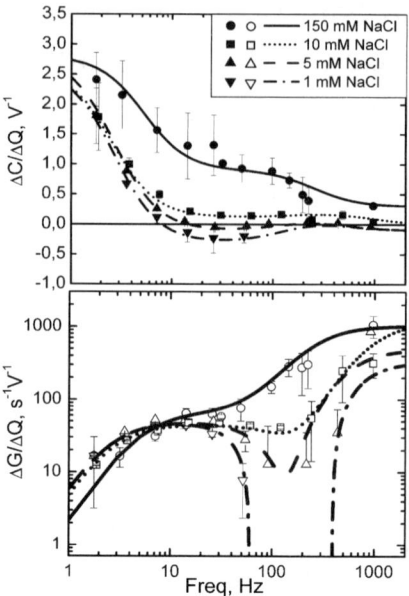

FIGURE 1. Frequency dependencies of capacitance and conductance increments, normalized to the net charge transferred through the membrane after release of ATP from caged-ATP. The solution contained various Na^+ concentrations (as indicated) and 30 mM imidazole, 10 mM $MgCl_2$, and 1 mM EDTA, pH 6.5. The lines were plotted according to the Eqs. (1) and (2) using following parameters: **150 mM NaCl:** $C_0 = 1.9$ V^{-1}, $\omega_0 = 33$ s^{-1}, $C_1 = 0.6$ V^{-1}, $\omega_1 = 1600$ s^{-1}, $C_{lim} = 0.3$ V^{-1}; **10 mM NaCl:** $C_0 = 2.5$ V^{-1}, $\omega_0 = 16$ s^{-1}, $C_1 = 0.2$ V^{-1}, $\omega_1 = 6000$ s^{-1} $C_2 = -0.08$ V^{-1}, $\omega_2 = 1000$ s^{-1}; **5 mM NaCl:** $C_0 = 2.9$ V^{-1}, $\omega_0 = 17$ s^{-1}, $C_1 = 0.2$ V^{-1}, $\omega_1 = 3000$ s^{-1}, $C_{lim} = -0.1$ V^{-1}, $C_2 = -0.2$ V^{-1}, $\omega_2 = 800$ s^{-1}; **1 mM NaCl:** $C_0 = 2.9$ V^{-1}, $\omega_0 = 17$ s^{-1}, $C_1 = 0.2$ V^{-1}, $\omega_1 = 3000$ s^{-1}, $C_{lim} = -0.1$ V^{-1}, $C_2 = -0.4$ V^{-1}, $\omega_2 = 800$ s^{-1}.

slowest step of Na^+ transport, the conformation transition E_1/E_2. The corner frequency coincides with the reciprocal time constant of an exponent decay of the falling phase of the ATP-induced current transient. The Lorentzian with a corner frequency ω_1 of about 2000 s^{-1} corresponds to the "intermediate" step, the release of the third Na^+ ion through an extracellular access channel. The constant term, C_{lim}, is assigned to the fast Na^+ ion release through an access channel; its rate could not be resolved so far.

The negative increments at low Na^+ may be explained by electrogenic Na^+ transport through a cytoplasmic access channel. This process is suppressed at saturating Na^+ concentrations. The negative changes depend on frequency, and this dependence can be described by an additional Lorentzian with negative amplitude, C_2. Its corner frequency, ω_2 (about 800 s^{-1}), was attributed to the rate of electrogenic Na^+ transport. The ratio of the "positive" and "negative" Lorentzian, C_0/C_2, is about 10, and the square root of this ratio, ~3, gives an estimate of the ratio of the depths of extracellular and cytoplasmic access channels. With the well-documented dielectric depth of the extracellular access channel of 0.7–0.75, the cytoplasmic depth is then calculated about 0.25.

ACKNOWLEDGMENTS

This work was supported financially by RFBR No. 01-04-49246 and INTAS (Project 01-0224)

REFERENCES

1. BORLINGHAUS, R., H.-J. APELL & P. LÄUGER. 1987. Fast charge translocations associated with partial reactions of the Na,K-Pump: I. Current and voltage transients after photochemical release of ATP. J. Membr. Biol. **97**: 161–178.
2. SOKOLOV, V.S., K.V. PAVLOV, K.N. DZHANDZHUGAZYAN & E. BAMBERG. 1992. Capacitance and conductivity changes during Na^+,K^+-ATPase action in model membrane. Biol. Membr. **6**: 1263–1272.
3. SOKOLOV, V.S., S.M. STUKOLOV, A.S. DARMOSTUK & H.-J. APELL. 1998. Influence of sodium concentration on changes of membrane capacitance associated with the electrogenic ion transport by the Na,K-ATPase. Eur. Biophys. J. **27**: 605–617.
4. SOKOLOV, V.S., A.G. AYUAN & H.-J. APELL. 2001. Assignment of charge movements to electrogenic reaction steps of the Na,K-ATPase by analysis of salt effects on the kinetics of charge movements. Eur. Biophys. J. **30**: 515–527.

Isolation of $(\alpha\beta)_4$-Tetraprotomer Having Half-of-Sites ATP Binding from Solubilized Dog Kidney Na^+/K^+-ATPase

YUTARO HAYASHI, NOBUKO SHINJI, YOSHIKAZU TAHARA, EMI HAGIWARA, HITOSHI TAKENAKA

First Department of Biochemistry, Kyorin University School of Medicine, Mitaka, Tokyo 181-8611, Japan

KEYWORDS: tetraprotomer; ATP binding; solubilized Na^+/K^+-ATPase

The M_r and ATPase activity of the oligomeric components of solubilized Na^+/K^+-ATPase (dog kidney) has been simultaneously determined to indicate that the $\alpha\beta$ protomer (P, M_r:1.56 × 10^5), a possible minimum enzyme unit, less likely formed a functional Na^+/K^+-ATPase in the intact membrane.[1] We later found that solubilization in potassium acetate (CH_3COOK) increased oligomers higher than $(\alpha\beta)_2$-diprotomer (D, 3.02 × 10^5). We here deduced from the M_r determined by a HPGC-LALLS analysis[1] that the higher oligomer (H) was $(\alpha\beta)_4$-tetraprotomer (T).

IDENTIFYING $(\alpha\beta)_4$-TETRAPROTOMER

The enzymes solubilized in $C_{12}E_8$ and 0.1 M CH_3COOK were chromatographed using a novel column (0.78ϕ × 30 cm, TSKgel G3000SWxL + TSKgel G4000SWxL, 1:1 w/w, Tosoh Co.) and the basic elution buffer (0.2 mg/mL $C_{12}E_8$, and [in mM] 90 NaCl, 10 KCl, 4 $MgCl_2$, 1 EDTA, 10 imidazole and 13 Hepes [pH7.0]) plus 60 µg/ml PS at 0°C. H showed 0.296 ± 0.004 mL/g as the specific refractive index increment (dn/dc_p) and 5.96 ± 0.12 × 105 ($n = 3$) as M_r in the HPGC-LALLS analysis. When the fraction of H thus isolated was rechromatographed by the above method, what emerged was a single protein peak with dn/dc_p of 0.258 mL/g and M_r of 6.14 × 10^5. Since H, D and P could be interchangeable, H would be an $(\alpha\beta)_4$-tetraprotomer, T, which ranged from 38–58% of all the oligomers.

Address for correspondence: Yutaro Hayashi, First Department of Biochemistry, Kyorin University School of Medicine, Mitaka, Tokyo 181-8611, Japan. Voice: +81-422-76-7651; fax: +81-422-76-7650.

yutahaya@kyorin-u.ac.jp

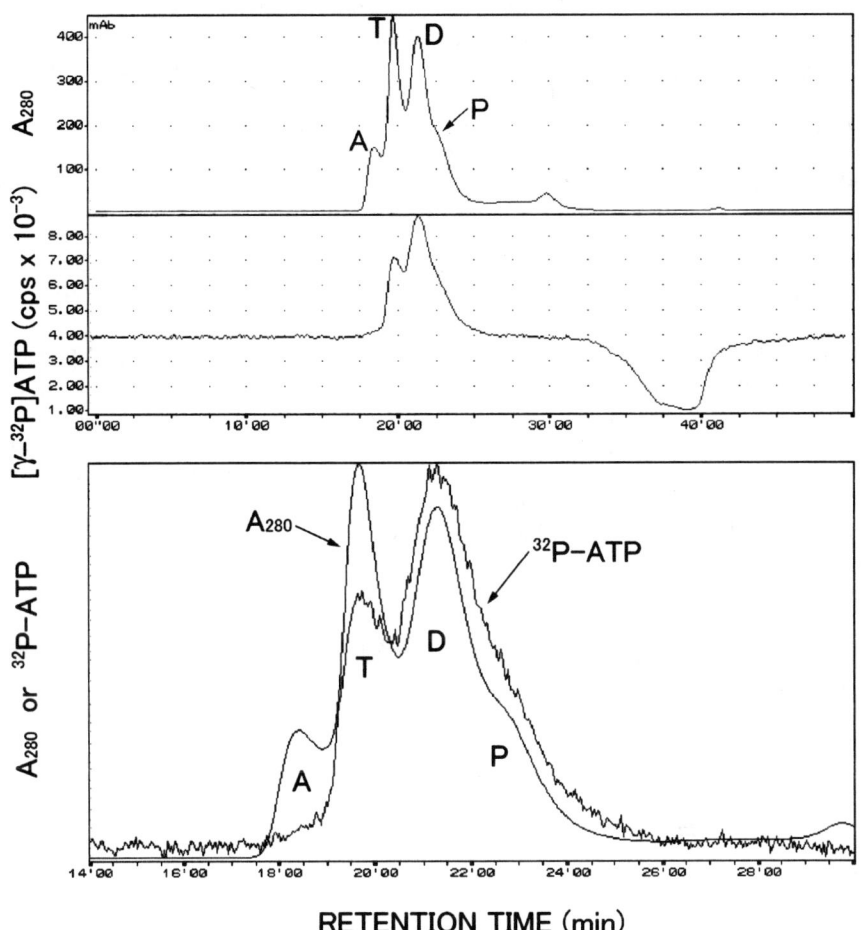

FIGURE 1. ATP-binding of T, D, and P detected by the Hummel-Dreyer method in 1.0 μM [γ-^{32}P] ATP at 0°C. *Top panel:* absorbance at 280 nm (*upper*) and a radioactivity (*lower*). *Bottom panel:* The two elution patterns in the top panel are superimposed.

ATPase ACTIVITIES OF T, D, AND P

The solubilized enzymes were chromatographed at 0°C using the conventional TSKgel G3000SW$_{XL}$ column and the basic elution buffer plus 80 μg/mL PS and 1 mM ATP. The effluent was warmed to 25.0°C immediately after eluting out of the column-oven, by which the respective oligomers started ATP hydrolysis. After the flow for 33 s at 25°C, the effluent was fractionated into microtubes chilled at 0°C at the 15-s interval, and the amount of P_i was determined. The ATPase activities of T, D, and P thus estimated were 8.17 ± 0.72, 11.4 ± 1.0, and 13.0 ± 1.5 ($n = 3$) U·mg^{-1}

protein, respectively. The ATPase activity of the intact enzyme used in the present experiment was 40.7 U·mg^{-1} protein at 37°C under the optimum conditions and 5.79 U·mg^{-1} protein in the elution buffer at 25°C. Accordingly, the activities of T, D, and P are inferred to be 57.4 ± 5.1, 80.1 ± 7.0, and 91.4 ± 10.5 U·mg^{-1} protein, respectively, under the optimum conditions at 37°C. The ATPase activity of T was substantially equal to the highest value reported for the membrane-bound enzyme. It is thus likely that the Na$^+$/K$^+$-ATPase in the membrane is at T form, and that D and P hydrolyze ATP without coupling to Na$^+$/K$^+$-transport.

AMOUNT OF ATP-BINDING WITH HIGH AFFINITY TO T, D, OR P

[γ-^{32}P] ATP bound to the oligomers was measured by the Hummel-Dreyer's method with the conventional column and the elution buffer containing 0.2 mg/mL C$_{12}$E$_8$, 60 μg/mL PS, 1 μM [γ-^{32}P] ATP, and (in mM) 100 NaCl, 1 EDTA, 10 imidazole and 13 Hepes (pH 7.0) at 0°C (FIG. 1). The amount of ATP bound was 0.40 ± 0.01, 0.71 ± 0.03, and 0.74 ± 0.02 ($n = 6$) mol ATP/mol protomer in T, D, and P, respectively. Therefore, the number of binding sites per protomer was 0.5, 1.0 and 1.0 for T, D, and P, respectively, which reversibly changed in a manner dependent on its oligomeric structure.

REFERENCE

1. HAYASHI, Y., et al. 1989. Biochim. Biophys. Acta **983:** 217–229.

ATPase Activity and Oligomerization of Solubilized Na^+/K^+-ATPase Maintained by Synthetic Phosphatidylserine

NOBUKO SHINJI,[a] YOSHIKAZU TAHARA,[a] EMI HAGIWARA,[a] TAKAYUKI KOBAYASHI,[a] KUNIHIRO MIMURA,[b] HITOSHI TAKENAKA,[a] AND YUTARO HAYASHI[a]

[a]*First Department of Biochemistry, Kyorin University School of Medicine, Mitaka, Tokyo 181-8611, Japan*

[b]*Department of Medical Technology, Toyo Public Health College, Shibuya, Tokyo 151-0071, Japan*

KEYWORDS: phosphatidylserine; oligomeric structure; tetraprotomer; solubilized Na^+/K^+-ATPase

When the membrane-bound Na^+/K^+-ATPase (dog kidney) was solubilized with $C_{12}E_8$ and subjected to gel chromatography at 0°C, natural phosphatidylserine (PS; bovine brain) should be added to the elution buffer to isolate enzymatically active oligomers, viz., $(\alpha\beta)_2$-diprotomer (D) and $\alpha\beta$-protomer (P).[1] We recently isolated active $(\alpha\beta)_4$-tetraprotomer (T) in an amount comparable to active D and P from the enzyme solubilized in the presence of CH_3COOK.[2] We here compare effects of the natural PS and four kinds of synthetic PS with various lengths of acyl chains (Avanti Polar Lipids, Inc.) on distribution of oligomeric forms and restoration of their ATPase activities.

ATPase ACTIVITIES OF OLIGOMERS VERSUS SPECIES OF PSs

The membrane-bound enzymes were solubilized with $C_{12}E_8$ in the presence of either 0.1 M CH_3COOK or $NaNO_3$, which was respectively designated as the CH_3COOK- or $NaNO_3$-solubilized enzyme, and was subjected to chromatography in a TSKgel G3000SWxL column (7.8ϕ × 300 mm, Tosoh Co.) with the basic elution buffer (0.2 μg/mL $C_{12}E_8$, and [in mM] 90 NaCl, 10 KCl, 4 $MgCl_2$, 1 EDTA, 10 imidazole and 13 Hepes [pH 7.0]) at 0°C. The elution patterns of solubilized enzymes

Address for correspondence: Yutaro Hayashi, First Department of Biochemistry, Kyorin University School of Medicine, Mitaka, Tokyo 181-8611, Japan. Voice: +81-422-76-7651; fax: +81-422-76-7650.
yutahaya@kyorin-u.ac.jp

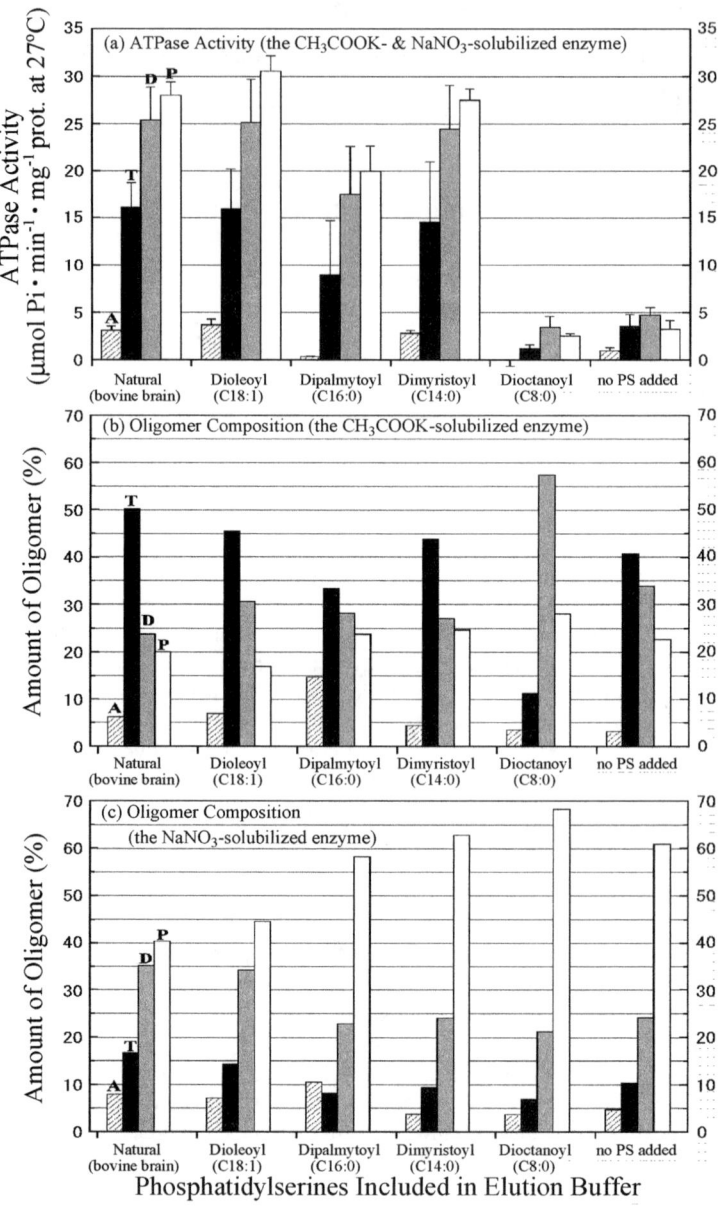

FIGURE 1. Effects of acyl moieties in PS on (**a**) the ATPase activity and (**b, c**) the distribution of the oligomers of the solubilized enzymes. (**a**) The mean ATPase activities of the oligomers obtained from the CH_3COOK- and $NaNO_3$-solubilized enzyme. (**b**) The amount of oligomers in the CH_3COOK- and (**c**) that in the $NaNO_3$-solubilized enzyme. *Hatched* column, A; *filled*, T; *shadowed*, D; *open*, P.

were distinctive in the amount of T and P (FIGS. 1b and c). The ATPase activities of the oligomers were estimated at 27 ± 1°C during chromatography with the basic elution buffer plus 2 mM ATP and 80 µg/mL dioleoyl PS (C18:1), as described elsewhere.[2] The CH_3COOK- and the $NaNO_3$-solubilized enzymes showed similar ATPase activities of oligomers: both showed that the mean values for T, D, and P were 16.0 ± 4.2, 25.1 ± 4.6, and 30.5 ± 1.6 µmol $P_i \cdot min^{-1}\, mg^{-1}$ protein ($n = 4$), respectively. The profiles of the ATPase activities thus obtained were similar to those obtained with natural PS, and were also substantially similar to those obtained in the presence of dipalmytoyl (C16:0) or dimyristoyl PS (C14:0) (FIG. 1a). In contrast, the activities without PS or with dioctanoyl PS (C8:0) were both less than 22% of those with dioleoyl PS. The ATPase activity exhibited by any oligomer was ouabain sensitive.

EFFECTS OF SYNTHETIC PS ON THE COMPOSITION OF OLIGOMERS

When the CH_3COOK-solubilized enzymes were chromatographed in the basic elution buffer plus 80 µg/mL various kinds of PS, the relative amounts of aggregate (A), T, D, and P reached 6, 50, 24 and 20%, respectively. As shown in FIGURE 1b, the composition of oligomers was very similar to that obtained in the presence of the dioleoyl, dipalmytoyl, or dimyristoyl PSs and also to the composition without added PS. However, the composition obtained in the presence of dioctanoyl PS was different from that described above. When the $NaNO_3$-solubilized enzymes were applied to the similar analyses, P was most abundant, but T was least in all cases (FIG. 1c). The results indicated that $NaNO_3$ diminished the ability of the synthetic PSs and/or residual endogenous phospholipids to keep P and D associated to form higher oligomers. In conclusion, the synthetic PSs bearing the acyl chains longer than C_{14} were found to mimic the natural PS to restore the ATPase activity and to hold the oligomeric structure as $(\alpha\beta)_4$-tetraprotomer in solubilized Na^+/K^+-ATPase, which was contrary to the case with dioctanoyl PS (C8:0).

REFERENCES

1. HAYASHI, Y., et al. 1989. Biochim. Biophys. Acta **983**: 217–229.
2. HAYASHI, Y., N. SHINJI, Y. TAHARA, et al. 2003. Isolation of $(\alpha\beta)_4$-tetraprotomer having half-of-sites ATP binding from solubilized dog kidney Na^+/K^+-ATPase. This volume.

Cation Requirement for Nucleotide Binding to Na,K-ATPase

MIKAEL ESMANN AND NATALYA U. FEDOSOVA

Department of Biophysics, University of Aarhus, DK-8000, Aarhus C, Denmark

KEYWORDS: nucleotide binding; Na,K-ATPase; cation binding

This poster summary deals with the cation requirement for induction of the protein conformation allowing high-affinity binding of nucleotide. We find that Na^+ and ionized Tris and imidazol are equally effective, and that 20–40 mM cation is sufficient for maximal affinity for ADP.

Pig kidney microsomal membranes were prepared according to the method of Klodos et al.,[1] treated with SDS,[2] and purified by differential centrifugation to a specific activity of 28 μmol per mg protein per min at 37°C.[1] Equilibrium binding of ADP was measured in double-labeling filtration experiments essentially as described.[3] In brief: Na,K-ATPase in 1 mM histidine neutralized with 0.045 mM CDTA to pH 7.0 (20°C) and chloride salts of Na^+, $Tris^+$, $choline^+$, $imidazol^+$ or *N*-methylglucamine were mixed with radiolabeled ADP. One milliliter of this suspension (usually 0.24–0.28 mg protein/mL) was loaded on two stacked Millipore HAWP 0.45-μm filters, and the bound and free ADP concentrations were calculated from the radioactivity on the upper filter. The maximal nucleotide binding capacity was about 2.8 nmol/mg protein.

In a medium with very low ionic strength (no added salt), binding of ADP to Na,K-ATPase is negligible at concentrations up to 20 μM ADP (FIG. 1). The implication is that the enzyme under these conditions is in a conformation with an inaccessible nucleotide site. One possibility is that this is the native, unliganded state of the enzyme. Another possibility is that a K^+ contamination in the μM range—inherent to Na,K-ATPase preparations—is sufficient to induce an E_2 form with low affinity for nucleotide.

Addition of NaCl to the incubation medium increases the binding of ADP markedly (FIG. 1). The equilibrium dissociation constant for ADP decreases from 3.4 μM at 5 mM NaCl to 0.4 μM at 35 mM NaCl. The maximal binding is approximately 2.8 nmol/mg protein at all NaCl concentrations tested. At concentrations of NaCl above 50 mM, the affinity for ADP decreases progressively. This is due to an ionic strength

---

Address for correspondence: Dr. Mikael Esmann, Department of Biophysics, University of Aarhus, Ole Worms Allé 185, DK-8000, Aarhus C, Denmark. Voice: +45-8942-2930; fax: +45-8612-9599.

me@biophys.au.dk.

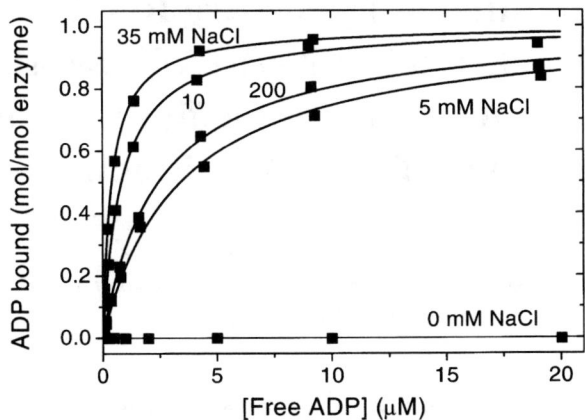

FIGURE 1. Equilibrium binding at 20°C of ^{14}C-ADP to Na,K-ATPase at different concentrations of NaCl. The amount of ^{14}C-ADP bound is measured with the filtration technique.[3] The lines represent single hyperbolic functions with a maximal binding of 1 mol per mol Na,K-ATPase. Dissociation constants were 3.41 μM (at 5 mM NaCl), 0.83 μM (at 10 mM NaCl), 0.40 μM (at 35 mM NaCl), and 2.45 μM (at 200 mM NaCl).

effect on the interaction between the negatively charged ADP and positively charged nucleotide binding site.[3] The cations of Tris, imidazol, choline, and N-methylglucamine were all found to induce high-affinity ADP-binding in a fashion similar to that of Na$^+$ (data not shown). The smallest dissociation constants in the presence of Tris and imidazol were similar to that in Na$^+$, and in optimal concentrations of choline or N-methylglucamine the dissociation constants were about 1 μM, twofold higher than for Na$^+$.

The similarity of the concentration dependence of the dissociation constant for ADP on Na$^+$, Tris, and imidazol in the 5–50 mM range (not shown) suggests a relatively nonspecific Debye screening of charges on the protein. This screening leads to a folding of the nucleotide domain in the E$_1$ state with an increased affinity for ADP.

REFERENCES

1. KLODOS, I., M. ESMANN & R.L. POST. 2002. Large-scale preparation of sodium-potassium ATPase from kidney outer medulla. Kidney Int. **62:** 2097–2100.
2. JØRGENSEN, P.L. 1975. Purification and characterization of (Na$^+$+K$^+$)-ATPase. III. Purification from the outer medulla of mammalian kidney after selective removal of membrane components by sodium dodecylsulphate. Biochim. Biophys. Acta **356:** 36–52.
3. FEDOSOVA, N.U., P. CHAMPEIL & M. ESMANN. 2002. Nucleotide binding to Na,K-ATPase: the role of electrostatic interactions. Biochemistry **41:** 1267–1273.

Single Mutation of Lys or Arg Residue in ATP Binding Pocket in Rat Na/K-ATPase Alpha-1 Subunit Induces Different Affinity Change in High- and Low-Affinity ATP Binding

TOSHIAKI IMAGAWA, SATOMI TERAMACHI, AND KAZUYA TANIGUCHI

Biological Chemistry, Division of Chemistry, Graduate School of Science, Hokkaido University, Sapporo 060-0810, Japan

KEYWORDS: Na/K-ATPase; ATP binding site; mutagenesis

INTRODUCTION

Na/K-ATPase belongs to the P-type ATPase group, which forms an acid-stable phophoenzyme (EP) during ATP hydrolysis. It is generally accepted that at least high and low ATP binding sites are present in the P-type ATPase. The issue of whether two different ATP binding sites reside simultaneously on the same Na/K-ATPase α-subunit, or are on different α-subunits with different conformational states, remains controversial.

X-ray crystallography revealed that the cytoplasmic headpiece of the Ca-ATPase contains one ATP binding pocket containing F487, K492, K515, and R560 (F475, K480, K501, and R544 in pig Na/K-ATPase).[1] In order to investigate the relationship between the high- and low-affinity ATP binding sites, we constructed ten mutants carrying a single amino acid substitution in the side chain that might possibly serve as ligands for ATP, and evaluated the consequence for high- and low-affinity ATP effects of each mutation.[2]

RESULTS AND DISCUSSION

HeLa cells transfected with the expression vector encoding the six mutants, F475Y, K480A/E, K501A/E, and R544A, resulted in the appearance of the cells that survived in the presence of 10 μM ouabain. Ouabain-resistant cell lines, via transfection with the four mutants, F475A/S/D or R544E, could not be obtained, suggesting that these mutations impaired the Na/K-ATPase activity to an extent that the cells were not able to survive.

Address for correspondence: Dr. Toshiaki Imagawa, Biological Chemistry, Division of Chemistry, Graduate School of Science, Hokkaido University, Kita-Ku, Kita 10 Nishi 8, Sapporo 060-0810, Japan. Voice: +81-11-706-2698; fax: +81-11-736-2074.
toshi@sci.hokudai.ac.jp

TABLE 1. ATP concentration dependence of EP formation, Na/K-ATPase activity, and inhibition of K-pNNPase activity[a]

	Wild type	F475Y	K480A	K480E	K501A	K501E	R544A
a	$K_{0.5}^{h}$ (0.026 μM)	4.8	1.0	**19**	2.0	**8.5**	**58.9**
b	$K_{0.5}^{l}$ (0.69 mM)	<5.8	1.1	<5.8	1.1	1.6	2.0
c	$K_{i,0.5}$ (0.36 mM)	8.4	1.5	20.7	0.9	6.8	4.0
d	$K_{0.5}^{l}/K_{0.5}^{h}$ (2.7 × 10^4)	<1.2	1.1	<0.3	0.5	0.2	**0.03**
e	$K_{i,0.5}/K_{0.5}^{h}$ (1.3 × 10^4)	1.7	1.5	1.1	0.7	0.8	**0.07**

NOTE: The value of the wild-type enzyme (the original parameters obtained are shown in parentheses) was assumed to be 1 and the relative value was compared to those of mutants.
[a]Measured as described previously.[2,4]

The apparent affinities for ATP were estimated by measuring high-affinity ATP-dependent phosphorylation ($K_{0.5}^{h}$: TABLE 1, a) and low-affinity activation of Na/K-ATPase ($K_{0.5}^{l}$: TABLE 1, b) or low-affinity ATP inhibition of K-pNPPase ($K_{i,0.5}$: TABLE 1, c). The data show that each single amino acid substitution, such as F475Y, K480E, K501E, and R544A, reduced the apparent affinity of the high- and low-affinity ATP effects to different extents (TABLE 1, d and e).

These results suggest that each α-chain contains only one ATP binding pocket, which can change its conformation to accept ATP with a high and low affinity, and that R544 and possibly K501 are more important for achieving the high-affinity ATP binding, while F475 and possibly K480 are more important for achieving the low-affinity ATP binding. The present finding is consistent with the view that Na/K-ATPase functions out of phase as a diprotomer, $(\alpha\beta)_2$, or a much higher oligomer, $(\alpha\beta)_4$.[3]

ACKNOWLEDGMENTS

This work was supported in part by grants-in-aid for Scientific Research (10308028 to K.T) and the International Scientific Research Program (10044048 to K.T.) from the Ministry of Education, Science, Sports and Culture of Japan.

REFERENCES

1. TOYOSHIMA, C., et al. 2000. Crystal structure of the calcium pump of sarcoplasmic reticulum at 2.6 A resolution. Nature **405**: 647–655.
2. TERAMACHI, S., et al. 2002. Replacement of several single amino acid side chains exposed to the inside of the ATP binding pocket induces different extents of affinity change in the high and low affinity ATP binding sites of rat Na/K-ATPase. J. Biol. Chem. **277**: 37394–37400.
3. TANIGUCHI, K., et al. 2001. The oligomeric nature of Na/K-transport ATPase. J. Biochem. **129**: 335–342.
4. IMAGAWA, T., et al. 1998. Does binding of ouabain to human alpha1-subunit of Na$^+$, K$^+$-ATPase affect the ATPase activity of adjacent rat alpha1-subunit? Jpn. J. Pharmacol. **76**: 415–423.

Localization of Catalytic Active Sites in the Large Cytoplasmic Domain of Na^+/K^+-ATPase

RITA KRUMSCHEID,[a] KLÁRA SUÁNKOVÁ,[b] RÜDIGER ETTRICH,[c] JAN TEISINGER,[b] EVEN AMLER,[b] AND WILHELM SCHONER[a]

[a]*Institute of Biochemistry and Endocrinology, Justus-Liebig-University, D-35392 Giessen, Germany*

[b]*Institute of Physiology, Czech Academy of Sciences, CZ-142 20 Prague 4, Czech Republic*

[c]*Institute of Physical Biology, University of South Bohemia, CZ-373 33 Nové Hrady, Czech Republic*

KEYWORDS: H_4H_5 loop; TNP-ATP; phosphatase activity

The controversy still continues whether Na^+/K^+-ATPase works as an oligomer with coexisting high- and low-affinity ATP sites, or as an $\alpha\beta$ monomer with consecutively existing high- and low-affinity ATP sites.[1,2] Recently the three-dimensional structure of the large cytoplasmic H_4-H_5-loop (L354-I777), containing the ATP site, became available by molecular modeling.[3] This method allowed us to identify in GST-H_4-H_5-loop fusion proteins, of variable length, the location of the ATP site(s), its intrinsic phosphatase activity, and to verify whether the loss of interacting amino acids or the change of the three-dimensional structure is responsible for any altered biological function.

EXPERIMENTAL METHODS

Constructs of the H_4-H_5-loop were obtained by insertion of stop codons between L354 and I777. Expression of the GST-fusion proteins containing this information in a pGEX-2T vector occurred in *E. coli*. TNP-ATP binding was performed by fluorometry in 20 mM TRIS/HCl, pH 7.8 at 37°C as described earlier.[2] Phosphatase activity was measured in the GST-fusion proteins in a variant of the method of Tran and Farley.[4] Molecular modeling was performed as described earlier.[3]

Address for correspondence: Dr. Wilhelm Schoner, Institute of Biochemistry and Endocrinology, Justus-Liebig-University, D-35392 Giessen, Germany. Voice: +49-641-9938170; fax: +49-641-9938179.

Schoner@vetmed.uni-giessen.de

Comparison of Phosphatase activity and TNP-ATP affinity

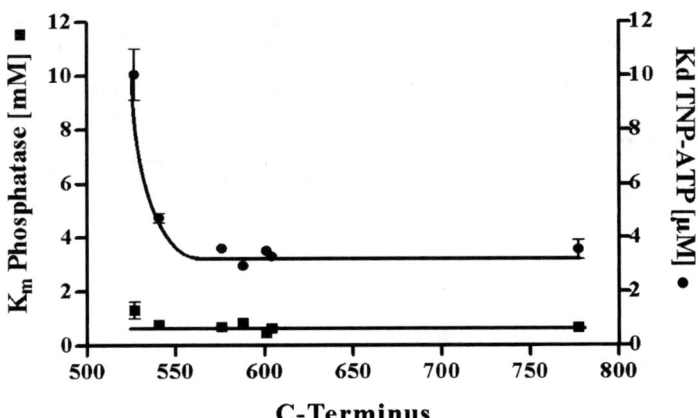

FIGURE 1. Comparison of the phosphatase activity (K_M) and TNP-ATP affinity (K_d) of GST-H$_4$-H$_5$ fusion products of variable length starting N-terminally at L354.

RESULTS AND DISCUSSION

FIGURE 1 shows that recognition of TNP-ATP, but not its phosphatase activity, is lost when the C-terminal truncation of the H$_4$-H$_5$-loop exceeds C549. This truncation does not remove any amino acids interacting directly with ATP in the intact recognition site in its open conformation.[3] Molecular modeling revealed that a change of protein stability of the ATP binding pocket is responsible for the loss of TNP-ATP recognition. Phosphatase activity was not affected by truncation, but was greatly enhanced when D369 was available as a phosphorylation site (mean V_{max} activity of all constructs containing D369: 11.4 ± 1 nmol/h·mg (± SEM); D369-deficient construct: 4.1 ± 2 nmol/h·mg; $P < 0.05$). Hence the N domain itself contains phosphatase activity, but transphosphorylation to D369 located on the P domain may increase it. Additionally, C549 was formerly identified as part of the low-affinity ATP site in the E$_2$ conformation.[2] C549 is, according to molecular modeling, close to the ATP site.[3] Since only one ATP site exists in the H$_4$-H$_5$-loop,[3] the former evaluation of Foerster distance[2] must be interpreted in favor of an oligomeric mechanism of Na$^+$/K$^+$-activated ATP hydrolysis.[1,2]

ACKNOWLEDGMENTS

This work was supported by the German and Czech Governments through Grants No. TSR-088-97 and CZE 00/033, by Grants No 309/02/1479, No. 204/01/0254 and No. 204/01/1001 of the Grant Agency of the Czech Republic, and by the research plan of the Faculty of Science (MSM 113100001).

REFERENCES

1. TANIGUCHI, K., S. KAYA, K. ABE & S. MÅRDH. 2001. The oligomeric nature of Na^+/K^+-ATPase. J. Biochem. **129:** 335–342.
2. LINNERTZ, H., H. KOST, P. OBSIL, et al. 1998. Erythrosin 5'-isothiocyanate labels Cys^{549} as part of the low-affinity binding site of Na^+/K^+-ATPase. FEBS Lett. **441:** 103–105.
3. ETTRICH, R., M. MELICHERCIK, J. TEISINGER, et al. 2001. Three-dimensional structure of the large cytoplasmatic H_4-H_5-loop of Na^+/K^+-ATPase deduced by restraint-based comparative modeling shows only one ATP-binding site. J. Mol. Model. **7:** 184–192.
4. TRAN, C.M. & R.A. FARLEY. 1999. Catalytic activity of an isolated domain of Na,K-ATPase expressed in *Escherichia coli*. Biophys. J. **77:** 258–266.

Calorimetry of Na,K-ATPase

M. STOLZ,[a] E. LEWITZKI,[a] E. SCHICK,[a] M. MUTZ,[b] AND E. GRELL[a]

[a]*Max-Planck-Institute of Biophysics, 60596 Frankfurt, Germany*
[b]*Novartis Pharma AG, 4002 Basel, Switzerland*

KEYWORDS: calorimetry; thermal denaturation; ouabain binding; nucleotide binding; DSC; ITC

In order to characterize the overall subunit interaction and the thermal stability of purified Na,K-ATPase isolated from pig kidney and dogfish (*Squalus acanthias*) rectal gland,[1] a differential scanning calorimetry (DSC) study is carried out (FIG. 1a). With regard to ligand binding, the interaction stoichiometry (number of active sites n per protomer) and the affinity of ouabain and nucleotide binding are investigated by titration calorimetry at 25°C. Similar to our earlier results,[2] the DSC thermogram of the pig kidney enzyme shows a single, very narrow, and almost symmetric endothermic denaturation transition at 57.0°C and a full width at half-maximum of 3.5°C. This is indicative of a uniform denaturation process of high cooperativity involving all subunits. Our result differs markedly from a recent investigation and thus questions the relevance of the interpretation postulated therein.[3] The thermogram of the dogfish enzyme shows a lower transition temperature (49.5°C) and a much broader transition range than for the kidney enzyme. In addition, it exhibits a well-separated pretransition at 37.5°C. The pretransition is tentatively assigned to the γ-like subunit of the dogfish enzyme. The broader transition can be an expression of less ordered subunit associations. The transition temperature of the micellar dogfish enzyme ($C_{12}E_8$ solubilized) is 8°C lower than that of the membrane-bound state. This can be due to an increased water access of the protein in the micellar state. All overall ΔH values are around 18 MJ/mol.

Calorimetric titrations with ouabain have enabled a precise determination of the binding stoichiometry, leading to values of n around 0.55.[1] In the presence of Mg^{2+} and P_i, equilibrium constants around 10^8 M^{-1} and ΔH and ΔS values around −90 kJ/mol and −150 J/K mol, respectively, are obtained for both enzymes. This unfavorable entropic contribution is assumed to be mainly due to an increased interaction between water and the protein as a consequence of a conformational transition.

High-affinity ADP and ATP binding to the dogfish enzyme (FIG. 1b) in 10 mM imidazole/HCl, 0.25 mM CDTA, 25% glycerol of pH 7.5, is observed only in the presence of NaCl or $MgCl_2$ or at high ionic strength. The resulting stoichiometric

Address for correspondence: E. Grell, Max-Planck-Institute of Biophysics, 60596 Frankfurt, Germany. Voice: +49-69-6303-290; fax: +49-69-6303-346.
ernst.grell@mpibp-frankfurt.mpg.de

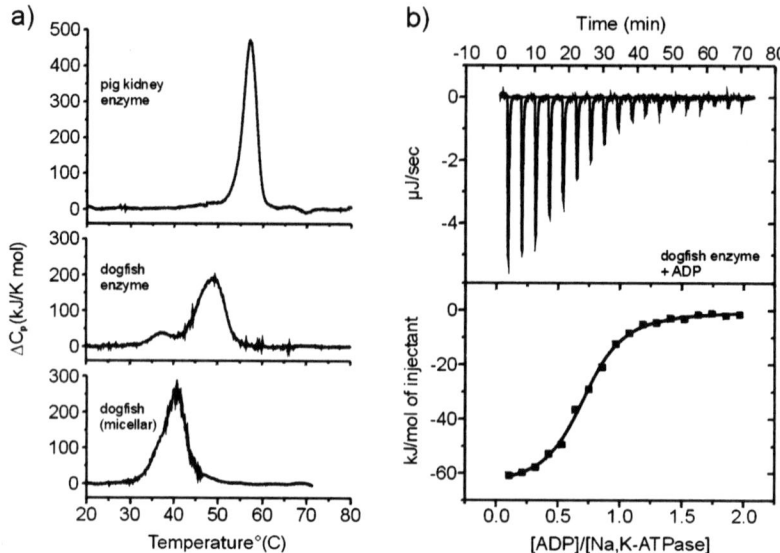

FIGURE 1. (a) DSC thermograms in 30 mM histidine/HCl, 0.1 mM EDTA, 25% glycerol pH 6.8 and (b) ADP binding titration (see text).

coefficients are smaller than 1, around 0.65. The affinities at low ionic strength are around 10^6 M^{-1} for both nucleotides and both enzymes. Our values are lower than those reported recently for the pig kidney enzyme.[4] The presence of Mg^{2+} does not lead to an increased affinity. The exothermic binding process is characterized by a larger ΔH value for ATP (–80 kJ/mol) than for ADP (–65 kJ/mol) in the presence of 3 mM NaCl, probably indicative of an increased number of protein interactions for the triphosphate compared to the diphosphate. The values of the enthalpy changes decrease with increasing ionic strength and are not sensitive to a CDTA concentration increase. The entropy changes upon ATP and ADP binding are large and negative (of the order of –120 J/K mol) at low ionic strength. This can again imply that the protein offers more access to water molecules in the bound state, for example, as a consequence of a structural change.

REFERENCES

1. GRELL, E., E. SCHICK & E. LEWITZKI. 2001. Membrane receptor calorimetry: cardiac glycoside interaction with Na,K-ATPase. Thermochim. Acta **380:** 245–254.
2. GRELL, E., M. MUTZ & E. MARTI. 1997. Membrane-bound Na$^+$,K$^+$-ATPase as the cardiac glycoside receptor: a thermochemical characterization. J. Therm. Anal. **49:** 1–9.
3. GRINBERG, A.V., N.M. GEVONDYAN, et al. 2001. The thermal unfolding and domain structure of Na$^+$/K$^+$-exchanging ATPase. Eur. J. Biochem. **268:** 5027–5036.
4. FEDOSOVA, N., P. CHAMPEIL & M. ESMANN. 2002. Nucleotide binding to Na,K-ATPase: the role of electrostatic interactions. Biochemistry **41:** 1267–1273 (and citations herein).

Expression of Na,K-ATPase in *P. pastoris*: Fe^{2+}-Catalyzed Cleavage of the Recombinant Enzyme

DAVID STRUGATSKY, RIVKA GOLDSHLEGER, EITAN BIBI, AND STEVEN J. D. KARLISH

Department of Biological Chemistry, Weizmann Institute of Science, Rehovoth 76100, Israel

KEYWORDS: Fe-catalyzed cleavage; recombinant Na,K-ATPase; *P. pastoris*

Specific oxidative cleavage of renal Na,K-ATPase catalyzed by Fe^{2+} ions or the ATP-Fe^{2+} complex provides information on spatial organization of the protein in E$_1$ and E$_2$ conformations and the ATP-Mg site.[1,2] We have expressed Na,K-ATPase in *Pichia pastoris* in order to study the consequences of amino acid substitution on cleavage patterns.

The methylotrophic yeast, *P. pastoris*, was chosen as a host organism due to its ability to grow to high cell density and to provide a large quantity of membranes. In addition, the yeast lacks endogenous Na,K-ATPase activity, and exogenous expression of the pump is not required for cell growth. These features allow analysis of mutants that inactivate enzymatic activity.[3,4] The α1 and β1 subunits of porcine Na,K-ATPase were cloned under regulation of the alcohol oxidase 1 (AOX1) promoter in an integrative pHIL-D2 yeast–bacterial shuttle vector, which was transfected into a *P. pastoris* protease-deficient strain (SMD1165). Replacement of the endogenous AOX1 gene with the expressed gene leads to a methanol utilization slow phenotype (Mut^s). Functional expression was optimized and conditions for expression of Na,K-ATPase at ca. 40 g/L were obtained. About 0.5 g of crude membranes with specific ouabain binding capacity of ~30 pmol/mg protein and Na,K-ATPase turnover rates of 7000–9000 min^{-1} at 37°C can be prepared from 3 L of culture.

Membranes containing the recombinant Na,K-ATPase were subjected to specific Fe^{2+}- or ATP-Fe^{2+}-catalyzed oxidative cleavage.[1,2] Specific cleavage fragments observed for renal Na,K-ATPase, including differences in E$_1$ and E$_2$ conformations, were observed, although the yields of fragments were lower for the yeast enzyme, apparently due to a lower stability. Mutants H286Q, D369N, and D369A have been expressed. Preliminary Fe^{2+}- and ATP-Fe^{2+}-catalyzed cleavage experiments provide

---

Address for correspondence: Steven J. D. Karlish, Department of Biological Chemistry, Weizmann Institute of Science, Rehovoth 76100, Israel. Voice: +972-8-934-2278; fax: +972-8-934-4118.
steven.karlish@weizmann.ac.il

evidence for two Fe^{2+} sites—site 1 in the cytoplasmic domains and site 2 at the membrane–water interface near M3 and M1—consistent with previous conclusions.[2] The D369N substitution greatly reduces the affinity and efficiency of ATP-Fe^{2+} in mediating cleavage in the P (^{712}VNDS) and N domains (near ^{440}VAGDA). This provides evidence for an interaction of ^{369}D with the Mg^{2+}-binding aspartate (^{710}D), consistent with an inference based on double mutant cycles,[4] and may also indicate a role for ^{443}D in Mg^{2+} coordination. *P. pastoris* expressing the Na,K-ATPase appears to be a promising system for such structural studies.

ACKNOWLEDGMENTS

This work was supported by the Israel Science Foundation (No. 15/00-1).

REFERENCES

1. GOLDSHLEGER, R. & S.J.D. KARLISH. 1997. Fe-catalyzed cleavage of the alpha subunit of Na/K-ATPase: evidence for conformation-sensitive interactions between cytoplasmic domains. Proc. Natl. Acad. Sci. USA **94**(18): 9596–9601.
2. PATCHORNIK, G., R. GOLDSHLEGER & S.J.D. KARLISH. 2000. The complex ATP-Fe(2+) serves as a specific affinity cleavage reagent in ATP-Mg(2+) sites of Na,K-ATPase: altered ligation of Fe(2+) (Mg(2+)) ions accompanies the E(1)→E(2) conformational change. Proc. Natl. Acad. Sci. USA **97**(22): 11954–11959.
3. PEDERSEN, P.A., J.H. RASMUSSEN & P.L. JORGENSEN. 1996. Consequences of mutations to the phosphorylation site of the alpha-subunit of Na,K-ATPase for ATP binding and E1-E2 conformational equilibrium. Biochemistry **35**: 16085–16093.
4. PEDERSEN, P.A., J.R. JORGENSEN & P.L. JORGENSEN. 2001. Importance of conserved α-subunit segment 709GDGVND for Mg2+ binding, phosphorylation, and energy transduction in Na,K-ATPase. J. Biol. Chem. **275**: 37588–37595.

The Mechanism of Na-K Interaction on Na,K-ATPase

CLAUDIA DONNET AND KATHLEEN J. SWEADNER

Laboratory of Membrane Biology, Massachusetts General Hospital, Charlestown, Massachusetts 02129, USA

KEYWORDS: kinetics; competition; K_{Na}; Na-K interaction; mechanism

There is much prior evidence that K^+ in high concentrations acts as an inhibitor of Na,K-ATPase activity through a shift in the apparent affinity of Na,K-ATPase for Na^+ (K_{Na}). Because cytoplasmic K^+ is generally high, this is important for understanding pump properties in physiological conditions. Differences in sensitivity to K^+ contribute to tissue-specific (and isoform-independent) differences in apparent Na^+ affinity[1] and to the effect of the γ subunit on Na,K-ATPase properties.[2,3] Much prior evidence shows a linear relationship between K_{Na} and $[K^+]$, consistent with a simple kinetic model for Na^+-K^+ competition. While investigating the roles of γ in NRK-52E cells, however, we observed that K_{Na} did not relate linearly to $[K^+]$, with or without γ. Differences in ligand concentrations compared to the literature were a possible reason for the discrepancy. This was investigated with purified rat kidney Na,K-ATPase.

Na,K-ATPase activity was measured as a function of Na^+ concentration in the presence of different concentrations of K^+. Ionic strength was kept constant by addition of choline when specified. The experiments were repeated, varying the concentrations of free Mg^{2+} and ATP and the kind of buffer used. Confirming previous reports, elevated K^+ decreased K_{Na} in all conditions. The data approximated a linear relationship between K_{Na} and $[K^+]$ only in conditions of 4 mM Mg^{2+}, 1 mM ATP with 1 mM free Mg^{2+}, but not with 3 mM Mg^{2+}, 3 mM ATP with <0.5 mM free Mg^{2+}. Furthermore, we observed effects of $[K^+]$ on V_{max} and on the degree of cooperativity for Na^+. These observations suggest that the K^+ effect on Na^+ activation is not a pure competitive effect. Models producing expressions for activity as a function of Na^+ and K^+ concentration were fitted to the whole set of data using two-variable regression, also not supporting simple competition. See FIGURE 1.

The following conclusions can be drawn: (a) Addition of choline to keep constant ionic strength affected the values of K_{Na}, but still produced nonlinear plots of K_{Na}

Address for correspondence: Claudia Donnet, Laboratory of Membrane Biology, 149 13th Street, #6118, Massachusetts General Hospital, Charlestown, MA 02129. Voice: 617-726-8560; fax: 617-726-5677.
  donnet@helix.mgh.harvard.edu

Ann. N.Y. Acad. Sci. 986: 249–251 (2003). © 2003 New York Academy of Sciences.

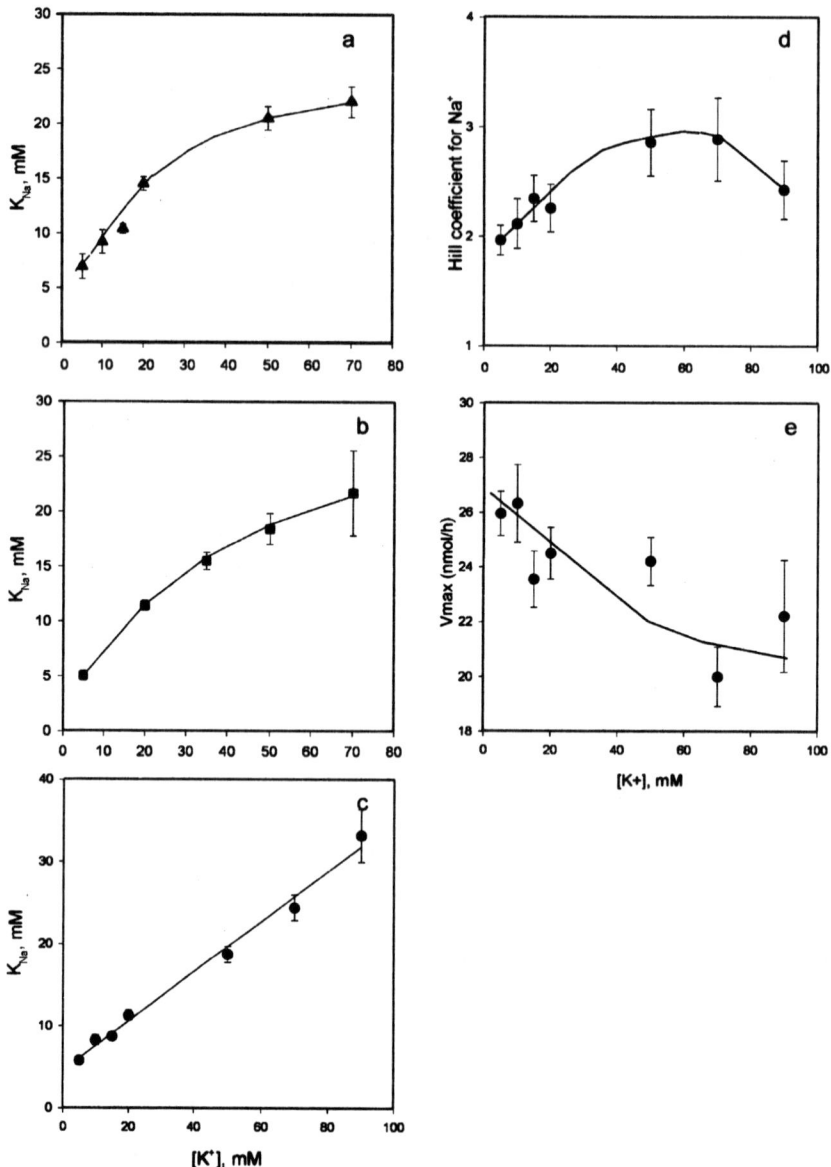

FIGURE 1. K^+ dependence of the kinetic parameters of Na^+ curves. *Left*: K_{Na} as a function of $[K^+]$ for activity measured in the presence of histidine as buffer, $[Mg^{2+}]_{free}$ = 0.3–0.5 mM, [EGTA] = 0 mM, and [ATP] = 3 mM (ionic strength varied with [NaCl] and [KCl]) (**a**); same conditions, but addition of choline to keep a constant ionic strength (**b**), or Tris as buffer, $[Mg^{2+}]_{free}$ = 1 mM, [EGTA] = 0.5 mM, [ATP] = 1 mM, and constant ionic strength kept by addition of choline (**c**). *Right*: The dependence of the Hill coefficient (**d**) and V_{max} (**e**) with $[K^+]$ under the same conditions as in panel c.

vs. [K$^+$] in histidine buffer containing low free Mg^{2+}. (b) Using higher Mg^{2+} concentration, Tris buffer and EGTA made K$_{Na}$ vs. [K$^+$] plots linear. (c) In addition to the effect on K$_{Na}$, we observed an effect of K$^+$ on V_{max} and in the degree of cooperativity, suggesting that the K$^+$ effect on Na$^+$ activation is not a pure competitive effect. Both sets of ligand conditions are possible in cells, so the complexity of the K$^+$ effect is predicted to have subtle consequences.

ACKNOWLEDGMENTS

This work was supported by NIH Grant No. HL36271.

REFERENCES

1. THERIEN, A.G. & R. BLOSTEIN. 1999. K$^+$/Na$^+$ antagonism at cytoplasmic sites of Na$^+$,K$^+$-ATPase: a tissue-specific mechanism of sodium pump regulation. Am. J. Physiol. **277**: C891–C898.
2. PU, H.X., F. CLUZEAUD, R. GOLDSHLEGER, *et al.* 2001. Functional role and immunocytochemical localization of the γ$_a$ and γ$_b$ forms of the Na,K-ATPase γ subunit. J. Biol. Chem. **276**: 20370–20378.
3. ARYSTARKHOVA, E., C. DONNET, N.K. ASINOVSKI & K.J. SWEADNER. 2002. Differential regulation of renal Na,K-ATPase by splice variants of the γ subunit. J. Biol. Chem. **277**: 10162–10172.

Salt Effects on the Kinetics of the Electrogenic Na⁺ Transport in the Na,K-ATPase

ARTEM G. AYUYAN,[a] VALERIJ S. SOKOLOV,[a] AND HANS-JÜRGEN APELL[b]

[a]*A. N. Frumkin Institute of Electrochemistry RAS, 117071 Moscow, Russia*
[b]*Department of Biology, University of Konstanz, 78457 Konstanz, Germany*

KEYWORDS: chaotropic ions; activation energy; transient currents

The kinetics of electrogenic Na⁺ transport through the Na,K-ATPase is decelerated in concentrated salt solutions. This effect is anion specific, and the effectiveness of the anions is I⁻ > Br⁻ > Cl⁻, which agrees with the Hofmeister series of chaotropic ions. It can be explained by an effect on the conformation transition, E_1-E_2.[1,2] Here, we show how different salts affect the apparent activation energy of the rate-limiting E_1-E_2 conformation transition. Experiments were performed with Na,K-ATPase–containing membrane fragments adsorbed to lipid membranes (BLMs). Transient currents were induced by ATP release from caged ATP[1] and were fitted by three exponentials. The rate constant k of the exponential reproducing the falling phase of the current represents the rate-limiting step.[1] According to the Arrhenius equation, $k = A \cdot \exp(-E_a/RT)$, the activation energy E_a and the preexponential term A were obtained from the temperature dependency. To study the salt effects on both sides of the protein, two types of experiments were carried out: (1) choline salts with different anions were added into the cell before adsorption of the membrane fragments to the BLM so that they could affect the Na,K-ATPase on both sides of the membrane; (2) salts were added after adsorption and could affect then only the cytoplasmic side.

Without iodide or bromide salts, the activation energy was 63 kJ/mol. As shown in FIGURE 1A, choline iodide decreased the activation energy E_a and the factor A. Comparable results were found with bromide (not shown). The decrease of rate constant k with ion concentration was dominated by A. This effect was more significant in experiments of type 1, which indicates that it is caused preferentially by action of the anions on the extracellular side.

As found in experiments with varying ATP concentration, the effect of iodide could not be explained by altered ATP-binding constants. Changes of k with temperature due to addition of the salts differed from those induced by a decreased amount of ATP released from caged ATP. At high ATP concentrations, the temperature dependence of k remained linear, regardless of the presence of choline salts. In contrast, at low ATP concentrations, it was nonlinear, with an activation energy E_a of 125 kJ/mol at temperatures above 21°C; at decreased temperatures, ln(k) bent to a lower value of $E_a \approx 25$ kJ/mol (FIG. 1B).

Address for correspondence: Artem G. Ayuyan, A. N. Frumkin Institute of Electrochemistry RAS, Leninsky pr. 31, 117071 Moscow, Russia. Voice/fax: +7-095-952-5582.
mennefer@netra.elchem.ac.ru

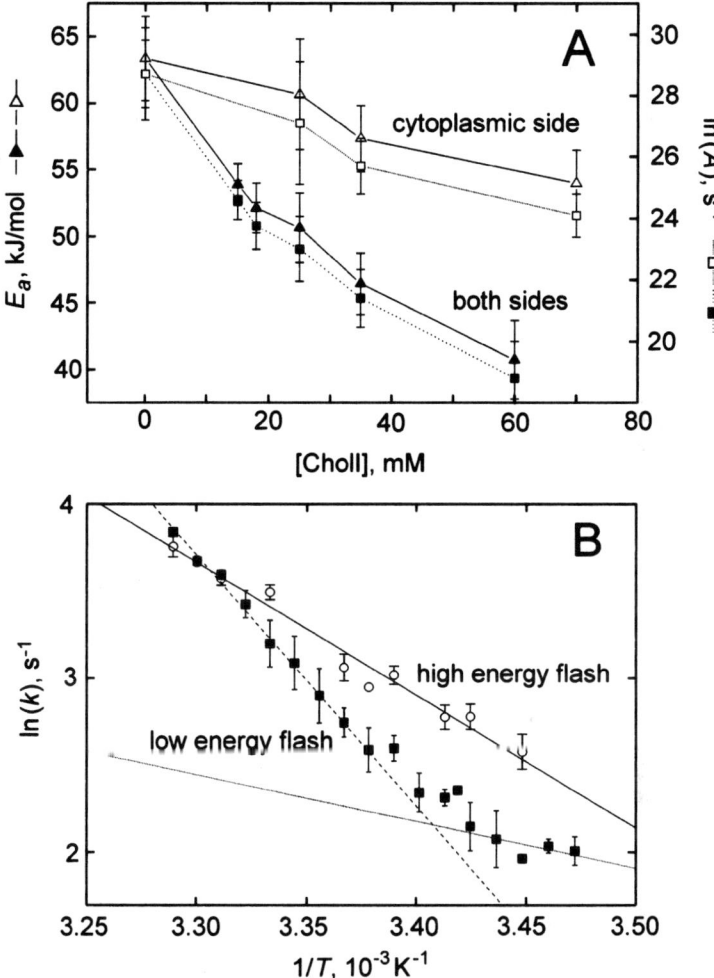

FIGURE 1. (A) Dependence of the activation energy E_a (triangles) and preexponential term A (squares) of the Arrhenius equation (see text) on choline iodide concentration obtained from type 1 (solid symbols) and type 2 (open symbols) experiments. BLMs were formed as described elsewhere.[1] (B) Arrhenius plot for different UV-flash intensities: 100% (circles) and 44% (squares) intensity. The solid, dashed, and light dotted lines correspond to activation energies of 63, 125, and 25 kJ/mol, respectively. Solutions are without choline salts.

ACKNOWLEDGMENTS

This work was supported financially by RFBR No. 01-04-49246 and INTAS (Project No. 01-0224).

REFERENCES

1. SOKOLOV, V.S. *et al.* 2001. Assignment of charge movements to electrogenic reaction steps of Na,K-ATPase by analysis of salt effects on the charge movements. Eur. Biophys. J. **30:** 515–527.
2. POST, R.L. & K. SUZUKI. 1991. A Hofmeister effect on the phosphoenzyme of Na,K-ATPase. *In* The Sodium Pump: Structure, Mechanism, and Regulation, pp. 201–209. Rockefeller University Press. New York.

Mutational Analysis of Ouabain Interaction with the M5–M6 Hairpin of Na,K-ATPase

L. Y. QIU, J. B. KOENDERINK, H. G. P. SWARTS, P. H. G. M. WILLEMS, AND J. J. H. H. M. DE PONT

Department of Biochemistry, Nijmegen Center for Molecular Life Sciences, 6500 HB Nijmegen, the Netherlands

KEYWORDS: Na,K-ATPase; ouabain; chimeric enzymes

Na,K-ATPase and gastric H,K-ATPase are two related enzymes that are responsible for active cation transport. Na,K-ATPase activity is specifically inhibited by ouabain, whereas gastric H,K-ATPase activity is not inhibited by this compound. The precise location of the ouabain-binding site in Na,K-ATPase is still unknown. In a previous study,[1] we demonstrated that the chimera HN34/56, which contained the α- and β-subunits of rat H,K-ATPase, but transmembrane hairpins M3–M4 and M5–M6 of rat Na,K-ATPase, could bind ouabain with high affinity. Chimeras containing either hairpin M3–M4 or M5–M6 of Na,K-ATPase did not bind ouabain. Transmembrane segments M5 and M6 are thought to be involved in the cation-binding pocket, in which movement of hairpin M5–M6 may play a role in active transport.[2] It has been proposed that binding of ouabain to this hairpin inhibits cation transport by immobilizing this transmembrane domain.[3] In the present study, we investigated which amino acids of the transmembrane M5 and M6 in Na,K-ATPase are involved in ouabain binding.

METHODS

There are 21 amino acids that are different between Na,K-ATPase and gastric H,K-ATPase in M5–M6 (FIG. 1). In this study, we divided them into seven groups and made mutations by replacing the Na,K-residues by their H,K-counterparts. *Xenopus* oocytes were injected with 10 ng of α-subunit and 2 ng of β-subunit cRNAs, and ouabain binding was determined.

Address for correspondence: L.Y. Qiu, Department of Biochemistry, Nijmegen Center for Molecular Life Sciences, P. O. Box 9101, 6500 HB Nijmegen, the Netherlands. Voice: +31-24-3614258; fax: +31-24-3616413.
L.Qiu@ncmls.kun.nl

Ann. N.Y. Acad. Sci. 986: 255–257 (2003). © 2003 New York Academy of Sciences.

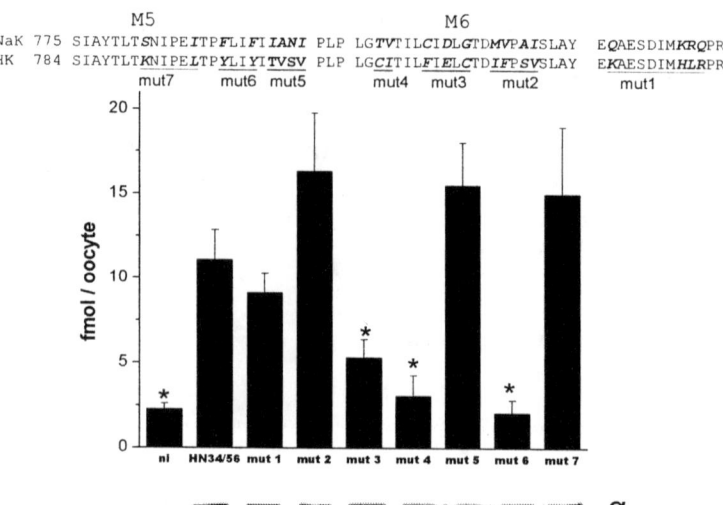

FIGURE 1. Expression of chimeric α-subunits in *Xenopus laevis* oocytes and [^{3}H]ouabain binding of mutated HN34/56 chimeras. The sequences of the M5/M6 regions of Na,K-ATPase and gastric H,K-ATPase and the seven mutations are indicated at the top. The α-subunits were detected with the polyclonal antibody 565–585. The enzymes were incubated at 21°C in the presence of 5.0 mM $MgCl_2$, 50 mM Tris–acetic acid (pH 7.0), 1.0 mM ATP, and 250 nM [^{3}H]ouabain. *Significantly ($p < 0.05$) different from HN34/56.

RESULTS AND DISCUSSION

To test the relative levels of expression of the mutated α-subunits in *Xenopus* oocytes, we performed Western blotting with total membrane preparations using a polyclonal antibody recognizing the M4–M5 loop of the gastric H,K-ATPase α-subunit. FIGURE 1 shows that the expression levels of the mutated α-subunits were similar to those of the chimera HN34/56.

To investigate which amino acids are involved in ouabain binding, we performed binding of [^{3}H]ouabain. FIGURE 1 illustrates the amount of binding of [^{3}H]ouabain to chimera HN34/56 and seven mutants (mut 1–7). HN34/56, mut 1, mut 2, mut 5, and mut 7 bound ouabain with high affinity, while mut 3, mut 4, and mut 6 did not bind ouabain. This indicates that, in the latter constructs, amino acids were mutated that are specific for ouabain binding to Na,K-ATPase. These three mutants still contain seven amino acids that are different between H,K-ATPase and Na,K-ATPase. We therefore conclude that 3–7 amino acids present in hairpin M5 and M6 of Na,K-ATPase might be important for ouabain binding. We will further investigate whether a chimera of H,K-ATPase, containing only M3–M4 and these 3–7 important amino acids of M5–M6 of Na,K-ATPase, still binds ouabain. If so, we will apply a similar approach to determine which amino acids of M3–M4 of Na,K-ATPase are also important for ouabain binding.

REFERENCES

1. KOENDERINK, J.B., H.P.H. HERMSEN, H.G.P. SWARTS et al. 2000. High-affinity ouabain binding by a chimeric gastric H^+,K^+-ATPase, containing transmembrane hairpins M3–M4 and M5–M6 at the α1-subunit of rat Na^+,K^+-ATPase. Proc. Natl. Acad. Sci. USA **97:** 11209–11214.
2. GATTO, C., S. LUTSENKO, J.M. SHIN, et al. 1999. Stabilization of the H,K-ATPase M5M6 membrane hairpin by K^+ ions: mechanistic significance for P_2-type ATPases. J. Biol. Chem. **274:** 13737–13740.
3. PALASIS, M., T.A. KUNTZWEILER, J.M. ARGUELLO & J.B. LINGREL. 1996. Ouabain interactions with the H5–H6 hairpin of the Na,K-ATPase reveal a possible inhibition mechanism via the cation binding domain. J. Biol. Chem. **271:** 14176–14182.

Role of the Isoform-Specific Region of the Na,K-ATPase Catalytic Subunit

MARIE-JOSÉE DURAN, SANDRINE V. PIERRE, DEBORAH L. CARR, AND THOMAS A. PRESSLEY

Department of Physiology, Texas Tech University Health Sciences Center, Lubbock, Texas, USA

KEYWORDS: Na,K-ATPase; isoform-specific region (ISR); PKC activation

INTRODUCTION

We have identified a region, the isoform-specific region (ISR), located in the major cytoplasmic loop of the Na,K-ATPase α catalytic subunit that greatly differs between the isoforms (FIG. 1). To evaluate the importance of this region, we constructed chimeras of the rodent α1 isoform in which the ISR was replaced.

METHODS

The α1HKα1 chimera was constructed by exchanging the rat α1 ISR with the corresponding sequence (TLEDPRDPRHL) from the rat gastric H,K-ATPase catalytic subunit. After transfection of the rat wild-type α1 and the α1HKα1 chimera into opossum kidney (OK) cells, selection of transfected cells was achieved using 3 µM ouabain, a concentration sufficient to kill nontransfected cells. All transfections produced viable colonies. Expression of the introduced sequences in OK cells was assessed by RT-PCR (data not shown). Enzymatic function was verified using ouabain-sensitive $^{86}Rb^+$ uptake assays. As Na,K-ATPase transport in OK cells is known to be increased by protein kinase C (PKC) stimulation,[1] we checked whether the ISR was involved in this process by treating the cells with phorbol myristate acetate (PMA).

RESULTS/DISCUSSION

Transfection of OK cells with the α1HKα1 chimera produced viable colonies, indicating that the α1 ISR is not essential for the overall enzymatic function. Replacement of the α1 ISR by the H,K-ATPase sequence abolished the PMA-induced increase in Na,K-ATPase transport observed with the wild-type α1 (FIG. 2). These results suggest that, although the ISR is not critical for overall enzymatic

---

Address for correspondence: Dr. Marie-Josée Duran, Department of Physiology, Texas Tech University Health Sciences Center, 3601, 4th Street, Lubbock, TX 79430. Voice: 806-743-4056; fax: 806-743-1512.

MarieJosee.Duran@ttuhsc.edu

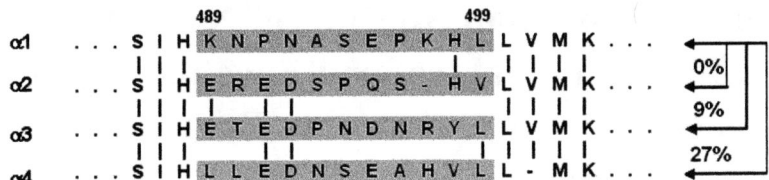

FIGURE 1. Alignment of the rat ISR (shaded) amino acid sequences. The percentage of homology between the α1 ISR and the other α isoform ISRs is also shown.

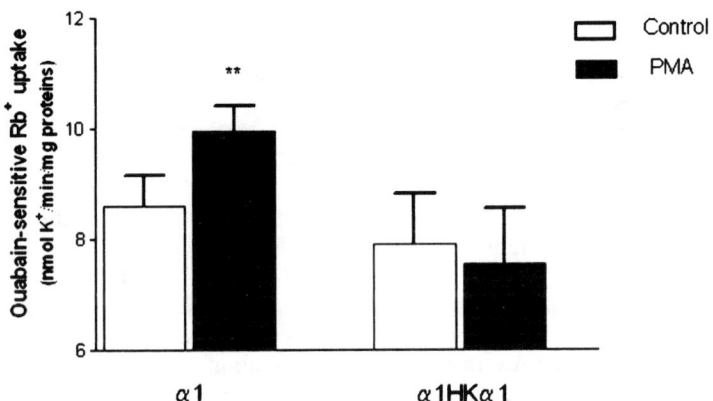

FIGURE 2. Ouabain-sensitive Rb$^+$ uptakes. ^{86}Rb$^+$ uptakes were performed as previously described.[2] **$p < 0.01$ vs. control.

function, the α1 ISR is implicated in PKC activation in OK cells. Our current hypothesis is that there is either a direct or indirect interaction between the phosphorylated serines (Ser 11 and 18) located in the N-terminus and the ISR of the Na,K-ATPase α1 isoform.

ACKNOWLEDGMENTS

Wild-type α1 cDNA was a gift from J. B. Lingrel (University of Cincinnati, Cincinnati, OH). This project was supported in part by a grant from the National Center for Research Resources [No. RR-19799].

REFERENCES

1. PEDEMONTE, C.H., T.A. PRESSLEY, M.F. LOKHANDWALA, et al. 1997. Regulation of Na,K-ATPase transport activity by protein kinase C. J. Membr. Biol. **155:** 219–227.
2. PIERRE, S.V., M-J. DURAN, D.L. CARR & T.A. PRESSLEY. 2002. Structure/function analysis of Na$^+$-K$^+$-ATPase central isoform-specific region: involvement in protein kinase C regulation. Am. J. Physiol. Renal Physiol. **283:** F1066-1074.

Structure/Function Analysis of Na,K-ATPase α1 and α2 Central Isoform-Specific Regions Reveals Their Involvement in Regulation by Protein Kinase C

S. V. PIERRE, M-J. DURAN, D. L. CARR, AND T. A. PRESSLEY

Department of Physiology, Texas Tech University Health Sciences Center, Lubbock, Texas, USA

KEYWORDS: α-isoforms; chimeras; PKC; potassium deocclusion

INTRODUCTION

The four isoforms of the Na,K-ATPase α-subunit are nearly identical, with the exception of the amino-terminus and a 10-residue region near the center of the molecule, the central isoform-specific region (ISR, K^{489}–L^{499}, FIG. 1). The specific functions served by the various isoforms may originate within these regions of structural divergence. The amino-terminus is clearly a source of isoform functional diversity, but the variability in kinetics and regulation of ion transport properties among species and isoforms suggests that additional regions may be involved.[1,2] We hypothesized that the central ISR is such a region. To examine its role, we constructed chimeric molecules in which the central ISRs of rat α1 and α2 isoforms were exchanged, and we characterized them for two properties known to differ dramatically among the isoforms—their K^+ deocclusion pattern and their response to protein kinase C (PKC) activation.

METHODS

Complementary DNA of rat ouabain-resistant α1, as well as α2 isoform modified to display α1-like ouabain-resistance, were used to obtain chimeras in which the central ISRs were exchanged. After stable transfection into ouabain-sensitive opossum kidney (OK) cells, recipient colonies were selected by exposure to ouabain concentrations sufficient to kill nontransfected cells. The expression and structure of the introduced forms were confirmed by direct detection of the exogenous polypeptides and mRNAs with specific probes (Western blotting and RT-PCR[3]). The rate

---

Address for correspondence: S.V. Pierre, Department of Physiology, Texas Tech University Health Sciences Center, 3601, 4th Street, Lubbock, TX 79430. Voice: 806-743-4056; fax: 806-743-1512.
Sandrine.Pierre@ttuhsc.edu.

Ann. N.Y. Acad. Sci. 986: 260–262 (2003). © 2003 New York Academy of Sciences.

of K^+ deocclusion was assessed at low ATP concentration (1 µM) in membranes isolated from transfected cells. In this condition, the response of Na^+-dependent ATP hydrolysis to varying concentrations of K^+ is a convenient and sensitive indication of the $E2(K) \rightarrow E1$ pathway of the Na^+,K^+-ATPase reaction.[1] Indeed, this part of the reaction becomes rate-limiting at low ATP concentration, and K^+ inhibits Na^+-ATPase activity of the α1 enzyme, whereas α2 is stimulated. Finally, the response to PKC was assessed by measuring Na,K-ATPase-mediated Rb^+ uptake in transfectants treated or not with the PKC activator, phorbol 12-myristate 13-acetate (PMA).

RESULTS/DISCUSSION

Comparisons of the chimeras with rat wild-type α1 and α2 isoforms expressed under the same conditions showed no difference in K^+ deocclusion kinetics. However, substitution of the α2 ISR into α1 doubled the increase in pump-mediated

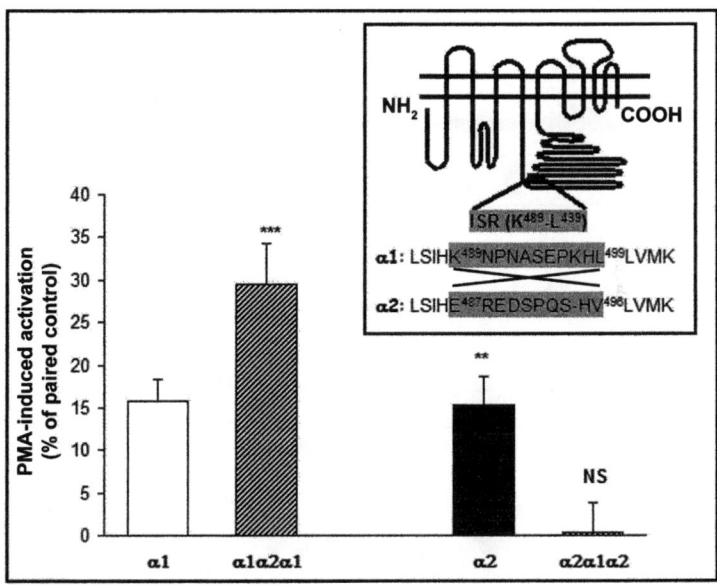

FIGURE 1. Effect of α1/α2 ISR exchange on PMA-dependent activation of cellular Na^+,K^+-ATPase-mediated Rb^+ transport. Na^+,K^+-ATPase-mediated transport was assayed in attached cells by measuring the ouabain-sensitive uptake of the K^+ congener, $^{86}Rb^+$. PKC activation was induced by a 5-min exposure of the cells to 10 µM PMA prior to the addition of Rb^+, and compared to paired control plates of cells exposed for 5 min to the same amount of vehicle alone (DMSO). Values are means ± SEM ($n = 6-11$) of flux activations induced by PMA exposure, expressed in percent of their paired controls (same transfection group, same passage, same day). Values were compared using two-tailed Student's paired t test— NS: nonsignificant; **$P < 0.01$; ***$P < 0.001$. The inset shows the location of the central ISR. The aligned amino acid sequences of the α1 and α2 ISR are shadowed.

transport elicited by exposure to the PKC agonist, PMA. In contrast, substitution of the α1 ISR into α2 eliminated the normal response seen with α2 (FIG. 1). These results suggest that the structure of the ISR may influence the overall response of the Na,K-ATPase to PKC, perhaps by altering the efficiency of recycling between the plasmalemma and intracellular membranes.

REFERENCES

1. DALY, S.E., L.K. LANE & R. BLOSTEIN. 1996. Structure/function analysis of the amino-terminal region of the α1 and α2 subunits of Na,K-ATPase. J. Biol. Chem. **271:** 23683–23689.
2. BLANCO, G. & R.W. MERCER. 1998. Isozymes of the Na-K-ATPase: heterogeneity in structure, diversity in function. Am. J. Physiol. **275:** F633–F650.
3. PIERRE, S.V. *et al*. 2002. Structure/function analysis of Na^+-K^+-ATPase central isoform-specific region: involvement in protein kinase C regulation. Am. J. Physiol. Renal Physiol. **283:** F1066–F1074.

Characterization of the Electrostatic Component of the Nucleotide Binding to Na,K-ATPase

N. U. FEDOSOVA,[a] P. CHAMPEIL,[b] AND M. ESMANN[a]

[a]*Department of Biophysics, University of Aarhus, DK-8000 Aarhus C, Denmark*

[b]*Section de Biophysique des Fonctions Membranaires (DBJC/CEA), Unité de Recherche Associée 2096 (CNRS) and LRA 17V (Université Paris-Sud), Centre d'Etudes de Saclay, 91191 Gif-sur-Yvette Cedex, France*

KEYWORDS: Na,K-ATPase; nucleotide binding; effective charge

The contribution of electrostatic forces to the interaction of Na,K-ATPase with adenine nucleotides was previously investigated by studying the effect of ionic strength on nucleotide binding.[1] The aim was to document, within the framework of the Debye-Hückel theory, the extent to which the long-range coulombic forces favor protein-ligand association, that is, to estimate the electrostatic attraction between the negatively charged nucleotide and the positively charged binding site during ligand approach. The reaction was assumed to be described by a simple binding scheme, E + A ↔ EA. Quantitative analysis of the K_d dependence on ionic strength showed that the product of the effective electrostatic charges on the ligand and the binding site was the same for all nucleotides (ATP, ADP, and MgADP), independent of the total charge of the nucleotide. The fact that dissociation rate constants (k_{off}) were different for the various ligands, and did not depend on ionic strength, as predicted by Debye-Hückel theory, demonstrated that the binding rate constants (k_{on}) were similar for all nucleotides tested, and therefore that association of nucleotides with Na,K-ATPase was governed by a partial charge rather than by the total charge of the nucleotide. Here, in time-resolved binding experiments, we directly determine the rates of association of the nucleotides and discuss them in relation to the dissociation rate constants and relative affinities of these nucleotides.

Na,K-ATPase was purified from pig kidney microsomal membranes to a specific activity of 28 µmoles per mg protein per min at 37°C. Equilibrium and time-resolved binding of nucleotides was measured in double-labeling filtration experiments,

Address for correspondence: N. U. Fedosova, Department of Biophysics, University of Aarhus, Ole Worms Allé 185, DK-8000 Aarhus C, Denmark. Voice: +45-8942-2933; fax: +45-8612-9599.

nf@biophys.au.dk

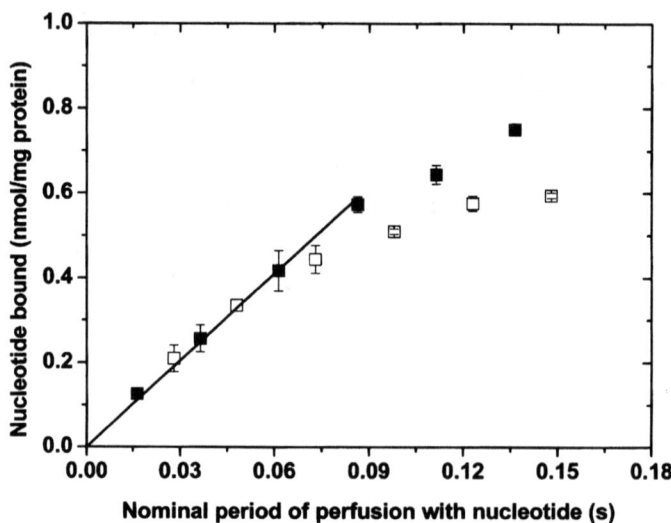

FIGURE 1. Initial velocity of nucleotide binding to Na,K-ATPase. Na,K-ATPase (2.8 nmol binding sites/mg protein; 0.01 mg per filter; wetting volume: 30 µL) in 35 mM Na^+ and 10 mM histidine (pH 7.0) was perfused at 20°C with 0.2 µM [^{32}P]ATP (*solid squares*) or [^{14}C]ADP (*open squares*) in the presence of 10 mM CDTA for various periods, and bound radioactivity was measured. Each data point represents the averages of three determinations. The slope of the straight line, 6.8 nmol/mg/s, gives the binding rate constant of about 12 $\mu M^{-1} \cdot s^{-1}$.

essentially as previously described.[1,2] For the latter experiments, a rapid filtration system was used (RFS-4, Bio-logic, France).

FIGURE 1 demonstrates that the initial velocity of binding is the same for ATP and ADP, that is, does not depend on the total charge of the nucleotide. Nevertheless, the equilibrium levels of nucleotide binding achieved after 4 s of perfusion are different (not shown in FIG. 1, but the tendency is already seen after 0.15 s of perfusion), in agreement with the difference in the K_d values observed for these nucleotides. The present data support the hypothesis that the contribution of the long-range coulombic forces to nucleotide-enzyme interaction does not depend on the total charge of the nucleotide and suggest that the short-range interactions determining the nucleotide dissociation rates make a major contribution to its binding affinity.

REFERENCES

1. FEDOSOVA, N.U., P. CHAMPEIL & M. ESMANN. 2002. Nucleotide binding to Na,K-ATPase: the role of electrostatic interactions. Biochemistry **41:** 1267–1273.
2. DUPONT, Y. 1984. A rapid-filtration technique for membrane fragments or immobilized enzymes: measurements of substrate binding or ion fluxes with a few-millisecond time resolution. Anal. Biochem. **142:** 504–510.

Independent Access of Fluorescein Isothiocyanate and Co(NH$_3$)$_4$ATP to Their Binding Sites on the Protomer of Na,K-ATPase

J. D. CAVIERES AND J. HADDOW

Department of Cell Physiology and Pharmacology, University of Leicester, Leicester LE1 9HN, United Kingdom

KEYWORDS: Na,K-ATPase; fluorescein isothiocyanate (FITC); Co(NH$_3$)$_4$ATP

INTRODUCTION

Following low-affinity binding, the substitution-inert analogue Co(NH$_3$)$_4$ATP inactivates E2 and overall reactions of Na,K-ATPase.[1–3] If E1 and overall reactions are obliterated with FITC, it is possible to inactivate the surviving K$^+$-phosphatase[2] with Co(NH$_3$)$_4$ATP. As both ligands can access all αβ protomers in the membrane enzyme,[3,4] this should mean that FITC and Co(NH$_3$)$_4$ATP bind at different sites on the α chain. However, although unlikely, it has been suggested that FITC and Co(NH$_3$)$_4$ATP might access a single pocket in the enzyme. If that was the case, one would predict that the presence of one bound ligand should decrease the binding affinity towards the other. We compared, therefore, the Co(NH$_3$)$_4$ATP affinity for the process of inactivation of the K$^+$ phosphatase activity of native and FITC-modified Na,K-ATPase in a single experiment.

METHODS

Na,K-ATPase was purified from pig outer medulla[5] and Co(NH$_3$)$_4$ATP was synthesized as described.[6] The FITC treatment was done at pH 9.2 and room temperature, until the Na,K-ATPase activity was around 1% (the K$^+$-phosphatase remaining at 94%). The Co(NH$_3$)$_4$ATP inactivation of the K$^+$-phosphatase was at 37°C and for up to 80 min, in 20 μL of 10 mM imidazole (pH 6.8). We arrested samples by diluting 2 μL into 1 mL containing (mM) KCl 150, MgCl$_2$ 6, imidazole (pH 7.2) 20, and 3-*O*-methylfluorescein phosphate 0.5. This was inside a cuvette in a U-3310 Hitachi spectrophotometer fitted with a six-cell changer thermostated to 37°C and magnetically stirred. After sampling mixtures at each of six Co(NH$_3$)$_4$ATP concentrations, the release of 3-*O*-methylfluorescein was followed at 475 nm over

Address for correspondence: Dr. J.D. Cavieres, Department of Cell Physiology and Pharmacology, University of Leicester, Leicester LE1 9HN, United Kingdom. Voice/fax: +44-116-252-3091.

jdc7@le.ac.uk

300 s, and the K^+-phosphatase activities calculated from linear slopes of 3-O-methylfluorescein release. This process was repeated at various inactivation intervals. For each Co(NH$_3$)$_4$ATP concentration, the fraction of the phosphatase activity left was plotted semilogarithmically against time to extract the inactivation rate constants.

RESULTS AND CONCLUSIONS

When the rate constants were plotted against Co(NH$_3$)$_4$ATP concentration, the curves were well fitted by Michaelis hyperbolae. The results for the native enzyme were $k_{\text{inact(max)}} = 0.0352 \pm 0.0027$ min^{-1} and $K_m = 0.454 \pm 0.113$ mM; for the FITC-treated enzyme, they were $k_{\text{inact(max)}} = 0.0285 \pm 0.0022$ min^{-1} and $K_m = 0.404 \pm 0.103$ mM. Evidently, occupation of the high-affinity ATP pocket by FITC did not hinder the low-affinity Co(NH$_3$)$_4$ATP binding that precedes the inactivation process.

It then seems reasonable to suppose that the bindings of FITC and Co(NH$_3$)$_4$ATP take place at some distance from each other and not within a common pocket. These may represent high-affinity and low-affinity ATP sites.

ACKNOWLEDGMENTS

This work was supported by research grants from The Wellcome Trust.

REFERENCES

1. SCHEINER-BOBIS, G., M. ESMANN & W. SCHONER. 1989. Shift to the Na$^+$ form of Na$^+$/K$^+$-transporting ATPase due to modification of the low-affinity site by Co(NH$_3$)$_4$ATP. Eur. J. Biochem. **183:** 173–178.
2. LINNERTZ, H., D. THÖNGES & W. SCHONER. 1995. Na$^+$/K$^+$-ATPase with a blocked E1ATP site still allows backdoor phosphorylation of the E2ATP site. Eur. J. Biochem. **232:** 420–424.
3. WARD, D.G., W. SCHONER & J.D. CAVIERES. 1994. Evidence for two distinct ATP sites in Na,K-ATPase purified from pig kidney. J. Physiol. **480:** 84P.
4. WARD, D.G. & J.D. CAVIERES. 1996. Binding of 2'(3')-O-(2,4,6-trinitrophenyl)ADP to soluble αβ protomers of Na,K-ATPase modified with fluorescein isothiocyanate: evidence for two distinct nucleotide sites. J. Biol. Chem. **271:** 12317–12321.
5. JORGENSEN, P.L. 1974. Purification and characterization of (Na$^+$ + K$^+$)-ATPase. III. Purification from the outer medulla of mammalian kidney after selective removal of membrane components by sodium dodecylsulphate. Biochim. Biophys. Acta **356:** 36–52.
6. CORNELIUS, R.D., P.A. HART & W.W. CLELAND. 1977. Phosphorus-31 NMR studies of complexes of adenosine triphosphate, adenosine diphosphate, tripolyphosphate, and pyrophosphate with cobalt(III) amines. Inorg. Chem. **16:** 2799–2805.

Importance of Thr^{214} in the Conserved TGES Sequence of the Na$^+$,K$^+$-ATPase for Vanadate Binding and Hydrolysis of E_2P

MADS TOUSTRUP-JENSEN AND BENTE VILSEN

Department of Physiology, University of Aarhus, DK-8000 Aarhus C, Denmark

KEYWORDS: Na$^+$,K$^+$-ATPase; mutagenesis; TGES

Thr^{214} of the conserved TGES sequence in the Na$^+$,K$^+$-ATPase is located in domain A thought to undergo large movements during the transport reaction (FIG. 1). It has been suggested that, in the E_2 and E_2P forms, domain A docks into domain P, making up a catalytic site that facilitates the nucleophilic attack by water on the phosphorus atom in the O–P bond.[1] Hence, the role of Thr^{214} was tested by mutagenesis.

RESULTS AND DISCUSSION

Titration of the Na$^+$-, K$^+$-, and ATP-dependence of ATPase activity of Thr^{214}→Ala showed changes in apparent affinities relative to the wild type, corresponding to a shift of the E_1–E_2 conformational equilibrium in favor of E_1. Analysis of the phosphoenzyme revealed a slightly reduced amount of E_2P, and the E_1P→E_2P transition rate was reduced compared with the wild type. Thr^{214}→Ala resembled the previously characterized domain A mutant Gly^{263}→Ala[2] with regard to displacement of the conformational equilibria in favor of the E_1 and E_1P forms, respectively, with the effects being less pronounced, however, for Thr^{214}→Ala.

Vanadate is a phosphate transition state analogue that stabilizes an E_2 conformation of the Na$^+$,K$^+$-ATPase, probably resembling the E_2–P or $E_2 \cdot P_i$ conformation. Thr^{214}→Ala displayed a conspicuous 151-fold reduction of the apparent vanadate affinity during catalytic turnover in the ATPase assay. This could not simply be explained by changes of the conformational equilibria, which were much less pronounced for Thr^{214}→Ala than for Gly^{263}→Ala.[2] Hence, the intrinsic vanadate affinity of the E_2 form was determined in an assay in which the enzyme was equilibrated with various vanadate concentrations in the presence of Rb$^+$ to accumulate the E_2(Rb$_2$) form, followed by phosphorylation to measure the vanadate free enzyme

Address for correspondence: Bente Vilsen, Department of Physiology, University of Aarhus, DK-8000 Aarhus C, Denmark. Voice: +45-8942-2832; fax: +45-8612-9065.
bv@fi.au.dk

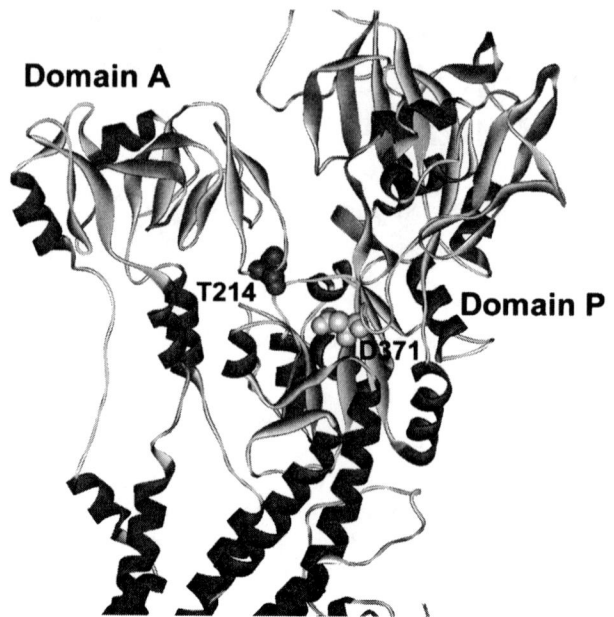

FIGURE 1. Location of Thr^{214} in the E_2 form according to the Ca^{2+}-ATPase structure.

fraction. This poise toward E_2 resulted in wild type–like vanadate affinity of Gly^{263}→Ala; that is, this mutation affects only the conformational equilibrium and not the intrinsic vanadate affinity of the E_2 form. By contrast, Thr^{214}→Ala showed a 4-fold reduced affinity, indicating that the E_2 form manifests *true* low vanadate affinity.

The reduced intrinsic vanadate affinity may be explained by a destabilization of the transition state in the hydrolysis of E_2P. Hence, the E_2P dephosphorylation rate[3] was studied. Relative to wild type, Thr^{214}→Ala showed a reduced E_2P dephosphorylation rate, independent of whether dephosphorylation was activated by Na$^+$ or K$^+$. This was caused by a reduction in V_{max} rather than a decrease in K$^+$ affinity at the activating E_2P sites. By contrast, the phosphorylation site in the E_1 form was unchanged since the phosphorylation rate of Thr^{214}→Ala with ATP was wild type like.

CONCLUSIONS

These results support the hypothesis that Thr^{214} is important for the ability of domain A to dock into domain P in the E_2 and E_2P conformations and for transition state stabilization and thus proper catalysis of E_2P dephosphorylation. A direct role of Thr^{214} in interaction with the phosphate in the transition state or with the catalytic Mg^{2+} ion, or a role in optimization of the structure of the catalytic site in the E_2P form, could be imagined.

REFERENCES

1. PATCHORNIK, G., R. GOLDSHLEGER & S.J.D. KARLISH. 2000. The complex ATP-Fe^{2+} serves as a specific affinity cleavage reagent in ATP-Mg^{2+} sites of Na,K-ATPase: altered ligation of Fe^{2+} (Mg^{2+}) ions accompanies the $E_1P \rightarrow E_2P$ conformational change. Proc. Natl. Acad. Sci. USA **97:** 11954–11959.
2. TOUSTRUP-JENSEN, M., M. HAUGE & B. VILSEN. 2001. Mutational effects on conformational changes of the dephospho- and phospho-forms of the Na^+,K^+-ATPase. Biochemistry **40:** 5521–5532.
3. TOUSTRUP-JENSEN, M. & B. VILSEN. 2002. Importance of Glu^{282} in transmembrane segment M3 of the Na^+,K^+-ATPase for control of cation interaction and conformational changes. J. Biol. Chem. **277:** 38607–38617.

Differential Inactivation of Na,K-ATPase by Erythrosin Isothiocyanate

M. TAYLOR, D. OWEN, A. TARIQ, AND J. D. CAVIERES

Department of Cell Physiology and Pharmacology, University of Leicester, Leicester LE1 9HN, United Kingdom

KEYWORDS: Na,K-ATPase; erythrosin isothiocyanate; ATP sites; inactivation

INTRODUCTION

ATP has dual activatory effects on Na,K-ATPase and we have argued[1] that this reflects the presence of two nucleotide sites per αβ protomer. ATP analogues have been used to investigate this situation in P-type ATPases and, out of those, erythrosin isothiocyanate (ErITC) seemingly inactivates SERCA by blocking low-affinity (but not high-affinity) ATP effects.[2] We wished to find out whether the same held true for Na,K-ATPase.

RESULTS AND CONCLUSIONS

FIGURE 1A shows an early experiment to study the inactivation by increasing ErITC concentrations at pH 9.2. As with SERCA,[3] the ErITC treatment appears much more effective when the samples are assayed at 2 mM ATP than at 10 or 1 µM ATP. However, these results changed when using a fresh vial of ErITC (FIG. 1B) and the inactivation appeared more extensive when examined at 1 µM ATP than at 2 mM ATP. In the absence of eosin, the hyperbolae could be fitted with $K_{0.5} = 16 \pm 3$ nM for 2 mM ATP and with $K_{0.5} = 7 \pm 2$ nM for 0.1 µM ATP. As we suspected that impurities were present in the old vial, the experiment was repeated using 10 µM eosin Y (a structurally similar ligand for ATP sites in Na,K-ATPase[4]) together with the ErITC. As can be seen from FIGURE 1B, the relative position of the two curves is reversed. In this case, $K_{0.5} = 36 \pm 6$ nM for 2 mM ATP and 63 ± 11 nM for 1 µM ATP; that is, $K_{0.5}$ for ErITC increases twofold when assaying at 2 mM ATP and eightfold when at 1 µM ATP (see arrows). In other words, the apparent competition by a second ATP-site ligand mimics the situation in FIGURE 1A. The change in the ErITC inactivation pattern can then be interpreted on the basis of two nucleotide sites per αβ protomer being differentially affected by the interaction of ErITC and its degradation products at more than one site.

Address for correspondence: Dr. M. Taylor, Department of Cell Physiology and Pharmacology, University of Leicester, Leicester LE1 9HN, United Kingdom. Voice: +44-116-2523085.
mt54@le.ac.uk

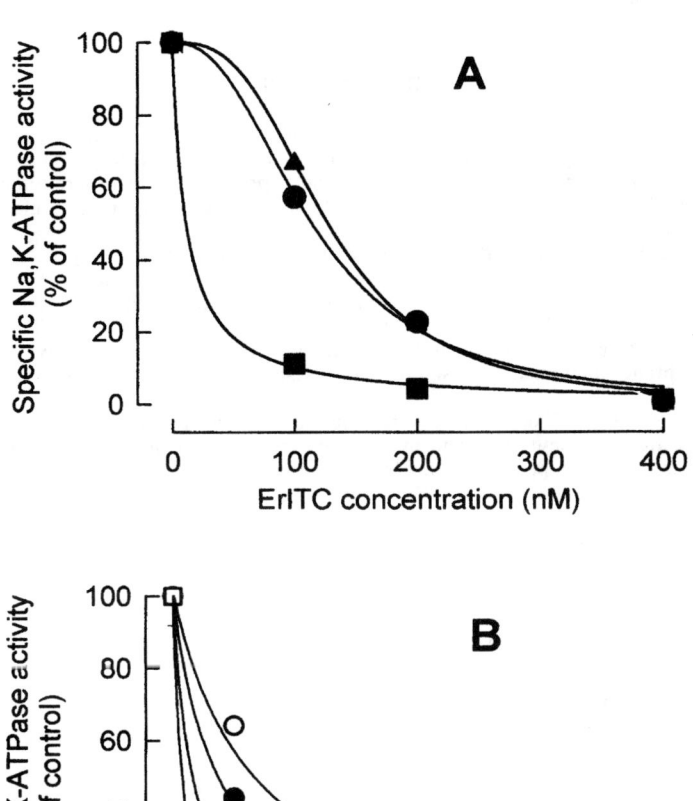

FIGURE 1. Differential inactivation by ErITC. **(A)** Samples were inactivated at pH 9.0 at increasing ErITC concentrations, diluted 500-fold, and their Na,K-ATPase activities determined with [γ^{32}P]-ATP at 2 mM (*squares*), 10 µM (*triangles*), and 0.1 µM (*circles*). **(B)** ErITC treatment in the presence (*squares*) or the absence (*circles*) of 10 µM eosin Y. After stopping and two washes at 425,000g, the Na,K-ATPase activity was assayed at 2 mM ATP (*open symbols*) and 1 µM ATP (*solid symbols*). Errors within symbols.

ACKNOWLEDGMENTS

This work was supported by research grants from The Wellcome Trust and by a European Union Grant to Young Researchers (to M. Taylor).

REFERENCES

1. WARD, D.G. & J.D. CAVIERES. 1998. Affinity labelling of two nucleotide sites on Na,K-ATPase using 2′(3′)-O-(2,4,6-trinitrophenyl)8-azidoadenosine 5′-[α-^{32}P]-diphosphate (TNP-8N$_3$-[α-^{32}P]ADP) as a photoactivatable probe: label incorporation before and after blocking the high affinity ATP site with fluorescein isothiocyanate. J. Biol. Chem. **273:** 33759–33765.
2. HUANG, S.H., S. NEGASH & T.C. SQUIER. 1998. Erythrosin isothiocyanate selectively labels lysine(464) within an ATP-protectable binding site on the Ca-ATPase in skeletal sarcoplasmic reticulum membranes. Biochemistry **37:** 6949–6957.
3. JORGENSEN, P.L. 1974. Purification and characterization of (Na$^+$ + K$^+$)–ATPase III: purification from the outer medulla of mammalian kidney after selective removal of membrane components by sodium dodecylsulphate. Biochim. Biophys. Acta **356:** 36–52.
4. BURDON, D., D.G. WARD et al. 2000. Photoinactivation of Na,K-ATPase by 2N$_3$-ATP: protection by nucleotides and the puzzling effect of eosin Y. In Na/K-ATPase and Related ATPases, pp. 417–420. Elsevier. Amsterdam/New York.

Mutational Analysis of the Interactions of the Alpha and Beta Subunits of the Na,K-ATPase

M. D. LAUGHERY, S. McLOUD, AND J. H. KAPLAN

Department of Biochemistry and Molecular Biology, Oregon Health and Sciences University, Portland, Oregon 97201-3011, USA

KEYWORDS: Na,K-ATPase; α subunit; β subunit

Functional Na,K-ATPase is a heterodimer of an α and a β subunit. The α subunit, with 10 transmembrane (TM) segments and an ATP-hydrolyzing domain, is primarily responsible for the ion transport activities of the Na,K-ATPase. However, the α is not stable and has no functional activity in the absence of β. β is a single TM protein with a small intracellular N-terminal region and a larger extracellular C-terminal domain. Previous work has suggested that β fulfills a chaperone role by aiding the folding and membrane insertion of α, as well as delivering the heterodimer to the plasma membrane (PM). Specifically, the extracellular region of β just beyond the TM segment has been implicated in interacting with α,[1] and the 10 most C-terminal residues have been deemed essential for heterodimer trafficking to the PM.[2,3]

We have utilized the baculovirus expression system to investigate assembly and trafficking of wild-type sheep α1 and mutant sheep β1. By separating the ER, Golgi, and PM compartments, it was previously demonstrated that the α subunit expressed alone does not traffic to the PM; however, when α and β are coexpressed, both subunits are targeted to the PM of High Five insect cells.[4] Through coexpression of wild-type α and mutant β followed by coimmunoprecipitation and fractionation analysis, the regions of β essential for assembly with α and those important for trafficking can be resolved. The methods used have been previously described.[4–6] A summary of trafficking and assembly results are presented in TABLE 1.

The removal of one of the three conserved disulfide bridges by substitution of both Cys residues to Ala (C126A/C149A, C157A/C176A, or C213A/C276A) resulted in β subunits capable of assembling with α. However, they are retained in the ER. The retention of these β subunits could be a response to inappropriate folding, to the lack of a normal trafficking signal, or to the accessibility of a specific ER retention signal.

Address for correspondence: M.D. Laughery, Department of Biochemistry and Molecular Biology, Oregon Health and Sciences University, 3181 SW Sam Jackson Park Road, Portland, OR 97201-3011. Voice: 503-494-1516; fax: 503-494-8393.
laughery@ohsu.edu

TABLE 1. Summary of β mutant characteristics

β mutant	α trafficking	β trafficking	Association	ATPase activity
C126A, C149A	ER retained	ER retained	α + β	Not significant
C157A, C176A	ER retained	ER retained	α + β	Not significant
C213A, C276A	ER retained	ER retained	α + β	Not significant
Δ125 flag	PM targeted	PM targeted	α + β	Not significant
Δ157 flag	ER retained	ER retained	α + β	Not significant
Δ212	ER retained	ER retained	α + β	Not significant

To further characterize trafficking and assembly requirements, C-terminal truncation mutants were created that removed the regions C-terminal to one (Δ212), two (Δ157Flag), or all three (Δ125Flag) disulfide bridges. A C-terminal flag tag was added when necessary for antibody detection. All of these mutants contained the sequence of β identified by yeast two hybrid as important for association with α.[1] The Δ212 and Δ157Flag mutants were retained in the ER, as might be expected if the C-terminal 10 residues are necessary for trafficking.[2,3] However, the Δ125Flag, which also lacks these residues, was able to traffic and assemble with the α subunit, in contrast to findings by others.[2,3]

REFERENCES

1. COLONNA, T.E., L. HUYNH & D.M. FAMBROUGH. 1997. Subunit interactions in the Na,K-ATPase explored with the yeast two-hybrid system. J. Biol. Chem. **272**(19): 12366–12372.
2. BEGGAH, A.T. *et al.* 1993. Hydrophobic C-terminal amino acids in the beta-subunit are involved in assembly with the alpha-subunit of Na,K-ATPase. Biochemistry **32**(51): 14117–14124.
3. HAMRICK, M., K.J. RENAUD & D.M. FAMBROUGH. 1993. Assembly of the extracellular domain of the Na,K-ATPase beta subunit with the alpha subunit: analysis of beta subunit chimeras and carboxyl-terminal deletions. J. Biol. Chem. **268**(32): 24367–24373.
4. GATTO, C., S.M. MCLOUD & J.H. KAPLAN. 2001. Heterologous expression of Na(+)-K(+)-ATPase in insect cells: intracellular distribution of pump subunits. Am. J. Physiol. Cell. Physiol. **281**(3): C982–C992.
5. HU, Y.K. & J.H. KAPLAN. 2000. Site-directed chemical labeling of extracellular loops in a membrane protein: the topology of the Na,K-ATPase alpha-subunit. J. Biol. Chem. **275**(25): 19185–19191.
6. HU, Y.K., J.F. EISSES & J.H. KAPLAN. 2000. Expression of an active Na,K-ATPase with an alpha-subunit lacking all twenty-three native cysteine residues. J. Biol. Chem. **275**(39): 30734–30739.

Mechanism of Phosphoryl Group Transfer

L. D. FALLER,[a] A. K. NAGY,[a] D. J. KANE,[a,b] AND R. A. FARLEY[b]

[a]*Departments of Medicine and Physiological Sciences, University of California at Los Angeles, Los Angeles, California, USA*

[b]*Department of Physiology and Biophysics, University of Southern California, Los Angeles, California, USA*

KEYWORDS: P-type pumps; HAD superfamily; ^{18}O exchange; site-directed mutagenesis

P-type pumps couple the energy chemically stored in ATP to ion transport by catalyzing transfer of the γ-phosphoryl group to water in two Mg^{2+}-dependent steps separated by protein conformational changes. The reaction occurs via nucleophilic attack of the acceptor, which is an aspartyl side chain in the first step and water in the second step, upon phosphorus. The haloacid dehydrogenase (HAD)–fold of the phosphorylation domain[1] predicts that amino acids in the conserved K...TGDGVND sequence participate in the catalytic mechanism.

Combining site-directed mutagenesis with measurements of stable oxygen isotope (^{18}O) exchange between P_i and water is a uniquely powerful method for identifying functional amino acids because ^{18}O exchange is catalyzed by a single step in the reaction cycle of P-type pumps. However, the interpretation of ^{18}O-exchange experiments depends upon the formal chemical mechanism, and currently different reaction mechanisms are claimed for the calcium (random binding) and sodium (ordered binding) pumps. Therefore, we repeated ^{18}O-exchange experiments that showed that the E_2 conformation of Na,K-ATPase binds Mg^{2+} before P_i[2] with purified Ca-ATPase. The Ca-ATPase results were remarkably similar to those obtained earlier with the sodium pump. The implication of ordered binding of Mg^{2+} before P_i to the E_2 conformations of both the calcium and sodium pumps for the molecular mechanism is that P-type pumps form a ternary enzyme·metal·phosphate complex.

The intrinsic Mg^{2+} dissociation constant to reform apoenzyme (K_{Mg}) and an apparent P_i dissociation constant $[K'_P/(1 + K_{H2O})]$ that depends upon the equilibrium constant (K_{H2O}) between covalently and noncovalently bound P_i can be estimated by fitting the rate equation for ordered binding to measurements of the ^{18}O-exchange rate as a function of the free Mg^{2+} and P_i concentrations. The intrinsic P_i dissociation constant from metalloenzyme (K'_P) is overestimated by about a factor of two because $K_{H2O} \approx 1$. TABLE 1 compares results obtained for mutations of the two

---

Address for correspondence: L.D. Faller, Departments of Medicine and Physiological Sciences, UCLA, Los Angeles, CA 90032. Voice: 310-268-3896; fax: 310-268-4963.
lfaller@ucla.edu

TABLE 1. Summary of exchange parameters

Expressed protein	$K_{0.5}$ (P_i)[a] (mM)	$K_{0.5}$ (Mg^{2+})[b] (mM)	K_{Mg} (mM)	$K'_P/(1 + K_{H2O})$ (mM)	P_c
WT	1.5 ± 0.2				0.20 ± 0.02
WT	2.9 ± 1.6	0.45 ± 0.16	0.9 ± 0.1	2.0 ± 1.0	0.22 ± 0.04
WT	1.62 ± 0.20	0.26 ± 0.05	0.7 ± 0.1	1.2 ± 0.1	0.26 ± 0.05
D717E	8.2 ± 4.2	2.4 ± 1.1	4.3 ± 1.3	2.6 ± 0.8	0.18 ± 0.02
D717N	1.9 ± 1.0	1.5 ± 0.8	**15 ± 4**	**0.22 ± 0.07**	0.29 ± 0.02
D717S	—	—	—		
D721N,S	—	—	—		

[a] $[Mg^{2+}] = 2$ mM.
[b] $[P_i] = 2$ mM.

aspartates in the K...TGDGVND sequence of human α1 Na,K-ATPase with wild-type exchange parameters. The dashes indicate that ^{18}O exchange could not be detected, and bold type denotes parameter estimates that differed significantly from wild-type values.

No exchange could be detected when D721, which coordinates the attacking water molecule in the mechanism[3] based on the crystal structure of the Mg^{2+}-dependent HAD superfamily member phosphoserine phosphatase (PSP), was changed. The results for mutant D717N dramatically illustrate the importance of deriving the rate equation and evaluating intrinsic constants. A roughly 15-fold decrease in affinity for Mg^{2+} is offset by a nearly 10-fold increase in affinity for P_i, so the half-maximum concentrations for titrations with Mg^{2+} or P_i resemble wild-type values. An earlier report by Jorgensen and collaborators[4] that the upstream aspartate in the K...TGDGVND sequence does not bind Mg^{2+} in the E_2 conformation of Na,K-ATPase was based on little, if any, shift in the half-maximum Mg^{2+} concentration for Mg^{2+} and P_i (or vanadate)–dependent ouabain binding to Na,K-ATPase when the upstream aspartate was changed to alanine. However, a dramatic shift in $K_{0.5}$ (Mg^{2+}) is not predicted by the results for D717N shown in the table, which are consistent with direct coordination of Mg^{2+} by D717 in a ternary complex with P_i as observed in the crystal structure of PSP.[3] Tighter binding of P_i to D717N is explained by ordered binding to metalloenzyme with higher positive charge density on the metal when the negative charge on one of the ligands coordinating Mg^{2+} is neutralized.

ACKNOWLEDGMENTS

This work was supported by NIH Grant No. DK52802 (to L. D. Faller and R. A. Farley) and a VA Merit Review (to L. D. Faller).

REFERENCES

1. TOYOSHIMA, C., M. NAKASAKO & H. OGAWA. 2000. Crystal structure of the calcium pump of sarcoplasmic reticulum at 2.6 Å resolution. Nature **405:** 647–655.
2. KASHO, V.N. *et al.* 1997. A proposal for the Mg^{2+} binding site of P-type ion motive ATPases and the mechanism of phosphoryl group transfer. Biochemistry **36:** 8045–8052.
3. WANG, W. *et al.* 2001. Crystal structure of phosphoserine phosphatase from *Methanococcus jannaschii*, a hyperthermophile, at 1.8 Å resolution. Structure **9:** 65–71.
4. PEDERSEN, P.A., J.R. JORGENSEN & P.L. JORGENSEN. 2000. Importance of conserved α-subunit segment ^{709}GDGVND for Mg^{2+} binding, phosphorylation, and energy transduction in Na,K-ATPase. J. Biol. Chem. **275:** 37588–37595.

Oligomeric Structure of P-Type ATPases Observed by Single Molecule Detection Technique

SHUNJI KAYA,[a] KAZUHIRO ABE,[a] KAZUYA TANIGUCHI,[a] MICHIO YAZAWA,[a] TSUYOSHI KATOH,[a] MAHITO KIKUMOTO,[b] KAZUHIRO OIWA,[b] AND YUTARO HAYASHI[c]

[a]*Division of Chemistry, Graduate School of Science, Hokkaido University, Sapporo 060-0810, Japan*

[b]*Kansai Advanced Research Center, Communications Research Laboratory, Kobe 651-2492, Japan*

[c]*Department of Biochemistry, School of Medicine, Kyorin University, Mitaka, Tokyo 181-8611, Japan*

KEYWORDS: single molecule imaging; fluorescence microscopy; oligomeric enzyme

The oligomericity of Na/K-ATPases such as diprotomeric and tetraprotomeric forms has been demonstrated by biochemical and physicochemical methods including molecular weight measurements by low-angle laser scattering photometry coupled with high-performance gel chromatography[1] and by rotary-shadowed electron microscopy.[2] It is noteworthy that techniques for the imaging of single molecules have been reported. We report here on attempts to directly detect Na/K-ATPase and H/K-ATPase molecules by total internal reflection microscopy (TIRFM).

RESULTS AND DISCUSSION

Lysine residues near the ATP binding pocket of Na/K- and H/K-ATPase, prepared from pig kidney and stomach, respectively, were labeled with fluorescein isothiocyanate (FITC). It has been shown that the solubilization of Na/K-ATPase by $C_{12}E_8$ results in a mixture of protomer, diprotomer, and higher oligomer(s) as evidenced by high-performance gel chromatography.[1] The observation of separated fraction by rotary-shadowed electron microscopy suggests that each fraction contains protomeric, diprotomeric, and tetraprotomeric Na/K-ATPase molecules.[2] To characterize the

Address for correspondence: Shunji Kaya, Division of Chemistry, Graduate School of Science, Hokkaido University N-10, W-8, Sapporo 060-0810, Japan. Voice/fax: +81-11-736-2074.
kayan@sci.hokudai.ac.jp

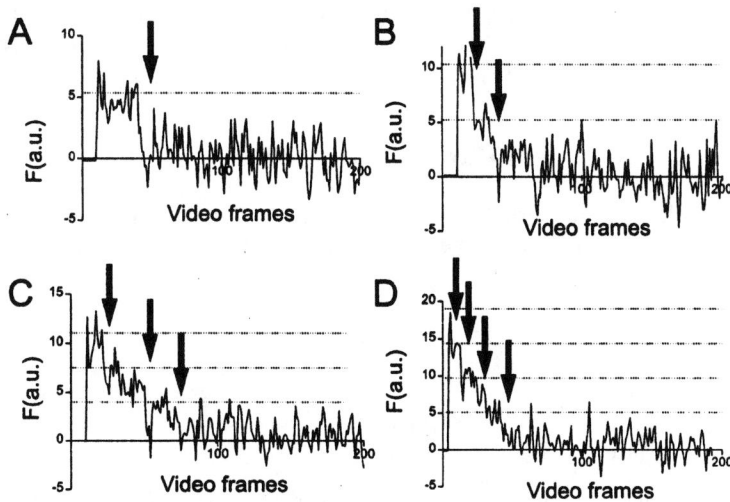

FIGURE 1. Quantized photobleaching of fluorescent molecules. Time course of the change in fluorescence intensity of FITC–Na/K-ATPase was recorded at video rate. Photobleaching occurred at the times indicated by the arrows. Quantized photobleaching was observed in **(A)** one, **(B)** two, **(C)** three, and **(D)** four steps.

oligomericity of Na/K- and H/K-ATPase, solubilized materials were examined by TIRFM. The TIRFM experiment was performed essentially as described previously.[3]

Fluorescence images (spots) of a single fluorophore attached to protein, bound to a glass surface, were observed in both preparations. The numbers of spots observed were proportional to the amount of enzyme used and, when unlabeled enzyme was used, the number was negligible. This suggests that each spot observed represents an FITC-labeled molecule attached to the protein molecule. The fluorescent image disappeared in a few seconds, indicating that photobleaching of the FITC occurs by laser irradiation. The change in fluorescence intensity of each spot was monitored immediately after laser irradiation. Video-rate images were recorded on videotape. The fluorescence images were digitally captured and the intensity change was then analyzed. The photobleaching of FITC molecules appeared to be quantized. When the initial intensity of the spot was high, a multistepwise decrease in fluorescence was observed. A typical time course of the change in fluorescence intensity is shown in FIGURE 1. About 250 independent spots observed from Na/K-ATPase were analyzed, and the value of the initial intensity was determined. The data obtained were fitted to the sum of four Gaussian functions. The population of 1, 2, 3, and 4 units of the initial intensity was determined to be 54%, 29%, 7%, and 10%, respectively. Considering the degree of FITC-labeling (around 90%), 3 and 4 units of population can be attributed to tetraprotomeric species. Based on these data, we conclude that the solubilized Na/K- and H/K-ATPase preparations contain mainly protomeric, diprotomeric, and tetraprotomeric forms of ATPase molecules. These results represent the first direct observation of a single P-type ATPase molecule in an aqueous solution of H/K- and Na/K-ATPase, thus demonstrating the existence of a tetraprotomeric enzyme form.

ACKNOWLEDGMENTS

This work was supported by Grants-in-Aid for Scientific Research (Nos. 13142201 and 13680703) from the Ministry of Education, Science, and Culture of Japan.

REFERENCES

1. MIMURA, K., H. MATSUI, T. TAKAGI & Y. HAYASHI. 1993. Change in oligomeric structure of solubilized $Na^+/K^{(+)}$-ATPase induced by octaethylene glycol dodecyl ether, phosphatidylserine, and ATP. Biochim. Biophys. Acta **1145:** 63–74.
2. YOKOYAMA, T., S. KAYA, K. ABE, et al. 1999. Acid-labile ATP and/or ADP/P(i) binding to the tetraprotomeric form of Na/K-ATPase accompanying catalytic phosphorylation-dephosphorylation cycle. J. Biol. Chem. **274:** 31792–31796.
3. OIWA, K., J.F. ECCLESTON, M. ANSON, et al. 2000. Comparative single-molecule and ensemble myosin enzymology: sulfoindocyanine ATP and ADP derivatives. Biophys. J. **78:** 3048–3071.

K^+ Induced Simultaneous Liberation of Two Moles of P_i, One from One Mole of EP and the Other from EATP, of Oligomeric H/K-ATPase from Pig Stomach

KAZUHIRO ABE, SHUNJI KAYA, TOSHIAKI IMAGAWA, AND KAZUYA TANIGUCHI

Department of Biological Chemistry, Division of Chemistry, Graduate School of Science, Hokkaido University, Sapporo 060-0810, Japan

KEYWORDS: H/K-ATPase; phosphoenzyme (EP); enzyme-bound ATP (EATP)

Gastric H/K-ATPase maintains half-site phosphorylation and half-site ATP binding during the H^+-ATPase reaction in the presence of high concentrations of ATP.[1]

To investigate the fate of enzyme-bound ATP (EATP), the enzyme was first incubated with various concentrations of [γ-^{32}P]ATP under turnover conditions. After 10 s, an excess of nonradioactive ATP was added, and the amount of breakdown of phosphoenzyme $E^{32}P$ (ΔEP) and that of the liberation of $^{32}P_i$ (ΔP_i) were determined. The ratio, $\Delta P_i/\Delta EP$, increased from 1 to ~2 with increasing concentrations of ATP (FIG. 1, $K_{1/2}$ = 0.16 mM). The simultaneous addition of K^+ with nonradioactive ATP also increased the ratio from 1 to ~1.7 ($K_{1/2}$ = 0.13 mM) with increasing initial concentrations of ATP (FIG. 1). These data suggest that both EP and EATP liberate stoichiometric amounts of P_i from EP and EATP.

When the reaction was chased with CDTA to chelate Mg^{2+}, the ratio remained constant: ~1 (FIG. 1). This suggests that Mg^{2+} is required for P_i liberation from EATP.

These data represent the first direct evidence, for the case of a P-type ATPase, in which two moles of P_i are simultaneously liberated from one mole of EP for half of the enzyme molecules and from one mole of EATP for the other half, during ATP hydrolysis, in either the absence or the presence of K^+.

The present data[1–5] as well as the recent finding by Hayashi *et al.*[6] that Na/K-ATPase activity of the tetraprotomer was around half of that of the diprotomer and protomer strongly indicate that the tetraprotomer is a functional unit of Na/K- and

---

Address for correspondence: Kazuhiro Abe, Department of Biological Chemistry, Division of Chemistry, Graduate School of Science, Hokkaido University, Sapporo 060-0810, Japan. Voice: 81-11-706-3483; fax: 81-11-736-2074.
ikkei@sci.hokudai.ac.jp

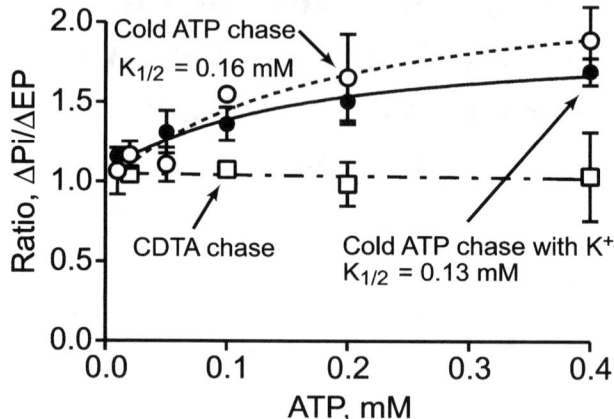

FIGURE 1. Dependence of ATP concentration on the ratio, $\Delta P_i/\Delta EP$, in the absence or presence of K^+. An excess of nonradioactive ATP, with (*closed circles*) or without (*open circles*) K^+, or CDTA (*open squares*) was added to the H/K-ATPase phosphorylated with increasing concentrations of $[\gamma\text{-}^{32}P]ATP$. Ratios of the amount of $^{32}P_i$ liberation (ΔP_i) to the amount of $E^{32}P$ (ΔEP) were obtained.

H/K-ATPase in the membrane. Diprotomeric and protomeric enzyme forms may be uncoupled forms of the enzyme with a higher specific activity.

ACKNOWLEDGMENTS

This work was supported in part by Grants-in-Aid for Scientific Research (Nos. 13142201 and 13680703) from the Ministry of Education, Science, Sports, and Culture of Japan.

REFERENCES

1. ABE, K. *et al.* 2002. Gastric H/K-ATPase liberates two moles of Pi from one mole of phosphoenzyme formed from high-affinity ATP binding site and one mole of enzyme-bound ATP at low-affinity site during cross-talk between catalytic subunits. Biochemistry **41:** 2438–2445.
2. TANIGUCHI, K. *et al.* 2001. The oligomeric nature of Na/K-transport ATPase. J. Biochem. **129:** 335–342.
3. SHIN, J.M. & G. SACHS. 1996. Dimerization of the gastric H^+,K^+-ATPase. J. Biol. Chem. **271:** 1904–1908.
4. KAYA, S. *et al.* 2003. Oligomeric structure of P-type ATPases observed by single molecule detection technique. This volume.
5. IMAGAWA, T. *et al.* 2003. Single mutation of Lys or Arg residue in ATP binding pocket in rat Na/K-ATPase alpha-1 subunit induces different affinity change in the high and low affinity ATP binding. This volume.
6. HAYASHI, Y. *et al.* 2003. Isolation of $(\alpha\beta)_4$-tetraprotomer having half-of-sites ATP binding from solubilized dog kidney Na^+/K^+-ATPase. This volume.

An Improved Method to Measure the Interactions of P-Type ATPases with the Lipidic Environment

VALERIA LEVI, JUAN P. F. C. ROSSI, ANA M. VILLAMIL GIRALDO, PABLO R. CASTELLO, AND F. LUIS GONZÁLEZ FLECHA

Instituto de Química y Fisicoquímica Biológicas, Departamento de Química Biológica, Facultad de Farmacia y Bioquímica, Universidad de Buenos Aires, Buenos Aires, Argentina

KEYWORDS: Na/K-ATPase; phosphatidylethanolamine; protein-amphiphile interactions; energy transfer

INTRODUCTION

Phospholipid-membrane protein interactions have great importance when optimizing conditions of protein reconstitution. It is well known that structure and function of P-ATPases are closely related to the characteristics of the lipids surrounding the transmembrane domain. We report here a method for measuring these interactions based on energy transfer between tryptophan residues of the protein and the fluorescent probe, 1-hexadecanoyl-2-(1-pyrenedecanoyl)-*sn*-glycero-3-phosphocholine (HPPC).

When HPPC is added to the micelle-solubilized enzyme, the probe molecules replace amphiphiles in contact with the transmembrane surface. In this condition, protein fluorescence decreases as a consequence of energy transfer between Trp and the pyrene moiety of HPPC. This decrease is related to the fraction of the transmembrane surface covered with HPPC (θ_{HPPC}) by

$$E_{app} = 1 - (I_{d,a}/I_d) = E \cdot \theta_{HPPC} \quad (1)$$

where $I_{d,a}$ and I_d are the fluorescence intensities of the protein in the presence and the absence of the acceptor. E_{app} and E are the apparent and the Förster energy transfer efficiencies, respectively.

θ_{HPPC} is determined by the detergent and phospholipid mole fractions in the micelles and by the affinities of these amphiphiles for the transmembrane domain of the protein.[1] Thus, the decrease of the fluorescence due to HPPC can be related to the micellar phase composition as follows:

Address for correspondence: F. Luis González Flecha, Departamento de Química Biológica–IQUIFIB, Facultad de Farmacia y Bioquímica, Universidad de Buenos Aires, Junín 956, C1113AAD Buenos Aires, Argentina. Voice: +54-11-4964-8289; fax: +54-11-4962-5457.
lgf@qb.ffyb.uba.ar

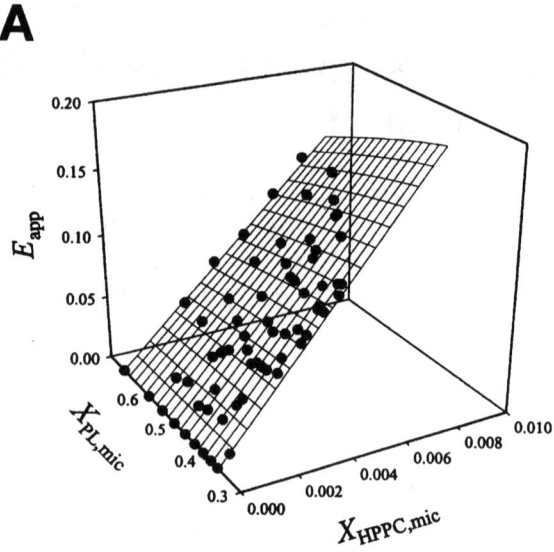

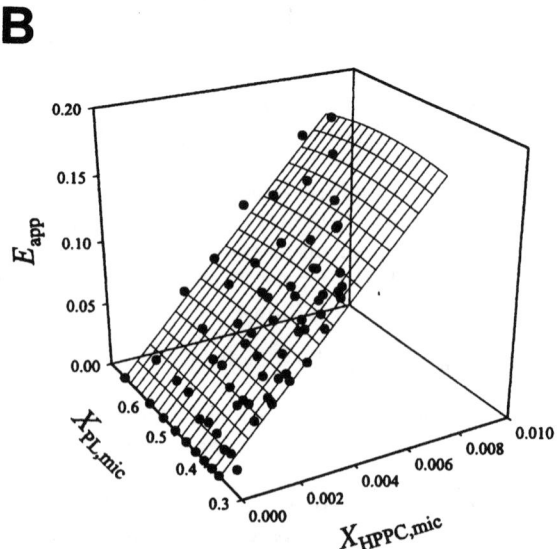

FIGURE 1. Dependence of the Na/K-ATPase fluorescence emission on the mole fractions of HPPC and $C_{12}E_{10}$. A mixture of DMPC:DMPE (mole ratio 1:1) was incubated with 200 mM mannitol (**A**) or glucose (**B**). Partially purified Na/K-ATPase was reconstituted in buffer I with 200 mM glucose, 140 mM $C_{12}E_{10}$, and the PL samples up to 280 µM of DMPC + DMPE. The enzyme was supplemented with $C_{12}E_{10}$ up to mole fractions within the range of 0.4 to 0.7. The emission intensity ($I_{d,a}$) was measured after adding increasing quantities of HPPC and mixing for 1 min. The surface is the graphical representation of equation 2 with the best-fit $K_{ex,PL}$ values.

$$E_{app} \cong X_{HPPC,mic} \cdot (\xi/[K_{ex,PL} \cdot X_{PL,mic}{}^{\beta(PL)} + X_{D,mic}{}^{\beta(D)}]) \qquad (2)$$

where $X_{i,mic}$ is the mole fraction of the i-amphiphile in the micelles and ξ is a constant for each system. β and $K_{ex,PL}$ are the stoichiometric coefficient and the displacement constant for the exchange between detergent (D) and phospholipid (PL) on the protein transmembrane surface, respectively.

For systems composed of more than one PL species, $K_{ex,PL}$ is a weighted average over the displacement constant of the phospholipids (Levi et al., unpublished).

EXPERIMENTAL PROCEDURES

Na/K-ATPase partially purified from pig kidney[2] was a gift from R. C. Rossi. Fragmented membranes were washed with 130 mM NaCl, 20 mM KCl, 1 mM DTT, 1 mM EGTA, and 10 mM Tris HCl, pH 7.4, at 25°C (buffer I). Proteins were solubilized by adding 1 mg of $C_{12}E_{10}$ per mL of the enzyme suspension (0.6 mg protein/mL), incubating for 10 min at 0°C, and centrifuging for 10 min at 14,000g. Spectroscopic data were collected as described previously.[1,3] Glycation of phosphatidylethanolamine was performed according to Ravandi et al.[4] Control experiments were carried out by incubating the lipids with mannitol instead of glucose.

RESULTS AND DISCUSSION

Na/K-ATPase was reconstituted in micelles composed of dimyristoyl phosphatidylcholine (DMPC), dimyristoyl phosphatidylethanolamine (DMPE), and the detergent poly(oxyethylene)-10-lauryl ether ($C_{12}E_{10}$). To assay the effect of phospholipid chemical modification on $K_{ex,PL}$, the enzyme was also reconstituted in a system identical to that previously described, except that the phospholipid mixture was incubated with glucose to obtain glycated phosphatidylethanolamine.[4] Na/K-ATPase fluorescence intensity was determined as a function of HPPC mole fraction and the composition of the micellar system (FIG. 1). Equation 2 was fitted to the experimental data obtaining a $K_{ex,PL}$ value of 1.7 ± 0.3 in the control experiment and 0.9 ± 0.1 for the glucose-treated PE. As observed, the $K_{ex,PL}$ value obtained for the PL mixture without the glucose treatment is greater than one, indicating that the average affinity for the Na/K-ATPase is higher than that of the detergent. In addition, $K_{ex,PL}$ significantly decreases after PL incubation, indicating a preferential adsorption of PE over glucose-treated PE to the transmembrane surface of the Na/K-ATPase.

CONCLUSIONS

We applied a simple energy transfer method to measure the affinity of amphiphiles for the transmembrane surface of P-ATPases. This method presents the advantage of using a probe that senses the competition among unlabeled amphiphiles and thus it does not contribute to $K_{ex,PL}$.

ACKNOWLEDGMENTS

This work was supported by UBA, CONICET, and Fundación Antorchas (Argentina).

REFERENCES

1. LEVI, V., J.P. ROSSI, M.M. ECHARTE, et al. 2000. Thermal stability of the plasma membrane calcium pump: quantitative analysis of its dependence on lipid-protein interactions. J. Membr. Biol. **173:** 215–225.
2. JENSEN, J., J.G. NORBY & P. OTTOLENGHI. 1984. Binding of sodium and potassium to the sodium pump of pig kidney evaluated from nucleotide-binding behaviour. J. Physiol. **346:** 219–241.
3. LEVI, V., J.P. ROSSI, P.R. CASTELLO & F.L. GONZÁLEZ FLECHA. 2002. Structural significance of the plasma membrane calcium pump oligomerization. Biophys. J. **82:** 437–446.
4. RAVANDI, A., A. KUKSIS & N.A. SHAIKH. 1999. Glycated phosphatidylethanolamine promotes macrophage uptake of low density lipoprotein and accumulation of cholesteryl esters and triacylglycerols. J. Biol. Chem. **274:** 16494–16500.

Extracellularly Applied Br-TITU Inhibits the Na^+/K^+ Pump by Interacting with Tryptophan at the Entrance to the Cation Sites

G. A. YUDOWSKI,[a] M. BAR SHIMON,[b] R. M. GONZÁLEZ-LEBRERO,[c]
R. C. ROSSI,[c] P. J. GARRAHAN,[c] S. J. D. KARLISH,[d] AND L. BEAUGÉ[a]

[a]*Laboratorio de Biofisica, Instituto M. y M. Ferreyra,
5000 Córdoba, Argentina*

[b]*Volcani Institute, Beit Dagon, Israel*

[c]*Departamento de Química Biológica, Facultad de Farmacia y Bioquímica, IQUIFIB, Universidad de Buenos Aires, Buenos Aires, Argentina*

[d]*Weizmann Institute of Science, Rehovot, Israel*

KEYWORDS: cation-tryptophan interaction; Br-TITU; Na^+/K^+-ATPase entrance port; tryptophan fluorescence

Aryl isothiouronium derivatives, among them Br-TITU, were designed as molecular probes for the cation binding sites of the Na^+,K^+ pump. Previous work showed that Br-TITU binds to renal Na^+,K^+-ATPase with two different affinities: <1 µM on the inside and around 50 µM, presumably on the outside.[1] Here, we describe properties and kinetic characteristics of Br-TITU as a cation antagonist at the extracellular surface of the Na^+,K^+-ATPase.

Br-TITU AS A CATION ANTAGONIST AT THE EXTRACELLULAR SURFACE

To explore if Br-TITU had an external effect on cation transport, we used the Na^+-Na^+ exchange model described previously.[2] For this purpose, we loaded red cells with ^{22}Na in Na^+ medium, pH 7.5. ^{22}Na ouabain-sensitive efflux was measured in the supernatant on the incubated red cells. These experiments were performed at three different external Na^+ concentrations (75, 100, 150 mM) in the absence and presence of three different concentrations of Br-TITU (30, 60, 100 µM). The Dixon plot of the data indicates that Br-TITU inhibits the ouabain-sensitive Na^+ efflux in a noncompetitive way: $K_{0.5} = 40$ µM.

Address for correspondence: G.A. Yudowski, Laboratorio de Biofisica, Instituto M. y M. Ferreyra, 5000 Córdoba, Argentina. Voice: +54-351-468-1465; fax: +54-351-469-5163.
gyudowski@immf.uncor.edu

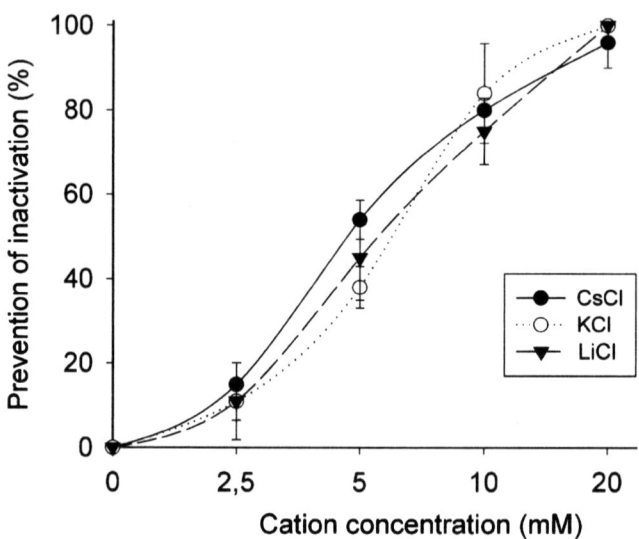

FIGURE 1. Cation-dependent inhibition of irreversible inactivation by Br-TITU. Human red blood cells loaded with ^{22}Na were preincubated 30 s in KCl (○), CsCl (●), or LiCl (▼) medium supplemented with 140 mM choline-Cl and 10 mM Tris-HCl, pH 9, with 200 µM Br-TITU. After preincubation, cells were washed in cold 140 mM choline-Cl and 10 mM Tris-HCl, pH 7.5, and ^{22}Na efflux was performed for 1 h in 140 mM NaCl and 10 mM Tris-HCl, pH 7.5, 37°C, in the absence and presence of 10^{-4} M ouabain.

IRREVERSIBLE INACTIVATION OF Na$^+$ EFFLUX BY Br-TITU AT ALKALI pH

An interesting characteristic of these aryl isothiouronium compounds is that, when incubated with the Na$^+$,K$^+$-ATPase at pH 9, they irreversibly inactivate the enzyme.[1] We have found recently that this irreversible inactivation occurs at the extracellular surface. After a short preincubation of Br-TITU with red cells at alkali pH, a large fraction of the ouabain-sensitive [^{22}Na]Na$^+$ efflux, measured at pH 7.5 in (Na$^+$+K$^+$)–free medium, was irreversibly inhibited. When presented during the preincubation, K$^+$, Cs$^+$, and Li$^+$ antagonized that inhibition, acting with similar apparent affinity; this is important for they display quite different affinities in their external activation of the Na$^+$ pump (FIG. 1). The prevention of this effect was not observed when Na$^+$ or choline-Cl was presented as a cation in the preincubation medium.

TRYPTOPHAN QUENCHING BY Br-TITU

Tryptophan fluorescence measurements from partially purified pig kidney Na$^+$/K$^+$-ATPase were used to explore Br-TITU binding. We observed that Br-TITU irreversibly quenches Trp fluorescence at pH 9 and this quenching is prevented by

K^+. This quenching occurs regardless of the conformational state of the pump and it is not prevented by Na^+.

Taken together, these data suggest that Br-TITU may bind and react with Trp residue(s) at an extracellular cation entrance port of the Na^+/K^+-ATPase, confirming a transit region through which the external activating cations must go through to get access to the extracellular transport sites. In addition, these results would show for the first time a cation-phi interaction on the aromatic side chain of tryptophan in a protein.

ACKNOWLEDGMENTS

This work was supported by a joint grant to P. J. Garrahan, L. Beaugé, and S. J. D. Karlish from Fundacion Antorchas.

REFERENCES

1. HOVING, S., M. BAR SHIMON, J.J. TIJMES, et al. 1995. Novel aromatic isothiouronium derivatives which act as high affinity competitive antagonists of alkali metal cations on Na/K-ATPase. J. Biol. Chem. **270:** 29788–29793.
2. GARRAHAN, P.J. & I.M. GLYNN. 1967. The behaviour of the sodium pump in red cells in the absence of external potassium. J. Physiol. **192:** 159–174.

Interactions between Cations and Na,K-ATPase Membranes Studied with Solid-State NMR

LOUISE ODGAARD JAKOBSEN,[a] NIELS CHR. NIELSEN,[b] AND MIKAEL ESMANN[a]

[a]*Department of Biophysics, University of Aarhus, DK-8000 Aarhus C, Denmark*

[b]*Institute of Molecular and Structural Biology, University of Aarhus, DK-8000 Aarhus C, Denmark*

KEYWORDS: cation binding; solid-state NMR

To fully understand the cation transport mechanism of Na,K-ATPase, it is important to obtain local structural information about the binding sites. With solid-state NMR methodology, it is possible to monitor the surroundings, that is, the local chemical environment, of NMR-sensitive nuclei. Such nuclei include, among others, the cations ^{87}Rb$^+$, ^{133}Cs$^+$, and ^{205}Tl$^+$, which can act as K$^+$-substitutes in the E$_2$-occluded Na,K-ATPase state.

From solid-state NMR spectra, several kinds of information can be extracted. Separation of the NMR signals according to different chemical environments for the nuclei is important. Stoichiometry of the cations in the different components can then be determined from the intensities of the resonances. Anisotropic interactions contain information about electronic environments, leading to structural information. Finally, chemical shifts and line widths may indicate an exchange process between different environments.

METHODS

Na,K-ATPase from shark rectal glands was prepared as described earlier.[1] NMR samples were prepared by centrifugation of the membranes, homogenization, and addition of small volumes of concentrated Cs$^+$ solutions. The final enzyme concentration is ~30 mg/mL, corresponding to a concentration of occlusion sites of about 0.18 mM. ^{133}Cs spectra were recorded at 52.4 MHz using a Varian Inova 400 (9.4 T)

Address for correspondence: Louise Odgaard Jakobsen, Department of Biophysics, University of Aarhus, Ole Worms Alle 185, DK-8000 Aarhus C, Denmark. Voice: +45-8942-2934; fax: +45-8612-9599.

lo@biophys.au.dk

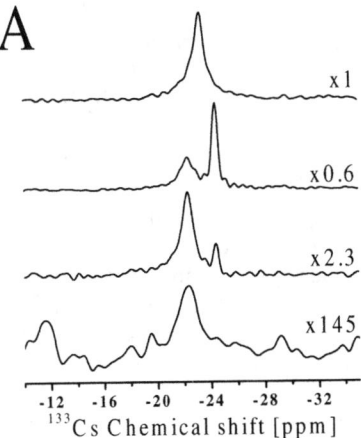

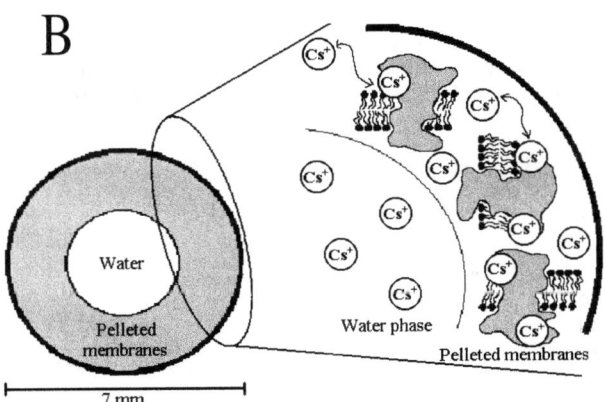

FIGURE 1. *Panel A* shows ^{133}Cs NMR spectra of Cs$^+$ in equilibrium with shark rectal gland Na,K-ATPase membranes: (1) [Cs$^+$] = 10 mM, no spinning; (2) [Cs$^+$] = 10 mM, with spinning; (3) re-recording of sample 2 after removal of the aqueous phase; (4) [Cs$^+$] = 0.1 mM. This spectrum was recorded in 72 h. *Panel B* shows a model of the NMR rotor seen from above and the phase separation of the membrane sample.

NMR spectrometer using a homebuilt double resonance magic angle spinning probe. The spinning speed was 2200 Hz and the temperature was 20°C.

A spectrum of Na,K-ATPase membranes with 10 mM Cs$^+$ present shows only a single peak: FIGURE 1A (top line). Better resolved spectra are achieved by magic angle spinning of the sample, whereby anisotropic interactions are averaged. The ^{133}Cs spectrum of such a sample (FIG. 1A, second line) shows two clearly separated peaks. The spinning separates the sample into a membrane-containing phase ("pelleted membranes") along the rotor wall and a water phase in the middle, as outlined in FIGURE 1B. Removal of the water phase after spinning for 6 h, followed by recording of a new spectrum of the same sample (FIG. 1A, third line), leads to identification of the origin of the peaks. The resonance at about −24.6 ppm corresponds to the water phase (very reduced in the third spectrum), and the membrane phase ^{133}Cs resonates at −22.5 ppm.

The membrane phase conceivably includes Cs^+ specifically bound at the occlusion sites, nonspecifically bound Cs^+, as well as aqueous Cs^+ in the residual water trapped between the membranes, and all three types of Cs^+ are assumed to contribute to the resonance from the membranous ^{133}Cs at −22.5 ppm.

Figure 1A (*bottom line*) shows a spectrum with 0.1 mM Cs^+ in Na,K-ATPase membranes. At this concentration the majority of the Cs^+ are expected to be occluded. The spectrum has a single resonance at −22.5 ppm, but a low signal/noise ratio. It is thus not possible with the present NMR experiments with Cs^+ to identify and characterize the occlusion sites in Na,K-ATPase in detail. However, the results reported here do form a basis for analysis of the interaction of Na,K-ATPase with ^{87}Rb, as well as the more NMR-sensitive nucleus ^{205}Tl (in progress).

REFERENCE

1. SKOU, J.C. & M. ESMANN. 1979. Preparation of membrane-bound and solubilized $(Na^+ + K^+)$–ATPase from rectal glands of *Squalus acanthias*: the effect of preparative procedures on purity, specific, and molar activity. Biochim. Biophys. Acta **567**: 436–444.

Interaction between ATP and the Na/K-ATPase from Duck Supraorbital Salt Glands

PROMOD R. PRATAP,[a] NATALIE OLDEN-STAHL,[a] OANA DEDIU,[b] AND G. ULRICH NIENHAUS[b,c]

[a]*Department of Physics and Astronomy, University of North Carolina at Greensboro, Greensboro, North Carolina 27402-6170, USA*

[b]*Abteilung Biophysik, Universität Ulm, Ulm, Germany*

[c]*Department of Physics, University of Illinois at Urbana-Champaign, Urbana, Illinois, USA*

KEYWORDS: Na/K-ATPase mechanism; enzyme-substrate interaction kinetics

INTRODUCTION

ATP binding to P-type ATPases results in transitions from α-helix to β-sheet.[1-3] The dynamics of these transitions were examined in Na^+/K^+-ATPase isolated from duck supraorbital salt glands[4] using FTIR, ATR-FTIR, and fluorescence spectroscopy.

Since ATPase activity depends on the presence of Na^+, K^+, and Mg^{2+}, experiments were designed to examine the effects of these ions on ATP-induced infrared and fluorescence signals. Results using ATR-FTIR spectroscopy indicate that, in the presence of Na, Na + K, or Mg, the $K_{0.5}$ value for ATP-induced changes in infrared absorbance was in the μM range, which was an order of magnitude higher than the $K_{0.5}$ for these changes in the absence of these. This difference was not observed with enzyme labeled with the fluorescent cyanine dye, Cy3.

MATERIALS AND METHODS

For ATR-FTIR experiments, protein (5 mg/mL) was suspended in D_2O buffer (20 mM HEPES, 79 mM salt, 1 mM EDTA, pD 7.5). Experiments were performed as described by Fringeli *et al.*[5] For fluorescence experiments, enzyme was labeled with Cy3-maleimide, with a modification of the protocol used to prepare IAF-labeled enzyme.[6] Emission wavelength was 571 nm; excitation wavelength was 547 nm.

RESULTS

Absorbance decrease at 1655 cm^{-1} (ΔA) was indicative of a transition from α-helix to β-sheet. ΔA was approximately 5% of total absorbance. These changes were measured as a function of ATP concentration under various ionic conditions and were fitted with a single rectangular hyperbola. Fits yielded $K_{0.5}$ of 2.1 µM

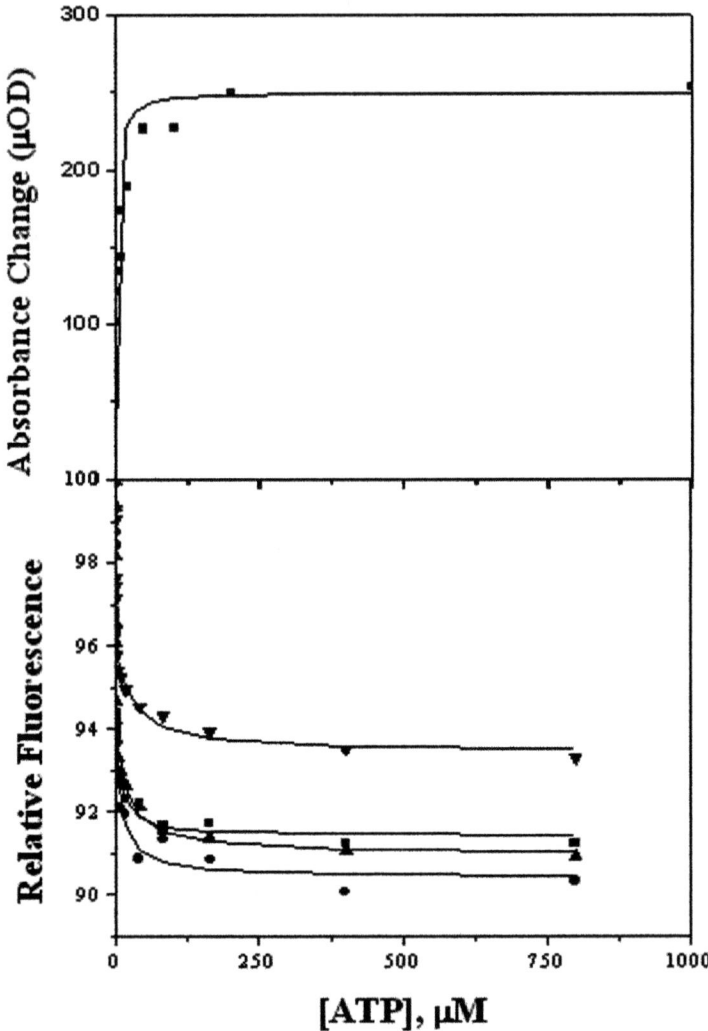

FIGURE 1. (**Top**) ΔA as a function of ATP concentration, with the enzyme in buffer with 74 mM Na$^+$ and 5 mM K$^+$. The solid line is a fit to a single hyperbola. (**Bottom**) ATP-induced change in the fluorescence of Cy3-labeled enzyme as a function of ATP concentration in Na$^+$ + K$^+$ (▼), Na$^+$ (▲), choline (■), and choline + Mg^{2+} (●) buffers. Solid lines are fits to a sum of two rectangular hyperbolae. Total salt concentration was 150 mM.

($Na^+ + K^+$ buffer), 4.2 µM (Na^+ buffer), 3.1 µM (choline buffer), and 0.1 µM (choline + Mg^{2+} buffer). See FIGURE 1, top.

ATP-induced fluorescence change of Cy3-labeled protein was measured under similar conditions (FIG. 1, bottom). The data were fitted with a sum of two hyperbolae. The apparent dissociation constants were approximately 50 nM and 16 µM under all ionic conditions.

DISCUSSION

The relative change in infrared absorbance was at least an order of magnitude higher with duck enzyme than with Na^+/K^+-ATPase from kidney and with other P-type ATPases.[1–3] This increased signal was also seen in transmission infrared experiments where absorbance change was induced by uncaging ATP from DMB-caged ATP (data not shown). This higher signal is probably due to the increased activity of the duck enzyme.

The half-maximal concentration of ATP required for the infrared signal was 20-fold or 30-fold lower in the absence of cations that bind to the enzyme (Na^+, K^+, and Mg^{2+}). This finding was reproducible, but not consistent with earlier findings with Na^+/K^+-ATPase from other sources.[7]

Experiments with Cy3 enzyme indicated that the signal was heterogeneous. This heterogeneity could be due to multiple labeling sites for Cy3 or shifts in the equilibrium between the E_1 and E_2 conformations. The nature of Cy3 labeling will be investigated further.

The results presented here are consistent with the measured changes being downstream from the ATP binding step in the enzyme reaction cycle.

REFERENCES

1. CHETVERIN, A.B. & E.V. BRAZHNIKOV. 1985. Do sodium and potassium forms of Na,K-ATPase differ in their secondary structure? J. Biol. Chem. **260:** 7817–7819.
2. BARTH, A. & W. MÄNTELE. 1998. ATP-induced phosphorylation of the sarcoplasmic reticulum Ca^{2+} ATPase: molecular interpretation of infrared difference spectra. Biophys. J. **75:** 538–544.
3. SCHEIRLINCKX, F. et al. 2001. Monitoring of secondary and tertiary structure changes in the gastric H^+/K^+-ATPase by infrared spectroscopy. Eur. J. Biochem. **268:** 3644–3653.
4. MARTIN, D.W. & J.R. SACHS. 1999. Preparation of Na^+,K^+-ATPase with near maximal specific activity and phosphorylation capacity: evidence that the reaction mechanism involves all of the sites. Biochemistry **38:** 7485–7497.
5. FRINGELI, U.P. et al. 1989. Polarized infrared absorption of Na^+/K^+-ATPase studied by attenuated total reflection spectroscopy. Biochim. Biophys. Acta **984:** 301–312.
6. KAPAKOS, J.G. & M. STEINBERG. 1982. Fluorescent labeling of ($Na^+ + K^+$)–ATPase by 5-iodoacetamidofluorescein. Biochim. Biophys. Acta **693:** 493–496.
7. FEDOSOVA, N.U., P. CHAMPEIL & M. ESMANN. 2002. Nucleotide binding to Na,K-ATPase: the role of electrostatic interactions. Biochemistry **41:** 1267–1273.

Three-Dimensional Structure-Activity Relationship Modeling of Digoxin Inhibition and Docking to Na^+,K^+-ATPase

W. JAMES BALL, JR.,[a] CAROL D. FARR,[a] STEFAN PAULA,[a] SUSAN M. KEENAN,[b] ROBERT K. DELISLE,[c] AND WILLIAM J. WELSH[b]

[a]*University of Cincinnati College of Medicine, Cincinnati, Ohio 45267, USA*

[b]*University of Medicine and Dentistry of New Jersey–Robert Wood Johnson Medical School, Piscataway, New Jersey 08845, USA*

[c]*Accelrys, Princeton, New Jersey 08543, USA*

KEYWORDS: digoxin docking; structural modeling; 3D-QSAR; Na^+-pump structure

The inotropic response of the heart to the therapeutic administration of cardiotonic steroids such as digoxin is well known, but the structural basis for digitalis binding and inhibition of Na^+,K^+-ATPase function is unknown. In this study, we used comparative molecular field analysis (CoMFA) to generate a 3D-quantitative structure-activity relationship (3D-QSAR) model that identifies the structural elements of digoxin involved in its inhibition of the Na^+-pump. Further, we used computational methods to model the transmembrane (TM) and extracellular domains of the enzyme's α1-subunit and to locate the putative site of digoxin binding.

MATERIALS AND METHODS

The inhibition of purified lamb kidney Na^+,K^+-ATPase activity by cardiotonic and hormonal steroids was determined using a spectrophotometric assay.[1] The molecular models of the structures of the compounds tested and the 3D-QSAR contour plots were generated using the Sybyl 6.6 (Tripos, Inc.) software on a Silicon Graphics workstation.[1] The modeling of the Na^+-pump α1 was done using the structure of an E_1 conformation of the SERCA1 Ca^{2+}-ATPase as determined by Toyoshima *et al.*[2] and sequence-structure threading methods and energy-minimizing calculations (Threader, U. College London; and InsightII, Accelrys). The docking of cardioactive compounds to the modeled surface of the enzyme was done using Gold 1.1 (Cambridge Cryst. Data Centre).

Address for correspondence: W. James Ball, Jr., University of Cincinnati College of Medicine, Cincinnati, OH 45267. Voice: 513-558-2388.
William.Ball@uc.edu

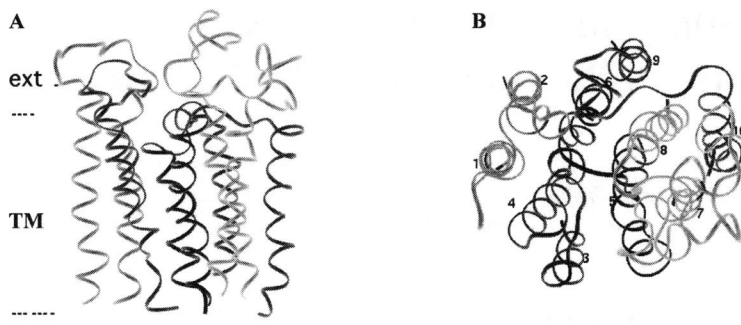

FIGURE 1. Threading/energy-minimized ribbon model of the TM and extracellular loops of sheep α1-subunit: **(A)** cross-sectional view; **(B)** extracellular-side view.

RESULTS AND DISCUSSION

The most potent inhibitors of the 47 compounds tested were the bufadienolides. The least potent compounds were chlormadinone and 17β-hydroxy-progesterone, while 12 other hormones tested had no significant effect. Interestingly, the active steroids showed a narrow range in their K_i values and, generally, the sugar moieties had little influence on inhibitory potency. Next, we constructed a statistically predictive 3D-QSAR model that indicated that electrostatic field contributions (60%) dominate over the steric field contributions (40%) towards digoxin binding.[1] The contour model also indicated that the receptor's binding site interactions encompass ~20 Å and bracket the lactone ring and α-sugar ends of digoxin. Our results suggest that digoxin binds the extracellular portion of α1 rather than within the membrane as proposed by Repke et al.[3] This work was extended by carrying out a "threading" sequence alignment of the lamb α1 to the SERCA1 Ca^{2+}-ATPase. The 3D-coordinates of the 10 transmembrane and connecting loop domains were assigned using InsightII, and the model was energy-minimized through a series of steps with the hydrogens, side chains, backbone atoms, and entire molecule allowed freedom of movement. FIGURES 1A and 1B, respectively, give ribbon representational views of the model. Digoxin and selected cardiotonic compounds were docked to the extracellular surface of α and a consensus binding orientation was achieved. Our results indicate that computational modeling techniques can locate a region of α1 that has physical properties consistent with available drug-binding data.

REFERENCES

1. FARR, C., C. BURD, M. TABET, et al. 2002. 3D quantitative structure-activity relationship study of the inhibition of Na,K-ATPase by cardiotonic steroids. Biochemistry **41:** 1137–1148.
2. TOYOSHIMA, C., H. NOKASAKO & H. OGAWA. 2000. Crystal structure of the calcium pump of sarcoplasmic reticulum at 2.6 Å resolution. Nature **404:** 647–655.
3. REPKE, K., K. SWEADNER, J. WEILAND, et al. 1996. In search of ideal inotropic steroids: recent progress. Prog. Drug Res. **47:** 9–52.

A Parallel Study of Eosin-Fluorescence Change and Rb^+ Occlusion in the Na^+/K^+-ATPase

M. R. MONTES, R. M. GONZÁLEZ-LEBRERO, P. J. GARRAHAN, AND R. C. ROSSI

Instituto de Química y Fisicoquímica Biológicas, Facultad de Farmacia y Bioquímica, Universidad de Buenos Aires, Buenos Aires, Argentina

KEYWORDS: eosin; conformational transition; Rb^+ occlusion

INTRODUCTION

Conformational changes in Na^+/K^+-ATPase have been studied using Eosin-Y (eosin).[1] This fluorescent probe binds with high affinity to the E_1 form of the enzyme (K_d = 0.2–0.5 µM), leading to an increase in fluorescence signal; in contrast, in media with K^+, fluorescence is low.[1] In our laboratory, we have developed a method that allows us to measure occlusion of the K^+-congener rubidium,[2] presumably testing the E_2 form of the enzyme. The aim of this work is a comparative characterization of the kinetics of the $E_1 \leftrightarrow E_2$ conformational transition, using both eosin-fluorescence change and Rb^+-occlusion experiments. Since eosin is known to bind noncovalently to the enzyme at the high-affinity binding site of ATP, we first explored how eosin affects the interaction between Rb^+ and the Na^+/K^+-ATPase.

MATERIALS AND METHODS

Na^+/K^+-ATPase was partially purified from pig kidney as described by Jensen *et al.*[3] All reactions were performed at 25°C in media containing 25 mM Imidazole-HCl (pH 7.4 at 25°C) and 0.25 mM EDTA. Occluded Rb^+ and changes in eosin fluorescence were measured according to Rossi *et al.*[2] and to Skou and Esmann,[1] respectively.

RESULTS AND DISCUSSION

Interactions of Eosin with the Rb^+-Containing Enzyme

We measured the amount of occluded Rb^+ (Rb_{occ}) in equilibrium as a function of [eosin] and [Rb^+]. We found that, at constant [Rb^+], eosin decreased Rb_{occ} to a non-

---

Address for correspondence: M.R. Montes, Instituto de Química y Fisicoquímica Biológicas, Facultad de Farmacia y Bioquímica, Universidad de Buenos Aires, C1113AAD Buenos Aires, Argentina. Voice: +5411-4964-5506; fax: +5411-4962-5457.
mmontes@qb.ffyb.uba.ar

TABLE 1. Effects of eosin and ATP on the initial rate of ^{86}Rb$^+$ deocclusion (v_d)

Condition	v_d [nmol (mg prot)$^{-1}$ s^{-1}]
Control	0.14 ± 0.02
100 µM eosin + 2 mM MgCl$_2$	4.40 ± 0.82
100 µM ATP + 2 mM MgCl$_2$	11.20 ± 3.88
1 mM eosin + 100 µM ATP + 2 mM MgCl$_2$	4.68 ± 0.40

NOTE: Reaction was started by mixing Na$^+$/K$^+$-ATPase (250 µg protein/mL) equilibrated with 100 µM ^{86}Rb$^+$ with enough deocclusion medium containing (final concentration) 100 µM Rb$^+$ (control) to cause a 20-fold isotopic dilution. Deocclusion medium also contained other ligands as indicated.

zero value that depended on [Rb$^+$]. Nonlinear regression analysis indicates that two eosins, rather than one, would bind both to the free and to the Rb$^+$-containing enzyme. In transient experiments, we found that eosin increased the rate of Rb$^+$ deocclusion. As this effect is similar to that known for ATP, exerted at a low-affinity binding site in E_2,[2] we investigated whether eosin binds to the same site. Therefore, we measured the time course of breakdown of occluded Rb$^+$ in the presence of eosin and/or ATP, and we calculated the initial rate of deocclusion (v_d). TABLE 1 shows that, in media containing 2 mM MgCl$_2$, the presence of either 100 µM eosin or ATP increased v_d, with ATP being 2- to 3-fold more effective than eosin. However, when 1 mM eosin and 100 µM ATP were added together, the value of v_d decreased to that obtained with 100 µM eosin, strongly suggesting that both ligands compete for the same site from which they promote deocclusion. Thus, eosin seems to bind not only to the high-affinity site for ATP,[1] but also to the low-affinity one, supporting the idea of eosin acting as an ATP analogue.

Comparative Study of the Kinetics of the $E_1 \leftrightarrow E_2$ Transitions

In parallel experiments, we measured the time courses of eosin-fluorescence change and occlusion of ^{86}Rb$^+$ upon addition of Rb$^+$ at different concentrations to the enzyme suspension. Fitted curves for occlusion and fluorescence change were normalized to 100% according to their maximal change and are shown in FIGURE 1. The tracings are double-exponential functions of time, with a fast and a slow component. It is easy to see that, at a given [Rb$^+$], both Rb$^+$ occlusion and fluorescence change show comparable time courses. However, as [Rb$^+$] increases, the curves for the fluorescence change tended to be somewhat faster than that observed for Rb$^+$ occlusion, a difference that was more apparent for the slow component of the curves. Although these results indicate that both eosin-fluorescence change and Rb$^+$ occlusion are strongly correlated phenomena, the small differences observed could reflect the fact that some stages in the conformational change that involve eosin release occur before occlusion phenomena.

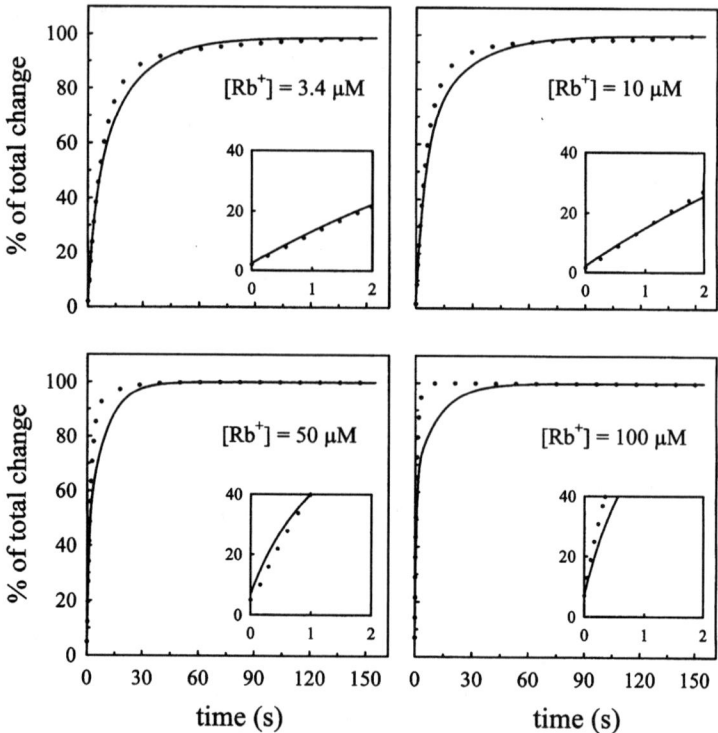

FIGURE 1. Normalized fitted curves for fluorescence change (*dotted line*) and Rb$^+$ occlusion (*solid line*). All reaction media contained 100 µg protein/mL of enzyme and 0.4 µM eosin. The insets show the initial part of the curves.

ACKNOWLEDGMENTS

This work was supported by grants from Agencia Nacional de Promoción Científica y Tecnológica, Universidad de Buenos Aires, Fundación Antorchas, and Consejo Nacional de Investigaciones Científicas y Técnicas.

REFERENCES

1. SKOU, J.C. & M. ESMANN. 1981. Eosin, a fluorescent probe of ATP binding to the (Na$^+$ + K$^+$)–ATPase. Biochim. Biophys. Acta **647:** 232–240.
2. ROSSI, R.C., S.B. KAUFMAN, R.M. GONZÁLEZ-LEBRERO, *et al.* 1999. An attachment for non-destructive, fast quenching of samples in rapid mixing experiments. Anal. Biochem. **270:** 276–285.
3. JENSEN, J., J.G. NØRBY & P. OTTOLENGHI. 1984. Binding of sodium and potassium to the sodium pump of pig kidney evaluated from nucleotide-binding behaviour. J. Physiol. **346:** 219–241.

The Sidedness of the Direct Route of Occlusion of K^+ in the Na^+/K^+-ATPase

RODOLFO M. GONZÁLEZ-LEBRERO, SERGIO B. KAUFMAN,
PATRICIO J. GARRAHAN, AND ROLANDO C. ROSSI

Instituto de Química y Fisicoquímica Biológicas and Departamento de Química Biológica, Facultad de Farmacia y Bioquímica, Universidad de Buenos Aires, Buenos Aires, Argentina

KEYWORDS: sidedness; occlusion; direct route

INTRODUCTION

Occlusion of up to two K^+ (or its congeners) per Na^+/K^+-ATPase molecule takes place either in media lacking Na^+, Mg^{2+}, and ATP (direct route) or during ATPase activity, after K^+ has promoted dephosphorylation of the enzyme (physiological route). In the latter condition, K^+ occlusion takes place through the extracellular sites of the enzyme. It is usually assumed that the direct route leads to K^+ occlusion through intracellular sites. However, there is no direct evidence to support this claim and some experimental results seem to disagree with this view.[1] We have reported results indicating that the direct route can lead to the sequential occlusion of one or two Rb^+ according to a single-file mechanism.[2] Similar mechanisms were proposed elsewhere to explain results obtained in the presence of MgP_i.[1,3] The sequential occlusion of the two Rb^+ can be envisioned as taking place in a narrow channel with gated entrances, with one to each side of the membrane, as shown in the following scheme:

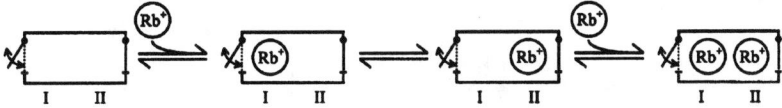

Occlusion occurs only when both gates are closed. We assume that, under the conditions of our experiments, one of the gates (that in position I) opens much more frequently than the other so that almost all the exchange of Rb^+ will take place through this gate. For Rb^+ to become occluded, it must bind to a site in position I and jump to a deeper position, II, before a second Rb^+ can be bound and occluded. Release of Rb^+ follows the reverse pathway.

Address for correspondence: Rodolfo M. González-Lebrero, Instituto de Química y Fisicoquímica Biológicas and Departamento de Química Biológica, Facultad de Farmacia y Bioquímica, Universidad de Buenos Aires, Junín 956, C1113AAD Buenos Aires, Argentina. Voice: +5411-4964-5506; fax: +5411-4962-5457.
gonlebre@mail.retina.ar

Using the assumption that the binding of MgP_i to the enzyme promotes deocclusion toward the extracellular gate[4] to assign a sidedness to positions I and II, we here provide experimental evidence to answer whether the occlusion of K^+ in the Na^+/K^+-ATPase via the direct route takes place through intracellular or extracellular sites of the enzyme.

RESULTS AND DISCUSSION

Since, at high [Rb^+], Rb^+ in position I exchanges with the media much faster than that in position II, we can label selectively the cation in either position using [^{86}Rb]Rb^+. To label Rb^+ in position II, the enzyme was first equilibrated in media

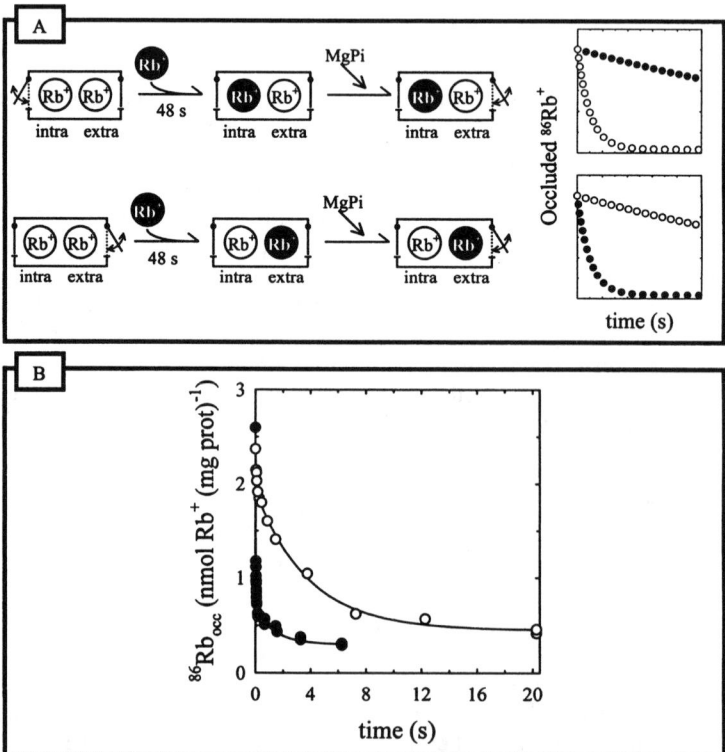

FIGURE 1. (**A**) Simulations of the release of $^{86}Rb^+$ (see main text). (**B**) Experimental time courses of release of occluded $^{86}Rb^+$ in media with 3 mM $MgCl_2$, 8 mM P_i, and 1 mM unlabeled Rb^+, when $^{86}Rb^+$ occupied position I (●) or II (○) of the occlusion sites. Continuous lines are plots of a single exponential function of time for the best fitting values of their rate coefficients, which were 35.8 ± 1.5 s^{-1} (●) and 0.263 ± 0.023 s^{-1} (○). Occluded Rb^+ was measured as described by Rossi et al.[5] All experiments were performed at 25°C in a rapid mixing apparatus using partially purified ATPase from pig kidney[6] in media containing 25 mM Imidazole-HCl (pH 7.4 at 25°C) and 0.25 mM EDTA.

with high [^{86}Rb$^+$] and then suspended for 48 s in a solution containing unlabeled Rb$^+$. Conversely, to label Rb$^+$ in position I, equilibration was reached in media with unlabeled Rb$^+$, and the 48-s incubation was carried out in media with ^{86}Rb$^+$. Preliminary experiments (not shown) indicated that incubation for 48 s was sufficient to allow the replacement of about 95% of the occluded Rb$^+$ in position I and only 3% of that in position II.

FIGURE 1A shows the expected results for the deocclusion of Rb$^+$ in MgP$_i$ from the enzyme states labeled in either position, I or II, assuming the following two possible alternatives for the sidedness of Rb$^+$ occlusion by the direct route: (i) if occlusion took place through intracellular sites, then ^{86}Rb$^+$ release in MgP$_i$ would be slow if ^{86}Rb$^+$ were in position I and fast if it were in position II (upper part of FIG. 1A); (ii) if the Rb$^+$ occlusion took place through extracellular sites, then the rate of release of ^{86}Rb$^+$ would be fast for the Rb$^+$ in position I and slow for the Rb$^+$ in position II (lower part of FIG. 1A). Experimental results are shown in FIGURE 1B. It can be seen that, when the Rb$^+$ occluded in position I is labeled, its rate of release is higher than when the labeled Rb$^+$ is in position II.

These results indicate that, if the above scheme is valid, Rb$^+$ becomes occluded and deoccluded by the direct route through the same pathway as that used in the presence of MgP$_i$, probably using the entrance facing the extracellular side of the membrane.

ACKNOWLEDGMENTS

This work was supported by grants from Fundación Antorchas, Agencia Nacional de Promoción Científica y Tecnológica, Consejo Nacional de Investigaciones Científicas y Técnicas, and Universidad de Buenos Aires, Argentina.

REFERENCES

1. FORBUSH III, B. 1987. Rapid release of ^{42}K or ^{86}Rb from two distinct transport sites on the Na,K-pump in the presence of Pi or vanadate. J. Biol. Chem. **262:** 11116–11127.
2. GONZÁLEZ-LEBRERO, R.M., S.B. KAUFMAN, M.R. MONTES, et al. 2002. The occlusion of Rb$^+$ in the Na$^+$,K$^+$-ATPase. I. The identity of occluded states formed by the physiological or the direct routes: occlusion/deocclusion kinetics through the direct route. J. Biol. Chem. **277:** 5910–5921.
3. GLYNN, I.M., J.L. HOWLAND & D.E. RICHARDS. 1985. Evidence for the ordered release of rubidium ions occluded within the Na,K-ATPase of mammalian kidney. J. Physiol. **368:** 453–469.
4. FORBUSH III, B. 1988. Occluded ions and Na,K-ATPase. In The Na$^+$,K$^+$-Pump. Part A: Molecular Aspects, pp. 229–248. Alan R. Liss. New York.
5. ROSSI, R.C., S.B. KAUFMAN, R.M. GONZÁLEZ-LEBRERO et al. 1999. An attachment for non-destructive, fast quenching of samples in rapid mixing experiments. Anal. Biochem. **270:** 276–285.
6. JENSEN, J., J.G. NØRBY & P. OTTOLENGHI. 1984. Binding of sodium and potassium to the sodium pump of pig kidney evaluated from nucleotide-binding behaviour. J. Physiol. **346:** 219–241.

The Muscle-Specific βm Protein Is Functionally Different from Other Members of the X,K-ATPase β-Subunit Family

NIKOLAY B. PESTOV,[a,b] GILLES CRAMBERT,[c] HAO ZHAO,[b] TATYANA V. KORNEENKO,[a] MIKHAIL I. SHAKHPARONOV,[a] KÄTHI GEERING,[c] AND NIKOLAI N. MODYANOV[b]

[a]*Shemyakin-Ovchinnikov Institute of Bioorganic Chemistry, Moscow 117997, Russia*
[b]*Medical College of Ohio, Toledo, Ohio 43614, USA*
[c]*University of Lausanne, Lausanne CH-1005, Switzerland*

KEYWORDS: X,K-ATPase β-subunit family; skeletal muscle; heart

Recently, we have identified the fifth member of the mammalian X,K-ATPase β-subunit gene family, ATP1B4.[1] The human, rat, and pig genes were found to be expressed in skeletal muscle and at a lower level in heart. From the point of view of primary structure, the encoded protein, termed βm, is homologous to other β-subunits. However, experimental characterization of βm revealed a number of properties unique among the other members of the structural family.[2–4] In contrast to mature forms of other known X,K-ATPase β-subunits, the carbohydrate moiety of βm is composed of short high-mannose or hybrid *N*-glycans.[2] This finding argued in favor of an intracellular location of βm in human skeletal muscle[2] and, indeed, fractionation of membranes from pig skeletal muscle demonstrated that the majority of the βm protein is concentrated not in the plasma membrane, but in the sarcoplasmic reticulum instead.[3] Subsequently, we have expressed βm in *Xenopus* oocytes[4] and showed that βm is stably expressed in the endoplasmic reticulum (ER) in its core-glycosylated, partially trimmed form. βm cannot associate with and stabilize Na,K-ATPase or gastric and nongastric H,K-ATPase α isoforms. Moreover, properties of deletion mutants and chimeras between βm and β1 indicate that this inability to interact with Na,K-ATPase α-subunit is conferred mostly by the βm C-terminal ectodomain.[4]

To test if the β-protein does or does not interact with the Na,K-ATPase α-subunits *in vivo*, in skeletal muscle membranes, immunoprecipitation with anti-βm antibodies has been employed. The immunoprecipitated material was probed with anti-βm and anti-Na,K-ATPase antibodies. As expected, the βm protein is readily detected in the material precipitated with anti-βm antibodies and is not detected when antibodies

Address for correspondence: Nikolay B. Pestov, Shemyakin-Ovchinnikov Institute of Bioorganic Chemistry, Moscow 117997, Russia. Voice: +7-095-3306556; fax: +7-095-3306456.
korn@mail.ibch.ru

against a nonrelated protein are used (anti-nongastric H,K-ATPase α-subunit that is known to be absent from skeletal muscle). Further, no band corresponding to the Na,K-ATPase α-subunit has been observed in both fractions when probed with an antibody specific for all Na,K-ATPase α-subunits (against C-terminal peptide KETYY).

From the data obtained, it is clear that βm is not able to strongly bind an α-subunit. However, it is reasonable to question whether the αβm interaction never occurs *in vivo*. Indeed, there are many hypothetical possibilities for such an interaction to exist. First, a hitherto unknown α-subunit isoform (or a noncanonical variant of a known isoform) may be a real partner of βm. Second, some protein modifications can change βm *in vivo* to make it potent for the αβ interaction. For example, a site-specific protease can cleave a βm domain that may be responsible for the inhibition of binding with the α-subunit. Third, in muscle cells, there may exist a triple interaction (α, β, and an unknown component that appears in the muscle cell only under some specific conditions, such as a developmental stage or a cellular stress). Finally, there is another possibility that βm, in fact, is not a β-subunit of the X,K-ATPase family, being rather a protein with a totally different function, perhaps a β-subunit of another P-type ATPase.

In conclusion, despite the fact that βm from a structural point of view obviously belongs to the protein family, well known as X,K-ATPase β-subunits, our results suggest that βm does not have a chaperone function for X,K-ATPase α-subunits, which is the primary role of X,K-ATPase β-subunits. It thus can be concluded that βm is not just another Na,K-ATPase β-subunit isoform (β4) and may have a totally different physiological function that is currently unknown. These observations emphasize the necessity to revise our understanding of the X,K-ATPase β-subunit family as a structural and functional entity and suggest that the physiological roles of its members are more diverse than predicted.

ACKNOWLEDGMENTS

This work was supported by grants from the National Institutes of Health (Nos. HL-36573 and GM-54997), the Swiss National Fund for Scientific Research (Nos. 31-53721.98 and 31-64793.01), and the Russian Foundation for Basic Research (No. 00-04-48153).

REFERENCES

1. PESTOV, N.B., G. ADAMS, M.I. SHAKHPARONOV & N.N. MODYANOV. 1999. Identification of a novel gene of the X,K-ATPase β-subunit family that is predominantly expressed in skeletal and heart muscles. FEBS Lett. **456:** 243–248.
2. PESTOV, N.B., T.V. KORNEENKO, H. ZHAO, *et al.* 2000. Immunochemical demonstration of a novel β-subunit isoform of X,K-ATPase in human skeletal muscle. Biochem. Biophys. Res. Commun. **277:** 430–435.
3. PESTOV, N.B., T.V. KORNEENKO, H. ZHAO, *et al.* 2001. The βm protein, a member of the X,K-ATPase β-subunits family, is located intracellularly in pig skeletal muscle. Arch. Biochem. Biophys. **396:** 80–88.
4. CRAMBERT, G., N.B. PESTOV, N.N. MODYANOV & K. GEERING. 2002. βm, a structural member of the X,K-ATPase β subunit family, resides in the ER and does not associate with any known X,K-ATPase α subunit. Biochemistry **41:** 6723–6733.

Influence of Intramembrane Electric Charge on H,K-ATPase

IRENA KLODOS

Department of Biophysics, University of Aarhus, DK-8000 Aarhus, Denmark

KEYWORDS: H,K-ATPase; intramembrane electric field; cation binding sites; alamethicin; lipophilic ions; TPP; TPB

Previous studies on Na,K-ATPase showed that the electric charge in the membrane affects all reaction steps connected with the access of Na^+ and K^+ to their extracellular binding sites.[1] We have now investigated the influence of membrane electric charge on interaction of H,K-ATPase with H^+ and K^+. All experiments were performed with a vesicular preparation of H,K-ATPase from hog stomach in the presence of a channel-forming peptide, alamethicin (1 mg/mg protein), to open the vesicles. Our studies showed that alamethicin, by ensuring access of ions and ATP to the intravesicular binding sites, exposes full H,K-ATPase activity [200–300 µmol ATP hydrolyzed/(mg protein·h)]. The accompanying pNPPase activity was not activated by permeabilization of the vesicles, which indicates that pNPP passes freely through the vesicle membrane and that access of cations to intravesicular binding sites does not play any role in the pNPPase activity.

The intramembrane electric field was modified, as previously,[1] by addition of the lipophilic ions, tetraphenylphosphonium (TPP^+) and tetraphenylboron (TPB^-), which partition into the membrane. The following reactions were studied: (1) ATP hydrolysis, reflecting the whole cycle; (2) EP formation, reflecting binding of extravesicular (intracellular) cations; (3) K-stimulated EP dephosphorylation, reflecting binding and transport through the membrane of intravesicular (extracellular) K^+; (4) K-stimulated pNPP hydrolysis, reflecting the last steps of the reaction cycle and, as shown in our studies, probably involving only binding of K^+ to the extravesicular sites. All experiments were performed in parallel at pH 6.0, 7.0, and 8.0. To avoid buffer ion effects on the reaction studied (which for H,K-ATPase are very significant), experiments were performed with the same buffer system containing 10 mM MES and 10 mM HEPES adjusted to the required pH. ATP hydrolysis was measured at 37°C with 5 mM $MgCl_2$ and 5 mM ATP. pNPP hydrolysis was measured at 37°C with 10 mM pNPP and 20 mM $MgCl_2$. Phosphorylation of the enzyme was performed at 20°C with 25 µM ATP and 5 mM $MgCl_2$. Dephosphorylation was induced by addition of 20 mM KCl plus 1 mM ATP.

Address for correspondence: Irena Klodos, Department of Biophysics, University of Aarhus, DK-8000 Aarhus, Denmark. Voice: +45-8942-2037; fax: +45-8612-9599.
ik@biophys.au.dk

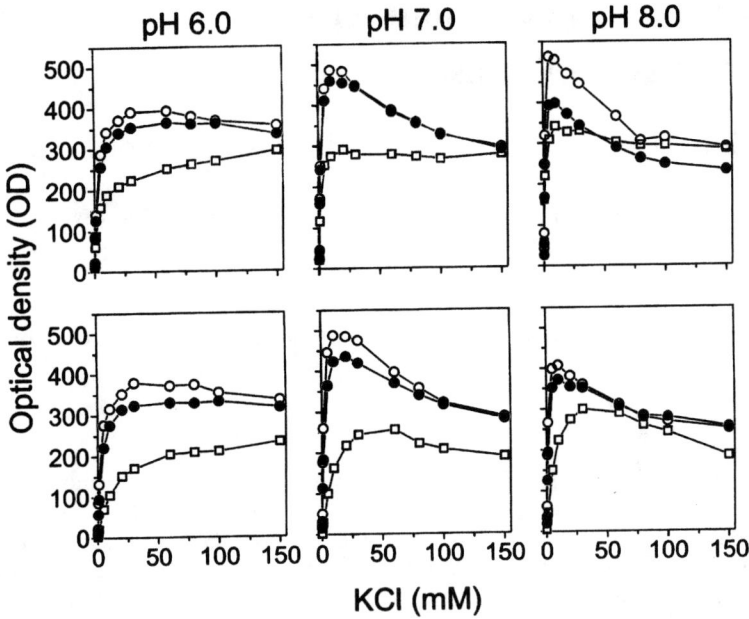

FIGURE 1. Effect of lipophilic ions on ATP hydrolysis by H,K-ATPase. The activity was measured at 37°C in the presence of 5 mM $MgCl_2$, 5 mM ATP, and 10 mM MES/10 mM HEPES adjusted to the pH shown in the figure. Symbols: (○) no lipophilic ion added. Upper panel: (●) 0.3 µM, (□) 5 µM TPB^- added. Lower panel: (●) 30 µM, (□) 300 µM TPP^+ added.

We found the hydrolysis rate of ATP by the enzyme to be significantly reduced by changes in the intramembrane electric field. Both TPP^+ and TPB^- inhibit the ATPase activity. Furthermore, high TPP^+ seemed to affect K^+-sensitivity (FIG. 1). However, the modification of the membrane charge did not affect the activity of pNPPase, the rate of formation of the phosphorylated acid-stable intermediates, nor the steady-state level of EP. Since both pNPPase activity and the phosphorylation appear to reflect binding of extravesicular cations, these results suggest that these sites are formed as shallow, low-field wells.

Further investigation of K-dependent dephosphorylation of EP, reflecting both binding and transport through the membrane of intravesicular K^+, is in progress.

REFERENCE

1. KLODOS, I., N.U. FEDOSOVA & L. PLESNER. 1995. Influence of intramembrane electric charge on Na,K-ATPase. J. Biol. Chem. **270**: 4244–4254.

The Role of Lys^{791} and Asn^{792} in Gastric H,K-ATPase

HERMAN G. P. SWARTS, PETER H. G. M. WILLEMS, JAN B. KOENDERINK, AND JAN JOEP H. H. M. DE PONT

Department of Biochemistry, Nijmegen Center for Molecular Life Sciences, 6500 HB Nijmegen, the Netherlands

KEYWORDS: H,K-ATPase; mutagenesis; conformational equilibrium

The Toyoshima model for Ca^{2+}-ATPase[1,2] visualizes the amino acids, present in transmembrane segments 4, 5, 6, and 8, involved in cation binding. It is likely that similar residues in gastric H,K-ATPase are involved in K$^+$ binding. Mutational studies on Asp^{824}, Glu^{820}, Glu^{795}, and Glu^{343} showed that these negatively charged amino acid residues are important for K$^+$ activation [De Pont et al., this volume]. Neutralization of the negative charge of residue Glu^{820} (M6) resulted in an enzyme with K$^+$-independent, but SCH 28080–sensitive ATPase activity, with a low vanadate sensitivity.[3,4] Mutants of Glu^{343} (M4), like E343D, could be phosphorylated, but dephosphorylation was blocked. Moreover, not the charge, but the side-chain geometry of the carbonyl residue of Glu^{795} (M5) was shown to be important for K$^+$-activated ATPase activity. In the present study, we investigated the role of two other polar amino acids in M5: Lys^{791} and Asn^{792}. Asn^{792} is conserved in P$_2$-type ATPases, while Lys^{791} is specific for H,K-ATPase (Ser in Na,K-ATPase and SR Ca^{2+}-ATPase). Several mutants, all or not combined with E820Q or E343D, were prepared and expressed in the baculovirus expression system.

All single, double, and triple mutants could be phosphorylated by ATP. The steady-state ATP-phosphorylation levels of the mutants, K791A, K791S, N792A, and N792Q, were nearly similar to that of the wild-type enzyme, whereas these levels were reduced when a negatively charged residue replaced Asn^{792} (N792D, N792E). The K791A, K791S, N792A, and N792Q mutants, however, showed only a low (K$^+$-stimulated) ATPase activity compared to the wild-type enzyme. The K$^+$-sensitivity of both the ATPase activity and the dephosphorylation reaction of these mutants was about 10 times reduced. The N792A-E820Q (but not the K791A-E820Q) double mutant showed a K$^+$-independent ATPase activity, indicating that Lys^{791} is also important for the K$^+$-independent hydrolysis of ATP.

Address for correspondence: Herman G.P. Swarts, Department of Biochemistry, Nijmegen Center for Molecular Life Sciences, 6500 HB Nijmegen, the Netherlands. Voice: +31-24-3614258; fax: +31-24-3616413.
 h.swarts@ncmls.kun.nl

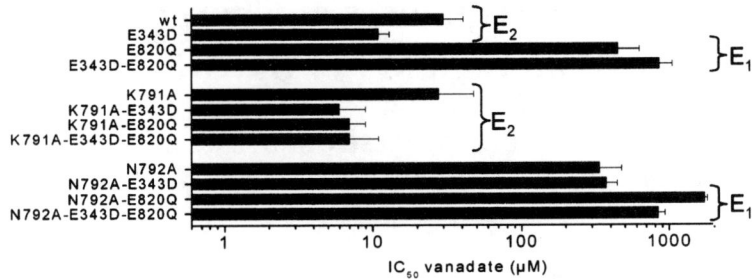

FIGURE 1. IC_{50} values of vanadate on the ATP-phosphorylation level of H,K-ATPase mutants.

The vanadate sensitivity of the ATP-phosphorylation (E-P) level was used as a tool to study the E_2/E_1 equilibrium. FIGURE 1 shows that the E-P levels of both the wild-type enzyme and the K^+-insensitive E343D mutant were lowered by very low vanadate concentrations ($IC_{50} \approx 30$ μM), whereas lowering of the E-P level of the K^+-independent mutant, E820Q, needed very high vanadate concentrations. This suggests a marked change in E_1/E_2 preference of the latter mutant. The vanadate sensitivity of the N792A mutant was about 10 times decreased, whereas the E_1/E_2 preference of the K791A mutant was similar to that of the wild type. If both K791A and E820Q were present, the E_1 preference of the E820Q mutant was overruled by the K791A mutation. The N792A mutation did not change the E_1 preference[5] of the E820Q mutation. These studies and those with triple mutants indicate that, regarding the E_1/E_2 preference, the order of dominance is

$$K791A > E820Q > N792A > E343D.$$

In summary, neither Lys^{791} nor Asn^{792} is essential for ATP-phosphorylation, whereas they are important for K^+-activation. In addition, the positive charge of Lys^{791} is crucial for the K^+-independent ATPase activity of the E820Q mutant and for its preference for the E_1 conformation.

REFERENCES

1. TOYOSHIMA, C., M. NAKASAKO, H. NOMURA & H. OGAWA. 2000. Crystal structure of the calcium pump of sarcoplasmic reticulum at 2.6 Å resolution. Nature **405:** 647–651.
2. TOYOSHIMA, C. & H. NOMURA. 2002. Structural changes in the calcium pump accompanying the dissociation of calcium. Nature **418:** 605–611.
3. SWARTS, H.G.P., H.P.H. HERMSEN, J.B. KOENDERINK, et al. 1998. Constitutive activation of gastric H^+,K^+-ATPase by a single mutation. EMBO J. **17:** 3029–3035.
4. HERMSEN, H.P.H., H.G.P. SWARTS, L. WASSINK, et al. 2001. Mimicking of K^+-activation by double mutation of glutamate-795 and glutamate-820 of gastric H^+,K^+-ATPase. Biochemistry **40:** 6527–6533.
5. SWARTS, H.G.P., J.B. KOENDERINK, H.P.H. HERMSEN, et al. 2001. K^+-independent gastric H^+,K^+-ATPase: dissociation of K^+-independent dephosphorylation and preference for the E_1-conformation by combined mutagenesis of transmembrane glutamate residues. J. Biol. Chem. **276:** 36909–36916.

Functional Consequences of Charge Reversals of Acidic Residues in M1 of the SR Ca-ATPase

ANJA PERNILLE EINHOLM, BENTE VILSEN, AND JENS PETER ANDERSEN

Department of Physiology, University of Aarhus, DK-8000 Aarhus C, Denmark

KEYWORDS: Ca-ATPase; mutagenesis; Ca^{2+} inlet; M1; Asp^{59}

A most conspicuous feature of transmembrane segment M1 of the SR Ca-ATPase is the presence of four negatively charged side chains near its N-terminus (Glu^{51}, Glu^{55}, Glu^{58}, and Asp^{59}). These residues have been suggested to be involved in the migration of Ca^{2+} between the cytoplasm and the intramembranous binding sites.[1] Most recently, the crystal structure of the Ca-ATPase in $E_2(TG)$ form,[2] showing that a water-accessible channel leads between M1 and M3 to Glu^{309} at Ca^{2+} binding site II, has attracted further attention to M1 as part of a possible entry pathway for Ca^{2+}. In relation to the specific cation selectivity of P-type ATPases, it is interesting that Glu^{55} and Glu^{58} in M1 are replaced by positive side chains in the Na,K-ATPase. Furthermore, the hydrogen bond between Glu^{58} and Glu^{309}, seen in the E_1Ca_2 crystal structure,[3] is interesting as it could be important in mediation of conformational effects upon Ca^{2+} binding.[1]

Previously, the above-mentioned acidic residues were mutated to alanines and/or glutamines with no effect on Ca^{2+} transport activity.[4] Such substitutions, however, are not expected to block the pathway. To test these ideas, we mutated Glu^{51}, Glu^{55}, Glu^{58}, and Asp^{59} individually to the positively charged and rather bulky arginine, expected to repel and sterically interfere with Ca^{2+} migration.

FIGURE 1 shows that the arginine substitutions of Glu^{51}, Glu^{55}, and Glu^{58} did not significantly alter the apparent affinity for Ca^{2+}. In contrast, replacement of Asp^{59} with arginine resulted in a more than 5-fold reduction of the apparent affinity for Ca^{2+} and in a 10-fold increase of the Ca^{2+} dissociation rate (see further discussion about Ca^{2+} dissociation in the article by J. P. Andersen *et al.* elsewhere in this volume).

Our results argue against a direct interaction of the side chains of Glu^{51}, Glu^{55}, and Glu^{58} of M1 with Ca^{2+} during its migration to and from the binding sites. Moreover, the fact that the substitution of Glu^{58} with arginine left the Ca^{2+} binding properties unaffected questions the existence of a close interaction between Glu^{58} and Glu^{309} in the native state of the pump. This apparent discrepancy with the

---

Address for correspondence: Jens Peter Andersen, Department of Physiology, Ole Worms Allé 160, DK-8000 Aarhus C, Denmark. Voice: +45-8942-2814; fax: +45-8612-9265.
jpa@fi.au.dk

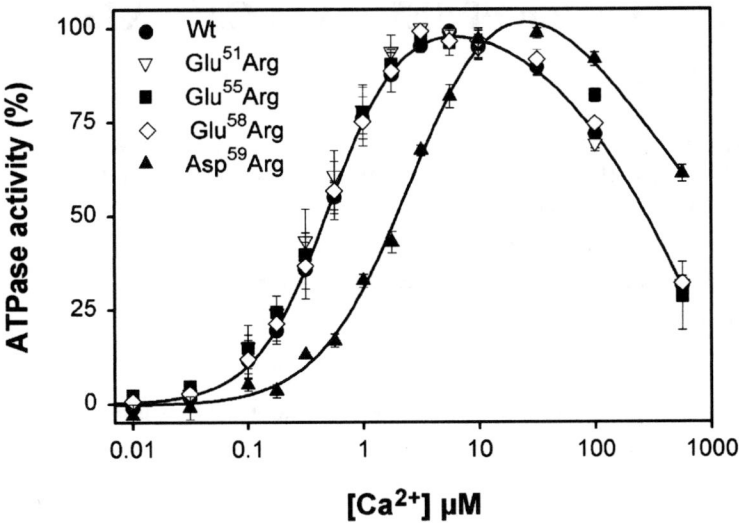

FIGURE 1. Ca^{2+} dependence of ATP hydrolysis at 37°C, pH 7.0.

published E_1Ca_2 crystal structure[3] may be explained by a high degree of mobility of the side chain of Glu^{58}.

Our results indicate, however, that the $Asp^{59} \rightarrow Arg$ mutation interferes with Ca^{2+} migration. This is in accordance with the location of the Ca^{2+} inlet between M1 and M3 suggested on the basis of the $E_2(TG)$ crystal structure.[2] Asp^{59} is more centrally located in relation to this inlet than Glu^{51}, Glu^{55}, and Glu^{58} because M1 bends at Asp^{59} during the E_1Ca_2 to $E_2(TG)$ transition. The effect of the $Asp^{59} \rightarrow Arg$ mutation may result either from direct steric and electrostatic interference with Ca^{2+} during its migration to and from the binding sites or from interference with the conformational change that bends M1 and opens the inlet.

REFERENCES

1. LEE, A.G. & J.M. EAST. 2001. What the structure of a calcium pump tells us about its mechanism. Biochem. J. **356:** 665–683.
2. TOYOSHIMA, C. & H. NOMURA. 2002. Structural changes in the calcium pump accompanying the dissociation of calcium. Nature **418:** 605–611.
3. TOYOSHIMA, C., M. NAKASAKO, H. NOMURA & H. OGAWA. 2000. Crystal structure of the calcium pump of sarcoplasmic reticulum at 2.6 Å resolution. Nature **405:** 647–655.
4. CLARKE, D.M., K. MARUYAMA, T.W. LOO, et al. 1989. Functional consequences of glutamate, aspartate, glutamine, and asparagine mutations in the stalk sector of the Ca^{2+}-ATPase of sarcoplasmic reticulum. J. Biol. Chem. **264:** 11246–11251.

Overexpression of SERCA1a Ca^{2+}-ATPase in Yeast

PIERRE FALSON,[a] GUILLAUME LENOIR,[a] THIERRY MENGUY,[a] FABIENNE CORRE,[a] CÉDRIC MONTIGNY,[a] PER A. PEDERSEN,[b] DENYSE THINÈS,[c] AND MARC LE MAIRE[a]

[a]*CEA Saclay, DSV/DBJC/SBFM, URA 2096 CNRS, 91191 Gif-sur-Yvette, France*

[b]*Biomembrane Center, 2100 Copenhagen OE, Denmark*

[c]*UCL, FYSA, 1348 Louvain-la-Neuve, Belgium*

KEYWORDS: SERCA1; Ca^{2+}-ATPase; expression; membrane protein; yeast

Much effort has been invested to express the SERCA1a Ca^{2+}-ATPase, a large membrane protein that transports Ca^{2+} across the sarcoplasmic reticulum membrane. Various expression systems have been set up with that goal, including COS cells[1,2] and *Spodopter fugiperda* Sf9[3] or Sf21[4] cells, via baculovirus and yeast,[5–7] all giving a relatively low amount of material.

Here, we have optimized a yeast expression system[8] for the production in yeast of high amounts of a C-terminal His-tagged Ca^{2+}-ATPase, using a galactose-regulated promoter. The expression was improved by optimizing the number of galactose inductions and modifying the host yeast to increase the amount of the Gal4p transcription factor, which controls the promoter behind which the SERCA gene was inserted. It was necessary to reduce the temperature of expression from 28°C to 18°C in order to enhance recovery of solubilized Ca^{2+}-ATPase, which could then be fully solubilized with L-α-lysophosphatidylcholine or by 50% with *n*-dodecyl-β-D-maltoside. FIGURE 1 displays an electron-microscope picture that shows that the expressed protein is located mainly in periendoplasmic membranes. A 4-L culture produced 100 mg of Ca^{2+}-ATPase, with 60 and 22 mg being pelleted with the heavy and light membrane fractions, respectively. The expressed Ca^{2+}-ATPase was further purified by metal affinity chromatography as described by Lenoir *et al.* elsewhere in this volume.[9]

Address for correspondence: Pierre Falson, CEA Saclay, DSV/DBJC/SBFM, URA 2096 CNRS, Bât 528, 91191 Gif-sur-Yvette, France. Voice: +33 1 6908 9882; fax: +33 1 6933 1351.
pierre.falson@cea.fr

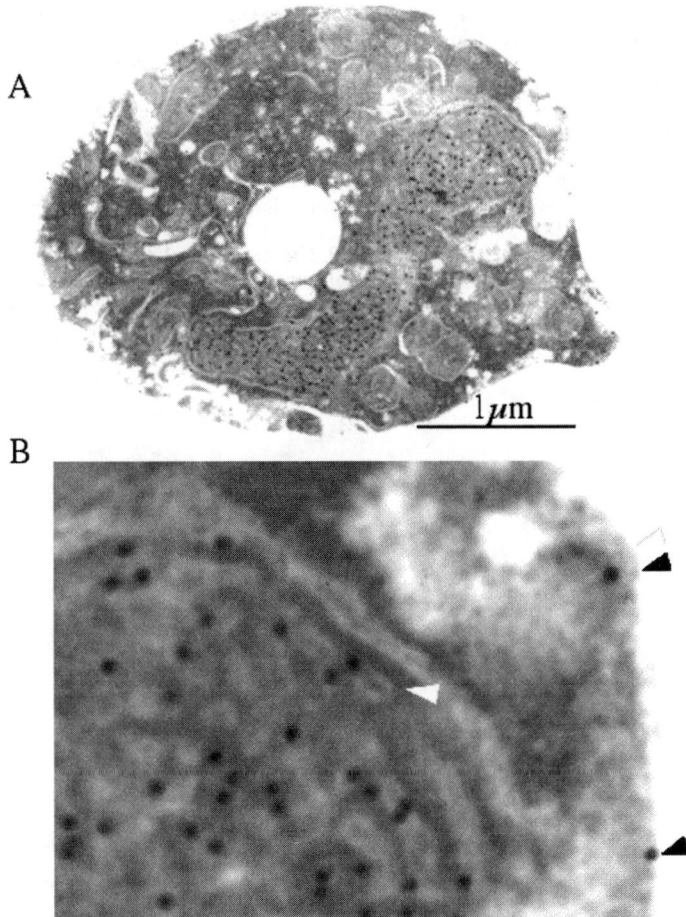

FIGURE 1. Localization of the Ca^{2+}-ATPase overexpressed in yeast. **(A)** W303.1b/Gal4 [pYeDP60-6HCSERCA1a] was grown in YPGE-2 medium for 36 h at 28°C and expression was then induced with galactose at time 0 and again 13 h later. Yeast cells were harvested after 19 h of expression, fixed and immunolabeled, and observed by electron microscopy. **(B)** Fourfold magnification of the upper-right corner of panel A.

REFERENCES

1. CLARKE, D.M., T.W. LOO, G. INESI & D.H. MACLENNAN. 1989. Location of high affinity Ca2+-binding sites within the predicted transmembrane domain of the sarcoplasmic reticulum Ca2+-ATPase. Nature **339:** 476–478.
2. ZHANG, Z., D. LEWIS, C. STROCK, *et al.* 2000. Detailed characterization of the cooperative mechanism of Ca(2+) binding and catalytic activation in the Ca(2+) transport (SERCA) ATPase. Biochemistry **39:** 8758–8767.
3. SKERJANC, I.S., T. TOYOFUKU, C. RICHARDSON & D.H. MACLENNAN. 1993. Mutation of glutamate 309 to glutamine alters one Ca(2+)-binding site in the Ca(2+)-ATPase of sarcoplasmic reticulum expressed in Sf9 cells. J. Biol. Chem. **268:** 15944–15950.

4. KARON, B.S., J.M. AUTRY, Y. SHI, *et al.* 1999. Different anesthetic sensitivities of skeletal and cardiac isoforms of the Ca-ATPase. Biochemistry **38:** 9301–9307.
5. CENTENO, F., S. DESCHAMPS, A.M. LOMPRE, *et al.* 1994. Expression of the sarcoplasmic reticulum Ca(2+)-ATPase in yeast. FEBS Lett. **354:** 117–122.
6. DEGAND, I., P. CATTY, E. TALLA, *et al.* 1999. Rabbit sarcoplasmic reticulum Ca(2+)-ATPase replaces yeast PMC1 and PMR1 Ca(2+)-ATPases for cell viability and calcineurin-dependent regulation of calcium tolerance. Mol. Microbiol. **31:** 545–556.
7. REIS, E.M., E. KURTENBACH, A.R. FERREIRA, *et al.* 1999. N-terminal chimeric constructs improve the expression of sarcoplasmic reticulum Ca(2+)-ATPase in yeast. Biochim. Biophys. Acta **1461:** 83–95.
8. POMPON, D., B. LOUERAT, A. BRONINE & P. URBAN. 1996. Yeast expression of animal and plant P450s in optimized redox environments. Methods Enzymol. **272:** 51–64.
9. LENOIR, G., T. MENGUY, F. CORRE, *et al.* 2002. Overproduction in yeast and rapid and efficient purification of the rabbit SERCA1a Ca(2+)-ATPase. Biochim. Biophys. Acta **1560:** 67–83.

TNP-8N$_3$-ADP Photoinactivation of the Phosphatase Activity of FITC-Modified SERCA

G. BARRIENTOS,[a] M. TAYLOR,[b] C. HIDALGO,[a] AND J. D. CAVIERES[b]

[a]*Departamento de Biología Celular y Molecular, ICBM, Facultad de Medicina, Universidad de Chile, Santiago, Chile*

[b]*Department of Cell Physiology and Pharmacology, University of Leicester, Leicester LE1 9HN, United Kingdom*

KEYWORDS: SERCA; Ca^{2+} pump; TNP-8N$_3$-ADP; photoinactivation; FITC; phosphatase

INTRODUCTION

SERCA is activated by ATP at micromolar, submillimolar, and millimolar concentrations.[1] At equilibrium, SERCA binds 1 mole of TNP-AMP, but 2 moles of TNP-ATP per mole of phosphorylation sites.[2] TNP-AMP can occupy the N-domain pocket in the 2.6-Å structure.[3] Deep into the pocket, the anchoring point[4] for FITC is found: Lys515.

The FITC modification suppresses the ATPase activity, but not E2 reactions like phosphatase activity and P$_i$ phosphorylation. Solubilization of FITC-modified SERCA with C$_{12}$E$_9$ does not increase its ATPase activity[4] as is found with Na,K-ATPase.[5] The implications are that all SERCA units bind FITC and that phosphatase substrate binding occurs elsewhere. We now show that TNP-8N$_3$-ADP can photoinactivate this surviving phosphatase activity.

METHODS

Rabbit skeletal muscle triads were prepared according to Hidalgo et al.[6] ATPase activities were determined at 3 mM [γ^{32}P]-ATP, and phosphatase activities at 10 mM p-nitrophenyl-phosphate,[5] both with and without 100 μM Ca^{2+}. The basic reaction medium contained 100 mM KCl, 20 mM Tris (pH 7.5), 5 mM MgCl$_2$, and 100 μM EGTA (solution A). FITC inactivation was done at room temperature, at 20 μM FITC and ca. 10 μM SERCA, in 50 mM Tris/Cl$^-$ (pH 9.0) plus 100 μM CaCl$_2$. TNP-

---

Address for correspondence: Dr. J.D. Cavieres, Department of Cell Physiology and Pharmacology, University of Leicester, Leicester LE1 9HN, United Kingdom. Voice: +44-116 2523091.

jdc7@le.ac.uk

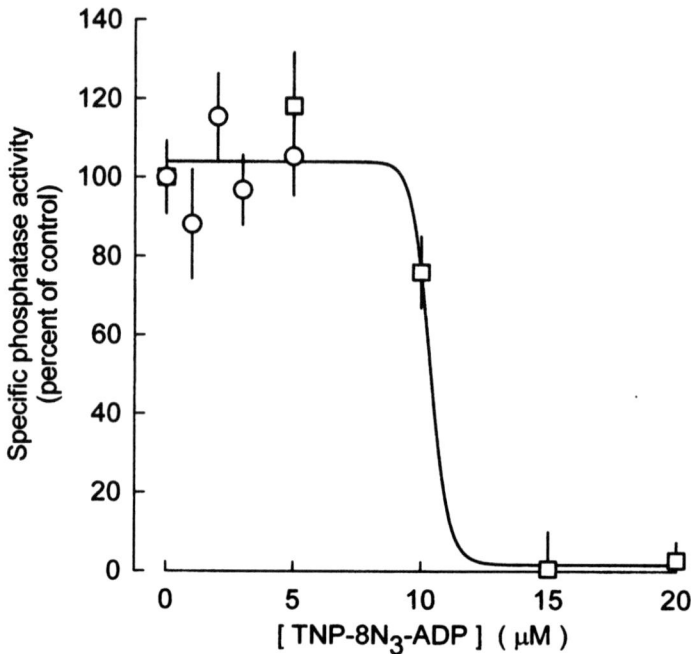

FIGURE 1. Photoinactivation of the Ca^{2+}-phosphatase activity of FITC-modified triads by TNP-8N$_3$-ATP.

8N$_3$-ADP was synthesized[7] and used for photoinactivation at 312 nm and at room temperature in solution A plus 200 µM CaCl$_2$. Protein was determined as described.[5]

RESULTS AND DISCUSSION

After 5 min with FITC, the Ca^{2+}-ATPase activity typically decreased to around 1%, while the Ca^{2+}-phosphatase activity remained at 85–90%. The results of two experiments to photoinactivate this Ca^{2+}-phosphatase activity at variable TNP-8N$_3$-ADP concentrations are shown in FIGURE 1. The lack of inactivation below 5 µM is probably due to high-affinity TNP-8N$_3$-ADP titration by other triad proteins.

If FITC fully blocks the ATP-binding pocket on SERCA's N-domain (where otherwise TNP-AMP binds with high affinity[2,3]), then TNP-8N$_3$-ADP should be found elsewhere on SERCA.

ACKNOWLEDGMENTS

This work was supported by research and travel grants from The Wellcome Trust (to J. D. Cavieres), by research grants from CONICYT-CHILE (to C. Hidalgo), and by a European Union Grant to Young Researchers (to G. Barrientos and M. Taylor).

REFERENCES

1. SCHATZMANN, H.J. 1989. The calcium pump of the surface membrane and of the sarcoplasmic reticulum. Annu. Rev. Physiol. **51:** 473–485.
2. SUSUKI, H., T. KUBOTA, K. KUBO & T. KANAZAWA. 1990. Existence of a low-affinity ATP-binding site in the unphosphorylated Ca^{2+}-ATPase of sarcoplasmic reticulum vesicles: evidence from binding of 2′,3′-(2,4,6-trinitrocyclohexadienylidene)-[^{3}H]AMP and -[^{3}H]ATP. Biochemistry **29:** 7040–7045.
3. TOYOSHIMA, CH., M. NAKASAKO, H. NOMURA & H. OGAWA. 2000. Crystal structure of the calcium pump of sarcoplasmic reticulum at 2.6 Å resolution. Nature **405:** 647–655.
4. MITCHINSON, C., A.F. WILDERSPIN, B.J. TRINAMMAN & N.M. GREEN. 1982. Identification of a labelled peptide after stoichiometric reaction of fluorescein isothiocyanate with the Ca^{2+}-dependent adenosine triphosphatase of sarcoplasmic reticulum. FEBS Lett. **146:** 87–92.
5. WARD, D.G. & J.D. CAVIERES. 1998. Photoinactivation of FITC-modified Na,K-ATPase by TNP-8N$_3$-ATP: abolition of E1 and E2 partial reactions by sequential block of high- and low-affinity nucleotide sites. J. Biol. Chem. **273:** 14277–14284.
6. HIDALGO, C., J. JORQUERA, V. TAPIA & P. DONOSO. 1993. Triads and transverse tubules isolated from skeletal muscle contain high levels of inositol 1,4,5-trisphosphate. J. Biol. Chem. **268:** 15111–15117.
7. SEEBREGTS, C.J. & D.B. MCINTOSH. 1989. 2′,3′-(2,4,6-Trinitrophenyl)-8-azido-adenosine mono-, di-, and triphosphates as photoaffinity probes of the Ca^{2+}-ATPase of sarcoplasmic reticulum. J. Biol. Chem. **264:** 2043–2052.

Ca^{2+} Occlusion of Sarcoplasmic Reticulum Ca^{2+}-ATPase by CrATP

BIRTE STÆHR JUUL AND JESPER V. MØLLER

Department of Biophysics, University of Aarhus, DK-8000 Aarhus C, Denmark

KEYWORDS: Ca^{2+}-ATPase; CrATP

The use of CrATP as a substitute of the physiological substrate, MgATP, has played a major role in the investigation of certain aspects of the transport mechanism of sarcoplasmic reticulum Ca^{2+}-ATPase: (1) to stabilize the ephemeral Ca^{2+}-occluded state to elucidate its characteristics and (2) to identify intramembrane residues involved in the high-affinity binding of the two transported Ca^{2+} by site-directed mutagenesis.[1] Despite intensive studies,[2,3] certain aspects of the reaction of Ca^{2+} with CrATP-complexed ATPase remain unclarified. This includes the existence of sequential or separate mechanisms for uptake and release of the two occluded calcium ions, and a definition of the access pathways to the intramembranous sites in the occluded state.

We initially found that, at 23°C, $^{45}Ca^{2+}$ added to the medium only exchanged with half of the unlabeled occluded Ca^{2+} over a period of 24 h, consistent with occlusion by a sequential and ordered mechanism. However, the outlook changed significantly when we changed to a procedure introduced by Coan et al.[3] in which the occlusion process is performed at 37°C to speed up reaction rates. In this case, occlusion was complete after 1–2 h when Ca^{2+}-ATPase was exposed to CrATP and $^{45}Ca^{2+}$ simultaneously (FIG. 1, ■). However, when the enzyme-CrATP complex with empty occluded sites was first formed in the presence of EGTA, subsequent $^{45}Ca^{2+}$ occlusion was clearly biphasic, reflecting stepwise uptake of the two occluded calcium ions (●). A biphasic pattern was also observed when $^{45}Ca^{2+}$ uptake was examined at steady state in exchange with fully occluded cold Ca^{2+} (★). This suggests that uptake of Ca^{2+} from the medium is not affected by the presence of Ca^{2+} in the occluded CrATP–Ca^{2+}-ATPase complex, indicating that exchange is a consequence of an infrequent opening of the cytosolic gate(s) by a Ca^{2+}-independent mechanism.

Further studies have revealed complementary features between release and uptake of occluded Ca^{2+} at 37°C. Under these conditions, one Ca^{2+} is relatively rapidly released without being affected by medium Ca^{2+} concentration (10^{-8}–10^{-3} M). The other occluded calcium ion is more slowly released and is bound with a very high

Address for correspondence: Birte Stæhr Juul, Department of Biophysics, University of Aarhus, DK-8000 Aarhus C, Denmark. Voice: +45-8942-2934; fax: +45-8612-9599.
bj@biophys.au.dk

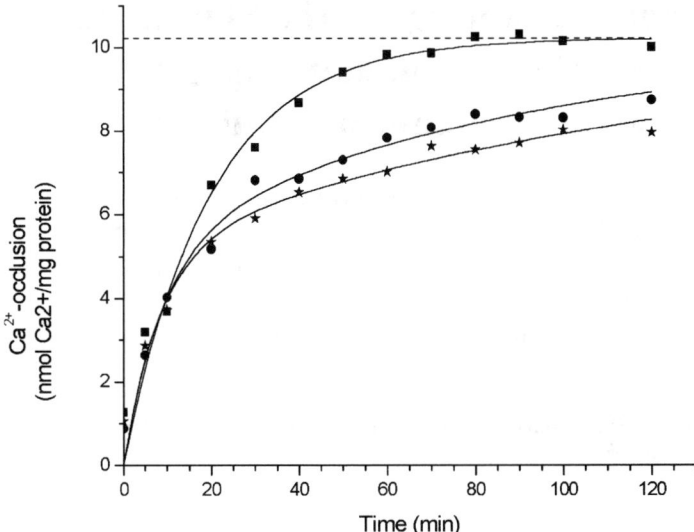

FIGURE 1. Occlusion of Ca^{2+} induced by complexation of Ca^{2+}-ATPase with CrATP at 37°C: (■) occlusion by purified Ca^{2+}-ATPase incubated with 0.6 mM $^{45}Ca^{2+}$, 0.5 mM EGTA, and 0.8 mM CrATP; (●) occlusion of Ca^{2+} by Ca^{2+}-ATPase preincubated with 0.5 mM EGTA and 0.8 mM CrATP for 90 min, before addition of 0.6 mM $^{45}Ca^{2+}$; (★) exchange of $^{45}Ca^{2+}$ added after preincubation with 0.6 mM cold Ca^{2+}, 0.5 mM EGTA, and 0.8 mM CrATP for 90 min, followed by addition of $^{45}Ca^{2+}$. All reactions took place in 50 mM Mes buffer (pH 7.0) at 37°C.

affinity ($K_d \approx 10^{-8}$ M). The release of Ca^{2+} from this site is only slightly dependent on the medium concentration of free Ca^{2+} and hence is independent of occupancy of the second site by Ca^{2+}. We conclude that uptake and release of Ca^{2+} from CrATP-reacted Ca^{2+}-ATPase basically occur at two independent sites with different affinities as previously suggested by Coan et al.[3] This would suggest the presence of different cytosolic gates for the two occluded calcium ions, but further work is needed to settle this question.

REFERENCES

1. VILSEN, B. & J.P. ANDERSEN. 1992. CrATP-induced Ca^{2+} occlusion mutants of sarcoplasmic reticulum. J. Biol. Chem. **267:** 25739–25743.
2. VILSEN, B. & J.P. ANDERSEN. 1992. Interdependence of Ca^{2+} occlusion sites in the unphosphorylated sarcoplasmic reticulum Ca^{2+}-ATPase complex with CrATP. J. Biol. Chem. **267:** 3539–3550.
3. COAN, C., J. JI-YING & J.A. AMARAL. 1994. Ca^{2+} binding to occluded sites in the CrATP-ATPase complex of sarcoplasmic reticulum: evidence for two independent high-affinity sites. Biochemistry **33:** 3712–3721.

A Model Accounting for the Simultaneous Transport of Calcium and Manganese in Sarcoplasmic Reticulum Membranes

DÉBORA ALEJANDRA GONZÁLEZ,[a] MARIANO ANÍBAL OSTUNI,[a,b] JEAN-JACQUES LACAPÈRE,[b] AND GUILLERMO LUIS ALONSO[a]

[a]*Cátedra de Biofísica, Facultad de Odontología, Universidad de Buenos Aires, Buenos Aires, Argentina*

[b]*Inserm U410, Faculté Xavier Bichat, 75870 Paris Cedex 18, France*

KEYWORDS: sarcoplasmic reticulum (SR); Ca^{2+}-ATPase; calcium; manganese; skeletal muscle

The ATPases in endosarcoplasmic reticulum membranes actively transport calcium to the reticular lumen. In skeletal muscle sarcoplasmic reticulum (SR), the Ca^{2+}-ATPase (SERCA 1 isoform) translocates up to two Ca^{2+} per one hydrolyzed ATP. It also transports Mn^{2+},[1] although this is not its physiological function. However, there are Ca,Mn-ATPases localized in the medial-Golgi compartment (PMR1, such as found in the yeast *Saccharomyces cerevisiae*) that could be important for regulating cytoplasmic Mn concentration.[2] Using isolated SR vesicles from rabbit skeletal muscle,[3] we investigated Ca-Mn competition, both for the binding to the Ca transport sites and for ATP-dependent uptake into the vesicles. We measured (^{45}Ca)Ca binding to the SR Ca-ATPase as a function of Mn concentration. Mn displaces Ca from the two high-affinity transport sites. The best curves were obtained simulating the following binding model, which includes the simultaneous binding of both cations:

$$E.Mn_2 \underset{K_2Mn}{\longleftrightarrow} E.Mn \underset{K_1Mn}{\longleftrightarrow} E \underset{K_1Ca}{\longleftrightarrow} E.Ca \underset{K_2Ca}{\longleftrightarrow} E.Ca_2$$

$$\updownarrow K_3Ca$$

$$E.Mn.Ca$$

where K_1Ca = 10 µM, K_2Ca = 0.1 µM, K_3Ca = 20 µM, K_1Mn = 0.5 mM, and K_2Mn = 20 mM.

Address for correspondence: Débora Alejandra González, Cátedra de Biofísica, Facultad de Odontología, Universidad de Buenos Aires, M.T. De Alvear 2142, (C1122AAH) Buenos Aires, Argentina. Voice/fax: +54-11-4964-1298.

debora@biofis.odon.uba.ar

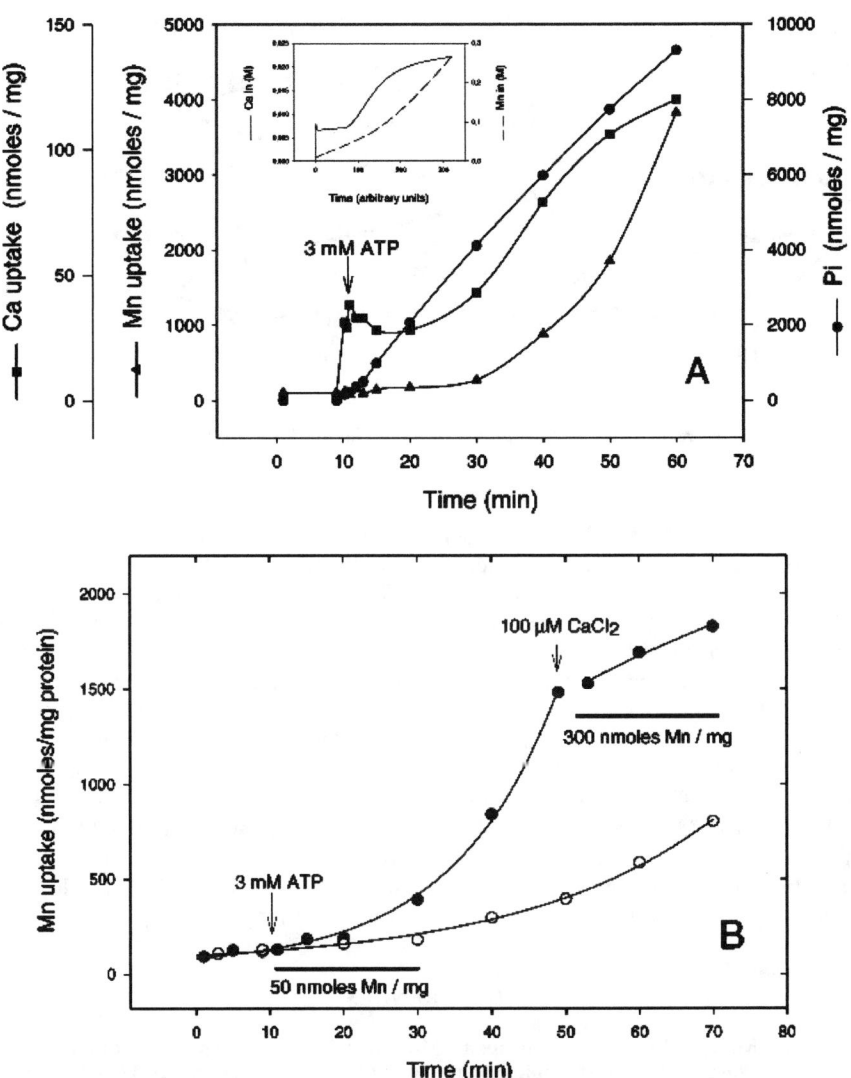

FIGURE 1. (A) Ca and Mn uptake and P_i production by SR vesicles in the presence of 0.1 M MOPS-Tris (pH 7.2), 0.1 M KCl, 10 mM (^{54}Mn)MnCl$_2$, and 20 µM added (^{45}Ca)CaCl$_2$. Inset: Numerical simulation of uptaken Ca and Mn according to the proposed model. (B) Mn uptake with (●) or without (○) 0.1 mM added CaCl$_2$. Horizontal bars indicate the time used to compare Mn accumulation in the presence of Ca in both curves.

Panel A in FIGURE 1 shows simultaneous Ca and Mn uptake by the SR Ca-ATPase as well as P_i production. Ca uptake during long periods shows a biphasic pattern. The second phase was associated with precipitation of calcium phosphate. Mn uptake exhibits an initial very low velocity followed by a progressive increase, reaching a high level of accumulated Mn. At long times, we always observed that uptaken Mn

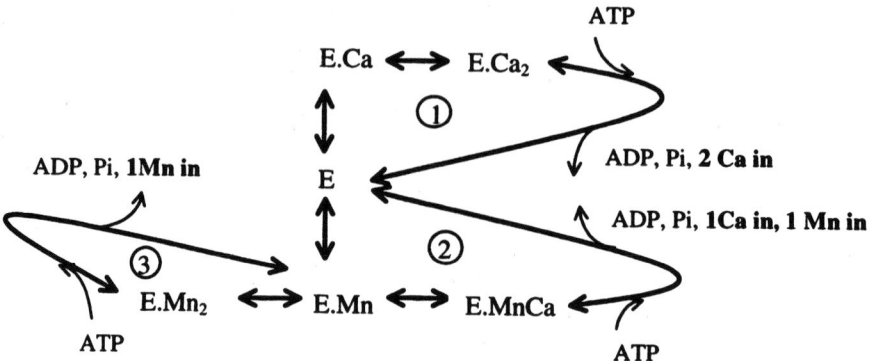

FIGURE 2. A model for the transport of Ca and Mn. See text for details.

is coupled 1:1 to ATP hydrolysis. Panel B shows that Mn uptake in the presence of added Ca is slower than in its absence, but here the late Ca addition does not reduce the rate of Mn uptake to the same extent. This result suggests a slow exchange between the E and E.Mn forms. The same conclusion can be deduced from Chiesi and Inesi's work.[1]

Taking into account all our results, we propose a model for the transport of Ca and Mn (FIG. 2). The initial high Ca uptake (cycle 1) is reduced when Mn drives the enzyme to cycle 2, which has a lower turnover. The vesicles lose Ca by diffusion until enough P_i is formed to precipitate internal Ca. A fraction of the molecules are retained in cycle 2 due to the very slow E.Mn dissociation. When cytoplasmic Ca is reduced by transport, Mn has more chance to bind to E.Mn and the rate of Mn uptake (through cycle 3) increases. We simulated numerically[4] the temporal evolution of Ca and Mn uptake, assigning kinetic constants to the model. Our results reproduce the pattern of the experimental curves (inset in panel A of FIG. 1).

REFERENCES

1. CHIESI, M. & G. INESI. 1980. Adenosine 5′-triphosphate dependent fluxes of manganese and hydrogen ions in sarcoplasmic reticulum vesicles. Biochemistry **19:** 2912–2918.
2. DÜRR, G. *et al.* 1998. The medial-Golgi ion pump Pmr1 supplies the yeast secretory pathway with Ca^{2+} and Mn^{2+} required for glycosylation, sorting, and endoplasmic reticulum–associated protein degradation. Mol. Biol. Cell **9:** 1149–1162.
3. CHAMPEIL, P. *et al.* 1985. Interaction of magnesium and inorganic phosphate with calcium-deprived sarcoplasmic reticulum adenosinetriphosphatase as reflected by organic solvent induced perturbation. Biochemistry **24:** 69–81.
4. HECHT, J.P., J.M. NIKONOV & G.L. ALONSO. 1990. A BASIC program for the numerical solution of the transient kinetics of complex biochemical models. Comput. Methods Programs Biomed. **33:** 13–20.

Interaction of an Aromatic Dibromo-Isothiouronium Derivative with the Ca-ATPase of Sarcoplasmic Reticulum

MERVYN C. BERMAN[a] AND STEVEN J. KARLISH[b]

[a]*Division of Chemical Pathology, University of Cape Town, Cape Town, South Africa*
[b]*Department of Biochemistry, Weizmann Institute, Rehovot, Israel*

KEYWORDS: sarcoplasmic reticulum; Ca^{2+}-ATPase; Br_2-TITU; E-P hydrolysis; TNP-ATP superfluorescence

Isothiouronium compounds act as high-affinity competitive antagonists for Na^+ and $K^+(Rb^+)$ cations of the renal Na^+/K^+-ATPase. The most potent inhibitor is 1,3-dibromo-2,4,6-tris(methylisothiouronium)benzene (Br_2-TITU). It binds to cation sites with a K_d of 0.35 µM.[1] The Ca^{2+}-ATPase of sarcoplasmic reticulum (SR) is known to have a K^+ site that activates E2-P hydrolysis with a K_d of 20–50 mM.[2–5]

Br_2-TITU inhibits both Ca^{2+} transport and Ca^{2+}-dependent ATPase with a $K_{0.5}$ of 10–30 µM. The effects of Br_2-TITU on total E-P were investigated with excess Ca^{2+} and ATP. During steady state at pH 6.8, total E-P increased from 0.5 to 5.5 nmol/mg. Inhibition was time-dependent, with a $t_{0.5}$ of ~3 minutes.

When turnover was inhibited by 100 µM Br_2-TITU by 80–90% activity, autofluorescence of endogenous Trp fluorescence was also quenched by EGTA, but it still responded to binding and release of Ca^{2+}. Inhibition of Trp fluorescence indicates that a global conformational change occurs and that E1/E2 transitions are retained.

The effects of Br_2-TITU on the exogenous fluophore, TNP-ATP, which monitors the environment of the ATP site and whose superfluorescence is related to the accumulation of $E2\text{-}P.2Ca^{2+}$, were investigated in both forward and back reactions of the Ca^{2+} pump. Br_2-TITU enhanced superfluorescence 2- to 5-fold under different conditions. Br_2-TITU thus favors accumulation of E2-P both in steady state (FIG. 1) and at equilibrium.

The kinetics of formation and decay of superfluorescence were studied by stopped flow. Br_2-TITU increased rate constants of the "on" reaction from ATP plus Ca^{2+} from 0.3 to 0.7 s^{-1}. "Off" reactions were decreased from 2.9 to 0.7 s^{-1} from ATP, and from 10.9 to 0.73 s^{-1} for P_i quench. The kinetics show that enhanced total E-P and E2-P are favored by increased rates of formation and decreased rates of E2-P decay.

Address for correspondence: Mervyn C. Berman, Division of Chemical Pathology, University of Cape Town, Cape Town, South Africa. Voice: +27-21-4066354; fax: +27-21-44884500.
mervyn@chempath.uct.ac.za

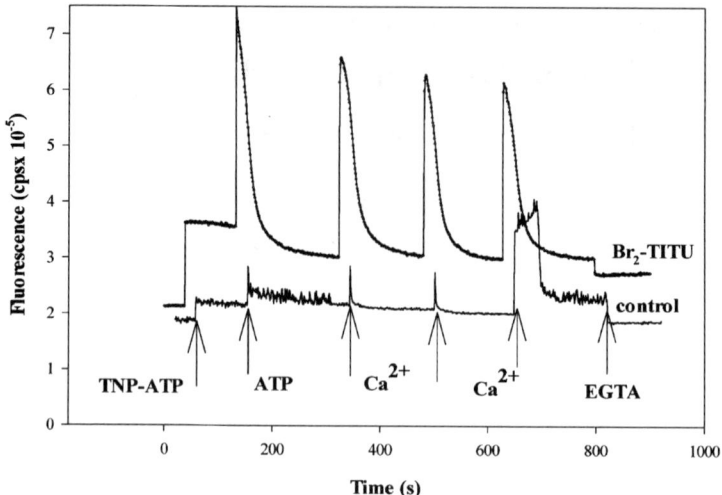

FIGURE 1. Effects of Br_2-TITU on TNP-ATP superfluorescence following pulsed additions of Ca^{2+}. SR vesicles were preincubated at 25°C for 20 min without and with 50 μM Br_2-TITU, following which fluorescence was recorded after 2.5 μM TNP-ATP, 2 mM ATP, and three pulses of 20 μM Ca^{2+} were added. The last addition of Ca^{2+} was increased to 100 μM Ca^{2+} in the control. EGTA, 2 mM, was added at the end of the experiment.

Inhibition of the Ca^{2+}-ATPase may be explained by a two-step reaction. Initially, Br_2-TITU could partition rapidly into the lipid annulus, and then more slowly alter the Ca^{2+}-ATPase conformation, leading to inhibition of E2-P hydrolysis.

Br_2-TITU should be a useful probe for inhibition of the Ca^{2+}-ATPase. It maximizes total E-P formation under conditions that are nonideal. The inhibitor may also be used to favor E2-P for structural studies.

REFERENCES

1. HOVING, S., M. BAR-SHIMON, J.J. TIJMES, et al. 1995. Novel aromatic isothiouronium derivatives which act as high affinity competitive antagonists of alkali metal cation on Na/K-ATPase. J. Biol. Chem. **270:** 29788–29793.
2. MCINTOSH, D.B. & P.D. BOYER. 1983. Adenosine 5′-triphosphate modulation of catalytic intermediates of calcium ion activated adenosinetriphosphatase of sarcoplasmic reticulum subsequent to enzyme phosphorylation. Biochemistry **22:** 2867–2875.
3. DAVIDSON, G. & M.C. BERMAN. 1987. Phosphoenzyme conformational states and nucleotide-binding site hydrophobicity following thiol modification of the Ca^{2+}-ATPase of sarcoplasmic reticulum from skeletal muscle. J. Biol. Chem. **262:** 7041–7046.
4. BISHOP, J.E., R.K. NAKOMOTO & G. INESI. 1986. Modulation of binding characteristics of a fluorescent derivative to the sarcoplasmic reticulum adenosinetriphosphatase. Biochemistry **25:** 696–703.
5. BERMAN, M.C. 1999. Regulation of Ca^{2+} transport by sarcoplasmic reticulum of Ca^{2+}-ATPase at limiting [Ca^{2+}]. Biochim. Biophys. Acta **1418:** 48–60.

Time-Resolved Partial Reactions of the SR Ca-ATPase Investigated with a Fluorescent Styryl Dye

CHRISTINE PEINELT AND HANS-JÜRGEN APELL

Department of Biology, University of Konstanz, 78457 Konstanz, Germany

KEYWORDS: SR Ca-ATPase; ion binding; kinetics; fluorescent styryl dye

The SR Ca-ATPase pumps two Ca^{2+} per ATP hydrolyzed into the lumen of the SR in exchange for two H^+. So far, only fluorescent styryl dyes may be used to detect electrogenic ion movements into and out of the binding sites of the pump.[1,2] Time-resolved kinetics can be analyzed by ATP-concentration jump experiments in which ATP is released from its inactive precursor caged-ATP by fast UV-laser flashes.[3] After release of ATP, a rising and a falling fluorescence intensity could be observed. The pump proceeds through its pump cycle and, as function of pH, it reaches a new steady state in a more or less protonated P-E_2 form (FIG. 1).

The time constant of the rising phase was 35 ± 5 ms. This constant was independent of substrate concentrations when $Ca^{2+} > 100$ nM (Ca^{2+} binding not limiting), when pH < 8 (at pH > 8, the ATP-release reaction is limiting; see FIG. 1), and when ATP > 20 µM (ATP binding not limiting).

The time constant of the falling phase had a minimum of 1.38 ± 0.29 s at pH 7, independent of the Ca^{2+} concentration. At higher pH, the H^+ concentration becomes a limiting factor and, at pH < 6.5, the affinity of the protein for H^+ is decreased.

The Arrhenius plots of the time constants of both fluorescence phases showed activation energies of 80–90 kJ/mol, comparable to the activation energy of E_1-E_2 transitions in the Na,K-ATPase. Therefore, the rate-limiting step of the rising phase is assumed to be a conformation transition included in the following scheme:[4]

$$(Ca_2)E_1\text{-}P \rightarrow P\text{-}E_2Ca_2 \rightarrow P\text{-}E_2{}^*Ca \rightarrow P\text{-}E_2Ca \rightarrow P\text{-}E_2.$$

On the basis of the very slow falling phase of the fluorescence, which reflects H^+ binding, the reaction sequence from the P-E_2 state onwards has to be supplemented by an additional step, assumed to be a slow conformational relaxation that precedes H^+ binding, such as $\cdots \rightarrow P\text{-}E_2{}^*s \rightarrow P\text{-}E_2 \rightarrow P\text{-}E_2H_2$.

Address for correspondence: Christine Peinelt, Department of Biology, University of Konstanz, Fach M635, 78457 Konstanz, Germany. Voice: +49-7531-882901; fax: +49-7531-883183.

Christine.Peinelt@uni-konstanz.de

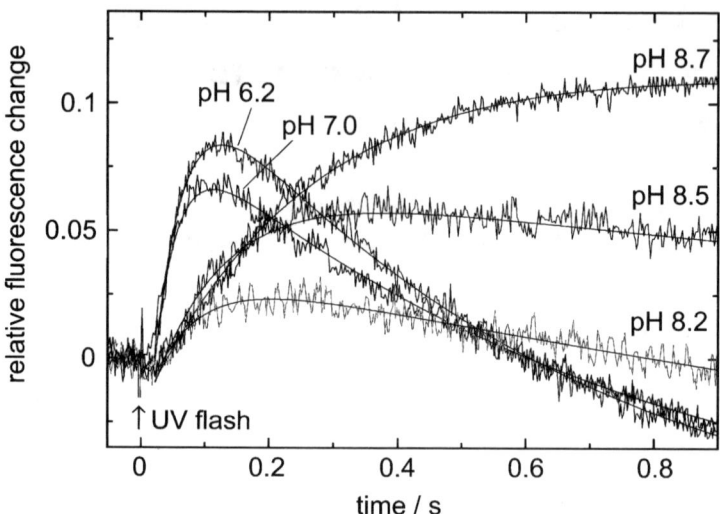

FIGURE 1. Fluorescence changes induced by the release of about 20 μM ATP. The UV flash occurred at time $t = 0$. The buffer contained 25 mM tricine, 50 mM KCl, 1 mM $MgCl_2$, 650 nM styryl dye 2BITC,[2] and 18 μg of SR Ca-ATPase. pH was adjusted according to the respective labels. The time course of the fluorescence was fitted with the sum of two exponentials. The constant level of fluorescence after times longer than 1 s was dependent on the steady state distribution into which the ion pumps relaxed after the ATP-concentration jump and was controlled by the amount of H^+ ions bound in the new steady state.

ACKNOWLEDGMENTS

This work was supported by the Deutsche Forschungsgemeinschaft (AP 45/4).

REFERENCES

1. BUTSCHER, C., M. ROUDNA & H-J. APELL. 1999. Electrogenic partial reactions of the SR-Ca-ATPase investigated by a fluorescence method. J. Membr. Biol. **168:** 169–181.
2. PEINELT, C. & H-J. APELL. 2002. Kinetics of the Ca^{2+}, H^+, and Mg^{2+} interaction with the ion-binding sites of the SR-Ca-ATPase. Biophys. J. **82:** 170–181.
3. HEYSE, S. *et al.* 1994. Partial reactions of the Na,K-ATPase: determination of rate constants. J. Gen. Physiol. **104:** 197–240.
4. INESI, G. & L. DE MEIS. 1989. Regulation of steady state filling in sarcoplasmic reticulum. J. Biol. Chem. **264:** 5929–5936.

Macrocyclic Carbon Suboxide Oligomers as Potent Inhibitors of the Na,K-ATPase

ROBERT STIMAC,[a] FRANZ KEREK,[b] AND HANS-JÜRGEN APELL[a]

[a]*Department of Biology, University of Konstanz, 78457 Konstanz, Germany*

[b]*MPI for Biochemistry, 82152 Martinsried, Germany*

KEYWORDS: macrocyclic carbon suboxide (MCS); sodium pump; inhibition; partial reactions

A new class of Na,K-ATPase inhibitors are macrocyclic carbon suboxide (MCS) factors, which are cyclo-oligomers of C_3O_2.[1] They inhibit the enzymatic activity with half-inhibiting concentrations of 10 nM, 100-fold lower than ouabain. In contrast to cardiac steroids, they hardly discriminate between rat and rabbit enzyme. Our preparation consists of structurally different MCS factors and their complex with Na^+ ions. The composition of this equilibrium mixture and its activity can be modulated by acids and bases.

Experiments with the fluorescent styryl dye, RH421, were performed to analyze the inhibitory effect on partial reactions of the pump cycle.[2] Na,K-ATPase was prepared from the outer medulla of rabbit kidneys. The buffer contained 30 mM imidazole, 5 mM $MgSO_4$, 1 mM EDTA (pH 6.95), 100 nM RH421, and 3 µg/mL Na,K-ATPase (at 20±0.5 °C). Specific fluorescence levels could be assigned to defined states in the pump cycle of the Na,K-ATPase.[3]

With RH421, the effects of MCS factors on various partial reactions of the Na,K-ATPase were studied (FIG. 1A). When adding MCS factors in states E_1, Na_3E_1, and $P-E_2$, subsequent fluorescence levels were reduced in amplitude, but fluorescence changes displayed the same directions (FIG. 1B). Although no enzymatic activity remained in the presence of 7 µM MCS, Na^+ and ATP additions still induced fluorescence changes. This suggests that the Na,K-ATPase can still bind Na^+ and be phosphorylated by ATP. K^+ was able to induce fluorescence changes from all states. The only case when fluorescence did not change dramatically was when adding MCS factors to $E_2(K_2)$.

In the presence of 1 mM K^+, saturating amounts of Na^+ are able to shift the enzyme into state Na_3E_1, and subsequent addition of ATP will induce phosphorylation and produce a fluorescence increase (FIG. 1C). Adding MCS factors to $E_2(K_2)$ (FIG. 1D) again reduces the Na^+ signal amplitude, but also decelerates significantly Na^+-binding kinetics. Subsequent addition of ATP indicates that, in the presence of MCS factors, the enzyme is not effectively being phosphorylated anymore.

Address for correspondence: Robert Stimac, Department of Biology, University of Konstanz, Fach M635, 78457 Konstanz, Germany. Voice: +49-7531-882901; fax: +49-7531-883183.
Robert.Stimac@uni-konstanz.de

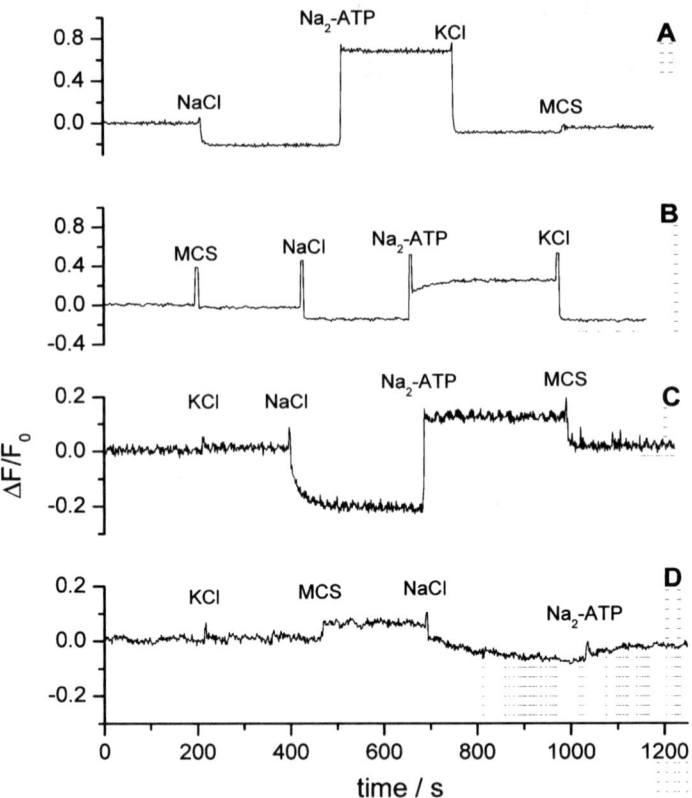

FIGURE 1. Substrate-induced partial reactions as detected by RH421 fluorescence experiments. Substrate additions are indicated: 50 mM NaCl, 100 μM Na$_2$ATP, 20 mM KCl (A, B) or 1 mM KCl (C, D), 7 μM MCS factors, and 50 μM ouabain. The effect of the MCS factors on the following protein conformation is shown: (A) $E_2(K_2)$, (B) E_1, (C) P-E_2/$E_2(K_2)$, and (D) $E_2(K_2)$.

From inhibition experiments with MCS factors and ouabain simultaneously, it was concluded that both inhibitors do not compete for the same binding site and that MCS factors inhibit the Na,K-ATPase reversibly in a state that follows in the Post-Albers cycle the ouabain-inhibited state.

In summary, MCS factors inhibit the Na,K-ATPase in an $E_2(K_2)$-like state in which enzyme phosphorylation is prevented. They do not compete with ouabain for the same binding site.

ACKNOWLEDGMENTS

This work was supported by the Deutsche Forschungsgemeinschaft (AP 45/8).

REFERENCES

1. KEREK, F. 2000. The structure of the digitalislike and natriuretic factors identified as macrocyclic derivatives of the inorganic carbon suboxide. Hypertens. Res. **23**(suppl.): S33–S38.
2. HEYSE, S. *et al.* 1994. Partial reactions of the Na,K-ATPase: determination of rate constants. J. Gen. Physiol. **104:** 197–240.
3. SCHNEEBERGER, A. & H-J. APELL. 1999. Ion selectivity of the cytoplasmic binding sites of the Na,K-ATPase: I. Sodium binding is associated with a conformational rearrangement. J. Membr. Biol. **168:** 221–228.

The Inherent Energy in SR Ca^{2+}-ATPase Is Convertible into Chemical Work

MAKOTO USHIMARU AND YOSHIHIRO FUKUSHIMA

Department of Chemistry, Kyorin University, School of Medicine, Mitaka, Tokyo, Japan

KEYWORDS: bioenergetics; energy conversion; Ca^{2+}-ATPase

For the ATP-ATPase system, the current principle of bioenergetics holds that the energy that can be converted into work by ATPase derives only from ATP and that the ATPase molecule is merely a device of energy conversion. However, is there any energy available for work in the ATPase molecule?[1] To answer this question, we have examined whether energy within the sarcoplasmic reticulum (SR) Ca^{2+}-ATPase molecule can be converted into chemical work during synthesis of the high-energy compound, ATP, from the low-energy compound, *p*-nitrophenylphosphate (pNPP):

$$E + pNPP \rightarrow EP + pNP,$$

$$EP + ADP \rightarrow ATP + E.$$

MATERIALS AND METHODS

SR microsomes (2 mg/mL) were incubated with 5 mM [^{32}P]-pNPP for 30 min at 20°C in 100 µL of a solution containing 10 mM $CaCl_2$, 5 µM A23187, 0.1 M KCl, 5 mM $MgCl_2$, and 10 mM HEPES (pH 8). After incubation, the reaction was cooled to about 0°C, and 10 µL of either 0.5 mM ATP or 1 mM ADP plus 0.5 mM ATP was added. All of the reactions were terminated by adding 0.6 mL of 0.6 M H_2SO_4 at the indicated times. The reaction solutions were centrifuged at 15,000 rpm for 5 min to separate protein precipitates from the reaction solution. Protein precipitates and supernatants were used for determining the amount of phosphoenzyme and ATP, respectively. The protein precipitate was filtered through Millipore membrane (pore size: 1.2 µm) and the radioactivity trapped on the membrane was determined by liquid scintillation counting. The supernatant was separated by an ion-pair reversed-phase HPLC column, which was equilibrated with 20 mM sodium phosphate (pH 7.0), 20% methanol, and 5 mM tetra-*n*-butylammonium as an ion-pair reagent.

Address for correspondence: Makoto Ushimaru, Department of Chemistry, Kyorin University, School of Medicine, 6-0-2 Shinkawa, Mitaka, Tokyo, Japan. Voice: +81-422-47-5511; fax: +81-422-44-1981.
ushimaru@kyorin-u.ac.jp

Nucleotides in the eluent were monitored by absorbance at 259 nm. The fraction that corresponded to ATP was collected and the amount of [^{32}P]-ATP was determined by liquid scintillation counting.

RESULTS AND DISCUSSION

ATP was synthesized from pNPP through the phosphoenzyme without a concentration gradient of Ca^{2+} across the SR membrane (FIG. 1). The amount of synthesized ATP was equal to the decrease in the amount of phosphoenzyme; in addition, the time course of ATP synthesis was symmetrical to that of phosphoenzyme decay (FIG. 1). These results indicate that ATP was synthesized from pNPP through the phosphoenzyme without an external source of energy and not through a direct exchange of P_i between pNPP and ATP.

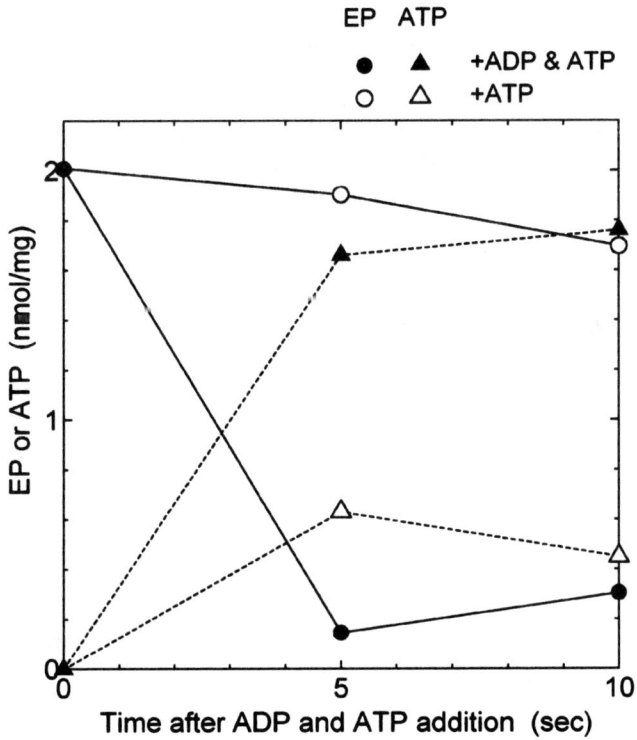

FIGURE 1. Time course of decay of phosphoenzyme and synthesis of ATP. The amount of [^{32}P]-phosphoenzyme remaining (*circles*) and the amount of [^{32}P]-ATP synthesized (*triangles*) at the indicated times after adding either 0.5 mM unlabeled ATP (*open symbols*) or 1 mM ADP plus 0.5 mM unlabeled ATP (*solid symbols*) to the reaction solution are shown.

The Gibbs' free-energy change of pNPP hydrolysis (4.5 kcal/mol) is about half of that of ATP (7 kcal/mol).[2] Because there was no energy input into the P_i exchange reaction, the enzyme molecule itself must have made up the deficiency in energy needed for ATP synthesis. Consequently, the ATPase molecule has inherent energy that can be converted into the chemical energy of ATP. These considerations have been already suggested by experiments carried out about 30 years ago in which ATP was synthesized from phosphoric acid through a Ca^{2+}-bound phosphoenzyme present on fragmented SR microsomes.[3,4]

The energy stored in Ca^{2+}-ATPase might be also converted to work needed for ion transport. If this is the case, then it raises certain questions. For example, if the ATPase molecule converts its own energy into work, then it must ultimately use up its stored energy through repetition of the catalytic cycle. Can this energy be recharged? If so, then how is this energy put back into the enzyme molecule?

REFERENCES

1. FUKUSHIMA, Y. & M. USHIMARU. 2002. General consideration of energy transduction by the ATPase-ATP system. Proc. Jpn. Acad. Ser. B **77:** 68–72.
2. INESI, G. 1971. *p*-Nitrophenyl phosphate hydrolysis and calcium ion transport in fragmented sarcoplasmic reticulum. Science **171:** 901–903.
3. DE MEIS, L. & M.G.C. CARVALHO. 1974. Role of the Ca^{2+} concentration gradient in the adenosine 5′–triphosphate-inorganic phosphate exchange catalyzed by sarcoplasmic reticulum. Biochemistry **13:** 5032–5038.
4. KNOWLES, A.F. & E. RACKER. 1975. Formation of adenosine triphosphatase from P_i and adenosine diphosphate by purified Ca^{2+}-adenosine triphosphatase. J. Biol. Chem. **250:** 1949–1951.

Purification of SERCA1a Ca^{2+}-ATPase Mutants Expressed in Yeast

GUILLAUME LENOIR, CÉDRIC MONTIGNY, MARC LE MAIRE, AND PIERRE FALSON

DSV/DBJC/SBFM and URA 2096 CNRS, CEA Saclay, 91191 Gif-sur-Yvette, France

KEYWORDS: yeast expression; membrane protein purification; Ca^{2+}-ATPase

The understanding of how membrane proteins work is often restricted because of the lack of sizable amounts of purified wild-type and mutant protein. All the systems used for heterologous expression of SERCA in COS cells or yeast produce limited amounts of SERCA1a. Nevertheless, site-directed mutagenesis studies on the sarcoplasmic reticulum (SR) Ca^{2+}-ATPase SERCA1a using such systems previously allowed identification of residues essential for the intramembranous binding of Ca^{2+}.[1] More recent studies have also highlighted the critical role for ATPase activation by Ca^{2+} of residues located in the extramembranous L6–7 loop.[2,3] A yeast expression system has been recently set up (see Ref. 4 and the report of Falson *et al.* in this volume) to produce large amounts of wild-type Ca^{2+}-ATPase. This system was used here to purify both the wild-type enzyme as well as three mutants with altered Ca^{2+}-binding residues (E309Q, E771Q, D800N) and a double mutant with an altered L6–7 loop (D813A/D818A).

A hexahistidine tag was introduced in the C-terminal of the SERCA protein, allowing the purification of the ATPase by Ni^{2+}-affinity chromatography. Starting from a membrane fraction containing 2% of expressed Ca^{2+}-ATPase, the Ca^{2+}-ATPase was first enriched to 4% by a stripping step carried out with 0.6 M KCl. Solubilization was subsequently performed using n-dodecyl-β-D-maltoside at a 3:1 to 5:1 detergent:protein ratio. The solubilized Ca^{2+}-ATPase was then submitted to Ni^{2+}-affinity chromatography (FIG. 1) and purified to about 50% homogeneity. Depending on the efficiency of solubilization, which was different for the various mutants, the amount of purified Ca^{2+}-ATPase finally obtained ranged from 0.5 to 1 mg per liter of yeast culture. The purified wild-type enzyme displayed the same K_m for Ca^{2+} and ATP as the native one, but a somewhat reduced specific ATPase activity of about 1.5 µmol ATP hydrolyzed/min/mg Ca^{2+}-ATPase, at pH 7.0 and 20°C. After removal of detergent with Bio-beads in the presence of phospholipids and 40% glycerol, it was stable and active for several months.[4] Once purified, ATPase with

Address for correspondence: Pierre Falson, DSV/DBJC/SBFM and URA 2096 CNRS, CEA Saclay, Bât 528, 91191 Gif-sur-Yvette, France. Voice: +33-1-6908-9882; fax: +33-1-6908-8139.
pierre.falson@cea.fr

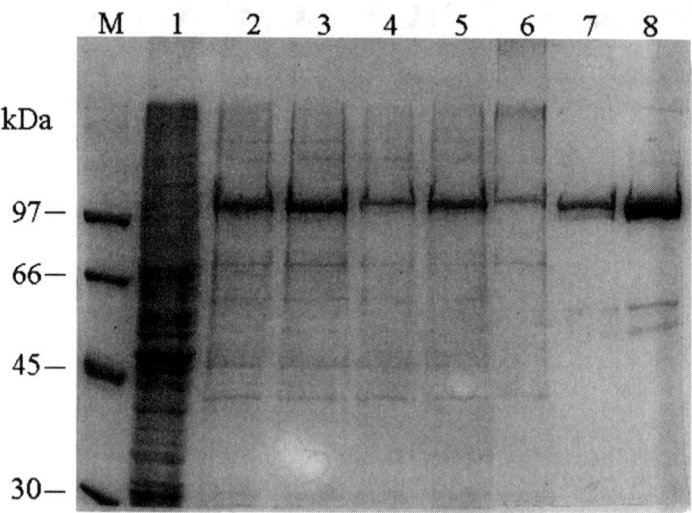

FIGURE 1. Coomassie blue–stained SDS-PAGE of the Ni^{2+}-NTA purified wild-type and mutated ATPases. Lane M corresponds to low molecular weight markers. Lane 1 corresponds to the initial KCl-stripped membranes. Lanes 2 to 6 correspond to purified wild-type ATPase or the E309Q, E771Q, D800N, or D813A/D818A mutant ATPases, respectively. Lanes 7 and 8 correspond to two different amounts of SR Ca^{2+}-ATPase, 1 and 3 µg, respectively.

mutated Ca^{2+}-binding sites (E309Q, E771Q, and D800N) displayed no noticeable Ca^{2+}-activated ATPase activity.

REFERENCES

1. CLARKE, D.M., T.W. LOO, G. INESI & D.H. MACLENNAN. 1989. Location of high affinity Ca^{2+}-binding sites within the predicted transmembrane domain of the sarcoplasmic reticulum Ca^{2+}-ATPase. Nature **339:** 476–478.
2. FALSON, P., T. MENGUY, F. CORRE, *et al.* 1997. The cytoplasmic loop between putative transmembrane segments 6 and 7 in sarcoplasmic reticulum Ca^{2+}-ATPase binds Ca^{2+} and is functionally important. J. Biol. Chem. **272:** 17258–17262.
3. MENGUY, T., F. CORRE, B. JUUL, *et al.* 2002. Involvement of the cytoplasmic loop L6–7 in the entry mechanism for transport of Ca^{2+} through the sarcoplasmic reticulum Ca^{2+}- ATPase. J. Biol. Chem. **277:** 13016–13028.

Mutational Analysis of Lys^{252} and Its Interaction with Loop 6–7 in the SR Ca^{2+}-ATPase

JOHANNES D. CLAUSEN AND JENS PETER ANDERSEN

Department of Physiology, University of Aarhus, Aarhus, Denmark

KEYWORDS: Ca^{2+}-ATPase; mutagenesis; lysine residue; Ca^{2+} dissociation; Ca^{2+} entrance

In a recent study from this laboratory,[1] it was found that the substitution of residues Lys^{252}LeuAspGlu^{255} in transmembrane segment M3 of the SR Ca^{2+}-ATPase with the corresponding residues of the Na^{+},K^{+}-ATPase, GluIleGluHis, resulted in a conspicuous increase of the rate of Ca^{2+} dissociation from the high-affinity binding sites of the E_1 form (i.e., toward the cytoplasmic side), as well as an increase of the rate of Ca^{2+} binding with associated conformational changes, suggesting a role for this segment in control of the gateway to the Ca^{2+} sites. Interestingly, the side chain of Lys^{252} seems to form hydrogen bonds with the two backbone carbonyls corresponding to Pro^{824} and Glu^{826} in the C-terminal part of the loop connecting the membrane-spanning segments M6 and M7 ("L6–7") in the Ca^{2+}-bound F_1 state (FIG. 1), but not in the thapsigargin-bound E_2 state. In this study, we investigated the importance of the proposed interaction between Lys^{252} and L6–7 in further detail by preparing a series of Ca^{2+}-ATPase point mutants with alterations to Lys^{252} in M3 and Pro^{824}, Lys^{825}, and Glu^{826} in L6–7.

The rate of Ca^{2+} dissociation from the cytoplasmically facing high-affinity Ca^{2+} sites was studied in rapid quench phosphorylation experiments by taking advantage of the requirement of the ATP phosphorylation reaction for Ca^{2+} occupation at both Ca^{2+} sites as previously described.[1] The Ca^{2+} dissociation rate of mutant Lys^{252}→Glu was conspicuously higher (5-fold) than that of the wild type, whereas more moderately increased rates (1.5- to 2-fold) were obtained with mutants Lys^{252}→Ala, Lys^{252}→Gln, Lys^{252}→Met, Lys^{252}→Arg, Pro^{824}→Leu, Lys^{825}→Leu, and Glu^{826}→Leu.

Furthermore, we studied the time course of Ca^{2+} binding with associated conformational changes in rapid quench phosphorylation experiments as previously described.[1] While phosphorylation initiated by simultaneous addition of Ca^{2+} and

Address for correspondence: Jens Peter Andersen, Department of Physiology, University of Aarhus, Ole Worms Allé 160, DK-8000 Aarhus C, Denmark. Voice: +45-8942-2814; fax: +45-8612-9065.

jpa@fi.au.dk

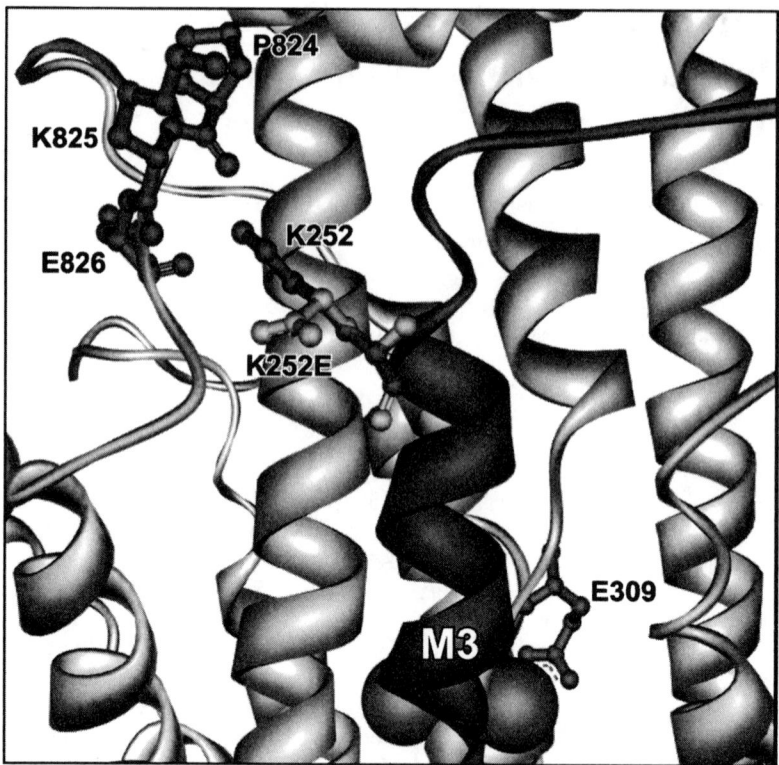

FIGURE 1. Structural organization of the region around Lys^{252}. Prepared by use of the atomic coordinates of the Ca^{2+}-ATPase in the E_1Ca$_2$ state, a mutant glutamate residue (*light gray*) has been inserted at position 252 on top of the wild-type lysine residue (*dark gray*).

ATP to Ca^{2+}-deprived wild-type enzyme is significantly slower than phosphorylation of wild-type enzyme preincubated with Ca^{2+}, there was not such a difference for mutant Lys$^{252}\rightarrow$Glu, with both k_{obs} values being close to that found for the wild-type enzyme preincubated with Ca^{2+}. This shows that the reactions involved in Ca^{2+} binding were no longer rate-limiting for Lys$^{252}\rightarrow$Glu. Hence, Ca^{2+} both dissociates and binds with a higher rate in Lys$^{252}\rightarrow$Glu as compared with the wild type.

These findings show that the point mutation Lys$^{252}\rightarrow$Glu has similar effects on Ca^{2+} migration as the previously analyzed chimeric mutation involving, in addition to Lys^{252}, three other residues.[1] The main reason for the conspicuous effects observed for Lys$^{252}\rightarrow$Glu seems to be the introduction of the negatively charged glutamate side chain (FIG. 1). The strong impact of the Lys$^{252}\rightarrow$Glu substitution on the rate of Ca^{2+} migration suggests electrostatic interaction with Ca^{2+} or electrostatic effects on the structure of the migration pathway or the gating process. Such effects could involve repulsion between the negative charge of the glutamate and the main chain carbonyls of Pro^{824} and Glu^{826} in L6–7.

REFERENCE

1. ANDERSEN, J.P. *et al*. 2001. Importance of transmembrane segment M3 of the sarcoplasmic reticulum Ca^{2+}-ATPase for control of the gateway to the Ca^{2+} sites. J. Biol. Chem. **276**: 23312–23321.

Phospholamban Inhibits Ca^{2+} Pump Oligomerization and Intersubunit Free Energy Exchange Leading to Activation of Cardiac Muscle SERCA2a

JAMES E. MAHANEY,[a] R. WAYNE ALBERS,[b] HOWARD KUTCHAI,[c] AND JEFFREY P. FROEHLICH[d]

[a]*Department of Biochemistry, West Virginia University, Morgantown, West Virginia 26506, USA*

[b]*National Institute of Neurological Disease and Stroke, National Institutes of Health, Bethesda, Maryland 20892, USA*

[c]*Department of Molecular Physiology and Biological Physics, University of Virginia, Charlottesville, Virginia 22908, USA*

[d]*National Institute on Aging, National Institutes of Health, Baltimore, Maryland 21224, USA*

KEYWORDS: cardiac muscle sarcoplasmic reticulum; phospholamban; SERCA2a; oligomerization; free energy exchange

Activation of cardiac muscle Ca^{2+}-ATPase (SERCA2a) by β_1-agonists results from PKA-dependent phosphorylation of phospholamban (PLB).[1] This increases the apparent Ca^{2+} affinity and V_{max} of SERCA2a, reflecting loss of regulatory control by PLB and enhanced efficiency of energy utilization by an unknown mechanism.

Pre-steady-state kinetic studies of native skeletal muscle SR Ca^{2+}-ATPase (SERCA1) demonstrate behavior consistent with an oligomeric pump, namely, acceleration of catalytic site dephosphorylation by ATP at a second high-affinity site,[2] a rapid phase of EGTA-induced E2P hydrolysis,[3] and deceleration of the fast E1P→E2P transition with stabilization of E1P in the steady state.[4] These effects are absent in native CSR Ca^{2+}-ATPase,[3] suggesting that SERCA2a in the presence of PLB is functionally monomeric. Saturation transfer EPR (ST-EPR) measurements of spin-labeled Ca^{2+}-ATPase in native CSR suggest that conditions relieving the effect of PLB promote oligomerization of SERCA2a.[5]

We tested whether the removal of PLB regulation converts SERCA2a to a functionally oligomeric state resembling SERCA1 using recombinant SERCA2a

Address for correspondence: James E. Mahaney, Department of Biochemistry, West Virginia University, Morgantown, WV 26506. Voice: 304-293-7756; fax: 304-293-6846.
jmahaney@hsc.wvu.edu

Ann. N.Y. Acad. Sci. 986: 338–340 (2003). © 2003 New York Academy of Sciences.

expressed with (+) or without (−) PLB in High Five insect cells (baculovirus). Quenched-flow mixing and ST-EPR were used to probe for intermolecular interactions in unlabeled and maleimide spin-labeled (MSL) SERCA2a.

RESULTS AND DISCUSSION

Spin-labeled SERCA2a coexpressed with PLB had a smaller rotational correlation time (63 μs) than either SERCA2a sans PLB (78 μs) or SERCA2a + phosphorylated PLB (97 μs), suggesting a more compact structure (monomer or compact dimer?) for SERCA2a + PLB. In rapid mixing experiments, SERCA2a + PLB behaved like native cardiac SR Ca^{2+}-ATPase,[3] showing a small (9%) steady-state fraction of E1P, the absence of a rapid phase of EGTA-induced E2P decay, and no stimulation by ATP (10–50 μM) of presteady EP overshoot decay. In contrast, SERCA2a sans PLB, like SERCA1, showed twofold stimulation of EP overshoot decay by ATP, a larger (31%) steady-state fraction of E1P, and the presence of a rapid

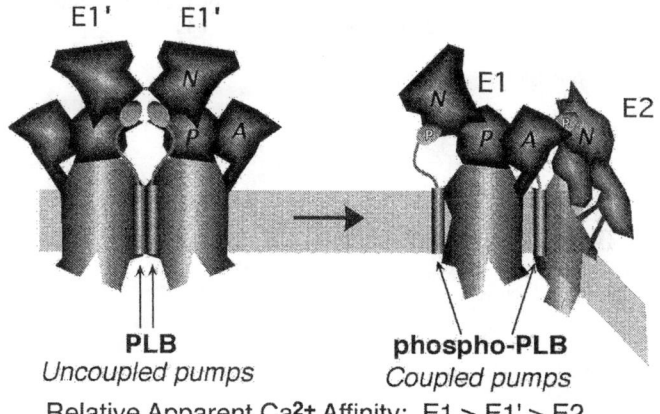

FIGURE 1. Model for the activation of SERCA2a. The E1 and E2 conformations of SERCA2a are based on the crystallographic data of Toyoshima et al.[6,7]

initial phase of EGTA-induced E2P decay. Thus, removal of PLB transforms expressed SERCA2a from a cardiac SR–like (functionally monomeric) enzyme to a skeletal SR–like (coupled oligomeric) enzyme.

Our model for SERCA2a coupling (FIG. 1), based on the SERCA1 crystal structures,[6,7] postulates interaction of the A domain of one pump with the N domain of another. PLB-regulated SERCA2a represented by the compact oligomeric starting state at the upper left is uncoupled. Asymmetric dimerization (2 E1′ → E1/E2) and conformational coupling subsequent to the removal of PLB regulation increase the apparent Ca^{2+} affinity and V_{max} by allowing an exchange of free energy between the subunits. The improved efficiency of operation comes from the utilization of chemical free energy available in the E1P → E2P transition that is dissipated in the uncoupled state as kinetic energy.

REFERENCES

1. TADA, M. 1992. Ann. N.Y. Acad. Sci. **671**: 92–103.
2. FROEHLICH, J.P. & E.W. TAYLOR. 1975. J. Biol. Chem. **250**: 2013–2021.
3. SUMIDA, M., T. WANG, A. SCHWARTZ, et al. 1980. J. Biol. Chem. **255**: 1497–1503.
4. FROEHLICH, J.P. & P.F. HELLER. 1985. Biochemistry **24**: 126–136.
5. NEGASH, S., L.T. CHEN, D.J. BIGELOW & T.C. SQUIER. 1996. Biochemistry **35**: 11247–11259.
6. TOYOSHIMA, C., M. NAKASAKO, H. NOMURA & H. OGAWA. 2000. Nature **405**: 647–655.
7. TOYOSHIMA, C. & H. NOMURA. 2002. Nature **418**: 605–611.

ATP Regulation of Calcium Binding in Ca^{2+}-ATPase Molecules of the Sarcoplasmic Reticulum

JUN NAKAMURA,[a] GENICHI TAJIMA,[a] AND CHIKARA SATO[b]

[a]*Department of Developmental Biology and Neurosciences, Graduate School of Life Sciences, Tohoku University, Sendai, Miyagi 980-8578, Japan*

[b]*Neuroscience Research Institute, National Institute of Advanced Industrial Science and Technology, Tsukuba 305-8568, Japan*

KEYWORDS: ATP; calcium binding; Ca^{2+}-ATPase

To know the effect of ATP on calcium binding of Ca^{2+}-ATPase of the sarcoplasmic reticulum (SR), calcium dependence of the Ca^{2+}-ATPase activity and the phosphorylation of the ATPase molecules was compared with that of the kinetic calcium binding in the absence of ATP. The results are discussed based on a model of two conformational variants (A and B forms) of the chemically equivalent ATPase molecules.[1]

The results show the existence of two types of regulatory sites of the enzyme molecules at which ATP binding improves the calcium binding performance of the molecules depending on the aggregation state of the molecules and pH: the two regulatory sites bind ATP at submillimolar (0.25 mM) and millimolar (5 mM) ATP, respectively. For example, in the SR membrane at pH 7.40, submillimolar ATP converted the calcium binding manner of the A form from noncooperative [Hill number (n_H) of ~1] to cooperative ($n_H \approx 2$), concurrent with a decrease in the apparent calcium affinity ($K_{0.5}$) from 2–6 to 0.1–0.3 µM. The binding of the A form became almost the same as that of the B form ($n_H \approx 2$, $K_{0.5} \approx 0.2$ µM), which was not affected by the ATP. Millimolar ATP further decreased the $K_{0.5}$ of the cooperative binding of the two forms to ~0.05 µM. The data are schematically represented in FIGURE 1. Here, for simplicity, the ATPase conformers that negatively cooperatively ($n_H < 1$), noncooperatively ($n_H \approx 1$), and positively cooperatively ($n_H \approx 2$) bind calcium ions are termed a, A, and B, respectively. It was also observed that ATP regulation exists in the detergent ($C_{12}E_8$)–solubilized, monomeric ATPase molecules that are in the E_2 state before calcium binding. The regulatory site seems to reside in the monomeric

Address for correspondence: Jun Nakamura, Department of Developmental Biology and Neurosciences, Graduate School of Life Sciences, Tohoku University, Sendai, Miyagi 980-8578, Japan. Voice: +81-22-217-6701; fax: +81-22-217-3683.
 jun-n@mail.cc.tohoku.ac.jp

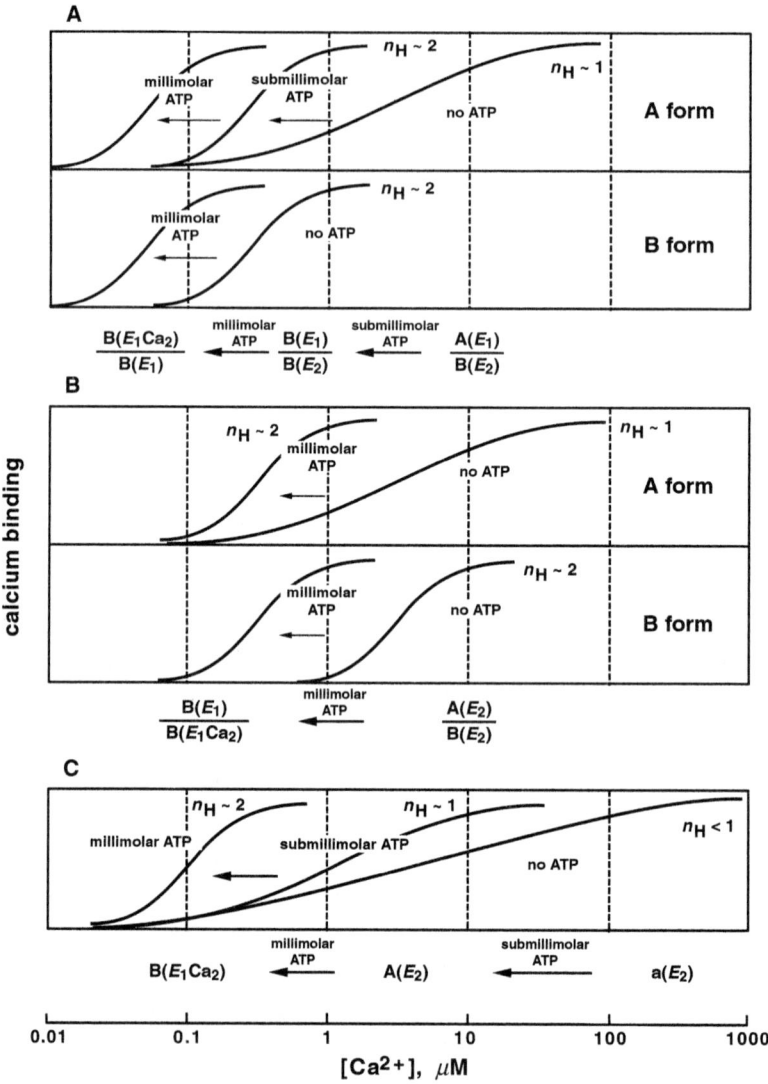

FIGURE 1. Schematic representation of the effect of ATP on calcium binding of the membranous Ca^{2+}-ATPase molecules at pH 7.40 **(A)** and 6.23 **(B)** and of the detergent-solubilized molecules at pH 7.40 **(C)**. The scheme below each panel shows the enzyme state (E_1 or E_2) of the ATPase molecules before and after the addition of calcium and the association state of the molecules under the different conditions of ATP concentration and pH (see text for details). (Excerpted from Ref. 2, with permission.)

form of the ATPase molecules in an alternate enzyme state of E_2 in the catalytic cycle. The E_2 state may be in equilibrium between two different E_2 states that have high and low affinity for ATP, respectively.[2]

ACKNOWLEDGMENTS

This work was supported in part by a grant from the Japan Biological Informatics Consortium.

REFERENCES

1. NAKAMURA, J. & T. FURUKOHRI. 1994. Two types of proton-modulated calcium binding in the sarcoplasmic reticulum Ca^{2+}-ATPase: I. A model of two different conformations of chemically equivalent ATPase molecules. J. Biol. Chem. **269:** 30818–30821.
2. NAKAMURA, J., G. TAJIMA, C. SATO, et al. 2002. Substrate regulation of calcium binding in Ca^{2+}-ATPase molecules of the sarcoplasmic reticulum: I. Effect of ATP. J. Biol. Chem. **277:** 24180–24190.

The Nature of the Low-Frequency Normal Modes of the E1Ca Form of the SERCA1 Ca^{2+}-ATPase

N. REUTER,[a] K. HINSEN,[b] AND J-J. LACAPÈRE[a]

[a]*U410 INSERM, F-75870 Paris Cédex 18, France*
[b]*CBM CNRS, F-45071 Orléans Cédex 2, France*

KEYWORDS: Ca^{2+}-ATPase; conformational changes; molecular modeling; normal modes

INTRODUCTION

The calcium ATPase is constituted of 3 cytoplasmic domains, named actuator (A), nucleotidic (N), and phosphorylation (P), and 10 transmembrane helices hosting the calcium binding sites. It is known that the cytoplasmic domains undergo large amplitude movements during the active transport of calcium ions.[1] We have used the atomic structure[2] of the E1Ca form of the Ca^{2+}-ATPase as a starting point for normal mode (NM) calculations. These computational methods are based on the hypothesis that the normal modes exhibiting the lowest frequencies describe the large amplitude movements in a protein and, as such, serve as a predictive tool for its conformational changes (for a review, see Ref. 3). The aim of this work is to determine the atomic displacements associated with the lowest frequency normal modes of the SERCA1 calcium pump and to understand how they can be related to functions of the pump.

METHODS

Since we are only interested in the motion of large domains, we used a simplified model[4] consisting of all $C\alpha$ atoms (994) of the 1EUL[2] atomic structure, and all the normal modes (2982) were calculated. All the calculations were done with the Molecular Modeling Toolkit (MMTK).[5] Below, we analyze the atomic displacements associated with the first three modes, that is, 7, 8, and 9 (modes 1 to 6 correspond to the global rotation and translation and thus their frequency is zero).

Address for correspondence: N. Reuter, U410 INSERM, 16, rue Henri Huchard, BP 416, F-75870 Paris Cédex 18, France. Voice: +33-1-44-85-61-32; fax: +33-1-42-28-87-65.
reuter@bichat.inserm.fr

RESULTS

The regions of the protein undergoing the movements with the largest amplitude along modes 7, 8, and 9 can be identified by the highest peaks in the plots in FIGURE 1. As predicted by EM[6] and X-ray[2] studies, the nucleotidic domain (N) shows high amplitude motions, which appear to be predominant in the two slowest modes. Modes 7 and 8 could describe the movements performed by the protein to bring the nucleotide binding site (in particular, K515) closer to the phosphorylation site (D351). Whereas modes 7 and 8 show predominantly movements of N, mode 9

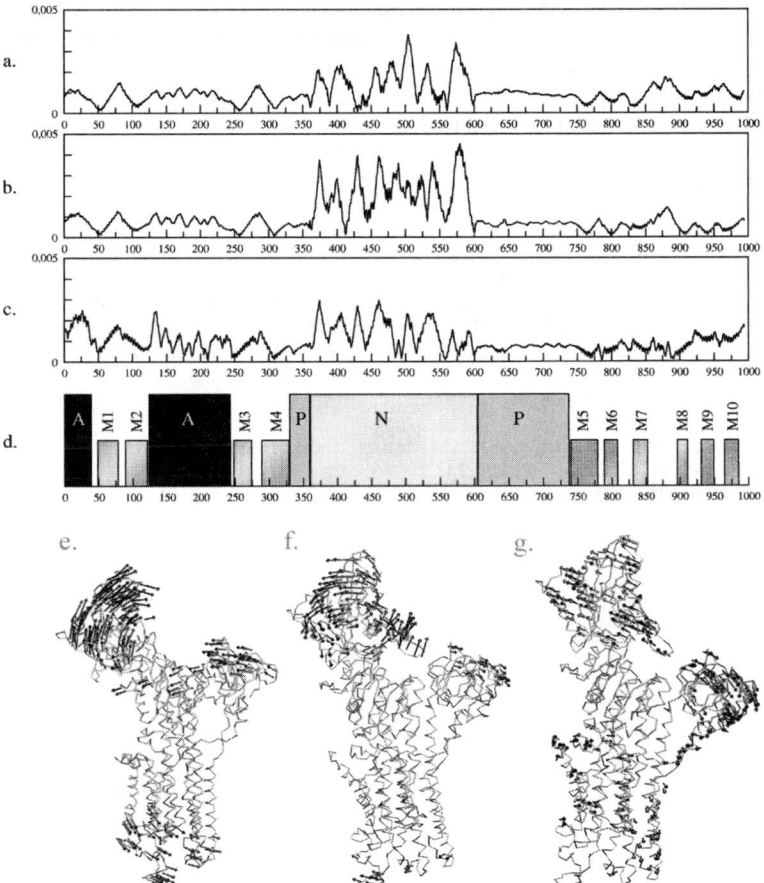

FIGURE 1. Atomic displacements associated with modes 7, 8, and 9: Normalized atomic displacements (plots **a** to **c**) (y-axis, in Å) with respect to the residue number in the sequence (x-axis), and vector field representation (plots **e** to **g**) associated with modes 7, 8, and 9, respectively. For the sake of clarity, only the vectors having a length greater than the average atomic displacement (calculated over all Cα atoms) are represented. Plot **d** sketches the correspondence between domain names and sequence numbers as defined in Ref. 1.

involves displacements of residues of both domains A and N with a comparable amplitude (FIG. 1, plot c) coupled to displacements of residues in the C-terminal helices. In this third mode, one can see cooperative movements of the cytoplasmic domains and some parts of the transmembrane segments.

CONCLUSIONS

Analysis of the three slowest modes of the E1Ca form of Ca^{2+}-ATPase shows features in good agreement with both the structural data available and functional features: (i) large movements of N and A, (ii) significant movements of the transmembrane helices, and (iii) coupled movements of the cytoplasmic and transmembrane domains. The latter item shows that movements of N and A can be transmitted from the cytoplasmic domains to the transmembrane part where the calcium binds. The observed displacement of N may be necessary for the formation of the active site in which ATP binding and further hydrolysis take place.

REFERENCES

1. LEE, A.G. & J.M. EAST. 2001. What the structure of a calcium pump tells us about its mechanism. Biochem. J. **356:** 665–683.
2. TOYOSHIMA, C. et al. 2000. Crystal structure of the calcium pump of sarcoplasmic reticulum at 2.6 Å resolution. Nature **405:** 647–655.
3. CORNELL, W.D. & S. LOUISE-MAY. 1998. Normal mode analysis. In Encyclopedia of Computational Chemistry, pp. 1904–1913. Wiley. Chichester/New York.
4. HINSEN, K. 1998. Analysis of domain motions by approximate normal mode calculations. Proteins **33:** 417–429.
5. HINSEN, K. 2000. The Molecular Modeling Toolkit: a new approach to molecular simulations. J. Comput. Chem. **21:** 79–85.
6. XU, C. et al. 2002. A structural model for the catalytic cycle of Ca^{2+}-ATPase. J. Mol. Biol. **316:** 201–211.

Protonation of the *Neurospora crassa* Plasma Membrane H^+-ATPase as a Function of pH Monitored by ATR-FTIR

O. RADRESA, V. RAUSSENS, J-M. RUYSSCHAERT, AND E. GOORMAGHTIGH

Structure and Function of Biological Membranes, Université Libre de Bruxelles, B1050 Brussels, Belgium

KEYWORDS: infrared spectroscopy; *Neurospora crassa*; H^+-ATPase; side chains

Plasma membrane H^+-ATPases from yeast and fungi exhibit an optimal activity at pH 6.0 and a bell-shaped dependence of the ATPase activity on intracellular pH.[1,2] Although it is well known that complexation by Mg^{2+} and nucleotides or vanadate is able to drive conformational changes in the *Neurospora* H^+-ATPase with only minor alteration of the secondary structure,[3,4] the structural effect of the ubiquitous H^+ ligand has drawn little interest so far. This contrasts with the many reports on the conformational effect of Ca^{2+} ions on the Ca^{2+}-ATPase. We report here on a novel approach for identifying structural changes in the reconstituted H^+-ATPase using the attenuated total reflection–Fourier transform infrared (ATR-FTIR) technique in the presence of a buffer flow running over the sample.

The H^+-ATPase was purified and reconstituted as described previously with a protein:lipid ratio equal to 1:5 (w/w).[5] Approximately 30 μg of the reconstituted enzyme was deposited under a nitrogen stream over a trapezoidal Ge crystal allowing 25 internal reflections. All spectra were recorded on a Bruker Equinox spectrometer with a resolution of 4 cm^{-1} and 254 scans/spectrum. Buffer composition is as follows: solution A is 3 mM Mes, 3 mM HEPES, pH 7.5; solution B is 3 mM Mes, 3 mM HEPES, pH x, with x decreasing from pH 7.5 to pH 5.5 by steps of 0.5 pH units. The experimental setup is shown in FIGURE 1a. In each cycle, solution A was flown over the sample (reference pH 7.5) for 3 min followed by 3 min at pH x.

The spectra recorded at each pH x were subtracted from the reference spectrum recorded just before with a scaling factor set at 1. The evolution of the difference spectra between 1800 cm^{-1} and 1300 cm^{-1} is shown in FIGURE 1b. Three main bands related to the protonation of either aspartic or glutamic amino acid side chains appear at 1402 cm^{-1}, 1575 cm^{-1}, and 1737 cm^{-1}. In the absence of the enzyme, the difference spectra did not display any of the features described above (not shown).

Address for correspondence: E. Goormaghtigh, ULB CP206/2, Bld. du Triomphe, B1050 Brussels, Belgium. Voice: +32-2-650-53-86; fax: +32-2-650-58-82.
egoor@ulb.ac.be

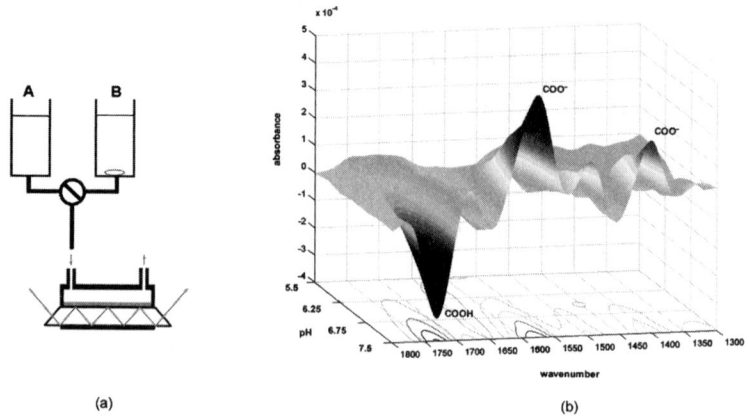

FIGURE 1. (a) Scheme of the experimental setup: two buffer solutions are separated by a computer-controlled valve delivering the solution in the flow cell where the IR spectra are recorded. (b) Evolution of the difference spectra between 1800 cm^{-1} and 1300 cm^{-1} revealing the protonation/deprotonation of acidic residues. (Solution A is at pH 7.5 and solution B is at pH x, with x varying from pH 7.5 to pH 5.5.)

Preliminary characterization indicates that the pK_a values of some of the acidic residue side chains are close to 6.0, distant from their value in solution, but very close to the optimal pH of the ATPase activity. Whether ionization of these acidic residues is related to the binding of the transported proton remains speculative. We are currently working at the identification of the structural modifications that may possibly accompany the observed protonation of the enzyme. We believe that, in combination with mutagenesis experiments, this approach should shed some light on the role of key residues involved in the pH dependence of several P-type ATPase partial reactions, and ultimately also to those of the residues directly involved in the transport reaction.

REFERENCES

1. BLANPAIN, J.P., M. RONJAT, P. SUPPLY, et al. 1992. The yeast plasma membrane H$^+$-ATPase: an essential change of conformation triggered by H$^+$. J. Biol. Chem. **267**: 3735–3740.
2. SCARBOROUGH, G.A. 1977. Properties of the *Neurospora crassa* plasma membrane H$^+$-ATPase. Arch. Biochem. Biophys. **180**: 384–393.
3. HENNESSEY, J.P., JR. & G.A. SCARBOROUGH. 1988. Secondary structure of the *Neurospora crassa* plasma membrane H$^+$-ATPase as estimated by circular dichroism. J. Biol. Chem. **263**: 3123–3130.
4. GOORMAGHTIGH, E., L. VIGNERON, G.A. SCARBOROUGH & J-M. RUYSSCHAERT. 1994. Tertiary conformational changes of the *Neurospora crassa* plasma membrane H$^+$-ATPase monitored by hydrogen/deuterium kinetics. J. Biol. Chem. **269**: 27409–27413.
5. VIGNERON, L., G.A. SCARBOROUGH, J-M. RUYSSCHAERT & E. GOORMAGHTIGH. 1995. Reconstitution of the *Neurospora crassa* plasma membrane H$^+$-adenosine triphosphatase. Biochim. Biophys. Acta **1236**: 95–104.

Mutagenic Study of Residues in Transmembrane Helix 4, 5, and 6 of the Plant Plasma Membrane P-Type H$^+$-ATPase

M. J. BUCH-PEDERSEN,[a] A. L. MØLLER,[a] AND M. G. PALMGREN[a,b]

[a]*Department of Plant Biology, The Royal Veterinary and Agricultural University, Thorvaldsensvej 40, DK-1871 Frederiksberg C, Denmark*

[b]*CNRS, Université Pierre et Marie Curie, Paris VI, 4, place Jussieu, F-75014 Paris, France*

KEYWORDS: plant plasma membrane; proton binding; H$^+$-ATPase

Plasma membrane H$^+$-ATPases are related to other P-type pumps such as Na$^+$/K$^+$- and Ca^{2+}-ATPases, but their mechanism of action is poorly known. In the sarcoplasmic reticulum Ca^{2+}-ATPase, two Ca^{2+} binding sites are formed by residues in transmembrane segments M4, M5, M6 and M8.[1] Likewise, proton coordinating residues in H$^+$-ATPases are likely to be formed in the membraneous part of the pump molecule. Arg655 and Asp684, situated in predicted transmembrane segments M5 and M6, respectively, are close to the predicted H$^+$ binding site in atomic models of plant[2] and fungal[3] P-type H$^+$-ATPases. The negatively charged Asp684 in M6 has previously been implicated in cation coordination in a plant H$^+$-ATPase.[4] The other, the positively charged Arg655 in M5, has not been assigned a role so far.

To explore the functional role of Arg655 and Asp684 in a plant H$^+$-ATPase, site-directed mutants at these two positions have been constructed in the *Arabidopsis* AHA2 plasma membrane H$^+$-ATPase and expressed in yeast. All the mutants were expressed as seemingly intact membrane proteins in the yeast host cells. The presence of a C-terminal-located histidine tag has allowed for efficient affinity purification of the mutated H$^+$-ATPases.

Surprisingly, all the tested Arg655 and Asp684 substitutions possess residual ATP hydrolytic activity (TABLE 1). In the catalytic model of P-type ATPases, the transported ion(s) needs to become bound to the E_1 conformation of the enzymes before transfer of the terminal phosphate from ATP to the ATPases can take place. Thus, our data suggest that Arg655 and Asp684 are not absolutely essential, neither for the initial proton-binding event, nor for the following $E_1 \rightarrow E_1P$ conformational transition. However, all tested Asp684 substitutions exhibit severe defects in the $E_1P \rightarrow E_2P$ transition, and detailed biochemical analysis has indeed suggested a role for

---

Address for correspondence: M.J. Buch-Pedersen, Department of Plant Biology, The Royal Veterinary and Agricultural University, Thorvaldsensvej 40, DK-1871 Frederiksberg C, Denmark. Voice: +45-3528-3327; fax: +45-3528-3365.
  mbp@kvl.dk

TABLE 1. H$^+$-ATPase activity of affinity-purified mutant AHA2 plasma membrane H$^+$-ATPases expressed in the yeast *Saccharomyces cerevisiae*

	ATP hydrolysis^a	H$^+$-pumping^b
WT	100 %	100 %
R655K	60 %	60 %
R655A	10 %	10 %
R655D	10 %	10 %
D684E	15 %	10 %
D684N	15 %	0 %
D684A	5 %	0 %
D684V	5 %	0 %
D684R	0.5 %	0 %

^{a}ATP hydrolytic activity of affinity-purified H$^+$-ATPases expressed relative to the ATP hydrolytic activity of the corresponding WT protein. 100% corresponds to 27 µmol P$_i$/min/mg protein.

^{b}Proton pumping measured in reconstituted vesicles. Proton pumping is expressed relative to WT proton pumping activity.

this residue in proton coordination in the catalytic cycle of the AHA2 H$^+$-ATPase (data not shown). Our mutational analysis of Asp684 single-point substituted enzymes suggests that the side chain of Asp684 is involved, but is not essential, for this initial binding (data not shown).

We are currently testing whether the carbonyl oxygens of M4 are indeed involved in proton coordination (data not shown). Thus, the transported ion could initially be positioned in a saddle on the unwound M4, allowing for hydrogen bonding with exposed carbonyl oxygen. Further in the catalytic cycle, coordination to the negatively charged Asp684 seems absolutely essential for any further progress in enzyme conformational changes. The role of Arg655 is more uncertain than the role of Asp684. An H$^+$-binding site with liganding groups from M4 and M6 would correspond to a Ca^{2+}-binding site II in SERCA1a.

ACKNOWLEDGMENTS

This work was supported by the European Union's Biotechnology Program and the Human Frontier Science Program Organization.

REFERENCES

1. TOYOSHIMA, C., M. NAKASAKO, H. NOMURA & H. OGAWA. 2000. Nature **405:** 647–655.
2. BUKRINSKY, J.T., M.J. BUCH-PEDERSEN, S. LARSEN & M.G. PALMGREN. 2001. FEBS Lett. **494:** 6–10.
3. KUHLBRANDT, W., J. ZEELEN & J. DIETRICH. 2002. Science **297:** 1692–1696.
4. BUCH-PEDERSEN, M.J., K. VENEMA, R. SERRANO & M.G. PALMGREN. 2000. J. Biol. Chem. **275:** 39167–39173.

Mutational Analysis of Charged Residues in the Putative KdpB-TM5 Domain of the Kdp-ATPase of *Escherichia coli*

MARC BRAMKAMP AND KARLHEINZ ALTENDORF

Universität Osnabrück, Fachbereich Biologie/Chemie, Abteilung Mikrobiologie, D–49069 Osnabrück

KEYWORDS: Kdp; potassium transport; mutational analysis

The P-type Kdp-ATPase is a high-affinity potassium transport system found in a variety of prokaryotes (http://biobase.dk/~axe/Patbase.html). The enzyme complex of *Escherichia coli* is composed of four subunits (KdpFABC), in which, in contrast to all other P-type ATPases known so far, ion transport (KdpA) and ATP hydrolysis (KdpB) are mediated by different subunits.[1] While the catalytic subunit KdpB shares all key features with other P-type ATPases,[2] KdpA shows some homologies to potassium channels, like KcsA.[3] Compared to other P-type ATPases the knowledge about transport mechanism and topology of the KdpFABC complex is rather scarce.

In order to elucidate the mechanism of ion transport initiated by ATP hydrolysis, phosphorylation and hence domain movement in KdpB, the two charged residues, Asp583 and Lys586 in the putative transmembrane helix 5 (TM5) (compare FIG. 1), which are highly conserved throughout the KdpB polypeptides of different species, were subjected to site-directed mutagenesis. We changed both residues to either a neutral, polar, or charged residue (D583A, D583S, D583N, D583K, D583E, K586A, K586T, K586D, K586R, and D583A:K586A). The mutations were introduced into a plasmid-encoded *kdpFABC* operon, which is under the control of the native promoter/operator region. Complementation tests with the *E. coli* strain TKW3205,[4] lacking all potassium transport systems, revealed that growth on low potassium concentrations was only restored by the conservative exchanges D583E and K586R, indicating the necessity of charges at these positions. The existence of a salt bridge can be ruled out, since the phenotype of the D583 mutants and the K586 mutants differ drastically. Elimination of the negative charge at position 583 leads to mutants unable to grow below 10 mM potassium, while elimination of the positive charge at position 586 leads to mutants, which could still grow at 0.5 mM potassium on solid agar plates. The double mutant D582A:K586A exhibits the same phenotype as the

---

Address for correspondence: Marc Bramkamp, University Osnabrueck, Department of Microbiology, Barbarastrasse 11, 49076 Osnabrueck, Germany. Voice: +49-541-9692867; fax: +49-541-9692870.
bramkamp@biologie.uni-osnabrueck.de

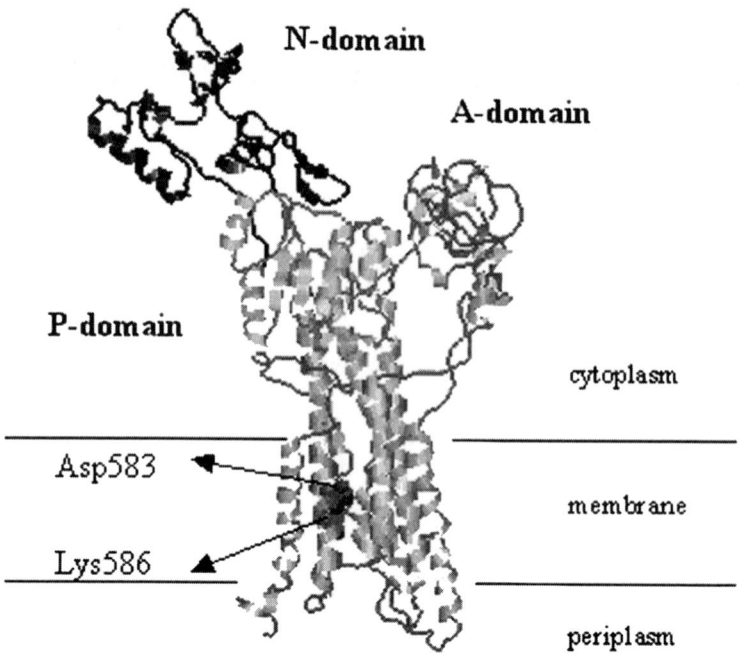

FIGURE 1. Model of KdpB according to the SERCA pump structure. The Ca^{2+}-ATPase structure (1EUL.pdb; according to Toyoshima and Nomura[5]) was used to model the KdpB subunit. Two charged residues, Asp583 and Lys586, are located in the membrane. These residues are highly conserved among the KdpB polypeptides of different origin.

D583 mutants, indicating the dominant effect of the negative charge. Purified Kdp-FABC complex carrying the mutation D583E shows K^+-stimulated ATPase activity similar to that of the wild type. However, the D583A complex exhibits maximal ATPase activity similar to the wild-type complex that is, however, no longer stimulated by K^+ ions. Reconstituted D583A complexes were not able to facilitate ion transport, although ATPase activity was close to 100% wild-type activity. Therefore, we conclude that elimination of the negative charge within TM5 leads to uncoupling of ATP hydrolysis and ion transport. Similar results were obtained for the complexes mutated at position K586. The ATPase activity of the K586R complex is stimulated by potassium ions, while the K586A complex is not.

In summary, it can be concluded that all mutations eliminating the charges at positions 583 and 586 lead to uncoupling of ATP hydrolysis and ion transport. Furthermore, these mutants seem to be shifted towards the E1 state, since all mutant complexes become more resistant towards *ortho*-vanadate. The charged residues within TM5 most likely form a strong dipole, which is involved in ion translocation. Movement of TM5 according to the predicted domain movements of P-type ATPases (E1 and E2 transition)[5] will probably move the dipole and might, in turn, have an effect on potassium ions occluded within the potassium channel-like protein KdpA.

The energy to push the K^+ ion against a concentration gradient through the selectivity filter of KdpA to the cytosolic side and the gating mechanism is most likely mediated by movements of the TM domains of KdpB. Charge movements, as already predicted by BLM measurements,[6,7] probably play a crucial role in this process.

REFERENCES

1. ALTENDORF, K. & W. EPSTEIN. 1996. The Kdp-ATPase of *Escherichia coli*. *In* Biomembranes (ATPases). Vol. 5. A.G. Lee, Ed.: 403–420. JAI Press. Greenwich and London.
2. ALTENDORF, K., M. GAßEL, W. PUPPE, *et al*. 1998. Structure and function of the Kdp-ATPase of *Escherichia coli*. Acta Physiol. Scand. **163**: 137–146.
3. DURELL, S.R., E.P. BAKKER & H.R. GUY. 2000. Does the KdpA subunit from the high affinity K(+)-translocating P-type KDP-ATPase have a structure similar to that of K(+) channels? Biophys. J. **78**: 188–199.
4. PUPPE, W., A. SIEBERS & K. ALTENDORF. 1992. The phosphorylation site of the Kdp-ATPase of *Escherichia coli*: site-directed mutagenesis of the aspartic acid residues 300 and 307 of the KdpB subunit. Mol. Microbiol. **6**: 3511–3520.
5. TOYOSHIMA, C. & H. NOMURA. 2002. Structural changes in the calcium pump accompanying the dissociation of calcium. Nature **418**: 605–611.
6. FENDLER, K., S. DRÖSE, K. ALTENDORF & E. BAMBERG. 1996. Electrogenic K^+ transport by the Kdp-ATPase of *Escherichia coli*. Biochemistry **35**: 8009–8017.
7. FENDLER, K., S. DRÖSE, W. EPSTEIN, *et al*. 1999. The Kdp-ATPase of *Escherichia coli* mediates an ATP-dependent, K^+-independent electrogenic partial reaction. Biochemistry **38**: 1850–1856.

Functional Roles of the α Isoforms of the Na,K-ATPase

JERRY LINGREL,[a] AMY MOSELEY,[a] IVA DOSTANIC,[a] MARC COUGNON,[a] SUIWEN HE,[b] PAUL JAMES,[c] ALISON WOO,[a] KYLE O'CONNOR,[a] AND JONATHAN NEUMANN[a]

[a]*Department of Molecular Genetics, Biochemistry, and Microbiology, University of Cincinnati, Cincinnati, Ohio 45267-0524, USA*

[b]*Department of Pharmacology and Cell Biophysics, University of Cincinnati, Cincinnati, Ohio 45267-0575, USA*

[c]*Department of Zoology, Miami University, Oxford, Ohio 45056, USA*

ABSTRACT: The Na,K-ATPase is composed of two subunits, α and β, and each subunit consists of multiple isoforms. In the case of α, four isoforms, α1, α2, α3, and α4 are present in mammalian cells. The distribution of these isoforms is tissue- and developmental-specific, suggesting that they may play specific roles, either during development or coupled to specific physiological processes. In order to understand the functional properties of each of these isoforms, we are using gene targeting, where animals are produced lacking either one copy or both copies of the corresponding gene or have a modified gene. To date, we have produced animals lacking the α1 and α2 isoform genes. Animals lacking both copies of the α1 isoform gene are not viable, while animals lacking both copies of the α2 isoform gene make it to birth, but are either born dead or die very soon after. In the case of animals lacking one copy of the α1 or α2 isoform gene, the animals survive and appear healthy. Heart and EDL muscle from animals lacking one copy of the α2 isoform exhibit an increase in force of contraction, while there is reduced force of contraction in both muscles from animals lacking one copy of the α1 isoform gene. These studies indicate that the α1 and α2 isoforms carry out different physiological roles. The α2 isoform appears to be involved in regulating Ca^{2+} transients involved in muscle contraction, while the α1 isoform probably plays a more generalized role. While we have not yet knocked out the α3 or α4 isoform genes, studies to date indicate that the α4 isoform is necessary to maintain sperm motility. It is thus possible that the α2, α3, and α4 isoforms are involved in specialized functions of various tissues, helping to explain their tissue- and developmental-specific regulation.

KEYWORDS: Na,K-ATPase; isoforms; sperm; muscle; heart; gene knockout

---

Address for correspondence: Jerry B. Lingrel, Ph.D., Department of Molecular Genetics, Biochemistry, and Microbiology, University of Cincinnati, P.O. Box 670524, Cincinnati, Ohio 45267-0524. Voice: 513-558-5324; fax: 513-558-1190.
Jerry.Lingrel@UC.edu

INTRODUCTION

Na,K-ATPase transports Na^+ out of the cell and K^+ in, using ATP as the energy source. The resulting gradient drives many transport processes through cotransporters and produces the electrogenic potential for nerve transmission and muscle contraction. The Na,K-ATPase is composed of two subunits, α and β, and in some tissues the enzyme is associated with another protein, γ.[1-4] Four isoforms of the α subunit occur in mammals, $\alpha 1$, $\alpha 2$, $\alpha 3$, and $\alpha 4$, and three isoforms of the β subunit have been identified, $\beta 1$, $\beta 2$, and $\beta 3$. The α isoforms are expressed in a tissue-specific and developmentally regulated manner.[5] The $\alpha 1$ isoform is expressed in every tissue, while the $\alpha 2$ isoform is limited largely to skeletal muscle, heart, and brain. The $\alpha 3$ isoform is expressed in brain and ovary, while the $\alpha 4$ isoform is expressed exclusively in sperm and its precursor cells.[6] The $\alpha 2$ isoform appears around birth in heart and skeletal muscle, but both the $\alpha 2$ and $\alpha 3$ isoforms are expressed earlier in brain development. The isoforms exhibit somewhat different properties in terms of cation binding, but it is unknown whether these differences are significant in terms of their overall physiological function.

In an attempt to understand the differential roles of the α isoforms, we have initiated a gene-targeting approach, where animals are produced lacking either one or both copies of the $\alpha 1$ or $\alpha 2$ isoform genes.[7] These animals provide the opportunity to examine Na,K-ATPase function with either reduced amounts or the absence of these two isoforms.

RESULTS AND DISCUSSION

Our approach for understanding the specific functional role of each of the α isoforms has employed gene-targeting studies. So far, this technology has been used to produce animals lacking either one copy or both copies of the $\alpha 1$ and $\alpha 2$ isoform genes in mice. In both cases, homologous recombination was used to both introduce a selectable marker and delete specific regions of the Na,K-ATPase required for function. Embryonic stem cells were generated carrying these modifications and these cells introduced into blastocysts of mice, which produced chimeric animals composed of cells originating from both wild type and targeted cells. Upon mating, some animals passed the modified genes through their germ cells, providing a permanent mouse line with the targeted allele carried in the genome. Mating animals with the disrupted $\alpha 1$ or $\alpha 2$ isoform on one chromosome allowed mice to be produced where the disruption occurred in both copies of the $\alpha 1$ or $\alpha 2$ isoform genes.

Animals lacking the $\alpha 1$ isoform gene do not produce offspring, and preliminary studies indicate that they do not develop past the blastocyst stage. This is not unexpected, as the $\alpha 1$ isoform is found in every tissue, and it is not unreasonable to think that as soon as the maternal Na,K-ATPase disappears the blastocyst will no longer survive. The $\alpha 2$ isoform, however, is not expressed during very early embryonic development, at least in the tissues studied to date, so it was expected that these animals would survive longer than the $\alpha 1$ isoform–deficient animals. Interestingly the $\alpha 2$ isoform–deficient animals developed to term but were either born dead, or died within the first few minutes following birth. There were no gross abnormalities associated with these animals, and histological studies revealed no differences between animals lacking the α2 isoform gene and wild type animals, with the exception of the lung, which did not appear to be fully inflated following birth. The α2 isoform was just beginning to be expressed in heart and skeletal muscle but was expressed at nearly maximal levels in the diaphragm muscle. Also, the α2 isoform was expressed in the brain at birth at levels 50–75% of those in the adult brain. It is possible that the animals could not breathe, due to a defect in either the ability of the diaphragm to contract or because of a lack of signal from the brain.

The animals lacking only one copy of either the α1 or α2 isoform gene are viable, mate, and produce healthy offspring. These animals thus provide a unique opportunity for studying the differential function of the α1 and α2 isoforms, as they produce approximately one-half the amount of α1 or α2 isoform protein, respectively. We have examined the function of two tissues, cardiac[7] and skeletal muscle,[8] in the animals lacking one copy of the α1 and α2 genes. Hearts from animals lacking one copy of the α2 isoform are hypercontractile, as would be expected based on studies of cardiac glycoside inhibition of the Na,K-ATPase. In this case, inhibition of the Na,K-ATPase causes an increase in force of contraction of the heart through the accumulation of intracellular Na^+, which increases intracellular Ca^{2+} through the Na-Ca exchanger. The higher intracellular Ca^{2+} levels results in larger Ca^{2+} transients, which in turn increases the force of contraction. However, when the contraction of the heart from animals lacking one copy of the α1 isoform gene is examined, an opposite phenotype is observed, that is, the hearts exhibit a hypocontractile phenotype. This is unexpected as it would have been assumed that partial loss of either the α1 or α2 isoform would cause an ionotrophic effect similar to inhibition of the Na,K-ATPase with cardiac glycosides. It is unknown why inhibition of the α1 isoform causes the heart to be hypocontractile, but we have ruled out Ca^{2+} overload as being responsible for this activity by varying extracellular Ca^{2+} in working heart preparations.

We have also examined the contractile properties of the extensor digitorum longus (EDL) muscle[8] in animals lacking one copy of either the α1 or α2 isoform gene. Similar to the heart, loss of one gene for either the α1 or α2 isoform gene reduces the levels of α1 and α2 isoform proteins by approximately half compared to wild type levels. EDL from animals lacking one copy of the α2 isoform gene exhibits an increase in the force of contraction, while those lacking one copy of the α1 isoform gene show a reduced force of contraction similar to the heart. This is true with both tetanic and twitch contractions. Relaxation is unaffected, which indicates that the difference in phenotype is not due to a defect in Ca^{2+} uptake following contraction.

In the case of heart muscle, approximately 80% of the α subunit is the α1 isoform, with 20% being the α2 isoform. This is reversed in EDL muscle, approximately 80% being α2 isoform and 20% the α1 isoform. Therefore, the differences in contractile properties do not appear to be related to the ratio of α1 or α2 isoform found in the tissue. The finding that two muscles, which vary in their amounts of α1 and α2 isoform, show a similar alteration in contraction suggests that a functional difference in the α1 and α2 isoforms represent a general phenomenon.

There are several explanations for these findings. First, the differences may result from the different kinetic properties between the two isoforms. It is also possible that the α1 and α2 isoforms are coupled to different biological processes in the cell. For example, it has been demonstrated that the α1 and α2 isoforms are differentially localized within the cell.[9,10] The α1 isoform appears to be uniformly distributed in

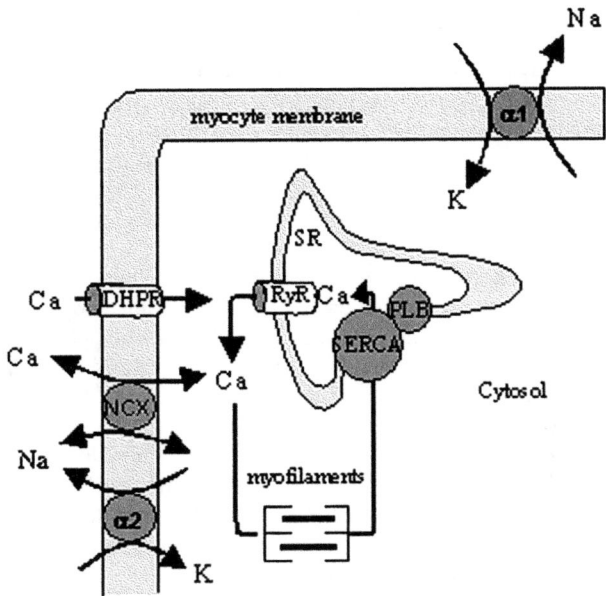

FIGURE 1. Model of differential activity of the Na,K-ATPase carrying the α1 or α2 isoform. The figure depicts the α2 Na,K-ATPase, being intimately associated with calcium handling, possibly through a physical separation from the α1 isoform of the Na,K-ATPase.

the plasma membrane, while the α2 isoform is limited to more specific areas in the plasma membrane, namely to regions overlying the endoplasmic reticulum. Also, the α2 isoform is enriched in t-tubules in skeletal muscle compared to the α1 isoform.[11] The Na–Ca exchanger also appears to colocalize in regions of the plasma membrane where the α2 isoform appears.[12] Therefore, it is possible that the α2 Na,K-ATPase and the Na–Ca exchanger are located in a region of the plasma membrane that regulates localized Ca^{2+} levels near the sarcoplasmic reticulum. Thus transient Ca^{2+} levels would be largely regulated by the α2 isoform of the Na,K-ATPase, while the α1 isoform of the Na,K-ATPase would carry out more generalized Na^+ and K^+ transport. A diagram showing this is presented in FIGURE 1. Additional studies are required to determine whether this model is correct.

The α4 isoform is found exclusively in testis and is present in mature sperm and their precursor cells.[13,14] While we have not developed mice lacking the α4 isoform, this isoform can be selectively inhibited in sperm because of its high affinity for ouabain compared to the low affinity of the α1 isoform, the only other isoform present. Ouabain at concentrations affecting only the α4 isoform completely inhibits sperm motility. We have further shown that the addition of nigericin, an H^+/K^+ ionophore, and monensin, an H^+/Na^+ ionophore, reinitiates ouabain-inhibited motility.[15] The K^+ ionophore, valinomycin, has no effect on the motility of ouabain-inhibited sperm. These inhibition studies suggest that H^+ transport is involved and that the α4 isoform may be necessary for the removal of intracellular H^+ from the sperm. It is known that sperm motility is highly sensitive to pH, and if hydrogen ions are allowed to accumulate, sperm motility is decreased. It is possible that the α4 isoform of the Na,K-ATPase drives the removal of H^+ from sperm through a Na/H exchanger. Both Na/H exchanger 1 and Na/H exchanger 5 are localized within the midregion of the sperm tail where the α4 isoform is also located. These studies, along with those above, may indicate that the α2, α3, and α4 isoforms function in a specialized role, possibly coupled with other ion exchangers, while the α1 isoform plays a more general transport function.

In addition to deleting genes, gene targeting can be used to produce animals where a gene has been modified in a particular way. To this end, we are developing mice that exhibit ouabain-resistant α2 and α3 isoforms. This will allow us to selectively inhibit the remaining isoform. In addition, these animals are likely to be useful in evaluating the role of endogenous ouabain. For example, if mice are produced with a low-affinity ouabain α2 isoform, then endogenous ouabain is unlikely to exhibit inhibitory activity on this isoform. If the resulting animals are compromised in some way, this will suggest than an endogenous ouabain may be responsible for the observed defect. On the other hand, if the animals are normal, the studies may suggest endogenous ouabain does not play an important role in regulating the α2 isoform of the Na,K-ATPase. This work is in progress.

SUMMARY

Using gene targeting, we have developed animals lacking either both copies or one copy of the α1 or α2 isoform of the Na,K-ATPase. The α1 isoform is required for animals to develop beyond the blastocyst stage, while animals lacking the α2 isoform develop to term, but are either born dead or die immediately after birth. Animals lacking one copy of the α1 or α2 isoform show different skeletal muscle and cardiac contraction phenotypes. With both tissues, reduction of one-half of the amount of α2 isoform results in an increased force of contraction, while a decreased force of contraction is observed in animals lacking one-half the α1 isoform. These studies indicate that these two isoforms play different functional roles, and it is possible that the α2 isoform is intimately involved in regulating Ca^{2+} levels related to muscle contraction, while the α1 isoform plays a more general transport role. The α4 isoform of the Na,K-ATPase is found exclusively in sperm and its precursor cells and is required for sperm motility. It is hypothesized that the Na,K-ATPase carrying the α4 isoform is responsible for removing H^+ from sperm. It is known that sperm are highly sensitive to intracellular pH, and it is possible that this isoform is coupled to the Na/H exchanger to carry out this function.

ACKNOWLEDGMENT

This work was supported by National Institutes of Health Grants R01 HL28573 and R01 HL66062.

REFERENCES

1. BEGUIN, P., X. WANG, D. FIRSOV, et al. 1997. The γ subunit is a specific component of the Na,K-ATPase and modulates its transport function. EMBO **16:** 4250–4260.
2. THERIEN, A.G., S.J. KARLISH & R. BLOSTEIN. 1999. Expression and functional role of the γ subunit of the Na,K-ATPase in mammalian cells. J. Biol. Chem. **274:** 12252–12256.
3. THERIEN, A.G., R. GOLDSHLEGER, S.J.D. KARLISH & R. BLOSTEIN. 1997. Tissue-specific distribution and modulatory role of the γ subunit of the Na,K-ATPase. J. Biol. Chem. **272:** 32628–32634.
4. ARYSTARKHOVA, E., R.K. WETZEL, N.K. ASINOVSKI & K.J. SWEADNER. 1999. The γ subunit modulates Na^+ and K^+ affinity of the renal Na,K-ATPase. J. Biol. Chem. **274:** 33183–33185.
5. ORLOWSKI, J. & J.B LINGREL. 1988. Tissue-specific and developmental regulation of rat Na,K-ATPase catalytic α isoform and β subunit mRNAs. J. Biol. Chem. **263:** 10436–10442.
6. WOO, A.L., P.F. JAMES & J.B LINGREL. 1999. Characterization of the fourth α isoform of the Na,K-ATPase. J. Membr. Biol. **169:** 39–44.
7. JAMES, P.F., I.L. GRUPP, G. GRUPP, et al. 1999. Identification of a specific role for the Na,K-ATPase α2 isoform as a regulator of calcium in the heart. Mol. Cell **3:** 555–563.
8. HE, S., D.A SHELLY, A.E. MOSELEY, et al. 2001. The α1 and α2 isoforms of Na,K-ATPase play different roles in skeletal muscle contractility. Am. J. Physiol. **281:** R917–R925.
9. JUHASZOVA, M. & M.P. BLAUSTEIN. 1997. Na^+ pump low and high affinity α subunit isoforms are differently distributed in cells. Proc. Natl. Acad. Sci. USA **94:** 1800–1805.
10. JUHASZOVA, M. & M.P. BLAUSTEIN. 1997. Distinct distribution of different Na^+ pump alpha subunit isoforms in plasmalemma. Physiological implications. Ann. N.Y. Acad. Sci. **834:** 524–536.
11. HUNDAL, H.S., D.L. MAXWELL, A. AHMED, et al. 1994. Subcellular distribution and immunocytochemical localization of Na,K-ATPase subunit isoforms in human skeletal muscle. Mol. Membr. Biol. **11:** 255–262.
12. JUHASZOVA, M., A. SHIMIZU, M.L. BORIN, et al. 1996. Localization of the Na^+-Ca^{2+} exchanger in vascular smooth muscle, and in neurons and astrocytes. Ann. N. Y. Acad. Sci. **779:** 318–335.
13. SHAMRAJ, O.I. & J.B LINGREL. 1994. A putative fourth Na^+,K^+-ATPase alpha-subunit gene is expressed in testis. Proc. Natl. Acad. Sci. USA **91:** 12952–12956.
14. WOO, A.L., P.F. JAMES & J.B LINGREL. 2000. Sperm motility is dependent on a unique isoform of the Na,K-ATPase. J. Biol. Chem. **275:** 20693–20699.
15. WOO, A.L., P.F. JAMES & J.B LINGREL. 2002. Roles of the Na,K-ATPase α4 isoform and the Na^+/H^+ exchanger in sperm motility. Mol. Reprod. Dev. **62:** 348–356.

Ion Pump–Interacting Proteins: Promising New Partners

PHILIPP PAGEL, ALESSANDRA ZATTI, TOHRU KIMURA, AMY DUFFIELD, VERONIQUE CHAUVET, VANATHY RAJENDRAN, AND MICHAEL J. CAPLAN

Department of Cellular and Molecular Physiology, Yale University School of Medicine, New Haven, Connecticut 06510, USA

ABSTRACT: The sorting and regulation of the Na,K and H,K-ATPases requires that the pump proteins must associate, at least transiently, with kinases, phosphatases, scaffolding molecules, and components of the cellular trafficking machinery. The identities of these interacting proteins and the nature of their associations with the pump polypeptides have yet to be elucidated. We have begun a series of yeast two-hybrid screens employing structurally defined segments of pump polypeptides as baits in order to gain insight into the nature and function of these interacting proteins.

KEYWORDS: ion pumps; interactions; tetraspan; polycystin; two hybrid

YEAST TWO-HYBRID SCREEN

The yeast two-hybrid system involves the expression in yeast of fusion proteins composed of portions of the GAL4 transcriptional activator.[1,2] In order for GAL4 to function, its DNA binding domain must be brought into close proximity with its transcriptional activation domain. A fusion protein composed of the polypeptide of interest coupled to the GAL4 DNA binding domain ("bait" construct) is coexpressed in yeast with fusion proteins that link the inserts of a cDNA library to the GAL4 transcriptional activator domain. If a cDNA encodes a polypeptide that interacts with the protein of interest, the two fusion proteins will form a complex that is then competent to activate the transcription of a selectable marker and/or a reporter gene under the control of a GAL4-sensitive promoter. This approach is especially well suited to identifying the participants in protein–protein interactions that may not be sufficiently strong to survive a biochemical isolation procedure.

We have used the yeast two-hybrid system to identify proteins that interact with the structurally defined cytoplasmic domains of the Na,K-ATPase α-subunit. This technique has also been applied to search for proteins that interact with the H,K-ATPase β-subunit N terminal cytoplasmic tail. Our studies have made use of the Matchmaker 3 system from Clontech, in which interaction between the GAL4 activation and DNA binding domains triggers transcription of both the HIS3 gene prod-

---

Address for correspondence: Michael J. Caplan, Department of Cellular and Molecular Physiology, Yale University School of Medicine, 333 Cedar Street, New Haven, CT 06510. Voice: 203-785-7316; fax: 203-785-4951.
Michael.caplan@yale.edu

uct and lacZ. Thus, colonies carrying interacting constructs can be rapidly selected on plates lacking histidine (-His), and β-galactosidase activity can be tested to rule out false positives. We have screened a commercially available human kidney cDNA library from Clontech. The poly A–primed kidney cDNAs are subcloned into the multiple cloning site of the pACT2 vector to create the necessary GAL4 activation domain fusion protein library. The bait constructs were transformed into yeast strain AH109, which has a lacZ gene under the control of the GAL4 promoter. To ensure that the bait constructs do not possess autonomous promoting activity, we assayed the singly transformed yeast for β-galactosidase activity.

Transformants expressing activation domain fusion proteins that interact with the bait construct were selected on His- plates and tested for lacZ expression as well, to rule out false positives. False positives due to autonomous promoting activity of activating domain fusion proteins were identified by genetically removing the binding domain construct and substituting empty binding domain plasmid or a binding domain plasmid encoding an unrelated fusion protein.

α-SUBUNIT BAITS

The high resolution structure of the Ca-ATPase reveals that this protein's cytoplasmic sequences fold to form three major structural domains, denoted P, N, and A[3] (FIG. 1). The structure of the Na,K-ATPase is very likely to recapitulate these features. The P domain includes two segments that are far from one another in the linear sequence and that collaborate to generate the region surrounding the phosphorylation site. The residues that are incorporated within the N loops together constitute the nucleotide-binding site. All of the sequence that participates in the formation of the P and N domains is contributed by the large cytoplasmic loop connecting transmembrane helices 4 and 5.

Recently, several groups have prepared bacterially expressed fusion proteins incorporating this entire cytoplasmic loop derived from the Na,K-ATPase α-subunit. Although isolated from contacts with any other portions of the pump proteins, these constructs appear to acquire a structure that is sufficiently similar to the normal folding pattern to permit them to exhibit limited pump-like abilities (Ref. 4 and W. Shoner, personal communication). The fusion proteins are able to bind ATP and to catalyze p-nitrophenylphosphatase activity. We would expect that these same pump sequences, when expressed in the context of a yeast two-hybrid bait fusion protein construct, would manifest the same degree of structural fidelity. Since the N and P domains appear to be responsible for pump-mediated ATP hydrolysis, it is likely that interacting proteins that exert influences on the pumps' enzymatic activities might associate with these domains. Furthermore, our sorting studies demonstrate that a conformational determinant embedded within this region seems to specify pump localization.[5] We chose, therefore, to prepare our first bait construct from the M4-M5 loop in order to search for potential partners of the N and P domains (top panel, FIG. 1). The pump sequence was subcloned into the pGBKT7 vector, in which the insert is fused to the carboxy terminus GAL4 DNA binding domain. As noted above, we have conducted a two-hybrid screen of a human kidney library using a version of this bait incorporating Na,K-ATPase sequence. The functional significance of the proteins identified through this effort are now in the process of being analyzed.

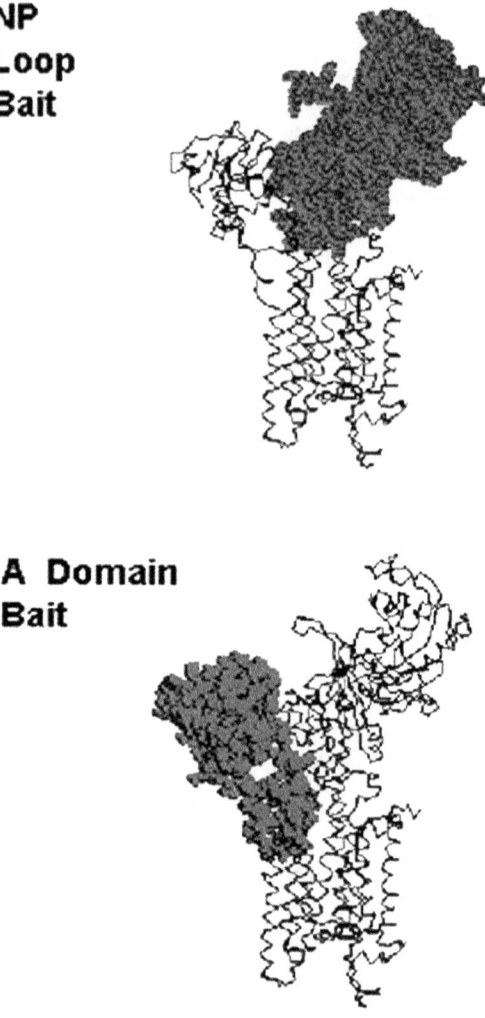

FIGURE 1. α-subunit baits for two-hybrid screens. The NP loop (*top panel*) and the A domain (*bottom panel*) are mapped onto the three-dimensional structure of the Ca-ATPase. The domains used in bait constructions are represented in gray spacefill.

The A domain of the Ca-ATPase is composed of the cytoplasmic N terminal tail arrayed along a surface of the rolled β strand structure created by the loop extending between transmembrane helices M2 and M3.[3] The extensive contacts connecting the N terminal tail to the M2-M3 loop strongly suggest that these two stretches of sequence are tightly bound to one another to form a single structural entity. Support for this hypothesis is found in predictions based on the EM structure of the Ca-ATPase in the E1 conformation. The A domain appears to undergo a dramatic rotation during the course of E1-E2 transition, during the course of which the interrelationship between the N terminal tail and the M2-M3 loop does not appear to be altered.[6,7] The A domain includes the putative PKC phosphorylation site,[8] as well as the putative binding site for phosphoinostide 3-kinase (PI3K).[9] Its extensive motion during the catalytic cycle has been suggested to play a role in the opening and closing of the cation binding pocket.[10] This structural domain, therefore, appears to be a likely substrate for interactions with proteins that regulate pump function.

We have performed two-hybrid screens using as bait a protein construct that should reproduce aspects of the A domain structure (FIG. 1, bottom panel). The α-subunit N terminal tail was attached through a linking sequence composed of (Gly-Gly-Gly-Gly-Ser)$_2$ to the sequence derived from the M2-M3 loop. The linking sequence was chosen based on its proven ability to form a flexible random coil[11] that will offer the N terminal tail sequence the freedom to participate in the energetically favorable associations that link it to the M2-M3 loop *in situ*. The pump sequence was subcloned into the pD134 vector, in which the insert is placed at the amino terminus of the GAL4 DNA binding domain. This orientation was expected to preserve the native topology of the pump sequences and allow them to fold to create a structure resembling the A domain if this assembly is sufficiently energetically favorable. In this manner it should have been possible to detect protein partners whose interactions with the A domain are dependent upon its native conformation. It is, of course, quite possible that this expectation is unjustified and the two portions of the bait construct do not interact appropriately. Even in this case, however, use of this construct in which the N terminus and the M2-M3 loop are arrayed in tandem should permit identification of those proteins capable of interacting with the individual linear sequences that constitute the A domain. Once again, the A domain bait construct incorporating Na,K-ATPase sequence was used in a screen against a human kidney cDNA library.

We have completed the yeast two-hybrid screens employing the large cytoplasmic loop containing the N and P domains as well as the A domain of the Na,K-ATPase α-subunit as baits. A screen using the N-terminal cytosolic tail of the H,K-ATPase β-subunit as bait has also been conducted. All of the candidates identified in the α-subunit screen were independently identified multiple times. Furthermore, they all meet criteria for specificity in the yeast two-hybrid assay. When expressed alone or in association with empty bait vector, they do not produce any detectable autoactivation. We have begun the process of characterizing the repertoire of interacting proteins detected in these screens and have selected the several polypeptides listed below for further analysis.

Na,K-ATPase α-SUBUNIT–CANDIDATE INTERACTING PROTEINS

Protein Phosphatase 2A Catalytic Subunit β

The Na,K-ATPase is regulated by PKC phosphorylation.[8,12–18] Furthermore, it has been suggested that recruitment of the molecules that may mediate pump endocytosis in response to dopamine stimulation may involve the activities of protein kinases and phosphatases.[19] In most cases the catalytic subunits of protein phosphatases associate with phosphoproteins via targeting subunits.[20] The targeting sub-

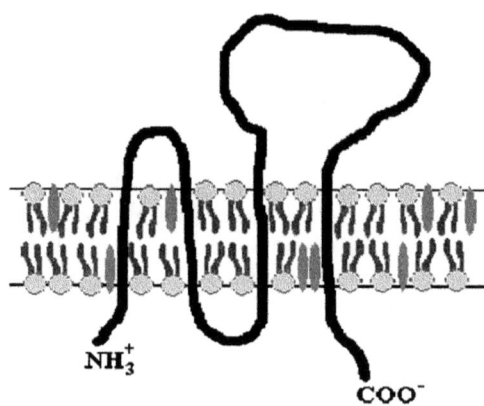

FIGURE 2. Generic structure of tetraspan proteins. Tetraspan proteins are thought to possess four transmembrane helices and short cytoplasmic N and C terminal tails. CD63 is heavily glycosylated on the larger second extracellular loop, whereas CD81 does not appear to carry any carbohydrates.

unit manifests a sequence motif that binds directly to the catalytic polypeptide. It has recently been shown, however, that substrate proteins can themselves express this catalytic subunit binding motif and thus bind directly to the catalytic subunits. This mechanism has been convincingly demonstrated for the Na,K,2Cl cotransporter,[21] suggesting that this interaction may also be physiologically relevant to both the regulation and the trafficking of the sodium pump as well.

CD81

The CD81 protein is a member of the tetraspan superfamily.[22] As their name implies, tetraspan proteins possess four membrane spanning domains (FIG. 2). Mammalian genomes appear to encode at least 28 distinct tetraspan sequences. All of the members of this family are relatively short (~250 amino acids), and all possess ~2 dozen conserved residues.[23] Tetraspan proteins are known to participate in a variety of complexes, with each other and with other membrane proteins, to form what has come to be called the "tetraspan web." Different tetraspan subclasses appear to exhibit distinct subcellular distributions. Like the Na,K-ATPase, CD81 is restricted to the basolateral plasmalemmal surfaces of epithelial cells.[24] Furthermore, while very little is known of the function of tetraspans, their ability to form complexes with a wide variety of membrane proteins suggests that they may participate in the formation of scaffolds and membrane domains that regulate signaling and sorting processes.

Neurabin-II/Spinophilin Binding Protein (rat lin10)

The *C. elegans* lin10 gene participates in a multiprotein complex involved in localizing the let 23 receptor to the basolateral surfaces of developing vulvar epithelial cells.[25] The initial positional cloning efforts resulted in misidentification of the lin10 coding sequence.[26] Before this error was discovered, however, a rat homologue of the misidentified gene product was cloned and named "rat lin10."[27] While rat lin10 exhibits no relationship to the protein subsequently shown to be responsible for lin10 activity, it turns out to be a fascinating protein that may link the sodium pump to a number of signaling and trafficking systems. A two-hybrid screen demonstrated that rat lin10 is associated with neurabin II/spinophilin.[28] This polypeptide is a targeting subunit for protein phosphatase I,[29] which has, in turn, been implicated in Na,K-ATPase regulation. Once again, therefore, our screen has identified an interaction that could bring a protein phosphatase into close proximity to the pump's own phosphorylation sites and the phosphorylation sites of other pump-associated proteins. Neurabin/spinophillin is also an actin-binding protein and may play a role in linking plasma membrane proteins, such as TGN38[30] and the D2 dopamine receptor,[31] to the cytoskeleton. The impact of D2 receptor stimulation on pump function in the renal proximal tubule[32] renders the prospect of a cytoskeletal scaffold linking the D2 dopamine receptor to the Na,K-ATPase especially interesting. Finally, neurabin/spinophillin has a PDZ domain and, like the Na,K-ATPase, is localized to lateral membranes in polarized epithelial cells.[33] Thus, the rat lin10 association may link the sodium pump to a variety of molecules involved in signal transduction and membrane protein localization.

SNARE-Associated Protein

The SNARE molecules are the principal transmembrane partners in the protein complex that mediates membrane fusion events.[34] The specificity and selectivity of fusion between intracellular vesicles and target membranes is determined, at least in part, by the complement of SNARE proteins that both the vesicular and target membranes possess.[35,36] Through an interaction with a SNARE-associated protein, the sodium pump may be sequestered into basolaterally directed vesicles during its initial biosynthetic sorting. Conversely, such an association might recruit the appropriate SNARE proteins into a forming carrier vesicle accumulating a cohort of sodium pumps. We expect, therefore, that this interaction could be relevant to Na,K-ATPase sorting and trafficking.

Polycystin 1

The C terminal tail of polycystin 1 was detected as an interacting protein in both the N,P loop and A domain screens. Mutation of the polycystin 1 (or PKD1) gene results in autosomal-dominant polycystic kidney disease.[37] Polycystin 1 is a massive membrane protein, with 11 predicted transmembrane spans and an extremely large N-terminal extracellular domain.[38] Its extracellular domain is thought to participate in the formation of intercellular contacts. The C-terminal cytoplasmic tail, which was identified in our screen as a putative Na,K-ATPase partner, has been shown to associate with a number of structural and signaling proteins, including β-catenin.[39,40] It has been suggested that Na,K-ATPase is mislocalized to the apical membrane in cyst lining epithelial cells from patients carrying polycystin 1 mutations.[41,42] An association between polycystin 1 and the sodium pump, therefore, could play a role in restricting the Na,K-ATPase to the basolateral domain or in modulating its activity in response to cues from neighboring cells.

H,K-ATPase β-SUBUNIT–INTERACTING PROTEIN

Under resting conditions, the gastric H,K-ATPase resides in an intracellular membranous storage compartment in its native gastric parietal cell.[43] Secretagogue stimulation of gastric acid production results in the fusion of these storage vesicles, which are referred to as tubulovesicular elements (TVE), with the apical plasma membrane. The cessation of gastric acid secretion requires that the H,K pump be endocytosed from the apical plasmalemma in a membrane internalization event that regenerates the TVE. We have previously shown that the cytoplasmic N terminal tail of the H,K-ATPase β-subunit contains a tyrosine-based endocytosis signal that is required for this regulated internalization.[44] Transgenic mice expressing H,K-ATPase β-subunit in which this signal has been disrupted are unable to remove H,K-ATPase from the cell surface and constitutively secrete gastric acid. We have now found that the β-subunit of the H,K-ATPase interacts directly and specifically with CD63. Like CD81, CD63 is a member of the tetraspan family and shares the same membrane topology (FIG. 2). While CD81 is localized to the basolateral plasma membrane in renal epithelial cells, CD63 is found in intracellular vesicular structures, such as endosomes, lysosomes, and secretory vesicles.[45] We find that CD63 is present in gastric TVE and interacts with the H,K-ATPase β-subunit in gastric tissue *in situ*. Furthermore, this interaction is required for the rapid endocytosis of the H,K-ATPase β-subunit protein. In the absence of CD63, the H,K-ATPase β-subunit accumulates at the plasmalemma, whereas H,K-ATPase β-subunit coexpressed with CD63 is rapidly endocytosed and accumulates in intracellular vesicles. These findings are especially interesting in light of the fact that CD63 binds to phosphoinositide 4-kinase, an enzyme whose activity has been shown to regulate a variety of membrane trafficking events.[46]

CONCLUSION

This short list of putative pump-interacting proteins is almost certainly far from complete. Our future efforts will focus on characterizing the physiological roles that these polypeptides may play in regulating pump distribution and function. We will also seek to expand the list of potential pump partners in a variety of cell types. It is only through this sort of analysis that we will come to better understand the pumps' quarternary structures as they exist in the plasma membranes of intact cells.

ACKNOWLEDGMENTS

We are grateful to all of the members of the Caplan Lab for their helpful ideas and discussions. This work was supported by NIH Grants GM42136 and DK17433.

REFERENCES

1. CHIEN, C.T. *et al.* 1991. The two-hybrid system: a method to identify and clone genes for proteins that interact with a protein of interest. Proc. Natl. Acad. Sci. USA **88:** 9578–9582.

2. FIELDS, S. & O. SONG. 1989. A novel genetic system to detect protein-protein interactions. Nature **340:** 245–246.
3. TOYOSHIMA, C. et al. 2000. Crystal structure of the calcium pump of sarcoplasmic reticulum at 2.6 A resolution. Nature **405:** 647–655.
4. GATTO, C., A.X. WANG & J.H. KAPLAN. 1998. The M4M5 cytoplasmic loop of the Na,K-ATPase, overexpressed in *Escherichia coli*, binds nucleoside triphosphates with the same selectivity as the intact native protein. J. Biol. Chem. **273:** 10578–10585.
5. DUNBAR, L.A., P. ARONSON & M.J. CAPLAN. 2000. A transmembrane segment determines the steady-state localization of an ion-transporting adenosine triphosphatase. J. Cell Biol. **148:** 769–778.
6. STOKES, D.L. et al. 1999. Comparison of H^+-ATPase and Ca^{2+}-ATPase suggests that a large conformational change initiates P-type ion pump reaction cycles. Curr. Biol. **9:** 672–679.
7. ZHANG, P. et al. 1998. Structure of the calcium pump from sarcoplasmic reticulum at 8-A resolution. Nature **392:** 835–839.
8. THERIEN, A.G. & R. BLOSTEIN. 2000. Mechanisms of sodium pump regulation. Am. J. Physiol. Cell Physiol. **279:** C541–566.
9. YUDOWSKI, G.A. et al. 2000. Phosphoinositide-3 kinase binds to a proline-rich motif in the Na^+, K^+-ATPase alpha subunit and regulates its trafficking. Proc. Natl. Acad. Sci. USA **97:** 6556–6561.
10. TOYOSHIMA, C. & H. NOMURA. 2002. Structural changes in the calcium pump accompanying the dissociation of calcium. Nature **418:** 605–611.
11. HUSTON, J.S. et al. 1988. Protein engineering of antibody binding sites: recovery of specific activity in an anti-digoxin single-chain Fv analogue produced in *Escherichia coli*. Proc. Natl. Acad. Sci. USA **85:** 5879–5883.
12. CHIBALIN, A.V. et al. 1999. Dopamine-induced endocytosis of Na^+,K^+-ATPase is initiated by phosphorylation of Ser-18 in the rat alpha subunit and is responsible for the decreased activity in epithelial cells. J. Biol. Chem. **274:** 1920–1927.
13. CHIBALIN, A.V. et al. 1998. Phosphorylation of the catalyic alpha-subunit constitutes a triggering signal for Na^+,K^+-ATPase endocytosis. J. Biol. Chem. **273:** 8814–8819.
14. CHIBALIN, A.V. et al. 1997. Receptor-mediated inhibition of renal Na(+)-K(+)-ATPase is associated with endocytosis of its alpha- and beta-subunits. Am. J. Physiol. **273:** C1458–1465.
15. OGIMOTO, G. et al. 2000. G protein-coupled receptors regulate Na^+,K^+-ATPase activity and endocytosis by modulating the recruitment of adaptor protein 2 and clathrin. Proc. Natl. Acad. Sci. USA **97:** 3242–3247.
16. CHIBALIN, A.V. et al. 1998. Phosphatidylinositol 3-kinase-mediated endocytosis of renal Na^+, K^+-ATPase alpha subunit in response to dopamine. Mol. Biol. Cell. **9:** 1209–1220.
17. FISONE, G. et al. 1995. Na+,K(+)-ATPase in the choroid plexus. Regulation by serotonin/protein kinase C pathway. J. Biol. Chem. **270:** 2427–2430.
18. LI, D. et al. 1998. Effects of okadaic acid, calyculin A, and PDBu on state of phosphorylation of rat renal Na^+-K^+-ATPase. Am. J. Physiol. **275:** F863–869.
19. YUDOWSKI, G.A., R. EFENDIEV, C.H. PEDEMONTE, et al. 2000. Dephosphorylation of dynamin by PP2a is critical for Na,K-ATPase endocytosis in response to dopamine receptor signals. J. Am. Soc. Nephrology **11:** 54a.
20. FAUX, M.C. & J.D. SCOTT. 1996. More on target with protein phosphorylation: conferring specificity by location. Trends Biochem. Sci. **21:** 312–315.
21. DARMAN, R.B., A. FLEMMER & B. FORBUSH. 2001. Modulation of ion transport by direct targeting of protein phosphatase type 1 to the Na-K-Cl cotransporter. J. Biol. Chem. **276:** 34359–34362.
22. LEVY, S., S.C. TODD & H.T. MAECKER. 1998. CD81 (TAPA-1): a molecule involved in signal transduction and cell adhesion in the immune system. Annu. Rev. Immunol. **16:** 89–109.
23. HEMLER, M.E. 2001. Specific tetraspanin functions. J. Cell Biol. **155:** 1103–1107.
24. OKOCHI, H. et al. 1999. Expression of tetraspans transmembrane family in the epithelium of the gastrointestinal tract. J. Clin. Gastroenterol. **29:** 63–67.

25. WHITFIELD, C.W. *et al.* 1999. Basolateral localization of the *Caenorhabditis elegans* epidermal growth factor receptor in epithelial cells by the PDZ protein LIN-10. Mol. Biol. Cell. **10:** 2087–2100.
26. KIM, S.K. & H.R. HORVITZ. 1990. The *Caenorhabditis elegans* gene lin-10 is broadly expressed while required specifically for the determination of vulval cell fates. Genes Dev. **4:** 357–371.
27. IDE, N. *et al.* 1998. Molecular cloning and characterization of rat lin-10. Biochem. Biophys. Res. Commun. **243:** 634–638.
28. IDE, N. *et al.* 1998. Interaction of rat lin-10 with brain-enriched F-actin-binding protein, neurabin-II/spinophilin. Biochem. Biophys. Res. Commun. **244:** 258–262.
29. HSIEH-WILSON, L.C. *et al.* 1999. Characterization of the neuronal targeting protein spinophilin and its interactions with protein phosphatase-1. Biochemistry **38:** 4365–4373.
30. ROUS, B.A. *et al.* 2002. Role of adaptor complex AP-3 in targeting wild-type and mutated CD63 to lysosomes. Mol. Biol. Cell. **13:** 1071–1082.
31. SMITH, F.D., G.S. OXFORD & S.L. MILGRAM. 1999. Association of the D2 dopamine receptor third cytoplasmic loop with spinophilin, a protein phosphatase-1-interacting protein. J. Biol. Chem. **274:** 19894–19900.
32. BERTORELLO, A. & A. APERIA. 1990. Inhibition of proximal tubule Na(+)-K(+)-ATPase activity requires simultaneous activation of DA1 and DA2 receptors. Am. J. Physiol. **259:** F924–928.
33. SATOH, A. *et al.* 1998. Neurabin-II/spinophilin. An actin filament-binding protein with one pdz domain localized at cadherin-based cell-cell adhesion sites. J. Biol. Chem. **273:** 3470–3475.
34. FERRO-NOVICK, S. & R. JAHN. 1994. Vesicle fusion from yeast to man. Nature **370:** 191–193.
35. BENNETT, M.K. 1995. SNAREs and the specificity of transport vesicle targeting. Curr. Opin. Cell Biol. **7:** 581–586.
36. GUO, W. *et al.* 2000. Protein complexes in transport vesicle targeting. Trends Cell Biol. **10:** 251–255.
37. TORRES, V.E. 1998. New insights into polycystic kidney disease and its treatment. Curr. Opin. Nephrol. Hypertens. **7:** 159–169.
38. HUGHES, J. *et al.* 1995. The polycystic kidney disease 1 (PKD1) gene encodes a novel protein with multiple cell recognition domains. Nat. Genet. **10:** 151–160.
39. KIM, E. *et al.* 1999. The polycystic kidney disease 1 gene product modulates Wnt signaling. J. Biol. Chem. **274:** 4947–4953.
40. HUAN, Y. & J. VAN ADELSBERG. 1999. Polycystin-1, the PKD1 gene product, is in a complex containing E-cadherin and the catenins. J. Clin. Invest. **104:** 1459–1468.
41. WILSON, P.D. *et al.* 1991. Reversed polarity of Na(+)-K(+)-ATPase: mislocation to apical plasma membranes in polycystic kidney disease epithelia. Am. J. Physiol. **260:** F420–430.
42. AVNER, E.D., W.E. SWEENEY JR. & W.J. NELSON. 1992. Abnormal sodium pump distribution during renal tubulogenesis in congenital murine polycystic kidney disease. Proc. Natl. Acad. Sci. USA **89:** 7447–7451.
43. HERSEY, S.J. & G. SACHS. 1995. Gastric acid secretion. Physiol. Rev. **75:** 155–189.
44. COURTOIS-COUTRY, N. *et al.* 1997. A tyrosine-based signal targets H/K-ATPase to a regulated compartment and is required for the cessation of gastric acid secretion. Cell **90:** 501–510.
45. KOBAYASHI, T. *et al.* 2000. The tetraspanin CD63/lamp3 cycles between endocytic and secretory compartments in human endothelial cells. Mol. Biol. Cell. **11:** 1829–1843.
46. YAUCH, R.L. & M.E. HEMLER. 2000. Specific interactions among transmembrane 4 superfamily (TM4SF) proteins and phosphoinositide 4-kinase. Biochem. J. **351:** 629–637.

Amino Acids in the TM4-TM5 Loop of Na,K-ATPase Are Important for Biosynthesis

JESPER R. JØRGENSEN, JENS HOUGHTON-LARSEN, METTE DORPH JACOBSEN, AND PER AMSTRUP PEDERSEN

Biomembrane Research Centre, August Krogh Institute, University of Copenhagen, 2100 Copenhagen OE, Denmark

ABSTRACT: The ten-transmembrane Na,K-ATPase α-subunit exposes very few amino acids to the extra membrane space except for an approximately 408 residue-long loop between transmembrane segments four and five. The present paper focuses on the role of this loop in biosynthesis of functional Na,K-ATPase. Expression of 39 mutations in this loop to phylogenetically conserved as well as nonconserved residues showed that only two could be expressed at 30°C. By contrast, only five could not be produced in a functional form at 15°C. A detailed analysis showed that a number of these mutants are temperature-sensitive folding mutants, as they induce the unfolded protein response at 30°C but not at 15°C. We used an algorithm to predict that residues ^{868}ENGFLIPIHLL^{878} in the L78 loop exposed to the endoplasmic reticulum lumen constitute the most likely BiP binding site. Correct folding of this sequence may be important in the endoplasmic reticulum quality control, as the same loop is responsible for the α-β-associations required to leave this compartment. On the basis of the Ca-ATPase crystal structure and the presented data, we propose a model to account for the role of the TM4-TM5 loop in Na,K-ATPase biosynthesis.

KEYWORDS: Na,K-ATPase; protein folding; yeast; heterologous expression; membrane proteins

The topology of polytopic membrane proteins is generally believed to be determined by signal anchor and stop transfer sequences.[1] *In vivo* studies in *Xenopus* oocytes using the reporter glycosylation assay have shown that TM1 and TM3 work as signal anchor sequences, and that TM2 and TM4 work as stop transfer sequences in biosynthesis of Na,K-ATPase.[2] Efficient membrane insertion of the four N-terminal membrane segments has also been demonstrated for the homologous sarcoplasmic reticulum Ca-ATPase[3] and H,K-ATPase.[4,5] Full-length Na,K-ATPase α-subunits are unstable in *Xenopus* oocytes without coexpression of the partner β-subunit and rapidly degraded. By contrast, C-terminally truncated α-subunit protein, including

---

Address for correspondence: Per Amstrup Pedersen, Biomembrane Research Centre, August Krogh Institute, University of Copenhagen, 2100 Copenhagen OE, Denmark. Voice: +45-35321667; fax: +45-35321567.
PApedersen@aki.ku.dk

TM1-TM4 and the large cytoplasmic domain, is retained in the endoplasmic reticulum (ER) but stable when expressed in excess of the endogenous β-subunit. The large cytoplasmic loop does not, therefore, per se expose any degradation signals during biosynthesis and must fold into a stable conformation, protecting it from proteolytic attack or prolonged association with chaperones.

Membrane insertion and folding of TM5 and TM6 are more complicated than for the four N-terminal transmembrane segments TM1-TM4. Presence of two helix-braking prolines and a charged residue in TM5 results in poor signal anchor activity.[2,6–8] Only 5% of the total TM5 population is inserted into the membrane when TM1-TM5 are expressed in *Xenopus* oocytes and analyzed by the reporter glycosylation assay.[2,6] Furthermore the truncated protein is unstable and degraded by the proteasome.[6] Since TM6 does not have any topogenic activity *in vitro*, posttranslational interactions between TM5 and TM6 may be important for membrane insertion of this pair.[2,6] Furthermore since mutations to residues TM5 and TM6 can change the topogenic activity of these segments, efficient membrane insertion may require some sequence-specific information. Membrane insertion of TM5 and TM6 does not, therefore, exclusively follow the signal anchor and stop transfer mechanism.

TM7 has only partial signal anchor activity due to the presence of helix-braking glycine and polar residues.[2] Like TM1-TM7, α-subunits that constitute TM1-TM8 are unstable when expressed alone, but coexpression with the β-subunit stabilizes the membrane insertion of the TM7-TM8 pair.[6] As the extracellular loop connecting TM7 and TM8 is essential for interactions with the β-subunit, this subunit may promote efficient membrane insertion of the TM7-TM8 membrane pair.

TM9 has poor membrane insertion properties because of the presence of two negatively charged and one polar residue, resulting in less overall hydrophobicity. Formation of the TM9-TM10 pair is mainly due to a strong stop transfer sequence in TM10.[2]

The crystal structure of the Ca-ATPase shows that TM5 is directly connected to the catalytic site.[9] The purpose of the present study has been to test the idea that the cytoplasmic protrusion guides membrane insertion of at least TM5 and subsequently promotes the efficient helix packing of the following transmembrane segments that are required for biosynthesis.

AMINO ACID SUBSTITUTIONS IN THE TM4-TM5 LOOP INTERFERE WITH α-SUBUNIT ACCUMULATION

In search of residues involved in nucleotide (ATP/ADP) binding or magnesium coordination,[10,11] we have generated a number of mutations in the α1 (TM4-TM5) loop of pig kidney Na,K-ATPase (TABLE 1). Some of these substitutions target phylogenetically conserved amino acids, while others are specific for the Na,K-ATPase. We initially screened for functional expression of these mutations in yeast after growth at temperatures between 30 and 35°C. Data in TABLE 1 show that the capacity for [^{3}H]-ouabain binding was between 0 and 0.5 pmol/mg for the mutants compared to an ouabain density of 8–15 pmol/mg for wild-type enzyme. Western blotting showed that only wild-type α-subunits accumulated at this temperature (data not shown). These mutants therefore seemed well suited for analyzing the involvement of the TM4-TM5 loop in α-subunit biosynthesis.

TABLE 1. Density of [^{3}H]-ouabain sites in membranes from yeast expressing mutant or wild-type Na,K-ATPase at 15°C or 30°C[a]

Allele	Expression at 30°C B_{max} (pmol/mg)	Expression at 15°C B_{max} (pmol/mg)
wt	8.7	12.7
D369A	9.4	ND
D369N	10.9	ND
R544K	ND	5.0
R544Q	ND	10.4
R544A	0	0
C549S	0	9.3
C549A	0.05	4.5
D555E	0.05	7.0
D555N	0	6.7
E556D	0	8.1
E556Q	0.09	3.5
D565E	0	9.7
D565N	0	5.6
D567E	0.18	13.8
D567N	0	7.3
D568N	0	7.1
D568E	0.14	10.8
D574N	0.08	6.4
D574E	0.006	8.7
D586N	0.02	0
D586E	0.10	5.5
R589Q	0	3.4
R589K	0.20	12.4
D594N	0.26	12.5
D594E	0.42	15.9
T708A	0	4.8
T708S	0	0.77
D710A	0	9.1
D710N	0	14.1
N713A	0.04	22
N713Q	0	13.9
D714A	0	0
D714N	0	0
S715A	0.6	4.6
S715T	0	3.5
K719A	0	8.3
K719R	0	11.2
K720A	0	4.5
K720R	0.006	10.2

[a]Membranes were isolated 48 hours after induction of Na,K-ATPase biosynthesis. ND, not determined.

MOST AMINO ACID SUBSTITUTIONS IN THE TM4-TM5 LOOP ACCUMULATE IN A TEMPERATURE-SENSITIVE WAY

A very successful approach in classical genetics has been to generate temperature-sensitive mutations that allow correct protein folding at the permissive temperature, but prevent folding at the restrictive temperature. We therefore analyzed whether the mutations depicted in TABLE 1 were temperature-sensitive mutations. The result of [^{3}H]-ouabain binding to membranes isolated from yeast strains expressing mutant or wild-type protein at 15°C is shown in TABLE 1. It can be seen that all mutants except R544A, D586N, T708S, D714A, and D714N accumulated to a level comparable to that of wild type. Western blotting to the same membranes showed that all proteins except R544A, D586N, and T708S accumulated at 15°C to approximately wild-type level (data not shown). The D714A and D714N mutants were the only ones to accumulate in a form that was unable to bind ouabain.[10]

TM4-TM5 MUTANT PROTEINS ARE RECOGNIZED BY THE ENDOPLASMIC RETICULUM QUALITY CONTROL AS MISFOLDED

The ER is responsible for quality control of secreted and membrane proteins. Only correctly folded proteins are allowed to leave the ER for their final destination. Presence of misfolded protein in the ER lumen or ER membrane induces a signal transduction pathway designated the unfolded protein response (UPR).[12] The molecular mechanism behind the unfolded protein response in yeast cells is depicted in FIGURE 1.

We expressed some of our mutants in the TM4-TM5 loop at 15°C and 35°C in a yeast strain carrying a chromosomally integrated URP reporter to see whether protein folding was compromised at the restrictive temperature. Data in FIGURE 2 show induction of the UPR with respect to expression temperature for the two representative mutants N713A and S715A. It can be seen that expression of these mutant proteins increased the unfolded protein response up to 34-fold compared to wild type. Expression of the other mutants in the ^{708}TGDGVNDSPALKK^{720} sequence also induced the unfolded protein response.[13]

THE L78 LOOP EXPOSES A POTENTIAL BiP BINDING SITE

Since mutations to the ^{708}TGDGVNDSPALKK^{720} segment in Na,K-ATPase induce the unfolded protein response, mutant protein must be recognized from the ER lumen. Therefore, the molecular mechanism underlying recognition and degradation of mutant proteins must involve more that recognition of degradation signals from the cytosolic side. However, only a few residues are exposed to the ER lumen during synthesis. Nevertheless, unassembled Na,K-ATPase α-subunits are coimmunoprecipitated with BiP.[14,15] The ability to bind BiP is a direct link to the unfolded protein response.[12] A central question is, therefore, how BiP recognizes and binds to Na,K-ATPase. BiP, with a peptide binding site of only seven amino acids, prefers an extended conformation of aromatic and hydrophobic residues in the binding cleft.[16,17]

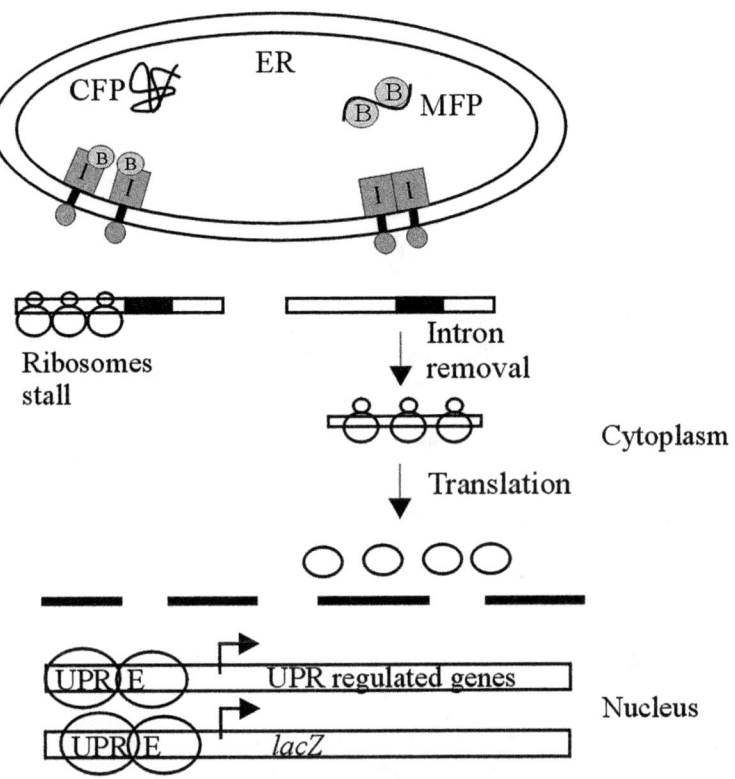

FIGURE 1. The molecular mechanism behind the unfolded protein response (UPR). In the presence of correctly folded protein (CFP) in the ER lumen or membrane, the molecular chaperone BiP (B) binds to the transmembrane Irep protein (I) and prevents it from dimerizing. In the presence of misfolded protein (MFP) in the ER, BiP is recruited from the Irep protein to the misfolded protein. As a consequence, Irep dimerizes and cross-phosphorylates. The dimerized Irep protein has RNase activity and removes an intron in the Hac1 mRNA. A t-RNA ligase joins the outermost ends of the digested Hac1 mRNA molecule, which is subsequently translated. The resulting Hac1p protein enters the nucleus and activates transcription from genes carrying an unfolded protein response element (UPRE) in their promoters. Our yeast reporter strain carries a fusion between a minimal promotor, the yeast UPRE, and the *E. coli* lacZ gene encoding β-galactosidase. Concomitant with induction of normal UPR genes, the yeast reporter strain will induce transcription of the lacZ gene. The level of β-galactosidase will subsequently report the degree of UPR induction.

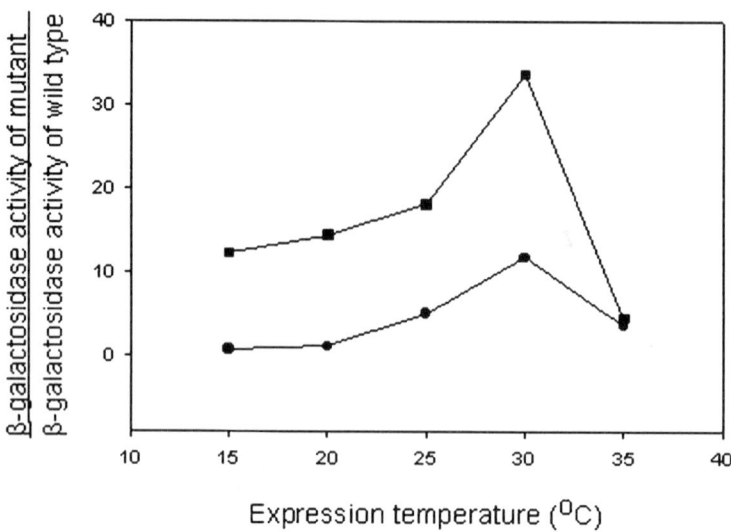

FIGURE 2. Induction of the unfolded protein response with respect to temperature. The induced unfolded protein response is given relative to wild type. Yeast cells grown at 30°C until $OD_{450} = 1$ were separated in 5 aliquots. After 30 minutes at the indicated temperature, Na,K-ATPase biosynthesis was induced by addition of 2% galactose. The cytoplasm was isolated after 48 hours and assayed for β-galactosidase activity. ■, S715A; ●, N713A.

However, hydrophobicity per se is not useful for prediction of BiP binding capacity. Therefore, affinity panning of phage libraries displaying random peptides was used to characterize BiP binding sequences from protein primary structures.[16] FIGURE 3 shows the BiP score of L78, the largest extracellular loop of the α-subunit. It can be seen that especially the ^{868}ENGFLIPIHLL^{878} sequence in the beginning of L78 is very likely to bind BiP. In addition several heptapeptides further downstream in L78 are potential BiP binding sites as well. None of the other loops facing the ER lumen are predicted to bind BiP (data not shown). When predicted BiP binding sequences are mapped onto the three-dimensional structure of the Fd antibody fragment, the majority of binding sites are located in contact points between the heavy and light chains.[18] Even though the contact site for the β-subunit of Na,K-ATPase[6,19] only shows moderate BiP binding capacity, association between subunits may be a key event that prevents rebinding of BiP and promotes further maturation. This is in line with the observation that BiP binding is reduced when the α-subunit associates with the β-subunit.[14] It is, therefore, proposed that mutations to the TM4-TM5 protrusion interfere with helix packing through TM5 and impede correct association with the β-subunit, leading to prolonged BiP binding to L78, ultimately resulting in UPR induction and protein degradation.

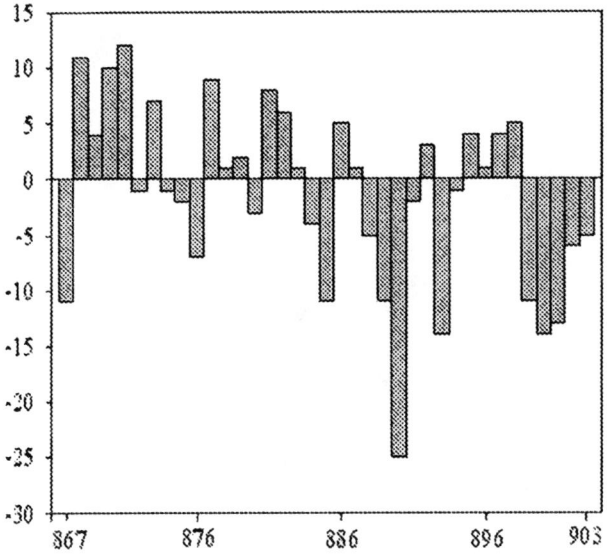

FIGURE 3. BiP score of the loop between TM7 and TM8 (L78). The most probable BiP binding site is shown in bold, while the primary structure required for β-subunit assembly is underlined. The algorithm scores amino acid sequences with a moving window of seven residues, corresponding to the binding cleft of BiP. Each amino acid is assigned an individual score, depending on the probability that BiP will bind to that particular hepta peptide.[16] Hepta peptides scoring higher than +10 will have an extremely high probability of BiP binding, while peptides with a score between +6 and +10 have 3 to 1 odds in favor of BiP binding. Scores from 0 to +5 have little predictive value, but hepta peptides with negative scores almost certainly never bind BiP. Hepta peptides are numbered according to the first amino acid.

A FOLDING MODEL FOR Na,K-ATPase

FIGURE 4 shows a model of the folding of wild-type and mutant α-subunit Na,K-ATPase at 35°C. In this model, membrane insertion of TM1-TM4 and folding of the large cytoplasmic domain up to the mutated residues is identical for wild-type and mutant protein. Subsequently, for wild-type protein, folding of the large cytoplasmic domain is completed, and the C-terminal transmembrane pairs inserted into the membrane. After association with the β-subunit, the stabilized heterodimer escapes the ER and is transported to the plasma membrane. In the case of mutant protein local folding until TM5 is disturbed, resulting in inefficient membrane insertion of the TM5-TM6 hairpin or interference with packing of C-terminal transmembrane helixes. As mutations in the TM4-TM5 cytoplasmic loop induce the unfolded protein response, it is most likely that at least TM5-TM8 to some extend are inserted into the membrane, but packed in a way that is recognized as misfolded.

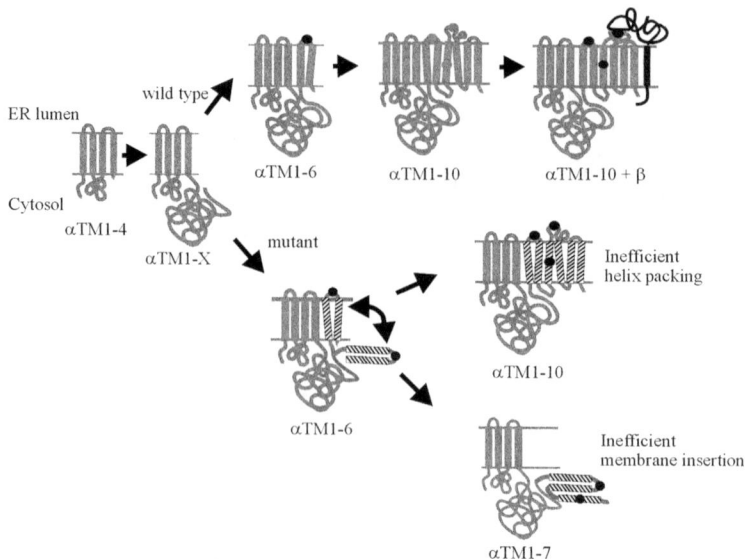

FIGURE 4. Folding of wild-type (*top panel*) and mutant (*lower panel*) α-subunit Na,K-ATPase at 35°C. Wild-type Na,K-ATPase is inserted into the ER membrane, associates with the β-subunit, and exits. By contrast, mutations to the cytoplasmic loop disturb packing of the C-terminal transmembrane helixes or cause inefficient membrane insertion. *Black dots* indicate the position of intrinsic degradation signals.[6] The β-subunit protein is shown in *black*.

It has been shown that inefficient membrane insertion of the C-terminal transmembrane pairs results in exposure of intrinsic degradation signals when the α-subunit is expressed alone.[6] Degradation signals are located in the loop between TM5 and TM6, and within TM7, and are exposed to the cytosol during biosynthesis. Furthermore, the loop between TM7 and TM8, interacting with the β-subunit, contains a degradation signal that is exposed to the ER lumen during biosynthesis, which could possibly be recognized by chaperones. All these degradation signals are masked only in the case of complete synthesis of the α-subunit, association with the β-subunit, and complete membrane insertion packing, as in the case of wild-type Na,K-ATPase.

CONCLUSIONS

Correct folding of the TM4-TM5 loop in the α1 subunit from pig kidney is necessary for biosynthesis of functional Na,K-ATPase. Misfolding of the cytoplasmic loop may interfere with insertion of TM5-TM10, as temperature-sensitive mutants are recognized as misfolded from the ER luminal side at the restrictive temperature. At least a part of the ER quality control seems to operate through BiP that has a potential recognition site in the L78 loop. TM5 may behave as a stick on a chain. Only optimal handling of the chain orients TM5 correctly toward the ER membrane.

REFERENCES

1. SAKAGUCHI, M. 1997. Mutational analysis of signal-anchor and stop-transfer sequences in membrane proteins. *In* Membrane Protein Assembly. R. Gunnar von Heijne, Ed.: 135–150. Landes. Austin, TX.
2. BEQUIN, P. *et al.* 1998. Membrane integration of α-subunits and β-subunit assembly. J. Biol. Chem. **273:** 24921–24931.
3. BAYLE, D. *et al.* 1997. In vitro translation analysis of integral membrane proteins. J. Recept. Signal Transduc. Res. **17:** 29–56.
4. BAMBERG, K. & G. SACHS. 1994. Topological analysis of H^+,K^+ATPase using in vitro translation. J. Biol. Chem. **269:** 16909–16919.
5. BEGGAH, A.T. *et al.* 1999. β-subunit assembly is essential for the correct packing and the stable membrane insertion of H,K-ATPase α-subunit. J. Biol. Chem. **274:** 8217–8223.
6. BEGUIN, P. *et al.* 2000. Endoplasmic reticumlum quality control of oligomeric membrane proteins: topogenic determinants involved in the degradation of the unassembled Na,K-ATPase α-subunit and its stabilisation by β-subunit assembly. Mol. Biol. Cell **11:** 1657–1672.
7. HOMAREDA, H. *et al.* 1989. Loacation of signal sequences for membrane insertion of the Na^+,K^+-ATPase alpha subunit. Mol. Cell. Biol. **9:** 5742–5745.
8. XIE, Y. & T. MORIMOTO. 1995. Four hydrophobic segments in the NH2-terminal third (H1-H4) of Na,K-ATPase α-subunit alternately initiate and halt membrane translocation of newly synthesized polypeptide. J. Biol. Chem. **270:** 11985–11991.
9. TOYOSHIMA, C. *et al.* 2000. Crystal structure of the calcium pump of sarcoplasmic reticulum at 2.6Å resolution. Nature **405:** 647–655.
10. PEDERSEN, P.A., J.R. JØRGENSEN & P.L. JORGENSEN. 2000. Importance of conserved α-subunit segment ^{709}TGDVND for Mg^{2+} binding, phosphorylation and energy transduction in Na,K-ATPase. J. Biol. Chem. **275:** 37588–37595.
11. JACOBSEN, M.D., P.A. PEDERSEN & P.L. JØRGENSEN. 2002. Importance of Na,K-ATPase residue α1-Arg^{544} in the segment Arg^{544}-Asp^{567} for high affinity binding of ATP, ADP, or MgATP. Biochemistry **41:** 1451–1456.
12. CHAPMAN, R., C. SIDRAUSKI & P. WALTER. 1998. Annu. Rev. Cell. Dev. Biol. **14:** 459–485.
13. JØRGENSEN, J.R. & P.A. PEDERSEN. 2001. Role of phylogenetically conserved amino acids in folding of Na,K-ATPase. Biochemistry **40:** 7301–7308.
14. BEGGAH, A.T. *et al.* 1996. Degradation and endoplasmic reticulum retention of unassembled α- and β-subunits of Na,K-ATPase correlate with interaction with BiP. J. Biol. Chem. **271:** 20895–20902.
15. BEGGAH, A.T. & K. GEERING. 1997. α- and β-subunits of Na,K-ATPase interact with BiP and calnexin. Ann. N. Y. Acad. Sci. **834:** 537–539.
16. BLOND-ELGUINDI, S. *et al.* 1993. Affinity panning of a library of peptides displayed on bacteriophages reveals the binding specificity of BiP. Cell **75:** 717–728.
17. FLYNN, G.C. *et al.* 1991. Peptide-binding specificity of the molecular chaperone BiP. Nature **353:** 726–730.
18. KNARR, G. *et al.* 1995. BiP binding sequences in antibodies. J. Biol. Chem. **270:** 27589–27594.
19. COLONNA, T.E., L. HUYNH & D.M. FAMBROUGH. 1997. Subunit interactions in the Na,K-ATPase with the two-hybrid system. J. Biol. Chem. **272:** 12366–12372.

Differential Degradation of the Na^+/K^+-ATPase Subunits in the Plasma Membrane

SHIGE H. YOSHIMURA AND KUNIO TAKEYASU

*Graduate School of Biostudies, Kyoto University,
Kitashirakawa-oiwake-cho, Sakyo-ku, Kyoto, 606-8502, Japan*

ABSTRACT: The cell-surface expression of the Na^+/K^+-ATPase is tightly regulated at the ER/Golgi transit level. The assembly of the α- and β-subunits is a prerequisite for the maturation of the enzyme and its exit from the ER. At present, the fate of the functional ATPase, once it reaches the plasma membrane, is obscure, although endocytosis, recycling, and degradation in the lysosome are known to be involved. In this study, we shall demonstrate the differential degradation of the Na^+/K^+-ATPase subunits in the plasma membrane: that is, lysosomal degradation of α-subunits and proteasomal degradation of β-subunits, and propose the possibility of the subunit dissociation in the plasma membrane.

KEYWORDS: subunit assembly; plasma membrane; degradation; lysosome; proteasome; ubiquitylation

INTRODUCTION

The functional expression of the Na^+/K^+-ATPase on the plasma membrane requires the assembly of α- and β-subunits in the endoplasmic reticulum (ER). The assembled enzyme can exit the ER and pass through the Golgi apparatus to finally reach to the plasma membrane. A number of studies have suggested that the β-subunit is indispensable for the structural and functional maturation of the enzyme,[1] as well as for its intracellular transport to the plasma membrane.[2] However, the functional involvement of the β-subunit in the catalytic cycle is still controversial.[3] It was indeed demonstrated in insect cells that the α-subunit had catalytic activity independent of the β-subunit.[4] There is also a series of experiments showing that the β-subunit is also involved in cell adhesion of neuronal cells.[5,6] Thus, an irresistible assumption would be that the state of the subunit assembly could dynamically fluctuate and the β-subunit may not be directly involved in the enzymatic function of the Na^+/K^+-ATPase in the plasma membrane. Here we tested this possibility by examining the fate of the α- and β-subunits in the plasma membrane. We pulse labeled the cell-surface proteins with membrane-impermeable biotinylation reagent and chased the fate of the α- and β-subunits (FIG. 1a).

Address for correspondence: Shige H. Yoshimura, Graduate School of Biostudies, Kyoto University, Kitashirakawa-oiwake-cho, Sakyo-ku, Kyoto, 606-8502, Japan. Voice: +81-75-753-6852; fax: +82-75-753-6852.
yoshimura@lif.kyoto-u.ac.jp

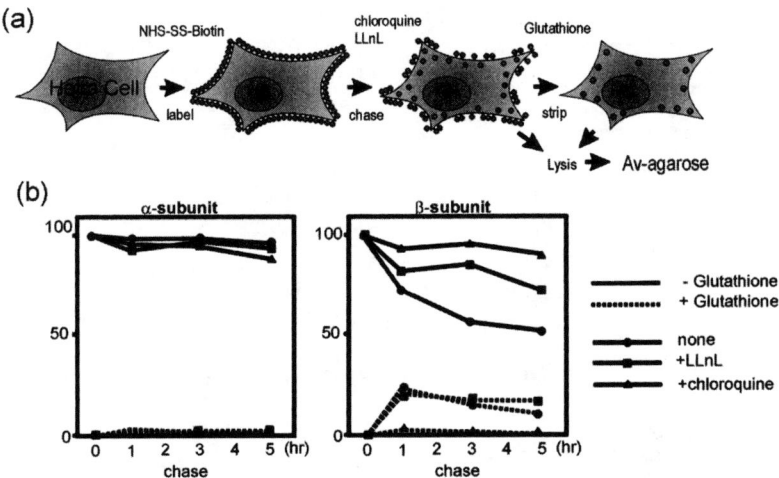

FIGURE 1. The α- and β-subunits undergo different degradation pathways. (**a**) Biotinylation of cell-surface proteins. HeLa cells were incubated with a membrane-impermeable biotinylation reagent (NHS-SS-Biotin) for 15 min on ice. The fates of the biotinylated proteins were chased in normal medium (DMEM) at 37°C. At the end of the chase period, the cells were either directly lysed or lysed after treatment with glutathione to cleave off the biotin moieties still remaining on the cell surface. The cell lysate was incubated with avidin-agarose for 1 hour at 4°C. Bound proteins were analyzed by SDS-PAGE followed by immunoblotting. (**b**) The different degradation rates of the cell-surface Na^+/K^+-ATPase subunits. The amount of the biotin-labeled α- and β-subunits were quantitated after 0, 1, 3, or 5 hours' chase. An inhibitor for the endosomal pathway (chloroquine) or for the proteasomal degradation (LLnL) was added in the chase medium.

RESULTS

The amount of the labeled Na^+/K^+-ATPase α-subunit in the plasma membrane was almost constant during the five-hour chase period (FIG. 1b, left), although it was decreased over the prolonged chase time via the lysosomal degradation pathway (data not shown). The internalization of the α-subunit was not detected within five hours. Thus, the α-subunit, once expressed on the cell surface, stably stays in the plasma membrane over several hours, in good agreement with the notion that the half-life of the α-subunit is about 24 hours in HeLa cells. When cells were chased in the presence of 1 mM ouabain, the α-subunit was swiftly internalized within one hour and remained for the next several hours without degradation (data not shown). This result also matches well the previous study in which the ouabain treatment induces the internalization of the enzyme.[7–9]

In contrast to the α-subunit, the β-subunit is swiftly internalized and degraded to about 70% within one hour in the absence of ouabain (FIG. 1b, right). After five hours, the total amount of the labeled β-subunit is reduced to about 50 percent. When ouabain (1 mM) is added to the chase medium, the internalization and degradation are slightly stimulated. Thus, in contrast to the α-subunit, both internalization and degradation are constantly occurring to the β-subunit.

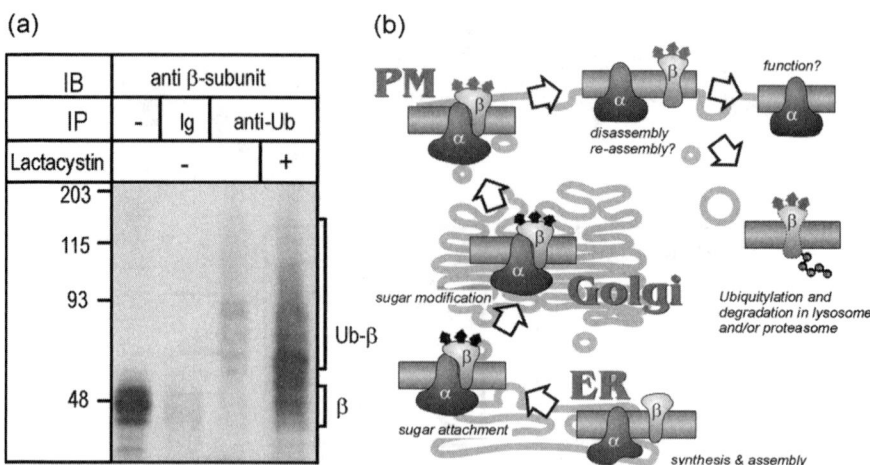

FIGURE 2. (a) The ubiquitylation of the β-subunit. HeLa cell lysate was immunoprecipitated with the antiubiquitin antibody and then analyzed by SDS-PAGE followed by immunoblot against the anti–β-subunit antibody. (b) The total life cycle of the Na^+/K^+-ATPase. The α- and β-subunits are assembled in the ER and go through the Golgi apparatus with a series of sugar modifications on the β-subunit. Once they reach the plasma membrane, both subunits are disassembled. The ubiquitylation of the β-subunit triggers the endocytosis and the following degradation process in lysosome and/or proteasome. The α-subunit remains in the plasma membrane and functions as an ion pump alone or together with a newly replaced β-subunit.

An inhibitor of the proteasomal degradation blocked the degradation of the β-subunit, but the endocytosis was still occurring (FIG. 1b, right). The endocytosis and the degradation of the β-subunit were completely blocked by chloroquine, an inhibitor of the endocytic pathway. These results indicate that the stability of the cell-surface β-subunit is regulated by the endosome/lysosome pathway and the proteasome-dependent protein degradation. The ubiquitylation of the β-subunit was detected directly by immunoprecipitation (FIG. 2a). The amount of the ubiquitylated β-subunit was increased when the proteasomal degradation was blocked. Thus, the ubiquitylation plays a key role in the regulation of the β-subunit turnover.

DISCUSSION

The function of the β-subunit has been thought to help in the correct folding of the α-subunit upon/after the insertion into the ER membrane. Only the α/β-assembled enzyme with a correct folding is allowed to exit the ER, follows a maturation process in the Golgi complex, and reaches the plasma membrane[1,10,11] (FIG. 2b). This role of the β-subunit predicts that the minimum unit of the functional enzyme in the plasma membrane is an α/β dimer. However, our data on the stability of the α- and β-subunits in the plasma membrane suggests that the majority of the α- and β-subunit complexes could be disassembled on the cell surface. Namely, after both subunits reach the plasma membrane, most of the complexes release the β-subunit, whereas the remaining α-subunits still possess the normal ouabain-binding activity. It is intriguing that the β-subunit is swiftly degraded by a ubiquitin-proteasome system, while the α-subunit stably stays on the cell surface for many hours. If the disassembled α-subunit stays as a single subunit on the cell surface, as suggested in FIG. 2b, the α-subunit may no longer require the β-subunit for its function. Indeed, some of the Na^+/K^+-ATPase α-subunits in *C. elegans* and *D. melanogaster* are expected not to assemble with the β-subunit.[12] Alternatively, the dissociated α-subunit may reassemble with the newly appeared β-subunit on the cell surface. It has been suggested that, in some cases, the β-subunit can reach the plasma membrane without the α-subunit.[13] This possible reassociation of the α- and β-subunit would suggest a new aspect of the subunit turnover on the cell surface.

REFERENCES

1. GEERING, K. 1991. The functional role of the beta-subunit in the maturation and intracellular transport of Na,K-ATPase. FEBS Lett. **285:** 189–193.
2. TAKEYASU, K. *et al.* 1989. Differential subunit and isoform expression involved in regulation of sodium pump in skeletal muscle. Curr. Topics Membr. Transport. **34:** 143–165.
3. YOSHIMURA, S.H. *et al.* 1998. The Na^+,K^+-ATPase carrying the carboxy-terminal Ca^{2+}/calmodulin binding domain of the Ca^{2+} pump has $2Na^+,2K^+$ stoichiometry and lost charge movement in Na^+/Na^+ exchange. FEBS Lett. **425:** 71–74.
4. BLANCO, G. *et al.* 1994. The alpha-subunit of the Na,K-ATPase has catalytic activity independent of the beta-subunit. J. Biol. Chem. **269:** 23420–23425.
5. SCHMALZING, G. *et al.* 1992. The adhesion molecule on glia (AMOG/beta 2) and alpha 1 subunits assemble to functional sodium pumps in *Xenopus* oocytes. J. Biol. Chem. **267:** 20212–20216.
6. GLOOR, S. *et al.* 1990. The adhesion molecule on glia (AMOG) is a homologue of the beta subunit of the Na,K-ATPase. J. Cell. Biol. **110:** 165–174.
7. NUNEZ-DURAN, H. *et al.* 1988. Ouabain uptake by endocytosis in isolated guinea pig atria. Am. J. Physiol. **255:** C479–485.
8. LAMB, J.F. & P. OGDEN. 1982. Internalization of ouabain and replacement of sodium pumps in the plasma membranes of HeLa cells following block with cardiac glycosides. Q. J. Exp. Physiol. **67:** 105–119.
9. COOK, J.S., E.H. TATE & C. SHAFFER. 1982. Uptake of [^{3}H]ouabain from the cell surface into the lysosomal compartment of HeLa cells. J. Cell. Physiol. **110:** 84–92.
10. FAMBROUGH, D.M., *et al.* 1994. Analysis of subunit assembly of the Na-K-ATPase. Am. J. Physiol. **266:** C579–589.
11. BEGUIN, P. *et al.* 1998. Membrane integration of Na,K-ATPase alpha-subunits and beta-subunit assembly. J. Biol. Chem. **273:** 24921–24931.
12. TAKEYASU, K. *et al.* 2001. P-type ATPase diversity and evolution: the origins of ouabain sensitivity and subunit assembly. Cell. Mol. Biol. **47:** 325–333.
13. BILLECOCQ, A. *et al.* 1997. 1,25-Dihydroxyvitamin D3 selectively induces increased expression of the Na,K-ATPase beta 1 subunit in avian myelomonocytic cells without a concomitant change in Na,K-ATPase activity. J. Cell. Physiol. **172:** 221–229.

FXYD Proteins as Regulators of the Na,K-ATPase in the Kidney

KATHLEEN J. SWEADNER, ELENA ARYSTARKHOVA, CLAUDIA DONNET, AND RANDALL K. WETZEL

Neuroscience Center, Massachusetts General Hospital, Charlestown, Massachusetts 02129, USA

ABSTRACT: The FXYD gene family has seven members in mammals and others in fish. Five of these (FXYD1, FXYD2, FXYD4, FXYD7, and PLMS from shark) have been shown to alter the activity of the Na,K-ATPase, as described by other papers in this volume. The gene structure of FXYD family members suggests assembly from protein domain modules and gene duplication. The γ subunit is unique in the family for having alternative splice variants in the coding region and can be posttranslationally modified with different final consequences for enzyme properties. The nonoverlapping distribution of γ and CHIF (FXYD4) in kidney helps to explain physiological differences in Na^+ affinity among nephron segments. We also detected phospholemman (FXYD1) in kidney. By immunofluorescence, it was found in extraglomerular mesangial cells (EM cells) of the juxtaglomerular apparatus and in the afferent arteriole. Contrary to many reports that only α1 and β1 are expressed in the kidney, we found that α2 and β2 are present, although not in any nephron segment. Both were detected in arterioles, and β2 was found in the EM cells. In contrast, α1, β1, and γ were found in adjacent macula densa. Phospholemman, α2, and β2 are proposed to have distinct roles in regulating the sodium pump in structures involved in tubuloglomerular feedback.

KEYWORDS: FXYD; gamma; phospholemman; genomic DNA; gene family

FXYD GENE FAMILY MEMBERS

The FXYD gene family in humans, mice, and rats has at least seven members.[1] To date, five members of the family have been implicated in modulation of Na,K-ATPase activity: the γ subunit[2,3] and CHIF[4,5] of kidney, PLMS of shark rectal gland,[6] phospholemman of heart and brain,[7,8] and the FXYD7 protein of brain.[9] The structure of the genes has been reported for FXYD1,[10] FXYD2,[11–13] and FXYD4.[14] It is interesting that, as small as these proteins are, they are encoded by genes with a minimum of six small exons. This contrasts with the genomic organization of phospholamban, the SERCA modulator, which has a single exon.

FIGURE 1 shows a diagram of the exon correspondence for FXYD1, FXYD2, and FXYD4. The shared sequence motif that identifies the family spans 35 amino acid

---

Address for correspondence: Kathleen J. Sweadner, Neuroscience Center, Massachusetts General Hospital, Charlestown, MA 02129. Voice: 617-726-8579; fax: 617-726-7526.
sweadner@helix.mgh.harvard.edu

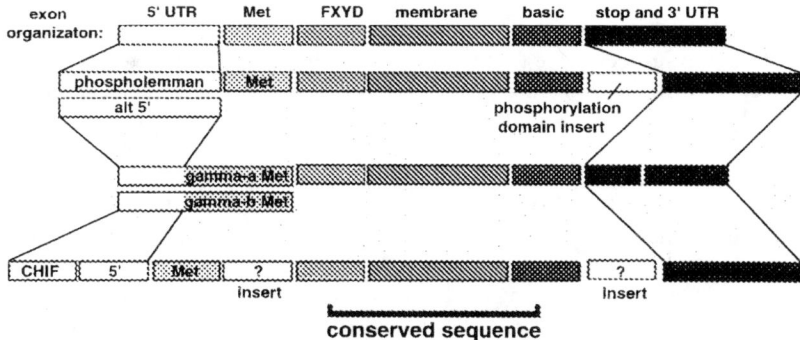

FIGURE 1. Genomic organization of FXYD family members. All of the FXYD genes studied to date are made up of a minimum of six exons. The central three form the conserved sequence, and the upstream and downstream exons generate structural diversity.

residues from PFxYD in the extracellular space to a stop transfer sequence with alternating basic residues and cysteines in the intracellular space. Curiously, it takes three exons to build this motif (although two of the exons are fused in mouse FXYD2). On the 5' side of the three-exon core, phospholemman has a minimum of two exons, one for the N-terminus and another (actually several alternative exons)[1] for the 5' untranslated sequence. Upstream of the core, the γ subunit gene has two (human) or three (mouse) alternative coding exons for the N-terminus, apparently with different promoters, and CHIF has four separate upstream exons, two just for the 5' untranslated sequence. Downstream of the core, γ has just one coding exon and an extra exon for the 3' untranslated sequence. Phospholemman and CHIF, in contrast, both have an extra coding exon before the last one, and in phospholemman this exon contains the phosphorylation domain. The organization suggests a shared origin for the three core exons, plus a varying number of exons on either side of them, generating sequences that make each gene product distinctive.

THE γ SUBUNIT AS A SODIUM PUMP MODULATOR

The isoforms of the Na,K-ATPase α and β subunits have been shown to affect functional properties by having different affinities for Na^+ or K^+, and different voltage sensitivity for the electrogenic step. Nonetheless, variations in the properties of Na,K-ATPases in different cell types have been observed that cannot be explained by variation of V_{max} or exchange of isoforms. For example, Na,K-ATPase prepared from whole kidney or renal medulla has predominantly α1 and β1 as detected with antibodies on Western blots. However, the Na,K-ATPase in microdissected nephron segments showed variation in properties. The first discovered was a variation in ouabain affinity,[15] but later Na^+ affinity was also found to differ, with a tendency for affinity to increase from proximal segment to collecting duct.[16–19] These observations stimulated a search for the expression of additional isoforms of α and β that might account for the functional properties of different segments. A number of labs reported finding only α1 and β1, though, mostly by methods entailing the detection of their mRNAs, including *in situ* hybridization.

We began to suspect a role for the γ subunit during attempts to determine whether the effects of different β subunits on Na$^+$ and K$^+$ affinity can be detected in established rat cell lines. The results were initially puzzling: a rat cell line with α1β1 (NRK-52E) had similar Na$^+$ affinity to one with α1β3 (C6) and to one with α1 and a mixture of β1 and β2 (L6). At the same time, however, all of the cell lines had higher Na$^+$ and K$^+$ affinities than a preparation purified from rat renal medulla. We observed the same discrepancy with other renal cell lines: pig LLC-PK1 cells compared to pig medullary enzyme, and canine MDCK cells compared to canine medullary enzyme. On SDS gels, using antibodies provided by R. Mercer and K. Geering, we found that the γ subunit could not be detected in the cell lines, something that was independently observed by Therien *et al.*[20] At almost the same time, Beguin *et al.*[21] and Therien *et al.*[20] published the first evidence that γ alters the properties of the Na,K-ATPase: its affinity and voltage sensitivity to extracellular K$^+$, and its activity and affinity for ATP.

Differences in affinity for Na$^+$ were the most significant effect seen in our initial experiments with cell lines, so we tested the hypothesis that the γ subunit was responsible by stably transfecting γa into NRK-52E, a rat renal cell line that does not normally express it. In multiple individual clones, expression of γ reduced affinity for Na$^+$ and K$^+$.[3] However, the experiments produced an unexpected bonus: individual clones differed in whether γ was fully posttranslationally modified, and the fully modified clones had reduced affinity for K$^+$, but not Na$^+$. It appeared that there were three potential forms of Na,K-ATPase: αβ, αβγ, and αβγ′, where the (′) is a modification that can change γ's effects.[3] Those conclusions were reached before we found that γ had a splice variant (γb, GenBank AF233060).[1,22] We have recently demonstrated that both splice variants can be modified and that the four structural variants show different combinations of alterations of Na$^+$ and K$^+$ affinity (E. Arystarkhova and R. Wetzel, this volume).[23]

The structure of the modification has so far resisted determination, but we have shown that mutations of serine and threonine residues in the N-terminus of γa block the modification, as detected by a shift in gel mobility after *in vitro* synthesis.[23] Such modification could be due to acylation or esterification. The modification also requires the addition of pancreatic microsomes during synthesis, suggesting that it is catalyzed by microsomal proteins and consistent with an intraluminal (extracellular) location.[4,23,24] Subsequently, mutation of oxygen-containing side chains in FXYD7 has also been shown to affect mobility after *in vitro* synthesis with microsomes, possibly indicating O-glycosylation.[9] In contrast, an extra γa band formed in HeLa, possibly an undissociated dimer, appears to require the C-terminus, but not the N-terminus.[25]

There is now agreement that changes in apparent Na$^+$ affinity are a major consequence of association with γ.[3,4,23,25,26] Evidence that modulation of Na$^+$ affinity is physiologically important is that the clones with reduced affinity also have a reduced growth rate.[23] There is less certainty about the significance of effects of γ on ATP affinity because it has been more obvious in some cell lines than others.[23,27] However, in HeLa, effects on ATP affinity were shown to depend on the cytoplasmic portion of the γ subunit.[26] Available evidence on what part of γ is responsible for effects on Na$^+$ and K$^+$ affinity does not yet give a clear picture. On the one hand, deletion or alanine substitution of the N-terminal spliced regions indicated that they are not required for alteration of ion affinity;[26] on the other hand, our experiments with γa and γb clones with different levels of posttranslational modification indicate that the extracellular splice region must be important.[23] A possible explanation is that the central portion of γ is essential for the basic effect, while the spliced region is a handle by which the effect can be modified.

FXYD PROTEINS IN THE JUXTAGLOMERULAR APPARATUS

It is already established that γ (FXYD2) and CHIF (FXYD4) have distinct distributions in the kidney.[3,25,28–30] Kidney mRNA for phospholemman (FXYD1) has been detected in Northern blots.[10] We and others have shown that phospholemman too associates with Na,K-ATPase[7,8] and thus we looked for its cellular distribution in the kidney. Phospholemman antibodies stained extraglomerular mesangial cells (EM cells, also known as lacis cells or Goormaghtigh cells), which form a pad under the macula densa and between it and the glomerular arterioles, at the vascular pole of the glomerulus (FIG. 2). Phospholemman was also found in the renin-producing cells and in the smooth muscle cells of the afferent arteriole and other cortical arterioles. Interestingly, these structures were not stained by antibodies against α1 or β1. Instead, we found that arterioles were stained by antibodies against α2 and β2, colocalized with phospholemman. This is the first time that alternative isoforms for α and β have been unambiguously identified in the kidney. In the EM cells, β2 colocalized with phospholemman, but we were unable to detect any α subunit. It has been reported before that little or no α can be detected in the regular mesangial cells of the glomerulus,[31] but α1 and β1 were detected in cultured mesangial cells;[32] thus, it is likely that α was present, but below the sensitivity of the antibodies used.

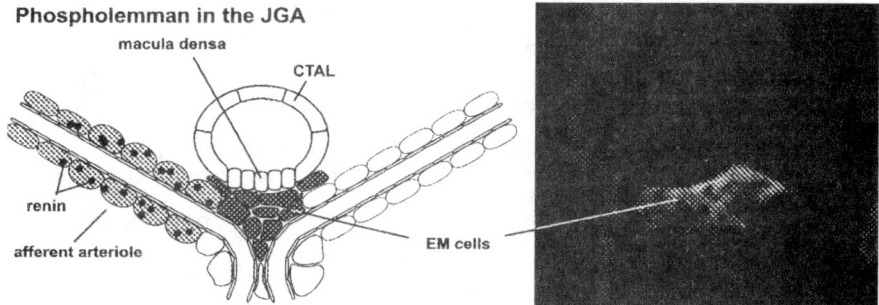

FIGURE 2. Phospholemman is in the juxtaglomerular apparatus (JGA) in EM cells and afferent arteriole. CTAL: cortical thick ascending limb. Macula densa is specialized for sensing luminal NaCl. The fluorescence image shows phospholemman-stained EM cells. Stained afferent arteriole, identified by the presence of renin granules (spots) proximal to the JGA, is not in the plane of this section, and the rest of the glomerulus is below the structures shown. Triple-label immunofluorescence was used to identify all of the JGA structures (not shown).

The macula densa, in contrast, expressed γ with α1 and β1, but not phospholemman. In rats, we did not reproduce findings reported for rabbits that the macula densa lacks any Na,K-ATPase.[33] The whole juxtaglomerular apparatus (macula densa, EM cells, afferent and efferent arterioles with renin-secreting granular cells) controls renal function by sensing the NaCl load in the lumen of the CTAL and causing both local and blood-borne, renin-mediated changes in the glomerular filtration rate to adjust urine output. Little is known about the function of the EM cells since they are difficult to study experimentally, but a role in ion movements and signal transduction is suggested by the presence of the $Na^+:K^+:2Cl^-$ transporter, NKCC1/BSC2. The presence of γ in macula densa and phospholemman in EM cells suggests an important role for the Na,K-ATPase in the NaCl sensor. Because phospholemman is phosphorylated by multiple protein kinases, it may be involved in signal transduction in the EM cells, in their intermediary position between tubule and vasculature.

REFERENCES

1. SWEADNER, K.J. & E. RAEL. 2000. The FXYD gene family of small ion transport regulators or channels: cDNA sequence, protein signature sequence, and expression. Genomics **68:** 41–56.
2. THERIEN, A.G., H.X. PU, S.J.D. KARLISH & R. BLOSTEIN. 2001. Molecular and functional studies of the gamma subunit of the sodium pump. J. Bioenerg. Biomembr. **33:** 407–414.
3. ARYSTARKHOVA, E., R.K. WETZEL, N.K. ASINOVSKI & K.J. SWEADNER. 1999. The γ subunit modulates Na^+ and K^+ affinity of the renal Na,K-ATPase. J. Biol. Chem. **274:** 33183–33185.
4. BEGUIN, P., G. CRAMBERT, S. GUENNOUN, et al. 2001. CHIF, a member of the FXYD protein family, is a regulator of Na,K-ATPase distinct from the γ subunit. EMBO J. **20:** 3993–4002.
5. GARTY, H., M. LINDZEN, R. SCANFANO, et al. 2002. A functional interaction between CHIF and Na,K-ATPase: implication for regulation by FXYD proteins. Am. J. Physiol. **283:** F607–F615.
6. MAHMMOUD, Y.A., H. VORUM & F. CORNELIUS. 2000. Identification of a phospholemman-like protein from shark rectal glands: evidence for indirect regulation of Na,K-ATPase by protein kinase C via a novel member of the FXYDY family. J. Biol. Chem. **275:** 35969–35977.
7. CRAMBERT, G., M. FUZESI, H. GARTY, et al. 2002. Phospholemman (FXYD1) associates with Na,K-ATPase and regulates its transport properties. Proc. Natl. Acad. Sci. USA **99:** 11476–11481.
8. FESCHENKO, M.S., C. DONNET, R.K. WETZEL, et al. 2003. Phospholemman, a single-span membrane protein is an accessory protein of Na,K-ATPase in cerebellum and choroid plexus. J. Neurosci. In press.
9. BEGUIN, P., G. CRAMBERT, F. MONNET-TSCHUDI, et al. 2002. FXYD7 is a brain-specific regulator of Na,K-ATPase α1-β isozymes. EMBO J. **21:** 3264–3273.
10. BOGAEV, R.C., L. JIA, Y.M. KOBAYASHI, et al. 2001. Gene structure and expression of phospholemman in mouse. Gene **271:** 69–79.
11. SWEADNER, K.J., R.K. WETZEL & E. ARYSTARKHOVA. 2000. Genomic organization of the human FXYD2 gene encoding the γ subunit of the Na,K-ATPase. Biochem. Biophys. Res. Commun. **279:** 196–201.
12. MEIJ, I.C., J.B. KOENDERINK, H. VAN BOKHOVEN, et al. 2000. Dominant isolated renal magnesium loss is caused by misrouting of the Na^+,K^+-ATPase γ subunit. Nat. Genet. **26:** 265–266.
13. JONES, D.H., M.C. GOLDING, K.J. BARR, et al. 2001. The mouse Na^+-K^+-ATPase γ-subunit gene (*Fxyd2*) encodes three developmentally regulated transcripts. Physiol. Genom. **6:** 129–135.

14. AIZMAN, R., C. ASHER, M. FUZESI, et al. 2002. Generation and phenotypic analysis of CHIF knockout mice. Am. J. Physiol. **283:** F569–F577.
15. DOUCET, A. & C. BARLET. 1986. Evidence for differences in the sensitivity to ouabain of NaK-ATPase along the nephrons of rabbit kidney. J. Biol. Chem. **261:** 993–995.
16. FERAILLE, E., M.L. CARRANZA, M. ROUSSELOT & H. FAVRE. 1994. Insulin enhances sodium sensitivity of Na,K-ATPase in isolated rat proximal convoluted tubule. Am. J. Physiol. **267:** F55–F62.
17. FERAILLE, E., M.L. CARRANZA, B. BUFFIN-MEYER, et al. 1995. Protein kinase C–dependent stimulation of Na,K-ATPase in rat proximal convoluted tubules. Am. J. Physiol. **268:** C1277–C1283.
18. BUFFIN-MEYER, B., S. MARSY, C. BARLET-BAS, et al. 1996. Regulation of renal Na^+,K^+-ATPase in rat thick ascending limb during K^+ depletion: evidence for modulation of Na^+ affinity. J. Physiol. **490:** 623–632.
19. BARLET-BAS, C., L. CHEVAL, C. KHADOURI, et al. 1990. Difference in the Na affinity of Na^+-K^+-ATPase along the rabbit nephron: modulation by K. Am. J. Physiol. **259:** F246–F250.
20. THERIEN, A.G., R. GOLDSHLEGER, S.J.D. KARLISH & R. BLOSTEIN. 1997. Tissue-specific distribution and modulatory role of the γ subunit of the Na,K-ATPase. J. Biol. Chem. **272:** 32628–32634.
21. BEGUIN, P., X. WANG, D. FIRSOV, et al. 1997. The γ subunit is a specific component of the Na,K-ATPase and modulates its transport function. EMBO J. **16:** 4250–4260.
22. SWEADNER, K.J., E. RAEL, R.K. WETZEL & E. ARYSTARKHOVA. 2000. Splice variants of the Na,K-ATPase gamma subunit. *In* Na/K-ATPase and Related ATPases, pp. 543–546. Elsevier. Amsterdam/New York.
23. ARYSTARKHOVA, E., C. DONNET, N.K. ASINOVSKI & K.J. SWEADNER. 2002. Differential regulation of renal Na,K-ATPase by splice variants of the γ subunit. J. Biol. Chem. **277:** 10162–10172.
24. MERCER, R.W., D. BIEMESDERFER, D.P. BLISS, JR., et al. 1993. Molecular cloning and immunological characterization of the γ polypeptide, a small protein associated with the Na,K-ATPase. J. Cell Biol. **121:** 579–586.
25. PU, H.X., F. CLUZEAUD, R. GOLDSHLEGER, et al. 2001. Functional role and immunocytochemical localization of the γa and γb forms of the Na,K-ATPase γ subunit. J. Biol. Chem. **276:** 20370–20378.
26. PU, H.X., R. SCANZANO & R. BLOSTEIN. 2002. Distinct regulatory effects of the Na,K-ATPase γ subunit. J. Biol. Chem. **277:** 20270–20276.
27. THERIEN, A.G., S.J.D. KARLISH & R. BLOSTEIN. 1999. Expression and functional role of the γ subunit of the Na,K-ATPase in mammalian cells. J. Biol. Chem. **274:** 12252–12256.
28. WETZEL, R.K. & K.J. SWEADNER. 2001. Immunocytochemical localization of the Na,K-ATPase α and γ subunits in the rat kidney. Am. J. Physiol. **281:** F531–F545.
29. SHI, H., R. LEVY-HOLZMAN, F. CLUZEAUD, et al. 2001. Membrane topology and immunolocalization of CHIF in kidney and intestine. Am. J. Physiol. **280:** F505–F512.
30. ARYSTARKHOVA, E., R.K. WETZEL & K.J. SWEADNER. 2002. Distribution and oligomeric association of splice forms of the Na,K-ATPase regulatory γ subunit in rat kidney. Am. J. Physiol. **282:** F393–F407.
31. KASHGARIAN, M., D. BIEMESDERFER, M. CAPLAN & B. FORBUSH III. 1985. Monoclonal antibody to Na,K-ATPase: immunocytochemical localization along nephron segments. Kidney Int. **28:** 899–913.
32. OHARA, T., U. IKEDA, S. MUTO, et al. 1993. Thyroid hormone stimulates Na^+-K^+-ATPase gene expression in cultured rat mesangial cells. Am. J. Physiol. **265:** F370–F376.
33. PETI-PETERDI, J., Z. BEBOK, J-Y. LAPOINTE & P.D. BELL. 2002. Novel regulation of cell [Na^+] in macula densa cells: apical Na^+ recycling by H-K-ATPase. Am. J. Physiol. **282:** F324–F329.

FXYD Proteins: New Tissue- and Isoform-Specific Regulators of Na,K-ATPase

KÄTHI GEERING,[a] PASCAL BÉGUIN,[a] HAIM GARTY,[b] STEVEN KARLISH,[b] MARIA FÜZESI,[b] JEAN-DANIEL HORISBERGER,[a] AND GILLES CRAMBERT[a]

[a]*Institute of Pharmacology and Toxicology, University of Lausanne, CH-1005 Lausanne, Switzerland*

[b]*Department of Biological Chemistry, Weizmann Institute of Science, Rehovot 76100, Israel*

ABSTRACT: The recently defined FXYD protein family contains seven members that are small, single-span membrane proteins characterized by a signature sequence containing an FXYD motif and three other conserved amino acid residues. Until recently, the functional role of FXYD proteins was largely unknown, with the exception of the γ subunit of Na,K-ATPase, which was shown to be a specific regulator of renal α1-β1 isozymes. We have investigated whether other members of the FXYD family may have a similar role as the γ subunit and have found that CHIF (corticosteroid hormone–induced factor, FXYD4), FXYD7, as well as phospholemman (FXYD1) specifically associate with Na,K-ATPase and preferentially with α1-β isozymes in native tissues, and produce distinct effects on the transport properties of Na,K-ATPase that are adapted to the physiological demands of the tissues in which they are expressed. These results provide evidence for a unique and novel mode of regulation of Na,K-ATPase by FXYD proteins that involves a tissue-specific expression of an auxiliary subunit of distinct Na,K-ATPase isozymes.

KEYWORDS: Na,K-ATPase transport modulators; FXYD proteins; *Xenopus* oocytes; renal Na$^+$ reabsorption; muscle contractility; neuronal excitability

INTRODUCTION

Considering the important physiological role of Na,K-ATPase, it is obvious that its expression and activity must be finely regulated and be adapted to changing physiological demands. A very basic requirement for Na,K-ATPase expression is the stoichiometric synthesis of its catalytic α subunit and its β subunit, which acts as a molecular chaperone that is necessary for the correct membrane insertion of the α subunit.[1] Four α and 3 β isoforms are expressed in a tissue-specific pattern and potentially can form 12 different isozymes with different transport properties.[2,3] Other mediators of Na,K-ATPase modulation include changes in the intracellular Na$^+$ concentration or peptide hormones and neurotransmitters that activate PKA or

---

Address for correspondence: Käthi Geering, Institute of Pharmacology and Toxicology, Rue du Bugnon 27, CH-1005 Lausanne, Switzerland. Voice: +041-21-692-54-10; fax: +041-21-692-53-55.
kaethi.geering@ipharm.unil.ch

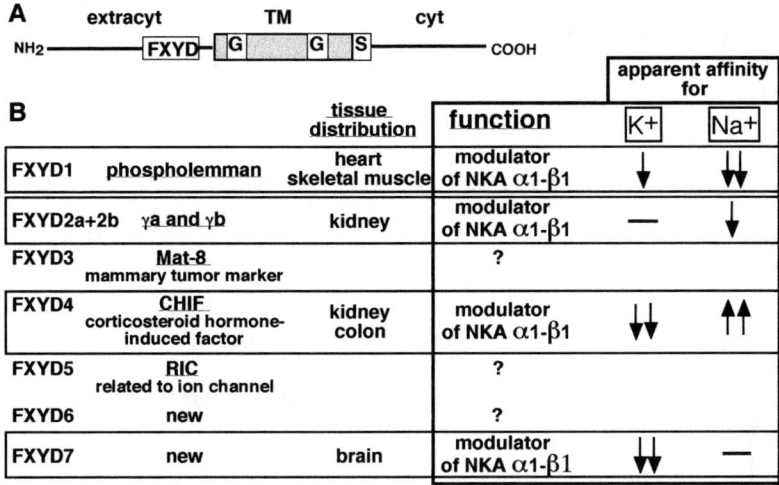

FIGURE 1. The family of FXYD proteins. **(A)** Shown is a linear model of a typical FXYD protein with 1 transmembrane (TM) segment, a signature sequence containing the FXYD motif and 3 other conserved amino acid residues, and a type-I membrane orientation resulting in the exposure of the N-terminus to the extracytoplasmic ("extracyt") side and of the C-terminus to the cytoplasmic ("cyt") side. **(B)** Tissue distribution and summary of the functional effects of different FXYD proteins on the apparent affinities for extracellular K^+ and intracellular Na^+ of Na,K-ATPase α1-β1 isozymes.

PKC and provoke phosphorylation of Na,K-ATPase that can modulate its distribution between the plasma membrane and intracellular stores.[4] Finally, long-term regulation, which ultimately leads to an increased number of Na,K-ATPase units at the cell surface, is mediated, for example, by aldosterone that affects α and β gene transcription.[5] Recently, we have elucidated still another, novel regulatory mechanism that involves isozyme- and tissue-specific interactions between Na,K-ATPase and small, single-span membrane proteins of the FXYD family.

The so-called FXYD protein family has recently been defined by Sweadner and Rael.[6] It contains 7 members that are characterized by 1 transmembrane domain and a signature sequence containing the FXYD motif and 3 other conserved residues (FIG. 1). Until recently, the function of these proteins was largely unknown, with the exception of the γ subunit of Na,K-ATPase,[7,8] which occurs as 2 splice variants, a and b,[6,9] and which we and others have identified as specific modulators of renal Na,K-ATPase α1-β1 isozymes.[10–13] Interestingly, the mutation of a conserved glycine residue in the γ subunit could recently be linked to cases of human primary hypomagnesemia probably indirectly caused by a dysregulation of Na,K-ATPase.[14]

Since several FXYD proteins induce ion-specific conductances when overexpressed in *Xenopus* oocytes, it was so far believed that proteins such as phospholemman (FXYD1),[15,16] Mat-8 (mammary tumor marker, FXYD3),[17] CHIF (corticosteroid hormone–induced factor, FXYD4),[18] or RIC (related to ion channel, FXYD5)[19] may function as channels themselves or may be modulators of ion channels. The biological relevance of these putative functions remains, however, unclear. We wondered whether one or the other or perhaps all of the FXYD proteins may have a similar function as the γ subunit and may act as tissue-specific modulators of Na,K-ATPase.

MATERIALS AND METHODS

All technical details can be found in several recent papers.[10,11,20,21]

RESULTS AND DISCUSSION

CHIF Is a Kidney- and Colon-Specific Modulator of Na,K-ATPase Favoring Na^+ Conservation

To test the hypothesis that other FXYD proteins than γ subunits may be modulators of Na,K-ATPase, we first concentrated on the protein CHIF, which, as the Na,K-ATPase and the γ subunit, is exclusively expressed in the basolateral membrane of epithelial cells, in this case of the kidney and the colon.[22]

CHIF shows more than 50% sequence similarity with γa and γb variants, which themselves differ only in the most N-terminal sequence. We determined that γ subunits as well as CHIF are type-I membrane proteins exposing the N-terminus to the extracytoplasmic side. CHIF, but not γ subunits, adopts this membrane orientation after cleavage of an N-terminal signal sequence.[11]

We first assessed whether CHIF could associate with Na,K-ATPase after coexpression in *Xenopus* oocytes. Similar to γ subunits, metabolically labeled CHIF can indeed be coimmunoprecipitated with an Na,K-ATPase α antibody, and this over prolonged chase periods. These stable interactions with Na,K-ATPase are specific since neither CHIF nor γ subunits associate with colonic H,K-ATPase, despite its close relation to Na,K-ATPase and a similar tissue distribution as CHIF.[11]

γ-Na,K-ATPase as well as CHIF-Na,K-ATPase complexes exist also in native tissues. Western blot analysis of kidney microsomes reveals the presence of γa and γb as well as of CHIF, and all 3 peptides can be coimmunoprecipitated with an Na,K-ATPase α antibody.[11]

To know whether CHIF modulates the transport properties of Na,K-ATPase, we first compared the K-activation of Na,K-pump currents in oocytes expressing Na,K-ATPase alone or together with γa or γb variants or with CHIF. The results show that CHIF significantly decreases the apparent affinity for extracellular K^+ of Na,K-pumps over a large range of membrane potentials, whereas γ variants only have a slight effect at physiologically relevant membrane potentials. Since intracellular Na^+ is the rate-limiting step of the Na,K-ATPase transport activity under physiological conditions, we also compared the effect of CHIF and γ variants on the Na^+ activation of Na,K-pumps. For these measurements, the renal epithelial Na^+ channel was coexpressed in oocytes, which permits control and calculation of intracellular Na^+ concentrations. Compared to Na,K-pumps expressed alone, Na,K-pumps coexpressed with CHIF indeed exhibit a significantly increased affinity for intracellular Na^+, whereas those coexpressed with γ variants have a decreased Na^+ affinity.[11] These differential effects of CHIF and γ subunits on the apparent Na$^+$ affinity of Na,K-ATPase are most likely of physiological relevance and reflect the different roles of Na,K-ATPase in the Na$^+$ reabsorption process in different segments of the renal tubule.

In the renal tubule, γ subunits are indeed mainly expressed in the thick ascending limb (TAL),[13] which reabsorbs as much as 25–40% of the filtered Na$^+$ load, whereas CHIF is present in the collecting duct (CD),[23] which is the ultimate site for electrolyte conservation. In the TAL, association of basolateral Na,K-pumps with the γ subunit, which reduces their apparent Na$^+$ affinity, is favorable since this permits efficient extrusion of cellular Na$^+$ even at a high cellular Na$^+$ load. On the other hand, in the CD, association of CHIF with Na,K-pumps, which increases their Na$^+$ affinity, is favorable since this permits efficient Na$^+$ reabsorption at low intracellular Na$^+$ concentrations. From the change in the apparent Na$^+$ affinity of Na,K-pumps associated with CHIF, it can indeed be estimated that, at physiologically low intracellular Na$^+$ concentrations, the transport rate of CHIF-associated Na,K-pumps is about 4 times higher than that of Na,K-pumps lacking CHIF.[11]

The collecting duct is a target site for aldosterone that can increase Na$^+$ reabsorption by 2- to 4-fold when Na$^+$ conservation is required. During the early phase of aldosterone action that mediates an increased apical Na$^+$ entry and produces small changes in intracellular Na$^+$ concentrations, Na$^+$ reabsorption should considerably improve in the presence of Na,K-pumps associated with CHIF. Moreover, in conditions of Na$^+$ deprivation, not only Na,K-ATPase, but also CHIF expression is increased.[22] This assures that CHIF does not become limiting for the formation of pumps with high Na$^+$ affinity.

Thus, in conclusion, among FXYD proteins, not only γ subunits, but also CHIF is a tissue-specific regulator of Na,K-ATPase. CHIF modulates the transport properties of Na,K-ATPase in an opposite way than γ subunits by increasing (rather than decreasing) its apparent Na$^+$ affinity. It is likely that this regulatory mechanism of Na,K-ATPase plays a crucial role in aldosterone-responsive tissues such as kidney and colon, which are responsible for the maintenance of body Na$^+$ and K$^+$ homeostasis.

FXYD7 Is a Brain-Specific Modulator of Na,K-ATPase α1-β Isozymes Favoring Neuronal Excitability

To further support the role of FXYD proteins in Na,K-ATPase modulation, we next analyzed FXYD7, a protein that has so far not been studied and that we found to be expressed exclusively in brain, both in neurons and in glial cells.[20]

Association of FXYD7 with Na,K-ATPase was again tested in *Xenopus* oocytes. Interestingly, after coexpression in oocytes, an FXYD7 antibody coimmunoprecipitated α1-β1, α2-β1, and α3-β1 isozymes, but not α-β2 isozymes, despite a similar expression of all Na,K-ATPase isozymes. These results indicate that FXYD7 can indeed associate with Na,K-ATPase isozymes, but only if they contain the β1 isoform.

By using α isoform-specific antibodies, we could also demonstrate Na,K-ATPase isozyme-specific association of FXYD7 in native brain tissue. Significantly, only α1 (but not α2 or α3) isoforms could be coimmunoprecipitated with an FXYD7 antibody in brain microsomes. Together with the results obtained in oocytes, these results indicate that, in brain, FXYD7 is specifically associated with Na,K-ATPase α1-β1 isozymes.

Interestingly, the functional effect of FXYD7 on α1-β1 isozymes differs from that of the γ subunit as well as from that of CHIF. FXYD7 increases the $K_{1/2}$ value for extracellular K$^+$ of α1-β1 isozymes by about 2-fold over a wide range of membrane potentials, whereas the $K_{1/2}$ value for intracellular Na$^+$ does not change.[20]

What then is the physiological relevance of α1-β1 isozymes with a low K$^+$ affinity in the nervous system? During neuronal activity, not only intracellular Na$^+$, but also extracellular K$^+$ concentrations increase significantly. To prevent K$^+$-mediated perturbations of the neuronal excitability, K$^+$ must be rapidly cleared from the extracellular space. Both the K$^+$ recovery during neuronal firing and the characteristic extracellular K$^+$ undershoot observed after neuronal firing are mediated by the Na,K-ATPase since both processes are abolished in the presence of ouabain.[24]

Little is known on the relative contribution of different Na,K-ATPase isozymes in the recovery of ion homeostasis during neuronal activity. It is likely that neuronal α3-β1 isozymes are responsible for efficient extrusion of excess intracellular Na$^+$ since they have a low intrinsic Na$^+$ affinity[3] and thus can increase their transport rate even when intracellular Na$^+$ is high. Based on the same reasoning, glial α2-β2 isozymes that have a low K$^+$ affinity[3] can be implicated in the recovery of extracellular K$^+$. However, to make the K$^+$ reuptake process efficient, it may be necessary that also the major Na,K-ATPase isozyme pool, namely the neuronal and glial α1-β1 complexes, must have a low K$^+$ affinity that they acquire by the association with FXYD7. Moreover, α1-β1 isozymes with a low K$^+$ affinity may also have a protective effect in preventing an excessive K$^+$ undershoot after sustained neuronal activity.

Phospholemman Is a Heart- and Skeletal Muscle–Specific Modulator of Na,K-ATPase Favoring Muscle Contractility

Of the remaining members of the FXYD family, we finally also analyzed FXYD1 or phospholemman. In contrast to most other family members, phospholemman has already been studied extensively and was found to be a main substrate for protein kinase A and C phosphorylation in heart and skeletal muscle,[16] to induce a hyperpolarization-activated Cl$^-$ conductance,[25] and to transport the zwitterionic amino acid taurine.[26] Based on these activities, it is believed that phospholemman plays a specific role in muscle contraction and cell volume regulation.

Significantly, phospholemman did also associate specifically with Na,K-ATPase after expression in oocytes and in fact with all α-β1 and α-β2 isozymes tested, but not with SERCA pumps.[21] Association of phospholemman with Na,K-ATPase probably occurs posttranslationally since the ratio between assembled phospholemman and α is at most 0.5 after a 6-h pulse, and only reaches a 1:1 stoichiometry after a 24-h chase.

Association of phospholemman with Na,K-ATPase could also be demonstrated in native tissue. In heart microsomes, phospholemman could be coimmunoprecipitated with an α1-specific antibody and to a much lesser extent with an α2 antibody; in skeletal muscle, α1 isoforms, but not α2 isoforms, could be coimmunoprecipitated with a phospholemman antibody.[21]

The functional analysis of phospholemman revealed only a small effect on the apparent K$^+$ affinity, but a substantial decrease in the apparent affinity for intracellular Na$^+$ of α1-β1 isozymes. This Na$^+$ effect is favorable in contractile tissues that need low Na$^+$ affinity Na,K-pumps that permit efficient extrusion of increased intracellular Na⁺ during action potentials and that hence ensure appropriate muscle contractility. As suggested by Na⁺ activation curves, only α1-β1 complexes associated with phospholemman can indeed increase their transport rate upon a rise in intracellular Na⁺, for example, up to 20 mM, whereas α1-β1 isozymes, which lack phospholemman, would transport already at a nearly maximal rate under these conditions.[21]

CONCLUSIONS

In conclusion, we have characterized 4 out of the 7 members of the FXYD family and our results suggest that most, if not all, of these proteins act as tissue-specific, regulatory subunits of Na,K-ATPase (FIG. 1).

Significantly, each of these auxiliary subunits has a distinct modulatory effect on the Na,K-ATPase transport function that is adapted to the specific physiological requirements of the tissue in which they are expressed. Moreover, FXYD proteins appear to modulate preferentially the function of the housekeeping, ubiquitous α1-β1 Na,K-ATPase isozymes even in excitable tissues that express other Na,K-ATPase isozymes with tissue-adapted transport properties. Altogether, these results highlight the exquisite complexity of the regulatory mechanisms of Na⁺ and K⁺ handling via the Na,K-ATPase that are necessary to ensure appropriate tissue functions such as renal Na⁺ reabsorption, muscle contraction, and neuronal excitability.

REFERENCES

1. GEERING, K. 2001. The functional role of β subunits in oligomeric P-type ATPases. J. Bioenerg. Biomembr. **33:** 425–438.
2. BLANCO, G. & R.W. MERCER. 1998. Isozymes of the Na-K-ATPase: heterogeneity in structure, diversity in function. Am. J. Physiol. **275:** F633–F650.
3. CRAMBERT, G., U. HASLER, A.T. BEGGAH, et al. 2000. Transport and pharmacological properties of nine different human Na,K-ATPase isozymes. J. Biol. Chem. **275:** 1976–1986.
4. THERIEN, A.G. & R. BLOSTEIN. 2000. Mechanisms of sodium pump regulation. Am. J. Physiol. **279:** C541–C566.
5. FÉRAILLE, E. & A. DOUCET. 2001. Sodium-potassium-adenosinetriphosphatase-dependent sodium transport in the kidney: hormonal control. Physiol. Rev. **81:** 345–418.
6. SWEADNER, K.J. & E. RAEL. 2000. The FXYD gene family of small ion transport regulators or channels: cDNA sequence, protein signature sequence, and expression. Genomics **68:** 41–56.
7. FORBUSH, B., III, J.H. KAPLAN & J.F. HOFFMAN. 1978. Characterization of a new photoaffinity derivative of ouabain: labeling of the large polypeptide and of a proteolipid component of the Na,K-ATPase. Biochemistry **17:** 3667–3676.
8. MERCER, R.W., D. BIEMESDERFER, D.P. BLISS, et al. 1993. Molecular cloning and immunological characterization of the γ-polypeptide, a small protein associated with the Na,K-ATPase. J. Cell Biol. **121:** 579–586.
9. KÜSTER, B., A. SHAINSKAYA, H.X. PU, et al. 2000. A new variant of the γ subunit of renal Na,K-ATPase: identification by mass spectrometry, antibody binding, and expression in cultured cells. J. Biol. Chem. **275:** 18441–18446.
10. BÉGUIN, P., X.Y. WANG, D. FIRSOV, et al. 1997. The γ subunit is a specific component of the Na,K-ATPase and modulates its transport function. EMBO J. **16:** 4250–4260.
11. BÉGUIN, P., G. CRAMBERT, S. GUENNOUN, et al. 2001. CHIF, a member of the FXYD protein family, is a regulator of Na,K-ATPase distinct from the γ-subunit. EMBO J. **20:** 3993–4002.

12. ARYSTARKHOVA, E., C. DONNET, N.K. ASINOVSKI & K.J. SWEADNER. 2002. Differential regulation of renal Na,K-ATPase by splice variants of the γ subunit. J. Biol. Chem. **277:** 10162–10172.
13. PU, H.X., R. SCANZANO & R. BLOSTEIN. 2002. Distinct regulatory effects of the Na,K-ATPase γ subunit. J. Biol. Chem. **277:** 20270–20276.
14. MEIJ, I.C., J.B. KOENDERINK, H. VAN BOKHOVEN, et al. 2000. Dominant isolated renal magnesium loss is caused by misrouting of the Na^+,K^+-ATPase γ-subunit. Nat. Genet. **26:** 265–266.
15. MOORMAN, J.R., C.J. PALMER, J.E. JOHN III, et al. 1992. Phospholemman expression induces a hyperpolarization-activated chloride current in *Xenopus* oocytes. J. Biol. Chem. **267:** 14551–14554.
16. PALMER, C.J., B.T. SCOTT & L.R. JONES. 1991. Purification and complete sequence determination of the major plasma membrane substrate for cAMP-dependent protein kinase and protein kinase C in myocardium. J. Biol. Chem. **266:** 11126–11130.
17. MORRISON, B.W., J.R. MOORMAN, G.C. KOWDLEY, et al. 1995. Mat-8, a novel phospholemman-like protein expressed in human breast tumors, induces a chloride conductance in *Xenopus* oocytes. J. Biol. Chem. **270:** 2176–2182.
18. ATTALI, B., H. LATTER, N. RACHAMIM & H. GARTY. 1995. A corticosteroid-induced gene expressing an "IsK-like" K^+ channel activity in *Xenopus* oocytes. Proc. Natl. Acad. Sci. USA **92:** 6092–6096.
19. FU, X. & M. KAMPS. 1997. E2a-Pbx1 induces aberrant expression of tissue-specific and developmentally regulated genes when expressed in NIH 3T3 fibroblasts. Mol. Cell. Biol. **17:** 1503–1512.
20. BÉGUIN, P., G. CRAMBERT, F. MONNET-TSCHUDI, et al. 2002. FXYD7 is a brain-specific regulator of Na,K-ATPase α1-β isozymes. EMBO J. **21:** 3264–3273.
21. CRAMBERT, G., M. FÜZESI, H. GARTY, et al. 2002. Phospholemman (FXYD1) associates with Na,K-ATPase and regulates its transport properties. Proc. Natl. Acad. Sci. USA In press.
22. SHI, H., R. LEVY-HOLZMAN, F. CLUZEAUD, et al. 2001. Membrane topology and immunolocalization of CHIF in kidney and intestine. Am. J. Physiol. **280:** F505–F512.
23. CAPURRO, C., N. COUTRY, J.P. BONVALET, et al. 1996. Cellular localization and regulation of CHIF in kidney and colon. Am. J. Physiol. **271:** C753–C762.
24. D'AMBROSIO, R., D.S. GORDON & H.R. WINN. 2002. Differential role of KIR channel and Na^+/K^+-pump in the regulation of extracellular K^+ in rat hippocampus. J. Neurophysiol. **87:** 87–102.
25. MOORMAN, J.R., S.J. ACKERMAN, G.C. KOWDLEY, et al. 1995. Unitary anion currents through phospholemman channel molecules. Nature **377:** 737–740.
26. KOWDLEY, G.C., S.J. ACKERMAN, Z.H. CHEN, et al. 1997. Anion, cation, and zwitterion selectivity of phospholemman channel molecules. Biophys. J. **72:** 141–145.

A Specific Functional Interaction between CHIF and Na,K-ATPase

Role of FXYD Proteins in the Cellular Regulation of the Pump

HAIM GARTY, MOSHIT LINDZEN, MARIA FÜZESI, ROMAN AIZMAN, RIVKA GOLDSHLEGER, CAROL ASHER, AND STEVEN J. D. KARLISH

Department of Biological Chemistry, Weizmann Institute of Science, Rehovot 76100, Israel

ABSTRACT: CHIF (corticosteroid hormone–induced factor) is a member of the FXYD family that shares ~50% homology with the γ subunit of Na,K-ATPase. It is expressed in renal collecting duct and distal colon, and is upregulated by Na^+ deprivation and high K^+ diet. Both CHIF and γ are coimmunoprecipitated by an anti-α subunit antibody, and α is immunoprecipitated by anti-γ and anti-CHIF antibodies. $^{86}Rb^+$ flux experiments in CHIF-transfected HeLa cells demonstrate that CHIF increases the affinity for cytoplasmic Na^+, but does not affect the affinity for extracellular K(Rb). A physiological role of CHIF in kidney function is further elucidated by the phenotypic analysis of CHIF knockout mice. Taken together with data by others, it appears that FXYD proteins are tissue-specific subunits or regulators of the Na,K-ATPase whose function is to adjust the pump kinetics to particular physiological needs.

KEYWORDS: CHIF; Na,K-ATPase; FXYD proteins; renal Na^+ absorption; renal K^+ secretion

INTRODUCTION

CHIF (corticosteroid hormone–induced factor) was cloned as an epithelial-specific aldosterone-induced mRNA.[1] Its deduced amino acid sequence predicts a single spanning transmembrane protein that shares significant homology with six other proteins whose function is poorly understood. Together they have been termed "the FXYD family" due to a common motif in their extracellular domain.[2] In addition to CHIF (FXYD4), the family includes phospholemman (FXYD1[3]), the γ subunit of Na,K-ATPase (FXYD2[4]), Mat-8 (FXYD3[5]), RIC (FXYD5[6]), and two additional proteins identified by searching the expressed sequence tag database (FXYD6 and 7[2]). Some of these proteins were reported to evoke channel activity in various expression systems.[1,5,7–10] However, the physiological relevance of these observations is yet unclear.

Address for correspondence: Haim Garty, Department of Biological Chemistry, Weizmann Institute of Science, Rehovot 76100, Israel. Voice: +972-8-934-2706; fax: +972-8-934-4177.
h.garty@weizmann.ac.il

Ann. N.Y. Acad. Sci. 986: 395–400 (2003). © 2003 New York Academy of Sciences.

Recent data by several laboratories have provided evidence that at least four of the FXYD proteins interact with the α subunit of the Na,K-ATPase (α) and alter the pump kinetics. The current paper summarizes data obtained in our laboratory indicating that CHIF is an epithelial-specific subunit of the pump that increases its affinity towards internal Na$^+$.

RESULTS

Previous studies have demonstrated that CHIF mRNA and protein are expressed only in kidney collecting duct and distal colon surface cells, and cannot be detected in a large variety of other epithelial and nonepithelial tissues. These include other fragments of the nephron and intestine, heart, brain, muscle, liver, skin, lung, uterus, testis, mammary glands, and salivary glands.[11,12] Like the γ subunit of Na,K-ATPase, (γ) CHIF resides in the basolateral membrane of the epithelial cells. However, distribution of the two FXYD proteins along the nephron appears to be mutually exclusive.[12] Expression of the two γ splice variants (γa and γb) starts at the medullary thick ascending limb (mTAL) and ends at the connecting tubules[13,14] (see, however, Pihakaski-Maunsbach et al., this volume). CHIF, on the other hand, is located only along the collecting duct (CD) and is present along the cortical (CCD), outer medullary (OMCD), and inner medullary (IMCD) collecting duct.

CHIF mRNA and protein are strongly induced by aldosterone, the major mineralocorticoid regulating Na$^+$ absorption and K$^+$ secretion in all vertebrates.[15] Similar effects are also seen by Na$^+$ deprivation that promotes aldosterone secretion.[12,16–18] An aldosterone-induced increase in CHIF mRNA is apparent only after >3 h, which classifies this effect as a "late" hormonal response.[15,16] Interestingly, the message and protein are differently affected by aldosterone (or Na$^+$ deprivation) and the effects also differ between the kidney and intestine. In the distal colon, CHIF mRNA is strongly elevated by either aldosterone or Na$^+$ deprivation, but such effects are not seen in kidney CD.[16,17] CHIF protein, on the other hand, is induced by these treatments in both the kidney and colon.[12,18] In addition to the upregulation by low Na$^+$ intake, CHIF is induced by K$^+$ loading through an aldosterone-independent mechanism.[12,17,18]

The fact that CHIF is localized in the basolateral membrane of epithelial cells, induced under conditions of an increased Na$^+$ absorption and/or K$^+$ secretion, and shares considerable homology with γ suggested that this protein too might be a regulator or subunit of the Na,K-ATPase. This assumption is strongly supported by two sets of data described in Reference 19: one is the specific coimmunoprecipitation of CHIF and α, and the other is functional effects of CHIF on the pump kinetics.

Coimmunoprecipitation of FXYD proteins and α has been done in colonocytes (CHIF) and kidney medulla (γ) solubilized in $C_{12}E_{10}$. Previous studies have demonstrated that this detergent preserves the native pump structure provided that the solubilizing buffer also includes Rb$^+$ plus ouabain or Na$^+$ plus oligomycin.[20] In agreement, we also found that the anti-α antibody can immunoprecipitate significant amounts of CHIF and γ only if the membrane solubilization buffer includes Rb$^+$ plus ouabain or Na$^+$ plus oligomycin.[19] It was further demonstrated that the anti-CHIF and anti-γ antibodies can immunoprecipitate α under these conditions. Thus, effective coprecipitation of either CHIF or γ with α requires preservation of the native pump structure, indicating that the interaction is indeed specific. Other studies summarized in reference 19 have demonstrated that the solubilized membranes do not contain mixed complexes of α with both CHIF and γ, or of γ with both γa and γb.

Functional effects of CHIF have been studied in HeLa cells overexpressing the rat α1 subunit of Na,K-ATPase. Cells were stably transfected with either CHIF or empty vector, resulting in matched cell lines that do or do not express this FXYD protein.[19] The pump activity could then be measured as ouabain-blockable $^{86}Rb^+$ uptake, and the pump's Na^+_{in} activation curve was evaluated by preincubating the cells with the Na^+ ionophore, monansin, in media containing different Na^+ concentrations. CHIF was found to decrease $K_{0.5}$ for internal Na^+ from 6.3 ± 2.0 mM to 1.9 ± 0.4 mM without significantly affecting V_{max} or the $K_{0.5}$ for external K^+. These effects are consistent with those reported by electrical recordings in *Xenopus* oocytes expressing rat α1β1 ± CHIF,[21] and are very different from the effects of γ measured either in transfected HeLa cells or in *Xenopus* oocytes.[14,22,23] The observed CHIF-induced increase in apparent Na^+ affinity predicts an up to 3.3-fold increase in pumping rate under limiting intracellular Na^+ activity. Such limitation is indeed expected to take place in the distal nephron under dietary Na^+ restriction.

To determine the role of CHIF in whole kidney physiology and look for other possible functions of this protein, we have generated and analyzed CHIF knockout mice.[18] The null mutated mice (–/–) were viable and, under normal conditions, not distinguishable from the matched wild-type population (+/+) in any of the parameters measured. These include glomerular filtration rate (GFR), water intake and urine volume, plasma and urine osmolarity, and plasma and urine electrolyte concentrations. In mice fed a high K^+ diet, two phenotypes associated with the null mutation of CHIF were apparent: the first was a ~25% increase in GFR and the second was an up to 2-fold increase in urine volume and water intake. Similar, but smaller effects were observed under Na^+ deprivation. These differences between +/+ and –/– mice were apparent under electrolyte stress, but not when diuresis was evoked by including sucrose in the drinking water and eliminating solid food. The data are consistent with an abnormality in water absorption that is secondary to a defect in electrolyte transport. We suggest that the deletion of CHIF does not affect overall electrolyte balance since the CD-specific inhibition is relatively small and well compensated by the increased GFR and transport in the more proximal nephron segments. The increased water excretion may therefore result from the increased GFR or a decrease in the driving force for water absorption across the IMCD. It was further shown that combining K^+ loading with inhibition of the $Na^+/K^+/Cl^-$ cotransporter by furosemide resulted in hyperkalemia and eventually lethality in the –/– (but not the +/+) mice. This may reflect either an excessive volume depletion or an inhibition of K^+ excretion (by the higher luminal K^+ in the CD) that becomes lethal in –/– mice.

In summary, the current data demonstrate the functional role of CHIF in the regulation of Na,K-ATPase activity and on whole body salt and water homeostasis.

DISCUSSION

Data summarized in this paper provide strong evidence for a functional role of CHIF in a kidney-specific regulation of the Na,K-ATPase and in whole body electrolyte metabolism. Effects on the pump kinetics have been reported for three other

FXYD proteins, that is, phospholemman,[24] γ,[14,22,23,25] and FXYD7.[26] Each of these has a distinct tissue distribution and different effects on the pump kinetics. Therefore, the emerging picture is that FXYD proteins are tissue-specific subunits or regulators of the Na,K-ATPase. They provide means to alter pump properties in one target tissue or cell type without affecting it elsewhere. As such, they may mediate regulation of the pump by dietary salt intake and mineralocorticoids (CHIF), PKA and PKC (phospholemman[27]), or anoxia and hyperosmotic stress (γ[22,28]). The above data do not exclude other cellular roles of these proteins. It is also possible that they have additional effects on the pump not manifested in the above expression systems. Such effects could involve additional interacting proteins and various agonists and antagonists.

A number of interesting questions remain to be answered. One is the structure-function relationships of these proteins and the identity of the residues interacting with α and mediating the functional effects. Do different FXYD with opposite functional effects interact with different domains on α or is the functional variability due to differences in FXYD sequences only? Some tools aimed at addressing these issues have been developed in our laboratory and are described in references 29 and 30. Another question is the possible regulation of the α/FXYD interactions. Some cells appear to contain two different FXYD, for example, γa and γb in the mTAL or both CHIF and γ in the IMCD. On the other hand, it appears that α does not interact simultaneously with two FXYD.[19] How is specificity determined in cells expressing more than one FXYD? Are there posttranslational modifications that control association/dissociation of the FXYD/α complex, or is the expression of these proteins the only mechanism by which regulation is evoked? Finally, one may wonder whether cells contain additional FXYD proteins not identified so far. Such additional family members may be present in a minor population of cells or expressed only under specific physiological conditions, and thus not represented in available databases.

CONCLUSIONS

Recent findings by a number of laboratories demonstrate that at least four of the FXYD proteins are tissue-specific regulatory subunits of the Na,K-ATPase. Thus, it is likely that all FXYD proteins are regulatory subunits that adjust the pump's kinetic properties to the specific needs of a particular tissue or physiological state. Many features of this unique regulatory mechanism await further study.

ACKNOWLEDGMENTS

Work in our laboratory was supported by research grants from the Minerva Foundation and the Israel Science Foundation. H. Garty is the incumbent of the Hella and Derrick Kleeman Chair for Biochemistry. S. J. D. Karlish is the incumbent of the William P. Smithburg Chair in Biochemistry.

REFERENCES

1. ATTALI, B., H. LATTER, N. RACHAMIM & H. GARTY. 1995. A corticosteroid-induced gene expressing an "IsK-like" K^+ channel activity in *Xenopus* oocytes. Proc. Natl. Acad. Sci. USA **92:** 6092–6096.
2. SWEADNER, K.J. & E. RAEL. 2000. The FXYD gene family of small ion transport regulators or channels: cDNA sequence, protein signature sequence, and expression. Genomics **68:** 41–56.
3. PALMER, C.J., B.T. SCOTT & L.R. JONES. 1991. Purification and complete sequence determination of the major plasma membrane substrate for cAMP-dependent protein kinase and protein kinase C in myocardium. J. Biol. Chem. **266:** 11126–11130.
4. MERCER, R.W., D. BIEMESDERFER, D.P. BLISS, *et al.* 1993. Molecular cloning and immunological characterization of the gamma polypeptide, a small protein associated with the Na,K-ATPase. J. Cell Biol. **121:** 579–586.
5. MORRISON, B.W., J.R. MOORMAN, G.C. KOWDLEY, *et al.* 1995. Mat-8, a novel phospholemman-like protein expressed in human breast tumors, induces a chloride conductance in *Xenopus* oocytes. J. Biol. Chem. **270:** 2176–2182.
6. FU, X. & M.P. KAMPS. 1997. E2a-Pbx1 induces aberrant expression of tissue-specific and developmentally regulated genes when expressed in NIH 3T3 fibroblasts. Mol. Cell. Biol. **17:** 1503–1512.
7. MOORMAN, J.R., C.J. PALMER, J.E. JOHN, *et al.* 1992. Phospholemman expression induces a hyperpolarization-activated chloride current in *Xenopus* oocytes. J. Biol. Chem. **267:** 14551–14554.
8. MOORMAN, J.R., S.J. ACKERMAN, G.C. KOWDLEY, *et al.* 1995. Unitary anion currents through phospholemman channel molecules. Nature **377:** 737–740.
9. MINOR, N.T., Q. SHA, C.G. NICHOLS & R.W. MERCER. 1998. The gamma subunit of the Na,K-ATPase induces cation channel activity. Proc. Natl. Acad. Sci. USA **95:** 6521–6525.
10. SHA, Q., K.L. LANSBERY, D. DISTEFANO, *et al.* 2001. Heterologous expression of the Na(+),K(+)-ATPase gamma subunit in *Xenopus* oocytes induces an endogenous, voltage-gated large diameter pore. J. Physiol. **535:** 407–417.
11. CAPURRO, C., J.P. BONVALET, B. ESCOUBET, *et al.* 1996. Cellular localization and regulation of CHIF in kidney and colon. Am. J. Physiol. Cell. Physiol. **271:** C753–C762.
12. SHI, H-K., R. LEVY-HOLZMAN, F. CLUZEAUD, *et al.* 2001. Membrane topology and immunolocalization of CHIF in kidney and intestine. Am. J. Physiol. Renal Physiol. **280:** F505–F515.
13. WETZEL, R.K. & K.J. SWEADNER. 2001. Immunocytochemical localization of Na-K-ATPase alpha- and gamma-subunits in rat kidney. Am. J. Physiol. Renal Physiol. **281:** F531–F545.
14. PU, H.X., F. CLUZEAUD, R. GOLDSHLEGER, *et al.* 2001. Functional role and immunocytochemical localization of the γa and γb forms of the Na,K-ATPase γ subunit. J. Biol. Chem. **276:** 20370–20378.
15. GARTY, H. 2000. Regulation of the epithelial Na^+ channel by aldosterone: open questions and emerging answers. Kidney Int. **57:** 1270–1276.
16. WALD, H., O. GOLDSTEIN, C. ASHER, *et al.* 1996. Aldosterone induction and epithelial distribution of CHIF. Am. J. Physiol. Renal Fluid Electrolyte Physiol. **271:** F322–F329.
17. WALD, H., M.M. POPOVTZER & H. GARTY. 1997. Differential regulation of CHIF expression by aldosterone and potassium. Am. J. Physiol. Renal Fluid Electrolyte Physiol. **272:** F617–F623.
18. AIZMAN, R., C. ASHER, M. FÜZESI, *et al.* 2002. Generation and phenotypic analysis of CHIF knockout mice. Am. J. Physiol. Renal Physiol. **283:** F569–F577.
19. GARTY, H., M. LINDZEN, R. SCANFANO, *et al.* 2002. A functional interaction between CHIF and Na,K-ATPase: implication for regulation by FXYD proteins. Am. J. Physiol. Renal Physiol. **283:** F607–F615.
20. CARRADUS, M. 2000. Ph.D. thesis approved by the Imperial College London University in December 2000.

21. BÉGUIN, P., G. CRAMBERT, S. GUENNOUN, et al. 2001. CHIF, a member of the FXYD protein family, is a regulator of Na,K-ATPase distinct from the gamma subunit. EMBO J. **20:** 3993–4002.
22. THERIEN, A.G., S.J. KARLISH & R. BLOSTEIN. 1999. Expression and functional role of the gamma subunit of the Na,K-ATPase in mammalian cells. J. Biol. Chem. **274:** 12252–12256.
23. BÉGUIN, P., X.Y. WANG, D. FIRSOV, et al. 1997. The gamma subunit is a specific component of the Na,K-ATPase and modulates its transport function. EMBO J. **16:** 4250–4260.
24. CRAMBERT, G., M. FÜZESI, H. GARTY, et al. 2002. Phospholemman (FXYD1) associates with Na,K-ATPase and regulates its transport properties. Proc. Natl. Acad. Sci. USA **99:** 11476–11481.
25. ARYSTARKHOVA, E., R.K. WETZEL, N.K. ASINOVSKI & K.J. SWEADNER. 1999. The gamma subunit modulates Na(+) and K(+) affinity of the renal Na,K-ATPase. J. Biol. Chem. **274:** 33183–33185.
26. BÉGUIN, P., G. CRAMBERT, F. MONNET-TSCHUDI, et al. 2002. FXYD7 is a brain-specific regulator of Na,K-ATPase alpha 1–beta isozymes. EMBO J. **21:** 3264–3273.
27. MAHMMOUD, Y.A., H. VORUM & F. CORNELIUS. 2000. Identification of a phospholemman-like protein from shark rectal glands: evidence for indirect regulation of Na,K-ATPase by protein kinase C via a novel member of the FXYDY family. J. Biol. Chem. **275:** 35969–35977.
28. CAPASSO, J.M., C. RIVARD & T. BERL. 2001. The expression of the gamma subunit of Na-K-ATPase is regulated by osmolality via C-terminal Jun kinase and phosphatidylinositol 3-kinase–dependent mechanisms. Proc. Natl. Acad. Sci. USA **98:** 13414–13419.
29. LINDZEN, M., R. AIZMAN, Y. LIFSHITZ, et al. 2003. Domains involved in the interactions between FXYD and the Na,K-ATPase. This volume.
30. FÜZESI, M., R. GOLDSHLEGER, H. GARTY & S.J.D. KARLISH. 2003. Defining the nature and sites of interaction between FXYD proteins and Na,K-ATPase. This volume.

Immunocytochemical Localization of Na,K-ATPase Gamma Subunit and CHIF in Inner Medulla of Rat Kidney

KAARINA PIHAKASKI-MAUNSBACH,[a,b] HENRIK VORUM,[c] ELSE-MERETE LØCKE,[a,b] HAIM GARTY,[d] STEVEN J. D. KARLISH,[d] AND ARVID B. MAUNSBACH[a,b]

[a]*The Water and Salt Research Center,* [b]*Department of Cell Biology, and* [c]*Department of Medical Biochemistry, University of Aarhus, DK-8000 Aarhus, Denmark*

[d]*Department of Biological Chemistry, Weizmann Institute of Science, Rehovoth 76100, Israel*

ABSTRACT: The γ subunit of Na,K-ATPase and CHIF both belong to the FXYD single-membrane-spanning protein family and have been suggested to have regulatory functions in kidney tubules. CHIF is known to be present in the collecting duct, and γ has been demonstrated in several segments of the rat kidney tubule, but never clearly in the inner medullary collecting duct (IMCD). Here, we demonstrate the cellular and subcellular localization of the γ subunit and CHIF in the IMCD in inner medulla by using Western blotting, laser-scanning confocal immunofluorescence, and immunoelectron microscopy. In the initial quarter of the IMCD (next to the outer medulla), antibodies against the C-terminal of γ as well as splice variant γa labeled the basolateral surface of intercalated cells (ICs), while principal cells (PCs) remained unlabeled. In the middle segment of the IMCD, all PCs exhibited distinct basolateral staining for the γC-terminal as well as γa and CHIF. Immunoelectron microscopy showed that the γC-terminal and CHIF were associated with the inner leaflet of the basolateral plasma membrane in the labeled cells. Immunoblotting demonstrated the presence of both the γC-terminal and γa in inner medullary tissue. However, splice variant γb was not detected in inner medulla by immunocytochemistry or immunoblotting. The present observations demonstrate that the Na,K-ATPase γ subunit and CHIF are strategically located in the inner medulla to participate in the fine-tuning of urine ion composition through the regulation of the Na,K-ATPase activity in the IMCD.

KEYWORDS: gamma subunit; CHIF; kidney inner medulla; collecting duct; intercalated cells; principal cells

Address for correspondence: Arvid B. Maunsbach, Department of Cell Biology, Institute of Anatomy, University of Aarhus, DK-8000 Aarhus, Denmark. Voice: +45-89423065; fax: +45-86128808.
maunsbach@ana.au.dk

INTRODUCTION

Na,K-ATPase is an integral membrane protein, responsible for the primary active transport of Na^+ and K^+ in animal cells. In most tissues, the enzyme consists of a catalytic α and a glycosylated β subunit; however, in kidney, there is also a small (~7 kDa) single-span membrane protein, the γ subunit, which is tightly associated with the αβ complex.[1,2] The γ subunit belongs to the FXYD protein family,[3] which contains several other small single-span membrane-spanning proteins, including CHIF (corticosteroid hormone–induced factor), which may also associate with the α subunit of Na,K-ATPase.[4,5] The γ subunit in the kidney has two splice variants, γa and γb, which differ at their extracellular N-termini, TELSANH and MDRWYL, respectively, but with an identical intracellular C-terminus.[6] The distribution of the γ subunit has been studied by immunofluorescence in kidneys of sheep[2] and rats.[7–9] It has been localized to basolateral membranes in most cortical and outer medullary tubules together with the α subunit, but never clearly in the inner medulla (IM). With respect to function, the γ subunit has been shown to modulate the properties of Na,K-ATPase by decreasing its apparent Na^+ affinity and increasing the apparent affinity for ATP.[7,10–13] CHIF protein and mRNA are present in the cortical and medullary collecting ducts, with the COOH tail facing the cell interior.[14,15] The functional properties of CHIF appear opposite from those of γ, and CHIF may modulate the Na,K-pump by increasing its apparent Na^+ affinity.[5,12,16]

In this study, we provide novel information by immunofluorescence and immunoelectron microscopy on the cellular and subcellular localizations of the Na,K-ATPase γ subunit and CHIF in the rat kidney IM as a basis for understanding the functional roles of γ and CHIF in the IMCD.

MATERIALS AND METHODS

Western Blotting

Membrane protein samples enriched in plasma membranes and intracellular vesicles were prepared from IM and the inner stripe of outer medulla (ISOM) of normal Wistar rat kidneys by centrifugation of a postmitochondrial fraction for 10 min at 133,000 g and separated by SDS-PAGE. Ten μg of prepared sample was loaded in each well, run on 10–20% gradient polyacrylamide minigels, and electrotransferred onto nitrocellulose membranes. Western blottings were made using polyclonal rabbit antibodies raised against the rat γC-terminal and the N-terminal splice variants, γa and γb.[6] Labeling was visualized with horseradish peroxidase–conjugated secondary antibodies diluted 1:5000 using the enhanced chemiluminescence system (Amersham International).

Immunocytochemistry

Kidneys from normal male Wistar rats were perfusion fixed with 4% paraformaldehyde and sagittal tissue blocks, including ISOM and IM embedded in paraffin. Sections were cut cold on a Leica RM2165 microtome at 2-μm thickness, and deparaffinized sections were preincubated in PBS containing 1% BSA or 0.1%

skimmed milk powder and 0.05 M glycine. They were then incubated for 1 h at room temperature with monoclonal antibodies against the α subunit of Na,K-ATPase, with polyclonal rabbit antibodies against the γC-terminal, the N-terminal splice variants γa and γb,[6] or CHIF.[5] The primary antibodies were detected with Alexa Fluor 546–labeled goat–anti-mouse IgG or with Alexa Fluor 488–labeled goat–anti-rabbit IgG and analyzed with a Leica TCS SL laser-scanning confocal microscope.

Immunoelectron Microscopy

Blocks from IM of perfusion-fixed kidneys were cryoprotected with sucrose and frozen in liquid N_2. The tissue was then either cryosectioned with a Leica-Reichert FCS cryoultramicrotome at −120°C or freeze-substituted in a Leica AFS freeze-substitution unit. The tissue was freeze-substituted in methanol containing 0.5% uranyl acetate at temperatures gradually rising from −90°C to −70°C, rinsed in pure methanol, infiltrated with Lowicryl HM20, and finally UV-polymerized at −45°C and at 0°C as described previously.[17] For immunoelectron microscopy, the ultrathin cryosections or the Lowicryl HM20 sections were labeled with the same antibodies as used for immunofluorescence and were detected with goat–anti-rabbit or goat–anti-mouse IgG conjugated to 10-nm colloidal gold particles. Sections were uranyl acetate–stained and examined in a FEI Morgagni 268D transmission electron microscope. Specificity of labeling was assessed by absorption controls or exclusion of primary antibody.

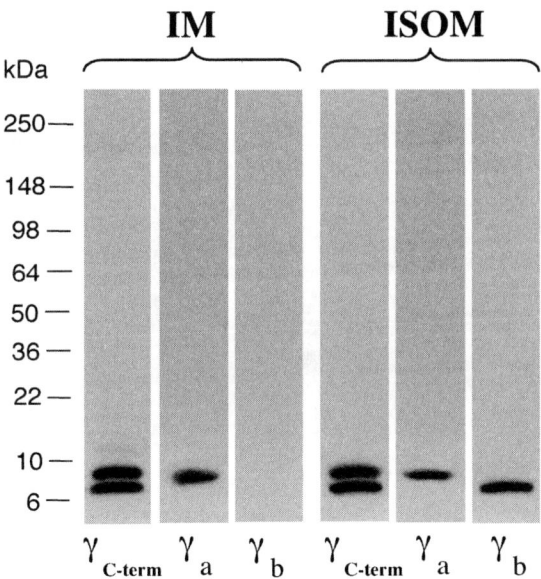

FIGURE 1. Western blot analysis of γ subunit of Na,K-ATPase in the IM and ISOM of normal rat kidney. Immunoblotting demonstrates that γC-terminal and γa, but not γb, are present in the IM.

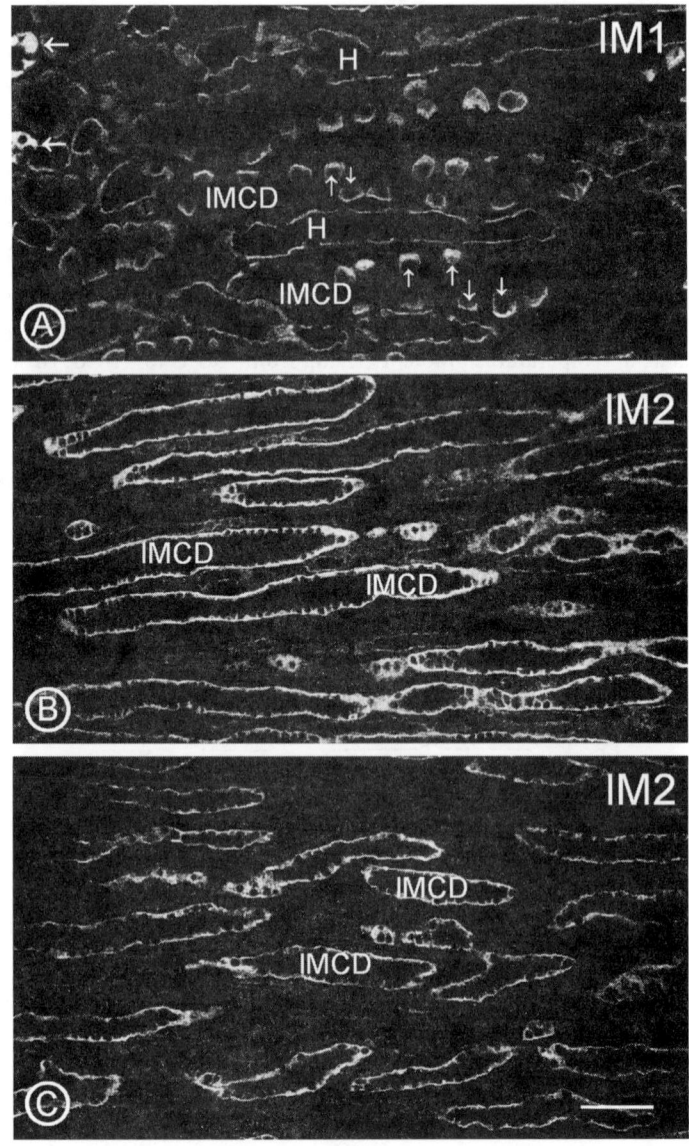

FIGURE 2. Immunofluorescence confocal microscopy of the initial portion of the inner medulla (IM1) showing γ subunit (C-terminal) in intercalated cells (*vertical arrows*) of collecting ducts (IMCD) and in thin limbs of Henle (H). *Horizontal arrows* indicate mTAL at the ISOM border. Note that the intensity of labeling in mTAL is much greater than in IM tubules **(A)**. Principal cells in collecting ducts in the middle portion of the inner medulla (IM2) are distinctly immunostained for the γ subunit, both the C-terminal **(B)** and the splice variant γa **(C)**. Bar in **C**, 50 μm.

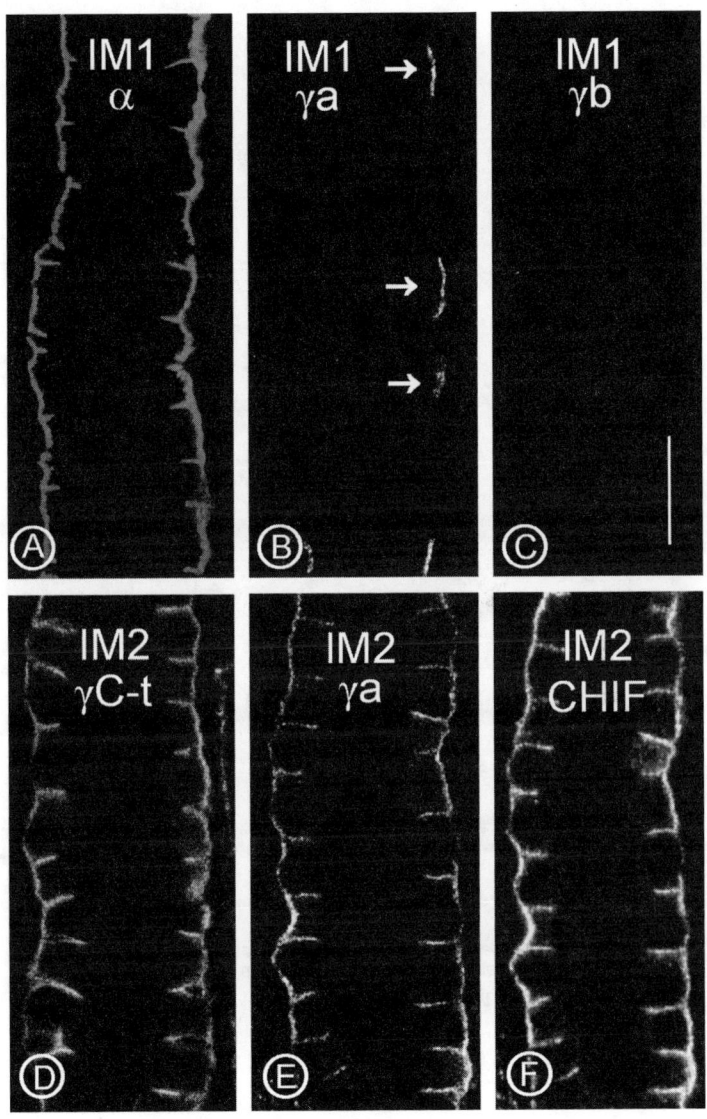

FIGURE 3. Collecting ducts in IM1 showing α subunit (**A**), γa (**B**), and absence of γb (**C**). *Arrows* in **B** point at intercalated cells. Principal cells in collecting ducts in the middle portion of the inner medulla (IM2) exhibit γ subunit (C-terminal) (**D**), γa (**E**), and CHIF (**F**). Bar, 20 μm.

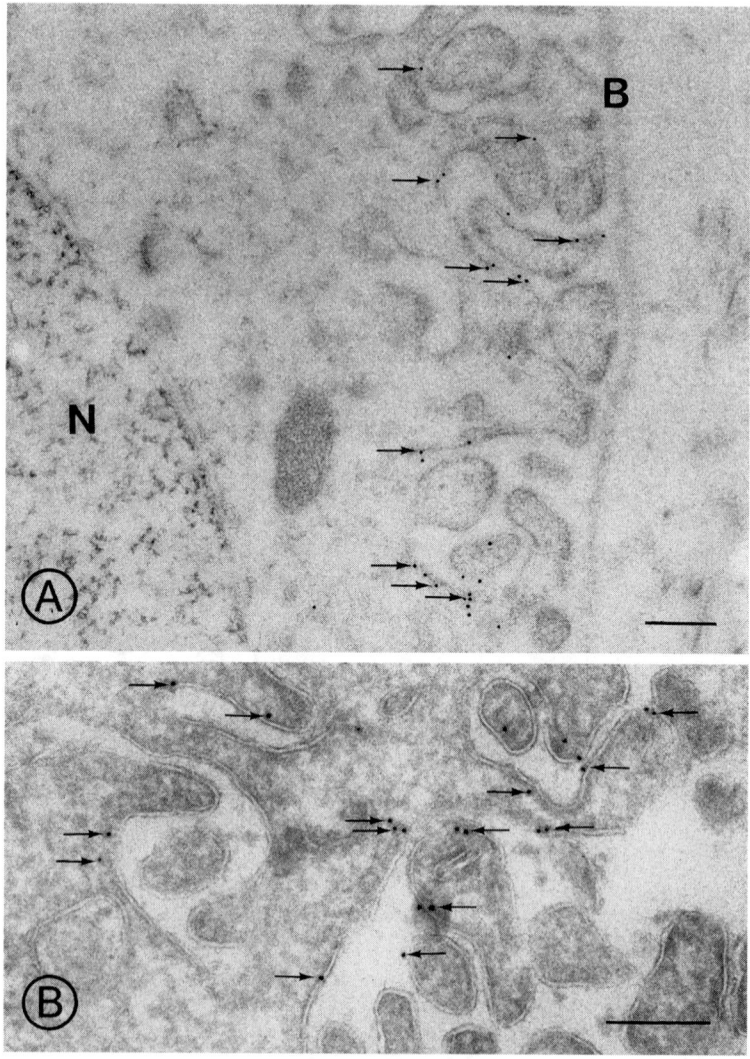

FIGURE 4. Immunoelectron microscope localization of γ subunit of Na,K-ATPase (**A**) and CHIF (**B**) in principal cells of collecting ducts in IM2 of normal rat kidney. Immunogold labels (*arrows*) are associated with the basolateral plasma membrane of the cells. Bar: 0.5 μm.

RESULTS

Immunoblotting was performed on membrane fractions from IM and ISOM. The γ subunit was present in both IM and ISOM as two main bands using anti-γ antibody recognizing the C-terminus of the protein (FIG. 1, lanes labeled γC-term). One single band was seen in both IM and ISOM for the γa splice variant. The γb splice variant was present in ISOM, but not in IM.

Immunofluorescence microscopy revealed the overall distributions of the different proteins in the IMCD (FIGS. 2 and 3) and immunoelectron microscopy clarified their subcellular locations (FIG. 4). In IM1, the initial quarter (about 25%) of IM next to outer medulla, antibodies against the γC-terminal (γC-term) and γa revealed strong basolateral labelings of intercalated cells (ICs), while adjacent principal cells (PCs) did not label (FIG. 2A). Thin limbs of Henle also stained for γa and C-terminal, and ascending thin limbs were often observed in direct continuity with thick ascending limbs (mTAL in FIG. 2A). PCs in IMCD in the middle portion (50%) of the IM (IM2) exhibited distinct basolateral immunofluorescence labeling for γC-terminal (FIG. 2B) as well as for γa (FIG. 2C). Notably, this labeling was less intense than the labeling of mTAL in the same section and with the same antibodies (FIG. 2A). The staining intensity for γC-terminal decreased slightly in IM3 (distal 25% of IM) towards the tip of the papilla. Absorption controls were negative.

The α subunit showed basolateral expression in all cells in IMCD. In ICs in IM1, the staining for γ colocalized with the α subunit (compare FIG. 3A and 3B). Notably, γb was not expressed at all in the IMCD (FIG. 3C), but showed strong staining in the medullary thick ascending limb (mTAL) in adjacent ISOM in similarity with γC-terminal (FIG. 2A, horizontal arrows) and γa.

Immunoelectron microscopy showed that the γC-terminal and γa were associated with the basolateral plasma membrane in ICs in IM1, in PCs in IM2 (FIG. 4A), and in IM3. The label was preferentially associated with the extensively folded basal regions of the plasma membrane, while those parts of the basal plasma membrane that directly abutted the basal lamina showed little or no label (FIG. 4A).

By immunofluorescence, CHIF showed a distinct basolateral localization in PCs (FIG. 3F). CHIF had the same expression pattern as γC-terminal and γa in IM2 (FIG. 3D and 3E), but not in IM1, where CHIF was only observed in PCs. Furthermore, immunoelectron microscopy showed that CHIF, in similarity with γ, was associated with the cytoplasmic face of the basolateral membrane (FIG. 4B), consistent with the suggested intracellular COOH tail of CHIF.[15]

DISCUSSION

Previous immunofluorescence studies have shown that the γ subunit of the Na,K-ATPase and its splice variants, γa and γb, have segment-specific distributions along the renal tubule in cortex and outer medulla.[2,6–9] However, the present results document that the γ subunit of Na,K-ATPase is present not only in the outer medulla, but also in the inner medulla of normal rat kidney, where it exhibits a pattern of localization that may be of physiological relevance. Thus, the γC-terminal and the γa splice variant are distinctly expressed in the middle and terminal portion of the inner medullary collecting duct, while γb is not observed at all in inner medulla. Furthermore, the γC-terminal and γa splice variant are present in the basolateral membranes of ICs in IM1, but not in adjacent PCs. The reason why labeling of γ in IM was observed in this investigation, but not reported in previous studies, is probably related to the high sensitivity and resolution obtained here with mild paraformaldehyde fixation and very thin sections both for immunofluorescence and immunoelectron microscopy. We did not detect the γb splice variant in inner medulla by immunofluorescence or Western blotting. However, the expression of γb in the IMCD may be related to the physiological conditions, since, interestingly, γb can be induced in cultured inner medullary cells following osmotic stress.[18] The α subunit, on the contrary, has a basolateral expression in all collecting duct cells in the rat inner medulla as previously observed.[14,19] In this investigation, the α and γ subunits colocalized in PCs in IM2 and IM3, while only α was observed in the PCs in IM1. The precise functions of the γ subunit in the kidney are still being investigated, but there is evidence that γ stabilizes E_1 conformation(s) of the enzyme and that anti-γ counteracts this effect.[12] Both γa and γb have similar effects on the Na,K-ATPase in increasing the affinity for ATP due to a shift in the enzyme's $E_1 \leftrightarrow E_2$ conformational equilibrium toward E_1. In addition, both γa and γb increase K^+ antagonism of cytoplasmic Na^+ activation.[20]

The present immunofluorescence observations of CHIF in the IMCD (FIG. 3F) are consistent with previous reports.[5,15] Additionally, our immunoelectron microscope observations (FIG. 4B) demonstrate that CHIF labeling is associated with the cytoplasmic face of the extensive basolateral membrane folds in the PCs. Different complexes between Na,K-ATPase and its subunits and CHIF, such as α/β/γa and α/β/CHIF, have been demonstrated by immunoprecipitation.[5,10] On the basis of the present immunofluorescence and immunoelectron microscope observations, which reveal the presence of both Na,K-ATPase γ subunit and CHIF in PCs in IM2 and IM3, additional complexes appear likely in the IMCD, such as α/β/γa in ICs in IM1 and in PCs in IM2 and IM3. The observation that PCs in IM1 did not stain for the γ subunit suggests that the composition of Na,K-ATPase in this segment is not α/β/γa, but either α/β/CHIF or only αβ. Complexes containing α/β/γa/CHIF have not been observed in immunoprecipitation experiments,[5] but the presence of both γ and CHIF in the basolateral membrane of PCs of IM2 and IM3 suggests that complexes such as α/β/γa and α/β/CHIF may exist simultaneously in these cells. It can be speculated that interactions between such complexes, or induced changes in their relative abundance, can influence the activity of the Na,K-ATPase.

In conclusion, the strategic localization of the Na,K-ATPase γ subunit as well as CHIF in the IMCD suggests that these FXYD proteins participate, individually or together, in the fine-tuning of ion excretion through regulation of the Na,K-ATPase activity in the IMCD. Further studies of the changes in the cellular localizations of Na,K-ATPase subunits and CHIF during diuresis, antidiuresis, and low/high Na/K intake may illuminate the significance of the characteristic localizations of the γ splice variants and the simultaneous presence of γ and CHIF in IM2 and IM3.

ACKNOWLEDGMENTS

This study was established and supported by the Danish National Research Foundation (Danmarks Grundforskningsfond), the Danish Medical Research Council, the Novo Nordisk Foundation, and the Karen Elise Jensen Foundation.

REFERENCES

1. FORBUSH, B., III, J.H. KAPLAN & J.F. HOFFMAN. 1978. Characterization of a new photoaffinity derivative of ouabain: labeling of the large polypeptide and of a proteolipid component of the Na,K-ATPase. Biochemistry 17: 3667–3676.

2. MERCER, R.W., D. BIEMESDERFER, D.P. BLISS, JR., et al. 1993. Molecular cloning and immunological characterization of the gamma polypeptide, a small protein associated with the Na,K-ATPase. J. Cell Biol. **121:** 579–586.
3. SWEADNER, K.J. & E. RAEL. 2000. The FXYD gene family of small ion transport regulators or channels: cDNA sequence, protein signature sequence, and expression. Genomics **68:** 41–56.
4. ATTALI, B., H. LATTER, N. RACHAMIM, et al. 1995. A corticosteroid-induced gene expressing an "IsK-like" K^+ channel activity in *Xenopus* oocytes. Proc. Natl. Acad. Sci. USA **92:** 6092–6096.
5. GARTY, H., M. LINDZEN, R. SCANZANO, et al. 2002. A functional interaction between CHIF and Na-K-ATPase: implication for regulation by FXYD proteins. Am. J. Physiol. Renal Physiol. **283:** F607–F615.
6. KÜSTER, B., A. SHAINSKAYA, H.X. PU, et al. 2000. A new variant of the γ subunit of renal Na,K-ATPase: identification by mass spectrometry, antibody binding, and expression in cultured cells. J. Biol. Chem. **275:** 18441–18446.
7. PU, H.X., F. CLUZEAUD, R. GOLDSHLEGER, et al. 2001. Functional role and immunocytochemical localization of the γa and γb forms of the Na,K-ATPase γ subunit. J. Biol. Chem. **276:** 20370–20378.
8. WETZEL, R.K. & K.J. SWEADNER. 2001. Immunocytochemical localization of Na-K-ATPase α and γ subunits in rat kidney. Am. J. Physiol. Renal Physiol. **281:** F531–F545.
9. ARYSTARKHOVA, E., R.K. WETZEL & K.J. SWEADNER. 2002. Distribution and oligomeric association of splice forms of Na(+)-K(+)-ATPase regulatory gamma-subunit in rat kidney. Am. J. Physiol. Renal Physiol. **282:** F393–F407.
10. BÉGUIN, P., X. WANG, D. FIRSOV, et al. 1997. The gamma subunit is a specific component of the Na,K-ATPase and modulates its transport function. EMBO J. **16:** 4250–4260.
11. THERIEN, A.G., S.J.D. KARLISH & R. BLOSTEIN. 1999. Expression and functional role of the γ subunit of the Na,K-ATPase in mammalian cells. J. Biol. Chem. **274:** 12252–12256.
12. THERIEN, A.G., R. GOLDSHLEGER, S.J.D. KARLISH, et al. 1997. Tissue-specific distribution and modulatory role of the gamma subunit of the Na,K-ATPase. J. Biol. Chem. **272:** 32628–32634.
13. ARYSTARKHOVA, E., R.K. WETZEL, N.K. ASINOVSKI, et al. 1999. The γ subunit modulates Na^+ and K^+ affinity of the renal Na,K-ATPase. J. Biol. Chem. **274:** 33183–33185.
14. CAPURRO, C., N. COUNTRY, J.P. BONVALET, et al. 1996. Cellular localization and regulation of CHIF in kidney and colon. Am. J. Physiol. **271:** C753–C762.
15. SHI, H., R. LEVY-HOLZMAN, F. CLUZEAUD, et al. 2001. Membrane topology and immuno-localization of CHIF in kidney and intestine. Am. J. Physiol. Renal Physiol. **280:** F505–F512.
16. BÉGUIN, P., G. CRAMBERT, S. GUENNOUN, et al. 2001. CHIF, a member of the FXYD protein family, is a regulator of Na,K-ATPase distinct from the gamma-subunit. EMBO J. **20:** 3993–4002.
17. MAUNSBACH, A.B. 1998. Immunolabeling and staining of ultrathin sections in biological electron microscopy. *In* A Laboratory Handbook, J.E. Celis, Ed.: 268–276. Academic Press. San Diego.
18. CAPASSO, J., C. RIVARD & T. BERL. 2001. The expression of the γ subunit of Na, K-ATPase is regulated by osmolarity via C-terminal Jun kinase and phosphatidylinositol 3-kinase-dependent mechanisms. Proc. Natl. Acad. Sci. USA **98:** 13414–13419.
19. SABOLIC, I., C.M. HERAK-KRAMBERGER, S. BRETON, et al. 1999. Na/K-ATPase in intercalated cells along the rat nephron revealed by antigen retrieval. J. Am. Soc. Nephrol. **10:** 913–922.
20. PU, H.X., R. SCANZANO & R. BLOSTEIN. 2002. Distinct regulatory effects of the Na,K-ATPase gamma subunit. J. Biol. Chem. **277:** 20270–20276.

Adaptation of Murine Inner Medullary Collecting Duct (IMCD3) Cell Cultures to Hypertonicity

JUAN M. CAPASSO, CHRISTOPHER J. RIVARD, LAURA M. ENOMOTO, AND TOMAS BERL

Department of Medicine, University of Colorado Health Sciences Center, Denver, Colorado 80262, USA

ABSTRACT: Recently, we have adapted IMCD3 cell cultures to survive under increasing hypertonic conditions (i.e., 600 and 900 mOsmol/kg H_2O). In adapted cells, ATPase activity is increased by one order of magnitude, while the expression of the α and β subunit is increased by a factor of 4 to 5 over controls (300 mOsmol/kg H_2O). Corresponding increases in mRNAs were also detected. The γ subunit has been described as being uniquely expressed in some areas of the kidney, but never in cell cultures (even those derived from kidney tissues). However, the γ subunit was detected at the protein and mRNA levels in the adapted IMCD3 cells. In contrast to the α and β subunits, the levels of γ protein and mRNA expression continue to increase as a function of the media ion concentration. We have also demonstrated that signaling pathways that upregulate the α, β, and γ subunits are very different. Increasing concentrations of the PI3 kinase inhibitor, LY294002, resulted in a dose-dependent reduction in the expression of the γ subunit, with total abolition at 10 μM. However, LY294002 had no significant effect on the expression of the α subunit. Inhibition of the JNK2 (but not of the JNK1) pathways by dominant negative transfections abolished the upregulation of the γ, but not the α subunit. Failure to upregulate the expression of the γ subunit was associated with a marked decrease in cell viability upon stress.

KEYWORDS: cell adaptation; signaling pathways; γ subunit regulation

Our laboratory has been involved in two main research areas: (a) cell adaptation to hypertonicity and (b) stress signaling pathways. In previous studies, we demonstrated that cell cultures derived from the kidney inner medulla, such as murine IMCD3, which survive *in vivo* at osmotic pressures as high as 3000 mOsmol/kg H_2O, do not survive *in vitro* more than 3 days under moderate osmotic stress (600 mOsmol/kg H_2O). While this survival time is considerably longer than in other kidney-derived cell cultures, such as M1, it does not permit chronic (long-term) studies to be carried out.

Address for correspondence: Juan M. Capasso, Department of Medicine, University of Colorado Health Sciences Center, Denver, CO 80262. Voice: 303-315-6723; fax: 303-315-4852.
juan.capasso@uchsc.edu

In order to circumvent this limitation, we have developed an adaptation protocol that allows kidney cell cultures to adapt and survive at 600–900 mOsmol/kg H_2O and beyond.[1] This protocol employs a stepwise adaptation of cell cultures to small increments in the medium tonicity, using NaCl. It is important to note that the protocol is not successful when other osmolytes such as mannitol or urea are employed. While the general appearances of control and adapted cell cultures are identical, under phase contrast microscopy, we have detected several differences at the physiological and biochemical levels, and additional investigations are currently under way. Examples of differences in cell physiology include the fact that adapted cells have a considerably longer doubling time than control cells, as well as higher requirements of energy sources.

Initial biochemical studies indicated that the inducible form of a critical heat shock protein, iHSP-70, is constitutively expressed in adapted cells, whereas it is totally absent in control cells.

The kinetics of activation of two MAP kinase pathways by NaCl increments, c-jun N-terminal kinase (JNK) and p-38, is blunted in adapted cells by a factor of four in comparison to control cells. This effect makes adapted cells markedly less sensitive and responsive to osmotic stress.

Another relevant finding is the upregulation in adapted cells of a fundamental enzyme in the kidney physiology, namely, Na/K-ATPase. In adapted cells, the mRNA levels of the α_1 subunit are increased severalfold as demonstrated by Northern blot analysis. Western blot analysis using specific antibodies revealed a fourfold increase in protein levels of both α_1 and β_1 subunits. This increase in protein level does not depend on the level of hyperosmolarity at which the cells were adapted (600 or 900 mOsmol/kg H_2O). The ouabain-sensitive ATPase activity is one order of magnitude higher in adapted cells as compared to control cultures. We have also shown that the upregulation of Na/K-ATPase subunits is critical for the resistance of the cells to hyperosmotic shock. Obviously, this dramatic increase in Na/K-ATPase activity could be only sustained by a parallel increase in ATP production. Using Northern blot analysis, we were able to demonstrate a large increase in mRNA levels of a key enzyme in the glycolytic pathway, namely, glyceraldehyde-3-phosphate dehydrogenase (GAPDH).

The γ subunit of the Na/K-ATPase is a member of the FXYD family of small regulatory proteins with a single transmembrane domain. In adults, it has been detected only in the kidney of a variety of mammals, but it has never been described in cell cultures of kidney or any other origin.[2] While the function of the γ subunit is quite complex, there is convincing evidence that its presence raises the affinity of Na/K-ATPase for ATP and decreases the affinity for sodium.[2,3]

Since the adapted cells are presumed to be under energetic stress, with possibly a limited availability of ATP, we evaluated these cells for expression of the γ subunit of Na/K-ATPase. A Northern blot analysis using a labeled oligonucleotide, mapping to the middle of the γ mRNA, revealed a strong positive signal with a length of 700 bases in cells adapted to 600 mOsmol/kg H_2O.[4] A much stronger signal was determined for cells adapted to 900 mOsmol/kg H_2O, while no signal was ever detected in control cells. Western blot analysis using an antibody raised against the common C-terminus of the γ protein reveals the same pattern. FIGURE 1 demonstrates the lack of γ subunit in control cells, but the concentration-dependent expression in cells adapted to 600 and 900 mOsmol/kg H_2O. This finding is in contrast with the up-

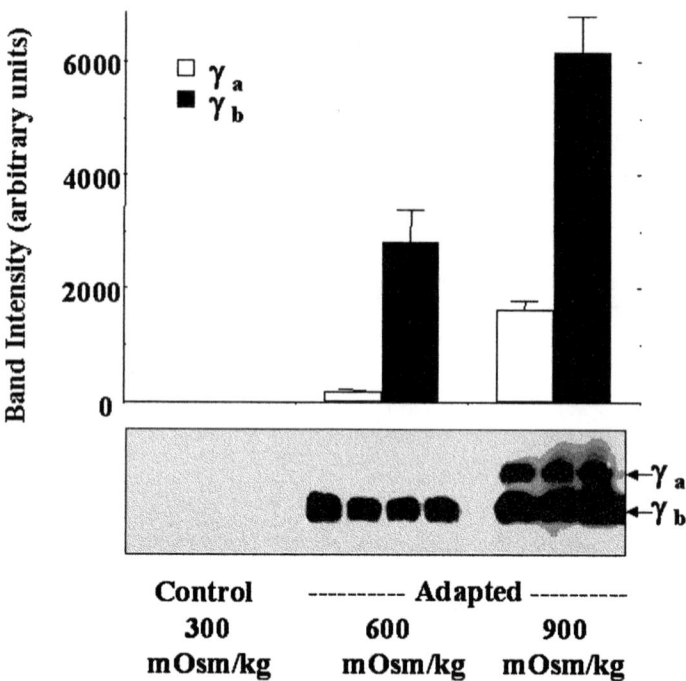

FIGURE 1. Cell lysates (100 μg of protein per lane) from IMCD3 control and adapted cells were analyzed by Western blot using γ subunit ATPase antibody. The identity of γ_b was corroborated with a specific antibody. The mean and SEM are from four different experiments at 300 and 600 mOsmol/kg H_2O and from three at 900 mOsmol/kg H_2O.

regulation of the α_1 and β_1 subunits of Na/K-ATPase that were determined to be independent of the osmolarity of the medium above 600 mOsmol/kg H_2O. It is obvious from the data in FIGURE 1 that two bands react with the antibody. These bands were subsequently identified as γ_a and γ_b splice variants using N-terminus-specific antibodies. Another interesting finding is the unequal expression of γ_a and γ_b. We have determined that γ_b is from 5 to 10 times more abundant than γ_a in adapted cells, in contrast to kidney tissues where they are present at similar levels. We have also demonstrated that the upregulation of the γ subunit is totally reversible. When adapted cells are returned to isotonic conditions, the γ subunit disappears with a half-time of about 17 h. While the upregulation of the γ subunit is dependent on tonicity, it is also osmolyte-dependent as neither mannitol nor urea is able to elicit this response at equivalent concentrations as with NaCl.

We have also been able to detect the phenomenon of adaptation to changes in osmolarity *in vivo*. Mice were overloaded with water by feeding them glucose solutions. The urine osmolarity dropped from 3000 to 500 mOsmol/kg H_2O and, concomitantly, the levels of γ_a dropped by 25% ($p < 0.02$) and γ_b by 41% ($p < 0.002$) in the inner medulla of the kidney, while no such variations were detected in the renal cortex.

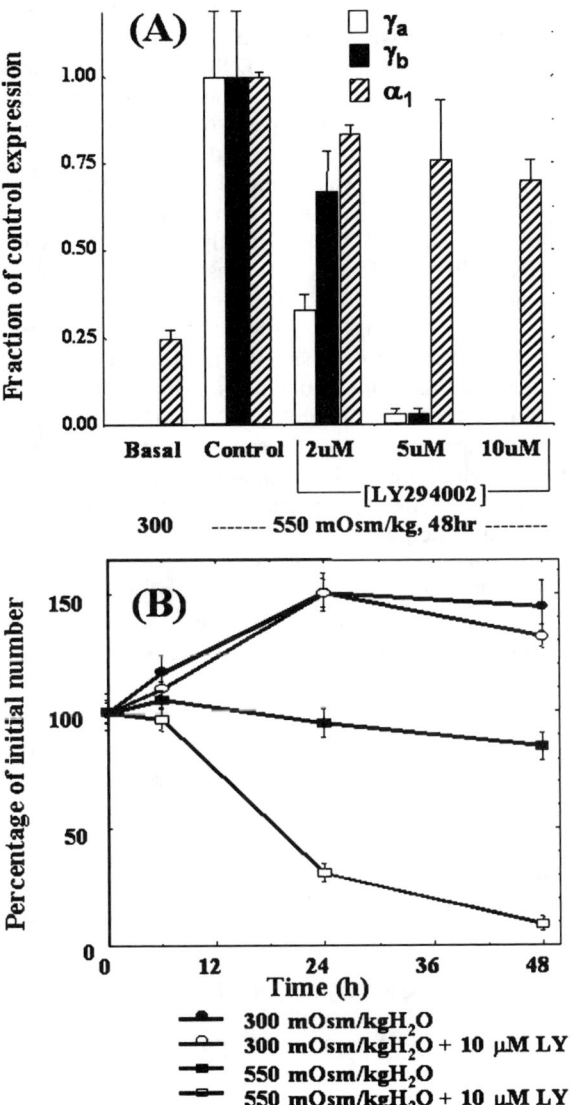

FIGURE 2. (A) Confluent IMCD3 cell cultures were kept for 24 h in low serum medium, incubated for 2 h with different concentrations of the inhibitor LY294002, and challenged with 250 mOsmol/kg H_2O of NaCl for 48 h. Cell lysates were prepared and analyzed by Western blot analysis ($n = 4$). (B) IMCD3 cells were grown in 24-well plates to confluence and treated as above with or without 10 μM LY294002 and 250 mOsmol/kg H_2O of NaCl. Cell survival was measured by the formation of formazan using the CellTiter 96 assay (Promega, Madison, WI). Results are the mean and SEM of six determinations.

Several stress signaling pathways could mediate the upregulation of the γ subunit of Na/K-ATPase, including the three MAP kinase pathways: ERK, p-38, and Jun kinase. We have investigated the participation of each pathway by employing specific inhibitors or dominant negative transfected cells. Neither the p-38 inhibitor, SB203580 (10 µM), nor the ERK inhibitor, PD98059 (20 µM), had any effect on the upregulation of α_1, γ_a, or γ_b subunits after an osmotic challenge with NaCl. Inhibition of the JNK1 pathway by dominant negative transfection (JNK1-APF) demonstrated the usual upregulation of the α and γ subunits upon osmotic stress. In contrast, when the JNK2 pathway was inhibited by the same method (JNK2-APF), whereas the upregulation of the α subunit was not affected, the upregulation of the γ subunit was totally abolished. More importantly, after 24 h of osmotic shock, the survival of the JNK1-APF-transfected cells was >80%, while the survival of the JNK2-APF-transfected cells was <20%.

The importance of the PI3 kinase signaling pathway on the upregulation of the γ protein was studied by employing the specific inhibitor, LY294002. The inhibitor reduced both the upregulation of the γ subunit as well as cell survival under osmotic stress. As shown in FIGURE 2A, the inhibitor LY294002 reduces the upregulation of both splice variants of the γ subunit in a dose-dependent manner, but it does not significantly affect the upregulation of the α subunit. FIGURE 2B shows that LY294002 at 10 µM does not have any effect on the survival of IMCD3 cells when they remain in isotonic medium. As anticipated, when cells are treated with LY294002 and are subjected to osmotic stress, approximately 30% of the cells survive after 24 h and less than 10% after 48 h.

In conclusion, we have demonstrated that the following events are necessary, but obviously not sufficient per se for the survival and/or adaptation of IMCD3 cells to osmotic stress:

(1) constitutive expression of the inducible HSP-70;
(2) upregulation of both the α_1 and β_1 subunits of the Na/K-ATPase;
(3) expression of the γ subunit of the Na/K-ATPase, especially the γ_b splice variant;
(4) desensitization of the JNK and p-38 MAP kinase pathways to increases in osmotic pressure;
(5) increase in ATP synthesis.

ACKNOWLEDGMENTS

This work was supported by Grant No. DK-19928 from the National Institutes of Health.

REFERENCES

1. CAPASSO, J.M., C.J. RIVARD & T. BERL. 2001. Long-term adaptation of renal cells to hypertonicity: role of MAP kinases and Na-K-ATPase. Am. J. Physiol. Renal Physiol. **280:** F768–F776.
2. THERIEN, A.G., R. GOLDSHLEGER, S.J.D. KARLISH & R. BLOSTEIN. 1997. Tissue-specific distribution and modulatory role of the gamma subunit of the Na,K-ATPase. J. Biol. Chem. **272:** 32628–32634.

3. ARYSTARKHOVA, E., R.K. WETZEL, N.K. ASINOVSKI & K.J. SWEADNER. 1999. The gamma subunit modulates Na(+) and K(+) affinity of the renal Na,K-ATPase. J. Biol. Chem. **274:** 33183–33185.
4. CAPASSO, J.M., C. RIVARD & T. BERL. 2001. The expression of the γ subunit of Na-K-ATPase is regulated by osmolality via C-terminal Jun kinase and phosphatidylinositol 3-kinase–dependent mechanisms. Proc. Natl. Acad. Sci. USA **98:** 13414–13419.

Gamma Structural Variants Differentially Regulate Na,K-ATPase Properties

ELENA ARYSTARKHOVA AND RANDALL K. WETZEL

Neuroscience Center, Massachusetts General Hospital, Charlestown, Massachusetts, and Harvard Medical School, Boston, Massachusetts, USA

ABSTRACT: Renal Na,K-ATPase is tightly bound to a small regulatory protein, the gamma subunit (FXYD2). In rat, it occurs in two splice variants with different N-termini. Immunolocalization on kidney sections revealed distinct distribution of the γ splice variants along the rat nephron. Where coexpressed, they coimmunoprecipitated with each other along with the alpha subunit, suggesting assembly in oligomeric complexes. Functional consequences of association with gamma were monitored in stably transfected NRK-52E cells. The outcome was that splice variants can differentially modulate the major intrinsic properties of the Na,K-ATPase under normal and stress-related conditions. The data imply an adaptive physiological mechanism of regulation of renal Na,K-ATPase through modulation of pump properties, gene expression, or both.

KEYWORDS: FXYD2; gamma subunit; Na^+ affinity; K^+ affinity; splice variants; oligomers

STRUCTURAL FORMS OF γ

The γ subunit has two variants generated by alternative splicing.[1] The calculated masses of the two splice variants in the rat, γa and γb, are very similar, so the proteins should comigrate on SDS gels. Nevertheless, they are always seen as a doublet where the larger species (γb) actually migrates faster. We demonstrated that γb also migrated faster than γa after *in vitro* biosynthesis.[2] When pancreatic microsomes were added to the reticulocyte lysates, the electrophoretic mobilities of both γa and γb were decreased detectably, albeit to different extents. Mutation of Ser and Thr residues blocked the posttranslational modification of γa. The conclusion was that the γ subunit may exist in four different structural forms: γa, γa′, γb, and γb′.

SEGMENT-SPECIFIC DISTRIBUTION OF THE γ SPLICE VARIANTS

By immunofluorescence, γ colocalized with α in many tubules, including medullary thick ascending limb (MTAL), distal convoluted tubules (DCT), connecting tubules (CNT), and proximal convoluted tubules (PCT), but γ was not found in

Address for correspondence: Elena Arystarkhova, Neuroscience Center, Massachusetts General Hospital, Charlestown, MA 02129. Voice: 617-726-8579; fax: 617-726-5677.
aristark@helix.mgh.harvard.edu

cortical thick ascending limb (CTAL) nor in cortical collecting duct (CCD).[3] Localization of the γ splice variants revealed that their distribution is strongly biased to different segments.[4] While both splice variants were found in MTAL in the inner stripe, γa staining was greatly reduced compared to γb in the outer stripe. The latter penetrated minimally into the cortex, but the rest of the CTAL was negative. Neither γ nor any other member of the FXYD family has yet been convincingly detected in the distal CTAL. The proportion of γb in cortex was seen to be distinctly lower than in medulla by blots, and γb was detected only in DCT and CNT by immunofluorescence. Conversely, we observed light, but uniform, γa stain in the PCT, correlating with the low abundance of the α1 subunit. No γb was detected in proximal tubules, either in kidney sections or in isolated proximal tubule cells.

OLIGOMERIC ASSEMBLY OF THE Na,K-ATPase

Immunoprecipitation revealed interaction between the splice variants of γ in rat kidney outer medulla membranes.[4] Specific antibodies against γa and γb coprecipitated not only their own antigens along with the α subunit, but each other's as well. The ratio among all three subunits (α1, γa, and γb) in the immunoprecipitates was invariant, even below the critical micelle concentration. Association of both γa and γb with dimers or tetramers of α is the most likely interpretation.

Specific association between the subunits was controlled by ligands that affect Na,K-ATPase conformation.[4] The best coprecipitation of γ splice variants and α was observed in buffers with NaCl, which favors the E1 conformation. Buffers with KCl, which favors E2, showed greatly reduced precipitation at the same ionic strength. Coprecipitation of γb with γa antibody was also reduced in the E2-P conformation induced by either $NaCl/ATP/Mg^{2+}$ or P_i/Mg^{2+}/ouabain. The data further support γ's functional integration into the Na,K-ATPase and suggest that stability of the γ interaction may depend on conformational rearrangements of the α (and/or β) subunits during enzyme turnover.

EXPRESSION OF γ IN NRK-52E CELLS AND IMPACT ON Na,K-ATPase ACTIVITY

Stable transfectants of γa and γb were generated in NRK-52E cells (normal rat kidney epithelial cell line) that do not express γ under normal conditions.[2,5] Two groups of clones were identified in both cases: γa was seen either as a doublet (γa* = γa and γa′) or as a single band with the electrophoretic mobility of the slower migrating species of the doublet (γa′). γb clones had either faster (γb) or slower migrating species (γb′). The γ splice variants evidently undergo posttranslational modification when expressed in some (but not all) NRK-52E clones.

The apparent affinities for Na^+, K^+, and ATP were all higher in wild-type NRK-52E cells than in rat kidney membranes. In transfectants, the apparent affinity for Na^+ was significantly decreased with γa*, γb, or γb′ clones (FIG. 1). Apparently, posttranslational modification of γb did not influence this property of the pump. Conversely, no change in apparent affinity for Na^+ was observed in the clones expressing fully modified γa′. The apparent affinity for K^+ was reduced in γa*, γa′,

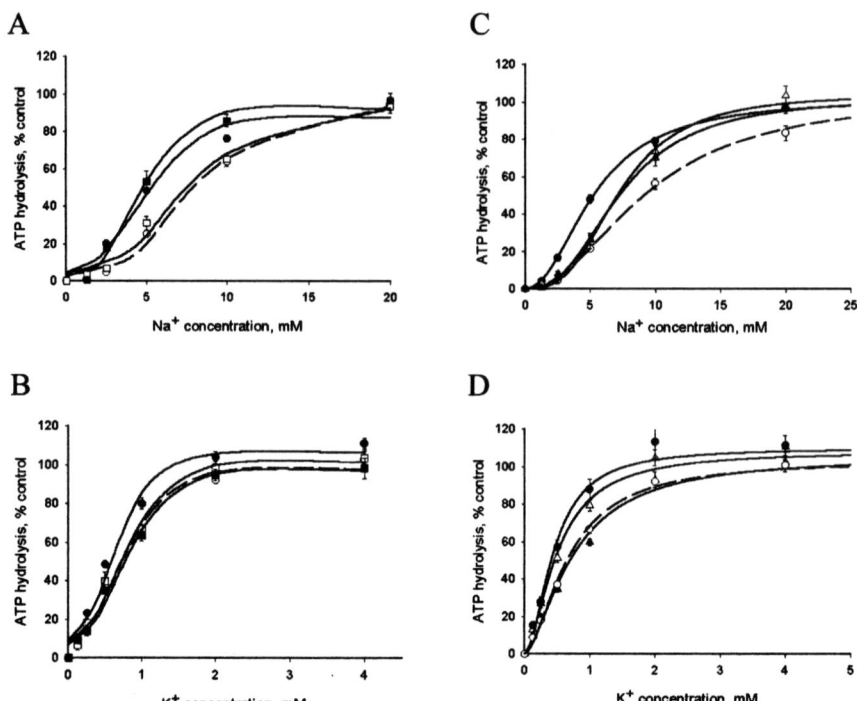

FIGURE 1. Differential effects of structural forms of γ on the Na,K-ATPase affinity for Na$^+$ and K$^+$ in transfected NRK-52E cells. Na,K-ATPase was partially purified from renal medulla (open circles), from mock-transfected NRK-52E cells (closed circles), or from stable clones expressing γa* (open squares), γa' (closed squares), γb (open triangles), or γb' (closed triangles). Apparent affinities for Na$^+$ (**A, C**) and K$^+$ (**B, D**) were measured *in vitro*, and the data were fitted to a cooperative model for ligand binding as described elsewhere.[2]

and γb' clones, but not in the γb clones (FIG. 1). Thus, the four types of clones encompassed every possible combination of Na$^+$ and/or K$^+$ affinity reduction.[2]

Transfection with γ did not make ATP affinity more like that of the kidney enzyme, however. Instead, it further increased the affinity for at least two of the molecular forms, γa' and γb, although the effect was very modest.[2]

Recently, we assessed Na,K-ATPase properties of γ knockout mice generated in the laboratory of G. Kidder (Jones, Li, Barr, Arystarkhova, Wetzel, Sweadner, Fong, and Kidder; manuscript in preparation). As predicted by the transfectants, Na,K-ATPase from kidney of wild-type animals had lower affinity for Na$^+$ than that from null mice. We also observed subtle differences in the K_m for ATP.

PHYSIOLOGICAL RELEVANCE OF γ EXPRESSION

In experiments with γ-transfected cells, we noticed a correlation between reduction in Na$^+$ affinity and the rate of cell proliferation. γa* transfectants had slower growth rate than mock-transfected cells or cells with fully modified γa'.[2] Analysis of the γb or γb′ transfectants revealed that their cell growth was similarly reduced compared to the mock-transfected cells. Since all three groups of slow-growing clones expressing γa*, γb, and γb′ possessed lower affinity for Na^+ than control or γa′ clones (FIG. 1), the data suggest that culture conditions create a selective pressure against high γ expression because of the reduction of Na,K-ATPase affinity for Na^+.

We have shown recently that γ expression can be induced in NRK-52E cells by exposure to hypertonic media (with NaCl or sucrose) or to ouabain.[6] In both cases, there was selective induction of γa. This upregulation of γ apparently entails post-transcriptional mechanisms affecting the rate of degradation.[6] Functionally, we found that γ-mediated effects on Na,K-ATPase may depend on the physiological status of the cell (Arystarkhova, Wetzel, and Sweadner; manuscript in preparation). Our initial data suggest that γ may play a dual role in regulation of the Na,K-ATPase in kidney by affecting either apparent affinity for Na^+ or the enzyme's V_{max} under normal or hyperosmotic conditions. Hypertonicity of the nephron varies greatly with depth and physiological status of the animal. Thus, the data may represent an important physiological link in understanding of mechanisms of normal and pathological alterations of sodium homeostasis in kidney.

REFERENCES

1. SWEADNER, K.J., R.K. WETZEL & E. ARYSTARKHOVA. 2000. Genomic organization of the human FXYD2 gene encoding the γ subunit of the Na,K-ATPase. Biochem. Biophys. Res. Commun. **279:** 196–201.
2. ARYSTARKHOVA, E., C. DONNET, N.K. ASINOVSKI & K.J. SWEADNER. 2002. Differential regulation of renal Na,K-ATPase by splice variants of the γ subunit. J. Biol. Chem. **277:** 10162–10172.
3. WETZEL, R.K. & K.J. SWEADNER. 2001. Immunocytochemical localization of the Na,K-ATPase α and γ subunits in the rat kidney. Am. J. Physiol. **281:** F531–F545.
4. ARYSTARKHOVA, E., R.K. WETZEL & K.J. SWEADNER. 2002. Distribution and oligomeric association of splice forms of the Na,K-ATPase regulatory γ subunit in rat kidney. Am. J. Physiol. **282:** F393–F407.
5. ARYSTARKHOVA, E., R.K. WETZEL, N.K. ASINOVSKI & K.J. SWEADNER. 1999. The γ subunit modulates Na^+ and K^+ affinity of the renal Na,K-ATPase. J. Biol. Chem. **274:** 33183–33185.
6. WETZEL, R.K., N.K. ASINOVSKI, E. ARYSTARKHOVA & K.J. SWEADNER. 2001. The regulatory gamma subunit of Na,K-ATPase can be upregulated in culture. J. Am. Soc. Nephrol. **12:** 43A.

Structure/Function Studies of the Gamma Subunit of the Na,K-ATPase

RHODA BLOSTEIN, HELEN X. PU, ROSEMARIE SCANZANO, AND ATHINA ZOUZOULAS

Departments of Medicine and Biochemistry, McGill University, Montreal, Canada

ABSTRACT: The Na,K-ATPase γ subunit is present primarily in kidney as two splice variants, γa and γb, which differ only at their extracellular N-termini. Two distinct effects of gamma are seen in biochemical Na,K-ATPase assays of mammalian (HeLa) cells transfected with γa or γb, namely, (i) a decrease in K'_{ATP} probably secondary to a shift in steady-state $E_1 \leftrightarrow E_2$ poise in favor of E_1 and (ii) an increase in cytoplasmic K^+/Na^+ antagonism seen as an increase in K'_{Na} at high K^+ concentration. Mutagenesis experiments involving alterations in extramembranous regions of γ indicate that different regions mediate the aforementioned distinct effects and that the effects appear to be long range. Studies of ouabain-sensitive fluxes with intact cells confirm the γ effects seen with membranes and also suggest an additional effect (increase) in apparent affinity for extracellular K^+. Alteration in gamma function was also evidenced in the behavior of a G41→R mutation within the transmembrane domain of gamma. G41R is associated with autosomal dominant renal magnesium wasting. Our studies show that this mutation in the γb variant retards trafficking of γ, but not αβ pumps, to the cell surface and abolishes functional effects of γ, consistent with the conclusion that the Mg^{2+} transport defect is secondary to loss of γ modulation of Na,K-ATPase function.

KEYWORDS: gamma subunit; structure/function

INTRODUCTION

The gamma subunit is a member of the FXYD family of small, single transmembrane proteins characterized by a FXYD motif in the N-terminus. There are at least seven family members and each is expressed in a tissue-specific fashion. (For a complete EST database search and analysis, see Ref. 14.) Although the existence of a small proteolipid in purified preparations of Na,K-ATPase was recognized over 30 years ago, it was Forbush and coworkers[1] who first demonstrated its specific association with the Na,K-ATPase purified from pig kidney. Almost 20 years later, Therien *et al.* showed that this "γ subunit" (FXYD2) is mainly a kidney-specific component of the Na,K-ATPase.[2]

Address for correspondence: Rhoda Blostein, Montreal General Hospital Research Institute, 1650 Cedar Avenue, Room L.11-132, Montreal, Quebec, Canada H3G 1A4. Voice: 514-934-1934, ext. 44501; fax: 514-934-8332.
Rhoda.Blostein@mcgill.ca

The γ subunit behaves as a kidney-specific modulator and affects apparent affinities for ATP[2,3] and cations.[4,5] In addition to γ, several other FXYD family members have been shown to modulate Na,K-ATPase function. These include phospholemman (PLM) or FXYD1 first described in dogs;[6] a PLM-like protein in sharks (PLMS[7]); FXYD4 or CHIF (channel inducing factor) first described in rats;[8] and FXYD7, a regulator of α1-β pumps in brain.[9] Recent reports describe functional properties of PLM,[10] CHIF,[11] and FXYD7.[9] There is evidence also for ion channel behavior or regulation of ion channels as seen, for example, with γ.[12] Details of the properties and cellular distribution of FXYD members other than gamma are described elsewhere in this volume.

Gamma appears as a doublet on SDS-PAGE. Mass spectrometry analysis of the rat enzyme indicates that the two species, γa (upper band, 7.184 kDa) and γb (lower band, 7.338 kDa), are splice variants such that the N-terminal residues, TELSANH, in γa are replaced by Ac-MDRWYL in γb.[13] These structures are identical to those predicted by an EST database search.[14] Additional bands seen on Western blots (γa' and γb') are presumably products of cell-specific posttranslational modifications.[13] Thus, in HEK, γa' runs faster than γa and, in HeLa, γb' runs slower than γb.

The functional basis for the presence of γa and γb remains largely unknown. Studies to date suggest that both have similar effects on Na,K-ATPase kinetics.[5] However, there are differences in their localization in the kidney: regions of overlapping as well as regions with distinct localizations have been clearly identified[5,15] and are discussed elsewhere in these proceedings.

DISTINCT EFFECTS OF GAMMA

The functional effects of γ described in earlier studies include (i) an increase in apparent affinity for ATP reflecting a shift in the steady-state E_1/E_2 distribution towards the E_1 conformation as shown by Therien et al.[2] and (ii) an increase in K^+/Na^+ competition at cytoplasmic Na^+ activation sites. Considering these effects in terms of the Albers-Post reaction mechanism [in particular, $E_2(K) \leftrightarrow E_1 + K^+$; $E_1 + ATP + Na^+ \leftrightarrow Na \cdot E_1P + ADP$], it is intriguing that the two effects of γ are paradoxally opposing. A higher affinity for K^+ at cytoplasmic Na^+ activation sites should shift the E_1/E_2 poise away from E_1, whereas a higher affinity for ATP should shift the poise away from E_2 towards E_1. The implications of these dichotomies are considered below.

More recently, we have used a mutagenesis approach to evaluate the structural basis for the distinct effects of γ on Na,K-ATPase kinetics. (For details, see Ref. 16.) Mutations of γa or γb were carried out in extramembranous N- and C-termini, as well as the transmembrane region (see Fig. 1). For the transmembrane region, the Glu41 → Arg substitution has been the focus of our experiments in view of its association with renal magnesium wasting.[17] Accordingly, analysis of the mutant γbG41R provides insight into the consequences of disruption of transmembrane structure and also addresses the question of altered γ-α/β interaction and targeting to the plasma membrane.

As described elsewhere,[5] mutant and control (empty vector) cDNA were transfected into rat α1-HeLa cells (stable clones obtained from J. B. Lingrel and maintained in micromolar ouabain to suppress activity of endogenous human Na,K-

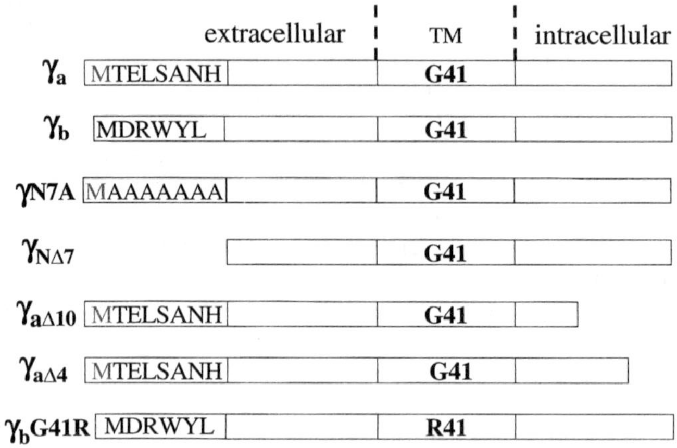

```
                    extracellular  |  TM  |  intracellular
γa    [MTELSANH  ]                [ G41 ]
γb    [MDRWYL   ]                 [ G41 ]
γN7A  [MAAAAAAA ]                 [ G41 ]
γNΔ7                              [ G41 ]
γaΔ10 [MTELSANH ]                 [ G41 ]
γaΔ4  [MTELSANH ]                 [ G41 ]
γbG41R[MDRWYL   ]                 [ R41 ]
```

FIGURE 1. Abbreviated scheme showing gamma mutants. Only variant- or mutant-distinct amino acid residues are shown.

TABLE 1. Summary of effects of γ mutants on Na,K-ATPase kinetics: comparison to mock-transfected cells

γ variant or mutant	K'_{ATP}	K'_{Na} (100 mM K$^+$)
γa	decreased[a]	increased[a]
γb	decreased[a]	increased[a]
γ-NΔ7	no change	increased
γ-N7A	decreased	increased
γa-CΔ4	no change	increased
γa-CΔ10	no change	increased
γbG41R	no change	no change

[a]Changes caused by wild-type γa and γb are a ~2-fold decrease in K'_{ATP} and a ~1.6-fold increase in K'_{Na}, respectively. For details, see Ref. 16.

ATPase, henceforth referred to as "α1-HeLa" cells). Membranes isolated from cells transfected with γa, γb, or mutant cDNA and from control (empty vector)–transfected control cells were used concurrently for Na,K-ATPase assays to determine apparent affinities for Na$^+$ measured at high (100 mM) K$^+$ concentration and for ATP as described previously.[5] The results summarized in TABLE 1 show that γbG41R replacement in the transmembrane region (but none of the extramembranous deletions) abolishes both effects of γ, namely, the decrease in K'_{ATP} and the increase in cytoplasmic K$^+$/Na$^+$ antagonism seen as an increase in K'_{Na} at high (100 mM) K$^+$

concentration. On the other hand, deletion of 10 and as few as 4 residues from the C-terminus abrogates the decrease in K'_{ATP} seen with both variants. An unexpected finding is that deletion of the variant-specific N-terminus also abolishes the decrease in K'_{ATP}.

The observation that none of the extramembranous mutants abolished both effects of γ indicates that all of these mutants associate with Na,K-ATPase $\alpha\beta$ dimers, consistent with the findings of Béguin et al.[18] showing that the FXYD motif is important for stable association. The finding that deletion of the N-terminus, like removal of the C-terminus (or addition of anti-C-terminal antibodies), abrogates the effect of γ on the $E_1 \leftrightarrow E_2$ conformational equilibrium points to long-range effects of γ-$\alpha\beta$ interactions on K'_{ATP}. Since the N-terminal deletion, but not N7A replacement, abrogates the K'_{ATP} effect, the γ effect to stabilize E_1 does not involve TELSANH-α or MDRWYL-α interactions, but rather the remainder of the chain.

FLUX STUDIES WITH INTACT CELLS

We have recently addressed the question of whether and to what extent the modulatory effects of γ deduced from the aforementioned experiments with isolated membrane preparations are relevant to γ regulation of the pump as it exists *in situ* in the intact mammalian cell. For example, it is plausible that the membrane isolation procedure disrupts or alters γ-α/β association; the lack of sidedness may also obscure kinetic behavior relevant to the intact cell. Accordingly, ouabain-sensitive $^{86}Rb^+(K^+)$ influx was measured in cell monolayers, with monensin added to maintain a constant Na^+ concentration as described previously,[19] with minor modifications.[11] Either Na^+ was varied and K^+ kept constant at a concentration sufficient to saturate extracellular K^+ sites or the cells were equilibrated with 20 mM Na^+ and extracellular K^+ was varied.

Apparent Cation Affinities

The γa-, γb-, or γbG41R-transfected α1-HeLa cells were analyzed concurrently with control mock-transfected cells. The results summarized in TABLE 2 show that both γ variants increase $K_{0.5(Na)}$. Whether the greater effect of γb compared to γa is due to greater expression or association with $\alpha\beta$ pumps remains to be determined. In contrast, a significant change in $K_{0.5(Na)}$ could not be detected with γbG41R. This behavior seen with intact (high intracellular K^+) cells is consistent with that seen in Na,K-ATPase assays using permeabilized membranes (TABLE 1).

In previous studies, we failed to detect a significant effect of γ on K^+ activation of Na,K-ATPase, except at suboptimal ATP concentration, under which condition an increase in K'_K was observed, presumably secondary to the γ-mediated decrease in K'_{ATP}.[20] In contrast, with intact cells (TABLE 2), a significant decrease in $K_{0.5(K)}$ is seen with either γa- or γb-transfected cells compared to mock- or γbG41R-transfected cells. This observation provides evidence of a heretofore undetected effect of γ. Current studies are under way to determine whether this γ effect denotes a higher affinity for K^+ binding to the transport site and/or a lower affinity for Na^+ binding to an allosteric inhibitory site (cf. Ref. 21).

TABLE 2. Effects of γ on apparent affinities for Na$^+$ and K$^+$: experiments with intact cells

Cells	$K_{0.5(Na)}$	$K_{0.5(K)}$
Control (mock)	7.89 ± 0.62	0.57 ± 0.02
γa	11.61 ± 0.76	0.38 ± 0.02
γb	17.32 ± 1.85	0.38 ± 0.03
γbG41R	7.43 ± 0.25	0.53 ± 0.01

NOTE: Values are averages ± SEM from at least four separate experiments. $K_{0.5(Na)}$ was determined from fitting the data to a simple Michaelis-Menten model, as in Ref. 11. $K_{0.5(K)}$ was determined using a cooperative model, with a mean value for the Hill coefficient ($n = 1.5$) obtained from all experiments.

Pump Flux versus ATP Concentration

The γ-mediated increase in affinity for ATP effected by γ transfection into cultured cells devoid of γ is applicable to the enzyme in renal tissue since a mirror image effect is seen following treatment of the renal enzyme with anti-γ. To understand whether this effect is relevant to the intact cell, we determined whether the relation between pump flux and varying cellular ATP concentration is modified by γ. This was tested by measuring ATP concentrations and ouabain-sensitive ^{86}Rb$^+$(K$^+$) influx in cells in which ATP concentration was varied by preincubation with antimycin A and varying amounts of glucose. Prior to the assay, two sets of triplicate samples were taken: one for ATP measurement and the other for ^{86}Rb$^+$(K$^+$) influx. In all experiments, pump rate increased continuously as a function of ATP concentration, up to at least 3 mM ATP; further, in 10 separate experiments (not shown) with control cells paired with either γa or γb, the increase was greater in the γ-transfected cells. Thus, the average fold increase in flux at 1 mM ATP (data taken from the fitted curves relating pump rate to ATP concentration) was 1.45 ± 0.11 (SEM). In 3 other paired experiments carried out with γbG41R, a difference from the paired controls was not detected.

G41R MUTATION ALTERS γ-αβ ASSOCIATION AND γ TRAFFICKING

In 2000, Meij *et al.* reported a point mutation in the γ subunit gene of Na,K-ATPase.[17] Their immunolocalization studies indicated misrouting of γ and retardation of α trafficking to the plasma membrane. Here, we describe the expression and behavior of γbG41R expressed in α1-HeLa cells. The charge alteration introduced by the G41R mutation reduces mobility of γb as seen in Western blots following perfluorooctanoate (PFO)–PAGE (see figure 1c in Ref. 16).

Coimmunoprecipitation experiments indicate defective association of this mutant with the α subunit. Thus, FIGURE 2 shows an immunoblot of α and γ before and after immunoprecipitation with anti-γ (C-terminal) antibodies in the presence of ouabain and Rb$^+$ as described by Garty *et al.*[11] The blot shows that α associates with γb, but not γbG41R. As summarized in TABLE 3, several observations support the conclusion that the γbG41R mutation causes defective processing and trafficking of γ (but not α) to the cell surface. One is the lack of posttranslational modification of γ (the upper

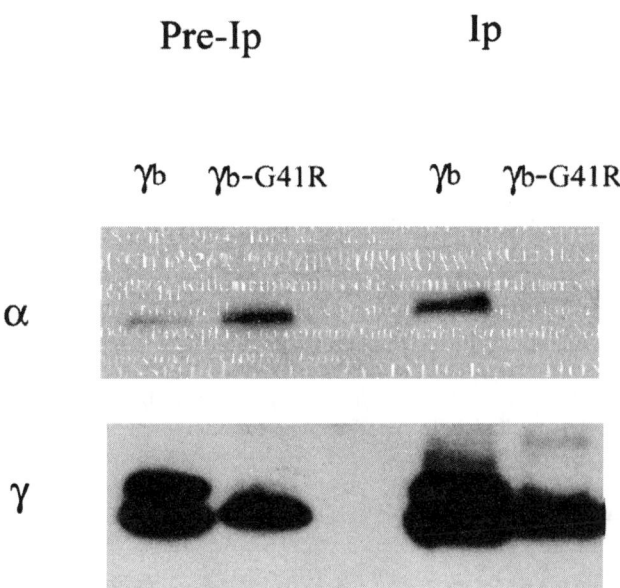

FIGURE 2. Coimmunoprecipitation of γb and γbG41R with alpha. Membranes were solubilized, as in Ref. 11, and then treated with anti-γ (C-terminal) antibodies.

TABLE 3. Assessment of γ and α trafficking to the plasma membrane

Parameter measured	γb		γbG41R	
Presence of posttranslationally modified γb′ in isolated clones	+		−	
Presence of γ in Golgi-rich subcellular fraction	(trace)		+	
Cell surface expression of α and γ determined by streptavidin isolation of biotinylated surface-exposed proteins	α +	γ +	α +	γ −

NOTE: For details, see Ref. 16.

γb′ band seen on SDS-PAGE) in each of eight separate clones of γbG41R cells compared to γb′ present in all (nine) clones of γb-transfected cells. A second is the high proportion of γbG41R compared to γb in Golgi-rich membranes, and relatively less in the plasma-rich fraction. The third is the absence of γbG41R, but not α, at the cell surface of γbG41R cells following isolation of surface biotinylated membrane proteins, which contrasts strikingly with the presence of both subunits at the surface of γb-transfected cells.

CONCLUSIONS

The experiments summarized in this monograph show clearly that effects of γ on K^+/Na^+ antagonism and on apparent ATP affinity are relevant to different regions of the gamma chain. The results also suggest "interplay" between these two opposing effects of γ whereby the γ-mediated increase in ATP affinity may counteract and hence minimize the "true" K^+/Na^+ antagonism and vice versa.

A physiological basis for the dual effects of γ is that it provides a fine-tuned, self-regulatory mechanism for balancing energy utilization and maintaining appropriate cation gradients across renal epithelial cells. Both variants are particularly abundant in the medullary thick ascending limb and, as argued elsewhere,[3] it is in this putatively anoxic region that an increased affinity for ATP would be important for maintaining pump activity; the moderate increase in K^+/Na^+ antagonism would counterbalance the ATP depletion and still maintain a suitably low intracellular Na^+ concentration. Recent studies by Garty *et al.* have shown that in certain regions with little, if any γ, in which the apparent affinity for Na^+ is higher (in particular, cortical and medullary collecting ducts), the renal pump is associated with CHIF.[11] CHIF has the opposite effect of γ on K'_{Na};[18] it increases the apparent Na^+ affinity at least two-fold, which these authors suggest may be critical for aldosterone-responsive tissues that have an important role in maintaining Na^+ and K^+ homeostasis.

An important role of γ in renal cation homeostasis secondary to its association with and modulation of Na,K-ATPase is demonstrated by our results showing the functional consequences of mutating Gly41 to Arg. This study provides evidence that the G41R substitution abrogates γ interaction with the $\alpha\beta$ pump, resulting in the failure of γ to traffic to the cell surface and to modulate pump kinetics. The former finding confirms the routing defect reported by Meij and coworkers.[17] In addition, our experiments demonstrate that $\alpha\beta$ pump trafficking, per se, is not notably affected, at least with the G41R mutation in the γb variant. Accordingly, the Mg^{2+} wasting seen in the dominant renal hypomagnesemia[17] appears to be secondary to abrogation of γ-mediated modulation of Na,K-ATPase kinetics. Although the exact mechanism for the effect of this mutation on Mg^{2+} wasting remains unknown, our findings suggest that a reduced apparent ATP affinity of $\alpha\beta$ pumps devoid of γ lowers pump activity and thus changes the electrochemical gradients of Na^+ and K^+, with secondary changes (reduction) in Mg^{2+} reabsorption. The foregoing interpretation is predicated, however, on the extent to which the effect of γ on K'_{ATP} is counteracted by its opposing effect on K^+/Na^+ antagonism in regions in which Mg^{2+} reabsorption is critical.

ACKNOWLEDGMENTS

This research was supported by grants from CIHR (No. MT-3876) and the Kidney Foundation of Canada.

REFERENCES

1. FORBUSH, B.D., J.H. KAPLAN & J.F. HOFFMAN. 1978. Characterization of a new photo-affinity derivative of ouabain: labeling of the large polypeptide and of a proteolipid component of the Na,K-ATPase. Biochemistry **17**: 3667–3676.

2. THERIEN, A.G., et al. 1997. Tissue-specific distribution and modulatory role of the gamma subunit of the Na,K-ATPase. J. Biol. Chem. **272**: 32628–32634.
3. THERIEN, A.G., et al. 2001. Molecular and functional studies of the gamma subunit of the sodium pump. J. Bioenerg. Biomembr. **33**: 407–414.
4. BÉGUIN, P., et al. 1997. The gamma subunit is a specific component of the Na,K-ATPase and modulates its transport function. EMBO J. **16**: 4250–4260.
5. PU, H.X., et al. 2001. Functional role and immunocytochemical localization of the gamma a and gamma b forms of the Na,K-ATPase gamma subunit. J. Biol. Chem. **276**: 20370–20378.
6. PALMER, C.J., B.T. SCOTT & L.R. JONES. 1991. Purification and complete sequence determination of the major plasma membrane substrate for cAMP-dependent protein kinase and protein kinase C in myocardium. J. Biol. Chem. **266**: 11126–11130.
7. CORNELIUS, F., Y.A. MAHMMOUD & H.R. CHRISTENSEN. 2001. Modulation of Na,K-ATPase by associated small transmembrane regulatory proteins and by lipids. J. Bioenerg. Biomembr. **33**: 415–423.
8. ATTALI, B., et al. 1995. A corticosteroid-induced gene expressing an "IsK-like" K^+ channel activity in *Xenopus* oocytes. Proc. Natl. Acad. Sci. USA **92**: 6092–6096.
9. BÉGUIN, P., et al. 2002. FXYD7 is a brain-specific regulator of Na,K-ATPase $\alpha 1$-$\beta 1$ isozymes. EMBO J. In press.
10. GEERING, K., P. Béguin, H. Garty, et al. 2003. FXYD proteins: new tissue- and isoform-specific regulators of Na,K-ATPase. This volume.
11. GARTY, H., et al. 2002. A specific functional interaction between CHIF and Na,K-ATPase: implication for regulation by FXYD proteins. Am. J. Physiol. In press.
12. SHA, Q., et al. 2001. Heterologous expression of the Na(+),K(+)-ATPase gamma subunit in *Xenopus* oocytes induces an endogenous, voltage-gated large diameter pore. J. Physiol. **535**: 407–417.
13. KUSTER, B., et al. 2000. A new variant of the gamma subunit of renal Na,K-ATPase: identification by mass spectrometry, antibody binding, and expression in cultured cells. J. Biol. Chem. **275**: 18441–18446.
14. SWEADNER, K.J. & E. RAEL. 2000. The FXYD gene family of small ion transport regulators or channels: cDNA sequence, protein signature sequence, and expression. Genomics **68**: 41–56.
15. ARYSTARKHOVA, E., R.K. WETZEL & K.J. SWEADNER. 2002. Distribution and oligomeric association of splice forms of Na(+)-K(+)-ATPase regulatory gamma-subunit in rat kidney. Am. J. Physiol. Renal Physiol. **282**: F393–F407.
16. PU, H.X., R. SCANZANO & R. BLOSTEIN. 2002. Distinct regulatory effects of the Na,K-ATPase subunit. J. Biol. Chem. **277**: 20270–20276.
17. MEIJ, I.C., et al. 2000. Dominant isolated renal magnesium loss is caused by misrouting of the Na(+),K(+)-ATPase gamma-subunit. Nat. Genet. **26**: 265–266.
18. BÉGUIN, P., et al. 2001. CHIF, a member of the FXYD protein family, is a regulator of Na,K-ATPase distinct from the gamma-subunit. EMBO J. **20**: 3993–4002.
19. MUNZER, J.S., et al. 1994. Tissue- and isoform-specific kinetic behavior of the Na,K-ATPase. J. Biol. Chem. **269**: 16668–16676.
20. THERIEN, A.G., et al. 2000. Structure/function studies of the gamma subunit of renal Na,K-ATPase. *In* Na/K-ATPase and Related ATPases, pp. 481–488. Elsevier. Amsterdam/New York.
21. BALSHAW, D.M., et al. 2000. Combined allosteric and competitive interaction between extracellular Na^+ and K^+ during ion transport by the $\alpha 1$, $\alpha 2$, and $\alpha 3$ isoforms of Na,K-ATPase. Biophys. J. **79**: 853–862.

Cell-Specific Expression of Three Members of the FXYD Family along the Renal Tubule

NICOLETTE FARMAN, MICHEL FAY, AND FRANÇOISE CLUZEAUD

INSERM U 478, Faculté de Médecine Xavier Bichat, 75870 Paris Cedex 18, France

ABSTRACT: The gamma subunit of Na/K/ATPase is a small membrane protein that shares homologies with other members of the FXYD family, like phospholemman and CHIF (corticosteroid hormone–induced factor). Both the gamma subunit and CHIF modulate sodium pump properties. The gamma subunit increases the apparent affinity of the pump for ATP and reduces its apparent affinity for sodium. CHIF, in contrast, augments its apparent affinity for sodium. Gamma subunit expression is essentially restricted to the kidney, with two main splice variants, γa and γb, which differ only at their extracellular N-termini. We have investigated in detail the cell-specific expression of the two splice variants of gamma within the kidney and compared it to that of CHIF. While both gamma variants affect catalytic properties of the pump (without detectable difference between a and b forms), their localization along the nephron is partially distinct. Both variants are coexpressed in the proximal tubule and in the medullary part of the thick ascending limb of Henle's loop (TAL). In contrast, their expression differs in the downstream tubular segments. Within the renal cortex, the sole gamma a variant was found in macula densa cells and in principal cells of the initial parts of the collecting duct. Gamma b is in the cortical part of the TAL. Outer and inner medullary collecting ducts lack detectable gamma expression. These latter nephron segments express CHIF, and no overlap between gamma and CHIF expression along the nephron was observed. Such distinct cell-specific expression argues for complementary roles to modulate Na/K/ATPase activity.

KEYWORDS: gamma subunit; sodium pump; Na/K/ATPase; splice variants; CHIF; renal tubule; immunolocalization

INTRODUCTION

In polarized epithelial cells, basolateral expression of the Na-K-ATPase is critical to ensure sodium absorption and potassium secretion. The Na/K/ATPase (sodium pump) is formed of two main subunits, α and β, and numerous data have accumulated on its function and regulation. More recently, a small protein, named γ, has been identified; it has been shown that it immunoprecipitates with α and β subunits and that it exerts a regulatory role on the sodium pump.[1–3] Curiously, its expression is not ubiquitous (in contrast with that of α and β subunits), but essentially restricted

---

Address for correspondence: Nicolette Farman, INSERM U 478, Faculté de Médecine Xavier Bichat, 16 rue Henri Huchard, BP 416, 75870 Paris Cedex 18, France. Voice: +331-44-85-63-23; fax: +331-42-29-16-44.
    farman@bichat.inserm.fr

Ann. N.Y. Acad. Sci. 986: 428–436 (2003). © 2003 New York Academy of Sciences.

to the kidney. Two splice variants of the γ subunit[4] have been identified (named γa and γb), sharing complete homology except for their extracellular last N-terminal amino acids (TELSANH and MDRWYL, respectively). The γ subunit also shares sequence homology with other proteins that play a role in the regulation of ion transporters; they form the FXYD family and, among them, CHIF (for corticosteroid hormone–induced factor) is also predominantly expressed in the kidney.[5–7]

Investigations to determine the roles of these putative regulatory proteins (γa, γb, and CHIF) include immunoprecipitation studies and expression in *Xenopus* oocytes or in transfected mammalian cells, together with the α and β subunits of the pump.[2,3,8,9] It has been shown that the γ subunit of Na-K-ATPase can modulate the activity of the sodium pump in two ways: (1) it increases the apparent affinity of ATP; (2) it increases the K^+ antagonism of cytoplasmic Na^+ activation, resulting in a reduction in the apparent affinity for sodium. Noticeably, these two effects are observed in the presence of γa as well as γb, with no difference in the modulatory effect of each of these two variants.[9] More recently, CHIF expression was also shown to modulate pump activity.[10] Interestingly, CHIF expression results in an increased affinity of the sodium pump for cytoplasmic sodium, an effect opposite to that yielded by the gamma subunit of Na/K/ATPase.

In order to get more insights into the functional relevance of these regulatory proteins, we have examined in detail their cell-specific expression along the renal tubule[7,9] and eventual regulation, in rats under different physiological status. These experiments consisted of immunolocalization of γ (or γa and γb) and CHIF, together with antibodies against proteins that identify cell types: anti-Tamm Horsfall protein (also named uromucoid) for the thick ascending limb of Henle's loop, and anti-aquaporin-2 (AQP2) for principal cells of the collecting duct and connecting tubule. In addition, cellular expression of the α subunit of Na-K-ATPase was also examined with anti-α monoclonal antibody (6H) and compared to that of gamma antibodies. By *in situ* hybridization experiments, the expression of mRNA encoding for the γ subunit was evaluated in the kidney of wild-type and CHIF-knockout mice; CHIF mRNA was also examined in rat kidney sections.

CELL-SPECIFIC EXPRESSION OF THE GAMMA SUBUNIT OF Na-K-ATPase IN THE KIDNEY

To answer this question, three different antibodies were used: one directed against the C-terminus of the γ subunit (γC-ter), common to both splice variants; one specific of the γa form; and one specific of the γb form.[9] Immunolocalization with γC-ter indicated evidence of several renal cell types expressing the γ subunit in their basolateral membrane. Cells with highest expression are those of the medullary part of the thick ascending limb of Henle's loop (mTAL); specific immunolabeling is also observed in the cortical part of the TAL (cTAL), in the proximal tubule, and in the collecting duct, but only in its initial cortical portion. Such expression pattern is in general accordance with data from Mercer *et al.*[1] An immunolocalization study was also performed recently by Wetzel and Sweadner,[11] which yielded results very close to ours, except for the lack of detection of the γ subunit in the cortical part of the TAL and in the cortical collecting duct in their report. Interestingly, they also documented γ expression in the distal tubule.

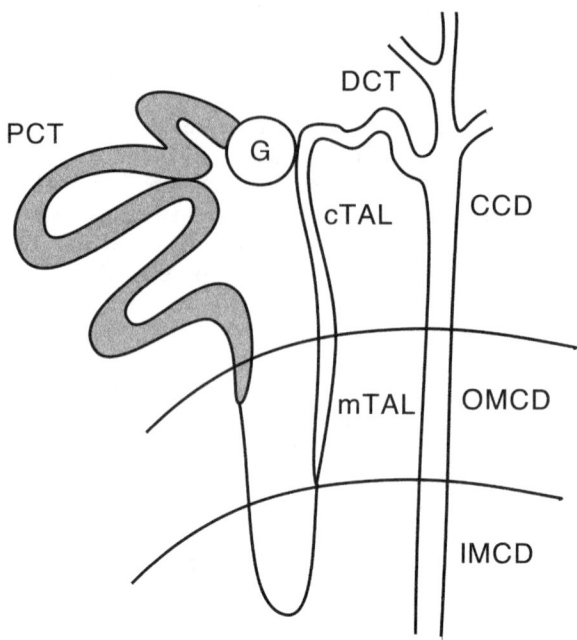

FIGURE 1. Schematic view of the nephron—G: glomerulus; PCT: proximal convoluted tubule; mTAL and cTAL: respectively, medullary and cortical parts of the thick ascending limb of Henle's loop; DCT: distal collecting tubule; CCD: cortical collecting tubule; OMCD: outer medullary collecting tubule; IMCD: inner medullary collecting tubule. The osmolarity of extracellular fluids is elevated in both medulla (containing mTAL and OMCD) and papilla (containing IMCD) as part of the mechanism of concentration of final urine. Such hyperosmolarity can reach very high values (2000–4000 mOsm/L) in some circumstances (water deprivation).

The identification of two splice variants, γa and γb, prompted us to examine whether and how they differ in their cell-specific expression along the nephron.[9] Results will be discussed in relation with the actual knowledge of the function of the epithelial cells where the gamma variants are expressed. FIGURE 1 is a schematic view of the nephron, illustrating the successive parts of the renal tubule, which will be referred to below.

Both splice variants have been identified in the basolateral membrane of the proximal tubule; this epithelium reabsorbs a large bulk of the sodium filtered by the glomerulus (about 50%), thus requiring high capacities of Na reabsorption.

In the mTAL, both γa and γb are coexpressed in the basolateral membrane; indeed, antibodies against each form colocalize with the expression of uromucoid. Although we ignore at the present time the relevance of γa versus γb expression, it should be reminded that immunoprecipitation experiments result either in αβγa complexes or in αβγb complexes. In other words, the αβ complexes do not associate with both variants of the gamma subunit (no αβγaγb complexes[9]). The mTAL has an important function, which is to reabsorb sodium chloride in excess of water, thus generating a sodium concentration gradient in the medulla. Sodium enters the water-impermeable TAL cells through the apical Na/K/2Cl cotransporter (NKCC2) and is extruded at the basolateral membrane by Na/K/ATPase. The number of pumps per mTAL cells is extremely high (several millions by cell), allowing very efficient transcellular sodium transport.[12] As a matter of fact, it has been estimated that transcellular sodium flux per min is 10-fold the intracellular content of the mTAL,[13] thus requiring very efficient pumping capacity of the cells. It should also be noted that the mTAL exerts NaCl reabsorption in very specific conditions: the extracellular environment is hyperosmotic and hypoxic. Adequate adaptation of sodium pump activity to such conditions may require the gamma subunit to yield efficient transepithelial sodium reabsorption.

At variance with the mTAL, the cTAL expresses only the γb form. Indeed, antibodies against γb only colocalize with uromucoid in the cTAL. Although the cells of the cTAL resemble those of the upstream mTAL, sodium fluxes through the cTAL are lower than through the mTAL; indeed, as tubular fluid flows along the TAL, active NaCl reabsorption reduces the concentration of sodium in the lumen of the tubule, thus reducing the driving force of Na reabsorption in the downstream tubule.

At the end of the TAL, the tubular fluid reaches the macula densa (MD), which consists of few specialized tubular cells considered as sodium chloride sensors.[14] We have documented the specific expression of the sole γa in the basolateral membrane of MD cells (while the preceding TAL cells are negative for γa). The MD of each tubule is in close contact with the glomerulus of the same nephron and with its afferent arteriole, containing renin-secreting cells. This unique spatial organization, named the juxtaglomerular apparatus, forms a functional unit with important regulatory roles of extracellular volumes and salt homeostasis. This occurs through the glomerulo-tubular feedback phenomenon and through control of renin release. The precise mechanisms that allow transduction by MD cells of the chemical information (NaCl concentration in the luminal fluid) into an integrated control of kidney function are still largely unknown. Expression of γa in the basolateral membrane of MD cells is striking, perhaps related to their Na sensor function.

We have not examined systematically gamma variant expression in the distal tubule, but a recent report of the group of Sweadner[15] provided evidence for expression of the γb form in these cells.

The gamma a variant (not b) was also observed in the initial portions of the cortical collecting duct (including the connecting tubule and initial part of the collecting tubule). It appears to be essentially in the basolateral membrane of principal cells (those that express AQP2 in their apical membrane). Sodium reabsorption by the collecting duct is quantitatively of limited magnitude as compared to the preceding nephron segments, but of major importance for the regulation of sodium reabsorption under the control of several hormones, in particular, aldosterone and vasopressin. Sodium enters the cells through the apical epithelial Na channel, ENaC; both hormones control ENaC activity in coordination with Na/K/ATPase. Aldosterone is required for the constitution of a reserve pool of sodium pumps, which can be recruited at the basolateral membrane when needed.[16] Both recruitment of pumps and activation of the enzyme occur when intracellular sodium or cell volume increases, for example, upon exposure to vasopressin.[17] It has been proposed that the gamma subunit of the pump may play a role in the recruitment of pumps, although no definitive evidence for such a role has been provided yet. We have also examined whether γa or γb immunofluorescence could vary upon changes in physiological status. However, no change in gamma immunostaining (intensity or cellular localization) was observed in collecting duct cells of kidneys from rats fed a low or a high sodium diet (i.e., with high or low plasma aldosterone concentration). These results argue against major regulation of gamma subunit expression in face of changes in sodium transport.

In all other collecting duct cells, γ expression was undetectable or only at very low levels.

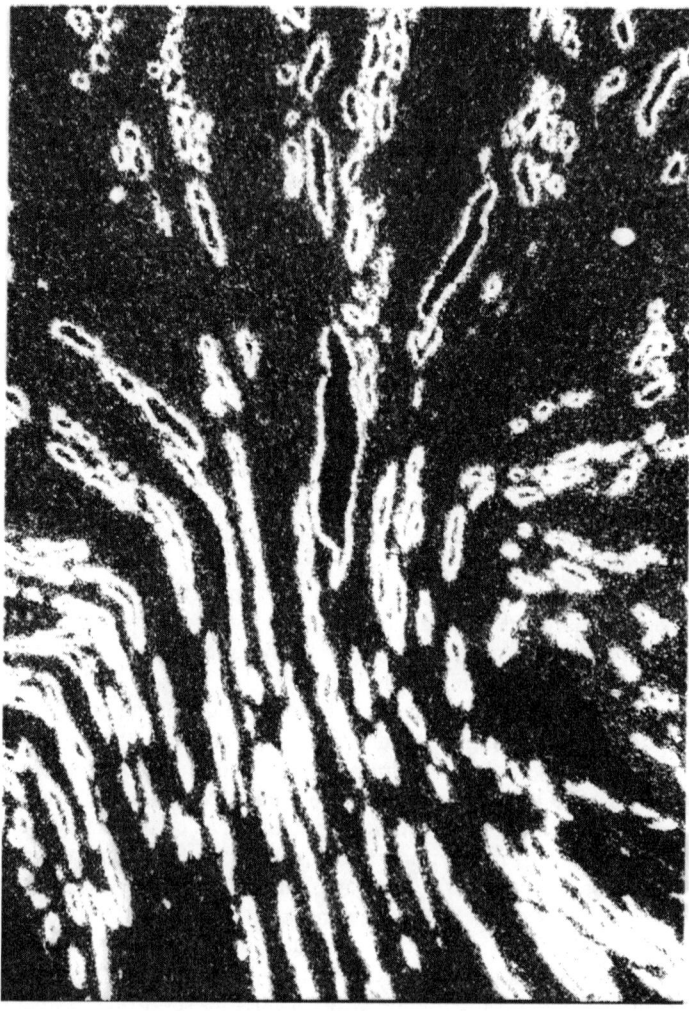

FIGURE 2. Expression of CHIF along the collecting duct. *In situ* hybridization shows CHIF mRNA expression all along the collecting duct, with an increasing intensity from cortex (upper part of photograph) to papilla (lower part of photograph).

EXPRESSION OF CHIF IN THE KIDNEY

Interestingly, collecting duct cells express CHIF, another member of the FXYD family.[5] As a matter of fact, CHIF is only in the colonic epithelium and in the kidney.[6] Within the kidney, its mRNA is localized all along the collecting duct, as illustrated in FIGURE 2. The cortical part of the collecting duct exhibits modest labeling, with progressive increase along the collecting duct, to reach strong levels over medullary and papillary parts of the collecting duct. This pattern of expression has also been documented at the protein level,[7] by immunofluorescence, using antibodies raised against a C-terminal peptide; in this study, CHIF was shown to be located in the basolateral membrane of principal cells of the collecting duct.

ARE CHIF AND GAMMA SUBUNIT EXPRESSED IN THE SAME KIDNEY CELLS?

Since both the gamma subunit of the sodium pump and CHIF were found in the cortical collecting duct (although at different locations), it appeared important to know whether they could be in the same cells. Double immunofluorescence experiments, with both anti-gamma subunit and anti-CHIF antibodies, revealed that this was not the case. It appeared clearly that the gamma subunit (the gamma a variant) is only in the initial parts of the collecting duct, with no overlap with CHIF, which became detectable only in the more distal portion of the cortical collecting tubule. Thus, these two modulators of the sodium pump do not colocalize. Such a finding is in accordance with their distinct opposite functional effects on the sodium affinity of Na/K/ATPase.

Another observation reinforces this notion. Mice with genetic invalidation of CHIF have been produced recently.[18] They appear normal and healthy, but high lethality occurs in CHIF-KO mice when challenged with high potassium diet and furosemide treatment. This observation suggests an impaired capacity to reabsorb sodium and fluid in their distal nephron tubular segments. We wanted to examine whether CHIF disruption could modify the renal expression of the gamma subunit of Na/K/ATPase, in view of putative compensation for the lack of CHIF. Kidneys from wild-type or CHIF-KO mice were hybridized with an RNA probe encoding for the gamma subunit (FIG. 3). In the absence of CHIF, we found no induction in gamma message in the cortex and medulla, and papillary collecting ducts remained negative for gamma expression. These observations suggest no compensatory role of the gamma subunit when the collecting duct lacks CHIF.

COMPARISON OF CHIF AND GAMMA SUBUNIT EXPRESSION SITES IN THE KIDNEY

FIGURE 4 summarizes our results on the expression pattern of three FXYD proteins along the renal tubule.[6,7,9] The gamma subunit is in the proximal tubule, the thick ascending limb of Henle's loop, the macula densa, and the initial parts of the collecting duct, with partially overlapping expression of its a and b splice variants. A complementary pattern of expression of CHIF was found as CHIF is all along the

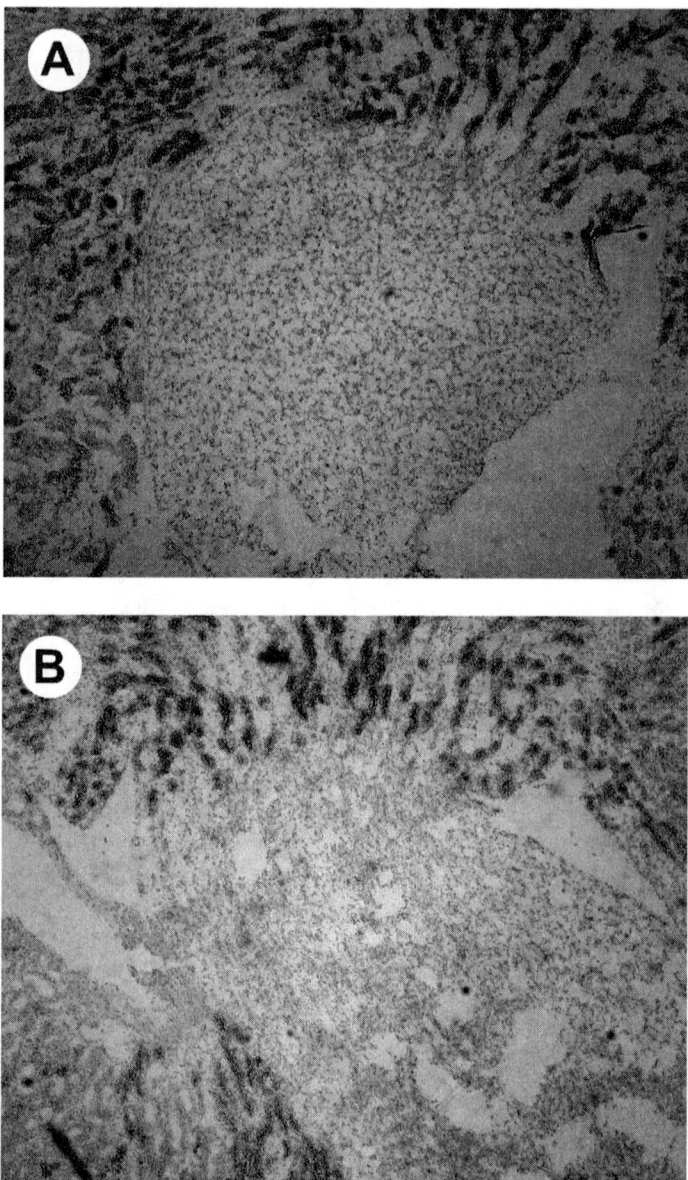

FIGURE 3. Gamma subunit expression in kidneys from wild-type (**A**) and CHIF-knockout (**B**) mice. *In situ* hybridization illustrates the pattern of gamma subunit mRNA expression at the medulla/papilla junction. mTAL exhibits high signal, while papillary tubule expresses background signal. No change in gamma subunit expression was apparent in CHIF-KO mice as compared to normal mice.

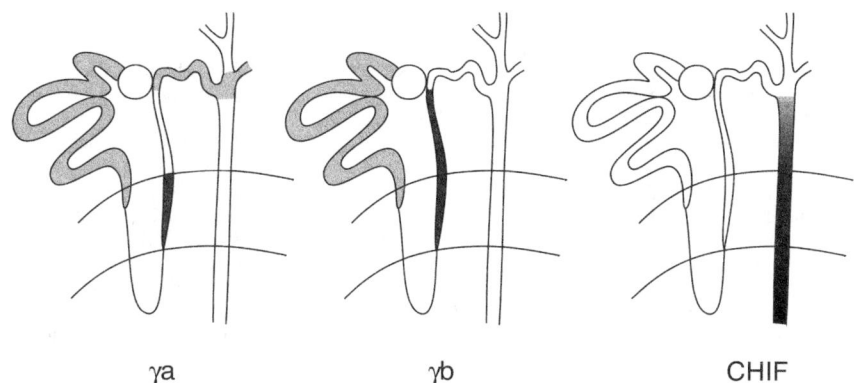

FIGURE 4. Summary of differential renal expression of the a and b variants of the gamma subunit and CHIF along the nephron.

collecting duct, with a gradient of expression starting in the cortex to increase all along the collecting duct, up to its papillary portion.

CONCLUSIONS

The question arises as to why three distinct regulators of the sodium pump are expressed in the kidney and how they modulate sodium reabsorption. Kidney cells are characterized by their regulatory role in sodium absorption, a function critical to body fluid and electrolyte homeostasis. These cells express the sodium pump in large amounts (which vary, however, between the different epithelia forming the nephron). Such excess of pumps is important to meet with the requirements of highly variable transcellular sodium fluxes and important variations in cell sodium entry. It is plausible to propose that cells with high transcellular sodium fluxes (i.e., proximal tubule, thick ascending limb of Henle's loop, macula densa cells, and initial parts of the distal nephron) may benefit from expression of the gamma subunit, which confers tolerance to relatively high intracellular sodium concentrations, by reducing Na/K/ATPase apparent affinity for cytoplasmic sodium. On the other hand, cells with lower transepithelial sodium reabsorptive fluxes (such as those of the collecting duct) need to adapt pump activity very precisely to environmental pressure (salt depletion, hormonal regulations, …) to ensure fine adjustments of sodium excretion. CHIF expression in collecting duct cells, by increasing sodium pump affinity for intracellular sodium, appears as an important factor to allow pump adaptations to such physiological requirements. We propose that CHIF may play a significant (and yet unexplored) role in the recruitment of the reserve pool of pumps in collecting duct cells. Future studies should help to elucidate these notions and will allow the evaluation of the respective roles of these FXYD proteins in renal Na/K/ATPase adaptative functions.

REFERENCES

1. MERCER, R.W., D. BIEMESDERFER, D.P. BLISS, Jr., *et al.* 1993. Molecular cloning and immunological characterization of the γ polypeptide, a small protein associated with the Na,K-ATPase. J. Cell Biol. **121:** 579–586.
2. THERIEN, A.G., R. GOLDSHLEGER, S.J.D. KARLISH & R. BLOSTEIN. 1997. Tissue-specific distribution and modulatory role of the γ subunit of the Na,K-ATPase. J. Biol. Chem. **272:** 32628–32634.
3. THERIEN, A.G., S.J.D. KARLISH & R. BLOSTEIN. 1999. Expression and functional role of the γ subunit of the Na,K-ATPase in mammalian cells. J. Biol. Chem. **274:** 12252–12256.
4. KUSTER, B., A. SHAINSKAYA, H.X. PU, *et al.* 2000. A new variant of the gamma subunit of renal Na,K-ATPase: identification by mass spectrometry, antibody binding, and expression in cultured cells. J. Biol. Chem. **275:** 18441–18446.
5. ATTALI, B., H. LATTER, N. RACHAMIM & H. GARTY. 1995. A corticosteroid-induced gene expressing an "IsK-like" K^+ channel activity in *Xenopus* oocytes. Proc. Natl. Acad. Sci. USA **92:** 6092–6096.
6. CAPURRO, C., N. COUTRY, J.P. BONVALET, *et al.* 1996. Cellular localization and regulations of CHIF in kidney and colon. Am. J. Physiol. **271:** C753–C762.
7. SHI, H., R. LEVY-HOLZMAN, *et al.* 2001. Membrane topology and immunolocalization of CHIF in kidney and intestine. Am. J. Physiol. **280:** F505–F512.
8. BÉGUIN, P., X. WANG, D. FIRSOV, *et al.* 1997. The γ subunit is a specific component of the Na,K-ATPase and modulates its transport function. EMBO J. **16:** 4250–4260.
9. PU, H.X., F. CLUZEAUD, R. GOLDSHLEGER, *et al.* 2001. Functional role and immunocytochemical localization of the gamma a and gamma b forms of the Na,K-ATPase gamma subunit. J. Biol. Chem. **276:** 20370–20378.
10. GARTY, H., M. LINDZEN, R. SCANZANO, *et al.* 2002. A specific functional interaction between CHIF and Na,K-ATPase: implication for regulation by FXYD proteins. Am. J. Physiol. **283:** F607–F615.
11. WETZEL, R.K. & K.J. SWEADNER. 2001. Immunocytochemical localization of Na-K-ATPase alpha- and gamma-subunits in rat kidney. Am. J. Physiol. **281:** F531–F545.
12. JORGENSEN, P.L. 1986. Structure, function, and regulation of Na,K-ATPase in the kidney. Kidney Int. **29:** 10–20.
13. CHEVAL, L. & A. DOUCET. 1990. Measurement of Na-K-ATPase-mediated rubidium influx in single segments of rat nephron. Am. J. Physiol. **259:** F111–F121.
14. SCHNERMANN, J. & J.P. BRIGGS. 1999. The macula densa is worth its salt. J. Clin. Invest. **104:** 1007–1009.
15. ARYSTARKHOVA, E., R.K. WETZEL & K.J. SWEADNER. 2002. Distribution and oligomeric association of splice forms of Na(+)-K(+)-ATPase regulatory gamma-subunit in rat kidney. Am. J. Physiol. **282:** F393–F407.
16. BLOT-CHABAUD, M., F. WANSTOK, J.P. BONVALET & N. FARMAN. 1990. Cell sodium-induced recruitment of Na(+)-K(+)-ATPase pumps in rabbit cortical collecting tubules is aldosterone-dependent. J. Biol. Chem. **265:** 11676–11681.
17. BLOT-CHABAUD, M., S. DJELIDI, N. COURTOIS-COUTRY, *et al.* 2001. Coordinate control of Na,K-ATPase mRNA expression by aldosterone, vasopressin, and cell sodium delivery in the cortical collecting duct. Cell. Mol. Biol. **47:** 247–253.
18. AIZMAN, R., C. ASHER, M. FUZESI, *et al.* 2002. Generation and phenotypic analysis of CHIF knockout mice. Am. J. Physiol. **283:** F569–F577.

Dominant Isolated Renal Magnesium Loss Is Caused by Misrouting of the Na$^+$,K$^+$-ATPase γ-Subunit

IWAN C. MEIJ,[a] JAN B. KOENDERINK,[b] JOKE C. DE JONG,[c] JAN JOEP H. H. M. DE PONT,[b] LEO A. H. MONNENS,[d] LAMBERT P. W. J. VAN DEN HEUVEL,[d] AND NINE V. A. M. KNOERS[a]

Departments of [a]Human Genetics, [b]Biochemistry, [c]Cell Physiology, and [d]Pediatrics, Institute of Cellular Signaling, University Medical Center Nijmegen, 6500 HB Nijmegen, the Netherlands

ABSTRACT: Hereditary primary hypomagnesemia comprises a clinically and genetically heterogeneous group of disorders in which hypomagnesemia is due to either renal or intestinal Mg^{2+} wasting. These disorders share the general symptoms of hypomagnesemia, tetany and epileptiformic convulsions, and often include secondary or associated disturbances in calcium excretion. In a large Dutch family with autosomal dominant renal hypomagnesemia, associated with hypocalciuria, we mapped the disease locus to a 5.6-cM region on chromosome 11q23. After candidate screening, we identified a heterozygous mutation in the *FXYD2* gene, encoding the Na$^+$,K$^+$-ATPase γ-subunit, cosegregating with the patients of this family, which was not found in 132 control chromosomes. The mutation leads to a G41R substitution, introducing a charged amino acid residue in the predicted transmembrane region of the γ-subunit protein. Expression studies in insect Sf9 and COS-1 cells showed that the mutant γ-subunit protein was incorrectly routed and accumulated in perinuclear structures. In addition to disturbed routing of the G41R mutant, Western blot analysis of *Xenopus* oocytes expressing wild-type or mutant γ-subunit showed mutant γ-subunit lacking a posttranslational modification. Finally, we investigated two individuals lacking one copy of the *FXYD2* gene and found their serum Mg^{2+} levels to be within the normal range. We conclude that the arrest of mutant γ-subunit in distinct intracellular structures is associated with aberrant posttranslational processing and that the G41R mutation causes dominant renal hypomagnesemia associated with hypocalciuria through a dominant negative mechanism.

KEYWORDS: γ-subunit; *FXYD2*; hypomagnesemia

Address for correspondence: Jan B. Koenderink, Department of Biochemistry (160), UMC Nijmegen, P. O. Box 9101, 6500 HB Nijmegen, the Netherlands. Voice: +31-24-3613517; fax: +31-24-3616413.

J.Koenderink@ncmls.kun.nl

INTESTINAL AND RENAL Mg^{2+} HOMEOSTASIS

Mg^{2+} is the fourth most abundant cation in the human body and the second most abundant intracellular cation and plays an essential role in a wide variety of biological activities.[1] Its concentration in the blood (normal: 0.75–1.4 mmol/L) is balanced by changes in urinary excretion of Mg^{2+} in response to alterations in uptake. The predominant site of Mg^{2+} uptake is the small bowel. On average, a net amount of approximately 100 mg of Mg^{2+} is absorbed by the small and large bowels each day.[2]

In the kidney, approximately 2500 mg of Mg^{2+} is filtered per day, 96% of which is reabsorbed along the nephron (FIG. 1, middle). Unlike Na$^+$ reabsorption, only a small fraction (5–15%) is reabsorbed in the proximal tubule. The major site of passive (paracellular) Mg^{2+} reabsorption is the thick ascending limb of Henle's loop (TAL), where 50–72% of the filtered Mg^{2+} is reabsorbed via the paracellular pathway (FIG. 1, left). Active (transcellular) reabsorption (FIG. 1, right) is thought to take place in the distal convoluted tubule (DCT) and accounts for 9–10% of the total filtered load.[3] There is ample evidence based on experiments in mouse DCT cells to assume that, in the DCT cells, both an apical Mg^{2+} channel and a basolateral Mg^{2+} extrusion system are involved in transepithelial Mg^{2+} transport (FIG. 1, right).[3,4] Renal Mg^{2+} handling is controlled by multiple hormones. The significance of hormonal influences of Mg^{2+} reabsorption has been extensively reviewed by de Rouffignac and Quamme[5,6] and by Dai et al.[7]

In serum, Mg^{2+} mainly exists in the free ionized form (55%). Approximately 34% is bound to proteins and the remaining 11% forms complexes with different agents such as phosphates and citrates. Of the total intracellular Mg^{2+}, less than 10% is

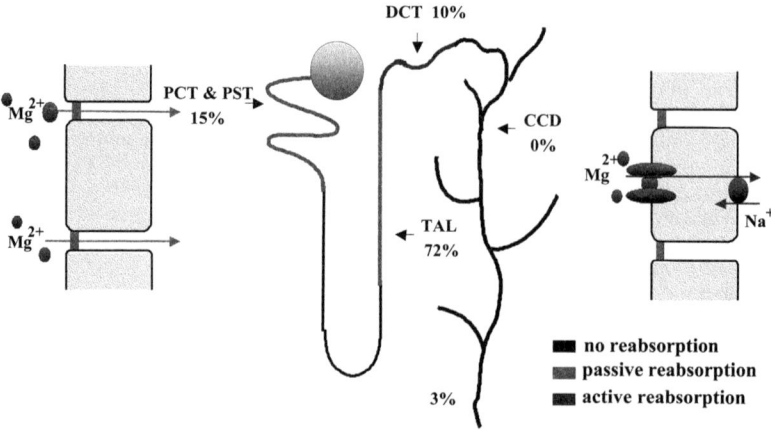

FIGURE 1. (*Middle*) Mg^{2+} reabsorption along the nephron: PCT = proximal convoluted tubule; PST = proximal straight tubule; TAL = thick ascending limb of Henle's loop; DCT = distal convoluted tubule; CNT = connecting tubule; CCD = cortical collecting duct. The percentage of filtered Mg^{2+} that is reabsorbed is indicated along the nephron as well as the final excretion of Mg^{2+}. (*Left*) Passive Mg^{2+} reabsorption via the paracellular pathway. (*Right*) A model for active Mg^{2+} reabsorption in the DCT through a putative apical Mg^{2+} channel and a putative basolateral Mg^{2+} extrusion mechanism.

present in the free ionized form, which is considered to be the metabolically active form.[8]

GENETICS OF Mg^{2+} HANDLING

Although Mg^{2+} plays an important role in many cellular enzymatic processes (phosphatases, ATPases, and RNA polymerases), it appeared very difficult to identify genes that are involved in Mg^{2+} homeostasis. This is due mainly to the lack of a suitable Mg^{2+} isotope, which could be used in functional assays. Rather, several studies of genetic Mg^{2+} wasting disorders have resulted in the identification of genes involved in Mg^{2+} reabsorption. Here, we will focus on the genetic disorder, hypomagnesemia, associated with hypocalciuria, in which we found the Na^+,K^+-ATPase γ-subunit to be involved.

Hypomagnesemia with Hypocalciuria

The existence of this disorder came to our attention more than 15 years ago, when two Dutch patients were submitted to hospital because of generalized convulsions. In these patients, serum Mg^{2+} was found to be as low as 0.39 mmol/L, without other plasma electrolyte abnormalities, including Ca^{2+} (both serum and normal ionized), Na^+, K^+, Cl^-, and bicarbonate. Blood pH was normal, and renin activity and plasma aldosterone levels were in the normal range.[8] The only abnormality found, in addition to hypomagnesemia, was lowered renal excretion of Ca^{2+}. Family members of these probands had low serum Mg^{2+} also, but lacked symptoms of Mg^{2+} depletion. Retention values of orally administered $^{28}Mg^{2+}$ and the effects of Mg^{2+} infusion on renal reabsorption showed that the defect must be located in the kidney. In these two families, the disorder was inherited as an autosomal dominant trait.[9]

Because of the phenotypic similarities with Gitelman syndrome, caused by mutations in *SLC12A3*, we speculate that the tubular defect in the patients with primary hypomagnesemia associated with hypocalciuria must be localized in the same nephron segment as the site of expression of *SLC12A3* (DCT), where active Mg^{2+} reabsorption is thought to take place. This hypothesis is sustained by studies in mice in which *SLC12A3* was knocked out since, in these mice, a Gitelman-like phenotype with hypomagnesemia, but without hypokalemia or metabolic alkalosis, was found.[10]

In order to identify the gene involved in hypomagnesemia/hypocalciuria, we performed linkage analysis on the genetic material of patients with this disorder. In a genome-wide linkage study, we first excluded a possible candidate region on chromosome 9q, encompassing the gene for intestinal hypomagnesemia with secondary hypocalcemia, and subsequently found linkage to markers on chromosome 11q23. Detailed haplotype analyses enabled us to limit the region of linkage to a 5.6-cM interval, in which an estimated 100 genes could be present. Furthermore, we identified a common haplotype segregating in both families, suggesting both their relationship through a common ancestor and the existence of a single, hypomagnesemia-causing mutation within them.[11] By screening of mapped genes and ESTs within the linkage region, we found an EST highly homologous to the rat Na^+,K^+-ATPase γ-subunit.

Na^+,K^+-ATPase γ-Subunit as a Candidate for Dominant Hypomagnesemia/Hypocalciuria

The γ-subunit of the Na^+,K^+-ATPase is a small, type-I transmembrane protein encoded by the *FXYD2* gene. Unlike the α- and β-subunits of Na^+,K^+-ATPase, the γ-subunit is not expressed in all cells. Rather, it is expressed in a tissue-specific pattern, most abundantly in kidney.[12,13] Although the highest expression is observed in kidney, the γ-subunit is not present in all nephron segments. In rat and mouse, γ-subunit expression was low in proximal tubule and macula densa and high in medullary thick ascending limb (mTAL) and DCT. Cortical collecting duct and superficial cortical TAL (cTAL) showed no γ-subunit expression,[14–16] whereas some expression was seen in deep cTAL.[16]

Since the discovery of the Na^+,K^+-ATPase γ-subunit,[17] its physiological function has been the subject of many *in vitro* studies. In spite of clear evidence for an ATPase-related function in experimental settings,[13,14,16,18–24] it remained uncertain whether or not the γ-subunit protein had a physiological role *in vivo*.

Altogether, this warranted further examination of the γ-subunit as a candidate for dominant hypomagnesemia. After sequencing, we found a mutation in the Na^+,K^+-ATPase γ-subunit in patients of both families with dominant renal hypomagnesemia. The G121→A mutation in the human *FXYD2* gene introduces a positively charged amino acid residue (G41R) into the predicted transmembrane region of the γ-subunit protein. In order to prove a functional effect of this mutation, we performed expression studies in several cell types. Heterologous expression of wild-type or mutant γ-subunit in insect Sf9 and COS-1 cells clearly showed that the mutant protein was retarded in distinct perinuclear structures, whereas the wild-type protein was routed to the plasma membrane.[25]

Similar to our previous observations in Sf9 and COS-1 cells, the mutant γ_a-subunit was not localized to the plasma membrane in *Xenopus laevis* oocytes (FIG. 2). In contrast, expression of the wild-type protein resulted in a bright plasma membrane staining in addition to intracellular staining. Western blot analysis of oocytes expressing wild-type or mutant γ_a-subunit revealed that the mutant γ_a-subunit lacked a posttranslational modification as compared to the wild type (FIG. 2). This was also seen by Pu *et al.*,[24] who introduced the G41R mutation into the γ_b-subunit splice variant. With centrifugal methods, these investigators concluded that the relative expression of the mutant protein was much higher in the Golgi apparatus compared to the wild-type protein, which is in agreement with our own immunological studies using an antibody against the Golgi-specific protein, RAB6B.[25] From these results, we conclude that the arrest of the G41R γ-subunit mutant takes place in the Golgi apparatus and is associated with aberrant posttranslational processing.

Haploinsufficiency versus Dominant-Negative

To obtain insight in the dominant mechanism of the disorder, we investigated two individuals with an 11q23.3-ter deletion.[26] FISH analysis with a probe against *FXYD2* revealed that these individuals lacked one copy of the *FXYD2* gene. If the lack of one copy of *FXYD2* would cause hypomagnesemia, then haploinsufficiency would most likely be the cause of hypomagnesemia in our patients as well. However, both individuals with 11q23.3-ter deletions had normal serum Mg^{2+} levels,

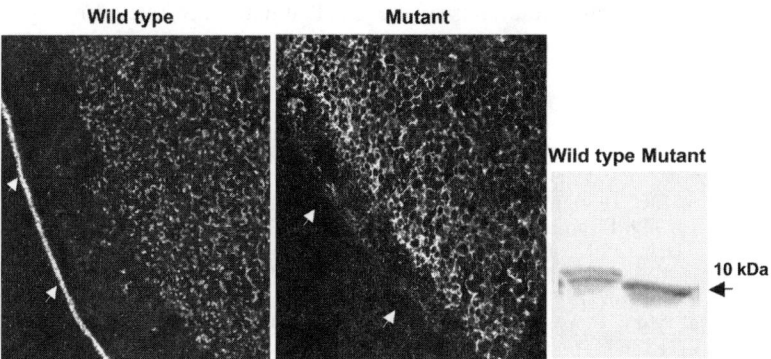

FIGURE 2. Confocal image of *Xenopus laevis* oocytes expressing wild-type (*left*) or mutant (*middle*) epitope-tagged γ-subunit protein. Oocytes expressing wild-type γ-subunit showed clear plasma membrane localization and an evenly distributed expression throughout the cytoplasm. Mutant protein was absent from the plasma membrane and accumulated in a thick band beneath the pigment layer. Western blot analysis (*right*) of *X. laevis* oocytes expressing epitope-tagged wild-type or mutant γ-subunit showed that the wild-type γ-subunit migrates as a doublet, whereas the upper band is missing in the lane expressing the G41R mutant. In each lane, the amount of protein equivalent to one oocyte was loaded.

excluding the possibility of a gene dosage effect and showing that, rather than haploinsufficiency, the presence of mutant γ-subunit causes hypomagnesemia. Our conclusion, therefore, is that the mutated γ-subunit protein actively interferes either directly or indirectly with normal Mg^{2+} reabsorption, consistent with a dominant-negative mechanism.

How Is the Na^+,K^+-ATPase γ-Subunit Involved in Hypomagnesemia?

Although several hypotheses have been proposed to explain the involvement of the Na^+,K^+-ATPase γ-subunit, the pathophysiology of autosomal dominant renal Mg^{2+} loss has not been resolved as yet. In our hands, preliminary immunoprecipitation studies (using the Na^+/oligomycin/$C_{12}E_{10}$ conditions according to Garty *et al.*[27]) indicate that the mutant γ-subunit protein is able to interact with αβ-complexes (not shown). This would support the hypothesis that the routing of not only the G41R γ-subunit, but of the whole Na^+,K^+-ATPase complex, is disturbed by the mutation in *FXYD2*.

A second explanation is that, in the presence of a mutated γ-subunit, the cell-specific fine regulation of Na^+,K^+-ATPase activity is no longer present, resulting in changes in intracellular ion composition and membrane potential that indirectly influence transepithelial Mg^{2+} transport. Alternatively, wild-type γ-subunit might have an additional function that is not known at present, such as regulating another ion channel on the basolateral membrane linked to Mg^{2+} reabsorption, which is abolished by the mutation.

Studies to investigate the effects of the G41R mutation in mice might provide more insight into the involvement of the Na$^+$,K$^+$-ATPase γ-subunit in renal Mg^{2+} handling in the near future.

REFERENCES

1. WACKER, W.E.C. 1980. Mg^{2+} and man. Harvard University Press. Cambridge, MA.
2. AGUS, Z.S. 1999. Hypomagnesemia. J. Am. Soc. Nephrol. **10:** 1616–1622.
3. QUAMME, G.A. 1997. Renal Mg^{2+} handling: new insights in understanding old problems. Kidney Int. **52:** 1180–1195.
4. RITCHIE, G., et al. 1996. Differentially expressed genes involved in regulation of epithelial Mg^{2+} transport identified by differential display [Abstr.]. J. Am. Soc. Nephrol. **7:** 1806.
5. DE ROUFFIGNAC, C. & G.A. QUAMME. 1994. Renal Mg^{2+} handling and its hormonal control. Physiol. Rev. **74:** 305–322.
6. QUAMME, G.A. & C. DE ROUFFIGNAC. 2000. Epithelial magnesium transport and regulation by the kidney. Front. Biosci. **5:** D694–D711.
7. DAI, L.J. et al. 2001. Mg^{2+} transport in the renal distal convoluted tubule. Physiol. Rev. **81:** 51–84.
8. FISELIER, T.J. et al. 1983. Levels of renin, angiotensin I and II, angiotensin-converting enzyme, and aldosterone in infancy and childhood. Eur. J. Pediatr. **141:** 3–7.
9. GEVEN, W.B. et al. 1987. Renal Mg^{2+} wasting in two families with autosomal dominant inheritance. Kidney Int. **31:** 1140–1144.
10. SCHULTHEIS, P.J. et al. 1998. Phenotype resembling Gitelman's syndrome in mice lacking the apical Na$^+$-Cl$^-$ cotransporter of the distal convoluted tubule. J. Biol. Chem. **273:** 29150–29155.
11. MEIJ, I.C. et al. 1999. Hereditary isolated renal magnesium loss maps to chromosome 11q23. Am. J. Hum. Genet. **64:** 180–188.
12. MERCER, R.W. et al. 1993. Molecular cloning and immunological characterization of the gamma polypeptide, a small protein associated with the Na,K-ATPase. J. Cell Biol. **121:** 579–586.
13. BÉGUIN, P. et al. 1997. The gamma subunit is a specific component of the Na,K-ATPase and modulates its transport function. EMBO J. **16:** 4250–4260.
14. ARYSTARKHOVA, E. et al. 1999. The gamma subunit modulates Na(+) and K(+) affinity of the renal Na,K-ATPase. J. Biol. Chem. **274:** 33183–33185.
15. WETZEL, R.K. & K.J. SWEADNER. 2001. Immunocytochemical localization of Na$^+$,K$^+$-ATPase alpha- and gamma-subunits in rat kidney. Am. J. Physiol. Renal Physiol. **281:** F531–F545.
16. ARYSTARKHOVA, E., R.K. WETZEL & K.J. SWEADNER. 2002. Distribution and oligomeric association of splice forms of Na$^+$,K$^+$-ATPase regulatory γ-subunit in rat kidney. Am. J. Physiol. Renal Physiol. **282:** F393–F407.
17. FORBUSH, B., III, J.H. KAPLAN & J.F. HOFFMAN. 1978. Characterization of a new photoaffinity derivative of ouabain: labeling of the large polypeptide and of a proteolipid component of the Na$^+$,K$^+$-ATPase. Biochemistry **17:** 3667–3676.
18. THERIEN, A.G. et al. 1997. Tissue-specific distribution and modulatory role of the γ subunit of the Na$^+$,K$^+$-ATPase. J. Biol. Chem. **272:** 32628–32634.
19. THERIEN, A.G., S.J. KARLISH & R. BLOSTEIN. 1999. Expression and functional role of the gamma subunit of the Na$^+$,K$^+$-ATPase in mammalian cells. J. Biol. Chem. **274:** 12252–12256.
20. THERIEN, A.G. et al. 2000. In Na$^+$,K$^+$-ATPase and Related Cation Pumps: 481–488. Elsevier. Amsterdam/New York.
21. ARYSTARKHOVA, E. et al. 2000. In Na$^+$,K$^+$-ATPase and Related Cation Pumps: 489–496. Elsevier. Amsterdam/New York.
22. ARYSTARKHOVA, E. et al. 2002. Differential regulation of renal Na,K-ATPase by splice variants of the gamma subunit. J. Biol. Chem. **277:** 10162–10172.

23. PU, H.X. *et al.* 2001. Functional role and immunocytochemical localization of the γ_a and γ_b forms of the Na$^+$,K$^+$-ATPase γ-subunit. J. Biol. Chem. **276:** 20370–20378.
24. PU, H.X., R. SCANZANO & R. BLOSTEIN. 2002. Distinct regulatory effects of the Na,K-ATPase gamma subunit. J. Biol. Chem. **277:** 20270–20276.
25. MEIJ, I.C. 2002. Gaining insights in renal magnesium handling. Ph.D. Thesis. University Medical Center Nijmegen, Nijmegen, the Netherlands.
26. MEIJ, I.C. *et al.* 2000. Dominant isolated renal magnesium loss is caused by misrouting of the Na$^+$,K$^+$-ATPase γ-subunit. Nat. Genet. **26:** 265–266.
27. GARTY, H. *et al.* 2002. A functional interaction between CHIF and Na-K-ATPase: implication for regulation by FXYD proteins. Am. J. Physiol. **283:** F607–F615.

FXYD7, the First Brain- and Isoform-Specific Regulator of Na,K-ATPase

Biosynthesis and Function of Its Posttranslational Modifications

GILLES CRAMBERT,[a] PASCAL BÉGUIN,[a] MARC ULDRY,[a] FLORIANNE MONNET-TSCHUDI,[b] JEAN-DANIEL HORISBERGER,[a] HAIM GARTY,[c] AND KÄTHI GEERING[a]

[a]*Institute of Pharmacology and Toxicology, University of Lausanne, CH-1005 Lausanne, Switzerland*

[b]*Institute of Physiology, University of Lausanne, CH-1005 Lausanne, Switzerland*

[c]*Department of Biological Chemistry, Weizmann Institute of Science, Rehovot 76100, Israel*

ABSTRACT: The FXYD protein family has recently been defined as a result of the search for homologues of the Na,K-ATPase γ subunit, CHIF, and phospholemman in EST and gene data banks. FXYD7 has been seen to have a role as a brain- and isozyme-specific regulator of Na/K-ATPase. In this study, the biosynthesis, membrane topology, nature, and role of the processing of FXYD7 are investigated.

KEYWORDS: FXYD protein; Na,K-pump; O-glycosylation; *Xenopus laevis* oocyte

INTRODUCTION

The FXYD protein family has recently been defined as a result of the search for homologues of the Na,K-ATPase γ subunit, CHIF, and phospholemman in EST and gene data banks.[1] This family contains seven members that display a common structure and an amino acid signature. In addition to the γ subunit of Na,K-ATPase (FXYD2) that we[2,3] and others[4,5] have defined as a specific regulator of the renal Na/K-ATPase, we demonstrated that CHIF[3] (FXYD4) and phospholemman[6] (FXYD1) are also able to associate with and modulate Na/K-ATPase activity. Moreover, we recently described the role of FXYD7 as a brain- and isozyme-specific regulator of Na/K-ATPase[7] (see also Geering *et al.* in this volume). For several FXYD proteins, it was shown that they are type-I membrane proteins, that they are co- or posttranslationally modified, and that they can reach the plasma membrane where they induce atypical ionic conductances. The nature of the co- or post-

---

Address for correspondence: Käthi Geering, Institute of Pharmacology and Toxicology, Rue du Bugnon 27, CH-1005 Lausanne, Switzerland. Voice: +041-21-692-54-10; fax: +041-21-692-53-55.
kaethi.geering@ipharm.unil.ch

translational modifications of γ subunits and CHIF is not known, whereas it is well established that phospholemman can be phosphorylated by protein kinase A and/or C.[8] Interestingly, Arystarkhova et al.[9] reported that the effects of the γ subunit on the Na,K-ATPase are influenced by its modifications. In this study, we investigate the biosynthesis, membrane topology, nature, and role of the processing of FXYD7.

MATERIALS AND METHODS

All technical details can be found in a recent publication.[7]

RESULTS AND DISCUSSION

Membrane Topology and Cell Surface Expression of FXYD7

Several FXYD proteins[10–13] were shown to induce ion conductances after expression in *Xenopus* oocytes and thus can reach the cell surface. The cell surface expression of FXYD7 was tested by anti-flag binding experiments on oocytes expressing FXYD7 containing a flag epitope at the N-terminus (N-Flag FXYD7). Iodinated anti-flag antibodies did not bind to noninjected oocytes nor to oocytes expressing wild-type FXYD7, but bound to oocytes expressing N-Flag FXYD7 (FIG. 1A). This result indicates that, similar to other FXYD proteins, FXYD7 is expressed at the cell surface in a type-I orientation, with its N-terminus exposed to the extracytoplasmic side. On the other hand, in contrast to other FXYD proteins, FXYD7 did not induce an ionic conductance neither during a rapid voltage step from −130 mV to +50 mV (FIG. 1B) nor after a long (4-s) depolarization (FIG. 1C) in the presence of K^+, Na^+, Cl^-, Ca^{2+}, and Mg^{2+} ions.

Processing of FXYD7

The processing of FXYD7 was followed by comparing the electrophoretic mobility of FXYD7 expressed in brain to that of FXYD7 expressed either *in vitro* in a reticulocyte lysate in the absence of microsomes or in *Xenopus* oocytes. As shown in FIGURE 1D (lane 1), *in vitro* translation of FXYD7 produced a protein with an apparent molecular mass of 14 kDa. When expressed in *Xenopus* oocytes (FIG. 1D, lane 2), the same 14-kDa protein was detected, indicating that FXYD7 has no cleavable signal sequence. In addition to the 14-kDa protein, we also observed a higher molecular mass doublet that most likely reflected co- or posttranslational modifications. In Western blots of brain microsomes, FXYD7 appeared as a single band corresponding to the lower band of the doublet observed in *Xenopus* oocytes (FIG. 1D, lane 3; compare lanes 2 and 3), indicating that FXYD7 is also processed in native tissue.

Computer-assisted analysis predicts that FXYD7 is O-glycosylated on three threonines in the extracytoplasmic N-terminus of the protein. To test this hypothesis, we mutated these threonines into alanines and tested the gel migration of the mutants expressed in *Xenopus* oocytes. As shown in FIGURE 1D (lane 4), after a short pulse with ^{35}S-methionine, FXYD7 appeared predominantly as a 14-kDa band and to a

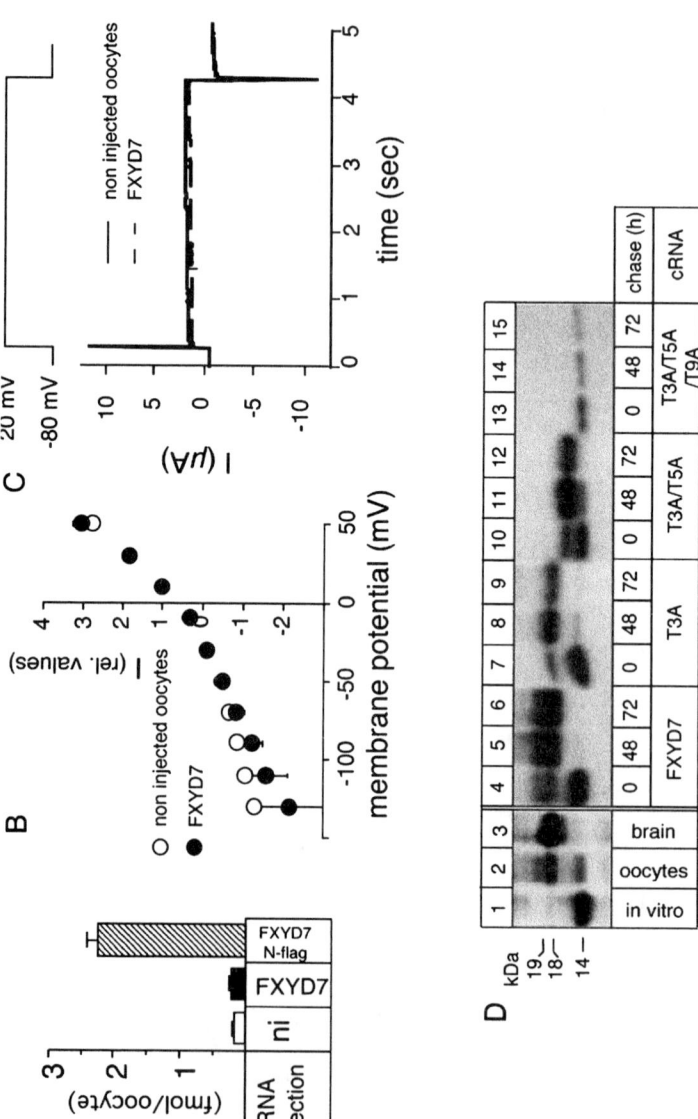

FIGURE 1. Cell surface expression and posttranslational modification of FXYD7. (**A**) Oocytes were injected or not (ni) with wild-type or N-flag FXYD7 cRNA (2 ng) and, after a 3-day incubation, were subjected to a radioimmunolabeling assay using [125I]-labeled anti-flag antibody (Sigma). (**B, C**) Three days after injection of FXYD7 cRNA (2 ng), current–voltage curves (from −130 mV to +50 mV; 200 ms each step) (**B**) and currents induced by a long depolarization (−80 mV to +20 mV for 4 s) (**C**) were recorded by a two-electrode voltage clamp. (**D**) FXYD7 expressed in reticulocyte lysate (lane 1), in oocytes (lane 2), and in brain microsomes (lane 3) was revealed with FXYD7 antibody in Western blot analysis. Oocytes were injected with wild-type or threonine mutant FXYD7 cRNA, metabolically labeled for 6 h, and subjected to 24- and 48-h chase periods (lanes 4–15). Microsomes were immunoprecipitated under denaturating conditions with an anti-FXYD7 antibody.

lesser extent as a doublet. However, after various chase periods (lanes 5–6), the 14-kDa band disappeared and the intensity of the doublet increased, indicating processing from a newly synthesized core to a posttranslationally modified protein. The triple mutant (FIG. 1D, lanes 13–15), in which all three threonines had been replaced by alanines, produced only the 14-kDa protein after a pulse period, which slowly disappeared during the chase. This result indicates that removal of the three threonines impedes posttranslational modifications and renders FXYD7 susceptible to degradation. The double mutant (T3A/T5A, FIG. 1, lanes 10–12) exhibited a 16-kDa band in both pulse and chase conditions, in addition to the 14-kDa band. This band, which is not found with the wild-type FXYD7, likely reflects posttranslational modification of one threonine (in this case, Thr 9). Finally, the replacement of one threonine by alanine led to the synthesis of the 14-kDa band in the pulse and of the lower band of the doublet (18 kDa) in both pulse and chase conditions (FIG. 1D, lanes 7–9), indicating that this band reflects the modification of two threonines. Together, these results show that the 14-kDa band corresponds to the core protein, whereas the lower and upper bands of the doublet are modified on two and three threonines, respectively. The presence of O-linked sugars was confirmed by testing the effect of neuraminidase and O-glycosidase on wild-type FXYD7 expressed in oocytes. Both enzymes were able to reduce the apparent molecular mass of the doublet, but not that of the core protein.[7]

Role of O-Glycosylation

FXYD7 is able to associate with the Na/K-ATPase and to decrease its K^+ affinity, but does not change its Na^+ affinity.[7] We investigated the role of O-glycosylation in these effects of FXYD7. After coexpression in oocytes, α1 and β1 subunits of the Na,K-pump could be coimmunoprecipitated either with wild-type FXYD7 or with the FXYD7 triple mutant (T3A/T5A/T9A).[7] Interestingly, the FXYD7 triple mutant, which was unstable when expressed alone (FIG. 1D, lanes 13–15), became stabilized through the association with the Na/K-ATPase.[7] On the other hand, the absence of O-glycosylation did not change the effect of FXYD7 on the cation affinity of Na,K-pumps.[7] All these results indicate that O-glycosylation of FXYD7 is not required for its effect on the Na,K-ATPase, but is necessary to stabilize the protein.

ACKNOWLEDGMENTS

This work was supported by grants from the Swiss National Fund for Scientific Research (Nos. 31-53721.98 and 31-64793.01) and the Roche Research Foundation.

REFERENCES

1. SWEADNER, K.J. & E. RAEL. 2000. The FXYD gene family of small ion transport regulators or channels: cDNA sequence, protein signature sequence, and expression. Genomics **68:** 41–56.
2. BÉGUIN, P., X. WANG, D. FIRSOV, et al. 1997. The γ subunit is a specific component of the Na,K-ATPase and modulates its transport function. EMBO J. **16:** 4250–4260.

3. BÉGUIN, P., G. CRAMBERT, S. GUENNOUN, et al. 2001. CHIF, a member of the FXYD protein family, is a regulator of Na,K-ATPase distinct from the γ subunit. EMBO J. **20:** 3993–4002.
4. ARYSTARKHOVA, E., R.K. WETZEL, N.K. ASINOVSKI, et al. 1999. The γ subunit modulates Na^+ and K^+ affinity of the renal Na,K-ATPase. J. Biol. Chem. **274:** 33183–33185.
5. THERIEN, A.G., S.J.D. KARLISH & R. BLOSTEIN. 1999. Expression and functional role of the γ subunit of the Na,K-ATPase in mammalian cells. J. Biol. Chem. **274:** 12252–12256.
6. CRAMBERT, G., M. FÜZESI, H. GARTY, et al. 2002. Phospholemman (FXYD1) associates with Na,K-ATPase and regulates its transport properties. Proc. Natl. Acad. Sci. USA **99:** 11476–11481.
7. BÉGUIN, P., G. CRAMBERT, F. MONNET-TSCHUDI, et al. 2002. FXYD7 is a brain-specific regulator of Na,K-ATPase α1-β isozymes. EMBO J. **21:** 3264–3273.
8. PALMER, C.J., B.T. SCOTT & L.R. JONES. 1991. Purification and complete sequence determination of the major plasma membrane substrate for cAMP-dependent protein kinase and protein kinase C in myocardium. J. Biol. Chem. **266:** 11126–11130.
9. ARYSTARKHOVA, E., C. DONNET, N.K. ASINOVSKI, et al. 2002. Differential regulation of renal Na,K-ATPase by splice variants of the gamma subunit. J. Biol. Chem. **277:** 10162–10172.
10. MOORMAN, J.R., C.J. PALMER, J.E. JOHN III, et al. 1992. Phospholemman expression induces a hyperpolarization-activated chloride current in *Xenopus* oocytes. J. Biol. Chem. **267:** 14551–14554.
11. MORRISON, B.W., J.R. MOORMAN, G.C. KOWDLEY, et al. 1995. Mat-8, a novel phospholemman-like protein expressed in human breast tumors, induces a chloride conductance in *Xenopus* oocytes. J. Biol. Chem. **270:** 2176–2182.
12. ATTALI, B., H. LATTER, N. RACHAMIM, et al. 1995. A corticosteroid-induced gene expressing an "IsK-like" K^+ channel activity in *Xenopus* oocytes. Proc. Natl. Acad. Sci. USA **92:** 6092–6096.
13. MOUNSEY, J.P., K.P. LU, M.K. PATEL, et al. 1999. Modulation of *Xenopus* oocyte–expressed phospholemman-induced ion currents by co-expression of protein kinases. Biochim. Biophys. Acta **1451:** 305–318.

Phosphorylation of the Na^+,K^+-ATPase in Skeletal Muscle

Potential Mechanism for Changes in Pump Cell-Surface Abundance and Activity

LUBNA AL-KHALILI, ANNA KROOK, AND ALEXANDER V. CHIBALIN

Department of Clinical Physiology, Karolinska Hospital, 171 76 Stockholm, Sweden, and Department of Physiology and Pharmacology, Karolinska Institute, 171 77 Stockholm, Sweden

ABSTRACT: In skeletal muscle, insulin stimulation leads to phosphorylation of Na^+,K^+-ATPase α-subunits on both serine/threonine and tyrosine residues, translocation of Na^+,K^+-ATPase molecules to the plasma membrane, and increased Na^+,K^+-ATPase activity. The molecular nature of the tyrosine kinase that phosphorylates Na^+,K^+-ATPase is not yet identified. *In vitro* phosphorylation experiments show that the α-subunit of Na^+,K^+-ATPase from skeletal muscle is a substrate for the tyrosine-specific protein kinase c-src. Tyrosine phosphorylation of the α-subunits of Na^+,K^+-ATPase may be an important mechanism for insulin-mediated regulation of Na^+,K^+-ATPase translocation and activity.

KEYWORDS: α-subunit; c-src tyrosine protein kinase; Na^+,K^+-ATPase; insulin; trafficking; phosphorylation; skeletal muscle

INTRODUCTION

Skeletal muscle contains one of the largest pools of Na^+,K^+-ATPase and therefore plays a central role in the extrarenal clearance of K^+ from the blood during ingestion or infusion of K^+.[1] Skeletal muscle is an important target tissue for insulin, a hormone that plays a major role in the control of both glucose transport and metabolism, as well as K^+ uptake. This latter effect of insulin is likely to be achieved through a stimulation of Na^+,K^+-ATPase in skeletal muscle and is not secondary to an increase in $[Na^+]_i$ via Na^+-H^+ antiporter stimulation.[2] Na-pump activation by insulin is PKC-dependent[3,4] and is partly mediated by an increase in cell surface appearance of Na^+,K^+-ATPase molecules. Isolation of skeletal muscle membrane fractions by differential centrifugations reveals that either injection of insulin *in vivo* in rats or *ex vivo* incubation of isolated rat skeletal muscle with insulin leads to an increase in $α_2$- and $β_1$-subunit abundance in the plasma membrane, with no change in $α_1$ and $β_2$

Address for correspondence: Alexander V. Chibalin, Ph.D., Section of Clinical Physiology, Integrative Physiology, Karolinska Institutet, von Eulers väg 4, 2 tr, SE-171 77 Stockholm, Sweden. Voice: +46-8-7287584; fax: +46-8-335436.
Alexander.Chibalin@kirurgi.ki.se

distribution.[4,5] One explanation for these data may be that the ubiquitously expressed Na^+,K^+-ATPase α_1-subunit isoform has more "housekeeping" functions, whereas the α_2-subunit isoform, which is predominantly expressed in adult skeletal muscle, is hormone-sensitive and responsible for most of the effects of insulin on Na^+,K^+-ATPase ion transport activity. However, alternative techniques of monitoring of Na^+,K^+-ATPase protein trafficking indicate that the α_1-subunit is also capable of mediating insulin-stimulated translocation to the plasma membrane; insulin promotes translocation of exofacially epitope-tagged rat Na^+,K^+-ATPase α_1-subunit to plasma membrane of HEK-293 cells.[6]

The molecular signaling mechanism by which insulin modulates Na^+,K^+-ATPase activity remains to be elucidated. We have recently shown that Na^+,K^+-ATPase α-subunits become phosphorylated on both serine/threonine and tyrosine residues in response to insulin in skeletal muscle via a PKC- and tyrosine kinase–dependent mechanism.[4] Tyrosine phosphorylation of the α-subunits of Na^+,K^+-ATPase may be an important mechanism of regulating insulin-mediated Na^+,K^+-ATPase activity and Na^+ and K^+ gradient.

RESULTS AND DISCUSSION

The molecular nature of the tyrosine kinase that phosphorylates Na^+,K^+-ATPase is not yet identified. Although the insulin receptor tyrosine kinase inhibitor, HNMPA-$(AM)_3$, completely blocked insulin-mediated phosphorylation of the Na^+,K^+-ATPase α-subunit in rat soleus muscle,[4] involvement of nonreceptor tyrosine kinases cannot be excluded.[7] Moreover, there is functional association between the insulin receptor and downstream effectors, including phosphatidylinositol 3-kinase and nonreceptor tyrosine kinase c-src.[8] Our results indicate that c-src phosphorylates the α-subunit of the Na^+,K^+-ATPase *in vitro* in skeletal muscle plasma membrane preparation (FIG. 1). Incorporation of [^{32}P] into Na^+,K^+-ATPase α-subunit occurred only in plasma membranes incubated in the presence of 0.2% Triton X-100. Similarly, other protein kinases such as PKA and PKG, which are capable of phosphorylating Na^+,K^+-ATPase α-subunit *in vivo* and in tissue homogenates, are also able to phosphorylate Na^+,K^+-ATPase α-subunit in membrane preparation only in the presence of detergent.[9,10] Thus, results from the *in vitro* phosphorylation experiment suggest that the α-subunit of Na^+,K^+-ATPase is a substrate for the nonreceptor tyrosine kinase c-src. Although Tyr-10 is a target for tyrosine kinases in the rat Na^+,K^+-ATPase α_1-subunit isoform,[11] the same residue is not likely to undergo phosphorylation in α_2-subunit, which is predominantly expressed in skeletal muscle. Tyr-10 is conserved and present in the α_2-subunit sequence; however, the amino acid environment of this residue is different. Rat α_2-subunit sequence comparison with the PhosphoBase database[12,13] predicts Tyr-537 as a most probable target for nonreceptor tyrosine kinases in rat Na^+,K^+-ATPase α_2-subunit. This tyrosine residue is located in the large cytoplasmic loop of Na^+,K^+-ATPase α-subunit and it is surrounded by the sequence AYMELG. One possible mechanism of Na-pump regulation by insulin could be through phosphorylation of tyrosine residues in the clathrin adaptor protein 2 (AP-2)–recognizing tyrosine-based endocytic motif, YMEL (YLEL), used for sorting proteins from the plasma membrane to endosomes.[14] It may arrest the formation of an endocytic complex of Na^+,K^+-ATPase α-subunit, AP-2, and clathrin, and may

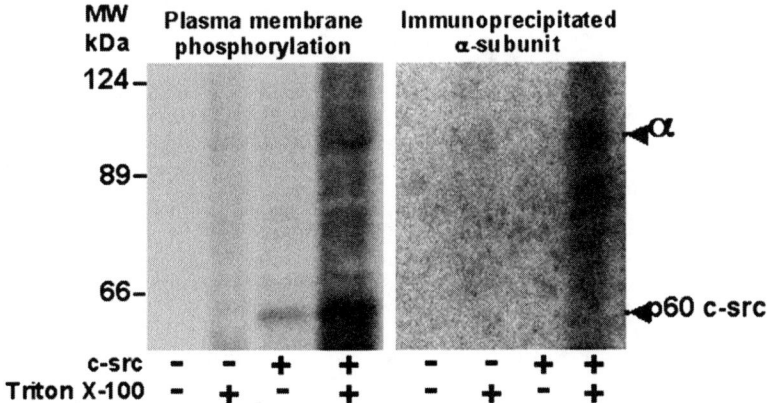

FIGURE 1. Phosphorylation of Na^+,K^+-ATPase α-subunit by c-src protein kinase. Rat skeletal muscle plasma membranes (100 μg) were incubated at 30°C in a final volume of 100 μL for 30 min in buffer containing 25 mM Tris-HCl (pH 7.2), 7.5 mM $MnCl_2$, 0.5 mM EGTA, 50 μM Na_3VO_4, 25 mM Mg-acetate supplemented with 100 μM [$\gamma^{32}P$]ATP (final activity: 5000 cpm/pmol, NEN) with (+) or without (−) 5 U of c-src protein kinase (Upstate Biotechnology, Lake Placid, NY), and 0.2% Triton X-100. Samples were analyzed by SDS-PAGE before and after immunoprecipitation with NK1 multisubunit-specific antibody.[5] Representative autoradiograms are shown.

prevent the Na^+,K^+-ATPase α-subunit from undergoing endocytosis from the plasma membrane, hence leading to an increased α-subunit abundance in plasma membrane due to constitutive exocytosis. Indeed, recent data indicate that the presence of Tyr-537 in such a motif within the Na^+,K^+-ATPase α_1-subunit is essential for AP-2 binding and clathrin-dependent endocytosis.[15,16]

In conclusion, the regulation of Na^+,K^+-ATPase translocation and activity is a complex multilevel process, and changes in α-subunit phosphorylation state require further study to be directly linked with changes in Na-pump activity in skeletal muscle. Phosphorylation is likely to play an important role in maintaining the intracellular distribution of Na^+,K^+-ATPase units and the regulation of Na^+ and K^+ gradients.

ACKNOWLEDGMENTS

We thank Juleen R. Zierath for helpful discussions and critical reading of the manuscript. This work was supported by grants from the Swedish Medical Research Council, the Novo-Nordisk Foundation, the Swedish Society of Medicine, the Thurings Foundation, and the Åke Wiberg Foundation.

REFERENCES

1. SWEENEY, G. & A. KLIP. 1998. Regulation of the Na,K-ATPase by insulin: why and how? Mol. Cell. Biochem. **182:** 121–133.
2. WEIL, E. et al. 1991. Mechanism of insulin-induced activation of Na,K-ATPase in isolated rat soleus muscle. Am. J. Physiol. **261:** C224–C230.
3. SAMPSON, S.R. et al. 1994. Role of protein kinase C in insulin activation of the Na-K pump in cultured skeletal muscle. Am. J. Physiol. **266:** C751–C758.
4. CHIBALIN, A.V. et al. 2001. Insulin- and glucose-induced phosphorylation of the Na^+,K^+-adenosine triphosphatase alpha-subunits in rat skeletal muscle. Endocrinology **142:** 3474–3482.
5. HUNDAL, H.S. et al. 1992. Insulin induces translocation of the alpha 2 and beta 1 subunits of the Na,K-ATPase from intracellular compartments to the plasma membrane in mammalian skeletal muscle. J. Biol. Chem. **267:** 5040–5043.
6. SWEENEY, G. et al. 2001. Insulin increases plasma membrane content and reduces phosphorylation of Na^+-K^+ pump alpha1-subunit in HEK-293 cells. Am. J. Physiol. **281:** C1797–C1803.
7. SUN, X.J. et al. 1996. The Fyn tyrosine kinase binds IRS-1 and forms a distinct signaling complex during insulin stimulation. J. Biol. Chem. **271:** 10583–10587.
8. GREY, A. et al. 2000. Evidence for a functional association between phosphatidylinositol 3-kinase and c-src in the spreading response of osteoclasts to colony-stimulating factor-1. Endocrinology **141:** 2129–2138.
9. CHIBALIN, A. et al. 1992. Phosphorylation of Na,K-ATPase alpha-subunits in microsomes and in homogenates of *Xenopus* oocytes resulting from the stimulation of protein kinase A and protein kinase C. J. Biol. Chem. **267:** 22378–22384.
10. FOTIS, H. et al. 1999. Phosphorylation of the alpha-subunits of the Na,K-ATPase from mammalian kidneys and *Xenopus* oocytes by cGMP-dependent protein kinase results in stimulation of ATPase activity. Eur. J. Biochem. **260:** 904–910.
11. FÉRAILLE, E. et al. 1999. Insulin-induced stimulation of Na,K-ATPase activity in kidney proximal tubule cells depends on phosphorylation of the alpha-subunit at Tyr-10. Mol. Biol. Cell **10:** 2847–2859.
12. KREEGIPUU, A. et al. 1998. Statistical analysis of protein kinase specificity determinants. FEBS Lett. **430:** 45–50.
13. KREEGIPUU, A. et al. 1999. PhosphoBase, a database of phosphorylation sites: release 2.0. Nucleic Acids Res. **27:** 237–239.
14. BOLL, W. et al. 1996. Sequence requirements for the recognition of tyrosine-based endocytic signals by clathrin AP-2 complexes. EMBO J. **15:** 5789–5795.
15. OGIMOTO, G. et al. 2000. G protein–coupled receptors regulate Na,K-ATPase activity and endocytosis by modulating the recruitment of adaptor protein 2 and clathrin. Proc. Natl. Acad. Sci. USA **97:** 3242–3247.
16. DONE, S.C. et al. 2002. Tyrosine 537 within the Na^+,K^+-ATPase alpha-subunit is essential for AP-2 binding and clathrin-dependent endocytosis. J. Biol. Chem. **277:** 17108–17111.

Physiological Functions of Plasma Membrane and Intracellular Ca^{2+} Pumps Revealed by Analysis of Null Mutants

GARY E. SHULL,[a] GBOLAHAN OKUNADE,[a] LYNNE H. LIU,[a] PETER KOZEL,[a] MUTHU PERIASAMY,[b] JOHN N. LORENZ,[c] AND VIKRAM PRASAD[a]

[a]*Department of Molecular Genetics, Biochemistry, and Microbiology, University of Cincinnati College of Medicine, Cincinnati, Ohio 45267, USA*

[b]*Department of Physiology and Cell Biology, Ohio State University, Columbus, Ohio 43210, USA*

[c]*Department of Molecular and Cellular Physiology, University of Cincinnati College of Medicine, Cincinnati, Ohio 45267, USA*

ABSTRACT: It is known that plasma membrane Ca^{2+}-transporting ATPases (PMCAs) extrude Ca^{2+} from the cell and that sarco(endo)plasmic reticulum Ca^{2+}-ATPases (SERCAs) and secretory pathway Ca^{2+}-ATPases (SPCAs) sequester Ca^{2+} in intracellular organelles; however, the specific physiological functions of individual isoforms are less well understood. This information is beginning to emerge from studies of mice and humans carrying null mutations in the corresponding genes. Mice with targeted or spontaneous mutations in plasma membrane Ca^{2+}-ATPase isoform 2 (PMCA2) are profoundly deaf and have a balance defect due to the loss of PMCA2 in sensory hair cells of the inner ear. In humans, mutations in *SERCA1* (*ATP2A1*) cause Brody disease, an impairment of skeletal muscle relaxation; loss of one copy of the *SERCA2* (*ATP2A2*) gene causes Darier disease, a skin disorder; and loss of one copy of the *SPCA1* (*ATP2C1*) gene causes Hailey-Hailey disease, another skin disorder. In the mouse, SERCA2 null mutants do not survive to birth, and heterozygous SERCA2 mutants have impaired cardiac performance and a high incidence of squamous cell cancers. SERCA3 null mutants survive and appear healthy, but endothelium-dependent relaxation of vascular smooth muscle is impaired and Ca^{2+} signaling is altered in pancreatic β cells. The diversity of phenotypes indicates that the various Ca^{2+}-transporting ATPase isoforms serve very different physiological functions.

KEYWORDS: embryonic stem cells; gene targeting; *ATP2A1*; *ATP2A2*; *ATP2A3*; *ATP2B1*; *ATP2B2*; *ATP2B3*; *ATP2B4*; *ATP2C1*

Address for correspondence: Gary E. Shull, Department of Molecular Genetics, Biochemistry, and Microbiology, University of Cincinnati College of Medicine, 231 Bethesda Avenue, ML 524, Cincinnati, OH 45267-0524. Voice: 513-558-0056; fax: 513-558-1885.
shullge@ucmail.uc.edu

INTRODUCTION

Data from molecular cloning studies and genome sequencing efforts suggest that mammalian tissues contain as many as 30–40 different Ca^{2+}-transporting ATPases encoded by alternatively spliced transcripts from at least nine genes. These include two secretory pathway Ca^{2+}-ATPase (SPCA) genes,[1,2] three sarco(endo)plasmic reticulum Ca^{2+}-ATPase (SERCA) genes,[3–7] and four plasma membrane Ca^{2+}-ATPase (PMCA) genes.[8–11] On the basis of sequence comparisons (FIG. 1), it is clear that the SERCA, PMCA, and SPCA pumps fall into three distinct classes of P-type ATPases, consistent with differences in their membrane locations and cellular functions.[2] It is well established that PMCAs extrude Ca^{2+} from the cell[12] and that SERCAs sequester Ca^{2+} in the sarcoplasmic or endoplasmic reticulum.[13] The membrane distribution and cellular functions of SPCAs are less well understood. The SPCAs exhibit a high degree of amino acid similarity to PMR1 (~50% identity), the yeast secretory pathway Ca^{2+} pump,[1,14–16] and to PACL (~35% identity with few gaps), a cyanobacterial Ca^{2+} pump.[17] The high degree of diversity among the Ca^{2+}-transporting ATPases suggests that the cellular functions of each class are very different and that specific isoforms may serve specialized physiological functions. Here, we discuss gene targeting and human genetics studies that provide new information about the physiological functions of these pumps.

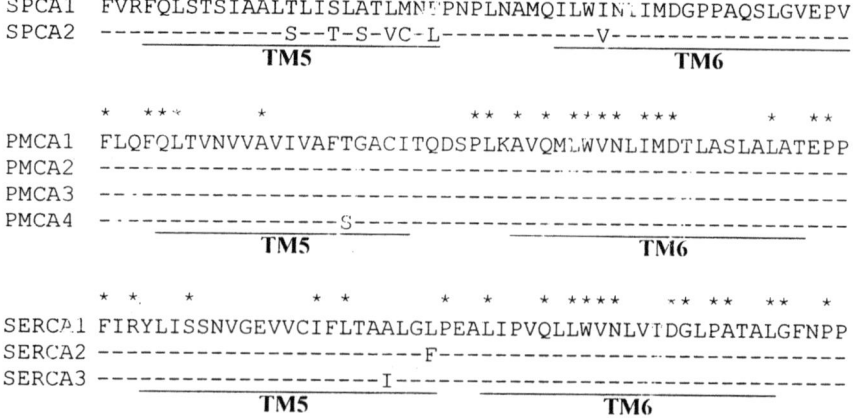

FIGURE 1. Mammalian Ca^{2+} pumps consist of three subfamilies of P-type ATPases and are encoded by nine distinct genes: comparisons of the amino acid sequences spanning transmembrane domains 5 and 6 (*underlined*; see Ref. 1) of the two SPCA, four PMCA, and three SERCA isoforms from rat. Homologues of the nine genes are also present in the human and mouse genomes. These transmembrane domains are relatively well conserved between members of the same subfamily (amino acid identity indicated by a *dash*), but are poorly conserved between the three different subfamilies (*asterisks* above PMCA and SERCA sequences indicate amino acid identity to SPCA).

SECRETORY PATHWAY Ca^{2+} PUMPS

Alternatively spliced C-terminal variants of mammalian SPCA1 were originally cloned from rat[1] and classified as possible Ca^{2+} pumps on the basis of their similarity to yeast PMR1. Later studies showed that *C. elegans* SPCA[18] and human SPCA1[2] were expressed in the Golgi and mediated the uptake of both Ca^{2+} and Mn^{2+}, consistent with similar studies of yeast PMR1.[19,20] A second SPCA isoform, termed SPCA2,[2] has been reported,[21] but nothing is currently known about its distribution or biological function.

Null mutations in a single copy of *ATP2C1*, the human gene encoding SPCA1, cause Hailey-Hailey disease, an autosomal dominant skin disorder.[22] Cultured keratinocytes from Hailey-Hailey patients have elevated intracellular Ca^{2+} and a reduction in the levels of Ca^{2+} that can be released by ionomycin.[22] Analysis of skin revealed that the extracellular Ca^{2+} gradient normally observed between basal and superficial epidermis[22] was perturbed. There have been several reports of squamous cell carcinomas in Hailey-Hailey disease.[23,24] In light of the squamous cell tumors in SERCA2-deficient mice,[25] discussed below, it is conceivable that SPCA1 haploinsufficiency also leads to a low incidence of skin cancer.

PLASMA MEMBRANE Ca^{2+} PUMPS

There are four distinct PMCA isoforms in mammalian tissues, termed PMCA1–4 (human gene symbols, *ATP2B1–4*). In our gene targeting studies, we have prepared knockout mice for PMCA1, PMCA2, and PMCA4. In the case of PMCA1, only heterozygous mutants survive, and these mice exhibit no readily apparent disease phenotype (Okunade and Shull, unpublished observations). PMCA4 homozygous mutants survive and appear healthy, but exhibit male infertility (Okunade and Shull, unpublished observations). Additional studies will be needed to characterize the PMCA1 and PMCA4 mutants.

Gene-targeted PMCA2 null mutant mice have a severe balance defect, and auditory brain stem responses show that they are profoundly deaf.[26] Histological analyses of the inner ear revealed a loss of otoconia, which may be responsible for the balance defect. Severe histopathology was also observed in the auditory system, but this alone is insufficient to explain the complete loss of hearing. PMCA2 mRNA is expressed at particularly high levels in outer hair cells,[27] and high levels of PMCA protein and activity have been demonstrated in stereocilia of sensory hair cells in the vestibular system.[28] There have been additional reports of mice with spontaneous mutations in the PMCA2 gene.[29,30] Immunocytochemical studies indicated that PMCA2 is expressed in the hair bundles of both inner and outer hair cells of the wild-type organ of Corti.[29,30] These data provide strong evidence that the profound deafness resulting from the loss of PMCA2 is due to a perturbation of hair cell function.

On some genetic backgrounds, PMCA2 heterozygous mutants exhibit a progressive loss of hearing as they age.[31] Because mice that exhibit an age-related hearing loss are more susceptible to noise-induced hearing loss,[32] it seemed possible that

PMCA2 mutations might cause noise-induced hearing loss. This was tested by subjecting heterozygous and wild-type mice to high noise intensities.[33] The results showed that heterozygous mutants were more susceptible to noise-induced hearing loss, thereby implicating PMCA2 as the first gene with a known protein product that causes a predisposition to noise-induced hearing loss.

SARCO(ENDO)PLASMIC RETICULUM Ca^{2+} PUMPS

Human genetics studies have shown that loss of SERCA1 causes an autosomal recessive form of the skeletal muscle disorder, Brody disease.[34,35] Consistent with absence of the major mechanism for Ca^{2+} uptake into the SR, these patients suffer from cramps and severely impaired relaxation of skeletal muscle during exercise. The physiological consequences of SERCA1-deficiency in Brody disease and the potential mechanisms that may provide partial compensation have been discussed in a recent review.[36]

When the mouse SERCA2 gene was inactivated by gene targeting, breeding of heterozygous mutants led to the birth of only wild-type and heterozygous mutant offspring.[37] Heterozygous mutants grew normally and appeared healthy; however, because reductions in SERCA2a mRNA, protein, and activity in cardiac muscle have been identified as possible causative factors in heart disease,[38] the effects of SERCA2 deficiency on cardiac performance were examined. Mean arterial pressure, left ventricular pressure, and positive dP/dt were significantly reduced, and negative dP/dt was significantly increased in 3- to 4-month-old heterozygous mutants.[37] SERCA2-deficient cardiac myocytes also exhibited reductions in contractility and relaxation, and SR Ca^{2+} stores and the amount of Ca^{2+} released during each contraction were reduced.[39] Some degree of compensation for the reduction in SERCA2a levels occurs by increased phosphorylation of phospholamban and by increased Na^+/Ca^{2+} exchange activity.[39]

In a particularly elegant study of the cardiac functions of SERCA2a, Wuytack's group prepared a mouse in which the splicing mechanism of the SERCA2 gene was modified to eliminate expression of SERCA2a.[40] SERCA2a null mutants exhibited an increased incidence of embryo lethality and cardiac malformations, but many survived to adulthood, indicating that SERCA2a was not essential for life. These mice had no SERCA2a expression in heart; however, SERCA2b levels were elevated to ~50% of the levels of SERCA2a normally observed in heart. Adult null mutants exhibited cardiac hypertrophy, and cardiac performance was impaired in a manner similar to that observed in mice lacking one copy of the SERCA2 gene.[37] Total Ca^{2+} sequestering activity was only slightly less in SERCA2a null mutants than it was in SERCA2 heterozygous mutants, but the apparent affinity for Ca^{2+} was significantly reduced.

In a study initially designed to test the hypothesis that a reduction in SERCA2a might lead to heart disease as the animals age, we made the unexpected observation that the mutant mice develop squamous cell tumors at a very high frequency.[25] These included hyperkeratinized squamous cell carcinomas and papillomas of the skin, oral mucosa, esophagus, and forestomach. Among the 14 pairs of wild-type and heterozygous mutant mice, none of the wild-type animals developed squamous cell tumors, but 13 of the knockouts were affected, with a total of 30 different tumors. These genetic data are the first clear evidence that perturbations of Ca^{2+} signaling and/or homeostasis can lead to cancer.

In response to agonist stimulation, a more rapid decay in Ca^{2+} transients of pancreatic acinar cells was observed in SERCA2 heterozygous mutants.[41] This surprising result was due to the upregulation of Ca^{2+} extrusion via PMCA activity, which compensated for the reduction in Ca^{2+} sequestering activity. Although there was a sharp reduction in the frequency of Ca^{2+} oscillations, agonist-stimulated exocytosis was not impaired, apparently due to the downregulation of synaptotagmin I, the low-affinity Ca^{2+} sensor, and upregulation of synaptotagmin III, the high-affinity Ca^{2+} sensor. Consequently, exocytosis in acinar cells from SERCA2-deficient mice had ~10-fold greater sensitivity to Ca^{2+} than that of wild-type cells.

In humans, null mutations in the SERCA2 (*ATP2A2*) gene cause Darier disease, an autosomal dominant skin disorder that arises in keratinized squamous epithelial cells.[42,43] Heart disease has not been reported in Darier patients,[44,45] and a recent study of 10 patients with Darier disease, using echocardiography,[46] revealed no evidence of heart disease. Although true Darier-like lesions have not been observed in mice,[25] the cell type in which tumors arise, keratinocytes, is the same. Possible mechanisms by which SERCA2 haploinsufficiency leads to Darier disease in humans and predisposes to squamous cell cancer in mice, which include alterations of gene expression, cell cycle regulation, and DNA repair, have been discussed.[25]

In contrast to SERCA2, SERCA3 has a limited cell-type distribution,[47] suggesting that it serves a specialized physiological or cellular function. Disruption of the SERCA3 gene resulted in the birth of viable null mutant mice in a normal Mendelian ratio.[48] The mutants grew normally, had no apparent disease phenotype, and were indistinguishable from wild-type mice in outward appearance and behavior.[48] So far, an *in vivo* physiological phenotype for SERCA3 null mutant mice has not been demonstrated. However, in response to acetylcholine, both endothelium-dependent relaxation of aortic rings and the Ca^{2+} transient in endothelial cells were reduced. Similar studies of tracheal samples revealed that epithelium-dependent relaxation of tracheal smooth muscle was altered.[49]

The possible role of SERCA3 in glucose homeostasis and insulin secretion was studied using the SERCA3-deficient mouse.[50] SERCA3 was shown to be coexpressed with SERCA2b in pancreatic β cells; however, plasma glucose and insulin levels were normal, indicating that this isoform is not essential for glucose homeostasis and insulin secretion. Following glucose stimulation, though, there were significant alterations in cytosolic Ca^{2+} oscillations in SERCA3-deficient β cells, and cytosolic Ca^{2+} transients in response to high K^+ were also altered. Surprisingly, the oscillations and transients exhibited an increased amplitude and decayed more rapidly in the absence of SERCA3. These results suggest that SERCA3 is active primarily when cytosolic Ca^{2+} is elevated, that it buffers the rise in Ca^{2+} during the upstroke phase, and that the pool it serves releases Ca^{2+} during the decay phase. There is evidence that SERCA3 missense mutations might contribute to type-II diabetes in humans.[51] Although the lack of an effect of SERCA3 ablation on glucose homeostasis or insulin secretion does not support this hypothesis, the disturbance of Ca^{2+} handling in pancreatic SERCA3-deficient β cells is consistent with the possibility that defective SERCA3 expression might serve as a contributing factor in the development of diabetes.

CONCLUSIONS

The mammalian Ca^{2+} pumps consist of three subfamilies of P-type ATPases (FIG. 1), with nine distinct genes, and many different splice variants. Phylogenetic analysis[2] shows a clear grouping of the members of each subfamily. SPCA1 and SPCA2 have homologues in lower eukaryotes[14,15] and even in bacteria,[17] suggesting that they may be more closely related to the earliest ancestral Ca^{2+} pump than the PMCAs and SERCAs. PMCAs are restricted to higher eukaryotes, although a related vacuolar Ca^{2+} pump, which lacks the C-terminal regulatory domains found in the four mammalian PMCAs, has been identified in yeast.[52] SERCAs, including the three mammalian isoforms, have been identified only in higher eukaryotes. With SPCAs expressed in the secretory pathway, PMCAs in the plasma membrane, and SERCAs in the sarcoplasmic or endoplasmic reticulum, it seems likely that at least part of the biological rationale for the extensive diversity among mammalian Ca^{2+}-transporting ATPases is the requirement for Ca^{2+} pumps with differing biochemical characteristics in different organellar or suborganellar compartments. The studies described here, involving analysis of gene-targeted mice and humans with null mutations in specific isoforms, support this hypothesis and are leading to a better understanding of the physiological and cellular functions of these pumps.

ACKNOWLEDGMENTS

Portions of this work were supported by NIH Grants HL61974 and HL64140.

REFERENCES

1. GUNTESKI-HAMBLIN, A., D.M. CLARKE & G.E. SHULL. 1992. Molecular cloning and tissue distribution of alternatively spliced mRNAs encoding possible mammalian homologs of the yeast secretory pathway calcium pump. Biochemistry **31:** 7600–7608.
2. TON, V.-K., D. MANDAL, et al. 2002. Functional expression in yeast of the human secretory pathway Ca^{2+},Mn^{2+}-ATPase defective in Hailey-Hailey disease. J. Biol. Chem. **277:** 6422–6427.
3. MACLENNAN, D.H., C.J. BRANDL, et al. 1985. Amino-acid sequence of a (Ca^{2+} + Mg^{2+})–dependent ATPase from rabbit muscle sarcoplasmic reticulum deduced from its cDNA sequence. Nature **316:** 696–700.
4. BRANDL, C.J., N.M. GREEN, et al. 1986. Two Ca^{2+}-ATPase genes: homologies and mechanistic implications of deduced amino acid sequences. Cell **44:** 597–607.
5. GUNTESKI-HAMBLIN, A., J. GREEB & G.E. SHULL. 1988. A novel Ca^{2+} pump expressed in brain, kidney, and stomach is encoded by an alternative transcript of the slow-twitch muscle sarcoplasmic reticulum Ca-ATPase gene. J. Biol. Chem. **263:** 15032–15040.
6. LYTTON, J. & D.H. MACLENNAN. 1988. Molecular cloning of cDNAs from human kidney coding for two alternatively spliced products of the cardiac Ca^{2+}-ATPase gene. J. Biol. Chem. **263:** 15024–15031.
7. BURK, S.E., J. LYTTON, et al. 1989. cDNA cloning, functional expression, and mRNA tissue distribution of a third organellar Ca^{2+} pump. J. Biol. Chem. **264:** 18561–18568.

8. SHULL, G.E. & J. GREEB. 1988. Molecular cloning of two isoforms of the plasma membrane Ca^{2+}-transporting ATPase from rat brain. J. Biol. Chem. **263:** 8646–8657.
9. GREEB, J. & G.E. SHULL. 1989. Molecular cloning of a third isoform of the calmodulin-sensitive plasma membrane Ca^{2+}-transporting ATPase that is expressed predominantly in brain and skeletal muscle. J. Biol. Chem. **264:** 18569–18576.
10. VERMA, A.K., A.G. FILOTEO, et al. 1988. Complete primary structure of a human plasma membrane Ca^{2+} pump. J. Biol. Chem. **263:** 14152–14159.
11. STREHLER, E.E., P. JAMES, et al. 1990. Peptide sequence analysis and molecular cloning reveal two calcium pump isoforms in the human erythrocyte membrane. J. Biol. Chem. **265:** 2835–2842.
12. CARAFOLI, E. 1994. Biogenesis: plasma membrane calcium ATPase—15 years of work on the purified enzyme. FASEB J. **13:** 993–1002.
13. MACLENNAN, D.H., W.J. RICE & N.M. GREEN. 1997. The mechanisms of Ca^{2+} transport by sarco(endo)plasmic reticulum Ca^{2+}-ATPases. J. Biol. Chem. **272:** 28815–28818.
14. RUDOLPH, H.K., A. ANTEBI, et al. 1989. The yeast secretory pathway is perturbed by mutations in PMR1, a member of a Ca^{2+} ATPase family. Cell **58:** 133–145.
15. ANTEBI, A. & G.R. FINK. 1992. The yeast Ca^{2+}-ATPase homologue, PMR1, is required for normal Golgi function and localizes in a novel Golgi-like distribution. Mol. Biol. Cell **3:** 633–654.
16. SORIN, A., G. ROSAS & R. RAO. 1997. PMR1, a Ca^{2+}-ATPase in yeast Golgi, has properties distinct from sarco/endoplasmic reticulum and plasma membrane calcium pumps. J. Biol. Chem. **272:** 9895–9901.
17. BERKELMAN, T., P. GARRET-ENGELE & N.E. HOFFMAN. 1994. The pacL gene of *Synechococcus* sp. strain PCC 7942 encodes a Ca^{2+}-transporting ATPase. J. Bacteriol. **176:** 4430–4436.
18. BAELEN, K.V., J. VANOEVELEN, et al. 2001. The Golgi PMR1 P-type ATPase of *Caenorhabditis elegans*. J. Biol. Chem. **276:** 10683–10691.
19. DURR, G., J. STRAYLE, et al. 1998. The medial-Golgi ion pump Pmr1 supplies the yeast secretory pathway with Ca^{2+} and Mn^{2+} required for glycosylation, sorting, and endoplasmic reticulum–associated protein degradation. Mol. Biol. Cell **9:** 1149–1162.
20. LAPINSKAS, P.J., K.W. CUNNINGHAM, et al. 1995. Mutations in PMR1 suppress oxidative damage in yeast cells lacking superoxide dismutase. Mol. Cell. Biol. **15:** 1382–1388.
21. ISHIKAWA, K., T. NAGASE, et al. 1998. Prediction of the coding sequences of unidentified human genes: X. The complete sequences of 100 new cDNA clones from brain which can code for large proteins *in vitro*. DNA Res. **5:** 169–176.
22. HU, Z., J.M. BONIFAS, et al. 2000. Mutations in *ATP2C1*, encoding a calcium pump, cause Hailey-Hailey disease. Nat. Genet. **24:** 61–65.
23. HOLST, V.A., K.P. FAIR, et al. 2000. Squamous cell carcinoma arising in Hailey-Hailey disease. J. Am. Acad. Dermatol. **43:** 368–371.
24. COCKAYNE, S.E., D.M. RASSL & S.E. THOMAS. 2000. Squamous cell carcinoma arising in Hailey-Hailey disease of the vulva. Br. J. Dermatol. **142:** 540–542.
25. LIU, L.H., G.P. BOIVIN, et al. 2001. Squamous cell tumors in mice heterozygous for the null allele of *Atp2a2*, encoding the sarco(endo)plasmic reticulum Ca^{2+}-ATPase isoform 2 Ca^{2+} pump. J. Biol. Chem. **276:** 26737–26740.
26. KOZEL, P.J., R.A. FRIEDMAN, et al. 1998. Balance and hearing deficits in mice with a null mutation in the gene encoding plasma membrane Ca^{2+}-ATPase isoform 2. J. Biol. Chem. **273:** 18693–18696.
27. FURUTA, H., L. LUO, et al. 1998. Evidence for differential regulation of calcium by outer versus inner hair cells: plasma membrane Ca-ATPase gene expression. Hear. Res. **123:** 10–26.
28. YAMOAH, E.N., E.A. LUMPKIN, et al. 1998. Plasma membrane Ca^{2+}-ATPase extrudes Ca^{2+} from hair cell stereocilia. J. Neurosci. **18:** 610–624.
29. STREET, V.A., J.W. MCKEE-JOHNSON, et al. 1998. Mutations in a plasma membrane Ca^{2+}-ATPase gene cause deafness in deaf waddler mice. Nat. Genet. **19:** 390–394.
30. TAKAHASHI, K. & K. KITAMURA. 1999. A point mutation in a plasma membrane Ca^{2+} ATPase gene causes deafness in Wriggle Mouse Sagami. Biochem. Biophys. Res. Commun. **261:** 773–778.

31. NOBEN-TRAUTH, K., Q.Y. ZHENG, et al. 1997. mdfw: a deafness susceptibility locus that interacts with deaf waddler (dfw). Genomics **44:** 266–272.
32. DAVIS, R.R., J.K. NEWLANDER, et al. 2001. Genetic basis for susceptibility to noise-induced hearing loss in mice. Hear. Res. **155:** 82–90.
33. KOZEL, P.J., R.R. DAVIS, et al. 2002. Deficiency in plasma membrane calcium ATPase isoform 2 increases susceptibility to noise-induced hearing loss in mice. Hear. Res. **164:** 231–239.
34. ODERMATT, A., P.E.M. TASCHNER, et al. 1996. Mutations in the gene encoding SERCA1, the fast-twitch skeletal muscle sarcoplasmic reticulum Ca^{2+}-ATPase, are associated with Brody disease. Nat. Genet. **14:** 191–194.
35. ODERMATT, A., K. BARTON, et al. 2000. The mutation of Pro789 to Leu reduces the activity of the fast-twitch skeletal muscle sarco(endo)plasmic reticulum Ca^{2+} ATPase (SERCA1) and is associated with Brody disease. Hum. Genet. **106:** 482–491.
36. MACLENNAN, D.H. 2000. Ca^{2+} signaling and muscle disease. Eur. J. Biochem. **267:** 5291–5297.
37. PERIASAMY, M., T.D. REED, et al. 1999. Impaired cardiac performance in heterozygous mice with a null mutation in the sarco(endo)plasmic reticulum Ca^{2+}-ATPase isoform 2 (SERCA2) gene. J. Biol. Chem. **274:** 2556–2562.
38. LOUKIANOV, E., Y. JI, et al. 1998. The sarco(endo)plasmic reticulum Ca^{2+} ATPase isoforms and their role in muscle physiology and pathology. Ann. N.Y. Acad. Sci. **853:** 251–259.
39. JI, Y., M.J. LALLI, et al. 2000. Disruption of a single copy of the Serca2 gene results in altered Ca^{2+} homeostasis and cardiomyocyte function. J. Biol. Chem. **275:** 38073–38080.
40. VER HEYEN, M., S. HEYMANS, et al. 2001. Replacement of the muscle-specific sarcoplasmic reticulum Ca^{2+}–ATPase isoform SERCA2a by the nonmuscle SERCA2b homologue causes mild concentric hypertrophy and impairs contraction-relaxation of the heart. Circ. Res. **89:** 838–846.
41. ZHAO, X-S., D.M. SHIN, et al. 2001. Plasticity and adaptation of Ca^{2+} signaling and Ca^{2+}-dependent exocytosis in Serca2$^{+/-}$ mice. EMBO J. **20:** 2680–2689.
42. SAKUNTABHAI, A., V. RUIZ-PEREZ, et al. 1999. Mutations in *ATP2A2*, encoding a Ca^{2+} pump, cause Darier disease. Nat. Genet. **21:** 271–277.
43. SAKUNTABHAI, A., S. BURGE, et al. 1999. Spectrum of novel *ATP2A2* mutations in patients with Darier's disease. Hum. Mol. Genet. **8:** 1611–1619.
44. MUNRO, C.S. 1992. The phenotype of Darier's disease: penetrance and expressivity in adults and children. Br. J. Dermatol. **127:** 126–130.
45. BURGE, S. 1994. Darier's disease—the clinical features and pathogenesis. Clin. Exp. Dermatol. **19:** 193–205.
46. TAVADIA, S., R.C. TAIT, et al. 2001. Platelet and cardiac function in Darier's disease. Clin. Exp. Dermatol. **26:** 696–699.
47. WUYTACK, F., L. DODE, et al. 1995. The SERCA3-type of organellar Ca pumps. Biosci. Rep. **15:** 299–306.
48. LIU, L.H., R.J. PAUL, et al. 1997. Defective endothelium-dependent relaxation of vascular smooth muscle and endothelial cell Ca^{2+} signaling in mice lacking sarco(endo)plasmic reticulum Ca^{2+}-ATPase isoform 3. J. Biol. Chem. **272:** 30538–30545.
49. KAO, J., C.N. FORTNER, et al. 1999. Ablation of the SERCA3 gene alters epithelium-dependent relaxation in mouse tracheal smooth muscle. Am. J. Physiol. **277:** 264–270.
50. ARREDOUANI, A., Y. GUIOT, et al. 2002. Serca3 ablation does not impair insulin secretion, but suggests distinct roles of different sarco-endoplasmic reticulum Ca^{2+} pumps in β-cells Ca^{2+} homeostasis. Diabetes. In press.
51. VARADI, A., L. LEBEL, et al. 1999. Sequence variants of the sarco(endo)plasmic reticulum Ca^{2+}-transport ATPase 3 gene (SERCA3) in Caucasian type II diabetic patients (UK prospective diabetes study 48). Diabetologia **42:** 1240–1243.
52. CUNNINGHAM, K.W. & G.R. FINK. 1994. Calcineurin-dependent growth control in *Saccharomyces cerevisiae* mutants lacking *PMC1*, a homolog of plasma membrane Ca^{2+} ATPases. J. Cell Biol. **124:** 351–363.

Characterization of PISP, a Novel Single-PDZ Protein That Binds to All Plasma Membrane Ca^{2+}-ATPase b-Splice Variants

GEOFFREY M. GOELLNER,[a] STEVEN J. DeMARCO,[b] AND EMANUEL E. STREHLER

Department of Biochemistry and Molecular Biology, Mayo Clinic, Rochester, Minnesota 55905, USA

ABSTRACT: Plasma membrane Ca^{2+} ATPases (PMCAs) maintain intracellular Ca^{2+} homeostasis and participate in the local regulation of Ca^{2+} signaling. Spatially separate demands for Ca^{2+} regulation require proper membrane targeting of PMCAs, but the mechanism of PMCA targeting is unknown. Using the PMCA2b carboxyl-terminal tail as yeast two-hybrid bait, we isolated a novel PDZ domain–containing protein from a human brain cDNA library. This protein, named PISP for PMCA-interacting single-PDZ protein, consists of 140 amino acids and contains little else besides a single PDZ domain. Pulldown experiments showed that PISP interacts with all PMCA b-splice forms. PISP was found to be ubiquitously expressed and, in MDCK cells, was present in a punctate pattern throughout the cytosol and at the basolateral membrane. When added to microsomal membranes expressing PMCA4b, PISP was unable to stimulate the PMCA-dependent ATPase activity. Our data suggest that PISP is a transiently interacting partner of the PMCA b-splice forms that may play a role in their sorting to or from the plasma membrane.

KEYWORDS: calcium pump; PDZ domain; plasma membrane Ca^{2+}-ATPase; PMCA

INTRODUCTION

The plasma membrane Ca^{2+} ATPases (PMCAs) are ubiquitous eukaryotic ion transporters responsible for the expulsion of Ca^{2+} from the cytosol.[1] Four genes encode human PMCA isoforms 1–4, and isoform complexity is further augmented by alternative RNA splicing.[2] PMCA isoforms and splice variants differ in their kinetic and regulatory properties, for example, in their sensitivity to calmodulin, which is the primary activator of the pumps.[3] The expression pattern of different

---

Address for correspondence: Emanuel E. Strehler, Department of Biochemistry and Molecular Biology, Mayo Clinic, 200 First Street S.W., Rochester, MN 55905. Voice: 507-284-9372; fax: 507-284-2384.
strehler.emanuel@mayo.edu
[a]Present address: Geoffrey M. Goellner, Department of Biochemistry, University of Utah School of Medicine, Salt Lake City, UT 84132.
[b]Present address: Steven J. DeMarco, Brain Research Institute, University of Zurich, CH-8057 Zurich, Switzerland.

Ann. N.Y. Acad. Sci. 986: 461–471 (2003). © 2003 New York Academy of Sciences.

PMCA isoforms and splice variants is tissue- and cell type–specific and developmentally regulated.[2] Even within a single cell, different PMCA isoforms may be expressed in a spatially distinct pattern.[4] However, the specific targeting mechanisms for different PMCA isoforms are unknown at present.

PMCA b-splice forms share the C-terminal amino acid sequence, ETSL/V, which fits the consensus for peptide ligands of type-I PDZ (PSD95/Dlg/ZO-1) domains.[5] PDZ domains are protein–protein interaction modules present in many proteins with diverse functions ranging from clustering receptors at the cell membrane to protein trafficking and signal transduction.[6] We[7–9] and others[10] have recently shown that PMCA b-splice forms bind to several PDZ proteins, including members of the membrane-associated guanylate kinase–like protein (MAGUK) family, as well as to NHERF2 and NO synthase-I. The significance of PMCA-PDZ protein interactions is not yet known, with the exception of the PMCA4b–NO synthase interaction, which appears to play a direct role in downregulating NO synthase activity.[10]

Here, we identify a novel PDZ protein from a yeast two-hybrid screen using the C-terminal tail of PMCA2b as bait. This protein, which we named PISP (for PMCA-interacting single-PDZ protein), binds to all PMCA b-splice forms via its single type-I PDZ domain and is ubiquitously expressed at low to moderate levels. When added to microsomal membranes expressing PMCA4b, PISP does not stimulate the PMCA-dependent ATPase activity, nor does it interfere with the calmodulin-dependent stimulation of the ATPase. Because of its ubiquitous coexpression with the PMCAs, PISP may be a common regulator of sorting of PMCA b-splice forms to or from the plasma membrane.

MATERIALS AND METHODS

Plasmid Constructs

The yeast two-hybrid bait construct pDB-CT2b has been described.[8] Expression plasmids for hPMCA1b, 2b, and 4b in pMM$_2$ have also been described.[8,11] An expression vector for rPMCA3b was reported by Enyedi et al.,[12] but contained a read-through of several amino acids at its C-terminus. Site-directed mutagenesis with the GeneEditor™ kit (Promega) was used to introduce a stop codon at the correct position, thus creating pMM$_2$-rPMCA3b. The cDNA for full-length PISP was PCR-amplified and cloned as BamHI-EcoRI fragment into pGEX-2TK (Amersham Biosciences) and pCMV-TAG2 (Stratagene).

Yeast Two-Hybrid Screening

The screening of a human brain cDNA library (BD Biosciences Clontech) using pDB-CT2b and the isolation of plasmids from positive yeast clones have been described.[8]

Protein Expression and GST Pulldowns

GST fusion proteins were expressed in E. coli BL21(DE3) and purified on glutathione-Sepharose (Sigma) as described.[8,13] For pulldown assays, fusion proteins were adjusted to ~0.5 mg/mL. COS-1 cells grown on 6-well plates were transfected with 2 μg DNA using Lipofectamine™ according to the manufacturer's instructions (LifeTechnologies). After ~48 h, the cells were rinsed and lysed, and the lysates collected as described.[8] Equal amounts of lysate from cells expressing PMCA1b, 2b, 3b, or 4b were incubated with Sepharose-bound GST or GST-PISP. Beads and lysate were rocked for 90 min at 4°C. Bound proteins were eluted from the washed beads, separated on 10% SDS-polyacrylamide gels, and transferred to nitrocellulose following standard Western blotting procedures.[13] The PMCAs were detected using the pan-PMCA antibody 5F10 diluted 1:2000.

Production of PISP-Specific Antibodies

Rabbit polyclonal antibodies against GST-PISP were raised at Cocalico (Reamstown, PA). Sera were heat-inactivated and anti-GST antibodies removed by GST-Sepharose chromatography (Amersham-Pharmacia). Anti-PISP antibodies (Ab737) in the GST-Sepharose flow-through were purifed using a GST-PISP fusion protein affinity column following established procedures.[13]

Northern Blot and Genomic Structure of PISP

A human multiple-tissue Northern blot (Clontech) was hybridized as described[14] using the ^{32}P-labeled PISP cDNA insert as probe. The structure of the PISP gene was determined by database analysis of the NCBI (http://www.ncbi.nlm.nih.gov) and UCSC (http://genome.ucsc.edu) human genome resources.

Tissue Expression and Cellular Localization of PISP

A multiple-tissue Western blot (Chemicon) was probed with affinity-purified anti-PISP antibody Ab737 (1.3 μg/mL). Madine Darby Canine Kidney (MDCK) cells grown on polylysine-covered glass coverslips were transfected with pCMV-PISP DNA using Lipofectamine 2000™ (Life Technologies). At 24–48 h after transfection, the cells were fixed for 15 min at room temperature in 3.5% paraformaldehyde in Dulbecco's PBS, permeabilized for 3 min with 0.2% Triton-X in PBS, and then washed, blocked, and incubated in primary (Ab737, 1 μg/mL) and secondary (anti-rabbit Alexa-594 1:800, Molecular Probes) antibodies as described.[8] Confocal micrographs were taken on a Zeiss LSM510 microscope.

Effect of PISP on PMCA Activity

ATP hydrolysis due to Ca^{2+} pump activity was measured by continuously recording the absorbance at 360 nm in a coupled enzyme assay as described,[15] using reconstituted human erythrocyte ghost membranes or microsomal preparations of baculovirus-infected Sf9 cells expressing hPMCA4b.[16] Baseline ATPase activity was determined, and the PMCA-dependent activity was measured after adding Ca^{2+} to 0.5 μM in the absence and presence of 235 nM calmodulin. The effects of GST alone (control) and of GST-PISP on PMCA activity were determined by adding the purified proteins to the assay either before or after calmodulin. The GST fusion proteins were added at 1:1 up to 10:1 stoichiometric ratios to calmodulin. Experiments were repeated at least twice, and data are expressed in arbitrary units as mean ± standard deviation.

RESULTS AND DISCUSSION

Isolation of a Novel Single-PDZ Protein by Yeast Two-Hybrid Screening

Screening a human brain cDNA library with a PMCA2b bait yielded five clones that specified PDZ proteins.[8,9] Among these, β2-syntrophin, SAP97/hDlg, and NHERF-2 (2 clones) corresponded to known PDZ proteins, whereas 1 encoded a novel protein that we named PISP for PMCA-interacting single-PDZ protein (FIG. 1A). The PISP cDNA sequence has been deposited from several cDNA profiling studies (accession nos. AF151061, AK024746, BC012996). Its open reading frame specifies a protein of 140 residues with a calculated mass of 16,131 Da. The protein consists of a single PDZ domain (residues 45–126) with only short N- and C-terminal extensions that show no similarity to other known domains. However, PISP has been highly conserved at least in mammals: the mouse and human homologues are 97% identical (FIG. 1A), suggesting a conserved function. The PDZ domain itself shows highest similarity (40–45% identity) to the single PDZ domain of *C. elegans* Lin-7[17] and its mammalian homologues (named Velis or Mals[18,19]), as well as to the PDZ domains of several members of the MAGUK family.[6] The PDZ domain of PISP corresponds to class I,[5] as expected for a PDZ domain binding the C-terminal ETSL/V sequence of the PMCAs.

Gene Structure of PISP

Using genome database mining tools, we determined the exon-intron structure of the human PISP gene (FIG. 1B). The gene on chromosome Xq12 contains seven exons spread over 3.3 kb of DNA. Strikingly, the 5' end (putative promoter region) of the PISP gene lies very close to and likely overlaps with the beginning of the gene for the kinesin family member, KIF4A. This large gene (over 130 kb) is transcribed in the opposite direction, with its first exon separated by less than 200 bp from exon 1 of the PISP gene (FIG. 1B). Such closely spaced head-to-head genes transcribed in opposite directions may be more common in eukaryotic genomes than previously thought and, in some cases, may be coordinately regulated and functionally related.[20]

PISP Is Ubiquitously Expressed and Localized in a Punctate Cytosolic Pattern and Apposed to the Membrane

A multiple-tissue Northern blot probed with the PISP cDNA revealed an approximately 1.4 kb mRNA in all tissues (FIG. 2A). The 1.4-kb transcript is expressed at relatively low levels, with slightly higher amounts in kidney and skeletal muscle. Using an affinity-purified antibody against PISP, we readily detected an approximately 17-kDa protein in all tissues on a multiple-tissue Western blot (FIG. 2B). This protein was most abundant in the kidney and also elevated in the liver. In addition, the anti-PISP antibody detected a band (sometimes a doublet) at about 22 kDa in some tissues (FIG. 2B). This band may correspond to a different, but closely related, PDZ protein or to an alternatively spliced and/or posttranslationally modified variant of PISP. Other mammalian proteins containing little else besides a single PDZ domain have been described, notably the Velis or Mals,[18,19] the Tax-interacting protein TIP-1,[21] and the activin receptor–interacting protein ARIP2.[22] However, none of these proteins is particularly closely related to PISP (except in the overall struc-

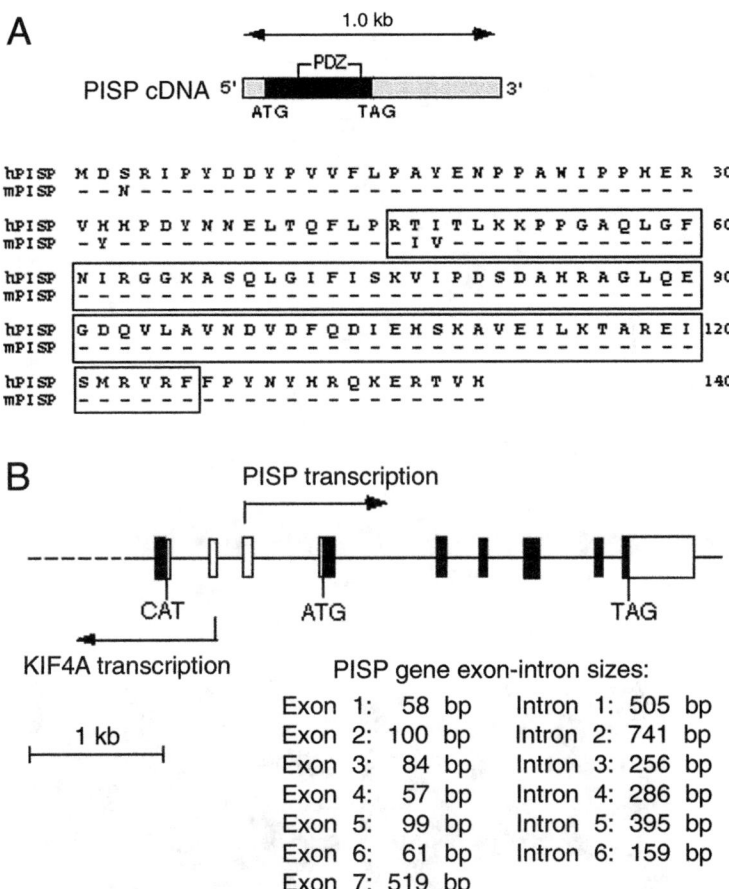

FIGURE 1. Identification of PISP. **(A)** *Top*: Structure of the PISP cDNA. 5′ and 3′ untranslated regions are shown as *gray boxes*; the coding region is in *black*. The translational start (ATG) and stop (TAG) codons are indicated, and the location of the single PDZ domain is shown between *brackets*. *Bottom*: Amino acid sequence alignment of human and mouse PISP (hPISP and mPISP). Sequence identity is indicated by hyphens in the mouse sequence; residues are numbered on the right. The PDZ domain is *boxed*. **(B)** Structure of the human PISP gene. The exon-intron structure of the human PISP gene and a part of the adjacent KIF4A gene on human chromosome Xq12 are shown to scale (*1-kb bar* on the bottom left). Protein-coding exons are shown as *black boxes*, exons specifying 5′ and 3′ untranslated sequences as *open boxes*, and introns are drawn as *lines*. The direction of transcription is indicated by *arrows*. Note that the PISP and KIF4A genes are transcribed from opposite strands. The start (ATG) and stop (TAG) codons of the PISP gene are indicated, as is the start codon (reverse complement of ATG) for the KIF4A gene. Exon and intron sizes of the PISP gene are listed on the bottom right.

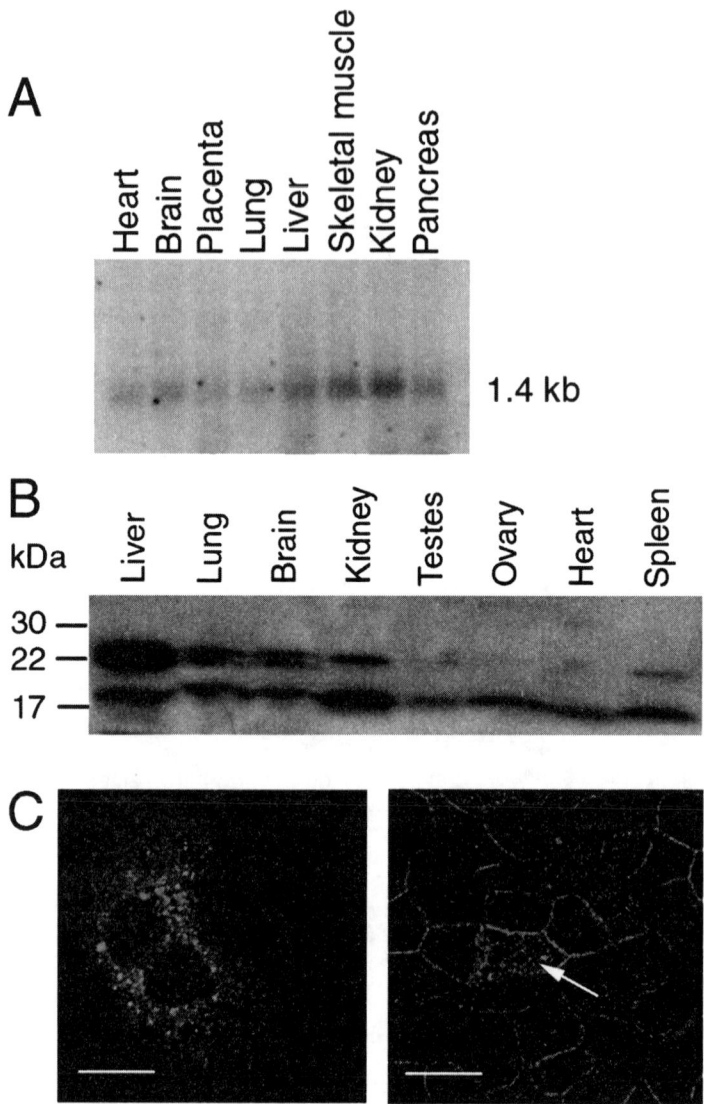

FIGURE 2. Tissue expression and cellular localization of PISP. **(A)** Multiple human tissue Northern blot hybridized with a full-length PISP cDNA probe. A band of about 1.4 kb is detected in all tissues (labeled on top of each lane). **(B)** Multiple human tissue Western blot probed with an affinity-purified polyclonal antibody against PISP. The position of molecular mass standards is given in kDa on the left; the tissues are identified on top of each lane. Note the presence of an approximately 17-kDa band corresponding to the predicted size of PISP in all tissues. Also note the presence of a second band (resolved as a doublet in lung and brain) in several tissues, most prominently in the liver. This is likely due to a cross-reacting protein related to PISP. **(C)** Full-length PISP was transiently expressed in MDCK

ture of the PDZ domain), and only Veli-3 and ARIP2 are close in M_r to the 22-kDa band detected by the anti-PISP antibody.

We determined the localization of PISP in MDCK cells by immunostaining. Because of the relatively low levels of endogenous PISP, we transfected the cells with recombinant PISP. In subconfluent, nonpolarized cells, PISP staining was predominantly cytosolic in a punctate or reticular pattern (FIG. 2C, left). A similar pattern for PISP staining was seen in polarized MDCK cells (FIG. 2C, right), although a significant amount of staining was also observed close to the lateral membrane known to be enriched in endogenous PMCAs.[8] Our data suggest that this single PDZ protein does not act as a "permanent" scaffold linking membrane proteins to the submembranous cytoskeleton or to each other. Rather, it may interact with its targets in transit through a cytosolic compartment or undergo transient interactions with the (endogenous) PMCAs at the basolateral membrane.

PISP Interacts with All PMCA b-Splice Forms

To test the specificity of the interaction of PISP with different PMCAs, we performed "pulldown" experiments. Expression vectors for PMCA1b, 2b, 3b, and 4b were transfected into COS-1 cells and the resulting lysate from these cells was presented to GST-PISP immobilized on glutathione-Sepharose. As shown in FIGURE 3, all PMCAs were able to bind to GST-PISP, whereas no interaction was seen with GST alone. As expected for PMCA-PDZ protein interactions, no binding was seen when GST-PISP was exposed to an a-splice form of the PMCA (PMCA4a) or to PMCA4b lacking its 6 C-terminal residues (data not shown). Thus, PISP appears to bind to all PMCA b-splice forms via its single PDZ domain.

Effect of PISP on PMCA Activity in Vitro

To test if PISP directly activates the PMCA, we measured the ATPase activity of the endogenous pump (predominantly PMCA4b) in reconstituted human erythrocyte membranes[15] or of recombinant PMCA4b expressed in Sf9 cell membranes.[16] FIGURE 4 shows that the addition of GST-PISP did not significantly stimulate the PMCA above its basal rate in the presence of 0.5 μM Ca^{2+}. By contrast, an equimolar amount of calmodulin (235 nM) fully activated the pump under these conditions, resulting in an almost 3-fold increase in PMCA activity.[15] Preincubation of microsomes with GST-PISP did not influence the calmodulin stimulation of the pump nor did this treatment lead to a synergistic stimulation of the pump beyond that obtained with calmodulin alone. The same result was seen when calmodulin was added to the microsomes prior to GST-PISP (FIG. 4). Essentially identical results were obtained using the negative control protein, GST (FIG. 4). Even when GST-PISP was added to the membranes in up to 10-fold excess (2.3 μM) over calmodulin, there was no

---

cells and localized by immunofluorescence using an affinity-purified anti-PISP antibody. In nonpolarized cells (*left*), PISP is present in a punctate pattern in a perinuclear intracellular compartment. In polarized cells (*right*), endogenous PISP and recombinant PISP overexpressed in a transfected cell (*arrow*) are present in a punctate pattern in the cytosol as well as apposed to the lateral cell membrane. Bars: 10 μm (*left*); 20 μm (*right*).

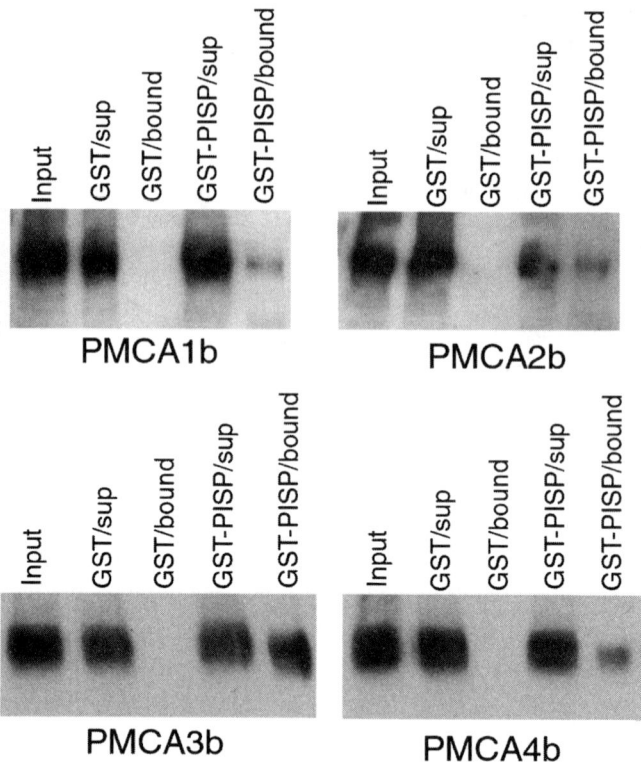

FIGURE 3. PISP binds to all PMCA b-splice forms. Aliquots of lysates from COS-1 cells transfected with full-length PMCA constructs (indicated below each panel) were incubated with glutathione-Sepharose containing equal amounts of GST alone or GST-PISP. After centrifugation, the supernatants were kept (sup), the pellets were washed, and the bound proteins eluted (bound). Aliquots of the starting lysates (25% of the total; input) and supernatants, as well as all of the bound protein, were separated by SDS-PAGE and transferred to nitrocellulose membrane. The membranes were probed with a monoclonal antibody against all PMCAs. Note that GST-PISP is able to pull down each of the four PMCAs, whereas GST does not pull down any of the pumps.

change in the extent of calmodulin-dependent pump stimulation (not shown). The affinity of the PMCA-PISP interaction is not known; however, the K_D of the interaction between a C-terminal peptide of PMCA4b and PDZ1+2 of hDlg/SAP97 is in the low nM range.[7] This is similar to the affinity of the PMCAs for calmodulin.[23] A GST-hDlg/SAP97 fusion protein containing all 3 PDZ domains also had no effect on the *in vitro* ATPase activity of the PMCAs and, like PISP, did not interfere with calmodulin stimulation of the pump (GMG and EES, unpublished observation). From these data, we conclude that PISP, despite being similar in size to calmodulin

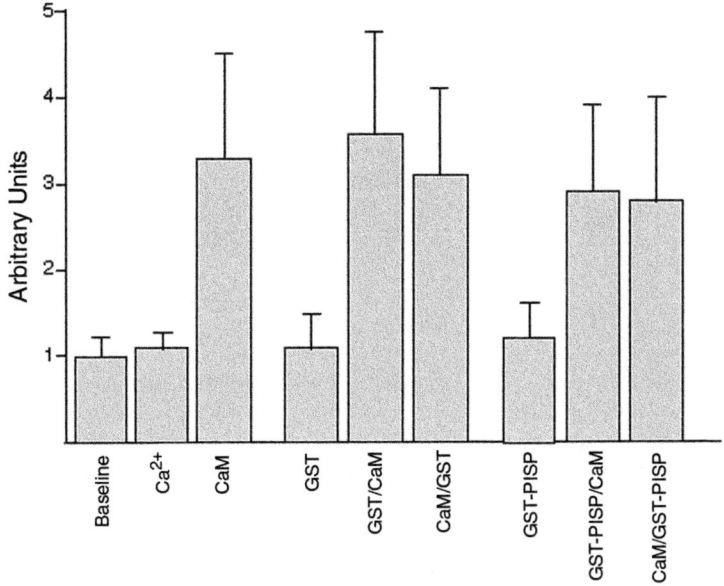

FIGURE 4. PISP does not stimulate the ATPase activity of the PMCA in microsomal membranes. Rates of PMCA-dependent ATP hydrolysis in human erythrocyte microsomal membranes were determined as described in MATERIALS AND METHODS and are shown in arbitrary units. Baseline, basal rate prior to adding Ca^{2+} and exogenous proteins; Ca^{2+}, rate after the addition of 0.5 µM free Ca^{2+}; CaM, addition of 235 nM calmodulin after adding Ca^{2+} causes a 3-fold increase in PMCA activity. GST, GST/CaM, and CaM/GST: Negative controls showing that GST alone (200 nM) has no effect and that adding GST before or after calmodulin does not inhibit PMCA stimulation. GST-PISP: GST-PISP (200 nM) does not stimulate PMCA activity on its own. GST-PISP/CaM: Preincubation of the membranes with GST-PISP (200 nM) does not prevent or enhance subsequent stimulation by calmodulin. CaM/GST-PISP: Adding GST-PISP after calmodulin is without effect on PMCA activity. All experiments were repeated at least twice and the mean values are shown. *Error bars* indicate the standard deviation.

and binding to the C-terminal tail of the PMCAs, does not affect PMCA activity under the conditions of our assay.

CONCLUSIONS

PISP is a novel, ubiquitously expressed single-PDZ domain protein that interacts with all PMCA b-splice forms. Because PMCAs (specifically, PMCA1b and PMCA4b) are also ubiquitously expressed, PISP may be a relevant interacting partner of the pumps in many cells. In contrast to multi-PDZ domain proteins such as the MAGUKs, PISP is unlikely to play a role as clustering and cytoskeletal anchoring protein for transmembrane transporters. Rather, its small size and cytoplasmic localization in a punctate and reticular pattern suggest that it may act as a "chaperone" for the PMCA b-splice forms in transit from the endoplasmic reticulum to the plasma membrane. Such a role would be reminiscent of the mammalian Lin-7 homologues (Velis) that are thought to be important for trafficking and sorting of membrane proteins in polarized cells.[19,24] Alternatively, a role for PISP in regulated endocytosis or signaling to or from the PMCAs could be envisaged. PISP could influence PMCA turnover by recruiting endocytic regulatory proteins to the pump, and the PISP-PMCA interaction itself could be regulated, for example, by phosphorylation of the PMCA C-terminal domain. Future work will hopefully clarify the role of this intriguing novel PDZ protein in the trafficking, endocytosis, or signaling of the PMCAs.

ACKNOWLEDGMENTS

We thank Ariel Caride (Mayo Clinic) for sharing materials and assisting with the ATPase activity measurement, and Krista Lehman, Anne Vrabel, and Michael Chicka for technical assistance. This work was supported by NIH Grant No. GM-58710. Geoffrey M. Goellner and Steven J. DeMarco contributed equally to this work.

REFERENCES

1. CARAFOLI, E. 1994. Biogenesis: plasma membrane calcium ATPase—15 years of work on the purified enzyme. FASEB J. **8:** 993–1002.
2. STREHLER, E.E. & D.A. ZACHARIAS. 2001. Role of alternative splicing in generating isoform diversity among plasma membrane calcium pumps. Physiol. Rev. **81:** 21–50.
3. PENNISTON, J.T. & A. ENYEDI. 1998. Modulation of the plasma membrane Ca^{2+} pump. J. Membr. Biol. **165:** 101–109.
4. DUMONT, R.A. et al. 2001. Plasma membrane Ca^{2+} ATPase isoform 2a is the PMCA of hair bundles. J. Neurosci. **21:** 5066–5078.
5. SONGYANG, Z. et al. 1997. Recognition of unique carboxyl-terminal motifs by distinct PDZ domains. Science **275:** 73–77.
6. HUNG, A.Y. & M. SHENG. 2002. PDZ domains: structural modules for protein complex assembly. J. Biol. Chem. **277:** 5699–5702.
7. KIM, E. et al. 1998. Plasma membrane Ca^{2+} ATPase isoform 4b binds to membrane-associated guanylate kinase (MAGUK) proteins via their PDZ (PSD-95/Dlg/ZO-1) domains. J. Biol. Chem. **273:** 1591–1595.
8. DEMARCO, S.J. & E.E. STREHLER. 2001. Plasma membrane Ca^{2+}-ATPase isoforms 2b and 4b interact promiscuously and selectively with members of the membrane-associated guanylate kinase family of PDZ (PSD-95/Dlg/ZO-1) domain–containing proteins. J. Biol. Chem. **276:** 21594–21600.
9. DEMARCO, S.J., M.C. CHICKA & E.E. STREHLER. 2002. Plasma membrane Ca^{2+} ATPase isoform 2b interacts preferentially with Na^+/H^+ exchanger regulatory factor 2 in apical plasma membranes. J. Biol. Chem. **277:** 10506–10511.
10. SCHUH, K. et al. 2001. The plasma membrane calmodulin-dependent calcium pump: a major regulator of nitric oxide synthase I. J. Cell Biol. **155:** 201–205.
11. ADAMO, H.P. et al. 1992. Overexpression of the erythrocyte plasma membrane Ca^{2+} pump in COS-1 cells. Biochem. J. **285:** 791–797.
12. ENYEDI, A. et al. 1997. Protein kinase C phosphorylates the "a" forms of plasma membrane Ca^{2+} pump isoforms 2 and 3 and prevents binding of calmodulin. J. Biol. Chem. **272:** 27525–27528.
13. AUSUBEL, F.M. et al., Eds. 1998. Current Protocols in Molecular Biology. Wiley. New York.
14. TOUTENHOOFD, S.L. et al. 1998. Characterization of the human CALM2 calmodulin gene and comparison of the transcriptional activity of CALM1, CALM2, and CALM3. Cell Calcium **23:** 323–338.

15. CARIDE, A.J et al. 1999. The rate of activation by calmodulin of isoform 4 of the plasma membrane Ca^{2+} pump is slow and is changed by alternative splicing. J. Biol. Chem. **274:** 35227–35232.
16. CARIDE, A.J. et al. 2001. The plasma membrane calcium pump displays memory of past calcium spikes: differences between isoforms 2b and 4b. J. Biol. Chem. **276:** 39797–39804.
17. SIMSKE, J.S. et al. 1996. LET-23 receptor localization by the cell junction protein LIN-7 during C. elegans vulval induction. Cell **85:** 195–204.
18. BUTZ, S., M. OKAMOTO & T.C. SÜDHOF. 1998. A tripartite protein complex with the potential to couple synaptic vesicle exocytosis to cell adhesion in brain. Cell **94:** 773–782.
19. JO, K. et al. 1999. Characterization of MALS/Velis-1, -2, and -3: a family of mammalian LIN-7 homologs enriched at brain synapses in association with the postsynaptic density-95/NMDA receptor postsynaptic complex. J. Neurosci. **19:** 4189–4199.
20. ADACHI, N. & M.R. LIEBER. 2002. Bidirectional gene organization: a common architectural feature of the human genome. Cell **109:** 807–809.
21. REYNAUD, C., S. FABRE & P. JALINOT. 2000. The PDZ protein TIP-1 interacts with the Rho effector rhotekin and is involved in Rho signaling to the serum response element. J. Biol. Chem. **275:** 33962–33968.
22. MATSUZAKI, T. et al. 2002. Regulation of endocytosis of activin type II receptors by a novel PDZ protein through Ral/Ral-binding protein 1–dependent pathway. J. Biol. Chem. **277:** 19008–19018.
23. ELWESS, N.L. et al. 1997. Plasma membrane Ca^{2+} pump isoforms 2a and 2b are unusually responsive to calmodulin and Ca^{2+}. J. Biol. Chem. **272:** 17981–17986.
24. BORG, J-P. et al. 1998. Identification of an evolutionarily conserved heterotrimeric protein complex involved in protein targeting. J. Biol. Chem. **273:** 31633–31636.

The Regulation of SERCA-Type Pumps by Phospholamban and Sarcolipin

DAVID H. MacLENNAN, MICHIO ASAHI, AND A. RUSSELL TUPLING

The Banting and Best Department of Medical Research, University of Toronto, Charles H. Best Institute, Toronto, Ontario, Canada M5G 1L6

ABSTRACT: Both sarcolipin (SLN) and phospholamban (PLN) lower the apparent affinity of either SERCA1a or SERCA2a for Ca^{2+}. Since SLN and PLN are coexpressed in the heart, interactions among these three proteins were investigated. When SERCA1a or SERCA2a were coexpressed in HEK-293 cells with both SLN and PLN, superinhibition resulted. The ability of SLN to elevate the content of PLN monomers accounts, at least in part, for the superinhibitory effects of SLN in the presence of PLN. To evaluate the role of SLN in skeletal muscle, SLN cDNA was injected directly into rat soleus muscle and force characteristics were analyzed. Overexpression of SLN resulted in significant reductions in both twitch and tetanic peak force amplitude and maximal rates of contraction and relaxation and increased fatigability with repeated electrical stimulation. Ca^{2+} uptake in muscle homogenates was impaired, suggesting that overexpression of SLN may reduce the sarcoplasmic reticulum Ca^{2+} store. SLN and PLN appear to bind to the same regulatory site in SERCA. However, in a ternary complex, PLN occupies the regulatory site and SLN binds to the exposed side of PLN and to SERCA.

KEYWORDS: Ca^{2+}-ATPase; sarcolipin; phospholamban; cardiomyopathy; regulatory molecules

INTRODUCTION

A major objective in our laboratory has been to define and characterize the sites of regulatory interaction between the cardiac sarco(endo)plasmic reticulum Ca^{2+}-ATPase (SERCA2a) and phospholamban (PLN). In its unphosphorylated state, PLN inhibits SERCA2a activity by lowering its apparent affinity for Ca^{2+}; when phosphorylated, for example, through adrenergic stimulation of the heart, PLN inhibitory function is reversed and Ca^{2+} pump activity is activated.[1] Thus, PLN is a key regulator of cardiac contractility through its effects on the function of SERCA2a. PLN-ablated mice[2,3] show high indices of cardiac contractility in the absence of catecholamine stimulation. Conversely, our studies of cardiac-specific overexpression of dominant, superinhibitory forms of PLN show that PLN superinhibition of cardiac

---

Address for correspondence: David H. MacLennan, The Banting and Best Department of Medical Research, University of Toronto, Charles H. Best Institute, 112 College Street, Toronto, ON, Canada M5G 1L6. Voice: 416-487-9729; fax: 416-487-8528.
david.maclennan@utoronto.ca

Ca^{2+} transport function leads to diminished contractility and even to cardiomyopathy.[4–6]

We and others have also investigated the interaction between SERCA and sarcolipin (SLN), a PLN homologue.[7–9] We showed that SLN inhibits SERCA1a just as PLN inhibits SERCA2a. Although SLN was a less effective inhibitor of SERCA1a than PLN, both proteins decreased the apparent affinity of SERCA for Ca^{2+}. Since this alteration in Ca^{2+} affinity can be measured in a concentration range in which Ca^{2+} signaling occurs, the phenomenon is likely to be relevant physiologically. To date, however, little is known about the physiological role of SLN. In this paper, we describe some of our recent investigations of the regulatory properties of SLN that lead us to believe that its regulatory role may be extensive not only in skeletal muscle where it was first discovered,[10] but also in heart.[7,11]

POTENTIAL ROLE FOR SLN IN CARDIAC CONTRACTILITY

PLN is expressed almost exclusively in cardiac, smooth, and slow-twitch skeletal muscles, while SLN is highly expressed in human fast-twitch skeletal muscles and in reduced amounts in cardiac and slow-twitch skeletal muscles.[7] In rats, however, SLN appears to be more highly expressed in cardiac muscle and to be expressed to

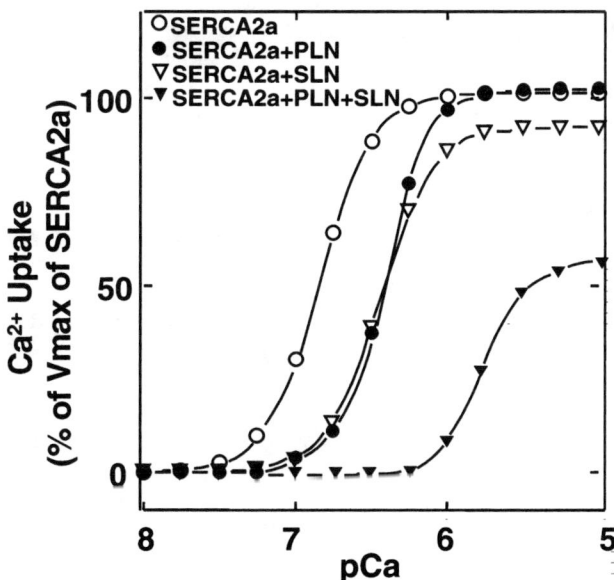

FIGURE 1. Effect of NF-SLN on the inhibition of SERCA2a by PLN. SERCA2a was expressed in HEK-293 cells alone or in the presence of NF-SLN, PLN, or NF-SLN and PLN. The Ca^{2+} dependence of Ca^{2+} transport activity in microsomal fractions was measured for SERCA2a alone (○) or in the presence of PLN (●), NF-SLN (▽), or both PLN and NF-SLN (▼). (Reproduced from Asahi et al.,[12] by permission from the American Society for Biochemistry & Molecular Biology.)

a more modest extent in fast-twitch skeletal muscles.[11] Since PLN and SLN are coexpressed in the heart, it was of interest to determine whether they interact as regulators of SERCA2a activity.

In early experiments, we noted that triple expression of SERCA1a with PLN and NF-SLN (SLN tagged at its N-terminus with the sequence, MDYKDDDDK, which introduces the FLAG with no effect on SLN function) led to superinhibition of SERCA1a activity by shifting Ca^{2+} affinity by nearly 1 pCa unit.[8] When reinvestigating this phenomenon[12] with both SERCA1a and SERCA2a, we found that NF-SLN was as effective in inhibiting SERCA2a as PLN, decreasing both V_{max} and apparent K_{Ca} (FIG. 1).

On the basis of the results of alanine-scanning mutagenesis of PLN,[13] we proposed that PLN can become superinhibitory by at least two mechanisms: the first, through an increase in the concentration of the PLN monomer, the inhibitory species, which would increase inhibition through mass action; the second, from an increased affinity of the mutant PLN for SERCA. Many mutations in the transmembrane domain of PLN become superinhibitory because they alter and thereby destabilize the "leucine zipper" that holds the pentameric form of PLN together, thereby increasing monomer content.[14] Other PLN mutations, exemplified by N27A,[5,15] are likely to become superinhibitory only because of their tighter association with SERCA in an inhibitory complex.

In our efforts to understand the mechanism by which a combination of PLN and NF-SLN might become superinhibitory, we found that the two proteins interact with each other in the absence of SERCA since antibodies against either protein coimmunoprecipitated at least 50% of the total amount of the other expressed protein. We also found that NF-SLN, PLN, and SERCA interact as a ternary complex that could be precipitated with antibodies against either regulatory protein. When we tried to determine the order of binding of the two regulatory molecules, we found that NF-SLN increased the amount of PLN that was precipitated when compared to controls where NF-SLN was not expressed with PLN and SERCA; however, PLN did not increase the amount of NF-SLN that was precipitated when compared to controls where PLN was not expressed with NF-SLN and SERCA.[12] Thus, the superinhibition seemed to be related to the inhibitory binding of PLN rather than NF-SLN.

We then found that the percentage of PLN in pentameric form is decreased when the amount of NF-SLN that was coexpressed with PLN was increased (FIG. 2). This is likely to occur because NF-SLN can interact directly with PLN monomers, preventing them from polymerizing into pentamers. Thus, the most likely explanation for the superinhibition of SERCA by a combination of PLN and NF-SLN is the mass action effect of the increased concentration of PLN monomers. However, the fact that V_{max} is also decreased suggests not only that superinhibition is due to an increase in the number of PLN monomers available to interact with the ATPase, but that the mechanism or the stoichiometry of interaction is changed. NF-SLN has little (if any) ability to form homo-oligomers,[9] making it unlikely that NF-SLN could become superinhibitory through depolymerization. In the normal human heart, SLN is moderately expressed;[7] however, if the expression of SLN should be elevated for any reason, the superinhibition that would result might have deleterious effects on cardiac contractility.

As an example, in heart failure, PLN and SERCA2a are both downregulated[16] so that Ca^{2+} stores are less effectively filled through the action of SERCA2a. As the

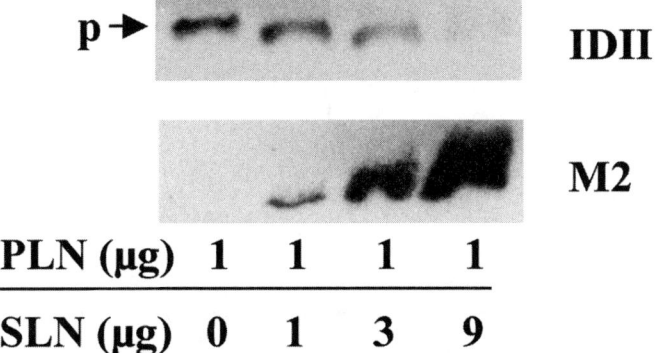

FIGURE 2. Prevention of PLN polymerization by NF-SLN. One μg of PLN was cotransfected into HEK-293 cells with 0, 1, 3, and 9 μg of NF-SLN. After 48 h, microsomal fractions were isolated and subjected to SDS-PAGE in 1.5% SDS in 15% acrylamide gels. The proteins were transferred to nitrocellulose membranes and stained with PLN antibody IDII or NF-SLN antibody M2 to illustrate the correlation between a decrease in PLN pentamer content and an increase in NF-SLN expression. (Reproduced from Asahi et al.,[12] by permission from the American Society for Biochemistry & Molecular Biology.)

Ca^{2+} store is depleted, the force of contraction is diminished. Under these conditions, the SLN to PLN ratio might change so that superinhibition might result, exacerbating the pathological condition. Thus, it is important to understand any potential that exists for the involvement of SLN in heart failure.

EVIDENCE FOR A ROLE OF SLN IN SKELETAL MUSCLE REGULATION

The physiological function of SLN has not been evaluated either in skeletal muscle or in cardiac muscle. Since SLN and PLN affect SERCA2a function in a similar fashion *in vitro*, it is likely that SLN and PLN would affect muscle contractility in a similar fashion. In order to explore the possibility that SLN can regulate SERCA2a activity and contractility in slow-twitch soleus muscle, just as PLN can regulate cardiac contractility, we expressed NF-SLN in rat soleus muscles from one hind limb by intramuscular injection and electrotransfer of rabbit NF-SLN cDNA.[17] The contralateral limb, injected with vector DNA only, served as a control.

Overexpression of NF-SLN was not very high. Relative to SERCA expression in soleus muscles, NF-SLN expression was only one-sixth of that which was optimal for regulation of a comparable amount of SERCA in HEK-293 cells. The effects of a small overexpression of NF-SLN in the presence of what must have been an optimal concentration of endogenous SLN were quite dramatic, however. The maximal Ca^{2+} transport activity in postnuclear homogenates was reduced by 31% from muscle expressing NF-SLN compared with control. The calculated K_{Ca} was unaffected by NF-SLN, but the absolute Ca^{2+} uptake rate was lower with NF-SLN expression, compared with control, over a range from low to high Ca^{2+} concentrations.

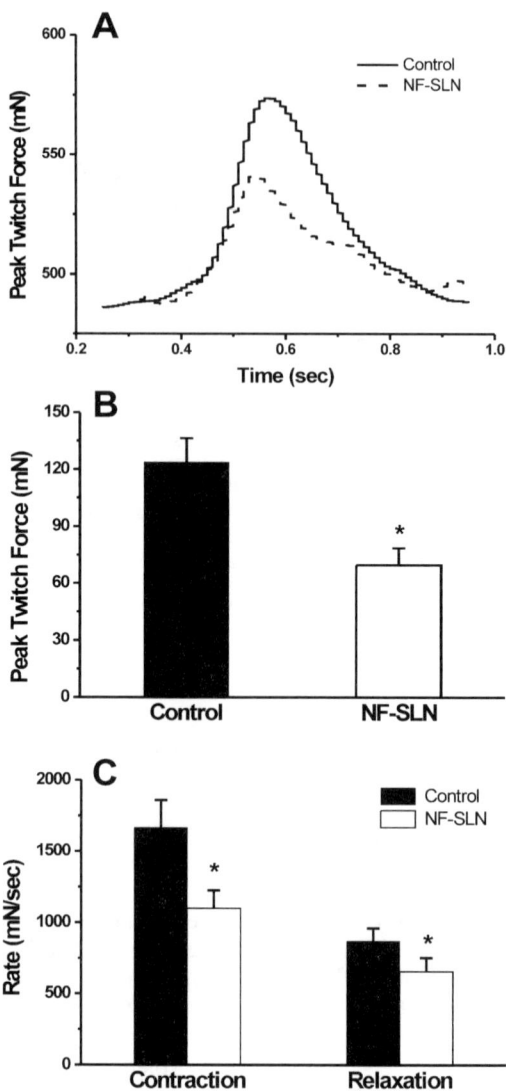

FIGURE 3. Effect of NF-SLN on soleus isometric twitch properties. Three days following injection and electrotransfer of NF-SLN into rat soleus, isometric twitch and tetanic properties were measured *in situ* and compared with twitch and tetanic properties assessed simultaneously in contralateral control muscles, which were injected with the expression vector pcDNA3.1. Twitch and tetanic contractile properties from three muscles that expressed the highest levels of NF-SLN and paired control muscles are shown. (**A**) A typical twitch force record for a soleus muscle that was injected with NF-SLN and a matched control from one animal. (**B**) Peak twitch amplitude (P_t) in control (■) and soleus muscles that expressed NF-SLN (□). (**C**) Maximal twitch rate of contraction ($+dF/dt$) and rate of relaxation ($-dF/dt$) in control (■) and soleus muscles that expressed NF-SLN (□). *$p < 0.05$ versus control. (Reproduced from Tupling *et al.*,[17] by permission from the American Society for Biochemistry & Molecular Biology.)

This decrease in Ca^{2+} uptake was reflected in muscle contractile performance, which was impaired following expression of NF-SLN in slow-twitch skeletal muscle (FIG. 3). In assessing both isometric and tetanic force properties in resting soleus, it was found that both the kinetics and amplitude of contraction were reduced with NF-SLN expression compared with control. Thus, NF-SLN expression in slow-twitch skeletal muscle was similar to a model of PLN overexpression in the heart, at least in terms of its effects on muscle contractility.

A lower rate of SERCA-mediated Ca^{2+} removal from the cytoplasm in the presence of NF-SLN would reduce the size of the Ca^{2+} store in the sarcoplasmic reticulum, and a reduced Ca^{2+} store could alter excitation-contraction coupling in skeletal muscle, accounting for the overall negative effect of NF-SLN on isometric twitch and tetanic contractile properties observed in our study. Since SLN is normally expressed in the heart, an obvious implication from this study and the study of Asahi et al.[12] is that overexpression of SLN in the heart would have the potential to impair cardiac contractile function. It is interesting to note, however, that the deleterious effects of SLN overexpression on muscle contractility did not depend on PLN expression since rat soleus does not express PLN.[17,18]

We also assessed the effects of NF-SLN expression on soleus susceptibility to fatigue, an important characteristic that distinguishes slow-twitch from fast-twitch fibers. Compared with control, resting tetanic force was 47% lower in soleus muscles that expressed NF-SLN. Nevertheless, our fatigue protocol indicated that there was a relatively greater loss of force and reduced rate of contraction in soleus muscles expressing NF-SLN. However, the relative change in relaxation rate was similar between soleus muscles that expressed NF-SLN and control. This suggested that there was no interaction between NF-SLN expression and fatigue, in relation to the slowing of relaxation that normally occurs with fatigue.[19] These results suggest that NF-SLN expression was only indirectly responsible for the observed differences in susceptibility to fatigue. In the relationship between force and pCa, there is a range over which small changes in pCa lead to large changes in force.[20] It is possible that resting soleus muscles expressing NF-SLN, as opposed to control, were already in that sensitive range where small reductions in Ca^{2+} release would lead to reductions in force. Since expression of NF-SLN in rat soleus resulted in a significant depression in muscle contractility and increased susceptibility to fatigue, it is clear that overexpression of SLN has the potential to impair skeletal muscle relaxation.

SITES OF INTERACTION BETWEEN SLN AND SERCA AND BETWEEN PLN AND SERCA

The cytosolic domains of PLN and SLN are very different in that PLN contains a highly conserved, 30-amino-acid, cytosolic sequence that contains sites for phosphorylation by several kinases,[1] while SLN has a poorly conserved, 7-amino-acid, cytosolic sequence that is not phosphorylated under normal conditions. A point of similarity is the conservation of $Asn(Ser)^4$ in the cytosolic sequence of SLN, which is homologous to $Asn(Lys)^{27}$ in PLN, and of Glu^7 in SLN, which corresponds with Asn^{30} in PLN. Of the 19 amino acids that make up the transmembrane sequence of SLN,[7,21] 8 are identical to amino acids in the corresponding positions in the 22-residue transmembrane sequence of PLN and 8 more are highly conserved as L, I, V, or M. SLN contains a highly conserved, hydrophilic, 5-amino-acid, C-terminal sequence, Arg-Ser-Tyr-Gln-Tyr[31], while the highly conserved, hydrophobic, 3-amino-acid, C-terminus of PLN is Met-Leu-Leu[52]. The C-terminal sequence of SLN may serve a different function than the C-terminus of PLN. It could, for example, be involved in the regulatory properties of SLN by interacting with SERCA or another luminal residue. Alternatively, it might serve as a targeting or ER/SR retention signal for SLN.

Analysis of the results of mutagenesis has identified specific amino acids in PLN likely to interact directly with specific amino acids in SERCA. Alanine-scanning mutagenesis of PLN has identified some 10 amino acids on one side of the transmembrane helix that interact with SERCA both functionally and physically, while another 10 amino acids on the other face of the helix are concerned with PLN oligomerization.[13] SERCA1a mutations, L321A, V795A, L892A, T805A, and F809A, diminish substantially the ability of both SLN and PLN to inhibit SERCA1a.[22] The published structures of SERCA1a in the E1Ca2 conformation,[23] which does not bind PLN, and the E2H conformation,[24] which does bind PLN, allow modeling of the interaction between PLN and SLN. As a starting point, it can be predicted that PLN lies in a groove formed between M2 and M9, which is closed in E1Ca2, but open in E2H, exposing the binding surfaces of M6 and M4 defined by mutagenesis. Such modeling highlights a number of probable interaction sites between amino acids in PLN domain II and M2, M4, M6, and M9 in SERCA1a.

Since SLN does not have a cytosolic phosphorylation site, it is likely that relief of SLN inhibition of SERCA1a occurs largely through elevation of Ca^{2+}. Since PLN and SLN are likely to fit into the same groove in SERCA1a, then the same conformational changes in SERCA1a would force SLN out of its inhibitory groove, alleviating the inhibition of SERCA1a by SLN.

While it is clear how PLN and SLN interact with SERCA when they are expressed singly with SERCA, as they are in some tissues, it is not so clear how they

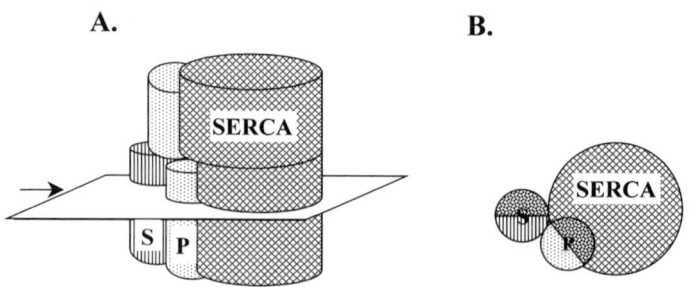

FIGURE 4. A model for the interaction of SLN with PLN and SERCA. The model suggests that PLN has the stronger affinity for the SERCA binding site when both PLN and SLN are expressed with SERCA. SLN, however, has a significant affinity for PLN and binds to the PLN-SERCA complex through its association with PLN rather than through a direct binding to SERCA. SLN increases the stability of the PLN-SERCA complex, suggesting that it may have additional interactions with the complex: **(A)** a side view; **(B)** a top view.

interact with SERCA when they are expressed together with SERCA. There is a robust interaction between NF-SLN and PLN that is strong enough to dissociate the PLN pentamer, presumably by inserting itself into the pentameric structure, thereby weakening it.[12] This might indicate that the SLN-PLN interaction has a higher affinity than the PLN-PLN interaction. Superinhibition resulting from triple expression of SLN, PLN, and SERCA seems to be related to the inhibitory binding of PLN rather than SLN. These observations imply that PLN, with higher affinity for SERCA1a, will occupy the binding site in SERCA1a, when PLN and NF-SLN are coexpressed with SERCA1a, and that NF-SLN will bind to the complex through its interaction with PLN. However, NF-SLN increases the stability of the PLN/SERCA complex, suggesting that it may have additional interactions with the complex. In FIGURE 4, these ideas are presented in a simple model of SERCA1a/PLN/SLN interaction.

ACKNOWLEDGMENTS

Original work described in this paper was supported by grants to D. H. MacLennan from the Canadian Institutes of Health Research and the Heart and Stroke Foundation of Ontario. R. Tupling was a postdoctoral Fellow of the Heart and Stroke Foundation of Canada.

REFERENCES

1. SIMMERMAN, H.K. & L.R. JONES. 1998. Phospholamban: protein structure, mechanism of action, and role in cardiac function. Physiol. Rev. **78:** 921–947.
2. LUO, W. *et al.* 1994. Targeted ablation of the phospholamban gene is associated with markedly enhanced myocardial contractility and loss of beta-agonist stimulation. Circ. Res. **75:** 401–409.
3. KOSS, K.L. & E.G. KRANIAS. 1996. Phospholamban: a prominent regulator of myocardial contractility. Circ. Res. **79:** 1059–1063.
4. ZVARITCH, E. *et al.* 2000. The transgenic expression of highly inhibitory monomeric forms of phospholamban in mouse heart impairs cardiac contractility. J. Biol. Chem. **275:** 14985–14991.
5. ZHAI, J. *et al.* 2000. Cardiac-specific overexpression of a superinhibitory pentameric phospholamban mutant enhances inhibition of cardiac function *in vivo*. J. Biol. Chem. **275:** 10538–10544.
6. HAGHIGHI, K. *et al.* 2001. Superinhibition of sarcoplasmic reticulum function by phospholamban induces cardiac contractile failure. J. Biol. Chem. **276:** 24145–24152.
7. ODERMATT, A. *et al.* 1997. Characterization of the gene encoding human sarcolipin (SLN), a proteolipid associated with SERCA1: absence of structural mutations in five patients with Brody disease. Genomics **45:** 541–553.
8. ODERMATT, A. *et al.* 1998. Sarcolipin regulates the activity of SERCA1, the fast-twitch skeletal muscle sarcoplasmic reticulum Ca^{2+}-ATPase. J. Biol. Chem. **273:** 12360–12369.
9. HELLSTERN, S. *et al.* 2001. Sarcolipin, the shorter homologue of phospholamban, forms oligomeric structures in detergent micelles and in liposomes. J. Biol. Chem. **276:** 30845–30852.
10. MACLENNAN, D.H. *et al.* 1972. Isolation of sarcoplasmic reticulum proteins. Cold Spring Harbor Symp. Quant. Biol. **37:** 469–478.
11. GAYAN-RAMIREZ, G. *et al.* 2000. Corticosteroids decrease mRNA levels of SERCA pumps, whereas they increase sarcolipin mRNA in the rat diaphragm. J. Physiol. **524:** 387–397.

12. ASAHI, M. *et al.* 2002. Sarcolipin inhibits polymerization of phospholamban to induce superinhibition of sarco(endo)plasmic reticulum Ca^{2+}-ATPases (SERCAs). J. Biol. Chem. **277:** 26725–26728.
13. KIMURA, Y. *et al.* 1997. Phospholamban inhibitory function is activated by depolymerization. J. Biol. Chem. **272:** 15061–15064.
14. SIMMERMAN, H.K. *et al.* 1996. A leucine zipper stabilizes the pentameric membrane domain of phospholamban and forms a coiled-coil pore structure. J. Biol. Chem. **271:** 5941–5946.
15. KIMURA, Y. *et al.* 1998. Phospholamban domain Ib mutations influence functional interactions with the Ca^{2+}-ATPase isoform of cardiac sarcoplasmic reticulum. J. Biol. Chem. **273:** 14238–14241.
16. LOUKIANOV, E. *et al.* 1998. Sarco(endo)plasmic reticulum Ca^{2+} ATPase isoforms and their role in muscle physiology and pathology. Ann. N. Y. Acad. Sci. **853:** 251–259.
17. TUPLING, A.R., M. ASAHI & D.H. MACLENNAN. 2002. Sarcolipin overexpression in rat slow twitch muscle inhibits sarcoplasmic reticulum Ca^{2+} uptake and impairs contractile function. J. Biol. Chem. **277:** 44740–44746.
18. DAMIANI, E., R. SACCHETTO & A. MARGRETH. 2000. Variation of phospholamban in slow-twitch muscle sarcoplasmic reticulum between mammalian species and a link to the substrate specificity of endogenous Ca^{2+}-calmodulin-dependent protein kinase. Biochim. Biophys. Acta **1464:** 231–241.
19. WESTERBLAD, H. & D.G. ALLEN. 1993. The contribution of $[Ca^{2+}]_i$ to the slowing of relaxation in fatigued single fibres from mouse skeletal muscle. J. Physiol. **468:** 729–740.
20. ALLEN, D.G., J. LANNERGREN & H. WESTERBLAD. 1995. Muscle cell function during prolonged activity: cellular mechanisms of fatigue. Exp. Physiol. **80:** 497–527.
21. WAWRZYNOW, A. *et al.* 1992. Sarcolipin, the "proteolipid" of skeletal muscle sarcoplasmic reticulum, is a unique, amphipathic, 31-residue peptide. Arch. Biochem. Biophys. **298:** 620–623.
22. ASAHI, M. *et al.* 1999. Transmembrane helix M6 in sarco(endo)plasmic reticulum Ca^{2+}-ATPase forms a functional interaction site with phospholamban: evidence for physical interactions at other sites. J. Biol. Chem. **274:** 32855–32862.
23. TOYOSHIMA, C. *et al.* 2000. Crystal structure of the calcium pump of sarcoplasmic reticulum at 2.6 A resolution. Nature **405:** 647–655.
24. TOYOSHIMA, C. & H. NOMURA. 2002. Structural changes in the calcium pump accompanying the dissociation of calcium. Nature **418:** 605–611.

The Thermogenic Function of the Sarcoplasmic Reticulum Ca^{2+}-ATPase of Normal and Hyperthyroid Rabbit

LEOPOLDO DE MEIS,[a] ANA PAULA ARRUDA, WAGNER S. DA-SILVA, MARCELO REIS, AND DENISE P. CARVALHO[b]

Departamento de Bioquímica Médica, Instituto de Ciências Biomédicas, Universidade Federal do Rio de Janeiro, Cidade Universitária, RJ, 21941 590, Brasil

[b]*Instituto de Biofísca, Universidade Federal do Rio de Janeiro, Cidade Universitária, RJ, 21941 590, Brasil*

ABSTRACT: After formation of a Ca^{2+} gradient, the amount of heat released during the hydrolysis of each mol of ATP cleaved (ΔH^{cal}) varies depending on the Ca^{2+}-ATPase isoform expressed by the muscle cell. In vesicles derived from the sarcoplasmic reticulum of white muscle (SERCA 1) most of the ATP cleaved is not coupled to Ca^{2+} transport, and the ΔH^{cal} varies between -20 and -22 kcal/mol. In contrast, in vesicles derived from red muscle (SERCA 2a) the hydrolysis of ATP is coupled with Ca^{2+} transport, and the ΔH^{cal} varies between -12 and -14 kcal/mol. Hyperthyroidism increases the rate of heat production by the Ca^{2+}-ATPase fourfold in white muscle and 40-fold in red muscle. In hyperthyroid rabbits, the amount of sarcoplasmic reticulum protein recovered from white and red muscle is four- to fivefold greater than that obtained from control rabbits. Hyperthyroid red muscle expresses SERCA 1, and the vesicles derived from these muscle hydrolyze ATP through a catalytic route that is not coupled to Ca^{2+} transport, thus increasing the amount of heat released during ATP hydrolysis, the ΔH^{cal} varying between -20 and -22 kcal/mol.

KEYWORDS: Ca^{2+}-ATPase; thermogenesis; Ca^{2+} transport; heat production; ATP hydrolysis; hyperthyroidism

INTRODUCTION

This work deals with two interconnected subjects: (1) the mechanism of energy interconversion by enzymes and (2) heat generation, a process that plays a key role in the regulation of several physiological processes, including body temperature, body weight, and cold acclimation.[1–3] Energy is released when ATP is hydrolyzed, and the various ATPases that catalyze ATP hydrolysis operate as specific energy transducers. The ATPases are able to convert the energy released into work (actomy-

Address for correspondence: Dr. Leopoldo de Meis, Departamento de Bioquímica Médica, Instituto de Ciências Biomédicas, Universidade Federal do Rio de Janeiro, Cidade Universitária, RJ, 21941 590, Brasil. Voice: +55-21-2270 1635; fax: +55-21-2270-8647.
demeis@bioqmed.ufrj.br

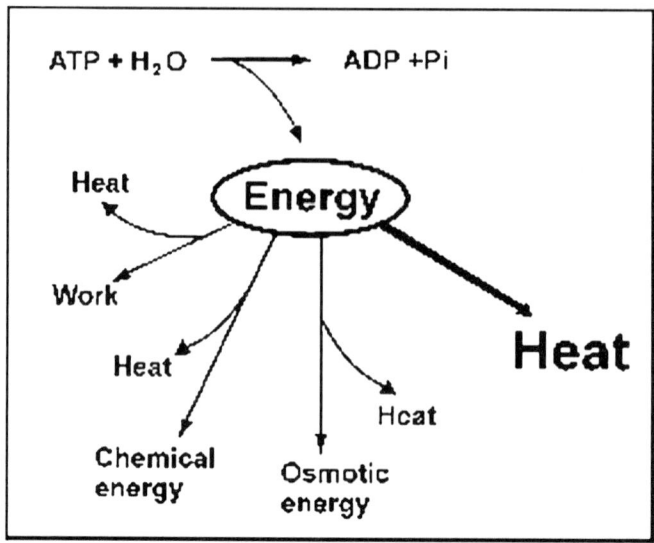

FIGURE 1. Conversion of the energy released during ATP hydrolysis.

osin) or other forms of energy, such as osmotic energy (active transport) or chemical energy (synthesis of new molecules) (FIG. 1). Usually, in these processes a part of the energy released is dissipated into the environment as heat.

The sarcoplasmic reticulum Ca^{2+}-ATPase can drive the hydrolysis of ATP through two different catalytic routes represented in FIGURE 2 as reaction 10 and reactions 1 to 6. Notice that only one of these routes is coupled to the translocation of Ca^{2+} through the membrane (reactions 1 to 6). The second route (reaction 10) does not promote the transport of Ca^{2+}; we refer to it as uncoupled ATPase activity.[3–6] In recent reports,[3,7–16] we observed that the amount of heat produced during ATP hydrolysis varies depending on the catalytic route through which the cleavage of ATP occurs. The heat produced during the hydrolysis of each mol of ATP (ΔH^{cal}) cleaved through the uncoupled route varies between 20 and 25 kcal. In this route all the energy derived from ATP hydrolysis seems to be converted into heat, with no energy used for Ca^{2+} transport. When processed through the coupled route, the heat produced decreases to the range of 8 to 12 kcal/mol ATP, because during cleavage a portion of the energy is used to translocate Ca^{2+} through the membrane (reactions 3 and 4 in FIG. 2) and only a fraction of the total amount of energy released is available to be converted into heat. During transport, the amount of ATP cleaved through the coupled and uncoupled route varies depending on the Ca^{2+}-ATPase isoform and the experimental conditions used. Thus, the value of the ΔH^{cal} of ATP hydrolysis measured varies depending on the balance between the two catalytic routes of the enzyme. The total amount of chemical energy released during ATP hydrolysis is always the same, but the fraction of the total energy converted into osmotic energy or heat can be modulated by the Ca^{2+}-ATPase isoform.

Different genes are known to encode the sarco/endoplasmic reticulum Ca^{2+}-ATPases (SERCA).[17,18] The SERCA 1 gene is expressed exclusively in white or fast skeletal muscle. The SERCA 2 gene is expressed in red or slow muscle (SERCA 2a) and in blood platelets (SERCA 2b). The physiological meaning of isoform diversity is not clear. The phenotype of white and red muscle and the transcription of SERCA 1 and SERCA 2a isoforms are regulated by the thyroid hormone 3,5,3'-triiodo L-thyronine (T_3).[19–21] In hypothyroid rats, white muscles are replaced by red muscles and the expression of SERCA 1 is decreased while SERCA 2a is overexpressed (TABLE 1). In hyperthyroid animals, the situation is inverted, white muscles are predominant, SERCA 1 is overexpressed, while SERCA 2a expression is repressed. The amount of Ca^{2+}-ATPase found in the sarcoplasmic reticulum of white muscles is larger than that found in red muscles.[19,22,23] The injection of T_3 in animals increases the amount of Ca^{2+}-ATPase found in the sarcoplasmic reticulum.[19,20] The data shown below indicate that (1) the ability to convert energy derived from ATP hydrolysis into heat varies depending on the Ca^{2+}-ATPase isoform used, and (2) in hyperthyroid rabbits there is a large increase in the amount of heat produced by the muscle Ca^{2+}-ATPase.

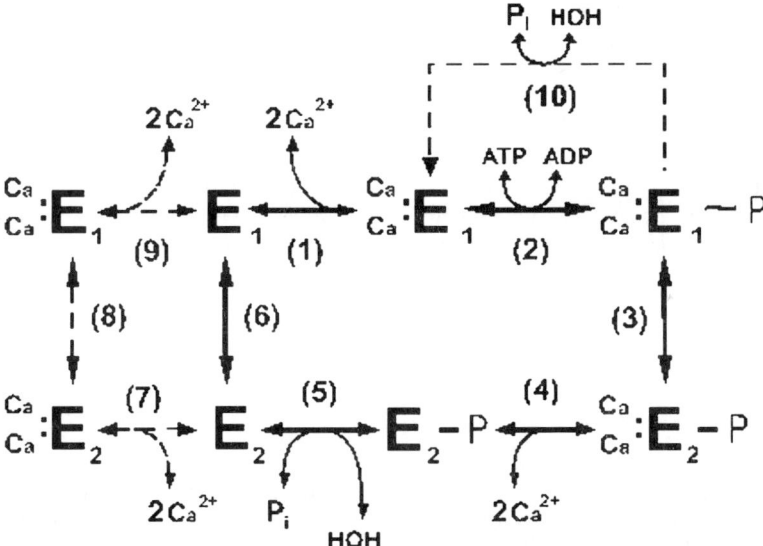

FIGURE 2. The catalytic cycle of the Ca^{2+}-ATPase. The sequence includes two distinct enzymes conformations, E_1 and E_2. The Ca^{2+} binding sites in the E_1 form face the external surface of the vesicle and have a high affinity for Ca^{2+} ($K_a = 10^{-6}$ M at pH 7). In the E_2 form the Ca^{2+} binding sites face the vesicle lumen and have a low affinity for Ca^{2+} ($K_a = 10^{-3}$ M at pH 7). The enzyme form E_1 is phosphorylated by ATP but not by P_i and, conversely, the enzyme form E_2 is phosphorylated by P_i but not by ATP. When the Ca^{2+} concentration on the two sites of the membrane is less than 50 µM, reaction 4 is irreversible and this forces the sequence to flow forward from reaction 1 to 6.[24–27] This is observed when leaky vesicles are used. With intact vesicles, the Ca^{2+} pumped by the ATPase is retained in the vesicle lumen. The high intravesicular Ca^{2+} concentration (~10 mM) permits the reversal of the catalytic cycle (reactions 5 to 1 backwards), during which a portion of the Ca^{2+} accumulated leaves the vesicles in a process coupled with the synthesis of ATP from ADP and P_i. For vesicles derived from white (fast) muscle (SERCA 1), the increase in the intravesicular Ca^{2+} concentration leads to ramifications of the catalytic cycle, the uncoupled Ca^{2+} efflux mediated by reactions 7 to 9, and the uncoupled ATPase activity[3–6] mediated by reaction 10.

WHITE MUSCLE (SERCA 1)

The amount of heat released during the hydrolysis of each ATP molecule varies depending on whether or not a transmembrane gradient is formed across the vesicle membrane.[3,7,9,12–16] In the absence of a Ca^{2+} gradient (leaky vesicles), between 10 and 12 kcal are released for each mol of ATP cleaved. In the presence of a Ca^{2+} gradient (intact vesicles), the heat released increases to the range of 20 to 24 kcal for each mole of ATP cleaved. This difference is promoted by the uncoupled Ca^{2+} efflux and by the uncoupled ATPase activity[3,11,13–15] represented by reactions 7 to 9 and reaction 10 in FIGURE 2. These two activities can be detected only when the vesicles are able to accumulate Ca^{2+} and a gradient is formed across the vesicle membrane.[3–6] Of the two activities, the uncoupled ATPase plays a predominant role in establishing the difference of heat measured in the presence and absence of a gradient.[3,13–15] In the presence of a gradient, a substantial amount of ATP is cleaved through the uncoupled catalytic route TABLE 2). When the uncoupled ATPase activity is inhibited, there is a decrease in the yield of heat produced during ATP cleavage,[7,9,13–15] thus abolishing the difference of ΔH^{cal} for ATP hydrolysis measured with intact and leaky vesicles. Both the coupled and uncoupled ATPase activities have the same Ca^{2+} dependence.[16] The uncoupled ATPase activity can be specifically inhibited by fluoride,[16] dimethyl sulfoxide,[15] low ATP and high ADP concentrations,[13] and decreasing assay medium temperature from 35°C to 25°C.[9,15] These conditions had no effect on the values of the ΔH^{cal} for ATP hydrolysis measured with leaky vesicles, because the uncoupled Ca^{2+} efflux and ATPase activity are only activated when the Ca^{2+} concentration in the vesicles lumen rises to the millimolar range.[4–6]

RED MUSCLE (SERCA 2A) AND BLOOD PLATELETS (SERCA 2B)

These vesicles accumulated less Ca^{2+} and hydrolyze less ATP than the vesicles derived from white muscles (FIGS. 3 and 4; TABLES 1 and 2). In the absence of a gradient, the ΔH^{cal} values for ATP hydrolysis measured with vesicles derived from both white and red muscles varies between −10 and −12 kcal/mol.[3,16] Formation of a gradient leads to an increase in the yield of heat produced during ATP hydrolysis in both cases, but in red muscle vesicles the heat released during ATP hydrolysis and the rate of the uncoupled ATPase were significantly smaller than those measured with vesicles derived from white muscle (TABLES 1 and 2; FIGS. 3 and 4). As was the case for red muscle, vesicles derived from blood platelets hydrolyze ATP and accumulate Ca^{2+} at a slow rate. The ΔH^{cal} of ATP hydrolysis of platelets varies between −10 and −12 kcal/mol both in the presence and absence of a gradient.[10] The predominant isoform found in red muscle is SERCA 2a, but a small fraction of the total Ca^{2+}-ATPase is SERCA 1.[16] Blood platelets do not express SERCA 1. Thus, the possibility remains that SERCA 2a, similar to SERCA 2b found in platelets, does not cleave ATP through the uncoupled route and the low uncoupled activity detected in TABLE 2 is derived from the small fraction of SERCA 1 expressed in red muscle.

TABLE 1. Ca^{2+}-ATPase activity and rate of heat production in white (fast) and red (slow) muscles

Vesicles	SRV* (mg/g muscle)	Ca^{2+}-ATPase (mmol/µg·min^{-1})	P_i (µmol/g muscle)	ΔH^{cal} (Kcal/mol)	mcal/g muscle
White muscle					
Control	0.36	1.62 ± 0.14 (7)[a]	0.58	−23.0 ± 2.5 (7)	13.41
Hyperthryroid	1.37	2.00 ± 0.11 (15)	2.74	−20.4 ± 1.2 (6)	55.90
Red muscle					
Control	0.17	0.23 ± 0.04 (11)[b]	0.04	−13.4 ± 0.78 (9)[c]	0.52
Hyperthryroid	0.92	1.11 ± 0.14 (5)	1.02	−21.3 ± 1.21 (3)	21.75

NOTE: Hyperthyroidism was induced by daily subcutaneous injection of L-thyroxine (200 µg/kg body weight) for 8 days, as previously reported.[20] Vesicles from control and hyperthyroid rabbits were prepared simultaneously in the same day using the same buffer solutions. Values are (*) average of two experiments or means ± SE of the number of experiments shown in parenthesis. Ca^{2+}-ATPase activity and heat production were measured as previously described.[13–16] The difference between the values of control and hyperthyroid were significant (t test) at the level of [a]$P < 0.05$, [b]$P < 0.001$, and [c]$P < 0.01$.

TABLE 2. Rates of coupled and uncoupled ATPase activity and heat release during ATP hydrolysis

| Vesicles | ΔH^{cal} (kcal/mol) | Ca^{2+}-ATPase activity (µmol/mg·min^{-1}) | | Ratio (b/a) |
		Coupled (a)	Uncoupled (b)	
White muscle				
Control	−23.0 ± 2.5 (7)	0.25 ± 0.03 (12)	1.07 ± 0.26 (6)[a]	4.28
Hyperthyroid	−20.4 ± 1.2 (6)	0.27 ± 0.02 (5)	1.80 ± 0.20 (5)	6.67
Red muscle				
Control	−13.4 ± 0.8 (9)	0.07 ± 0.01 (9)	0.08 ± 0.02 (7)[b]	1.07
Hyperthyroid	−21.3 ± 1.2 (3)	0.03 ± 0.01 (3)	1.39 ± 0.26 (3)	53.5

NOTE: Conditions were as in TABLE 1. The difference between the values of control and hyperthyroid were significant (t test) at the level of [a]$P < 0.05$ and [b]$P < 0.001$.

HYPERTHYROIDISM

There is an increase of oxygen consumption and heat production in hyperthyroid animals. The mechanism underlying this effect is unclear. In hyperthyroidism there is a simultaneous increase of SERCA 1 and a decrease of SERCA 2a expression.[20,21] In agreement with earlier reports,[19] the amount of sarcoplasmic reticulum protein recovered from hyperthyroid muscles was four- to fivefold greater than that obtained from control rabbits (TABLE 1). Western blot analysis revealed that in addition to SERCA 2a, there was a significant increase of SERCA 1 in red muscle vesicles. Vesicles from white muscle did not alter its pattern and expressed only SERCA 1 in both control and hyperthyroid rabbits (data not shown). The effect of hyperthyroidism in red muscles was more pronounced than that noted in white muscle (TABLES 1 and

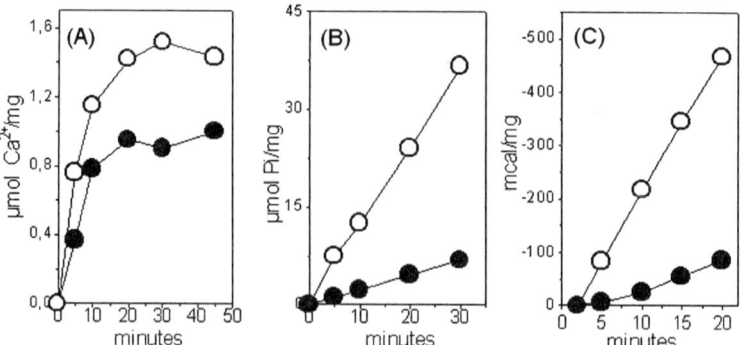

FIGURE 3. Ca^{2+} transport (**A**), ATP hydrolysis (**B**) and heat production (**C**) by vesicles derived from red muscles. The assay medium composition was 50 mM MOPS-Tris buffer (pH 7.0), 1 mM ATP, 2 mM $MgCl_2$, 100 μM $CaCl_2$, 100 mM EGTA, 10 mM P_i, 100 mM KCl, and 5 mM NaN_3. (●) Control and (○) hyperthyroid vesicles. The reaction was performed at 35°C and was started by the addition of vesicles, 10-μg protein/mL; Ca^{2+} uptake (**A**), Ca^{2+}-ATPase activity (**B**), and heat production (**C**) were measured as previously described.[13–16] The figure shows a typical experiment.

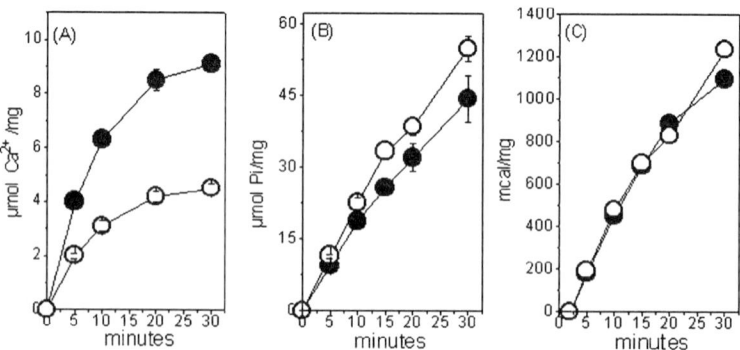

FIGURE 4. Ca^{2+} transport (**A**), ATP hydrolysis (**B**), and heat production (**C**) by vesicles derived from white muscles. (●) Control and (○) hyperthyroid vesicles. Assay medium and experimental conditions were as in FIGURE 3. Values are mean ± SE of 6 experiments.

2). Vesicles from hyperthyroid rabbit red muscle were able to accumulate more Ca^{2+}, hydrolyzed more ATP, and produced more heat (FIG. 3) than the vesicles of control animals. In addition, the ΔH^{cal} of ATP hydrolysis became more negative than that measured in control vesicles. This change was associated with a significant increase of the uncoupled ATPase activity (TABLE 2). In contrast to red muscle, the vesicles derived from hyperthyroid white muscle were able to accumulate less Ca^{2+} than the control vesicles (FIG. 4). This was accompanied by a small but significant increase of the uncoupled ATPase activity (TABLE 2). The data of TABLE 1 indicate that in hyperthyroidism the rate of heat production by the Ca^{2+}-ATPase increases by a factor of 4 in white muscle and by a factor of 40 in red muscle. This was calculated taking into account the amount of sarcoplasmic reticulum protein recovered from muscle homogenates, the rate of the Ca^{2+}-ATPase activity, and ΔH^{cal} value of ATP hydrolysis. In white muscle, the increase of heat production seems to be related solely to the increase in the amount of Ca^{2+}-ATPase available in the muscle. In red muscles, the large increase of heat production is related not only to the increase of Ca^{2+}-ATPase available, but mainly to the expression of SERCA 1, which is able to cleave ATP through the uncoupled catalytic route ATPase (TABLE 2).

When extended to humans, the data described in this report indicate that the total amount of heat dissipated daily by the Ca^{2+}-ATPase should vary depending on the ratio between white and red muscle. Hyperthyroidism enhances the contribution of the Ca^{2+}-ATPase to the total heat production of the body, the effect on red muscle being far more pronounced than that of white muscle. Skeletal muscle is by far the most abundant tissue of human body and accounts for more than 50% of the total oxygen consumption in a resting human being and up to 90% during very active muscular work. Thus, a small change in heat production in skeletal muscles reflects a significant modification in the total body heat production.

ACKNOWLEDGMENTS

This work was supported by grants from PRONEX—Financiadora de Estudos e Projetos (FINEP), Conselho Nacional de Desenvolvimento Científico e Tecnológico (CNPq), and by Fundação de Amparo à Pesquisa do Estado do Rio de Janeiro (FAPERJ).

REFERENCES

1. ROLFE, D.F.S. & G.C. BROWN. 1997. Cellular energy utilization and molecular origin of standard metabolic rate in mammals. Physiol. Rev. **77:** 731–758.
2. JANSKÝ, L. 1995. Humoral thermogenesis and its role in maintaining energy balance. Physiol. Rev. **75:** 237–259.
3. DE MEIS, L. 2002. J. Membr. Biol. **188:** 1–9.
4. YU, X. & G. INESI. 1995. Variable stoichiometric efficiency of Ca^{2+} and Sr^{2+} transport by the sarcoplasmic reticulum ATPase. J. Biol. Chem. **270:** 4361–4367.
5. FORTEA, M.I., F. SOLER & F. FERNANDEZ-BELDA. 2000. Insight into the uncoupling mechanism of sarcoplasmic reticulum ATPase using the phosphorylating substrate UTP. J. Biol. Chem. **275:** 12521–12529.
6. LOGAN-SMITH, M.J., P.J. LOCKYERS, J.M. EAST & A.G. LEE. 2001. Curcumin, a molecule that inhibits the Ca^{2+}-ATPase of sarcoplasmic reticulum but increases the rate of accumulation of Ca^{2+}. J. Biol. Chem. **276:** 46905–46911.
7. DE MEIS, L., M.L. BIANCONI & V.A. SUZANO. 1997. Control of energy fluxes by the sarcoplasmic reticulum Ca^{2+}-ATPase: ATP hydrolysis, ATP synthesis and heat production. FEBS Lett. **406:** 201–204.
8. DE MEIS, L. 1998. Control of heat produced during ATP hydrolysis by the sarcoplasmic reticulum Ca^{2+}-ATPase in the absence of a Ca^{2+} gradient. Biochem. Biophys. Res. Commun. **243:** 598–600.
9. DE MEIS, L. 1998. Control of heat production by the Ca^{2+}-ATPase of rabbit and trout sarcoplasmic reticulum. Am. J. Physiol. **274** (Cell Physiol. **43**)**:** C1738–C1744.

10. MITIDIERI, F. & L. DE MEIS. 1999. Ca^{2+} release and heat production by the endoplasmic reticulum Ca^{2+}-ATPase of blood platelets: effect of the platelets activating factor. J. Biol. Chem. **274:** 28344–28350.
11. DE MEIS, L. 2000. ATP synthesis and heat production during Ca^{2+} efflux by sarcoplasmic reticulum Ca^{2+}-ATPase. Biochem. Biophys. Res. Commun. **276:** 35–39.
12. DE MEIS, L. 2001. Role of the sarcoplasmic reticulum Ca^{2+}-ATPase on heat production and thermogenesis. Biosci. Rep. **21:**113–137.
13. DE MEIS, L. 2001. Uncoupled ATPase activity and heat production by the sarcoplasmic reticulum Ca^{2+}-ATPase. J. Biol. Chem. **276:** 25078–25087.
14. REIS, M., M. FARAGE, A.C. SOUZA & L. DE MEIS. 2001. Correlation between uncoupled ATPase hydrolysis and heat production by the sarcoplasmic reticulum Ca^{2+}-ATPase. J. Biol. Chem. **276:** 42793–42800.
15. BARATA, H. & L. DE MEIS. 2002. J. Biol. Chem. **277:** 16868–16872.
16. REIS, M., M. FARAGE & L. DE MEIS. 2002. Mol. Memb. Biol. In press.
17. MACLENNAN, D.H., C.J. BRANDL, B. KORCZAK & N.M. GREEN. 1985. Amino-acid sequence of Ca^{2+}, Mg^{2+}-dependent ATPase from rabbit muscle sarcoplasmic reticulum, deduced from its complementary DNA sequence. Nature **316:** 696–700.
18. LYTTON, J., M. WESTIN, S.E. BURK, et al. 1992. Functional comparisons between isoforms of the sarcoplasmic reticulum family of calcium pumps. J. Biol. Chem. **267:** 14483-14489.
19. NUNES, M.T., A.C. BIANCO, A. MIGALA, et al. 1985. Tyroxine induced transformation in sarcoplasmic reticulum of rabbit soleus and psoas muscles. Z. Naturforsch. **40c:** 726–734.
20. ARAI, M., K. OTSU, D.H. MACLENNAN, et al. 1991. Effect of thyroid-hormone on the expression of messenger RNA encoding sarcoplasmic-reticulum proteins. Circ. Res. **69:** 266–276.
21. VAN DER LINDEN G.C., W.S. SIMONIDES, A. MULLER, et al. 1996. Fiber specific regulation of Ca^{2+}-ATPase isoform expression by thyroid hormone in rat skeletal muscle. Am. J. Physiol. **271** (Cell Physiol. **40**): C1908–C1919.
22. DULHUNTY, A.F., M.R. BANYARD & C.C. MEDVECZKY. 1987. Distribution of calcium ATPase in the sarcoplasmic reticulum of fast- and slow-twitch muscles determined with monoclonal antibodies. J. Membr. Biol. **99:** 79–92.
23. WU, K., W. LEE, J. WEY, et al. 1995. Localization and quantification of endoplasmic reticulum Ca^{2+}-ATPase isoforms transcripts. Am. J. Physiol. **269** (Cell Physiol. **38**): C775–C784.
24. DE MEIS, L. & A.L. VIANNA. 1979. Energy interconversion by the Ca^{2+}-transport ATPase of sarcoplasmic reticulum. Annu. Rev. Biochem. **48:** 275–292.
25. TANFORD, C. 1984. Twenty questions concerning the reaction cycle of the sarcoplasmic reticulum calcium pump. CRC Crit. Rev. Biochem. **17:** 123–151.
26. INESI, G. 1985. Mechanism of Ca^{2+} transport. Annu. Rev. Physiol. **47:** 573–601.
27. DE MEIS, L. 1989. Role of water in the energy of hydrolysis of phosphate compounds—Energy transduction in biological membranes. Biochim. Biophys. Acta **973:** 333–349.

Na,K-ATPase as a Signal Transducer

OLEG AIZMAN AND ANITA APERIA

Department of Woman and Child Health, Karolinska Institutet,
Astrid Lindgren Children's Hospital, Stockholm, Sweden

ABSTRACT: Recent studies have indicated that Na,K-ATPase may, in addition to being the key regulator of intracellular Na^+ and K^+ concentration, act as a signal transducer. Despite extensive research, the biological role for ouabain, a natural ligand of Na,K-ATPase, is not well understood. We have reported that exposure of rat proximal tubular cells (RPTC) to doses of ouabain that inhibit the Na,K-ATPase activity by less than 50% (10 nM – 500 μM), will induce intracellular $[Ca^{2+}]_i$ oscillations and that this calcium signal leads to activation of the transcription factor NF-κB. The ouabain-induced calcium oscillations were blocked by an inhibitor of the IP_3 receptors but not by phospholipase C inhibitors nor by cellular depletion of IP_3, suggesting that the calcium signal is not due to phospholipase C–mediated IP_3 release. Fluorescence resonance energy transfer (FRET) studies suggested a close proximity between the Na,K-ATPase and IP_3 receptor. Our findings demonstrate a novel principle for calcium signaling via Na,K-ATPase.

KEYWORDS: Na,K-ATPase; Ca^{2+} oscillations; ouabain; IP_3 receptor; FRET; transcriptional regulation; NF-κB

INTRODUCTION

Several recent studies suggest that in addition to its key role as a regulator of cell ion homeostasis, Na,K-ATPase may act as a signal transducer and activator of gene transcription.[1–4] To further elucidate this role for Na,K-ATPase, we have investigated intracellular signaling pathways activated by the ouabain/Na,K-ATPase complex. Here we report that ouabain-bound Na,K-ATPase can induce intracellular calcium oscillation. The majority of results presented in this review article have been published previously by Aizman *et al.*[5] Calcium (Ca^{2+}) is involved in the regulation of such diverse cellular processes as gene transcription, cell adhesion, cell growth, proliferation, and apoptosis.[6,7] Intracellular Ca^{2+} $[Ca^{2+}]_i$ oscillation may be the most versatile of all intracellular signals, since the cell can decode differences in the amplitude and frequency of these oscillations and translate them into specific cellular responses.[8,9] Low-frequency Ca^{2+} oscillations will specifically activate the transcription factor NF-κB by triggering proteolysis of the inhibitory subunit, IκB.[10,11] NF-κB plays an important role in regulation of cell growth, proliferation, and apoptosis.[12]

Address for correspondence: Anita Aperia, Department of Women and Child Health, Karolinska Institutet, Astrid Lindgren Children's Hospital, Q2:09, S-171 76 Stockholm, Sweden. Voice: +468-51777326; fax: +468-51777328.

anita.aperia@kbh.ki.se

METHODS

The majority of experiments were performed on primary cultures of rat proximal tubule cells (PTC), obtained from kidneys of 20-day-old Sprague-Dawley rats. These cells grow as clusters after 2–3 days in culture. At that time, they have a well-preserved phenotype and display the typical morphology of epithelial cells from proximal tubule. They have, as is typical for all renal tubule cells, a high level of Na,K-ATPase. In addition, rat PTC are particularly well suited for Ca^{2+} measurements with Fura-2AM.[13] Rat Na,K-ATPase has a relatively low ouabain sensitivity and full inhibition of the enzyme requires millimolar concentrations of ouabain.[14] Changes in Ca^{2+} concentration induced by ouabain/Na,K-ATPase complex were monitored by ratiometric fluorescent microscopy with Ca^{2+}-sensitive fluorophore, Fura-2AM. Frequency analysis was applied to determine specificity of the Ca^{2+} signal. To investigate the biological relevance of induced Ca^{2+} signal, we studied activation of the Ca^{2+}-dependent transcriptional factor, NF-κB. The interaction of Na,K-ATPase with IP_3R and its role in the generation of Ca^{2+} signals were explored by co-immunoprecipitation and fluorescence resonance energy transfer (FRET) analysis.

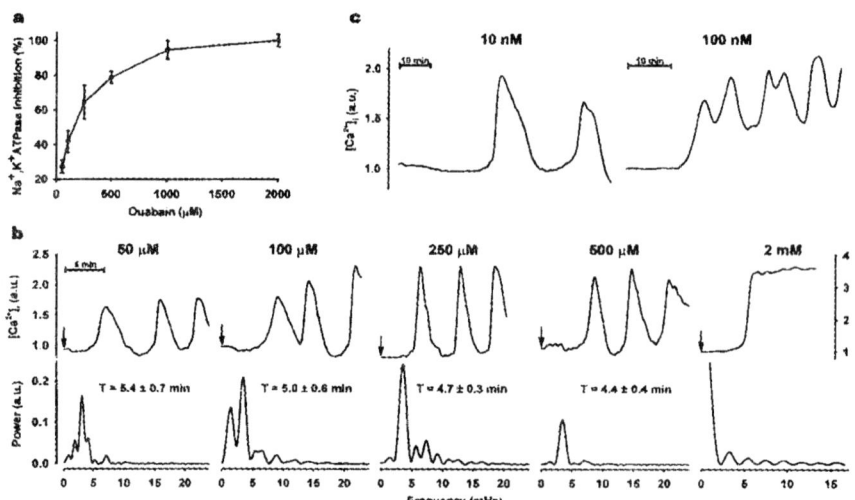

FIGURE 1. Effect of ouabain on $[Ca^{2+}]_i$ in primary culture of RPT cells. (**a**) Na,K-ATPase activity measured as ouabain-sensitive $^{86}Rb^+$ uptake (mean ± S.E.). (**b**) (*Upper panel*) Representative single cell $[Ca^{2+}]_i$ tracings in response to indicated ouabain concentrations. At time 0 (*arrow*), cells were exposed to ouabain concentrations ranging from 50 μM to 2 mM and recordings were made every 30 s. Arbitrary units (a.u.) represent ratio values corresponding to $[Ca^{2+}]_i$ changes. (*Lower panel*) Spectral analysis of ouabain-induced $[Ca^{2+}]_i$ oscillations. Each plot corresponds to the single cell recording above. $[Ca^{2+}]_i$ oscillation periodicity (T) of each ouabain concentration was calculated as mean ± S.E. from approximately 50 cells from at least three separate experiments. (**c**) Representative single cell $[Ca^{2+}]_i$ tracings observed in cells superfused for 3 hours at a slow rate (100 μL/min) with nanomolar ouabain. (From Aizman *et al.*[1] With permission from the National Academy of Sciences.)

RESULTS AND DISCUSSION

When PTC were exposed to ouabain in doses resulting in only partial Na,K-ATPase inhibition (10 nM–250 µM) (FIG. 1a), we observed slow, regular $[Ca^{2+}]_i$ oscillations (FIG. 1b). This response was detected in approximately one-third of the cells and was generally initiated in one cell at the periphery of a cell cluster and then propagated to neighboring cells. To determine to what extent the oscillations possessed an intrinsic regularity, we performed power spectrum analysis. Power spectrum analysis revealed a periodicity between 5.4 ± 0.7 min for 50 mM ouabain and 4.4 ± 0.4 min for 250 mM ouabain (FIG. 1b). The amplitude of the oscillations for all partially inhibitory ouabain doses was in the same range. A dose of 2 mM ouabain, which causes complete inhibition of rat Na,K-ATPase activity, did not cause oscillations, but resulted in a sustained increase in $[Ca^{2+}]_i$ (FIG. 1b).

It may be argued that Ca^{2+} oscillations were demonstrated in response to pharmacological doses of ouabain. *In vivo*, circulating levels of ouabain rarely exceed the picomolar-to-nanomolar range. However, it should be taken into account that ouabain binding to Na,K-ATPase is tight and long lasting—the "on-rate" for ouabain is ~20 times faster than the "off-rate."[15] Thus, per time unit, more ouabain molecules associate than dissociate with Na,K-ATPase. In tissue exposed to circulating blood, the number of Na,K-ATPase molecules occupied by ouabain will approach an equilibrium exponentially over time. In support of this, we demonstrated that when PTC cells were superfused with solutions containing ouabain in nanomolar range for more than an hour, Ca^{2+} oscillations were observed in 5% of the cells, while 30% responded at 250 µM ouabain (FIG. 1c). It should also be taken into account that the numbers of ouabain molecules bound to Na,K-ATPase must exceed a threshold to trigger a global cellular Ca^{2+} response.

To examine whether the Ca^{2+} oscillations were secondary to inhibition of Na,K-ATPase activity, we studied the effect of lowering extracellular K^+ concentration. Both ouabain and graded reduction of extracellular K^+ evoked similar, dose-dependent increases in intracellular Na ($[Na^+]_i$), indicating that both ouabain and low K^+ inhibited Na,K-ATPase activity to the same extent. However, lowering extracellular K^+ did not trigger calcium oscillations.[5]

The next set of studies was performed to elucidate the molecular mechanism of Ca^{2+} oscillations induced by the ouabain/Na,K-ATPase complex. In cells where the intracellular Ca^{2+} stores of the endoplasmic reticulum were depleted by preincubation with the sarco-endoplasmic reticulum ATPase (SERCA) pump inhibitor (20 µM cyclopiazonic acid), ouabain failed to induce Ca^{2+} oscillations. Regulated Ca^{2+} release from intracellular stores occurs via InsP$_3$Rs or via ryanodine receptors (RyR). Stimulation of RyR with a low dose of ryanodine (10 µM) did not have any effect on $[Ca^{2+}]_i$. Inhibition of RyR by a higher dose of ryanodine (100 µM) did not prevent ouabain-induced $[Ca^{2+}]_i$ oscillations.[5] From this we conclude that RyR are not involved in ouabain-induced Ca^{2+} oscillations and that functioning RyR are of little, if any, importance in RPT cells. In contrast, IP$_3$ receptor inhibition with membrane-permeable IP$_3$R inhibitor 2-aminoethoxydiphenyl borate (2-APB) completely abolished the oscillations,[5] indicating that Ca^{2+} release via IP$_3$ receptor played a major role in Ca^{2+} oscillation induced by ouabain. Although 2-APB is the most specific membrane-permeable IP$_3$R inhibitor available today, it can also affect Ca^{2+} release–

activated Ca^{2+} channels (CRAC).[16] $InsP_3Rs$ are, either directly or via an anchor protein, coupled to CRAC channels, the function of which is essential for the maintenance of Ca^{2+} oscillations.[17,18] We conclude from these studies that release of calcium from intracellular stores via IP_3R is an essential contribution to the ouabain/Na,K-ATPase–induced calcium oscillations.

We also found that calcium-free media abolished ouabain-induced Ca^{2+} oscillations. Influx of Ca^{2+} from the extracellular space may, at least partially, occur via L-type voltage gated Ca^{2+} (L-VGC) channels, which are to a limited extent expressed in RPT cells[19] Two inhibitors of L-VGC channels, nifedipine and verapamil, both abolished the ouabain-induced Ca^{2+} oscillations, suggesting that L-VGC channels are involved in the generation of Ca^{2+} oscillations. However, both these inhibitors also exhibit significant antioxidant activity[20] and there is some evidence that they may also inhibit CRAC channels.[21] Therefore, it is possible that the observed inhibition of Ca^{2+} oscillations by nifedipine and verapamil is not solely a result of their effect on L-VGCC.

Since Na,K-ATPase is an electrogenic pump, it cannot be excluded that, even in epithelial cells, partial inhibition of its activity by ouabain may lead to some membrane depolarization and therefore may activate L-VGCC. Depolarization of cell membrane by the depolarizing agent, 4-aminopyridine (4-AP) or high extracellular K^+, however, did not cause $[Ca^{2+}]_i$ oscillations.[5] Activation of L-VGCC by BayK 8644 (an L-type voltage-gated Ca^{2+} channel agonist; Sigma) causes a slow increase in intracellular Ca^{2+} but no oscillations. Taken together, these data show that activation of L-VGCC alone is not sufficient to trigger ouabain-induced Ca^{2+} oscillations.

The classical way to activate IP_3Rs involves increased generation of IP_3, triggered by ligand/G-protein coupled receptor interaction and PLC activation. Preincubation of cells with a PLC inhibitor, U73122, did not prevent ouabain-induced oscillations (unpublished observation). To test the efficiency of PLC inhibition, cells were also exposed to bradykinin, a well-known activator of phospholipase C (PLC) and $InsP_3$ production. Preincubation of RPT cells with a PLC inhibitor abolished bradykinin-induced Ca^{2+} transients. Taken together, these findings indicate that activation of PLC and subsequent increased generation of IP_3 are not required for induction of Ca^{2+} oscillations by ouabain/Na,K-ATPase complex.

A close spatial proximity between plasma membrane and endoplasmic reticulum has been demonstrated in renal epithelial cells.[22] It was therefore reasonable to hypothesize that IP_3R may interact with Na,K-ATPase. In ongoing studies we have found that Na,K-ATPase co-localizes with two subtypes of IP_3R (types 2 and 3). To further investigate the spatial relationship between Na,K-ATPase and $InsP_3R$ we are now using FRET, which provides resolution in the nanometer scale. The study is performed on COS cells, stably transfected with green fluorescent protein (GFP)-tagged rat Na,K-ATPase catalytic $\alpha 1$ subunit. Na,K-ATPase α-subunit fused to GFP on the NH_2-terminus (NKA-GFP) serves as a FRET donor. IP_3R types 2 and 3 labeled with Cy3-conjugated secondary antibody serve as FRET acceptor. According to the spectral properties, the combination of GFP and Cy3 allows detection of FRET at a distance up to approximately 12 nm.[23] So far we have observed a significant FRET effect between Na,K-ATPase and IP_3R types 2 and 3, indicating that Na,K-ATPase and $InsP_3R$ are separated by less than ~12 nm. Preincubation of cells with ouabain appears to enhance Na,K-ATPase/IP_3R FRET efficiency.

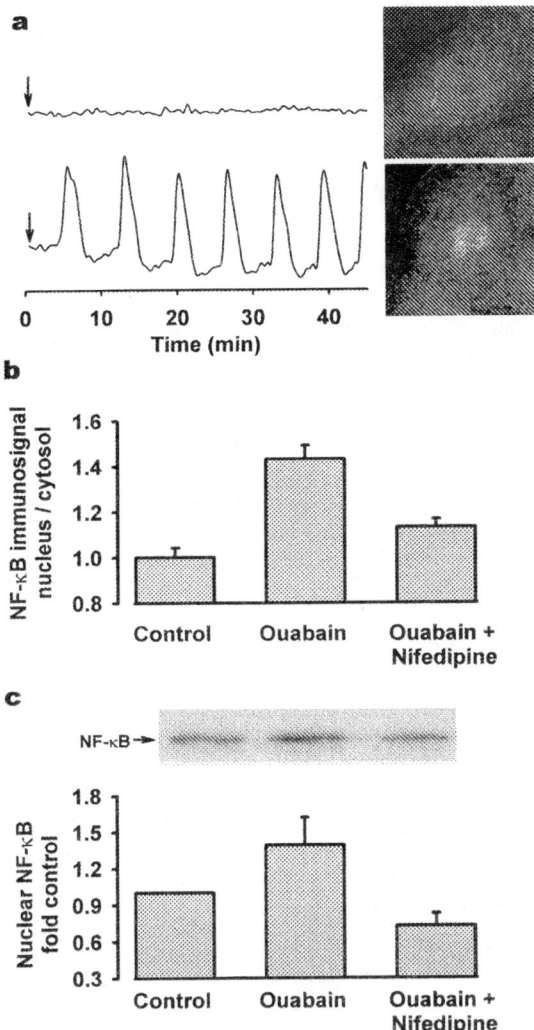

FIGURE 2. Effect of $[Ca^{2+}]_i$ oscillations on ouabain-induced NF-κB activation. (**a**) A cell cluster was treated with 250 μM ouabain (*arrow*) and individual cells were analyzed for both $[Ca^{2+}]_i$ and NF-κB immunofluorescence. *Upper panel* shows a typical nonoscillating $[Ca^{2+}]_i$ response (*left*) and its corresponding cellular NF-κB localization (*right*). *Lower panel* shows a typical oscillating $[Ca^{2+}]_i$ response (*left*) and its corresponding cellular NF-κB localization (*right*). (**b**) Semiquantitative analysis of NF-κB immunofluorescence signal showing translocation from cytosol to nucleus in cells exposed to 250 μM ouabain, in the absence or presence of 50 μM nifedipine. Values are mean ± S.E. 50–150 cells. Representative Western blot and densitometric analysis of 3–5 experiments showing changes in (**c**) nuclear NF-κB protein in cells exposed to 250 μM ouabain in the presence or absence of nifedipine. (From Aizman *et al.*[1] With permission from the National Academy of Sciences.)

Since both Na,K-ATPase and IP$_3$R are anchored by cytoskeleton proteins,[24,25] it was reasonable to expect that perturbation of the actin cytoskeleton network may influence the physical and/or functional Na,K-ATPase/IP$_3$R interaction. In fact, actin cytoskeleton hyperpolymerization or depolymerization by jasplakinolide (JP) or cytochalasin D (CytD), respectively,[26] abolished ouabain-induced Ca^{2+} oscillations.

We suggested a model for Ca^{2+} signaling triggered by ouabain/Na,K-ATPase complex. According to this model Na,K-ATPase and IP$_3$R interact with each other and this interaction requires an intact actin cytoskeleton. Ouabain, via allosteric changes in Na,K-ATPase, enhances the Na,K-ATPase/IP$_3$R interaction and triggers the frequency of the Ca^{2+} oscillations. Perturbations in actin cytoskeleton disrupt Na,K-ATPase/IP$_3$R interaction and thereby abolish Ca^{2+} oscillations. Ouabain-induced Ca^{2+} oscillations also require several permissive factors, such as Ca^{2+} influx via voltage-gated Ca^{2+} channels and/or CRAC channels.

The involvement of signaling cascades initiated by ouabain/Na,K-ATPase complex in the regulation of cell growth, proliferation and apoptosis have been previously suggested by several groups.[4,27,28] NF-κB is involved in the transcriptional regulation of many genes related to growth, differentiation, and apoptosis.[12] Additionally, NF-κB activation has been shown to be sensitive to and preferentially activated by slow Ca^{2+} oscillations.[10] This prompted us to study the effect of ouabain-induced Ca^{2+} oscillations on the activity of the transcription factor NF-κB. In unstimulated cells, NF-κB is predominantly located in the cytoplasm in association with inhibitory peptide IκB. Upon activation, NF-κB translocates to the nucleus.[12] The ratio between NF-κB immunosignal from the nucleus and from the cytosol has been semiquantitatively estimated (FIG. 2b). NF-κB nuclear staining was increased in cells that responded to ouabain with typical [Ca^{2+}]$_i$ oscillations (FIG. 2a). In cells where no effect of ouabain on [Ca^{2+}]$_i$ was detected, we did not observe any significant effect on the nuclear NF-κB immunosignals.[5] Subcellular fractionation and immunoblot studies the confirmed cytosolic–nuclear redistribution of NF-κB signals in all cells exposed to ouabain doses sufficient to trigger calcium oscillations (FIG. 2c). Interestingly, cells exposed to an ouabain concentration (2 mM) that caused a sustained increase in [Ca^{2+}]$_i$ exhibited a lesser degree of NF-κB activation.[5]

SUMMARY

We have demonstrated a novel role for Na,K-ATPase as a signal transducer involved in transcriptional regulation in mammalian cells. Taking into account the ubiquitous expression of Na,K-ATPase, it will be important to identify the cell-specific effects of this signaling. It will also be important to explore how the function of Na,K-ATPase as the key regulator of intracellular Na$^+$ and K$^+$ homeostasis may interrelate to its function as a signal transducer.

REFERENCES

1. KOMETIANI, P. *et al.* 1998. Multiple signal transduction pathways link Na$^+$/K$^+$-ATPase to growth-related genes in cardiac myocytes. The roles of Ras and mitogen-activated protein kinases. J. Biol. Chem. **273:** 15249–15256.

2. PENG, M., L. HUANG, Z. XIE, et al. 1996. Partial inhibition of Na$^+$/K$^+$-ATPase by ouabain induces the Ca^{2+}-dependent expressions of early-response genes in cardiac myocytes. J. Biol. Chem. **271:** 10372–10378.
3. TIAN, J., X. GONG & Z. XIE. 2001. Signal-transducing function of Na$^+$-K$^+$-ATPase is essential for ouabain's effect on [Ca^{2+}]$_i$ in rat cardiac myocytes. Am. J. Physiol. (Heart Circ. Physiol.) **281:** J1899–J1907.
4. XIE, Z. & A. ASKARI. 2002. Na(+)/K(+)-ATPase as a signal transducer. Eur. J. Biochem. **269:** 2434–2439.
5. AIZMAN, O., P. UHLEN, M. LAL, et al. 2001. Ouabain, a steroid hormone that signals with slow calcium oscillations. Proc. Natl. Acad. Sci. USA **98:** 13420–13424.
6. BERRIDGE, M.J., M.D. BOOTMAN & P. LIPP. 1998. Calcium—a life and death signal. Nature **395:** 645–648.
7. BERRIDGE, M.J., P. LIPP & M.D. BOOTMAN. 2000. The versatility and universality of calcium signalling. Nat. Rev. Mol. Cell Biol. **1:** 11–21.
8. BERRIDGE, M.J. 1997. The AM and FM of calcium signalling. Nature **386:** 759–760.
9. DE KONINCK, P. & H. SCHULMAN. 1998. Sensitivity of CaM kinase II to the frequency of Ca^{2+} oscillations. Science **279:** 227–230.
10. DOLMETSCH, R.E., K. XU & R.S. LEWIS. 1998. Calcium oscillations increase the efficiency and specificity of gene expression. Nature **392:** 933–936.
11. BAEUERLE, P.A. & D. BALTIMORE. 1988. I kappa B: a specific inhibitor of the NF-kappa B transcription factor. Science **242:** 540–546.
12. DELFINO, F. & W.H. WALKER. 1999. Hormonal regulation of the NF-kappaB signaling pathway. Mol. Cell. Endocrinol. **157:** 1–9.
13. UHLEN, P. et al. 2000. Alpha-haemolysin of uropathogenic E. coli induces Ca^{2+} oscillations in renal epithelial cells. Nature **405:** 694–697.
14. LINGREL, J.B. 1992. Na,K-ATPase: isoform structure, function, and expression. J. Bioenerg. Biomembr. **24:** 263–270.
15. KAWAMURA, A. et al. 2001. Biological implication of conformational flexibility in ouabain: observations with two ouabain phosphate isomers. Biochemistry **40:** 5835–5844.
16. PRAKRIYA, M. & R.S. LEWIS. 2001. Potentiation and inhibition of Ca(2+) release-activated Ca(2+) channels by 2-aminoethyldiphenyl borate (2-APB) occurs independently of IP(3) receptors. J. Physiol. **536:** 3–19.
17. MA, H.T., K. VENKATACHALAM, J.B. PARYS & D.L. GILL. 2002. Modification of store-operated channel coupling and inositol trisphosphate receptor function by 2-aminoethoxydiphenyl borate in DT40 lymphocytes. J. Biol. Chem. **277:** 6915–6922.
18. PUTNEY, J.W. 1999. "Kissin' cousins": intimate plasma membrane-ER interactions underlie capacitative calcium entry. Cell **99:** 5–8.
19. ZHAO, P.L. et al. 2002. Tubular and cellular localization of the cardiac L-type calcium channel in rat kidney. Kidney Int. **61:** 1393–1406.
20. MAK, I.T., J. ZHANG & W.B. WEGLICKI. 2002. Protective effects of dihydropyridine Ca-blockers against endothelial cell oxidative injury due to combined nitric oxide and superoxide. Pharmacol. Res. **45:** 27–33.
21. KRUTETSKAIA, Z.I., O.E. LEBEDEV, N.I. KRUTETSKAIA & T.V. PETROVA. 1997. [Organic and inorganic blockers of potential-dependent Ca^{2+} channels inhibit store-dependent entry of Ca^{2+} into rat peritoneal macrophages]. Tsitologiia **39:** 1131–1141.
22. ICHITANI, Y. et al. 2001. Cyclooxygenase-1 and cyclooxygenase-2 expression in rat kidney and adrenal gland after stimulation with systemic lipopolysaccharide: in situ hybridization and immunocytochemical studies. Cell Tissue Res. **303:** 235–252.
23. NG, T. et al. 1999. Imaging protein kinase Calpha activation in cells. Science **283:** 2085–2089.
24. BOURGUIGNON, L.Y., H. JIN, N. IIDA, et al. 1993. The involvement of ankyrin in the regulation of inositol 1,4,5-trisphosphate receptor-mediated internal Ca^{2+} release from Ca^{2+} storage vesicles in mouse T-lymphoma cells. J. Biol. Chem. **268:** 7290–7297.
25. DEVARAJAN, P., D.A. SCARAMUZZINO & J.S. MORROW. 1994. Ankyrin binds to two distinct cytoplasmic domains of Na,K-ATPase alpha subunit. Proc. Natl. Acad. Sci. USA **91:** 2965–2969.
26. PATTERSON, R.L., D.B. VAN ROSSUM & D.L. GILL. 1999. Store-operated Ca^{2+} entry: evidence for a secretion-like coupling model. Cell **98:** 487–499.

27. MCCONKEY, D.J., Y. LIN, L.K. NUTT, *et al.* 2000. Cardiac glycosides stimulate Ca^{2+} increases and apoptosis in androgen-independent, metastatic human prostate adenocarcinoma cells. Cancer Res. **60:** 3807–3812.
28. YEH, J.Y., W.J. HUANG, S.F. KAN & P.S. WANG. 2001. Inhibitory effects of digitalis on the proliferation of androgen dependent and independent prostate cancer cells. J. Urol. **166:** 1937–1942.

Molecular Mechanisms of Na/K-ATPase–Mediated Signal Transduction

ZIJIAN XIE

Department of Pharmacology and Medicine, Medical College of Ohio, Toledo, Ohio 43614, USA

ABSTRACT: Our recent work shows that in addition to pumping ions, Na/K-ATPase acts as a signal transducer. Binding of ouabain to Na/K-ATPase changes the interaction of the enzyme with neighboring membrane proteins and induces the formation of multiple signaling modules, resulting in activation of Src, transactivation of the EGF receptor (EGFR), and increased production of reactive oxygen species (ROS). Interaction of these signals leads to activation of several other cascades, including p42/44 and p38 MAPKs, phospholipase C, and protein kinase C isozymes, in a cell-specific manner. Ouabain also increases $[Ca^{2+}]_i$ and contractility, induces some of the early-response protooncogenes, and activates transcription factors AP-1 and NF-κB. Interplay among these pathways eventually results in changes in the expression of a number of growth-related genes and in cell growth. Significantly, inhibition of Src blocked many of the aforementioned ouabain-activated signaling pathways. Furthermore, Src binds to Na/K-ATPase directly and ouabain regulates the interaction between Src and the enzyme, resulting in Src activation. To address the possibility that the signaling Na/K-ATPase is concentrated in a separate pool on the plasma membrane, we have assessed interaction of the enzyme with caveolins. These studies indicated that Na/K-ATPase was concentrated in caveolae/rafts. In addition, caveolin-1 can be co-immunoprecipitated with Na/K-ATPase. Finally, we have shown that the signaling function of the enzyme is also pivotal to ouabain-induced nongenomic effects on cardiac myocytes.

KEYWORDS: Na/K-ATPase; ouabain; Src; caveolae; inter-receptor communication

INTRODUCTION

Na/K-ATPase was discovered as an energy transducing ion pump, and its pumping function and its regulation have been studied extensively since 1957.[1,2] Although early findings suggested that the enzyme also played a role in regulation of gene expression and cell growth,[3,4] only in recent years have studies been performed to investigate the molecular mechanisms by which this plasma membrane enzyme organizes signaling modules and regulates the functions of different proteins.[5-14] This work, done mostly on neonatal rat cardiac myocytes, shows that Na/K-ATPase has multiple signaling partners and that ouabain activates various signaling branches

---

Address for correspondence: Dr. Zijian Xie, Department of Pharmacology and Medicine, Medical College of Ohio, Toledo, OH 43614. Voice: 419-383-4182; fax: 419-383-2871.
zxie@mco.edu

Ann. N.Y. Acad. Sci. 986: 497–503 (2003). © 2003 New York Academy of Sciences.

to send messages to various intracellular organelles via regulation of interactions between the enzyme and its partners. Significantly, many of these findings have also been reported recently in cells other than cardiac myocytes.[9,13,15–19] In addition, Na/K-ATPase was found to be essential for dopamine-induced activation of phosphatidylinositol 3-kinase in rat kidney proximal tubular cells.[20] Realization that Na/K-ATPase is a signal transducer has prompted us to ask "What are the possible endogenous ligands for Na/K-ATPase?" In principle, there are at least three major potential classes of chemicals that could serve as Na/K-ATPase ligands under both physiological and pathological conditions. The first group of chemicals is the cardiac glycosides (e.g., endogenous ouabain or marinobufagin-like factors).[21–23] The second class will be those that may chemically modify the enzyme either reversibly or irreversibly. We propose that ROS may fit in this category.[24] The third will be alterations in the concentrations of intracellular sodium and extracellular potassium. This article reviews some of our recent mechanistic studies that are related to the initiation, compartmentalization, and biological consequences of Na/K-ATPase–mediated signal transduction.

Src AND Na/K-ATPase–MEDIATED SIGNAL TRANSDUCTION

Interaction of Na/K-ATPase with Src

Src family kinases are 52–62-kDa membrane-associated nonreceptor tyrosine kinases that are key regulators of various signal transduction pathways.[25,26] The kinase activity of Src is regulated by tyrosine phosphorylation.[26] Phosphorylation of Tyr^{418} activates the kinase activity of Src. On the other hand, phosphorylation of a conserved Tyr^{529} induces formation of an inactive conformation through intramolecular interaction between the SH2 domain and the C-terminus. Interaction of the SH3 domain with the linker between the kinase domain and the SH2 domain further keeps the enzyme in the inactive state. Therefore, competitive binding of a regulatory protein to either the SH2 or SH3 domain will destabilize the inactive conformation of Src, resulting in increases in phosphorylation of Tyr^{418} and the kinase activity. We showed previously that binding of ouabain to Na/K-ATPase activated Src. Since Na/K-ATPase is not a receptor tyrosine kinase, we proposed that ouabain might regulate the interaction between Src and Na/K-ATPase, resulting in Src activation. To test this hypothesis, we have performed the following three sets of experiments. First, we measured the effects of ouabain on tyrosine phosphorylation of Src, since either stimulation of Tyr^{418} phosphorylation via a conformation change or Tyr^{529} dephosphorylation by a phosphotyrosine phosphatase can stimulate Src activity. We found that in both A7r5 cells and LLC-PK1 cells ouabain stimulated Tyr^{418} phosphorylation, but had no effect on Tyr^{529} phosphorylation.[13] These data support our proposition that ouabain regulates the interaction between Na/K-ATPase and Src. Therefore, in the second set of experiments we immunoprecipitated α1 from both ouabain-treated and control LLC-PK1 cells, and probed for Src. We found that ouabain increased Src binding to the Na/K-ATPase signaling complex in a dose- and time-dependent manner. In addition, when Src was immunoprecipitated from the LLC-PK1 cells, ouabain clearly increased the co-precipitated α1 subunit of the enzyme.[13] These findings indicate that there is indeed an interaction between Na/K-

ATPase and Src in response to ouabain. Because Src could bind to Na/K-ATPase through a scaffold protein, we have recently performed a third set of experiments to dissect the nature of the Src interaction with the enzyme. GST-Src was expressed in *E. coli*, purified, and used to pull down 1% Triton X-100 solubilized pig kidney Na/K-ATPase. The experiments showed that the solubilized Na/K-ATPase bound to GST-Src in a dose-dependent manner. In short, the above findings clearly demonstrate that Src can interact with Na/K-ATPase directly and that ouabain regulates the interaction between Src and Na/K-ATPase.

Src Relays Ouabain Signal from Na/K-ATPase to the EGFR and Other Signaling Cascades

It is well established that receptor tyrosine kinases (RTKs) are central elements for cellular signal transduction.[25] In recent years there is a growing body of evidence that RTKs cross-communicate with other signaling systems to integrate the variety of extracellular stimuli into a limited number of signaling pathways. The activated EGFR, for example, has been identified as a critical element in the signal transduction networks of cytokines, H_2O_2, and those using G protein–coupled receptors.[25] Since Src family kinases can couple the receptors lacking intrinsic kinase activity to RTKs, we proposed that activation of Src by ouabain might serve as a mediator for Na/K-ATPase to communicate with the EGFR so that the extracellular ouabain signal can be transmitted to the Ras/MAPK cascade. Indeed, our recent work showed that activation of Src was essential for the ouabain-induced activation of p42/44 MAPKs, since ouabain failed to activate MAPKs in A7r5 cells that were pretreated with Src inhibitor PP2 and herbimycin A.[13] This was further supported by the experiments performed in both SYF and SYF + Src cells.[13] The SYF cells are derived from mouse embryos harboring functional null mutations in both alleles of the Src family kinases Src, Yes, and Fyn. The SYF + c-Src cells are the stable transfectants of the SYF cells that express c-Src. While ouabain activated p42/44 MAPKs in a dose-dependent manner in the SYF + c-Src cells, it failed to stimulate MAPKs in the SYF cells. Furthermore, we showed that inhibition of Src also blocked ouabain-, but not EGF-induced Src binding to the EGFR and subsequent EGFR tyrosine phosphorylation. Finally, we found that the transactivated EGFR was capable of recruiting and phosphorylating the adaptor protein Shc, resulting in increased binding of the Grb2/Sos complex to the activated EGFR receptors. This led to the recruitment of Ras and the subsequent activation of the Raf/MEK/MAPK cascade. Significantly, we showed that the EGFR, but not the platelet-derived growth factor receptor, was involved in ouabain-induced activation of MAPKs. Taken together, these new findings support the proposal that activation of Src is the initial critical step that relays the signal emanating from the interaction of ouabain with Na/K-ATPase to the EGFR.

CAVEOLAE AND COMPARTMENTALIZATION OF Na/K-ATPase

Caveolins are 21–24 kDa membrane-associated scaffold proteins and are major structural components of caveolae, which exist as flask-shaped vesicular invaginations of plasma membrane and are rich in cholesterol, glycosphingolipids, and sph-

Na/K ATPase α Subunits		Caveolin-binding Motif 100	Ouabain-binding Domain 120	Caveolin-binding Motif 990
P05023	A1A1_HUMAN	CRQLGGGSM	SIQAATEEEQN---DNLY	KPTWWECAFPYSLLI
P05024	A1A1_PIG	CRQLGGGSM	GIQAATEEEQN---DNLY	KPTWWECAFPYSLLI
P06685	A1A1_RAT	CRQLGGGSM	GIRSATEEEPN---DELY	KPTWWECAFPYSLLI
Q92123	A1A1_XENLA	CRQLGGGSM	GIQAAMEEEQN---DNLY	KPTWWECAFPYSLII
Q92030	A1A1_ANGAN	CRQLGGGSM	GIQAASEDEAN---DNLY	KPSQWECAFPYSLLI
P25489	A1A1_CATCO	CKQMGGGSM	GIIAAMEDEAN---DNLY	KPNAWECAFPYSLLI
P05025	AT1A_TORCA	CRQLGGGSI	GIQVATVDNEAN---DNLY	KPSAWECAFPYSLII
P13607	ATNA_DROME	CKNLGGGAM	SIQASTSEEEAD---DNLY	KLVQWEPAIPEALAI
Q27766	MTH1733_CTEFE	CKNLGGGAL	SIQASTVEEEAD---DNLY	KFVWWLPALPEMISI
P28774	AT1B_ARTSF	CKNLGGGAL	GIEASSGNEDMLK--DNLY	KINAWEPALPESFLI
P17326	AT1A_ARTSF	CKQLGGGQM	TMEKYKN-PDVLG--DNLY	KIWQWEPPMPESLLI
P35317	AT1A_HYDAT	CKQMGGGSM	GIRAVRD-TNENM--DELY	NFTWWLPGLPESLLI
Q27461	EAT-6_CAEEL	CKNLGGGAM	SVLYFTMEYESK---DNLY	RFSWWSCALPESILI
Q9W5Y2	CG17923_DROME	LKTMGGGAI	LIQLQTQHEEPD---DNLY	KFIQWWEIYAFPGLLI
T18833	CAEEL	LAGSIGGGNF	GMLLSMSDD-EEVPKDNMY	RLEIGELALPEAFFI
O45240	T23E1.2_CAEEL	LAGSIGGGNF	GMLLSMSDD-EEVPKDNMY	RLEIGELALPEAFFI
Q95024	IONA_DICDI	LGKCFTNFEMI	GLDRNQR--------VNLY	PGLEWAXPLPMIFCL
O43134	PAT1_BLAEM	LECLLALENF	GLDPVSN--------YANIY	NPLYLLIPFGVGFVL
O16436	C02E7.1_CAEEL	LRQFKNLLWI	IYDPSDL--------TNLC	PWQCWLVPIVVGVWI
O16331	C09H5.2_CAEEL	VRQFKNLLWV	IYDPTDA--------LNLY	PWECWLVPVIVGIWI
Q9W248	CG3701_DROME	LKSCESILGI	YLFATKTPDNGKVDPEFLV	ELHEGHELLTNCPFM

FIGURE 1. CBMs and ouabain binding domains in IIC Na,K-ATPase α subunits.

ingomyelin.[27] Three major groups of caveolins have been identified and named as caveolin-1, 2, and 3. The primary sequence of caveolin-1 contains three palmitoylation sites, a central hydropholic domain (residues 102–134) that anchors to membranes, an oligomerization domain (residues 61–101), and a scaffolding domain (residues 82–101).[27] Interaction between the oligomerization domains and the C-terminal domains results in formation of high molecular oligomers containing 14 to 16 caveolins, which is important for the scaffolding function of caveolins.[27] Interestingly, multiple proteins such as EGFR, Src, PKC, and Ras contain caveolin-binding motifs (CBM),[27] and interaction of caveolin-1 with these proteins plays an important role in clustering of these signaling proteins in the compartment of caveolae. Database searches indicate that both α and β subunits of mammalian Na/K-ATPase contain conserved CBM (e.g., $\Phi XX\Phi XXXX\Phi$ and $\Phi X\Phi XXXX\Phi$), where Φ represents an aromatic amino acid residue. Interestingly, the CBM in the α subunits was acquired quite early during evolution. It first appeared in one of the mutated Na/K-ATPases in *C. elegans,* and has been conserved since *Drosophila* (FIG. 1). Significantly, the appearance of the CBMs correlates well with the occurrence of the domain for ouabain binding. These findings led us to propose that caveolae may cluster the signaling Na/K-ATPase with its partners. Confocal imaging of immunostained LLC-PK1 cells showed that Na/K-ATPase co-localized with caveolin-1 on plasma membrane. Both density gradient fractionation experiments and immunoprecipitation using anti-cav-1–coated magnetic beads revealed that Na/K-ATPase was enriched in caveolae, together with Src and other signaling proteins.[28] To test whether the ouabain-activated signal transduction can originate from caveolae, we immunoprecipitated caveolin-1 from ouabain-treated LLC-PK1 cells, and probed for α1 and Src by Western blot analysis of the immunocomplex. We found that 100 nM ouabain (2 min exposure) increased the binding of both Na/K-ATPase and Src to caveolin-1. More importantly, ouabain stimulated tyrosine-phosphorylation of caveolin-1. Taken together, these findings clearly demonstrated that the ouabain-bound Na/K-ATPase was capable of recruiting and assembling of both Src and caveolin-1 into signaling modules in LLC-PK1 cells. In addition, these findings provide strong evidence that at least some of the signaling events of ouabain must be originated from caveolae in these cells.

BIOLOGICAL CONSEQUENCES OF THE SIGNAL TRANSDUCING FUNCTION OF Na/K-ATPase

Regulation of Cardiac Growth and Growth-Related Genes by Na/K-ATPase in Cardiac Myocytes

Several years ago we became interested in the role of Na/K-ATPase in the nonproliferative growth (hypertrophy) of cardiac myocytes. This stemmed from the growing realization that cardiac hypertrophy plays an important role in the development of heart failure. Our early studies demonstrated that exposure of the cultured cardiac myocytes to ouabain stimulated cardiac growth.[5–8] Like other hypertrophic stimuli, ouabain also regulated transcription of several hypertrophic marker genes in cardiac myocytes.[5–8] Clearly, Na/K-ATPase must now be considered as a potential signal transducer for hypertrophic growth in the heart along with other membrane receptors.

Regulation of $[Ca^{2+}]_i$ and Myocyte Contractility by Ouabain Involves p42/44 MAPKs and ROS

Since activation of PTKs preceded the increases in $[Ca^{2+}]_i$ in response to ouabain,[12] we have tested if the signal transducing function of Na/K-ATPase contributes to nongenomic effects of ouabain on myocytes. These studies showed the following:[12,14] First, inhibition of either Src or Ras abolished ouabain-induced increases in both $[Ca^{2+}]_i$ and contractility. Second, while activation of p42/44 MAPKs was required for ouabain-induced rise in $[Ca^{2+}]_i$, both MAPKs and ROS contributed to ouabain regulation of cardiac contraction.[14] Finally, ouabain stimulated mitochondrial ATP-sensitive K channel (mitoK$_{ATP}$) activity, resulting in increases in mitochondrial ROS production. Inhibition of mitoK$_{ATP}$ by 5-hydroxydecanoate significantly reduced ouabain-induced increases in contractility. Thus, ouabain regulates cardiac contractility via activation of at least two major pathways. Activation of p42/44 MAPKs and inhibition of the ion pumping function of the Na/K-ATPase by ouabain increased $[Ca^{2+}]_i$, whereas opening of mitoK$_{ATP}$ stimulated the production of ROS. Both $[Ca^{2+}]_i$ and ROS, in turn, worked in concert, resulting in increases in contractility in cardiac myocytes. Although most of the above studies need to be repeated in the isolated heart and whole animals, the significance of these studies should be noted. First, the above findings provide a new insight into the mechanism of ouabain action on contractility. Second, they point to a possibility that blocking of certain pathways may separate the therapeutic effect from the toxicity of ouabain on the heart.

Ouabain Protects the Perfused Rat Heart against Ischemia-Reperfusion Injury

Ischemia preconditioning describes an experimental treatment of the normal heart that reduces myocyte damage from a subsequent ischemia-reperfusion event. We were struck by the fact that elements of the ouabain signaling pathway, including

Src, PKC, and mitoK$_{ATP}$ overlap with those involved in the signaling pathway of preconditioning.[29] Accordingly, we carried out experiments on perfused rat hearts to determine whether pretreatment with ouabain is cardioprotective. Results from ten preparations indicate that 80 µM ouabain administered 15 min before 30-min ischemia is cardioprotective. These data show that ouabain protection is equal to protection by ischemia preconditioning, with respect to enzyme release during the first 10 min of reperfusion. To ensure that the cardioprotective effects of ouabain are not limited to these ouabain-insensitive rodents, we also measured whether ouabain protects rabbit cardiac myocytes from ischemia. Rabbit cardiac myocytes express ouabain-sensitive Na/K-ATPase, and 50% inhibition occurs when the cells are exposed to 1 µM ouabain. Using an *in vitro* ischemic model, we showed that preincubation of cardiac myocytes with 0.1 to 1 µM ouabain gave a dose-dependent protection of these cells from ischemia-induced cell death. Since digitalis glycosides are still widely used in the treatment of congestive heart failure, the above findings have significant clinical implications.

CONCLUSIONS AND FUTURE PERSPECTIVES

The work of the past few years has clearly demonstrated that Na/K-ATPase is an important signal transducer. Some of our recent work has begun to identify the various signaling partners of Na/K-ATPase and the domains that are involved in such interactions. It is important to emphasize that the above studies are not only important for us to understand the events regulated by the signaling Na/K-ATPase, but also the regulation of the enzyme as an energy transducer by different signaling pathways. Clearly, this new line of investigation will provide vast opportunities for significant expansion of research in the Na/K-ATPase field.

ACKNOWLEDGMENTS

This work was supported by National Institutes of Health Grants HL-36573, HL-63238, and HL-67963 by the National Heart, Lung and Blood Institute.

REFERENCES

1. SKOU, J.C. 1957. The influence of some cations on an adenosine triphosphatase from peripheral nerves. Biochim. Biophys. Acta **23:** 394–401.
2. LINGREL, J.B. & T. KUNTZWEILER. 1994. Na$^+$,K$^+$-ATPase. J. Biol. Chem. **269:** 19659–19662.
3. PRESSLEY, T.A. 1992. Ionic regulation of Na,K-ATPase expression. Semin. Nephrol. **12:** 67–71.
4. NAKAGAWA, Y., V. RIVERA & A.C. LARNER. 1992. A role for the Na$^+$/K$^+$-ATPase in the control of human c-fos and c-jun transcription. J. Biol. Chem. **267:** 8785–8788.
5. XIE, Z. 2001. Ouabain interaction with cardiac Na/K-ATPase reveals that the enzyme can act as a pump and a signal transducer. Cell. Mol. Biol. **47:** 383–390.
6. PENG, M., L. HUANG, Z. XIE, *et al.* 1996. Partial inhibition of Na$^+$/K$^+$-ATPase by ouabain induces the Ca^{2+}-dependent expressions of early-response genes in cardiac myocytes. J. Biol. Chem. **271:** 10372–10378.
7. HUANG, L., H. LI & Z. XIE. 1997. Ouabain-induced hypertrophy in cultured cardiac myocytes is accompanied by changes in expressions of several late response genes. J. Mol. Cell. Cardiol. **29:** 429–437.

8. KOMETIANI, P., J. LI, L. GNUDI, et al. 1998. Multiple signal transduction pathways link Na^+/K^+-ATPase to growth-related genes in cardiac myocytes: The roles of Ras and mitogen-activated protein kinases. J. Biol. Chem. **273:** 15249–15256.
9. XIE, Z., P. KOMETIANI, J. LIU, et al. 1999. Intracellular reactive oxygen species mediate the linkage of Na/K-ATPase to hypertrophy and its marker genes in cardiac myocytes. J. Biol. Chem. **274:** 19323–19328.
10. HAAS, M., A. ASKARI & Z. XIE. 2000. Involvement of Src and epidermal growth factor receptor in the signal transducing function of Na^+/K^+-ATPase. J. Biol. Chem. **275:** 27832–27837.
11. LIU, J., J. TIAN, M. HAAS, et al. 2000. Ouabain interaction with cardiac Na^+/K^+-ATPase initiates signal cascades independent of changes in intracellular Na^+ and Ca^{2+}. J. Biol. Chem. **275:** 27838–27844.
12. TIAN, J., X. GONG & Z. XIE. 2001. Signal-transducing function of Na^+/K^+-ATPase is essential for ouabain's effect on $[Ca^{2+}]_i$ in rat cardiac myocytes. Am. J. Physiol. **281:** H1899–H1907.
13. HAAS, M., H. WANG, J. TIAN & Z. XIE. 2002. Src-mediated interceptor cross-talk between the Na/K-ATPase and the EGF receptor relays the signal from ouabain to mitogen-activated protein kinases. J. Biol. Chem. **277:** 18694–18702.
14. TIAN, J., J. LIU, J.I. SHAPIRO, et al. 2002. Involvement of mitogen-activated protein kinases and reactive oxygen species in the inotropic action of ouabain on cardiac myocytes. A potential role for mitochondrial K_{ATP} channels. Mol. Cell. Biochem. In press.
15. ZHOU, X., G. JIANG, A. ZHAO, et al. 2001. Inhibition of Na,K-ATPase activates PI3 kinase and inhibits apoptosis in LLC-PK1 cells. Biochem. Biophys. Res. Commun. **285:** 46–51.
16. CONTRERAS, R.G., L. SHOSHANI, C. FLORES-MALDONADO, et al. 1999. Relationship between Na(+),K(+)-ATPase and cell attachment. J. Cell Sci. **112:** 4223–4232.
17. WATABE, M., N. KAWAZOE, Y. MASUDA, et al. 1997. Bcl-2 protein inhibits bufalin-induced apoptosis through inhibition of mitogen-activated protein kinase activation in human leukemia U937 cells. Cancer Res. **57:** 3097–3100.
18. AYDEMIR-KOKSOY, A., J. ABRAMOWITZ & J.C. ALLEN. 2001. Ouabain induced signaling and vascular smooth muscle cell proliferation. J. Biol. Chem. **276:** 46605–46611.
19. AIZMAN, O., P. UHLEN, M. LAL, et al. 2001. Ouabain, a steroid hormone that signals with slow calcium oscillations. Proc. Natl. Acad. Sci. USA **98:** 13420–13424.
20. YUDOWSKI, G.A., R. EFENDIEV, C.H. PEDEMONTE, et al. 2000. Phosphoinositide-3 kinase binds to a proline-rich motif in the Na^+,K^+-ATPase α subunit and regulates its trafficking. Proc. Natl. Acad. Sci. USA **97:** 6556–6561.
21. SCHONER, W. 2002. Endogenous cardiac glycosides, a new class of steroid hormones. Eur. J. Biochem. **269:** 2440–2448.
22. HAMLYN, J.M., M.P. BLAUSTEIN, S. BOVA, et al. 1991. Identification and characterization of a ouabain-like compound from human plasma. Proc. Natl. Acad. Sci. USA **81:** 6259–6263.
23. FEDOROVA, O.V., P.A. DORIS & A.Y. BAGROV. 1998. Endogenous marinobufagin-like factor in acute plasma volume expansion. Clin. Exp. Hypertens. **20:** 581–591.
24. LI, J., J. LIU, A. ASKARI & Z. XIE. 2000. Stimulation of hypertrophic growth in cardiac myocytes by reactive oxygen species. FASEB J. **14:** A1453.
25. PRENZEL, N., O.M. FISCHER, S. STREIT, et al. 2001. The epidermal growth factor receptor family as a central element for cellular signal transduction and diversification. Endocr.-relat. Cancer **8:** 11–31.
26. THOMAS, S.M. & J.S. BRUGGE. 1997. Cellular functions regulated by Src family kinases. Annu. Rev. Cell. Dev. Biol. **13:** 513–609.
27. ANDERSON, R.G. 1998. The caveolae membrane system. Annu. Rev. Biochem. **67:** 199–225.
28. LIU, L., K. MOHAMMADI, B. AYNAFSHAR, et al. 2003. Role of caveolae in the signal transducing function of cardiac Na^+/K^+-ATPase. Am. J. Physiol. In press.
29. GARLID, K.D., P. PAUCEK, V. YAROV-YAROVOY, et al. 1997. Cardioprotective effect of diazoxide and its interaction with mitochondrial ATP-sensitive K^+ channels. Possible mechanism of cadioprotection. Circ. Res. **81:** 1072–1082.

Low Concentrations of Ouabain Activate Vascular Smooth Muscle Cell Proliferation

JULIUS C. ALLEN, JOEL ABRAMOWITZ, AND ASLIHAN KOKSOY

Baylor College of Medicine, Houston, Texas 77030, USA

ABSTRACT: In vascular smooth muscle cells the sodium pump complex can act as an intracellular signal transducing complex activated by low ouabain concentrations, which inhibit sufficient pumps to activate a transduction cascade via transactivation of EGFR, but insufficient pumps to alter intracellular ions. Higher concentrations interfere with proliferation. This biphasic ouabain response occurs in human, canine, and rat VSMC at concentrations that reflect the differing ouabain affinities of the α1 isoforms of the three species. This supports the proposal that this effect occurs via ouabain binding to the α1 subunit of the Na pump. These data suggest a new transducing function of ouabain–Na pump interaction, distinct from the cellular ionic effects resulting from pump inhibition. This transducing function occurs at ouabain concentrations that do not perturb cytoplasmic ion content and requires specific localization of pumps to caveolae.

KEYWORDS: ouabain, Na,K-ATPase; VSMC proliferation; EGFR, ERK1/2

INTRODUCTION

Recently a number of laboratories have reported the observation that the Na pump may have a function in addition to its well-known control of ionic gradients. First shown by Askari and coworkers in rat cardiomyocytes,[1] lower concentrations of the cardiac glycoside ouabain activated a signaling pathway Src→EGFR→ERK1/2. This pathway was entirely separate from any inhibitory effect of ouabain on the pump.[2]

RESULTS AND DISCUSSION

When we studied this highly novel effect in canine vascular smooth muscle cells (VSMC), the cellular response to very low concentrations of ouabain was proliferation, as measured by both DNA synthesis and cell counts.[3]

Effects of low concentrations of ouabain have now been shown to occur in a variety of cell types: rat cardiac muscle cells;[1] kidney epithelial cells (Ca oscillations);[4] prostate smooth muscle cells (proliferation);[5] and in canine, human, and

Address for correspondence: Dr. Julius C. Allen, Department of Medicine, Section of Cardiovascular Sciences, Baylor College of Medicine, One Baylor Plaza, Houston, TX 77030. Voice: 713-798-4977.
  juliusa@bcm.tmc.edu

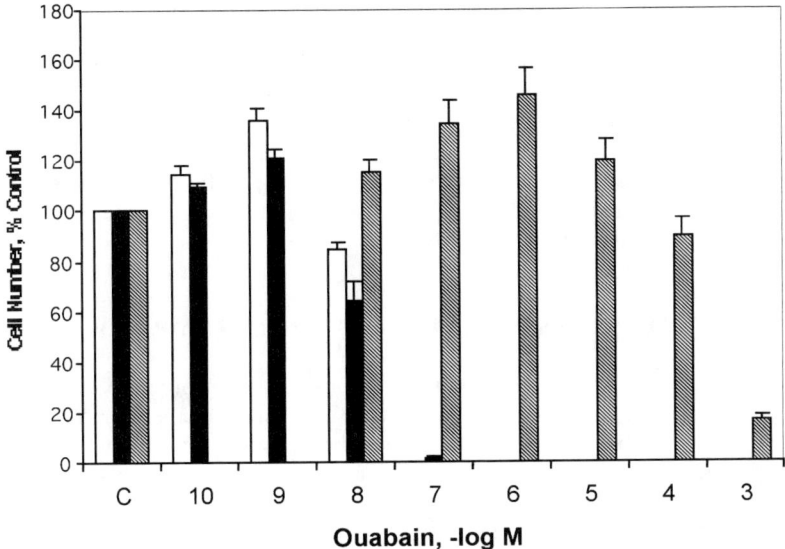

FIGURE 1. Ouabain dose-response effects on proliferation of canine VSMC (*open bars*), human VSMC (*closed bars*), and rat VSMC (A_7r_5 cells) (*hatched bars*). Cell incubation densities varied depending on the cell type. Cells were incubated in ouabain-free medium overnight, then ouabain was added at the indicated concentrations. Incubation was continued for 5 days. Cell numbers were counted on the fifth day.

rodent VSMC. A variety of different responses occur in response to this effect. We showed that 0.1–1.0 nM ouabain activated the same pathway (Src→EGFR→ERK1/2) in canine VSMC as did higher concentrations in rat cardiomyocytes, and caused cellular proliferation. We also showed that higher concentrations of ouabain inhibited the proliferation effect, that is, the effect was biphasic.

An explanation for these biphasic effects follows. At very low ouabain concentrations, ~1–2% of the pumps are inhibited. Any small increase in Na_i would immediately be eliminated by the remaining activity of the uninhibited pumps. Thus the effect on Na_i is small, transitory, and difficult to detect.

Because these proliferative effects occur at very low concentrations of ouabain, the usual ligand binding studies for determination of affinity are difficult to perform. Thus we have assumed that the α1 subunit is the receptor and have performed an ouabain dose-response curve on rat VSMC (A_7r_5 cells) known to contain only the α1 subunit. This subunit is three orders of magnitude less sensitive than the subunit in "sensitive" species. A comparison of these dose-response curves is shown in FIGURE 1.

The effect in both human and canine VSMC starts at 0.1 nM, peaks at 1 nM, and is inhibited at 10 nM. With the rat VSMC line A_7r_5, the proliferative effect begins at 0.1 µM, peaks at 1 µM (three orders of magnitude higher than the effective concentrations with the sensitive cells), followed by inhibition. These data are consistent with the hypothesis that ouabain is activating the signaling pathway by binding to the α1 subunit.

We propose that ouabain interacts with the receptor (α1 subunit) activating the proliferative pathway. While the pumps to which these low concentrations of ouabain bind are by definition inhibited, insufficient numbers are inhibited to effect a detectable change in Na_i. Furthermore, any increase in Na_i that does occur is reduced by the remaining uninhibited pumps. Thus at low ouabain concentrations with minimal increases in Na_i, the proliferative pathway is predominant and proliferation occurs in the absence of detectable pump inhibition. However, as the ouabain concentration is increased and more pumps are inhibited, an increase in Na_i occurs, which in turn inhibits the proliferative pathway.

FUNCTIONAL INTEGRATION OF THE Na PUMP IN VSMC

It might seem that the ionic control (responsive) and proliferative (regulatory) functions of the Na pump described herein are unrelated. We therefore propose that the Na pump distribution within the cell membrane is not uniform, and there are at least two locations of Na pumps within the membrane, one in the caveolae and the other in the bulk membrane. Such specific localization of a variety of proteins, including channel proteins and the sarcoplasmic reticulum Ca pump, occur within these specialized lipid rafts.[6–8] A major function of these caveolae can be defined by their content of the scaffolding protein, caveolin, of which there are at least three tissue-specific isoforms. This protein supports clustering of proteins that are not usually in close proximity when limited to the bulk cell membrane.

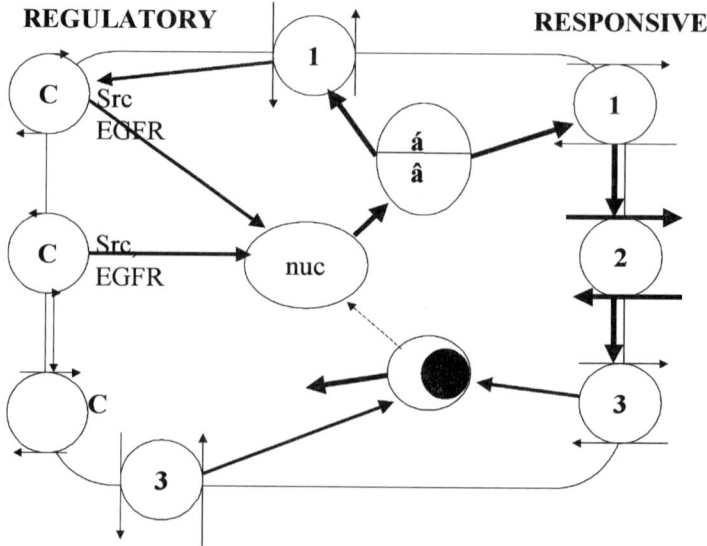

FIGURE 2. A model of Na pump membrane compartmentation and the potential interaction between regulatory and responsive functions. This figure is described in the text.

There are two specific types of functions of the Na pump—"responsive" and "regulatory." The responsive function of the pump occurs as it responds to physiologic or pharmacologic incidents and is represented by the right side of FIGURE 2. The pumps are designated 1, 2, and 3. These three pumps may have different ionic capacities, but are localized to the bulk membrane.

As pump 1 is moved into the membrane, it may proceed to become a functional pump 2, or it may move toward the caveolae. While the mechanism of activation of protein movement into caveolae is largely unknown, it has been suggested that the regulatory protein may be EGF, as it interacts with its receptor (EGFR). Carpenter[9] has suggested that the EGFR is a nexus for both trafficking and signaling. Such a concept has not yet been applied to the Na pump, but it is important to note that in the cardiac data from Xie and Askari,[1,2] and our VSMC data,[3] transactivation of EGFR is essential for the initiation of the cascade.

Thus pump 1 may become attached to caveolin "C" and continue to be further localized to the caveolae, where it then can be brought into close proximity with the signaling proteins Src and EGFR. It would then be able to respond to low concentrations of ouabain and thus activate a signaling cascade. This model suggests that specific localization of pump proteins to specific membrane sites can significantly modulate cellular functions.

PHYSIOLOGICAL RELEVANCE

The possible role of an endogenous digitalis-like factor (EDLF) compound in the etiology of hypertension has been discussed for some time, most recently in two short reviews.[10,11] The current effects of low concentrations of ouabain via MAPK activation on VSMC proliferation now offers an additional possible functional pathway. In a very recent paper,[12] it was shown that both basal- and angiotensin II–activated MAPK of aortic cells were elevated in 9-week spontaneously hypertensive rats (SHR) when compared to control. These observations in a model of hypertension, coupled with our observations of VSMC MAPK activation by low levels of ouabain certainly warrant attention and open the possibilities that the function of an EDLF need not be limited to an effect on ion transport.

REFERENCES

1. PENG, M., L. HUANG, W.-H HUANG & A. ASKARI. 1996. Partial inhibition of NaK ATPase by ouabain induces the Ca dependent expressions of early response genes in cardiac myocytes. J. Biol. Chem. **271:** 10372–10378.
2. HAAS, M., H. WANG, J. TIAN & Z. XIE. 2002. Src-mediated inter-receptor cross talk between the NaK ATPase and the EGFR relays the signal from ouabain to MAP kinases. J. Biol. Chem. **277:** 18694–18702.
3. KOKSOY, A.A., J. ABRAMOWITZ & J.C. ALLEN. 2001. Ouabain induced signaling and vascular smooth muscle cell proliferation. J. Biol. Chem. **276:** 46605–46611.
4. AIZMAN, O., P. UHLEN, M. LAL, et al. 2001. Ouabain, a steroid hormone that signals with low calcium oscillations. Proc. Natl. Acad. Sci. USA **98:** 13420–13424.
5. CHUEH, S.-C., J.-H. GUH, J. CHEN, et al. 2001. Dual effects of ouabain on the regulation of proliferation and apoptosis in human prostatic smooth muscle cells. J. Urol. **166:** 347–353.

6. XIE, Z. & A. ASKARI. 2002. Na,K ATPase as a signal transducer. Eur. J. Biochem. **269:** 2434–2439.
7. USHIO-FUKAI, L. N. MILENSKI, N. SANTANAM, *et al.* 2001. Cholesterol depletion inhibits EGFR transactivation by angiotensin II in vascular smooth muscle cells. J. Biol. Chem. **276:** 48269–48275.
8. DANIEL, E.E. 2002. The smooth muscle cell membrane and peripheral sarcoplasmic reticulum: their interactions are organized and may be crucial. J. Hypert. **21:** 367–370.
9. CARPENTER, G. 2000. The EGF receptor: a nexus for trafficking and signaling. Bioessays **22:** 697–707.
10. SCHONER, W. 2002. Endogenous cardiac glycosides, a new class of steroid hormones. Eur. J. Biochem. **269:** 2440-2448.
11. DMITRIEVA, R.I. & P.A. DORIS. 2002. Cardiotonic steroids: potential endogenous Na pump ligands with diverse function. Exp. Bio. Med. **227:** 561–569.
12. KUBO, T., T, IBUSUKI, S. CHIBA, *et al.* 2002. Altered MAPK activation in vascular smooth muscle cells from SHR. Clin. Exper. Physiol. Pharm. **29:** 537–543.

Regulation of Ca^{2+} Signaling by Na^+ Pump Alpha-2 Subunit Expression

VERA GOLOVINA,[a] HONG SONG, PAUL JAMES,[b] JERRY LINGREL,[c] AND MORDECAI BLAUSTEIN

Department of Physiology, University of Maryland School of Medicine, Baltimore, Maryland 21201, USA

[b]*Department of Zoology, Miami University, Oxford, Ohio 45056, USA*

[c]*Department of Molecular Genetics, Biochemistry and Microbiology, University of Cincinnati College of Medicine, Cincinnati, Ohio 45267, USA*

ABSTRACT: The role of the Na^+ pump α2 subunit in Ca^{2+} signaling was examined in primary cultured astrocytes from wild type (WT) mouse fetuses and those with a null mutation in one (Het) or both (KO) α2 genes. Cytosol Na^+ and Ca^{2+} concentrations ($[Na^+]_{CYT}$ and $[Ca^{2+}]_{CYT}$) were measured with benzofuran isophthalate (SBFI) and Fura-2. Het astrocytes express ≈50% of normal α2; KO cells express none. Resting $[Na^+]_{CYT}$ = 6.5 mM in WT, 6.8 mM in Het, and 8.0 mM in KO cells; 500 nM ouabain (which inhibits only α2) equalized $[Na^+]_{CYT}$ at 8 mM in all genotypes. Resting $[Ca^{2+}]_{CYT}$ = 132 nM in WT, 162 nM in Het and 196 nM in KO cells. Cyclopiazonic acid (CPA), which inhibits endoplasmic reticulum (ER) Ca^{2+} pumps, induces elevation of $[Ca^{2+}]_{CYT}$. These Ca^{2+} responses are augmented in Het and KO cells. This correlates with α2 Na^+ pump and Na^+/Ca^{2+} exchanger (NCX) localization in plasma membrane (PM) microdomains that overlie the ER. Selective reduction of α2 Na^+ pump activity apparently elevates *local* $[Na^+]$ and, via NCX, $[Ca^{2+}]$ in the tiny cytosolic space between the PM and ER. This augments adjacent ER Ca^{2+} stores and amplifies Ca^{2+} signaling without elevating bulk $[Na^+]_{CYT}$.

KEYWORDS: astrocytes; Fura-2; SBFI; Na,K-ATPase isoforms; transgenic mice

INTRODUCTION

Cytosolic Ca^{2+} signals regulate numerous physiological processes in all cells. To help elucidate how these signals are controlled, we studied the influence of Na^+ pump expression on resting $[Ca^{2+}]_{CYT}$ and Ca^{2+} signaling in primary cultured mouse cortical astrocytes.

Both a PM ATP-driven Ca^{2+} pump (PMCA) and a Na^+/Ca^{2+} exchanger (NCX) regulate resting $[Ca^{2+}]_{CYT}$ in astrocytes[1] as in other cell types. Most of the intracellular Ca^{2+} in quiescent cells is stored in the ER. By controlling $[Ca^{2+}]_{CYT}$, the

---

[a]Address for correspondence: Vera A. Golovina, Ph.D., Department of Physiology, University of Maryland School of Medicine, 655 W. Baltimore St., Baltimore, MD 21201. Voice: 410-706-4164; fax: 410-706-8341.

vgolovin@umaryland.edu

PMCA and NCX indirectly influence ER Ca^{2+} store content. During cell activity, much of the "signal Ca^{2+}" comes from the ER stores, although some also enters the cells through PM Ca^{2+}-permeable channels.

The NCX depends upon the PM Na^+ electrochemical gradient generated by the Na^+ pump. The Na^+ pump $\alpha 2$ isoform and NCX are confined to PM microdomains that overlie the "junctional" ER (jER) in astrocytes and other cells, whereas PMCA and the Na^+ pump $\alpha 1$ isoform are more uniformly distributed.[2] Thus, Na^+ pumps with $\alpha 2$ subunits, should regulate, via NCX, not only the local $[Ca^{2+}]$ in the junctional space, but also the jER $[Ca^{2+}]$ that plays a key role in Ca^{2+} signaling. To test this hypothesis, we measured "bulk" $[Na^+]_{CYT}$ and $[Ca^{2+}]_{CYT}$ in resting astrocytes, and the increase in $[Ca^{2+}]_{CYT}$ induced by blocking the ER Ca^{2+} pump (SERCA). Astrocytes express only $\alpha 1$ and $\alpha 2$ Na^+ pump isoforms.[2] Therefore, these parameters were studied in astrocytes from normal and $\alpha 2$ transgenic mice.[3]

METHODS

Mice with null mutations in one or both $\alpha 2$ genes (i.e., heterozygotes or Het = $\alpha 2^{+/-}$, and homozygote knockouts, KO = $\alpha 2^{-/-}$, respectively) were generated as described.[3] Wild type mice (WT = $\alpha 2^{+/+}$) were from the same litters. Het mice appear normal and develop normally into adults.[3] KO mice die very shortly after birth,[3] but the fetuses appear normal. Astrocytes were cultured from near-term fetuses,[4] and $[Ca^{2+}]_{CYT}$ and $[Na^+]_{CYT}$ were measured with Fura-2 and benzofuran isophthalate (SBFI), respectively.[4]

RESULTS AND DISCUSSION

Na^+ Pump α Subunit Isoform Expression

Astrocytes, which normally express both the $\alpha 1$ and $\alpha 2$ isoforms,[2] are able to grow normally when $\alpha 2$ is knocked out. Het astrocytes express $\approx 50\%$ of normal (WT) $\alpha 2$; KO cells do not express any $\alpha 2$. Expression of $\alpha 1$ is normal in Het and KO cells. The morphology of cells from the three genotypes is very similar.

Cytosolic Na^+ Concentrations

The "bulk" $[Na^+]_{CYT}$ was 6.5 ± 0.1 mM in quiescent WT astrocytes and 6.8 ± 0.4 mM in Het cells. $[Na^+]_{CYT}$ was, however, slightly elevated (8.0 ± 0.3 mM) in KO cells (FIG. 1A). These small effects might be attributed to a small reduction in total α subunit expression because $\alpha 2$ accounts for only $\approx 20\%$ of the total α subunit and an even smaller fraction of the total Na^+ pump flux in WT astrocytes.

The effects of 500 nM and 1 mM ouabain on $[Na^+]_{CYT}$ were compared. In rodents, 500 nM ouabain blocks only $\alpha 2$ ($IC_{50} = 20-100$ nM), whereas 1 mM blocks both isoforms ($\alpha 1$ $IC_{50} = 50-100$ mM).[5] Ouabain (500 nM, 10 min) increased $[Na^+]_{CYT}$ only to 8.0 mM in WT and Het cells. It did not affect $[Na^+]_{CYT}$ in KO cells. In contrast, 1 mM ouabain increased $[Na^+]_{CYT}$ by ≈ 2 mM/min in all three genotypes. Clearly, the elevated resting $[Na^+]_{CYT}$ in KO cells is due to reduced Na^+ extrusion

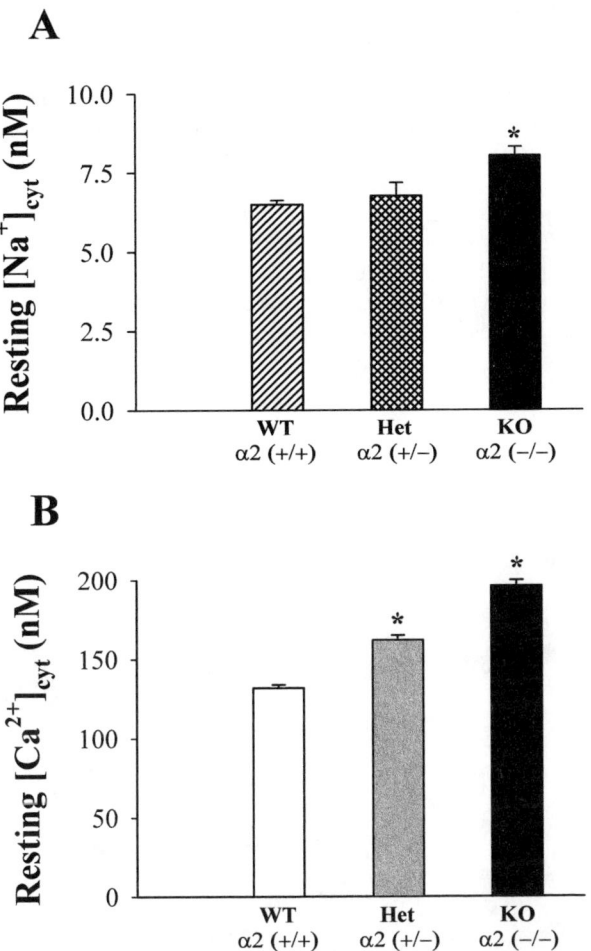

FIGURE 1. (**A**) Resting $[Na^+]_{CYT}$ in astrocytes from WT (69 cells), Het (40 cells), or KO (43 cells) fetuses. (**B**) Resting $[Ca^{2+}]_{CYT}$ data are from 670 WT cells, 244 Het cells, and 465 KO cells. Each bar shows data from 4–5 fetuses from two litters. *$p < 0.001$ vs. WT.

by Na^+ pumps with α2 subunits because the effect of α2-KO is mimicked by 500 nM ouabain. Thus, the α1 isoform is the primary determinant of $[Na^+]_{CYT}$, and α2 has only minor influence on bulk $[Na^+]_{CYT}$.

Resting $[Ca^{2+}]_{CYT}$ and Releasable ER Ca^{2+}

Resting $[Ca^{2+}]_{CYT}$ data are summarized in FIGURE 1B. The average resting $[Ca^{2+}]_{CYT}$ is 132 ± 2 nM in the WT astrocytes and is elevated by ≈23% in Het cells and by ≈49% in KO cells.

Storage of Ca^{2+} in the ER is governed by the ambient $[Ca^{2+}]_{CYT}$ and by SERCA. Thus, if resting $[Ca^{2+}]_{CYT}$ is elevated in cells from Het and KO mice, we would expect storage of (releasable) Ca^{2+} in the ER to be increased. Cyclopiazonic acid (CPA, 10 µM), which unloads ER Ca^{2+}, elevated $[Ca^{2+}]_{CYT}$ to 563 nM in WT cells, 695 nM in Het cells and 871 nM in KO mice incubated in the presence of 1.8 mM extracellular Ca^{2+}. Comparable data were obtained from cells incubated in Ca^{2+}-free medium.

While more pronounced effects were observed in KO astrocytes, the resting $[Ca^{2+}]_{CYT}$ and Ca^{2+} transient data from Het cells are noteworthy because $[Na^+]_{CYT}$ was not elevated (FIG. 1A). These effects on Ca^{2+} transients are attributable to inhibition of Na^+ pumps with α2 subunits because 100–500 nM ouabain has a comparable effect in astrocytes.[6]

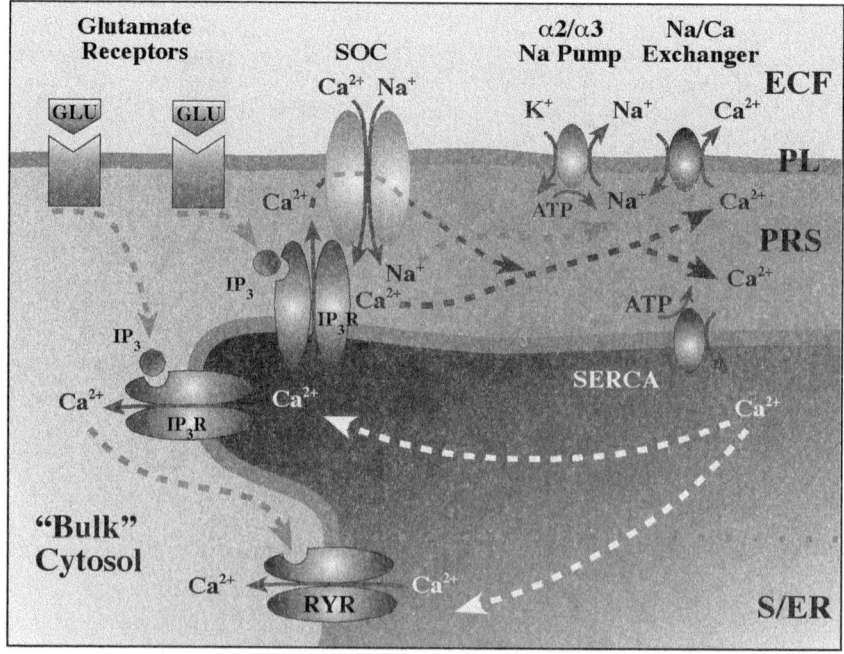

FIGURE 2. The PM-jER region showing key transporters involved in local control of jER Ca^{2+} stores and modulation of Ca^{2+} signals. The PM microdomain overlying jER contains α2/α3 Na^+ pumps, NCX, and store-operated channels (SOCs). The jER contains SERCA and inositol trisphosphate and ryanodine receptors (IP$_3$R, RYR). A tiny, "diffusion-restricted" cytosolic space (PRS) lies between the PM (PL) and jER. Shading indicates relative concentrations of Na^+ and/or Ca^{2+}. ECF = extracellular fluid; α1 Na^+ pumps are widely distributed in the PM, but may be excluded from these microdomains.

Reduced α2 Activity Augments Ca^{2+} Signaling by Regulating $[Na^+]$ in a Sub-PM Cytosolic Compartment

Na^+ pump inhibition and reduction of the PM $[Na^+]$ gradient, for example, by ouabain, promotes Ca^{2+} entry via NCX and augments Ca^{2+} signaling in most cell types.[7] This does not occur when NCX expression is blocked.[6]

How does reduced Na^+ pump α2/α3 subunit activity modify Ca^{2+} homeostasis without raising bulk $[Na^+]_{CYT}$ (in α2 Hets)? We must assume that these Na^+ pumps regulate the local $[Na^+]$ and, via NCX, $[Ca^{2+}]$ in a distinct subcompartment of the cytosol. Na^+ and Ca^{2+} diffusion between this PM-reticular space (PRS in FIG. 2) and bulk cytosol must be markedly restricted.[1] Such a compartment also explains how the low Na^+ affinity α2 and α3 isoforms function while the "housekeeping" high Na^+ affinity α1 isoform maintains bulk $[Na^+]_{CYT}$ below 10 mM.

Immunocytochemistry[2,8] reveals that PM microdomains overlying jER contain Na^+ pumps with α2 or α3 (but not α1) subunits, and NCX. These proteins and those in the jER are all involved in Ca^{2+} signaling. In contrast, other regions of the PM are rich in PMCA and Na^+ pumps with α1 subunits.[8] This structural and functional coupling, depicted in FIGURE 2, can be used to explain the augmented Ca^{2+} transients observed in Het and KO astrocytes, and in WT astrocytes treated with low-dose ouabain.[6] These structural units (PLasmERosomes) obviously play a key role in regulating Ca^{2+} signaling. Therefore, it seems appropriate to refer to them as "Ca^{2+} signaling complexes."

ACKNOWLEDGMENTS

Supported by the National Institutes of Health, the American Heart Association, and Miami University.

REFERENCES

1. BLAUSTEIN, M.P. & V. GOLOVINA. 2001. Structural complexity and functional diversity of endoplasmic reticulum Ca^{2+} stores. Trends Neurosci. **24:** 602–608.
2. JUHASZOVA, M. & M. P. BLAUSTEIN. 1997. Na^+ pump low and high ouabain affinity alpha subunit isoforms are differently distributed in cells. Proc. Natl. Acad. Sci. USA **94:** 1800–1805.
3. JAMES, P.F., et al. 1999. Identification of a specific role for the Na,K-ATPase α2 isoform as a regulator of calcium in the heart. Mol. Cell **3:** 555–563.
4. GOLOVINA, V.A., et al. 1996. Modulation of two functionally distinct Ca^{2+} stores in astrocytes: role of the plasmalemmal Na/Ca exchanger. Glia **16:** 296–305.
5. BLANCO, G. & R.W. MERCER. 1998. Isozymes of the Na,K-ATPase: heterogeneity in structure, diversity in function. Am. J. Physiol. **275:** F633–F650.
6. BLAUSTEIN, M.P., et al. 2002. Na/Ca exchanger and PMCA localization in neurons and astrocytes: functional implications. Ann. N. Y. Acad. Sci. **976:** 356–366.
7. BLAUSTEIN, M.P. & W.J. LEDERER. 1999. Sodium/calcium exchange: its physiological implications. Physiol. Rev. **79:** 763–854.
8. JUHASZOVA, M. & M.P. BLAUSTEIN. 1997. Distinct distribution of different Na^+ pump alpha subunit isoforms in plasmalemma. Physiological implications. Ann. N. Y. Acad. Sci. **34:** 524–536.

Visualization of Na,K-ATPase Interacting Proteins Using FRET Technique

PER UHLÉN

Department of Woman and Child Health, Karolinska Intitutet,
Astrid Lindgren Children's Hospital, S-171 76, Stockholm, Sweden

ABSTRACT: Signaling transduction mediated by protein aggregates within specific microdomains has been receiving increased attention. We previously showed that Na,K-ATPase, partially inhibited by ouabain, induces intracellular calcium (Ca^{2+}) oscillations which involve Ca^{2+} release from the endoplasmic reticulum (ER). Plasma membrane bound Na,K-ATPase and proteins in the ER are in close proximity to each other, and signal transduction may occur via a physical interaction or a microdomain. To study these signaling pathways and intricate microenvironments, sophisticated methods are required. One way to detect molecular interactions in the nanometer scale (1−10 nm) is fluorescence resonance energy transfer (FRET). Thus, FRET provides vital insight into the action of Na,K-ATPase to trigger intracellular signaling events.

KEYWORDS: Na,K-ATPase; fluorescence resonance energy transfer; FRET; signaling microdomain; protein–protein interaction

INTRODUCTION

It is becoming increasingly recognized that the compartmentalization of proteins within specific microdomains of the cell is essential to the appropriate functioning of many signal transduction pathways. For example, G proteins tend to exist not randomly on the plasma membrane, but rather in concentrated specialized distinct microdomains.[1] Previous studies have suggested that physical docking of intracellular ER Ca^{2+} stores with the plasma membrane activates Ca^{2+} release-activated Ca^{2+} (CRAC) channels through physical interaction between CRAC channels and the inositol-1,4,5-trisphosphate receptor ($InsP_3R$) in ER.[2] Also, evidence has been provided describing the presence of a spatially restricted signaling microdomain containing the B_2 bradykinin receptor and the $InsP_3R$.[3] To investigate the close proximity between proteins within microdomains, complex cell imaging techniques and analysis methods are necessary.

Address for correspondence: Dr. Per Uhlén, Department of Woman and Child Health, Karolinska Institutet, Astrid Lindgren Children's Hospital Q2:09, S-171 76 Stockholm, Sweden. Voice: +46-8-51777340; fax: +46-8-51777328.
per.uhlen@kbh.ki.se

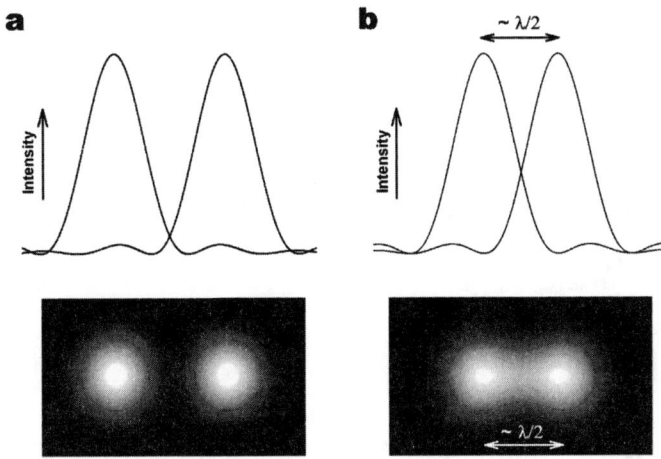

FIGURE 1. The limit of resolution refers to the smallest separation of two point sources of negligible diameter that can be distinguished as separate. (*Upper panels*) densitometer tracings taken along a line passing through the geometric "Gaussian" image points. (*Lower panel*) illustrate how point-objects emerge in a microscope. (**a**) Two easily resolved points. (**b**) Two objects separated by less than ~$\lambda/2$ will smear together as one and will show up as co-localized in the microscope.

IMMUNOCYTOCHEMICAL STUDIES

Immunocytochemical studies with labeled proteins have been used for decades to determine colocalization between proteins. However, cytochemical studies provide an indication of association, not interaction, and precise determination of proximity is limited by the resolution of the optical system. The smallest separation between two point-objects while still allowing them to be resolved is referred to as the limit of resolution. According to Rayleigh's criterion, two point-objects are said to be just resolved when the center of one Airy disk falls on the first minimum of the other Airy disk (FIG. 1). The Airy disk radius is given by

$$r_{\text{Airy}} = \frac{1.22\lambda}{2NA} \sim \frac{\lambda}{2},$$

where λ is the wavelength of light and *NA* is the numerical aperture of the objective. Thus, the maximal spatial resolution is approximately equal to half the excitation wavelength (typically >250 nm).

FRET

The FRET technique offers the unique possibility of monitoring molecules separated by 1–10 nm. Since sizes of proteins are in the nanometer scale, FRET can be used to extend the resolution of a fluorescence microscope to determine protein–

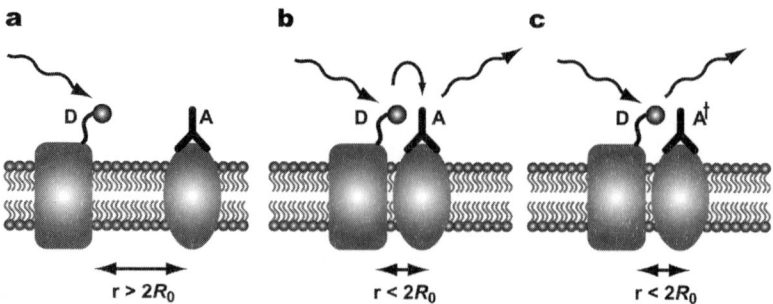

FIGURE 2. FRET can be used to show that colocalized proteins interact. Here, two membrane proteins are labeled, one with GFP and the other with an antibody directly tagged with a fluorophore. With favorable characteristics, these fluorophores can be used as a FRET donor–acceptor (D–A) pair. (**a**) No FRET will occur if the proteins are separated by distances greater than $2R_0$. However, if the distance is less than $\sim\lambda/2$, the proteins will appear as colocalized. (**b**) If the proteins interact and are separated by distances less than $2R_0$, FRET will occur. (**c**) The extent of donor quenching due to FRET can be quantified by eliminating the acceptor by photobleaching ($A^\dagger$). Fluorescence from D will increase after acceptor photobleaching.

protein interactions. FRET reflects the transfer of energy from an excited state of donor fluorophore (D) to a nearby acceptor fluorophore (A) (FIG. 2).[4,5] The energy transfer efficiency, E, is defined by

$$E = \frac{1}{1 - (r/R_0)^6}$$

where r is the distance separating the D–A pair and R_0 (the Förster radius) is the distance at which 50% energy transfer takes place (typically 2–6 nm).[6] Since the FRET effect decreases as the sixth power of the separation distance between the donor and acceptor, FRET occurs when the donor and acceptor are in the range $0.5R_0 < r < 1.5R_0$. Experimentally, FRET can be detected in several ways. In photobleaching-FRET either the donor[7] or acceptor[8] is photobleached. Images are recorded before and after photobleaching of a region within the sample. Thereafter, the ratio between before and after photobleaching is calculated to quantify the rate of energy transfer.

Intermolecular energy transfer critically depends on the stoichiometry of acceptors to donors and can vary with transfection efficiencies. Most favorable conditions occur when all the donors are paired with an acceptor, because unpaired protein adds noise to the signal. FRET experiments yield information about spatial location and exact timing of interaction. This information is not available from conventional detection systems such as coimmunoprecitation, cross-linking, yeast two-hybrid, phage display, or unlinked noncomplementing mutant detection.[9] The introduction of the green fluorescent protein (GFP) has greatly facilitated the ability to exploit FRET-based imaging microscopy *in vivo*.[10]

Na,K-ATPase AND FRET

In ongoing studies, performed by our group, we have examined protein interactions in cells stably transfected with GFP-tagged Na,K-ATPase. We recently reported an additional role for Na,K-ATPase as a signal transducer inducing intracellular Ca^{2+} oscillations following partial occupancy by its endogenous ligand, ouabain.[11] Studies of the structure–function relationship of the catalytic subunit of Na,K-ATPase have indicated that the N-terminus is the most flexible part of the molecule and the domain is most likely to interact with other molecules.[12] Hence, we used GFP fused to the N-terminus of the Na,K-ATPase α1-subunit (Na,K-ATPase-GFP) to serve as the FRET donor. Ca^{2+} oscillations are known to occur as interplay between different Ca^{2+} transporters.[13] The $InsP_3R$, which releases Ca^{2+} from intracellular ER Ca^{2+} stores, plays an essential role in this signaling event.[14] Histochemical studies using electron microscopy have demonstrated a close proximity between the plasma membrane and ER.[15] We were interested in investigating the signaling pathway between Na,K-ATPase and $InsP_3R$. As FRET acceptor, a Cy3-labeled $InsP_3R$ antibody, located in the C-terminus, was used ($InsP_3R$-Cy3). The fluorophores, GFP and Cy3, form a Förster D–A pair[16] with $R_0 = 60$ Å, which allows detection of energy transfer at distances up to ~12 nm. FRET efficiency was quantified as the release of donor quenching after irreversible photobleaching of the acceptor in a plasma membrane region of the cell. Calculations of the energy transfer between Na,K-ATPase-GFP and $InsP_3R$-Cy3 appear to give enhanced Na,K-ATPase-GFP fluorescent intensity following photobleaching. Thus, our findings suggest that Na,K-ATPase and $InsP_3R$ are separated by not more than ~12 nm and may indicate a signal transduction pathway between Na,K-ATPase and $InsP_3R$.

REFERENCES

1. NEUBIG, R.R. 1994. Membrane organization in G-protein mechanisms. FASEB J. **8:** 939–946.
2. MA, H.T. et al. 2000. Requirement of the inositol trisphosphate receptor for activation of store-operated Ca^{2+} channels. Science **287:** 1647–1651.
3. DELMAS, P. et al. 2002. Signaling microdomains define the specificity of receptor-mediated InsP3 pathways in neurons. Neuron **34:** 209–220.
4. FÖRSTER, T. 1946. Energiewanderung und Fluoteszenz. Naturwissenschaften **6:** 166–175.
5. FÖRSTER, T. 1948. Intermolecular energy migration and fluorescence. Ann. Phys. (Leipzig) **2:** 55–75.
6. WU, P. & L. BRAND. 1994. Resonance energy transfer: methods and applications. Anal. Biochem. **218:** 1–13.
7. JOVIN, T.M. & D.J. ARNDT-JOVIN. 1989. Luminescence digital imaging microscopy. Annu. Rev. Biophys. Biophys. Chem. **18:** 271–308.
8. MIYAWAKI, A. & R.Y. TSIEN. 2000. Monitoring protein conformations and interactions by fluorescence resonance energy transfer between mutants of green fluorescent protein. Methods Enzymol. **327:** 472–500.
9. MENDELSOHN, A.R. & R. BRENT. 1999. Protein interaction methods—toward an endgame. Science **284:** 1948–1950.
10. BASTIAENS, P.I. & R. PEPPERKOK. 2000. Observing proteins in their natural habitat: the living cell. Trends Biochem. Sci. **25:** 631–637.
11. AIZMAN, O. et al. 2001. Ouabain, a steroid hormone that signals with slow calcium oscillations. Proc. Natl. Acad. Sci. USA **98:** 13420–13424.

12. SEGALL, L., L.K. LANE & R. BLOSTEIN. 2002. New insights into the role of the N terminus in conformational transitions of the Na,K-ATPase. J. Biol. Chem. **277:** 35202–35209.
13. BERRIDGE, M.J., P. LIPP & M.D. BOOTMAN. 2000. The versatility and universality of calcium signalling. Nat. Rev. Mol. Cell Biol. **1:** 11–21.
14. MIKOSHIBA, K. 1997. The InsP3 receptor and intracellular Ca^{2+} signaling. Curr. Opin. Neurobiol. **7:** 339–345.
15. MAUNSBACH, A.B. 1981. Ultrastructure of epithelia as related to models of iso-osmotic transport. *In* Physiology of Non-excitable Cells, Vol. 3. J. Salánki, Ed.: 81–88. Akadémiai Kiadó. Budapest.
16. NG, T. *et al.* 1999. Imaging protein kinase Calpha activation in cells. Science **283:** 2085–2089.

Na/K-ATPase Regulates Intracellular ROS Level in Cerebellum Neurons

ALEXANDER BOLDYREV,[a] ELENA BULYGINA, MARIA YUNEVA, AND WILHELM SCHONER[b]

M.V. Lomonosov Moscow State University, 119992 Moscow, Russia
[b]*Justus Liebig University, D-35932 Giessen, Germany*

KEYWORDS: reactive oxygen species (ROS); flow cytometric analysis; cerebellum granule cells; glutamate receptors; NMDA; protein kinases

The interrelation between NMDA-receptors and Na/K-ATPase in cerebellum granule cells prepared from 10–12-day-old rats was studied by measuring Na/K-ATPase and reactive oxygen species (ROS) produced after treatment of neurons with NMDA. Preparation and loading of cerebellum granule cells with the fluorescent dyes for ROS were described elsewhere.[1] For ROS measurement, the neurons were loaded with dihydrorhodamine 123 (DHR) or di-(acetoxymethyl) ester of carboxy-H_2DCF-DA (DCF) using stationary fluorescence technique (Perkin-Elmer Spectrometer LS50B) or FACStar (Becton Dickinson) flow cytometry analysis. Viability of the cells was determined with propidium iodide (PI). Unspecific actions of ligands on Na/K-ATPase were evaluated from their effects tested on disintegrated cells without preincubation. In the figure only specific (receptor-mediated) effects of ligands are presented. Na/K-ATPase was measured in parallel samples after disintegration of neuronal membranes by freezing–thawing.[1]

Activation of the neurons by increased concentrations of NMDA (0.1–1 mM) leads to a progressive inhibition of Na/K-ATPase to about half of its control activity. This inhibition is prevented by 10 µM MK-801 (which inhibits NMDA-dependent ionic channels) and is abolished by 1 mM cysteine when added to NMDA-exposed cells. NMDA induces a progressive increase in intracellular ROS. Suppression of the NMDA-induced increase in ROS by MK-801 was also noticed. Inhibition of Na/K-ATPase is maximal at 0.5 mM NMDA and a further increase in NMDA concentration results in an increase of ROS, but not in ATPase inhibition. This means that only part of Na/K-ATPase in cerebellum neurons is sensitive to ROS. Because of high sensitivity of ATPase to ROS and its ability to be restored by cysteine, this portion of the enzyme presumably can be identified as ($\alpha 2 + \alpha 3$) isoforms, whereas the rest of the activity can be attributed to the $\alpha 1$ isoform.

[a]Address for correspondence: A. Boldyrev, M.V. Lomonosov Moscow State University, 119992 Moscow, Russia. Voice/fax: +7-095-939-1398.
aaboldyrev@mail.ru

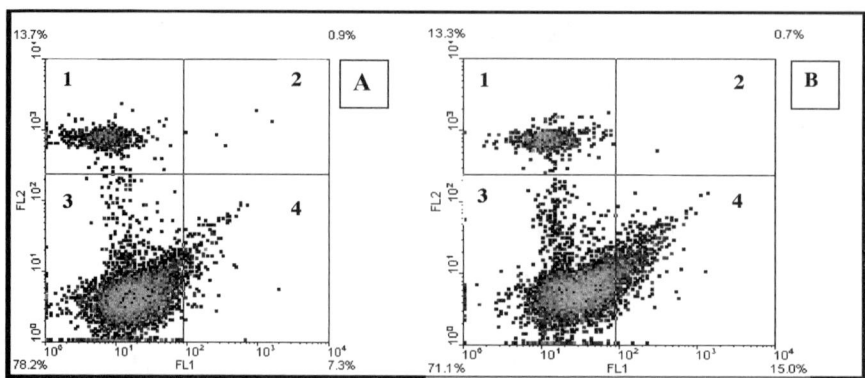

FIGURE 1. Ouabain increases ROS production by cerebellum neurons. Distributions of the cells between necrotic (PI-labeled, quadrants 1 and 2) and viable (DCF-labeled, quadrants 3 and 4) subpopulations in (**A**) control and (**B**) ouabain-treated (100 µM during 30 min) samples are shown. At the angles of each plot, the percent of the cells in the subsequent quadrant is shown. The FL1 axis corresponds to DCF and the FL2 axis to PI fluorescence.

Recently, an increase of the ROS levels by ouabain inhibition of Na/K-ATPase in cardiac cells has been reported.[2] In our experiments ouabain stimulated ROS generation in a dose-dependent manner starting at 50 µM. This effect is attenuated both by D-AP5 (which prevents NMDA binding) and by MK-801 (which inactivates the NMDA-linked ionic channels). More detailed analysis of the phenomenon was done by flow cytometry. FIGURE 1 demonstrates that 30-min incubation with 100 µM ouabain resulted in a 1.75-fold increase (from 20.1 ± 2.4 to 35.3 ± 2.8 units) in mean fluorescence of viable neurons and simultaneous selection of the neurons with enormously high ROS fluorescence (FIG. 1A and 1B, quadrant 4). At the same time, the portion of necrotic (PI-labeled) cells was unchanged (FIG. 1, quadrants 1 and 2). Exposure of the cells to NMDA, which resulted in increase in mean fluorescence was also accompanied by an increase in the amount of the neurons in the fourth quadrant, but the effects of NMDA and ouabain were not doubled.

The data presented demonstrate that ouabain-sensitive ($\alpha 2 + \alpha 3$) isoforms of neuronal Na/K-ATPase, which contributes about 50% of the total sodium pump activity, is the target for ROS inhibition, and the ouabain-insensitive $\alpha 1$ isoform may itself regulate its production.

ACKNOWLEDGMENTS

This work was supported by Deutscher Akademischer Austausch Dienst (DAAD), Germany (S-325) and by the Russian Foundation for Basic Research (RFBR), Russia (99-04-49420).

REFERENCES

1. Bulygina, E. *et al.* 2002. Activation of glutamate receptors inhibits Na/K-ATPase of cerebellum granule cells. Biochem. Mosc. **67:** 1209–1214.
2. Xie, Z. *et al.* 1999. Intracellular reactive oxygen species mediate the linkage of Na/K-ATPase to hypertrophy and its marker genes in cardiac myocytes. J. Biol. Chem. **274:** 19323–19328.

Intermolecular Interaction between Na^+/K^+-ATPase α Subunit and Glycogen Phosphorylase

KUNIO TAKEYASU, TSUBASA KAWASE, AND SHIGE H. YOSHIMURA

Graduate School of Biostudies, Kyoto University, Kyoto, 606-8502, Japan

Keywords: protein–protein interaction; functional regulation; glycolysis; glycogen phosphorylase

The Na^+/K^+-ATPase exhibits an evolutionary conservation[1] and requires an assembly of at least two subunits via an interaction between the extracellular regions.[2] In addition to the well-characterized interaction between the α and β subunits of the Na^+/K^+-ATPase, there also exist interactions between the subunits and other cytoplasmic proteins. First, the α subunit forms a homo-dimer under certain conditions.[3,4] Second, in cells, the α subunit is anchored to a distinct place via an interaction with ankyrin.[5,6] Third, the α subunit also interacts with various types of protein kinases.[7,8] Fourth, the α subunit binds certain chemicals at its cytoplasmic domains.[9,10] Here we report the fifth interaction between the α subunit and the glycogen phosphorylase.

The carboxy-terminal domain (Asp_{1000}-Tyr_{1021}) of the chicken Na^+/K^+-ATPase α subunit was expressed in bacteria as a GST-fusion protein and coupled to the glutathione beads. The beads were then incubated with the cell extract prepared from the chicken kidney. Bound proteins were analyzed by SDS-PAGE (FIG. 1a). A protein with the molecular weight of around 100 kDa was identified (arrowhead) and isolated followed by a CNBr treatment. Peptide sequencing of the obtained fragments demonstrated that all the fragments analyzed were derived from the glycogen phosphorylase.

A polyclonal antibody against the phosphorylase expressed in bacteria was raised and used to examine the subcellular localization of the antigen (FIG. 1b). Colocalization of the Na^+/K^+-ATPase and the phosphorylase on the plasma membrane was evident. A functional interaction between the Na^+/K^+-ATPase and the glycogen phosphorylase was also demonstrated by the Na^+/K^+-ATPase activity assays in the presence and absence of the phosphorylase. The Na^+/K^+-ATPase activity was repressed in the presence of the phosphorylase (FIG. 1c).

Address for correspondence: Dr. Kunio Takeyasu, Graduate School of Biostudies, Kyoto University, Kitashirakawa-oiwake-cho, Sakyo-ku, Kyoto, 606-8502, Japan. Voice: +81-75-753-6852; fax: +82-75-753-6852.
takeyasu@lif.kyoto-u.ac.jp

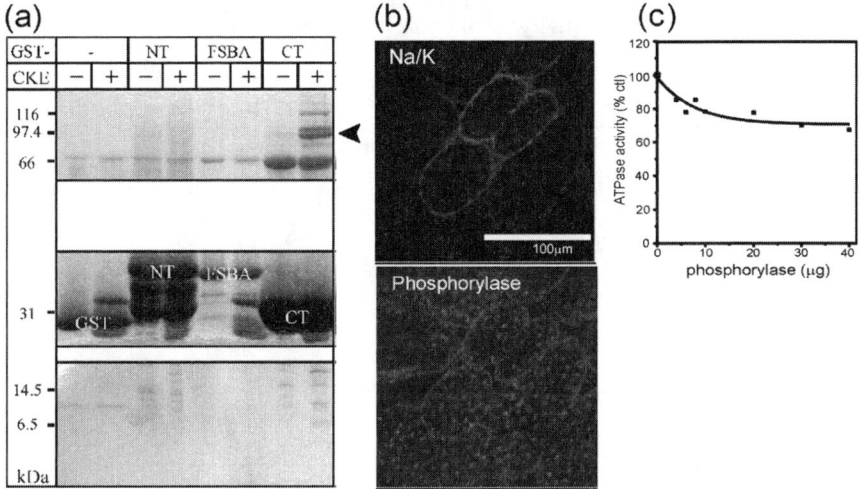

FIGURE 1. (a) Pull-down assay with GST-fused carboxyl-terminal region of the α subunit. The amino-terminal domain (NT), the FSBA binding domain (FSBA), or the carboxyl-terminal domain (CT) of the Na^+/K^+-ATPase α subunit was expressed as a GST-fusion protein in bacteria. Chicken kidney extract (CKE) was incubated with the GST-fusion protein coupled to the glutathione beads. Bound proteins were analyzed by SDS-PAGE and CBB staining. (b) Colocalization of the Na^+/K^+-ATPase and glycogen phosphorylase in skeletal muscle. A section of the chicken skeletal muscle was immunostained by anti-Na^+/K^+-ATPase β-subunit (*upper panel*) and anti-glycogen phosphorylase antibodies (*lower panel*). (c) The glycogen phosphorylase downregulates the Na^+/K^+-ATPase activity. The Na^+/K^+-ATPase activity was measured in the presence of an increasing amount of phosphorylase b (0~40 μg).

REFERENCES

1. TAKEYASU, K. et al. 2001. P-type ATPase diversity and evolution: the origins of ouabain sensitivity and subunit assembly. Cell. Mol. Biol. **47:** 325–333.
2. FAMBROUGH, D.M. et al. 1994. Analysis of subunit assembly of the Na-K-ATPase. Am. J. Physiol. **266:** C579–C589.
3. SARVAZYAN, N.A., N.N. MODYANOV & A. ASKARI. 1995. Intersubunit and intrasubunit contact regions of Na^+/K^+-ATPase revealed by controlled proteolysis and chemical cross-linking. J. Biol. Chem. **270:** 26528–26532.
4. KOSTER, J.C., G. BLANCO & R.W. MERCER. 1995. A cytoplasmic region of the Na,K-ATPase alpha-subunit is necessary for specific alpha/alpha association. J. Biol. Chem. **270:** 14332–14339.
5. NELSON, W.J. & P.J. VESHNOCK. 1987. Ankyrin binding to (Na^++K^+)ATPase and implications for the organization of membrane domains in polarized cells. Nature **328:** 533–536.
6. DEVARAJAN, P., D.A. SCARAMUZZINO & J.S. MORROW. 1994. Ankyrin binds to two distinct cytoplasmic domains of Na,K-ATPase alpha subunit. Proc. Natl. Acad. Sci. USA **91:** 2965–2969.

7. BERTORELLO, A.M. et al. 1991. Phosphorylation of the catalytic subunit of Na^+,K^+-ATPase inhibits the activity of the enzyme. Proc. Natl. Acad. Sci. USA **88:** 11359–11362.
8. HORIUCHI, A. et al. 1993. D1A dopamine receptor stimulation inhibits Na^+/K^+-ATPase activity through protein kinase A. Mol. Pharmacol. **43:** 281–285.
9. OHTA, T., K. NAGANO & M. YOSHIDA. 1986. The active site structure of Na^+/K^+-transporting ATPase: location of the 5'-(p-fluorosulfonyl)benzoyladenosine binding site and soluble peptides released by trypsin. Proc. Natl. Acad. Sci. USA **83:** 2071–2075.
10. HOMAREDA, H., T. ISHII & K. TAKEYASU. 2000. Binding domain of oligomycin on Na^+,K^+-ATPase. Eur. J. Pharmacol. **400:** 177–183.

Molecular Activity of Na$^+$,K$^+$-ATPase Relates to the Packing of Membrane Lipids

PAUL L. ELSE,[a,b] BEN J. WU,[a,b] L. H. STORLIEN,[d] AND A. J. HULBERT[c]

[a]*Metabolic Research Centre,* [b]*Department of Biomedical Science,*
[c]*Department of Biological Science, University of Wollongong,
Wollongong, NSW 2522 Australia*

[d]*AstraZeneca, R&D, Mondal, SE 431.83, Sweden*

KEYWORDS: molar activity; fatty acids; sodium pump; pressure

INTRODUCTION

Metabolism varies several hundredfold between animals yet a few common processes associated with membranes consume most of the energy and account for similar proportions of metabolism. On the basis of these observations we hypothesize that variation in animal metabolism is due to membranes setting the overall pace of a small number of common processes.[1] We consider that membranes drive changes in metabolism through changes in lipid composition, which in turn change fundamental membrane properties that influence molecular activities (i.e., the rate each enzyme molecule turns over substrate) of membrane-bound proteins. To examine this idea, packing characteristics of isolated lipids from preparations with different sodium pump molecular activities were measured in monolayers at surface pressures similar to those in membranes.

METHODS

Microsomes from toad (*Bufo marinus*) and rat (*Rattus norvegicus*) kidney and brain were used in all experiments. Sodium pump molecular activities varied from 1420–7194 ATP/min measured at 37°C. Compositional analyses were performed as previously described.[2] Molecular packing of extracted lipid/phospholipid as monolayers was measured as surface pressure/average molecular area isotherms by the Langmuir trough technique ($N=24$).

Address for correspondence: Dr. Paul Else, Department of Biomedical Science, University of Wollongong, Northfields Ave, Wollongong, NSW 2512, Australia. Voice: +61-2-42213496; fax: +61-2-42214096.

pelse@uow.edu.au

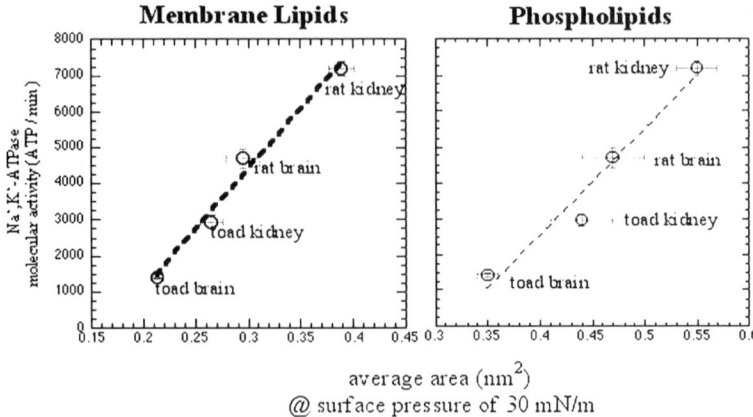

FIGURE 1. Molecular activity of sodium pumps from rat and toad kidney and brain plotted against the average area of lipid and phospholipid molecules at a monolayer pressure of 30 mN/m ($N=6 \pm$ SEM for each point).

RESULTS AND DISCUSSION

The area occupied by lipid (i.e., includes cholesterol) and phospholipid molecules from kidney and brain of rat and toad were determined from monolayers at a surface pressures of 30 mN/m (measurements at 10 and 20 mN/m showed similar results) and correlated with the molecular activities of sodium pumps measured from the same preparations at 37°C. The significant positive relationships found between the surface area occupied by lipid ($r=0.92$) and phospholipid ($r=0.82$) molecules and sodium pump molecular activity suggests that lipid and phospholipid molecules, which maintain large areas of influence, are linked to increased protein activity (FIG. 1). A functional link is supported by work[2] showing changes in sodium pump molecular activity following membrane "swapping." Composition analyses suggest fatty acyl composition as the most likely cause of activity differences with several fatty acids and indices showing significant correlation with pump activity. The work supports the notion of changes in membrane composition, in turn, changing the activity of membrane proteins with similar consequences for metabolism.

REFERENCES

1. HULBERT, A.J. & P.L. ELSE. 1999. Membranes as possible pacemakers of metabolism. J. Theor. Biol. **199:** 257–274.
2. ELSE, P.L. & B.J. WU. 1999. What role for membranes in determining the higher sodium pump molecular activity of mammals compared to ectotherms? J. Comp. Physiol. B **169:** 296–302.

Proteins Binding to α1β1 Isozyme of Na,K-ATPase

NATALIYA DOLGOVA, NATALIYA MAST, OLGA AKIMOVA,
ALEXANDER RUBTSOV, AND OLGA LOPINA

*Department of Biochemistry, School of Biology, Moscow State University,
119899 Moscow, Russia*

KEYWORDS: protein–protein interactions; immunoprecipitation; protein overlay

Na,K-ATPase interacts with various proteins, some of which have now been identified (γ-subunit, phospholemman, ankyrin, phosphoinositide-3 kinase). In this study we identified proteins that bind to Na,K-ATPase from duck salt glands (DSG) and outer medulla of pig kidney (OMPK).

METHODS

Immunoprecipitation

Na,K-ATPase was purified as described.[1,2] Antibodies against DSG Na,K-ATPase were produced in rabbits. Immunoprecipitation was done in the presence of Triton X-100 (1%). The precipitate was subjected to SDS-PAGE. The DSG homogenate was incubated with cAMP and [γ-^{32}P]ATP. After SDS-PAGE, proteins were identified by autoradiography.

Protein Overlay

Cytosol fractions of DSG and OMPK were subjected to SDS-PAGE. Proteins were transferred to nitrocellulose, treated with Na,K-ATPase, anti-Na,K-ATPase antibodies, then with secondary antibodies conjugated with horseradish peroxidase.

Department of Biochemistry, School of Biology, Moscow State University, 119899 Moscow, Russia. Voice: +7-095-9394434; fax: +7-095-9393955.
nv_dol@mail.ru

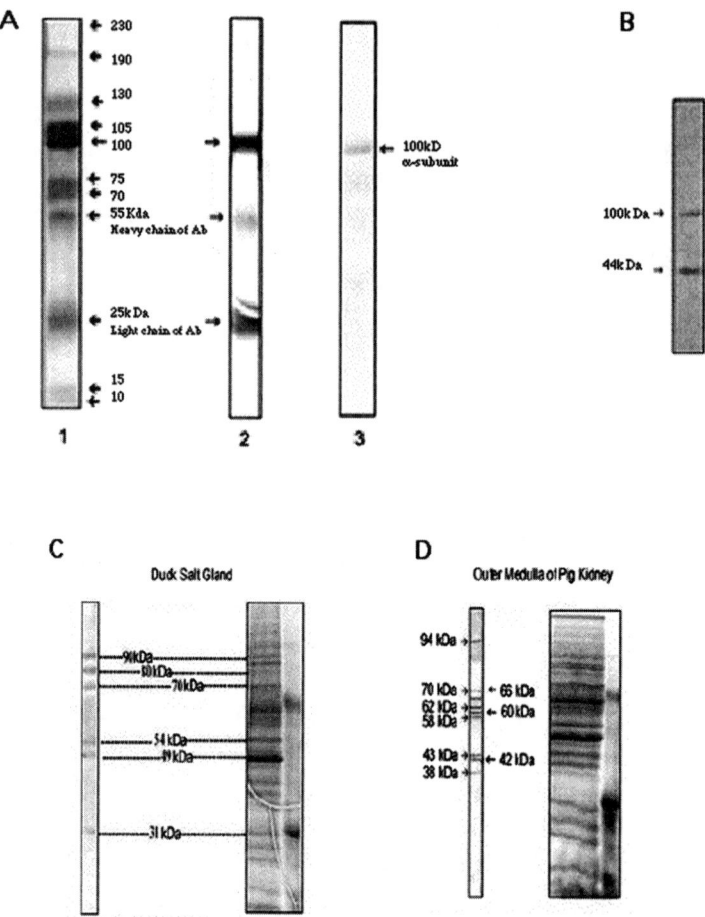

FIGURE 1. (**A** and **B**) Immunoprecipitation of Na,K-ATPase from OMPK homogenate. (**A**) Gel stained with Coomassie blue after SDS-PAGE of precipitate obtained in the presence of either 1% Triton X-100 (lane 1) or 1% Triton X-100 + 3.7% SDS (lane 2). In lane 3, proteins located on lane 1 stained with anti-Na,K-ATPase antibodies. (**B**) Autoradiography of gel after SDS-PAGE of immunoprecipitate obtained from DSG homogenate in the presence of cAMP and [γ-^{32}P]ATP]. (**C** and **D**) Cytosol proteins of DSG (**C**) and OMPK (**D**) interacting with Na,K-ATPase revealed by a protein overlay method. (**C** & **D**, *on the right*) gels stained with Coomassie blue after SDS-PAGE of cytosol fractions; (**C** & **D**, *on the left*) staining of the same proteins as described in METHODS.

RESULTS

The results of Na,K-ATPase immunoprecipitation from OMPK homogenate are presented in FIGURE 1 (A and B). If precipitation was performed in the presence of Triton X-100 (lane 1), 11–12 proteins were detected in the precipitate. If SDS was added, only three proteins were found (lane 2). Two of these bands, with molecular masses (M_r) 55 and 25 kDa, are heavy and light chains of immunoglobulins. They were stained with secondary antibodies and were also detected when precipitation was carried out without homogenate. The protein with M_r 100 kDa is the Na,K-ATPase α subunit. It was stained with anti-Na,K-ATPase antibodies (lane 3). Similar results were obtained after immunoprecipitation of Na,K-ATPase from DSG homogenate. If immunoprecipitation was conducted after addition of cAMP and [γ-^{32}P]ATP, two labeled proteins (100 and 44 kDa) were found (FIG. 1B). The 100-kDa protein appears to be the Na,K-ATPase α subunit, which is known to be a target for phosphorylation by PKA.[3]

The results of protein overlay are also presented in FIGURE 1 (C and D). Six cytosol proteins of DSG with M_r in the range 31–90 kDa bound to purified Na,K-ATPase (FIG. 1C). About ten cytosol proteins of OMPK (M_r 38–94 kDa) were found to interact with Na,K-ATPase (FIG. 1D). The data suggest that both the Triton X-100–soluble fraction and the cytosol fraction DSG and OMPK contain different proteins interacting with Na,K-ATPase.

ACKNOWLEDGMENTS

This study was supported by Grants No 01-04-48237 from the Russian Foundation for Basic Research (RFBR) and 01-0224 from the International Association for the Promotion of Cooperation with Scientists from the New Independent States of the former Soviet Union (INTAS).

REFERENCES

1. SMITH, T.W. 1988. Purification of Na$^+$,K$^+$-ATPase from the supraorbital salt gland of the duck. Methods Enzymol. **156:** 46–48.
2. JORGENSEN, P.L. 1988. Purification of Na$^+$,K$^+$-ATPase: enzyme sources, preparative problems, and preparation from mammalian kidney. Methods Enzymol. **156:** 29–43.
3. MURTAZINA, D.A., et al. 2001. Phosphorylation of the α-subunit of Na,K-ATPase from duck salt glands by cAMP-dependent protein kinase inhibits the enzyme activity. Biochemistry (Moscow) **66:** 1066–1077.

Domains Involved in the Interactions between FXYD and Na,K-ATPase

MOSHIT LINDZEN, ROMAN AIZMAN, YAEL LIFSHITZ, MARIA FÜZESI, STEVEN J. D. KARLISH, AND HAIM GARTY

Department of Biological Chemistry, The Weizmann Institute of Science, Rehovot 76100, Israel

KEYWORDS: FXYD proteins; Na,K ATPase; co-immunoprecipitation; function–structure relationships

CHIF (FXYD4) and γ (FXYD2) are kidney-specific members of the FXYD family, which interact with the α subunit of Na,K-ATPase (α) and affect the pump kinetics.[1,2] We have characterized the association of these proteins and α in transfected HeLa cells overexpressing the α1 subunit of the rat Na,K-ATPase. Both CHIF and γ could be specifically immunoprecipitated by anti-α antibodies from $C_{12}E_{10}$ solubilized cells, providing that the native structure of the pump was maintained by including Rb^+ and ouabain in the cell solubilizing buffer. Co-immunoprecipitation efficiency and stability of the two FXYD/α complexes were quite different. The anti-α antibody immunoprecipitated ~5% of total amount of γ expressed in the cells, but only ~1.5% of the total amount of CHIF. Moreover, the α/γ complex was stable up to 1.5 mg/ml $C_{12}E_{10}$. The α/CHIF complex on the other hand, was much more sensitive to the $C_{12}E_{10}$ concentration. These differences have been used to identify FXYD domains and residues involved in the interaction with α. Both CHIF and γ are single-span transmembrane (TM) proteins with an extracellular N terminal and an intracellular C tail. Chimera, in which one of the three domains comes from one protein and the rest from the other, were constructed, expressed in HeLa cells, and assayed for co-immunoprecipitation with α. It was found that chimera whose TM domain was derived from CHIF formed complexes with low stability in $C_{12}E_{10}$ and with poor co-immunoprecipitation efficiency. Chimera whose TM domain was derived from γ were efficiently immunoprecipitated and the α/chimera complex were stable at high concentrations of $C_{12}E_{10}$. Chimera having the extracellular domain of γa or γb and the TM domain of CHIF failed to translate protein even though the transfected cells expressed FXYD mRNA at levels comparable to those of CHIF or γ.

Amino acids involved in the α/FXYD interaction were further analyzed by mutating conserved residues that differ in CHIF and γ. One such area is the amino acid

Address for correspondence: Haim Garty, Department Biological Chemistry, Weizmann Institute of Science, Rehovoth, 76100, Israel. Voice: +972-8-934-2706; fax: +972-8-934-4177.
h.garty@weizmann.ac.il

triplet ^{54}AMA located in the TM domain of CHIF near the membrane/cytoplasm interface. Mutating this triplet to the corresponding amino acids in γ (LIL), resulted in an α/FXYD complex whose stability in $C_{12}E_{10}$ was as high as that of γ. The mutation did not alter the effect of CHIF on the pump kinetics measured as ouabain-blockable $^{86}Rb^+$ fluxes. Thus, hydrophobic interactions that involve the inner face of the membrane play a key role in the α/CHIF interaction, but do not participate in the functional effect of this FXYD protein.

ACKNOWLEDGMENTS

This work was supported by the Minerva Foundation and the Weizmann Institute Renal Research Fund.

REFERENCES

1. Pu, H.X. et al. 2001. Functional role and immunocytochemical localization of the γa and γb forms of the Na,K-ATPase γ subunit. J. Biol. Chem. **276:** 20370–20378.
2. Garty, H. et al. 2002. A functional interaction between CHIF and Na,K-ATPase. Implication for regulation by FXYD proteins. Am. J. Physiol. Renal Physiol. **283:** F607–F615.

Defining the Nature and Sites of Interaction between FXYD Proteins and Na,K-ATPase

MARIA FÜZESI, RIVKA GOLDSHLEGER, HAIM GARTY, AND STEVEN J. D. KARLISH

Department of Biological Chemistry, Weizmann Institute of Science, Rehovoth, 76100, Israel

KEYWORDS: FXYD proteins; Na,K-ATPase; coimmunoprecipitation; cross-linking

The FXYD family of single trans-membrane segment proteins appear to act as tissue-specific regulators of the Na,K-ATPase.

Assembly of the γ, corticosteroid hormone–induced factor (CHIF), and phospholemman (PLM) with the α subunit of the Na,K-ATPase was studied in specific coimmunoprecipitation experiments, using rat kidney medullary membranes, colonic membranes, and bovine sarcolemma. The coimmunoprecipitation is optimal after solubilization with $C_{12}E_{10}$ in media containing Na^+ plus oligomycin or Rb^+ plus ouabain. In these conditions, the native Na,K-ATPase protein structure is preserved. The order of sensitivity to the ligand conditions was PLM > γ > CHIF and may reflect the strength of the α-FXYD interaction. In kidney tubules, both γ splice variants (γa and γb) and CHIF are expressed in a nephron segment–specific manner.[1,2] Coimmunoprecipitation experiments demonstrate $\alpha/\beta/\gamma$a, $\alpha/\beta/\gamma$b, or α/β/CHIF complexes and exclude the existence of mixed complexes, such as $\alpha/\beta/\gamma$a/γb or $\alpha/\beta/\gamma$/CHIF.[3]

Covalent cross-linking was initiated in order to identify positions of α, β, and γ subunit interactions. Using pig kidney Na,K-ATPase, EDC (1-ethyl-3-(dimethylaminopropyl)-carbodiimide) and DST (disuccinimidyl tartrate) form β-γ cross-links, while o-PDM (N,N'-o-phenylene dimaleimide), DB (dibromobimane), and DST form α-γ cross-links. The α-γ cross-link is amplified in E1(Na) conformation compared to E2(Rb), consistent with the known effect of γ to stabilize E1 conformation.[1] 19 kDa membranes were used to further locate the interacting sites of γ-α and γ-β. 19 kDa membranes are produced by extensive tryptic digestion of the renal Na,K-ATPase and contain membrane-embedded fragments of the α subunit, a partially cleaved β subunit, and an intact γ subunit.[4] DST and EDC form three main cross-linked products recognizing anti-γ. Two bands run parallel with the intact and cleaved β subunits (γ-β, γ-β50). In the case of DST, a pair of bands, and with EDC,

Address for correspondence: Steven J.D. Karlish, Department of Biological Chemistry, Weizmann Institute of Science, Rehovoth, 76100, Israel. Voice: +972-8-934-2278; fax: +972-8-934-4118.

steven.karlish@weizmann.ac.il

a single band of ~23 kDa (γ-β16), are also stained with anti-β16, which recognizes a tryptic fragment of β containing the transmembrane domain. Overall the results indicate more than one site of interaction between γ and β subunits. PNGase F digestion of the DST- and EDC-treated 19 kDa membranes caused a shift in the relative mobility of the γ-β and γ-β50 cross-linked products, resulting in two bands of ~45 and ~29 kDa. This result proves a γ-β cross-linking in the extracellular domains. So far, further definition of γ-α cross-links has not been achieved.

ACKNOWLEDGMENT

This work was supported by the Minerva Foundation.

REFERENCES

1. Pu, H.X. et al. 2001. Functional role and immunocytochemical localization of the gamma a and gamma b forms of the Na,K-ATPase gamma subunit. 2001. J. Biol. Chem. **276:** 20370–20378.
2. Shi, H.-K. et al. 2001. Membrane topology and immunolocalization of CHIF in kidney and intestine. Am. J. Physiol. Renal Physiol. **280:** F505–F515.
3. Garty, H. et al. 2002. A functional interaction between CHIF and Na,K-ATPase. Implication for regulation by FXYD proteins. Am. J. Physiol. In press.
4. Capasso, J. et al. 1992. Extensive digestion of Na+,K(+)-ATPase by specific and nonspecific proteases with preservation of cation occlusion sites. J. Biol. Chem. **267:** 1150–1158.

NO Regulation of Na,K-ATPase

Nitric Oxide Regulation of the Na,K-ATPase in Physiological and Pathological States

DORETTE Z. ELLIS AND KATHLEEN J. SWEADNER

Neuroscience Center, Massachusetts General Hospital, Charlestown, Massachusetts 02129, USA

KEYWORDS: Na,K-ATPase; nitric oxide (NO); amyotrophic lateral sclerosis

In actively firing neurons, the Na,K-ATPase utilizes up to 50% of the cells' ATP. A compromised gradient alters membrane potential and affects a number of transport and cotransport processes that provide the cells with nutrients or that regulate intracellular concentrations of ions implicated in specialized neuronal functions such as synaptic potential.

Amyotrophic lateral sclerosis (ALS) is a fatal neurological disease with pathological features that include degeneration of motor neurons in the spinal cord. Approximately 20% of the familial cases are linked to mutations in the Cu/Zn superoxide dismutase gene (SOD1). SOD1 catalyzes the conversion of $O_2^{\bullet}$ to hydrogen peroxide and water. Motor neuron degeneration could be caused by alterations of free radical homeostasis and subsequent oxidative toxicity. Several theories for the molecular defects of ALS have been proposed, including an aberrant gain in SOD1 function. For example, mutant SOD1 may have enhanced ability to interact with hydrogen peroxide to form hydroxyl radicals.

We previously reported a biochemical pathway through which carbachol/nitric oxide (NO), soluble guanylate cyclase (sGC), and cGMP inhibit Na,K-ATPase activity.[1,2] This same pathway is shared by oxygen free radicals and forms a convergence of extracellular signals that regulate Na,K-ATPase activity. Because NO is a free radical and because free radical homeostasis is altered in ALS, we determined how these changes affect Na,K-ATPase. Transgenic mice containing mutant human SOD1, their nontransgenic littermates, and transgenic mice containing overexpressed normal human SOD1 were used.

FIGURE 1A shows that the NO donor, DETA-NO, inhibited Na,K-ATPase activity in spinal cord tissue slices from nontransgenic control animals. The specific sGC inhibitor, ODQ (1 µM), abolished the DETA-NO effect, confirming the involvement of sGC. In transgenic SOD1 overexpressors (FIG. 1B) and transgenic mutant SOD1

Address for correspondence: Dorette Z. Ellis, Neuroscience Center, Massachusetts General Hospital, CNY 149-6118, Charlestown, MA 02129. Voice: 617-726-8560.
ellis@helix.mgh.harvard.edu

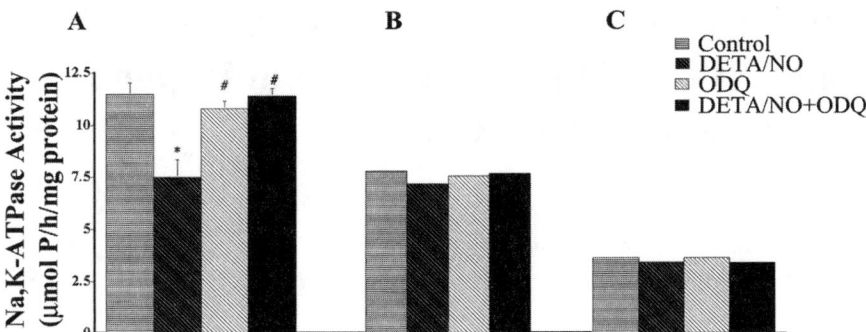

FIGURE 1. DETA-NO-induced inhibition of ouabain-sensitive Na,K-ATPase was abolished in spinal cord tissue slices from transgenic mutant SOD1 mice and transgenic normal human SOD1 overexpressors. Values represent the mean ± SEM for experiments on 3 animals done in triplicate. *Significantly different from the control at $p < 0.05$ (by ANOVA and Fisher's PLSD). #Significantly different from DETA-NO-treated samples at $p < 0.05$ (by ANOVA and Fisher's PLSD).

mice (FIG. 1C), the Na,K-ATPase activity was significantly decreased and the inhibition by DETA-NO was completely abolished. Western blot analysis quantitatively assessed protein levels and demonstrated that there were significant decreases in the α subunit in transgenic mutant SOD1 mice, while there were no changes in β subunits.[3] Global losses of all Na,K-ATPase α subunits in spinal cord from transgenic mutant SOD1 mice were observed using confocal immunofluorescence. The losses in activity exceeded the loss of protein, which would suggest the involvement of oxidative damage to the Na,K-ATPase.

REFERENCES

1. ELLIS, D.Z., J.A. NATHANSON, J. RABE & K.J. SWEADNER. 2001. Carbachol and nitric oxide inhibition of Na,K-ATPase activity in bovine ciliary process. Invest. Ophthalmol. Visual Sci. **42:** 2625–2631.
2. ELLIS, D.Z., J.A. NATHANSON & K.J. SWEADNER. 2000. Carbachol inhibits Na,K-ATPase activity in choroid plexus via stimulation of the NO/cGMP pathway. Am. J. Physiol. Cell. Physiol. **279:** C1685–C1693.
3. ELLIS, D.Z., J. RABE & K.J. SWEADNER. 2003. Global loss of Na,K-ATPase and its nitric oxide-mediated regulation in a transgenic mouse model of amyotrophic lateral sclerosis. J. Neurosci. **23:** 43–51.

Functional Expression of the α4 Isoform of the Na,K-ATPase in Both Diploid and Haploid Germ Cells of Male Rats

GUSTAVO BLANCO

Department of Molecular and Integrative Physiology, University of Kansas Medical Center, Kansas City, Kansas 66160, USA

KEYWORDS: Na,K-ATPase; isoforms; α4; germ cells; testes

INTRODUCTION

The α4 polypeptide is the Na,K-ATPase α isoform with the most restricted pattern of expression. The subunit is selectively expressed in rat testes, where it predominates in spermatozoa.[1–3] Besides this particular localization, α4 exhibits enzymatic properties that are unique. In the rat, the sensitivity of α4 to ouabain makes it possible to distinguish it from α1.[1,3] Here, ouabain inhibition profiles of Na,K-ATPase were used to explore the functional expression of α4 in rat testes and in germ cells isolated from the gonad at different stages of maturation.

METHODS

Testes were obtained from 7-day-old, 18-day-old, and adult Sprague-Dawley rats. Spermatogonia, spermatocytes, and spermatids were isolated from rat testes by differential sedimentation velocity, using counterflow elutriation.[4] Spermatozoa were obtained from rat epididymis. All samples were homogenized in 250 mM sucrose, 25 mM imidazole (pH 7.4), and 0.1 mM EGTA, and the lysates used after centrifugation at $1000 \times g$ for 10 min. Ouabain dose-response curves were performed in medium containing 120 mM NaCl, 30 mM KCl, 3 mM $MgCl_2$, 0.2 mM EGTA, 2.5 mM sodium azide, 30 mM Tris-HCl (pH 7.4), 3 mM cold ATP, 0.2 µCi $\gamma[^{32}P]$-ATP, and different ouabain concentrations. Curve fitting was performed using least-square nonlinear regression.[1]

Address for correspondence: Gustavo Blanco, Department of Molecular and Integrative Physiology, University of Kansas Medical Center, 3901 Rainbow Boulevard, Kansas City, KS 66160. Voice: 913-588-7405.
gblanco@kumc.edu.

TABLE 1. Inhibition constants (K_i) values and relative amounts of Na,K-ATPase α1 and α4 isoforms from testes of rat at different ages and from isolated germ cells

	α1 Isoform		α4 Isoform	
	K_i (M)	Percentage	K_i	Percentage
Rat testes				
7 day old	$5.3 \pm 5.5 \times 10^{-5}$	86.6 ± 1.9	$1.4 \pm 3.0 \times 10^{-9}$	13.4 ± 2.8
18 day old	$7.2 \pm 2.2 \times 10^{-5}$	72.4 ± 1.8	$3.2 \pm 2.8 \times 10^{-9}$	27.6 ± 2.9
adult	$1.5 \pm 1.3 \times 10^{-4}$	45.0 ± 1.4	$1.6 \pm 1.0 \times 10^{-9}$	55.0 ± 2.8
Spermatogonia	$1.3 \pm 0.6 \times 10^{-4}$	52.9 ± 3.9	$7.5 \pm 2.7 \times 10^{-9}$	47.1 ± 5.7
Spermatocytes	$5.1 \pm 0.9 \times 10^{-5}$	52.7 ± 1.7	$7.3 \pm 2.9 \times 10^{-9}$	47.3 ± 2.3
Spermatids	$2.7 \pm 1.3 \times 10^{-5}$	42.3 ± 3.7	$2.4 \pm 2.8 \times 10^{-9}$	57.5 ± 5.6
Spermatozoa	$1.3 \pm 0.6 \times 10^{-5}$	32.6 ± 2.6	$2.8 \pm 2.0 \times 10^{-9}$	67.4 ± 3.4

NOTE: Isoform K_i values and percentile amounts were calculated from dose-response curves for the ouabain inhibition of Na,K-ATPase activity. Values represent the mean ± SEM.

RESULTS AND DISCUSSION

The high sensitivity of the α4 isoform was used to study the functional expression of the polypeptide during male germ cell development. This was first determined in rat testes at different ages. Three time points in the sexual maturation of rats were chosen: day 7 after birth, when the testes contain a majority of germ cells at the stage of spermatogonia; day 18, when the gametes have differentiated into spermatocytes; and finally, adult animals, in which cells at all stages of spermatogenesis are found. At all ages, ouabain dose-response curves exhibited a biphasic pattern (data not shown), with K_i indicating the presence of the α1 and α4 isoforms (TABLE 1). As shown, the relative amount of α4 doubled by day 18, and became almost half of the total Na,K-ATPase at adulthood. The maximal expression of α4 in spermatozoa agrees with previous results, indicating a drastic reduction in activity of the isoform in testes from azoospermic mice.[1] To follow the activity of the α4 isoform more directly, ouabain inhibition profiles of Na,K-ATPase were performed in the isolated gametes at different stages of differentiation. As shown in TABLE 1, all samples exhibited inhibition constants compatible with the activity of the α1 and α4 isoforms. Gamete differentiation was accompanied with a progressive augment in the percentile amounts of α4, to peak in spermatozoa, where the isoform comprises two thirds of the Na,K-ATPase of the cells.

SUMMARY

Altogether these results indicate that the activity of the Na,K-ATPase α4 isoform is present in both the diploid and haploid germ cells of testes, and that its function is upregulated with germ cell maturation. Therefore, α4 function may not only be physiologically important in mature spermatozoa,[5] but may also play a role throughout differentiation of the male gametes.

ACKNOWLEDGMENT

Supported by National Science Foundation grant MCB-9982901.

REFERENCES

1. BLANCO, G., G. SÁNCHEZ, R.J. MELTON, *et al.* 2000. The α4 isoform of the Na,K-ATPase is expressed in the germ cells of the testes. J. Histochem. Cytochem. **48:** 1023–1032.
2. BLANCO, G., J.R. MELTON, G. SANCHEZ & R.W. MERCER. 1999. Functional characterization of a testes-specific-subunit isoform of the Na,K-ATPase. Biochemistry **38:** 13661–13669.
3. WOO, A.L., P.F. JAMES & J.B. LINGREL. 1999. Characterization of the fourth isoform of the Na,K-ATPase. J. Membr. Biol. **169:** 39–44.
4. GRABSKE, R.J., S. LAKE, B.L. GLEDHILL & M.L. MEISTRICH. 1975. Centrifugal elutriation: separation of spermatogenic cells on the basis of sedimentation velocity. J. Cell. Physiol. **86:** 177–190.
5. WOO, A.L., P.F. JAMES & J.B. LINGREL. 2000. Sperm motility is dependent on a unique isoform of the Na,K-ATPase. J. Biol. Chem. **275:** 20693–20699.

Responses at the Translational Level to Heterologous Expression of the Na,K-ATPase

LOTTE STEFFENSEN AND PER AMSTRUP PEDERSEN

August Krogh Institute, University of Copenhagen, Universitetsparken 13, 2100 Copenhagen OE, Denmark

KEYWORDS: translational regulation; yeast; Na,K-ATPase; membrane protein

Although proteomic analysis of sequenced eukaryotic genomes has revealed that more that one third of all genes are predicted to encode membrane proteins[1] the three-dimensional structure is only known for approximately 30 membrane proteins including three of eukaryotic origin.[2] The primary reason for the lack of three-dimensional structures for eukaryotic membrane proteins is that membrane proteins cannot be purified due to their low density in native tissue. Furthermore, the expression systems developed for soluble proteins are unable to supply the densities and quantities of membrane proteins required for purification on the milligram scale. We have therefore initiated a study to determine the responses to heterologous membrane protein production at the translation level.

Phosphorylation of the α subunit of eukaryotic initiation factor 2α (eIF-2α) is a well-characterized mechanism regulating translation initiation in response to different kinds of stress. In the yeast *Saccharomyces cerevisiae* different forms of stress[3,4] induce phosphorylation of eIF-2α by the Gcn2p kinase, leading to translation of *GCN4* mRNA. *GCN4* mRNA carries four small open reading frames upstream from the Gcn4p coding region. The four small open reading frames prevent translation initiation at the Gcn4p coding region in presence of nonphosphorylated eIF-2α. Phosphorylation of eIF-2α causes the ribosome to bypass open reading frames 2–4 and initiate translation of the Gcn4p open reading frame with high efficiency. To monitor transcription of the *GCN4* gene and translation of the *GCN4* mRNA we have used two reporter plasmids.[5]

Purified membrane- and cytosolic fractions from yeast cells were analyzed for [^{3}H]-ouabain binding and β-galactosidase activity, respectively. The results presented in TABLE 1 show that functional Na,K pumps were produced and localized to the yeast membranes. The data also show that induction of Na,K-ATPase biosynthesis resulted in an eightfold increase in the translation initiation rate of *GCN4* mRNA

Address for correspondence: Per Amstrup Pedersen, August Krogh Institute, University of Copenhagen, Universitetsparken 13, 2100 Copenhagen OE, Denmark. Voice: +45-35321667; fax: +45-35321567.

PAPedersen@aki.ku.dk

TABLE 1. Galactosidase activity in purified cytoplasmic fractions from yeast strains carrying either the translational *GCN4-lacZ* reporter or the transcriptional *GCN4$^{\Delta uORF}$-lacZ* reporter

Reporter Plasmid	Expression Plasmid	Gcn4p-lacZ Enzyme Activity (nmol/min·mg protein)	[^{3}H]-ouabain Binding (pmol/mg protein)
pPAP2915/□□□□LacZ	$\alpha_1\beta_1$	70	12.5
pPAP2983/ ××××LacZ	$\alpha_1\beta_1$	1189	13.1
pPAP2915/□□□□LacZ	empty vector	9	0
pPAP2983/ ××××LacZ	empty vector	956	0

NOTE: β-galactosidase activity in purified cytoplasmic fractions from yeast strains carrying either the translational *GCN4-lacZ* reporter (illustrated with *open boxes*) or the transcriptional *GCN4D$^{\Delta uORF}$-lacZ* reporter (illustrated with *crosses*) and the indicated expression plasmid. [^{3}H]-oaubain binding was determined to isolated membrane fractions as described previously.[6]

while *GCN4* transcription only showed a modest 24% increase. This shows that the transcription factor Gcn4p must play a hitherto unknown role in the physiological response to heterologous membrane protein expression.

In conclusion, we have found a new role for the Gcn4 transcription factor in responding to the stress imposed on the host cell by expressing a heterologous protein. It is our intention to identify genes that are upregulated or downregulated as a host response to heterologous protein expression. The tools for these studies will be cDNA chips. It is our hope that knock-out of some of the genes regulated by Gcn4p will increase the expression level of heterologous membrane proteins.

REFERENCES

1. CHERVITZ, A. *et al.* 1998 Comparison of the complete protein sets of worm and yeast: orthology and divergence. Science **282:** 2022–2028.
2. CAMPBELL, I.D. 2002. The march of structural biology. Nature Reviews Mol. Cell. Biol. **3:** 377–3813.
3. GOOSSENS, A., T.E. DEVER, A, PASCUAL-AHUIR & R. SERRANE. 2001. The protein kinase Gcn2p mediates sodium toxicity in yeast. J. Biol. Chem. **276:** 30753–30760.
4. YANG, R., S.A. WEK & R.C. WEK. 2000. Glucose limitation induces GCN4 translation by activation of Gcn2 protein kinase. Mol. Cell. Biol. **20:** 2706–2717.
5. HINNEBUSCH, A.G. 1985. A hierarchy of trans-acting factors modulates translation of an activator of amino acid biosynthetic genes in *Saccharomyces cerevisiae*. Mol. Cell. Biol. **5:** 2349–2360.
6. PEDERSEN, P.A, J.H. RASMUSSEN & P.L. JORGENSEN. 1996. Expression in high yield of pig alpha 1 beta 1 Na,K-ATPase and inactive mutants D369N and D807N in *Saccharomyces cerevisiae*. J. Biol. Chem. **271:** 2514–2522.

Protein Kinase C Phosphorylation Directed at Novel C-Terminal Sites in Na,K-ATPase

ANDERS KRÜGER, YASSER A. MAHMMOUD, AND FLEMMING CORNELIUS

Department of Biophysics, University of Aarhus, Denmark

KEYWORDS: Na,K-ATPase phosphorylation; PKA/PKC crosstalk

Recently a novel class of protein kinase C (PKC) sites has been characterized in Na,K-ATPase from shark rectal glands.[1] These sites are located close to the inner face of the plasma membrane and, like the PKA site, they become exposed to PKC after detergent treatment. The sites are located in the C-terminal part of the α subunit at the proximal part of M5/M6 and in the L89 loop. The latter is only 4 amino acids upstream from the PKA site (FIG. 1, left). In the present study we reconstituted N-terminal truncated Na,K-ATPase from shark rectal gland in lipid vesicles and demonstrated the phosphorylatability of these novel sites in the absence of detergent, as seen in the autoradiogram from the phosphorylation of α, and of the 19-kDa fragment after proteolytic fingerprinting (FIG. 1A, middle). The close proximity of the PKA and PKC site in the L89 loop could indicate that this loop is a motif for crosstalk between the two signaling pathways. The right panel of FIGURE 1A showing proteolytic fingerprinting of reconstituted truncated enzyme indicates that this might be so: The first lane (1) shows phosphorylation of the 19-kDa band by PKA. Lane 2 shows the same experiment, but after subsequent PKC phosphorylation. Lane 3 shows what happens if the order of phosphorylations is reversed. Here, PKC phosphorylation apparently completely inhibits subsequent phosphorylation by PKA.

Next, we investigated the K^+-activation of N-terminal truncated Na,K-ATPase before and after PKC phosphorylation in the presence of the detergent $C_{12}E_8$. As seen from FIGURE 1B (left panel), N-terminal truncation in itself inhibited the enzyme, as previously shown.[2] However, K^+-activation (and Na^+-activation, not shown) of truncated enzyme at saturating ATP showed no effect of C-terminal PKC phosphorylation, whereas at 10 µM ATP (where K^+ deocclusion is rate-limiting), a small but significant inhibition by ~20% ($P=0.0001$) is observed (FIG. 1B, right panel).

In conclusion, PKC phosphorylation of the C-terminal PKC phosphorylation sites close to the inner face of the membrane are only accessible in the presence of detergent in shark membrane bound Na,K-ATPase, but become accessible after re-

---

Address for correspondence: Flemming Cornelius, Department of Biophysics, University of Aarhus, Aarhus, Denmark. Voice: +45-89422926; fax: +45-86129599.
fc@biophys.au.dk

Ann. N.Y. Acad. Sci. 986: 541–542 (2003). © 2003 New York Academy of Sciences.

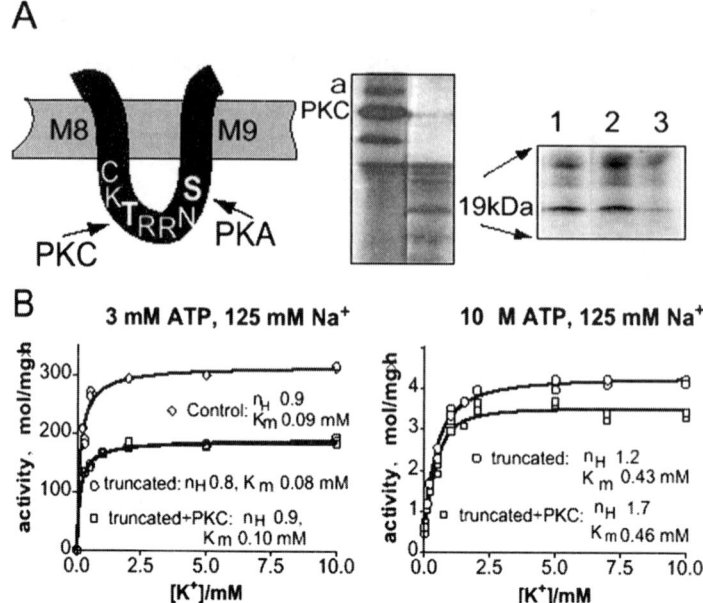

FIGURE 1. (A) Location (*left*) and phosphorylation after reconstitution of PKC sites within the 19-kDa fragment (*middle*). The order of a combined phosphorylation of PKA and PKC is shown to the *right*. (B) K^+-activation at saturating and subsaturating ATP concentration before and after PKC phosphorylation of truncated enzyme in the presence of 20 mM $C_{12}E_8$. Also shown is a control before truncation.

constitution. Combined phosphorylation assay of the PKC/PKA sites in the L89 loop indicates that this is a functional motif for crosstalk between the two protein kinase signaling pathways. A small but significant inhibition of the Na,K-ATPase activity was observed by C-terminal PKC phosphorylation, but only at limiting ATP, indicating interaction of PKC phosphorylation with the K^+ sites. This is in accordance with the close spatial location of the novel PKC sites to the K^+ binding pocket.

REFERENCES

1. MAHMMOUD, Y.A. & F. CORNELIUS. 2002. Protein kinase C phosphorylation of purified Na,K-ATPase: C-terminal phosphorylation sites at the α- and γ-subunits close to the inner face of the plasma membrane. Biophys. J. **82:** 1907–1019.
2. JORGENSEN, P.L. & J.H. COLLINS. 1986. Tryptic and chymotryptic cleavage sites in sequence of α-subunit of $(Na^+ + K^+)$-ATPase from outer medulla of mammalian kidney. Biochim. Biophys. Acta **860:** 570–576.

Modification of the PKC Phosphorylation Site Ser-23 of the Rat α1 Subunit

L. A. VASILETS,[a,b] A. SPIELMANN,[c] AND W. SCHWARZ[c]

[b]*Department of Physiology, Martin-Luther-University, Magdeburgerstrasse 12, D-06112 Halle/Saale, Germany*

[c]*Max-Planck-Institute for Biophysics, D-60596 Frankfurt/Main, Germany*

KEYWORDS: PKC; phosphorylation mutants; regulation; voltage clamp; Na^+,K^+-ATPase

Phosphorylation of the catalytic subunit of the Na^+,K^+-ATPase by protein kinases plays an important role in regulation of cation transport. The most extensively studied protein kinase, PKC, exhibits its functional effects on cellular and molecular levels, particularly by (1) changing the number of the active Na^+,K^+-ATPase molecules in the plasma membranes by activation of endocytosis, and by (2) phosphorylation of the α subunit and direct modulation of transport activity (reviewed in Vasilets,[1] for example). Which of the two mechanisms dominates depends on cell type and on phosphorylation conditions. Often direct regulation of the transport cycle may be overlooked because of the complexity of the PKC effects during activation of intracellular signaling. In addition, the transient character of the PKC effects restricts detailed investigation of transport properties of phosphorylated Na^+,K^+-ATPase. To investigate effects of PKC-mediated phosphorylation on transport cycle, we constructed mutants that mimic the continuously phosphorylated state at Ser-23. This was achieved by substitution of Ser-23 of the rat α1 subunit by negatively charged Glu or Asp. Previously, we demonstrated that the acidic replacement mimics PKC-mediated inhibition of cation transport by the Na^+,K^+-ATPase.[2] In addition, substitution of Ser-23 by Glu increased cellular expression of the α subunit, as judged from immunoblots of oocyte homogenates. However, the possibility could not be excluded that cellular expression of the modified α subunits differs from that in the surface membrane. To understand whether targeting of functionally active pumps in the surface membrane is also enhanced due to modification at Ser-23, we performed voltage-clamp analysis of rat α1 wild-type (WT) and S23E- and S23D-mutant Na^+,K^+ pumps expressed in *Xenopus* oocytes. FIGURE 1A shows transient currents generated by the expressed pump variants in Na^+/Na^+ exchange mode in response to

---

[a]Address for correspondence: Larisa Vasilets, Department of Physiology, Martin-Luther-University, Magdeburgerstrasse 12, D-06112 Halle/Saale, Germany. Voice: +49-345-5574740; fax: +49-345-5574140.
Larisa.vasilets@medizin.uni-halle.de

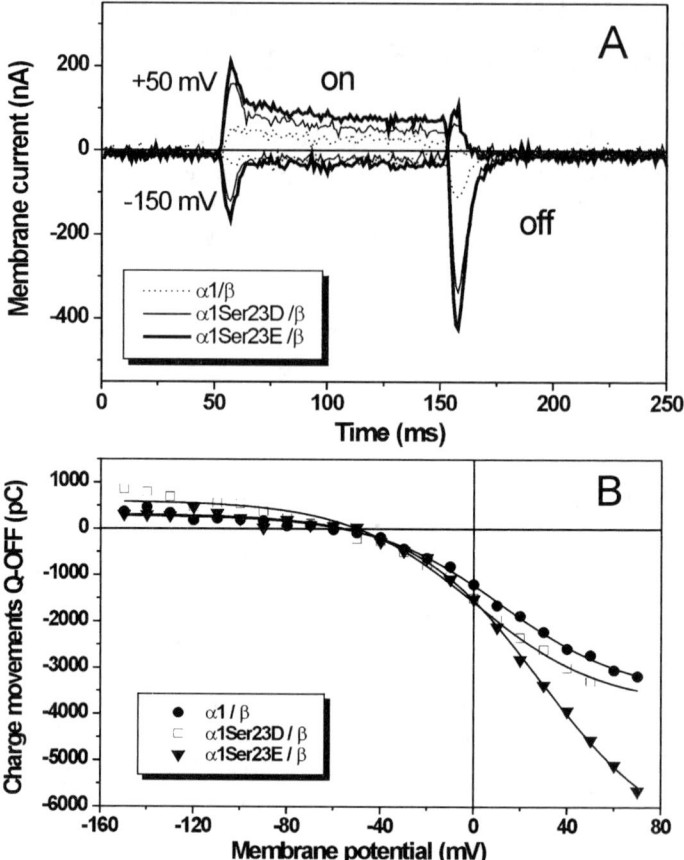

FIGURE 1. (**A**) Ouabain-sensitive transient currents for voltage jumps from -60 to $+50$ mV and from -60 to -150 mV for the WT pump and mutants. (**B**) Voltage dependencies of transiently moved charges during off-response described by the Fermi equation:

$$Q(V) = Q_{min} + Q_{tot} \{1/[1 + \exp(z_q(V - V_m)F/RT)]\}$$

selected potential steps. FIGURE 1B shows voltage dependencies of the transiently moved charges obtained by integration of transient currents. From the voltage dependence of the transient currents in Na^+/Na^+ exchange mode, the effective charge $z_{eff}e$ moved by a single pump molecule and the total charge transported by the entire population of the ATPase in the cell membrane Q_{tot} have been determined. The number of functioning $\alpha1/\beta$ complexes in the oocyte membrane has been determined from the ratio $N=Q_{tot}/z_{eff}e$, and was 2.6×10^{10} for the WT and 4.8×10^{10} and 3×10^{10} for S23E and S23D mutants, respectively (increase by factors 1.9 and 1.1). Western blot analysis showed that the total number of expressed S23E and S23D pumps was by factors of 2.3 and 1.3 higher than the number of WT pumps. Thus calculation of the number of active pump complexes from charge movement and from Western blot analysis yields quantitatively similar results with respect to an increase in expression of the phosphorylation mutants. The transport rate determined from the ratio of steady-state currents to total charge $k=I_{ss}/Q_{tot}$ was 24.6 s^{-1} for the WT and was reduced to 10 s^{-1} and 19 s^{-1} for the S23E and S23D mutants, respectively.

The results demonstrate that substitution of Ser-23 with negatively charged residues, which mimics binding of the phosphoryl group to the PKC phosphorylation site, (1) leads to inhibition of transport rate of the Na$^+$,K$^+$-ATPase and (2) enhances cellular expression of the α subunits as well as targeting of the active Na$^+$,K$^+$-ATPase complexes to the surface membrane.

REFERENCES

1. VASILETS, L.A. 2002. Mechanisms of short-term regulation of the Na$^+$/K$^+$-ATPase by protein kinases. Biol. Membr. **19:** 77–82.
2. VASILETS, L.A., R. POSTINA & S.N. KIRICHENKO. 1999. Mutation of Ser-23 of the α1 subunit of the rat Na$^+$/K$^+$-ATPase to negatively charged amino acid residues mimic the functional effect of PKC-mediated phosphorylation. FEBS Lett. **455:** 8–12.

Expression of a Na,K-ATPase-EGFP Chimera in COS Cells

Can Internalization Explain PKA- or PKC-Mediated Inhibition of ^{86}Rb Uptake?

BO KRISTENSEN, SVEND BIRKELUND,[a] AND PETER LETH JORGENSEN

Biomembrane Center, The August Krogh Institute, University of Copenhagen, Universitetsparken 13, DK-2100 Copenhagen OE, Denmark

[a]*Department of Medical Microbiology and Immunology, The Bartholin Building, University of Aarhus, Wilhelm Meyers Allé, Building 240, DK-8000 Aarhus C, Denmark*

KEYWORDS: Na,K-ATPase-EGFP chimera; protein kinases A and C; recycling endosomes

Protein kinases A (PKA) and C (PKC) are known regulators of Na,K-ATPase activity. Phosphorylation of the α1 subunit is important for the short- and long-term regulation of Na,K-ATPase activity and may alter intrinsic enzymatic properties including maximum velocity, E1-E2 equilibrium, and Na$^+$ affinity.[1] Alternately, fast changes in cellular Na,K-pump activity following hormone or protein kinase activation can be caused by internalization of pump molecules into endosomes.[2]

To visualize putative effects of PKA and PKC on Na,K-ATPase intracellular trafficking, enhanced green fluorescent protein (EGFP) was fused to the C-terminal of rat Na,K-ATPase α1 subunit. The chimera (Na,K-EGFP) was stably expressed in COS-1 cells using puromycin and ouabain selection procedures.[3]

In control cells, Na,K-EGFP was sparsely present (9%) in the recycling endosome (RE) compartment envisioned with Texas-red transferrin. Activation of PKA or PKC increased the amount of Na,K-EGFP in REs fourfold and decreased enzyme activity 22% (^{86}Rb-uptake assays). Simultaneous activation of PKA and PKC had almost additive effects on the presence of Na,K-EGFP in REs (sixfold over control) and inhibition of enzyme activity (40%).[3] The number of different vesicles per 100 μm^2 of cell section area is shown in FIGURE 1. Since the average surface area of a RE containing Na,K-EGFP is 1.9 μm^2, the PKA+PKC–mediated increase in density of these vesicles (1.0 per 100 μm^2, FIG. 1) correspond to internalization of only 2 to 4% of the total plasma membrane area. Internalization of Na,K-ATPase can therefore only account for the 40% inhibitory effects of PKA+PKC on enzyme activity[3] if the relative concentration of Na,K-EGFP in REs is 10–20-fold higher than in plasma

---

Address for correspondence: Peter Leth Jorgensen, Biomembrane Center, The August Krogh Institute, University of Copenhagen, Universitetsparken 13, DK-2100 Copenhagen OE, Denmark. Voice: +45-35-32-16-70; fax: +45-35-32-15-67.
pljorgensen@aki.ku.dk

Ann. N.Y. Acad. Sci. 986: 546–547 (2003). © 2003 New York Academy of Sciences.

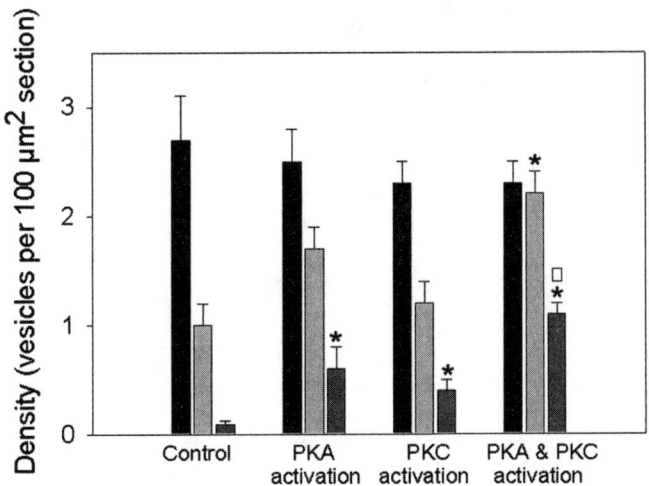

FIGURE 1. Quantitative analysis of internalization. The number of transferrin vesicles (*black bars*), Na,K-EGFP vesicles (*light gray bars*), and double labeled vesicles (*dark gray bars*) per 100 μm^2 of cell section area was determined. *$P < 0.05$ compared to control values. □$P < 0.05$ compared to either PKA (100 mM forskolin + 1 mM IBMX for 45 min) or PKC activation (100 nM phorbol 12-myristate 13-acetate for 45 min).

membranes. Though platelet-derived growth factor receptor densities are increased 5- to 15-fold by internalization,[4] the high density of Na,K-ATPase in basolateral kidney membranes (up to 5,000 per μm^2), makes large increments of enzyme densities less likely.

To analyze the effects of PKA and PKC on cell size, section areas of Na,K-EGFP–expressing cells were determined at different time intervals (0–75 min). Linear regression analysis of data showed that 45 min of kinase activation caused cell shrinkage: PKA, $-12 \pm 4\%$; PKC, $-15 \pm 1\%$; PKA and PKC, $-20 \pm 2\%$.

Several mechanisms exerting their effects in concert are therefore probably involved in the PKA and PKC mediated inhibition of Na,K-ATPase: (1) internalization of the enzyme, (2) general activity reduction caused by cell shrinkage, and (3) possible changes of enzyme properties like Na$^+$-affinity or E1-E2 equilibrium.[1]

REFERENCES

1. FERAILLE, E. & A. DOUCET. 2001. Sodium-potassium-adenosine triphosphatase-dependent sodium transport in the kidney: hormonal control. Physiol. Rev. **81:** 345–418.
2. CHIBALIN, A.V., J.R. ZIERATH, A.I. KATZ, et al. 1998. Phosphatidylinositol 3-kinase–mediated endocytosis of renal Na,K-ATPase alpha subunit in response to dopamine. Mol. Biol. Cell. **9:** 1209–1220.
3. KRISTENSEN, B., S. BIRKELUND & P.L. JORGENSEN. 2003. Trafficking of Na,K-ATPase fused to enhanced green fluorescent protein mediated by protein kinase A or C. J. Membr. Biol. **191:** 25–36.
4. ROSENFELD, M.E., D.F. BOWEN-POPE & R. ROSS. 1984. Platelet-derived growth factor: morphologic and biochemical studies of binding, internalization, and degradation. J. Cell. Physiol. **121:** 263–274.

PKA and PKC Phosphorylation of Gastric H,K-ATPase

YASSER AHMED MAHMMOUD AND FLEMMING CORNELIUS

Department of Biophysics, University of Aarhus, Aarhus, DK-8000 C Denmark

KEYWORDS: protein kinase; H,K-ATPase phosphorylation

In this study we demonstrate that incubation of H,K-ATPase membranes with purified protein kinase C (PKC) in a typical PKC phosphorylation mixture resulted in phosphorylation of the H,K-ATPase α-subunit with a stoichiometry of about 0.6 mol P_i/mol α-subunit. However, in the presence of detergent the phosphorylation stoichiometry doubled, indicating exposure of additional PKC sites. The α-subunit of H,K-ATPase was also found to be a substrate for protein kinase A (PKA), which phosphorylated the α-subunit to a stoichiometry of about 0.7 mol P_i/mole α-subunit, but only in the presence of detergent.

Truncation of the N-terminus by controlled proteolysis demonstrated that phosphorylation of native H,K-ATPase α-subunit by PKC is occurring exclusively at the N-terminus, which contains the conserved consensus PKC phosphorylation site ($KXS^{27}KKK$). It should be noted that in contrast to the Na,K-ATPase, N-terminal truncation of the H,K-ATPase α-subunit generated active, fully functional enzyme without significant differences in the affinity for K^+, or in the maximum hydrolytic activity (V_{max}). After truncation the enzyme could be phosphorylated by PKC only in the presence of detergent to a stoichiometry of about 0.4 mol P_i/mol α-subunit. In order to identify where the detergent-induced phosphorylation is located, we performed proteolytic fingerprinting, which demonstrated phosphorylation of the "19kDa" fragment containing Thr^{950} and a 12kDa fragment containing Ser^{786}. Both sites are located within conserved consensus PKC motifs. Thus, the protein kinase phosphorylation pattern of H,K-ATPase resembles the one found in Na,K-ATPase membrane preparations.[1]

In order to investigate whether PKC-phosphorylation at the conventional N-terminal site had any direct functional effects on the catalytic properties of H,K-ATPase, the K^+-activation of the enzyme was measured at saturating and non-saturating ATP concentrations. FIGURES 1A and 1B show the K^+-stimulated ATPase activity of untreated H,K-ATPase (control) compared to enzyme phosphorylated by PKC, measured at either 10 μM or 3 mM ATP, respectively. As seen, neither V_{max} nor the apparent K^+-affinity ($K_{0.5}$) was affected by PKC phosphorylation at low ATP concentration, whereas 70% activation was observed at 3 mM ATP, pH 7.2. In accor-

---

Address for correspondence: Dr. Yasser Ahmed Mahmmoud, Department of Biophysics, University of Aarhus, Aarhus, Denmark. Voice: +45 8942 2927; fax:+45 8612 9599.
e-mail:yam@biophys.au.dk

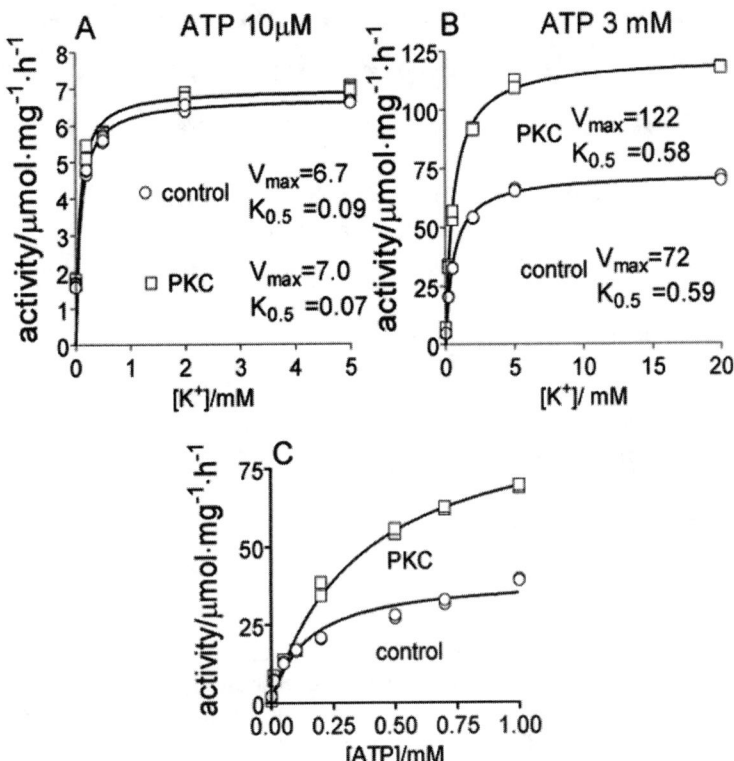

FIGURE 1. (A and B) K$^+$-activation curves at 10 μM and 3 mM ATP, respectively. (C) ATP substrate curves in control and after PKC phosphorylation.

dance, the ATP-activation curve demonstrated PKC activation at ATP concentrations > 20 μM, indicating that the effect was mainly on steps following the phosphoryl transfer step.

In conclusion, two new C-terminal PKC phoshorylation sites are demonstrated in gastric H,K-ATPase close to the inner face of the plasma membrane and only accessible to the kinase in the presence of detergent, just as found for Na,K-ATPase.[1] PKC phosphorylation at the conventional N-terminal Ser[27] resulted in significant stimulation of the H,K-ATPase activity at saturating ATP concentrations. The effect is probably mediated by a shift in the E_1/E_2 conformational equilibrium. We could not detect any FXYD proteins in the H,K-ATPase membranes, suggesting that N-terminal PKC phosphorylation is a direct conformation switching signal, and not an indirect signal initiated by phosphorylation of another small protein, as found in other P-type ATPases.

REFERENCE

1. MAHMMOUD, Y. A. & F. CORNELIUS. 2002. Biophys. J. **82:** 1907–1919.

Seasonal Changes of Ca-ATPase Activity in Skeletal Muscle Sarcoplasmic Reticulum of the Ground Squirrel *Spermophilus undulatus*

ALEXANDER S. KONDRASHEV-LUGOVSKII, ANNA N. MALYSHEVA, KENNETH B. STOREY,[a] OLGA D. LOPINA, AND ALEXANDER M. RUBTSOV

Department of Biochemistry, School of Biology, Lomonosov Moscow State University, 119992 Moscow, Russia

[a]*Department of Biology and Department of Chemistry, Carleton University, Ottawa K1S 5B6, Canada*

KEYWORDS: sarcoplasmic reticulum; Ca-ATPase activity; protein kinases; hibernation

Winter survival for many small animals can only be assured by hibernation, a torpor response to cold temperatures characterized by a strong metabolic rate depression and, as a result, reduction of body temperature to near ambient. Because winter hibernation consists of repeated bouts of torpor followed by brief arousals, skeletal muscles must readily regain their active state during these periodic awakenings. However, very little is known about mechanisms that adjust skeletal muscle contractile activity for this purpose. Here we studied seasonal changes of Ca-ATPase activity of sarcoplasmic reticulum (SR) from skeletal muscles of a typical hibernator, the ground squirrel *Spermophilus undulatus*.

METHODS

Membrane preparations of SR were obtained by differential centrifugation.[1] Enzyme kinetic parameters, protein composition of SR membranes, and *in vitro* phosphorylation of SR proteins by endogenous protein kinases were studied by standard methods as described previously.[1,2]

Address for correspondence: Alexander S. Kondrashev-Lugovskii, Department of Biochemistry, School of Biology, Lomonosov Moscow State University, 119992 Moscow, Russia. Voice: +7-095-9394434; fax: +7-095-9393955.
alexandr_ks@mail.ru

RESULTS

The activity of Ca-ATPase in SR preparations of the ground squirrel is decreased about twofold (from 5.2±0.4 to 2.4±0.3 µmol/min/mg SR protein) during the winter period independently of the physiological state of the animals (active or hibernating). This decrease of activity is connected with both ≈25% decrease of Ca-ATPase content in SR membranes and changes of its kinetic parameters: dependence of Ca-ATPase activity on ATP concentration is characterized by additional activation by high ATP concentrations (>500 µM) in SR of summer animals and is converted into a standard Michaelis curve at winter time. The content of Ca-binding proteins calsequestrin, sarcalumenin, and histidine-rich Ca-binding protein is decreased three- to fourfold, but their phosphorylation by endogenous protein kinases of SR preparations is increased about twofold during winter as well as phosphorylation of unidentified proteins with molecular masses 96 and 70 kDa. In SR of summer-active animals, the increased level of phosphorylation is detected for unidentified 44 kDa and 24 kDa proteins. Despite the fact that phosphorylation of Ca-ATPase protein by protein kinases was not detected, *in vitro* stimulation of the activity of endogenous protein kinases in SR preparations of winter ground squirrels restores both Ca-ATPase activity and enzyme kinetic parameters. It is suggested that the phosphorylation/dephosphorylation of unknown regulatory protein(s) plays a critical role in seasonal regulation of Ca-ATPase activity in skeletal muscle SR of ground squirrels.

ACKNOWLEDGMENTS

The research described in this publication was supported in part by Grant No. 01-04-48237 from the Russian Foundation for Basic Research and grant NSERC (Canada).

REFERENCES

1. MALYSHEVA, A.N., *et al.* 2001. Ca-ATPase activity and protein composition of sarcoplasmic reticulum membranes isolated from skeletal muscle of typical hibernator, the ground squirrel *Spermophilus undulatus*. Biosci. Rep. **21:** 831–838.
2. RUBTSOV, A.M. 2001. Hibernation: Protein adaptations. *In* Cell and Molecular Responses to Stress. Vol. 2. Protein Adaptations and Signal Transduction. K.B. Storey & J.M. Storey, Eds.: 57–71. Elsevier Science B.V. Amsterdam.

Acidic-Lipid Responsive Regions of the Plasma Membrane Ca^{2+} Pump

HUGO P. ADAMO, FELICITAS DE TEZANOS PINTO, LUIS M. BREDESTON, AND GERARDO R. CORRADI

Instituto de Química y Fisicoquímica Biológicas, Universidad de Buenos Aires, Junín 956 1113, Buenos Aires, Argentina

KEYWORDS: Ca^{2+} pump; Ca^{2+}-ATPase; PMCA; acidic lipids

The Ca^{2+} pump from plasma membranes (PMCA) is highly sensitive to the lipid environment and apparently is much more so than other P-type ATPases, in particular toward acidic lipids.[1] This unique response of the PMCA has been shown to be in part due to the fact that acidic lipids bind to the C-terminal autoinhibitory domain and mimic the activating action of calmodulin.[2] In addition, because acidic lipids enhance the Ca^{2+} sensitivity of the PMCA to a greater extent than calmodulin, the existence of an independent acidic-lipid responsive region has been suggested.[2,3] Studies of controlled proteolysis have shown that a lipid-like activation is attained after the cleavage of the PMCA in the segment connecting the A domain with the M3 transmembrane helix. In this position, the PMCA contains the insertion of a segment we have called A_L (residues 296–356) that is not present in the NaK-ATPase and SERCA pumps. We have recently reported that recombinant PMCAs with deletions involving residues 350–356 near the M3 helix are completely inactive, indicating that this region is critical for function.[4] In contrast, deletion of residues 296–349 resulted in a fully active pump with a high affinity for Ca^{2+}, characteristic of the activated form of the PMCA that would result from binding of acidic lipids at the lipid-binding site of region A_L.

In order to investigate further the effects of acidic lipids on the PMCA, we constructed a mutant designed d296-349ct120 lacking the C-terminal 120 amino acids containing the acidic lipid/calmodulin-binding site and the A_L region. In contrast with our previous studies, which were carried out using the microsomal fraction of transfected mammalian cells, we expressed d296-349ct120 in *Saccharomyces cerevisiae*, and the mutant protein was solubilized, purified, and reconstituted into liposomes. When reconstitution was made in liposomes containing only phosphatidylcholine the mutant was capable of hydrolyzing ATP, but it had a very low activity. The activity of the d296-349ct120 enzyme increased with the content

Address for correspondence: Hugo P. Adamo, IQUIFIB-Universidad de Buenos Aires, Junín 956 1113, Buenos Aires, Argentina. Voice: +54-1-4964-8289; fax +54-1-4962-5457.
hpadamo@qb.ffyb.uba.ar

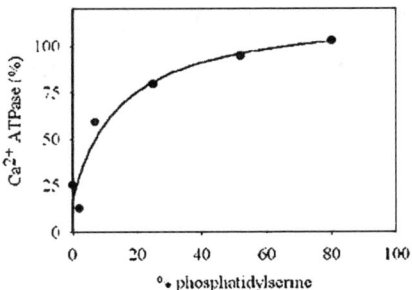

FIGURE 1. Stimulation of the PMCA mutant d296-349ct120 by acidic lipids. The recombinant d296-349ct120 enzyme was solubilized from yeast membranes and reconstituted into phosphatidylcholine liposomes containing the indicated proportion of phosphatidylserine. The Ca^{2+} ATPase activity was measured at 37°C for 30 min following the release of inorganic phosphate from $[\gamma\text{-}^{32}P]ATP$. The reaction was performed with 5 µg of reconstituted PMCA protein in 20 mM MOPS pH 7.1, 150 mM KCl, 5 mM $MgCl_2$, 3 mM ATP, 0.5 mM EGTA and $CaCl_2$ to give a free Ca^{2+} concentration of 50 µM.

of phosphatidylserine in the liposomes reaching at 25% of phosphatidylserine—a value about 3 times higher than in the absence of acidic lipid (FIG. 1). This result shows that despite the deletion of the proposed acidic-lipid responsive regions, the purified PMCA still needed high amounts of phosphatidylserine for optimal activity. Since the inner half of the plasma membrane in which the PMCA is embedded has approximately 12% of phosphatidylserine, the lower activity of the d296-349ct120 enzyme reconstituted in phosphatidylcholine liposomes is likely the consequence of the depletion of acidic lipids. Thus, the effect of acidic lipids on the PMCA is not exclusively dependent upon the N- and C-terminal regulatory regions and probably can be exerted directly by affecting the conformation of the transmembrane domain. While these results suggest that the dependence of the activity of the PMCA with acidic lipids is an intrinsic property of the enzyme, the occurrence of the A_L region could be envisioned as part of a mechanism, leading to a more sensitive regulation based on subtle changes in the amount or quality of the acidic lipids normally present in the plasma membrane.

REFERENCES

1. ENYEDI, A. et al. 1987. The maximal velocity and the calcium affinity of the red cell calcium pump may be regulated independently. J. Biol. Chem. **262:** 6425–6430.
2. FILOTEO, A.G. et al. 1992. The lipid-binding peptide from the plasma membrane Ca^{2+} pump binds calmodulin, and the primary calmodulin-binding domain interacts with lipid. J. Biol. Chem. **267:** 11800–11805.
3. ZVARITCH, E. et al. 1990. Mapping of functional domains in the plasma membrane Ca^{2+} pump using trypsin proteolysis. Biochemistry **29:** 8070–8076.
4. DE TEZANOS PINTO F. & H.P. ADAMO. 2002. Deletions in the acidic lipid-binding region of the plasma membrane Ca^{2+} pump. J. Biol. Chem. **277:** 12784–12789.

Short-Term Aldosterone Action on Na,K-ATPase Surface Expression

Role of Aldosterone-Induced SGK1?

FRANÇOIS VERREY, VANESSA SUMMA, DIRK HEITZMANN, DAVID MORDASINI,[a] ALAIN VANDEWALLE,[b] ERIC FÉRAILLE,[a] AND MARIJA ZECEVIC

Institute of Physiology, University of Zurich, CH-8057 Zurich, Switzerland

[a]*Division of Nephrology and Department of Pathology, University of Geneva, CH-1211 Geneva, Switzerland*

[b]*INSERM, Unité 478, Faculté de Médecine Xavier-Bichat, F-75870 Paris, France*

ABSTRACT: Aldosterone controls extracellular volume and blood pressure by regulating Na^+ reabsorption across epithelial cells of the aldosterone-sensitive distal nephron (ASDN). This effect is mediated by a coordinate action on the luminal channel ENaC (generally rate limiting) and the basolateral Na,K-ATPase. Long-term effects of aldosterone (starting within 3 to 6 hours and increasing over days) are mediated by the direct and indirect induction of stable elements of the Na^+ transport machinery (e.g., Na,K-ATPase α subunit), whereas short-term effects appear to be mediated by the upregulation of short-lived elements of the machinery (e.g., ENaC α subunit) and of regulatory proteins, such as the serum- and glucocorticoid-regulated kinase SGK1. We have recently shown that in cortical collecting duct (CCD) from adrenalectomized (ADX) rats, the increase in Na,K-ATPase activity (approximately threefold in 3 h), induced by a single aldosterone injection, can be fully accounted for by the increase in Na,K-ATPase cell-surface expression. Using the model cell line mpkCCD$_{cl4}$, we showed that the parallel increase in Na,K-ATPase function [assessed by Na^+ pump current (I_p) measurements] and cell-surface expression depends on transcription and translation, and that it is not secondary to a change in apical Na^+ influx. As a first approach to address the question whether the aldosterone-induced regulatory protein SGK1 might play a role in mediating Na,K-ATPase translocation, we have used the *Xenopus laevis* expression system. SGK1 coexpression indeed increased both the Na^+ pump current and the surface expression of pumps containing the rat α1 subunits. In summary, aldosterone controls Na^+ reabsorption in the short term not only by regulating the apical cell-surface expression of ENaC but also by coordinately acting on the basolateral cell-surface expression of the Na,K-ATPase. Results obtained in the *Xenopus* oocyte expression system suggest the possibility that this effect could be mediated in part by the aldosterone-induced kinase SGK1.

KEYWORDS: kidney cortical collecting duct; aldosterone-induced proteins; epithelial sodium channel; ENaC; sodium transport regulation

Address for correspondence: François Verrey, University of Zurich, Institute of Physiology, Winterthurerstrasse 190, CH-8057 Zurich, Switzerland. Voice: +41-1-635-5044; fax: +41-1-635-6814.
verrey@access.unizh.ch

Ann. N.Y. Acad. Sci. 986: 554–561 (2003). © 2003 New York Academy of Sciences.

INTRODUCTION

The main physiological role of the mineralocorticoid hormone aldosterone is to increase the extracellular volume in response to volume depletion (signaled by the renin-angiotensin system) and it does so mainly by stimulating the reabsorption of Na^+ from the tubular lumen of the kidney nephrons and of other organs, such as sweat glands and the distal colon. Aldosterone action is also connected to the regulation of K^+ homeostasis. On the one hand, high extracellular K^+ is a stimulus for aldosterone secretion and, on the other hand, the secretion of K^+ into the kidney tubule is directly linked to the aldosterone-regulated Na^+ reabsorption that creates the electrical driving force for K^+ secretion.[1]

The main control site of Na^+ reabsorption regulated by aldosterone is the apical Na^+ entry site, namely ENaC. The activity of this highly selective, amiloride-sensitive channel is rate limiting for Na^+ reabsorption provided that there is sufficient active Na,K-ATPase in the basolateral membrane to maintain a low intracellular Na^+ level (otherwise, ENaC activity would be decreased by feedback inhibition) and sufficient open K^+ channels to ensure an adequate membrane potential. Thus, despite the kinetic reserve of Na,K-ATPase (activation by intracellular Na^+ with positive cooperativity), a regulation of Na,K-ATPase that matches that of ENaC is necessary to maintain a stable intracellular Na^+ concentration, and therefore ENaC function, over a broad range of reabsorption rates.

The physiological response to aldosterone in transporting epithelia can be divided into short-term and long-term effects. The short-term aldosterone action (early effect) on Na^+ reabsorption and K^+ secretion can be observed as rapidly as 30 min after the beginning of an aldosterone treatment.[2,3] In the long term (late effect), aldosterone induces a more durable increase in the transport capacity of the target cells.[1]

In additional to the transcriptionally mediated effects described above, aldosterone has been shown over the past years to produce near-immediate cardiovascular effects whose physiological role remains enigmatic. These effects are mediated by a nontranscriptional mechanism (also called nongenomic or nonclassical) that must involve a receptor that has not yet been identified.[4]

SHORT-TERM REGULATION OF THE EPITHELIAL Na CHANNEL BY ALDOSTERONE

The amiloride-sensitive epithelial sodium channel (ENaC) expressed along the aldosterone-sensitive distal nephron (ASDN) is a main target of aldosterone. Our recent experiments on kidneys and isolated tubules of adrenalectomized rats that received a single dose of aldosterone showed that the expression of ENaC α subunit is regulated in the short term, at the mRNA and protein levels, whereas the β and γ subunits are expressed constitutionally.[5] Studies in cell cultures systems have also shown that the half-life of ENaC subunits is short.[6] Thus, the rapid up- and down-regulation of the total cellular amount of αβγENaC can play an important role in the short-term regulation of transepithelial Na^+ transport.

Interestingly, immunofluorescence images also revealed an effect of aldosterone on the subcellular localization of ENaC (all three subunits), which displays a gradient along the proximal-to-distal axis of the tubule. The single aldosterone injection performed in these experiments induced in the proximal segments of the ASDN a shift of the subunits from a punctuated distribution all over the cell cytosol to an apical localization. In contrast, the cellular distribution remained unchanged in later tubular segments. Observations made in long-term treated animals suggest that the length of the tubular segment in which the apical shift of ENaC takes place is regulated. Thus, we have hypothesized that (an) additional factor(s) that display(s) an axially heterogeneous distribution is/are required to induce this apical shift.

OLD OBSERVATIONS ON Na,K-ATPase REGULATION IN KIDNEY TARGET CELLS

The long-term treatment of animals with aldosterone has been shown to induce in the cortical collecting duct an increase in the amount of Na,K-ATPase mRNA (α subunit), protein, and activity (see original references in Verrey et al.[7]). This long-term regulation appears to depend also on other factors that play a permissive role, such as the luminal uptake of Na^+. Interestingly, it has been shown, using a cell culture system, that this slow induction (starting to be measurable at the protein level after 5 h) can be traced back to a rapid transcriptional activation that was measured as early as 15 min after hormone addition.[8,9] The long delay between the very early transcriptional response and the late accumulation of active pumps can be explained by the fact that the pumps are, in contrast to the channels, much more numerous and that they represent a stable component of the transport machinery (half-life of approximately one day) and that it thus takes time until an approximately threefold increase in pump production impacts on the total pool level. This slow time course is clearly not compatible with the functional short-term regulatory action described in the introduction. Already 20 years ago, some authors noticed that the activity of the rat cortical collecting duct Na,K-ATPase, measured as hydrolytic activity after cell permeabilization by freeze-thawing, slowly decreased following adrenalectomy, but rapidly returned to normal (1–3 h) upon aldosterone treatment.[10,11] It is only very recently that we have shown that this rapid reactivation of pumps appears to be mediated by a translocation of preexisting pumps to the surface of cells (see below). The fact that this effect was originally observed at the level of the Na,K-ATPase activity can be explained by the permeabilization technique used (freeze-thawing) that allows the substrates to diffuse into the cytosol but not into the interior of vesicles. The activity of pumps present in intracellular vesicles can be revealed by permeabilizing the membranes with detergents.[12]

EARLY ALDOSTERONE-REGULATED PROTEINS INVOLVED IN THE SHORT-TERM REGULATION

K-Ras was the first regulatory protein induced by aldosterone, identified in a target epithelium (A6 cells), that acts on ENaC function and surface expression in the *Xenopus laevis* oocyte expression system.[13,14] Later experiments performed in the laboratory of D. Eaton have suggested that the expression of K-Ras plays a rate-limiting role for Na^+ transport activation by aldosterone in A6 epithelia.[15] However, we have recently observed that K-Ras mRNA is not induced by aldosterone in mam-

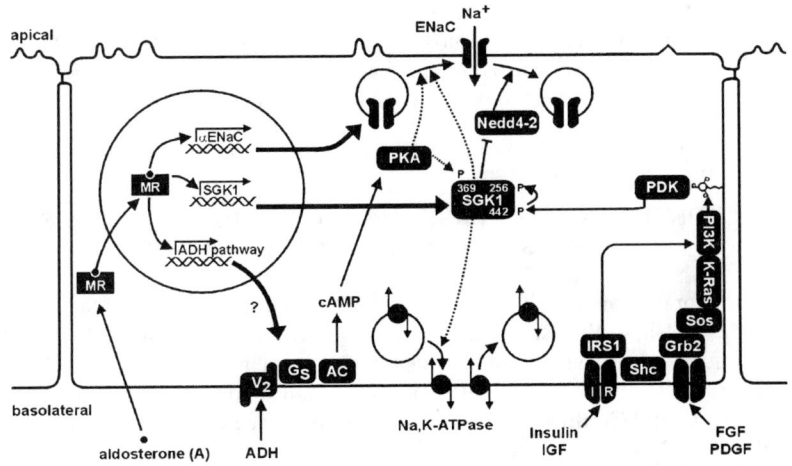

FIGURE 1. Schematic representation of regulatory pathways participating or modulating the aldosterone action in segment-specific cells of the aldosterone-sensitive distal nephron (ASDN). *Thick arrows* indicate transcriptionally mediated induction by aldosterone of gene products that affect Na$^+$ reabsorption. *Thin arrows* indicate regulatory pathways the importance of which appears to be established in the context of Na$^+$ reabsorption. *Stippled arrows* indicate where the precise site of impact or the existence of intermediate steps is not established.

malian ASDN (Zecevic and Verrey, unpublished observation). Nonetheless, this small G-protein might play, in some context, an important role in Na$^+$ transport regulation, since it controls a strategically important site of convergence within the regulatory network that controls SGK1 expression and activity (see below and FIG. 1).

The serum- and glucocorticoid-regulated kinase (SGK1) is an inducible serine/threonine protein kinase that was first identified in mammary tumor cells and subsequently in the hepatoma cell line HepG2 because of its induction by high osmolarity.[16,17] In a collaborative study with our laboratory, David Pearce identified SGK1 as an aldosterone-regulated protein that strongly stimulates ENaC cell-surface expression and function when expressed in *Xenopus* oocytes.[18] In a later study, we showed that SGK1 is indeed rapidly induced at the mRNA and protein levels in the segment-specific cells of the entire ASDN.[5,19]

Besides the fact that it is inducible at the mRNA/protein level, SGK1 was shown to be tightly controlled posttranslationally. Indeed, its activity depends in particular on its phosphorylation by 3-phosphoinositide-dependent protein kinase (PDK)[20,21] (FIG. 1). PDK1 activity in turn depends on the presence of the membrane lipid phosphatidylinositol(3,4,5)P3, the phosphorylation of which in position 3 is catalyzed by PI3-kinase. This latter enzyme is regulated by several inputs, in particular by insulin and IGF via the insulin receptor substrate-1 (IRS1) and also by Ras, and thus represents a site of convergence upstream of SGK1.

Downstream of SGK1, the ubiquitin ligase Nedd4-2 has been identified as a phosphorylation target.[22] Interestingly, the coexpression of this protein with ENaC in *Xenopus* oocytes was already well known to downregulate ENaC surface localization.[23] Thus, this cascade represents the first continuous link to be proposed that leads from transcriptional regulation by aldosterone (SGK1) to the activation of ENaC. However, this cascade has not yet been demonstrated in an epithelium, and some data generated by others appear for the time being contradictory with this proposed mechanism.[6,15]

ALDOSTERONE INCREASES Na,K-ATPase CELL-SURFACE EXPRESSION IN THE SHORT TERM

In a collaborative study with Eric Féraille, we have recently studied the impact of aldosterone on Na,K-ATPase function and cell-surface expression both in cortical collecting ducts of adrenalectomized rats (same treatment as above for the study of ENaC and SGK1) and in a cell culture system (mouse cortical collecting duct cell line mpkCCD$_{cl4}$).[24] Confirming the previous findings (see above), the hydrolytic activity of the Na,K-ATPase measured in freeze-thawed cortical collecting ducts was increased three- to fourfold when measured 3 h after an aldosterone injection. Western blot analysis revealed that the total Na,K-ATPase α subunit pool was not increased significantly during the same time period. In contrast, the cell-surface labeled Na,K-ATPase α subunit was increased nearly sixfold. Thus, the increase in Na,K-ATPase activity can be fully accounted for by an increase in the amount of pumps expressed at the cell surface.

A similar, but quantitatively smaller effect of aldosterone on the Na,K-ATPase cell-surface expression was observed in mpkCCD cells that was parallel to the increase in pump function measured as pump current in epithelia permeabilized apically to monovalent ions using amphotericin B. Using inhibitors in this *in vitro* system, we could show that the effect of aldosterone on Na,K-ATPase surface expression/function depends on ongoing transcription and translation, but not on the increase in apical Na$^+$ influx mediated by aldosterone (FIG. 2).

Thus, it appears that aldosterone acts on Na$^+$ in the short term not only by acting on the cell-surface expression of apical ENaC but also by acting on that of the basolateral Na,K-ATPase.

To test whether the short-term regulation of the sodium pump by aldosterone depends on specific features of the α1 subunit, we generated mpkCCD cell lines expressing the human Na,K-ATPase α subunits isoforms α1, α2, or α3 (Summa and Verrey, in preparation). An ouabain-sensitive current generated by pumps containing the exogenous human α1 and α2 subunit could be measured, whereas no additional cardiotonic-sensitive current was found in α3 subunit–expressing cells. This exogenous pump current was increased by an aldosterone treatment only in the case of the α1 subunit isoform expressing cells, indicating that isoform-specific sequences are required for that regulation.

POSSIBLE ROLE OF SGK ON Na,K-ATPase REGULATION

To address the question of whether SGK1 might play a role in the regulation of the Na,K-ATPase cell-surface localization, we coexpressed both proteins in *Xenopus* oocytes. Wild-type SGK1 appeared to increase the number of pumps that contain the

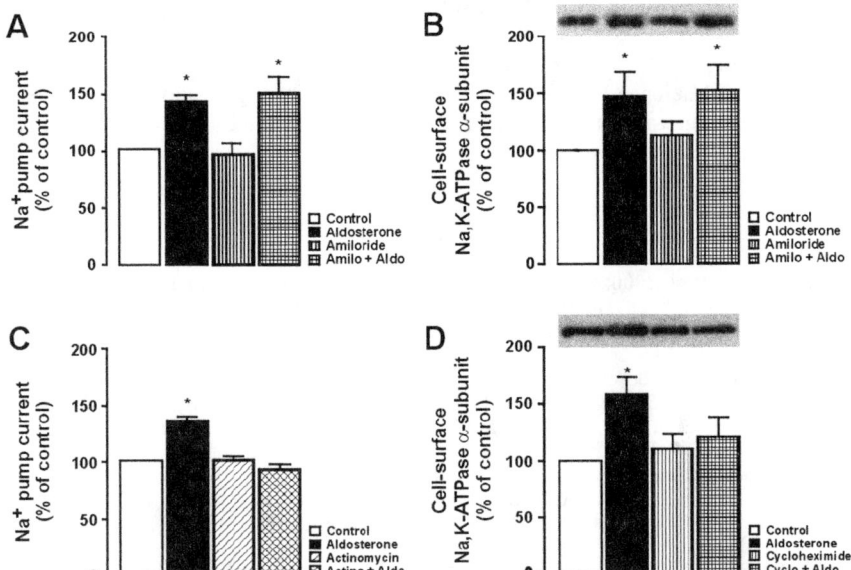

FIGURE 2. The short-term stimulatory effect of aldosterone on Na,K-ATPase pump current and cell surface expression does not depend on apical Na^+ influx but on ongoing transcription and translation. Experiments were performed on mpkCCD$_{cl4}$ epithelia grown on filters. (**A** and **B**) Blocking Na^+ influx via ENaC with amiloride during the aldosterone treatment (2 and 3 h, respectively) does not modify baseline nor aldosterone-induced Na,K-ATPase pump activity and/or cell-surface expression. (**C** and **D**) Inhibition of transcription (actinomycin D) or translation (cycloheximide) during an aldosterone treatment (2 and 3 h, respectively) does not modify the baseline nor the aldosterone-induced Na,K-ATPase pump activity and cell-surface expression. *Bars* represent percentage of controls. (Modified from Summa *et al.*[24])

exogenous α subunit at the cell surface and also the pump current carried by these pumps. In contrast, SGK1 mutant at the level of its catalytic site had no effect (Zecevic, Heitzmann, and Verrey, in preparation).

CONCLUSIONS

Taken together, these data suggest that aldosterone acts in the short term on the cell-surface expression of the basolateral Na,K-ATPase in a manner similar to its action onto apical ENaC and that in both cases SGK1 plays an essential role on the pathway that leads from the transcriptional regulation to the functional response. Whereas it appears that Nedd4-2 represents a link between SGK1 and ENaC, no mediator lying between SGK1 and the Na,K-ATPase has been proposed as yet. Further experiments within an epithelial system will be required to address this question.

ACKNOWLEDGMENTS

The authors thank Christian Gasser for artwork. This work was supported by the Swiss National Science Foundation Grants 31-59141.99 to F.V. and 31-50830.99 to E.F.; by the Hartmann-Müller Stiftung in Zürich; and by the Carlos and Elsie de Reuter Foundation in Geneva.

REFERENCES

1. VERREY, F. et al. 2000. Control of Na^+ transport by aldosterone. In The Kidney, Physiology and Pathophysiology. 3rd ed. D.W. Seldin and G. Giebiesch, Eds.: 1441–1471. Lippincott, Williams & Wilkins. Philadelphia.
2. HORISBERGER, J.D. & J. DIEZI. 1983. Effects of mineralocorticoids on Na^+ and K^+ excretion in the adrenalectomized rat. Am. J. Physiol. **245:** F89–F99.
3. EL MERNISSI, G. & A. DOUCET. 1983. Short-term effect of aldosterone on renal sodium transport and tubular Na-K-ATPase in the rat. Pflueger's Arch. **339:** 139–146.
4. LOSEL, R.M. et al. 2002. Nongenomic effects of aldosterone: cellular aspects and clinical implications. Steroids **67:** 493–498.
5. LOFFING, J. et al. 2001. Aldosterone induces rapid apical translocation of ENaC in early portion of renal collecting system: possible role of SGK. Am. J. Physiol. (Renal Physiol.) **280:** F675–F682.
6. DE LA ROSA, D.A., H. LI & C.M. CANESSA. 2002. Effects of aldosterone on biosynthesis, traffic, and functional expression of epithelial sodium channels in A6 cells. J. Gen. Physiol. **119:** 427–442.
7. VERREY, F., J. BERON & B. SPINDLER. 1996. Corticosteroid regulation of renal Na,K-ATPase. Mineral Electrolyte Metab. **22:** 279–292.
8. VERREY, F., J.P. KRAEHENBUHL & B.C. ROSSIER. 1989. Aldosterone induces a rapid increase in the rate of Na,K-ATPase gene transcription in cultured kidney cells. Mol. Endocrinol. **3:** 1369–1376.
9. BERON, J. & F. VERREY. 1994. Aldosterone induces early activation and late accumulation of Na-K-ATPase at surface of A6 cells. Am. J. Physiol. **266:** C1278–C1290.
10. HORSTER, M., H. SCHMID & U. SCHMIDT. 1980. Aldosterone in vitro restores nephron Na-K-ATPase of distal segments from adrenalectomized rabbits. Pfluegers Arch. **384:** 203–206.
11. PETTY, K.J., J.P. KOKKO & D. MARVER. 1981. Secondary effect of aldosterone on Na-K-ATPase activity in the rabbit cortical collecting tubule. J. Clin. Invest. **68:** 1514–1521.
12. GONIN, S. et al. 2001. Cyclic AMP increases cell surface expression of functional Na,K-ATPase units in mammalian cortical collecting duct principal cells. Mol. Biol. Cell **12:** 255–264.
13. SPINDLER, B. et al. 1997. Characterization of early aldosterone-induced RNAs identified in A6 kidney epithelia. Pflugers Arch. **434:** 323–331.
14. MASTROBERARDINO, L. et al. 1998. Ras pathway activates epithelial Na^+ channel and decreases its surface expression in Xenopus oocytes. Mol. Biol. Cell **9:** 3417–3427.
15. STOCKAND, J.D. 2002. New ideas about aldosterone signaling in epithelia. Am. J. Physiol. (Renal Physiol.) **282:** F559–F576.
16. WEBSTER, M.K. et al. 1993. Characterization of sgk, a novel member of the serine/threonine protein kinase gene family which is transcriptionally induced by glucocorticoids and serum. Mol. Cell. Biol. **13:** 2031–2040.
17. WALDEGGER, S. et al. 1997. Cloning and characterization of a putative human serine/threonine protein kinase transcriptionally modified during anisotonic and isotonic alterations of cell volume. Proc. Natl. Acad. Sci. USA **94:** 4440–4445.
18. CHEN, S. et al. 1999. Epithelial sodium channel regulated by aldosterone-induced protein sgk. Proc. Natl. Acad. Sci. USA **96:** 2514–2519.
19. LOFFING, J., et. al 2001. Mediators of aldosterone action in the renal tubule. Curr. Opin. Nephrol. Hypertens. **10:** 667–675.

20. PARK, J. et al. 1999. Serum and glucocorticoid-inducible kinase (SGK) is a target of the PI 3-kinase-stimulated signaling pathway. EMBO J. **18:** 3024–3033.
21. KOBAYASHI, T. & P. COHEN. 1999. Activation of serum- and glucocorticoid-regulated protein kinase by agonists that activate phosphatidylinositide 3-kinase is mediated by 3-phosphoinositide-dependent protein kinase-1 PDK1 and PDK2. Biochem. J. **339:** 319–328.
22. DEBONNEVILLE, C. et al. 2001. Phosphorylation of Nedd4-2 by Sgk1 regulates epithelial Na^+ channel cell surface expression. EMBO J. **20:** 7052–7059.
23. KAMYNINA, E. et al. 2000. A novel mouse Nedd4 protein suppresses the activity of the epithelial Na^+ channel. FASEB J. **15:** 204–214.
24. SUMMA, V. et al. 2001. Short term effect of aldosterone on Na,K-ATPase cell surface expression in kidney collecting duct cells. J. Biol. Chem. **276:** 47087–47093.

Renal Tubule Sodium Transporter Abundance Profiling in Rat Kidney

Response to Aldosterone and Variations in NaCl Intake

MARK A. KNEPPER, GHEUN-HO KIM,[a] AND SHYAMA MASILAMANI

Laboratory of Kidney and Electrolyte Metabolism, National Heart, Lung, and Blood Institute, National Institutes of Health, Bethesda, Maryland 20892-1603, USA

[a]*Department of Internal Medicine, Hallym University Hangang Sacred Heart Hospital, Seoul, Korea*

ABSTRACT: Based on extensive physiological study of sodium transport mechanisms along the renal tubule, complementary DNAs for all of the major transporters and channels responsible for renal tubular sodium reabsorption have been cloned over the past decade. There is now a comprehensive set of cDNA and antibody probes that can be used to investigate physiological mechanisms on a molecular level. Using rabbit polyclonal antibodies to all of the major renal Na transport proteins, we have developed profiling methods allowing comprehensive, integrated analysis of sodium transporter protein abundance changes along the renal tubule in response to physiological and pathophysiological perturbations. Here, we review some of our recent findings with this approach, focusing on renal responses to aldosterone and to variations in NaCl intake.

KEYWORDS: hypertension; aldosterone; vasopressin; ENaC

INTRODUCTION

The regulation of blood pressure and extracellular fluid volume is critically dependent on the control of renal sodium chloride excretion. The work of Guyton and his followers established a critical role for the kidney in the maintenance of normal blood pressure.[1,2] The rate of renal sodium excretion is determined by the relative rates of glomerular filtration of sodium ions by the glomerulus and net reabsorption of sodium ions along the renal tubule. Renal tubule sodium reabsorption is accomplished in the various renal tubule segments largely by active transport mechanisms that depend on specific Na transporter proteins (FIG. 1). In general, the basolateral component of sodium transport in each renal tubule segment is dependent on the same transporter, the Na,K-ATPase (α1-β1 complex). However, the apical component of sodium transport is mediated by a different transporter in each segment (FIG. 1). The properties of these apical transport proteins are summarized in TABLE 1. In

Address for correspondence: Mark A. Knepper, M.D. Ph.D., National Institutes of Health, Bldg. 10, Room 6N260, 10 Center Drive, MSC 1603, Bethesda, Maryland, 20892-1603, USA. Voice: 301-496-3064; fax 301-402-1443.
knep@helix.nih.gov

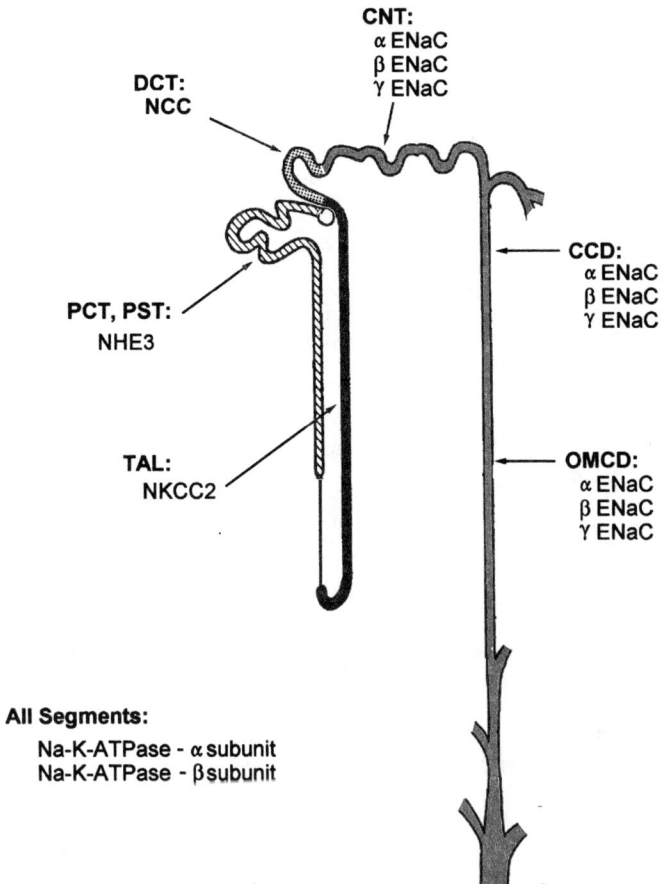

FIGURE 1. Diagram of renal tubule enumerating the major Na transport proteins in each segment. (See TABLE 1 for terminology.)

general, renal tubule sodium reabsorption is accomplished by coordinate regulation of these apical transporters and the basolateral Na,K-ATPase. Short-term regulation typically occurs as a result of post-translational modifications or trafficking of these proteins. In addition, long-term regulation of each of these transporters occurs via changes in transporter protein abundances. We have developed techniques for the integrated study of long-term regulation of all of the major Na transporter proteins, an approach that has been referred to as either "renal tubule sodium transporter abundance profiling" or "antibody-based targeted proteomics."[3,4] The concept is to study all of the major renal Na transporter proteins (as well as the major aquaporin water channels) simultaneously in intact animals so as to understand the integrated response to various physiological and pathophysiological perturbations associated with altered sodium ion balance in the intact rat or mouse. In this short review, we

TABLE 1. Major apical sodium transporter and sodium channel proteins in kidney

Name	Identification	Location	Amino acids (number)	Actual MW (kDa)[a]	Antibody Reference
NHE3	Type 3 Na-H exchanger	Proximal tubule	831	84	(21)
NKCC2	Type 2 Na-K-2Cl cotransporter	Thick ascending limb	1095	163	(31)
NCC	Na-Cl cotransporter	Distal convoluted tubule	1002	165	(7)
ENaC-α	α subunit of epithelial Na channel	Connecting tubule Collecting duct	699	86	(5)
ENaC-β	β subunit of epithelial Na channel	Connecting tubule Collecting duct	640	88	(5)
ENaC-γ	γ subunit of epithelial Na channel	Connecting tubule Collecting duct	649	85	(5)

[a]Actual molecular weight determined by immunoblotting.

summarize pertinent results from a series of studies done in our laboratory in which we have used this approach to study the roles of aldosterone and variations in dietary NaCl intake in renal sodium transport regulation.

RENAL TUBULE SODIUM TRANSPORTER ABUNDANCE PROFILING OF RESPONSE TO DIETARY NaCl RESTRICTION

The chief approach that we have utilized has been to develop high-quality polyclonal antibodies to each of the major renal sodium transporter proteins (TABLE 1), and to use these antibodies to screen renal homogenates for transporters whose abundances are up- or down-regulated in response to a given physiological manipulation. The approach generally employs both semiquantitative immunoblotting and immunocytochemistry. An example of the type of data obtained with semiquantitative immunoblotting is shown in FIGURE 2. This figure shows renal tubule sodium transporter abundance changes that occur in the rat kidney in response to dietary sodium restriction (a decrease in sodium chloride intake from 2 mmoles NaCl per 200 g of body weight per day to 0.1 mmoles NaCl per 200 g of body weight per day). As can be seen, dietary NaCl restriction only altered the expression of NaCl transporters expressed in the renal tubule segments beyond the macula densa (the site of tubuloglomerular feedback), namely the distal convoluted tubule, connecting tubule, and collecting duct. There was marked up-regulation of the thiazide-sensitive Na-Cl cotransporter (NCC) (expressed in the distal convoluted tubule) and of the α subunit of the amiloride-sensitive epithelial sodium channel (ENaC) (expressed in the connecting tubule and collecting duct). Interestingly, there was a significant decrease in the abundance of the β subunit of ENaC, and a marked qualitative change in γ ENaC consisting of the appearance of a lower molecular weight form, postulated to be due

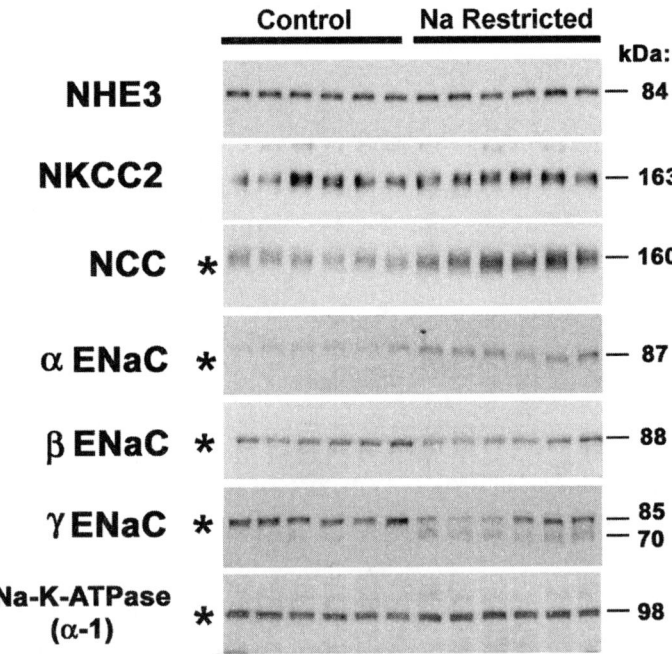

FIGURE 2. Immunoblots demonstrating changes in Na transporter protein abundances in response to long-term dietary NaCl restriction. Control rats received 2 mmoles of NaCl per 200 g of body weight per day. Na-restricted rats received 0.02 mmoles of NaCl per 200 g of body weight per day. *Asterisks* mark transporters whose abundances were significantly changed. (See TABLE 1 for terminology.)

to a physiological proteolytic cleavage of the external loop of the γ subunit.[5] Accompanying these changes, we found by immunocytochemistry that dietary NaCl restriction was associated with a shift in the cellular distribution of ENaC in connecting tubule and collecting duct principal cells from a broad intracellular distribution to the apical region of the cell.[5] Similar changes in ENaC distribution have been described by Loffing et al.[6] Interestingly, immunoblotting also demonstrated a small but statistically significant increase in the abundance of the α-1 subunit of the Na,K-ATPase in response to dietary NaCl restriction to 122 ± 6 percent of control (FIG. 2), although this technique cannot identify the specific segments in which the increase occurs.

PROFILING OF RESPONSES TO DIETARY NaCl RESTRICTION, ALDOSTERONE ADMINISTRATION, AND MINERALOCORTICOID RECEPTOR BLOCKADE

FIGURE 3 summarizes densitometry data from four separate renal tubule profiling studies quantifying responses to dietary NaCl restriction, aldosterone administration, administration of the mineralocorticoid antagonist spironolactone, and in-

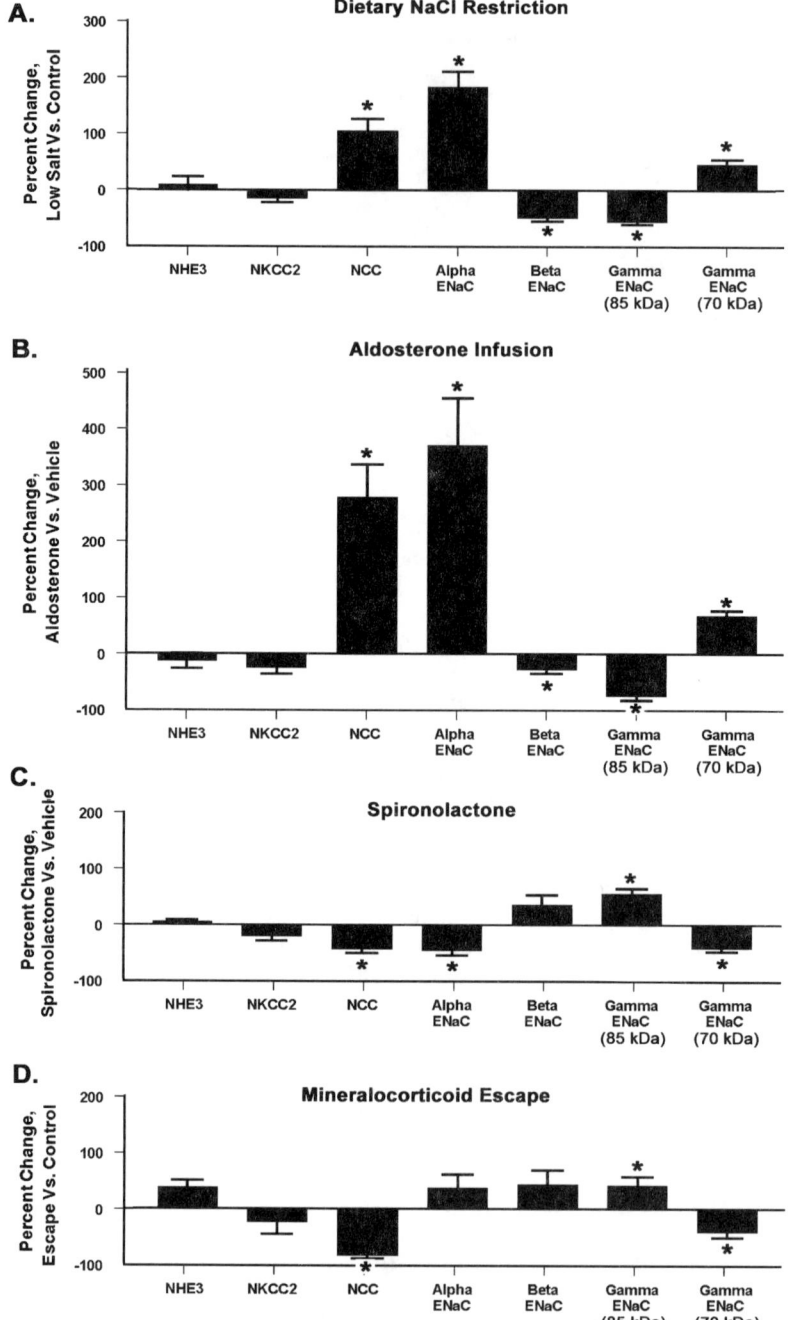

FIGURE 3. *See following page for legend.*

creased NaCl intake in the presence of high levels of circulating aldosterone (mineralocorticoid escape) in kidney homogenates. As can be seen, the profile for aldosterone infusion[5,7] is very similar to that for dietary NaCl restriction,[5,8] suggesting that the changes seen in response to dietary NaCl restriction could be due in part to the associated rise in circulating aldosterone.

Although ENaC has long been viewed as a target for regulation by aldosterone, the role of the thiazide-sensitive NCC as an aldosterone-responsive sodium transporter was not widely accepted prior to these studies. The strong increase in NCC abundance in response to either dietary NaCl restriction (FIG. 3A) or aldosterone infusion (FIG. 3B), coupled with the finding that spironolactone (a mineralocorticoid antagonist) decreases NCC abundance (FIG. 3C),[9] however, demonstrates that NCC is indeed a mineralocorticoid-responsive protein. Interestingly, the increase in NCC protein in response to dietary NaCl restriction was not associated with a measurable change in NCC mRNA, suggesting that the effect of aldosterone is not mediated by transcriptional regulation of the NCC gene, but rather may be due to effects on NCC translation or protein half-life.[8]

As shown in FIGURE 3 (A and B), of the three subunits of ENaC, only the α subunit increases in protein abundance in response to dietary NaCl restriction or aldosterone infusion. We have demonstrated that spironolactone administration totally blocks the increase α-ENaC abundance in response to dietary NaCl restriction[9] and that the increase in α-ENaC protein is associated with an increase in α-ENaC mRNA abundance,[8] supporting the view that the α-ENaC gene is a target for transcriptional regulation by the mineralocorticoid receptor.

In contrast to α-ENaC, the protein abundance of β-ENaC is decreased in response to dietary NaCl restriction or aldosterone administration (FIG. 3). This result is in accord with the view that the synthesis of α-ENaC protein is rate limiting for assembly of the ENaC complex[10] and that the idea that ENaC complex maturation and translocation to the apical plasma membrane may be inevitably followed by ENaC complex endocytosis and degradation, thus decreasing ENaC subunit protein half-life.[11]

The γ subunit of ENaC undergoes a complex qualitative change in response to dietary NaCl restriction (FIG. 2) or aldosterone administration.[5] These perturbations are associated with the appearance of a 70 kDa γ-ENaC band with a concomitant decrease in the main 85 kDa band. This change in molecular weight has been postulated to be due to a physiological proteolytic cleavage of γ-ENaC, although this possibility had not yet been successfully addressed. The molecular weight of the

---

FIGURE 3. Results of renal tubule sodium transporter abundance profiling experiments in rats. (**A**) Dietary NaCl restriction.[5,8] See FIGURE 2 caption for conditions. (**B**) Aldosterone infusion.[5] Rats were infused with aldosterone subcutaneously by osmotic minipump at a rate of 200 μg per day; controls were infused with vehicle. All rats received 1 mmole NaCl per 200 g of body weight per day in food. (**C**) Spironolactone administration.[9] All rats received a restricted amount of NaCl (0.02 mmole NaCl per 200 g of body weight per day). Experimental rats received high-dose spironolactone dissolved in olive oil and mixed with food in an amount sufficient to give 0.35 mg spironolactone/g body weight/day. Control rats received only olive oil. (**D**) Mineralocorticoid escape.[17] All rats received aldosterone infusion by osmotic minipump at a rate of 200 μg per day. Control rats received a low NaCl intake (0.02 mmole NaCl per 200 g body weight per day), while experimental rats received a high NaCl intake (2 mmoles of NaCl per 200 g body weight per day).

lower band shifts to 52 kDa after deglycosylation,[5] suggesting that the putative proteolytic cleavage would have to occur in the early part of the extracellular loop of ENaC to explain the size shift. In studies examining the early stages of the response to dietary NaCl restriction, the molecular weight shift of γ-ENaC was the only one of the responses seen in FIGURE 2 to be manifest within 15 hours of NaCl restriction.[12] Patch-clamp observations in split-open cortical collecting ducts established that there was a marked increase in amiloride-sensitive Na channel activity at that time, raising the possibility that the molecular weight shift plays a role in the activation or trafficking of ENaC. As shown in FIGURE 3C, spironolactone partially reversed the molecular weight shift of γ-ENaC seen in response to dietary NaCl restriction, suggesting that the molecular weight shift is a response to mineralocorticoid receptor activation.

Although spironolactone was effective in blocking two of the responses to dietary NaCl restriction in the collecting duct (blocked increase in α-ENaC abundance and molecular weight shift of γ-ENaC, FIG. 3C), a third effect was not blocked by spironolactone, namely the redistribution of γ-ENaC from a broad intracellular distribution to the apical region of the cell.[9] Although apical redistribution of ENaC can clearly seen in response to aldosterone,[13] it appears possible that a similar redistribution could occur in an aldosterone-independent manner, perhaps triggered by phosphorylation of the regulatory kinase sgk1.[14,15]

FIGURE 3(D) shows the changes in apical Na transporter protein abundances in mineralocorticoid escape. Mineralocorticoid escape (a manifestation of pressure natriuresis[16]) is the process whereby the kidney is able to overcome the Na-retaining effects of aldosterone in states of mineralocorticoid excess in order to excrete ingested NaCl and maintain salt balance. Increased excretion of NaCl in mineralocorticoid escape is associated with a large decrease in the abundance of the thiazide-sensitive cotransporter, NCC.[17] As illustrated in FIGURE 3(D), however, mineralocorticoid escape does not simply result from a reversal of the primary effects of aldosterone, since the abundance of the α subunit of ENaC was not suppressed. Nevertheless, there was a significant degree of reversal of the 85 to 70 kDa shift in the molecular weight of the γ-subunit of ENaC normally associated with high circulating levels of aldosterone, indicating that a role for ENaC in the mineralocorticoid escape phenomenon and pressure natriuresis cannot be ruled out.

ACKNOWLEDGMENTS

The work described herein was supported by the intramural budget of the National Heart Lung and Blood Institute (National Institutes of Health, project no. Z01-HL-01282-KE to M.A. Knepper).

REFERENCES

1. GUYTON, A.C. 1991. Blood pressure control: Special role of the kidneys and body fluids. Science **252:** 1813–1816.
2. LIFTON, R.P., A.G. GHARAVI & D.S. GELLER. 2001. Molecular mechanisms of human hypertension. Cell **104:** 545–556.
3. KNEPPER, M.A. & S. MASILAMANI. 2001. Targeted proteomics in the kidney using ensembles of antibodies. Acta Physiol. Scand. **173:** 11–21.

4. KNEPPER, M.A. 2002. Proteomics and the kidney. J. Am. Soc. Nephrol. **13:** 1398–1408.
5. MASILAMANI, S., G.-H. KIM, C. MITCHELL, *et al.* 1999. Aldosterone-mediated regulation of ENaC α, β, and γ subunit proteins in rat kidney. J. Clin. Invest. **104:** R19–R23.
6. LOFFING, J., L. PIETRI, F. AREGGER, *et al.* 2000. Differential subcellular localization of ENaC subunits in mouse kidney in response to high- and low-Na diets. Am. J. Physiol. Renal Physiol. **279:** F252–F258.
7. KIM, G.-H., S. MASILAMANI, R. TURNER, *et al.* 1998. The thiazide-sensitive Na-Cl cotransporter is an aldosterone-induced protein. Proc. Natl. Acad. Sci USA **95:** 14552–14557.
8. MASILAMANI, S., X.-Y. WANG, G.-H. KIM, *et al.* 2002. Time course of renal Na-K-ATPase, NHE3, NKCC2, NCC, and ENaC abundance changes with dietary NaCl restriction. Am. J. Physiol. Renal Physiol. In press.
9. NIELSEN, J., T.-H. KWON, S. MASILAMANI, *et al.* 2002. Sodium transporter abundance profiling in kidney: effect of spironolactone. Am. J. Physiol. Renal Physiol. **283:** 923–933.
10. MAY, A., A. PUOTI, H.P. GAEGGELER, *et al.* 1997. Early effect of aldosterone on the rate of synthesis of the epithelial sodium channel alpha subunit in A6 renal cells. J. Am. Soc. Nephrol. **8:** 1813–1822.
11. DE LA ROSA, D.A., H. LI & C.M. CANESSA. 2002. Effects of aldosterone on biosynthesis, traffic, and functional expression of epithelial sodium channels in a6 cells. J. Gen. Physiol. **119:** 427–442.
12. FRINDT, G., S. MASILAMANI, M.A. KNEPPER & L.G. PALMER. 2001. Activation of epithelial Na channels during short-term Na deprivation. Am. J. Physiol. Renal Physiol. **280:** F112–F118.
13. LOFFING, J., M. ZECEVIC, E. FERAILLE, *et al.* 2001. Aldosterone induces rapid apical translocation of ENaC in early portion of renal collecting system: possible role of SGK. Am. J. Physiol. Renal Physiol. **280:** F675–682.
14. WANG, J., P. BARBRY, A.C. MAIYAR, *et al.* 2001. SGK integrates insulin and mineralocorticoid regulation of epithelial sodium transport. Am. J. Physiol. Renal Physiol. **280:** F303–F313.
15. FALETTI, C.J., N. PERROTTI, S.I. TAYLOR & B.L. BLAZER-YOST. 2002. sgk: an essential convergence point for peptide and steroid hormone regulation of ENaC-mediated Na^+ transport. Am. J. Physiol. Cell Physiol. **282:** C494–C500.
16. HALL, J.E., J.P. GRANGER, M.J. SMITH & A.J. PREMEN. 1984. Role of renal hemodynamics and arterial pressure in aldosterone "escape." Hypertension **6:** I183–I192.
17. WANG, X.-Y., S. MASILAMANI, J. NIELSEN, *et al.* 2001. The renal thiazide-sensitive Na-Cl cotransporter as mediator of the aldosterone-escape phenomenon. J. Clin. Invest. **108:** 215–222.

Mechanism of Control of Na,K-ATPase in Principal Cells of the Mammalian Collecting Duct

ERIC FÉRAILLE, DAVID MORDASINI, SANDRINE GONIN,
GEORGES DESCHÊNES,[a] MANLIO VINCIGUERRA, ALAIN DOUCET,[b]
ALAIN VANDEWALLE,[c] VANESSA SUMMA,[d] FRANÇOIS VERREY,[d] AND
PIERRE-YVES MARTIN

*Division de Néphrologie, Fondation pour Recherches Médicales,
CH-1211 Genève 4, Switzerland*

[a]*Service de Néphrologie Pédiatrique, Hôpital Armand Trousseau,
F-75571 Paris Cedex 12, France*

[b]*Centre National de la Recherche Scientifique URA 1859, Centre d'Etudes de Saclay,
F-91191 Gif sur Yvette, France*

[c]*Institut National de la Santé et de la Recherche Médicale U478,
Faculté de Médecine Xavier Bichat, F-75870 Paris Cedex 18, France*

[d]*Institute of Physiology, University of Zürich, CH-8057 Zürich, Switzerland*

ABSTRACT: The collecting duct is the site of final Na reabsorption according to Na balance requirements. Using isolated rat cortical collecting ducts (CCD) and mpkCCD$_{cl4}$ cells, a mouse cortical collecting duct cell line, we have studied the physiological control of Na,K-ATPase, the key enzyme that energizes Na reabsorption. Aldosterone, a major regulator of Na transport by the collecting duct, stimulates Na,K-ATPase activity through both recruitment of intracellular pumps and increased total amounts of Na pump subunits. This effect is observed after a lag time of 1 hour and is independent of Na entry through ENaC, but requires *de novo* transcription and translation. Vasopressin and cAMP, its second messenger, stimulate Na,K-ATPase activity within minutes through translocation of Na pumps from a brefeldin A–sensitive intracellular pool to the plasma membrane. Dysregulation of collecting duct Na,K-ATPase activity is at least in part responsible of the Na retention observed in nephritic syndrome. In this setting, Na,K-ATPase activity and subunit synthesis are specifically increased in CCD. In conclusion, aldosterone, vasopressin, and intracellular Na control the cell surface expression of Na,K-ATPase and translocation from intracellular stores is a major mechanism of regulation of Na,K-ATPase activity in collecting duct principal cells.

KEYWORDS: Na,K-ATPase; collecting duct; aldosterone; vasopressin; sodium; nephrotic syndrome

Address for correspondence: Eric Féraille, Division de Néphrologie, Fondation pour Recherches Médicales, 64 Ave de la Roseraie, CH-1211 Genève 4, Switzerland. Voice: +41-22-382-38-37; fax: +41-22-347-59-79.
 Eric.Feraille@medecine.unige.ch

Mammalian kidneys play a major role in the homeostasis of body fluid compartments. Despite large quantitative and qualitative variations in dietary intake of solutes and water, the kidneys are able to maintain the composition and the volume of extracellular and intracellular compartments within a narrow range. Another important function of the kidney is the clearing of waste products, such as urea and creatinine, generated by the metabolism. This clearing function requires filtration of large amounts of fluids to ensure elimination of waste products and to maintain low circulating concentrations of these substances. Consequently, the human kidneys generate daily close to 180 liters of ultrafiltrate containing 25 moles of Na. Ultrafiltrate is generated during the passage of blood into the glomeruli. The reabsorption process taking place along kidney tubules results in the daily generation of 1–2 liters of final urine containing 1 to 5% of the filtered Na load. This tremendous work is accomplished by successive nephron segments, which exhibit specific functional properties and hormonal control.[1] Quantitatively, the proximal tubule is the most important segment since it reabsorbs about 70% of filtered water and Na. The loop of Henle next reabsorbs close to 20% and 15% of filtered water and Na, respectively. The distal convoluted tubule is impermeable to water and reabsorbs close to 5% of filtered Na. The connecting tubule and collecting duct are responsible for the fine-tuning of

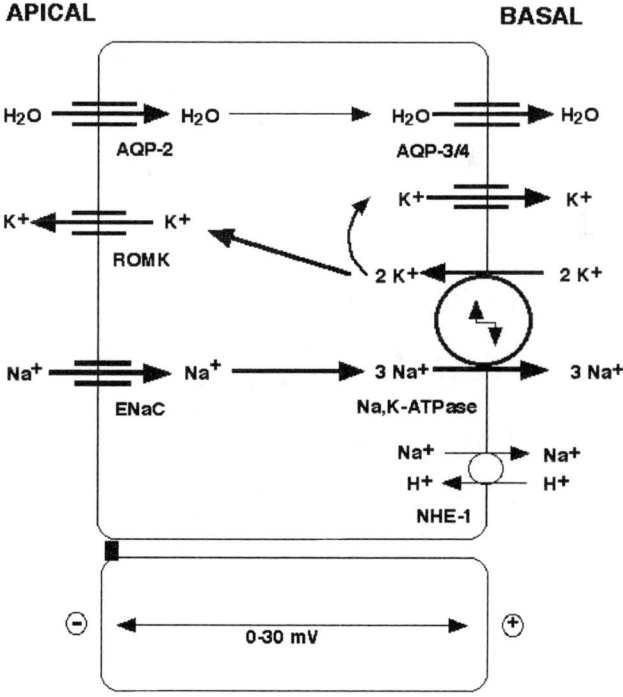

FIGURE 1. Cellular mechanism of sodium, potassium, and water transport in principal cells of the collecting duct. Arrows indicate net fluxes of water and ions. The names of the currently cloned transporters are mentioned.

water and Na reabsorption, their function being tightly controlled by hormones and non-hormonal factors in order to meet water and Na balance requirements.

Connecting tubules and collecting ducts are made up of several cell types but only principal cells are responsible for Na and water reabsorption, whereas intercalated cells are involved in acid-base secretion. The function of principal cells requires the presence of specific membrane transporters able to drive unidirectional Na and water fluxes. A schematic representation of the functional organization of the collecting duct principal cell is depicted by FIGURE 1. Water enters the luminal (apical) side via aquaporin 2 water channels and leaves the cell through basolateral aquaporin 3 and 4 water channels. The apical Na entry pathway is accounted for by the epithelial Na channel (ENaC) and intracellular Na is then extruded by basolateral Na,K-ATPase, which provides the driving force for Na reabsorption.

The major factors contributing that control Na,K-ATPase activity in collecting duct principal cells will be briefly analyzed. In addition, we will briefly discuss the role of Na,K-ATPase in Na retention in nephrotic syndrome as well as the mechanism of its dysregulation in this setting.

CONTROL OF Na,K-ATPase ACTIVITY BY ALDOSTERONE

The mineralocorticoid hormone aldosterone is a potent stimulator of Na reabsorption by the collecting duct. Infusion of aldosterone to adrenalectomized and glucocorticoid-supplemented rats decreases urinary Na excretion after a 30–60 min lag period.[2] Stimulation of Na reabsorption by aldosterone is mediated by coordinated activation of ENaC and Na,K-ATPase. In isolated mammalian collecting ducts as well as in cellular models of aldosterone-responsive epithelia,[3-5] aldosterone increases the density of active ENaC in apical membranes. The mechanism underlying the upregulation of number of active ENaC at the apical membrane is not yet fully elucidated. Recruitment of intracellular channels and *de novo* synthesis of ENaC subunits both contribute to the stimulatory effect of aldosterone.[6] However, activation of silent channels already present in the membrane should be also taken into consideration.

Stimulation of Na,K-ATPase activity by aldosterone has been demonstrated two decades ago. Infusion of aldosterone to adrenalectomized rats increased specific ouabain binding and Na,K-ATPase V_{max} within 1–3 hours in isolated cortical collecting ducts (CCD).[7,8] We have recently investigated the mechanism by which aldosterone stimulates Na,K-ATPase in isolated rat CCDs and in mpkCCD$_{cl4}$ cells,[9] a model of mammalian collecting duct principal cells. Indeed, mpkCCD$_{cl4}$ cells, derived from mouse CCD, develop into a tight epithelium and exhibit the major properties of collecting duct principal cells including stimulation of electrogenic Na transport by aldosterone[5] and vasopressin.[10] FIGURE 2 shows that intravenous infusion of aldosterone to adrenalectomized rats induced nearly a threefold increase in Na,K-ATPase activity measured at V_{max} in isolated CCDs. This stimulation was associated with a dramatic increase in Na,K-ATPase cell surface expression measured by Western blot performed after biotinylation and streptavidin precipitation of cell surface proteins (FIG. 2, B). This large increase in cell surface expression of Na,K-ATPase was associated with a small increase in the total cellular pool of Na,K-ATPase subunits. Similar results were obtained in mpkCCD$_{cl4}$ cells which displayed

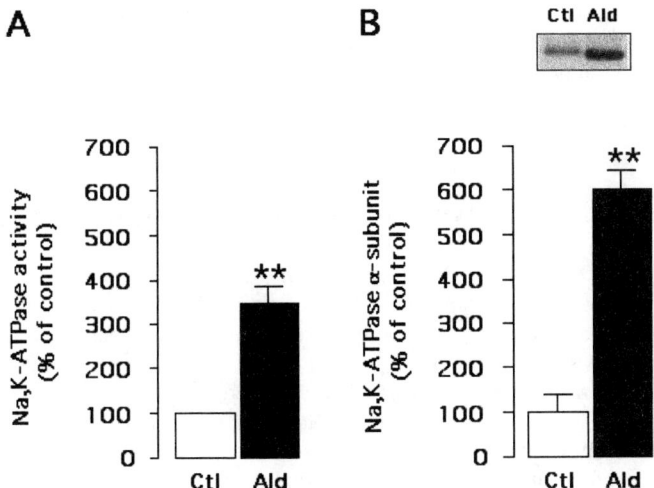

FIGURE 2. Effect of aldosterone (single injection, 3 h) on Na,K-ATPase activity (V_{max}) and cell surface expression in CCDs of adrenalectomized rats. (**A**) Effect of aldosterone on Na,K-ATPase activity. The activity of Na,K-ATPase was determined under V_{max} conditions on freeze/thawing permeabilized CCDs of control (Ctl) and aldosterone-injected (Ald) rats. The results, expressed as percentages of controls, are means ± SE from nine independent experiments. **$P<0.01$. (**B**) Effect of aldosterone on cell surface expression of Na,K-ATPase. Cell surface proteins from microdissected CCDs were biotinylated, solubilized, and precipitated by streptavidin-agarose beads, and the Na,K-ATPase α-subunit was detected by Western blotting. A representative immunoblot is show in the upper panel, and the bars in the lower panel represent the densitometric quantification values, expressed as percentages of control from seven independent experiments. **$P<0.01$.

an increase in maximal Na pump current and cell surface expression of Na,K-ATPase after 2 hours incubation in the presence of aldosterone. On the other hand, Western blotting on total mpkCCD$_{cl4}$ cell extracts revealed that the total pool of Na,K-ATPase increased later and to a lesser extent than cell surface expression. Altogether, these results indicate that short-term aldosterone stimulates Na,K-ATPase activity through an increase in cell surface expression of Na pumps. This effect most likely relies on the translocation of an intracellular reservoir of Na pumps to the plasma membrane and is independent of increased abundance of total Na,K-ATPase subunits. The presence of an intracellular pool of Na pumps in kidney cells is strongly suggested by cell fractionation[11] and cell-surface labeling studies.[12] However, localization and identification of the aldosterone-responsive intracellular pool of Na pumps remains to be investigated by morphological methods. On the other hand, the mechanism by which aldosterone increases the total cellular pool of Na,K-ATPase also remains to be determined. Results obtained in amphibian A6 cells, a widely used model of aldosterone-responsive epithelial cells, suggest that aldosterone increases Na,K-ATPase α- and β-subunit mRNA abundance.[13]

The short-term increase of Na,K-ATPase activity and cell surface expression in response to aldosterone observed in isolated rats CCDs and mpkCCD$_{cl4}$ cells is independent of apical Na entry through ENaC and requires *de novo* transcription and translation.[9,14] These results indicate that aldosterone-induced recruitment of Na pumps is not mediated by an increase in intracellular Na concentration brought about by ENaC stimulation and is most likely mediated by one of several short-term aldosterone-induced proteins.

CONTROL OF Na,K-ATPase ACTIVITY BY VASOPRESSIN/cAMP

Vasopressin, coupled to cAMP generation through V2 receptor binding, is another major regulator of Na transport in the collecting duct. *In vitro* microperfusion studies performed in rat CCD have demonstrated that vasopressin and cell-permeant cAMP analogues stimulate Na reabsorption within minutes.[15,16] Stimulation of Na reabsorption by vasopressin and cAMP is mediated by a coordinated activation of ENaC and Na,K-ATPase. Stimulation of Na transport in isolated rat CCDs in response to vasopressin/cAMP is associated with depolarization of the apical membrane[17] and an increase in apical Na conductance.[18] ENaC stimulation by vasopressin has been confirmed by patch-clamp experiments performed on the apical

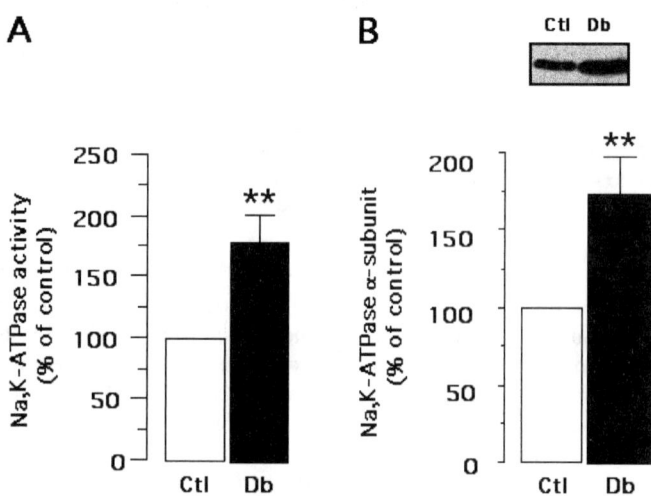

FIGURE 3. Effect of dibutyryl cAMP (db-cAMP, 15 min) on Na,K-ATPase activity (V_{max}) and cell surface expression in isolated rat CCDs. (**A**) Effect of db-cAMP on Na,K-ATPase activity. The activity of Na,K-ATPase was determined under V_{max} conditions on freeze/thawing permeabilized CCDs after incubation in the absence (Ctl) or presence of db-cAMP (Db). The results, expressed as percentages of controls, are means ± SE from five independent experiments. **$P<0.01$. (**B**) Effect of db-cAMP on cell surface expression of Na,K-ATPase. Cell surface proteins from microdissected CCDs were biotinylated, solubilized, and precipitated by streptavidin-agarose beads, and the Na,K-ATPase α-subunit was detected by Western blotting. A representative immunoblot is show in the upper panel, and the bars in the lower panel represent the densitometric quantification values, expressed as percentages of control from twelve independent experiments. **$P<0.01$.

membrane of isolated rat CCDs.[18] In amphibian A6 cells, vasopressin and cAMP analogues increased the apical membrane density of active ENaC in a brefeldin A–sensitive manner.[19] These results suggest that vasopressin induces translocation of ENaC from a Golgi-derived intracellular pool.

Although stimulation of Na,K-ATPase is a prerequisite for an increased Na reabsorption, initial studies reported an inhibition of Na,K-ATPase activity in response to vasopressin and cAMP analogues.[20] However, recent studies from our laboratories have shown that inhibition of Na,K-ATPase in response to cAMP relies on artefactual activation of phospholipase A2 by metabolic stress.[21] Indeed, we have recently shown that vasopressin (Georges Deschênes, personal communication) and cAMP analogues[12] induced a twofold stimulation of both transport and maximal hydrolytic activities of Na,K-ATPase in well oxygenated isolated rat CCDs. This effect was observed within minutes and was associated with a proportional increase in Na,K-ATPase cell surface expression (FIG. 3). Yet the total cellular pool of Na,K-ATPase remained unchanged. Similar results were obtained in mpkCCD$_{cl4}$ cells. Several lines of experimental evidence strongly suggest that cAMP recruits Na pumps from an intracellular reservoir to the plasma membrane. Firstly, saponin permeabilization of isolated rat CCDs increased Na,K-ATPase activity by twofold and abolished the stimulatory effect of cAMP. This observation suggests that, in contrast with classical freeze/thawing permeabilization, saponin permeabilization allowed the measurement of total Na,K-ATPase activity. Secondly, the effect of cAMP was prevented by reducing the incubation temperature to 20°C, by buffering intracellular calcium with BAPTA or by brefeldin A preincubation, both of which alter membrane trafficking from the Golgi apparatus and from specialized endosomal compartments. Finally, cAMP did not alter the internalization rate of Na,K-ATPase. The localization and identity of the cAMP-reponsive pool of Na,K-ATPase and its relation to an aldosterone-controlled reservoir remain to be determined.

DYSREGULATION OF Na,K-ATPase ACTIVITY IN NEPHROTIC SYNDROME

Nephrotic syndrome is characterized by massive albuminuria associated with hypoalbuminemia. In humans, nephrotic syndrome is either a primary disease without glomerular morphologic alterations or a consequence of inflammatory or deposit glomerular diseases. Nephrotic syndrome is associated with avid renal Na retention leading to extracellular fluid expansion and edema. Early *in vivo* micropuncture experiments performed in the unilateral model of puromycin aminonucleoside (PAN)-induced nephrotic syndrome in rats have shown that the collecting duct is a major site of Na retention.[22] This observation was recently confirmed by *in vitro* microperfusion experiments on isolated CCDs from PAN nephrotic rats.[23] Na retention in nephrotic syndrome is at least in part explained by increased Na,K-ATPase activity, which provides the driving force for tubular Na reabsorption. Indeed, Na,K-ATPase activity was specifically increased in CCD of several rat models of nephrotic syndrome, such as PAN and adriamycin nephrosis, Heyman nephritis, or mercury chloride–induced glomerulonephritis,[24,25] and correlated closely to decreased urinary Na excretion.[25] Moreover, experiments performed in adrenalectomized rats and in Brattleboro rats, which exhibit spontaneous knockout of the vasopressin gene, have

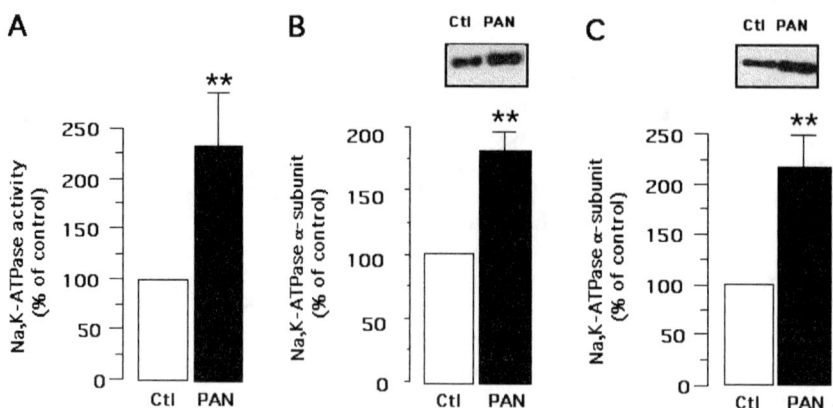

FIGURE 4. Na,K-ATPase hydrolytic activity (V_{max}) and total and cell surface expression of the Na,K-ATPase α-subunit in CCDs of normal and PAN nephrotic rats. (**A**) Na,K-ATPase activity. The activity of Na,K-ATPase was determined under V_{max} conditions on freeze/thawing permeabilized CCDs of control (Ctl) and PAN nephrotic (PAN) rats. The results, expressed as percentages of controls, are means ± SE from seven independent experiments. **$P<0.01$. (**B**) Total expression of Na,K-ATPase. Total proteins from microdissected CCDs were solubilized and the Na,K-ATPase α-subunit was detected by Western blotting. A representative immunoblot is show in the upper panel, and the bars in the lower panel represent the densitometric quantification values, expressed as percentages of control from six independent experiments. **$P<0.01$. (**C**) Cell surface expression of Na,K-ATPase. Cell surface proteins from microdissected CCDs were biotinylated, solubilized, and precipitated by streptavidin-agarose beads, and the Na,K-ATPase α-subunit was detected by Western blotting. A representative immunoblot is show in the upper panel, and the bars in the lower panel represent the densitometric quantification values, expressed as percentages of control from twenty independent experiments. **$P<0.01$.

established that increased Na,K-ATPase expression and Na retention are independent of aldosterone and vasopressin in PAN nephrotic rats.[24,25]

We have recently investigated the mechanism of increased Na,K-ATPase activity in CCD of PAN nephrotic rats.[26] As depicted by FIGURE 4, the twofold increase in maximal Na,K-ATPase hydrolytic activity was associated with a proportional increase of cell surface and total Na,K-ATPase. Indirect immunofluorescence imaging showed that increased Na,K-ATPase expression was restricted to CCD principal cells and that the enzyme was properly located at the basolateral pole of principal cells. These results suggest that increased Na,K-ATPase activity results from an increase in Na,K-ATPase subunits synthesis and delivery of newly synthesized Na pumps to the plasma membrane in PAN nephrotic rat CCD. This interpretation was further supported by the observed increase of the number of mRNA encoding Na,K-ATPase α1- and β1-subunits.

In addition to increased expression levels, cAMP-induced recruitment of Na pumps to the plasma membrane is abolished in PAN nephrotic rat CCDs. This unresponsiveness is not related to absence of an intracellular pool of Na pumps, since the size of this reservoir is increased in nephrotic rat CCDs. Nor can it be explained by defective vasopressin V_2 receptor/adenylyl cyclase system, since its functional expression is not altered in nephrotic rats. Finally, it cannot be attributed to prestimulation of the protein kinase A pathway since H89, a specific inhibitor of protein kinase A, did not alter Na,K-ATPase activity. These observations indicate that newly synthesized Na,K-ATPase units are targeted to the plasma membrane without transiting through the cAMP-responsive pool and that membrane trafficking is altered in PAN nephrotic syndrome.

CONCLUSIONS

Throughout this review, we have summarized the recent results from our laboratories focusing on the regulatory mechanisms that control Na,K-ATPase activity in collecting duct principal cells. We showed that apical Na entry through ENaC, and basolateral Na exit through Na,K-ATPase, are coordinately controlled by aldosterone and vasopressin, the major hormones that regulate Na reabsorption in the collecting duct. In addition we have provided evidence indicating that increased Na,K-ATPase activity secondary to increased synthesis of Na,K-ATPase subunits in the CCD is involved in Na retention in nephrotic syndrome. Increasing pieces of experimental evidence indicate that recruitment of Na,K-ATPase from an intracellular pool to the plasma membrane is a key regulatory mechanism shared by aldosterone and vasopressin. However, the localization and identification of an intracellular reservoir of Na pumps remains to be investigated by morphological methods. Finally, our results indicate that Na,K-ATPase can be deliverered to the plasma membrane either directly through the biosynthetic pathway or in response to a hormonal stimulus that recruits Na pumps from an intracellular pool.

REFERENCES

1. FÉRAILLE, E. & A. DOUCET. 2001. Sodium-potassium-adenosinetriphosphatase-dependent sodium transport in the kidney: Hormonal control. Physiol. Rev. **81:** 345–418.
2. HORISBERGER, J.D. & J. DIEZI. 1983. Effects of mineralocorticoids on Na^+ and K^+ excretion in the adrenalectomized rat. Am. J. Physiol. **254:** F89–F99.
3. PACHA, J., G. FRINDT, L. ANTONIAN, et al. 1993. Regulation of Na channels of the rat cortical collecting tubule by aldosterone. J. Gen. Physiol. **102:** 25–42.
4. ALVAREZ DE LA ROSA, D., H. LI & C.M. CANESSA. 2002. Effects of aldosterone on biosynthesis, traffic, and functional expression of epithelial sodium channels in A6 cells. J. Gen. Physiol. **119:** 427–442.
5. BENS, M., V. VALLET, F. CLUZEAUD, et al. 1999. Corticosteroid-dependent sodium transport in a novel immortalized mouse collecting duct principal cell line. J. Am. Soc. Nephrol. **10:** 923–934.
6. LOFFING, J., M. ZECEVIC, E. FÉRAILLE, et al. 2001. Aldosterone induces rapid apical translocation of ENaC in early portion of renal collecting system: possible role of SGK. Am. J. Physiol. **280:** F675–F682.
7. EL MERNISSI, G. & A. DOUCET. 1983. Short-term effect of aldosterone on sodium transport and tubular Na-K-ATPase in the rat. Pflügers. Arch. **399:** 139–146.
8. EL MERNISSI, G. & A. DOUCET. 1984. Specific activity of Na-K-ATPase after adrenalectomy and hormone replacement along the rabbit nephron. Pflügers. Arch. **402:** 258–263.
9. SUMMA, V., D. MORDASINI, F. ROGER, et al. 2001. Short term effect of aldosterone on Na,K-ATPase cell surface expression in kidney collecting duct cells. J. Biol. Chem. **276:** 47087–47093.

10. VANDEWALLE, A., M. BENS & J.-P. DUONG VAN HUYEN. 1999. Immortalized kidney epithelial cells as tools for hormonally regulated ion transport studies. Curr. Opin. Nephrol. Hypertens. **8:** 581–587.
11. CHIBALIN, A.V., A.I. KATZ, P.-O. BERGGREN & A.M. BERTORELLO. 1997. Receptor-mediated inhibition of renal Na^+-K^+-ATPase is associated with endocytosis of its α- and β-subunits. Am. J. Physiol. **273:** C1458–C1465.
12. GONIN, S., G. DESCHÊNES, F. ROGER, et al. 2001. Cyclic AMP increases cell surface expression of functional Na,K-ATPase units in mammalian cortical collecting duct principal cells. Mol. Biol. Cell **13:** 255–264.
13. VERREY, F. E. SCHAERER, P. ZOERKLER, et al. 1987. Regulation of Na^+,K^+-ATPase mRNAs, protein synthesis, and sodium transport in cultured kidney cells. J. Cell Biol. **104:** 1231–1237.
14. BARLET-BAS, C., C. KHADOURI, S. MARSY & A. DOUCET. 1988. Sodium-independent in vitro induction of Na^+,K^+-ATPase by aldosterone in renal target cells: permissive effect of triiodothyronine. Proc. Natl. Acad. Sci. USA **85:** 1707–1711.
15. HAWK, C.T., L.H. KUDO, A.J. ROUCH & J.A. SCHAFER. 1993. Inhibition by epinephrine of AVP- and cAMP-stimulated Na and water transport in Dahl rat CCD. Am. J. Physiol. **265:** F449–F460.
16. HAWK, C.T., L. LI & J. A. SCHAFER. 1996. AVP and aldosterone at physiological concentrations have synergistic effects on Na^+ transport in rat CCD. Kidney Int. **50:** S35–S41.
17. SCHAFER, J.A. & S.L. TROUTMAN. 1990. Vasopressin and mineralocorticoids increase apical membrane driving force for K^+ secretion in rat CCD. Am. J. Physiol. **258:** F823–F831.
18. SCHAFER, J.A. & S.L. TROUTMAN. 1990. cAMP mediates the increase in apical membrane Na^+ conductance produced in rat CCD by vasopressin. Am. J. Physiol. **259:** F823–F831.
19. KLEYMAN, T.R., S.A. ERNST & B. COUPAYE-GERARD. 1994. Arginine vasopressin and forskolin regulate apical cell surface expression of epithelial Na channels in A6 cells. Am. J. Physiol. **266:** F506–F511.
20. SATOH, T., H.T. COHEN & A.I. KATZ. 1992. Intracellular signaling in the regulation of renal Na-K-ATPase. J. Clin. Invest. **89:** 1496–1500.
21. KIROYTCHEVA, M., L. CHEVAL, M.L. CARRANZA, et al. 1999. Effect of cAMP on the activity and the phosphorylation of Na^+,K^+-ATPase in rat thick ascending limb of Henle. Kidney Int. **55:** 1819–1831.
22. ICHIKAWA, I., H.G. RENNKE, J.R. HOYER, et al. 1983. Role for intrarenal mechanisms in the impaired salt excretion of experimental nephrotic syndrome. J Clin Invest **71:** 91–103.
23. DESCHÊNES, G., M. WITTNER, A. DI STEFANO, et al. 2001. Collecting duct is a site of sodium retention in PAN nephrosis: A rationale for amiloride therapy. J. Am. Soc. Nephrol. **12:** 598–601.
24. VOGT, B. & H. FAVRE. 1991. Na^+,K^+-ATPase activity and hormones in single nephron segments from nephrotic rats. Clin. Sci. **80:** 599–604.
25. DESCHÊNES, G. & A. DOUCET. 2000. Collecting duct Na^+/K^+-ATPase activity is correlated with urinary sodium excretion in rat nephrotic syndromes. J. Am. Soc. Nephrol. **11:** 604–615.
26. DESCHÊNES, G., S. GONIN, E. ZOLTY, et al. Increased synthesis and AVP unresponsiveness of Na,K-ATPase in collecting duct from nephrotic rats. J. Am. Soc. Nephrol. **12:** 2241–2252.

Themes in Ion Pump Regulation

F. CORNELIUS AND Y. A. MAHMMOUD

Department of Biophysics, University of Aarhus, DK-8000, Århus C, Denmark

ABSTRACT: The Na,K-ATPase is an ion pump present in the plasma membrane. It is of vital importance for cell and body homeostasis and as such is under strict hormonal control. The molecular basis for the acute regulation of Na,K-ATPase is multisite phosphorylation by protein kinases that can alter its behavior. This includes direct effects on the Na,K-ATPase activity, regulation by membrane trafficking, and even dynamic regulation of interaction with regulatory proteins. In shark Na,K-ATPase, the latter includes functional interaction with a small hydrophobic protein of the FXYD protein family, phospholemman-like protein from shark, PLMS. This article summarizes our recent work on the mechanisms involved in the acute regulation of the Na,K-ATPase studied using a plasma membrane preparation from shark salt glands as an epithelial transport model.

KEYWORDS: FXYD proteins; PLMS; protein kinase; multisite phosphorylation; leucine zipper

MULTISITE PHOSPHORYLATION

The shark rectal gland mediates Cl^- secretion by secondary active transport via the basolateral $Na^+:K^+:2Cl^-$ cotransporter deriving the energy from the Na^+ gradient established by basolateral Na,K-ATPase. The shark rectal glands therefore are a rich source for Na,K-ATPase, which can be isolated as membrane fragments after homogenization and differential centrifugation. We have used this preparation for several years and its functional properties are well characterized.

The Na,K-ATPase is a substrate for phosphorylation by several protein kinases including both serine/threonine kinases and tyrosine kinases. Functional effects after phosphorylation by the cAMP-dependent protein kinase A (PKA) and the Ca^+/phospholipid-dependent protein kinase C (PKC) have been most intensely studied (reviewed by Therien and Blostein[1]). A PKA phosphorylation site, Ser-938, located in the C-terminal part of the α subunit in the L89 loop is universally conserved in all isoforms and is phosphorylated *in vitro* to stoichiometric values only in the presence of detergents, or after reconstitution.[2] PKC phosphorylation occurs at a low stoichiometry at a conserved Ser-11 in the N terminus, except in rat α1 and α3 isoforms in which an additional PKC phosphorylations site at Ser-18 is present. In rat, phosphorylation of these sites is involved in the regulation of Na,K-ATPase trafficking.[3]

Address for correspondence: F. Cornelius, Department of Biophysics, University of Aarhus, DK-8000, Århus C, Denmark. Voice: +45-89422926; fax: +45-86129599.
fc@biophys.au.dk

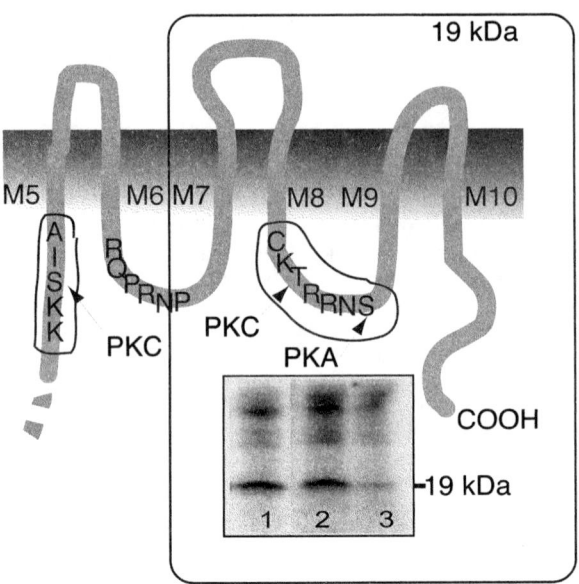

FIGURE 1. Location of novel PKC phosphorylation sites in Na,K-ATPase. Crosstalk between the PKA and PKC sites is shown in the inset depicting an autoradiogram showing the 19-kDa band of N-terminal truncated enzyme after phosphorylation by PKA (lane 1), PKA followed by PKC (lane 2), or PKC followed by PKA (lane 3).

Like other Na,K-ATPases, the shark enzyme is phosphorylated by PKA in the C-terminal part and by PKC in the N terminus at the conventional sites, mentioned above. PKA phosphorylation of reconstituted shark Na,K-ATPase stimulates the pump directly.[2] PKC phosphorylation of shark Na,K-ATPase is at a low stoichiometry and may be primarily involved in regulation of protein–protein interaction as described below.

Recently, we observed that after controlled N-terminal truncation that removes the N-terminal PKC site the shark Na,K-ATPase was phosphorylated by PKC in stoichiometric amount if detergents were present. By proteolytic fingerprinting, we identified a new class of conserved PKC phosphorylation sites in shark Na,K-ATPase close to the inner face of the plasma membrane.[4] As seen in FIGURE 1, these sites are located in the proximal part of the M5M6 hairpin and in the L89 loop and are conserved in all isoforms, indicating a physiological role. As with the PKA site the new PKC phosphorylation sites are accessible *in vitro* after reconstitution into liposomes in the absence of detergents. Preliminary results using detergent-treated shark Na,K-ATPase, which is N-terminally truncated to remove the conventional N-terminal PKC site, indicate that PKC phosphorylation decreases K^+ activation at low ATP concentration, where K^+ deocclusion becomes rate limiting.[5] This would be understandable from the close spatial position of these phosphorylation sites and the presumed location of the K^+-binding pocket within transmembrane segments M4, M5, M6, and M8.[6]

Consider also the close position of one of the new PKC sites only four amino acids upstream of the PKA site in the L89 loop. This could indicate that the L89 loop might function as a motif for crosstalk between the PKA and PKC signaling pathways. That this seems to be true is indicated by the observations that PKA phosphorylation at Ser-938 within the 19-kDa fragment is almost completely prevented by PKC phosphorylation at the nearby Thr-934 (see FIG. 1, *inset*).[5] Indeed, some molecular devices must be present for direction and timing of the phosphorylation events, but we do not know which yet. The need for detergents to expose the phosphorylation sites in the L89 loop *in vitro* may indicate that the cell exposure of these sites is itself regulated. Indeed, the transmembrane segments M5M6 and M8M9 around the novel PKC sites have been shown to be very flexible, and in the L67 loop a SH3-binding motif (+XPXXP) is, in fact, conserved.

Multisite phosphorylation of Na,K-ATPase can enable any effect, being regulation of membrane trafficking, direct effects on catalytic activity, or regulation of protein–protein interaction, to operate in the same protein by regulation and timing of the phosphorylation events, lending an enormous capacity for amplification, integration, and processing of signals in the cell.

REGULATION OF SHARK Na,K-ATPase BY FXYD PROTEIN

At the 9th International Conference on the Na/K-ATPase and Related ATPases in Japan in 1999, we reported the identification of a new small regulatory protein specifically associated with shark Na,K-ATPase.[7] The protein was termed phospholemman-like protein from shark, PLMS (Swiss-Prot accession no. P82542), because of its close homology with phospholemman, PLM, the major protein kinase substrate in the myocardium.[8] It belongs to a family of small hydrophobic proteins that was subsequently characterized at the gene level and termed the FXYD protein family.[9] One of the members is the γ subunit that modulates kidney Na,K-ATPase,[1,10] whereas the others were unrelated to any other proteins and were of unknown function. We suggested that this group of proteins all served as potential regulators of Na,K-ATPase.[7,11] This meeting demonstrated that in addition to γ and PLMS this is the case for at least three more members of this family, namely, PLM, CHIF, and FXYD7. The latter has not yet been characterized at the protein level.

Thus, members of the FXYD family act as tissue-specific regulators of Na,K-ATPase, but the mechanism for this regulation seems to be variable. We have studied the PLMS/α interaction in the shark rectal gland membranes and compared it with the γ interaction with pig kidney Na,K-ATPase.[7] A major difference seems to be that PLMS, like PLM, contains a C-terminal multisite phosphorylation motif, which is absent in other FXYD proteins. This enables the association of PLMS to be dynamically regulated via protein kinase phosphorylation. Our recent work on the mechanisms involved in the acute regulation of shark Na,K-ATPase is summarized, and attempts to derive a more general model for the regulation are described.

Structure and Localization of PLMS

In collaboration with G. Cramb's group at St. Andrews, we recently have cloned and sequenced PLMS from *Squalus acanthias* (manuscript in preparation). The partial sequence previously published (P82542) comprising the extracellular and membrane domains is confirmed, revealing in addition the cytoplasmic domain containing several threonine and serine residues as potential PKA and PKC target sites. This is in accord with previous results demonstrating that PLMS is a substrate for both PKA and PKC.[7]

In the close relative *Scyliorhinos canicula*, we also identified a 15-kDa protein that was a substrate for PKC. The partial sequence including the N-terminal 29 amino acids demonstrates that it is an FXYD homologue to PLMS in *Squalus*: VNE-PADXXAR<u>FTYD</u>XYGLXVVGLIVAAVL. A partial N-terminal sequence of a 13-kDa protein from *S. canicula* recently published[12] as a PLMS homologue without the FXYD motif thus is likely to be a hemoglobin contaminant.

PLMS is found to be coexpressed and coimmunoprecipitated with shark Na,K-ATPase demonstrating the specific association with the shark enzyme.[7] In collaboration with A. B. Maunsbach we recently have probed the cellular localization of Na,K-ATPase and PLMS using immunogold-labeled antibodies to α and PLMS (manuscript in preparation). The results demonstrate that both proteins are located to basolateral membranes, in accord with the epithelial transport model for the shark rectal gland.

Functional Effects of PLMS/α Interaction

In principle, the interaction between Na,K-ATPase and PLMS could be extracellular, transmembrane, or intracellular. In the shark membrane preparation, we measured functional effects using either detergents at concentrations below cmc to break hydrophobic interactions in the membrane, or PKC phosphorylation and controlled proteolysis of PLMS to disrupt cytoplasmic interactions, or extracellular-directed antibody to probe extracellular interactions.

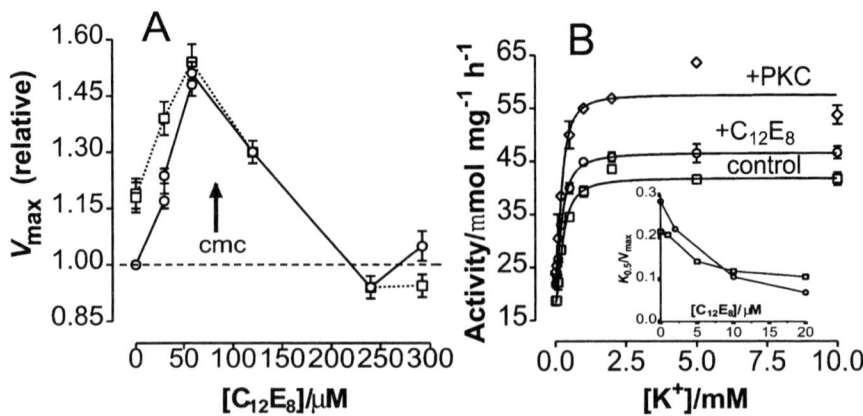

FIGURE 2. Panel A shows maximal hydrolytic activity (V_{max}) of enzyme incubated with increasing concentration of $C_{12}E_8$ either without or after PKC phosphorylation. Panel B shows K^+ activation at 10 μM ATP in controls, with 20 μM $C_{12}E_8$, or with 20 μM $C_{12}E_8$ after PKC phosphorylation. Based on data from Mahmmoud et al.[7]

FIGURE 2A demonstrates that sub-cmc concentrations of the detergent $C_{12}E_8$ activates shark Na,K-ATPase significantly. The hydrolytic activity measured at optimal conditions (V_{max}) increases by greater than 50%. At suboptimal $C_{12}E_8$ concentrations, the maximum hydrolytic activity is increased further on PKC phosphorylation, indicating additive activation of two processes when either the transmembrane hydrophobic interaction or the interaction at the cytoplasmic PKC phosphorylation site are interrupted. The same conclusion can be reached from Figure 2B, with the K^+ activation at 10 µM ATP. With 20 µM $C_{12}E_8$, the increased activity is potentated by PKC phosphorylation. However, the ratio $K_{0.5}/V_{max}$, which is a unique signifier for ligand interaction with the substrate, decreases equally in the two cases, indicating that the increased K^+-binding affinity is primarily ascribed to $C_{12}E_8$, that is, to the effect of interrupting the hydrophobic interaction of PLMS with the α subunit in the membrane phase (FIG. 2B, inset).

To further investigate the functional effects of interaction between the α subunit and the C-terminal phosphorylation domain of PLMS, we defined conditions in which the phosphorylation domain of PLMS, which is preceded by several trypsin cleavage sites, could be removed without affecting the α subunit. This proved to be possible by incubating enzyme at a very low protein to trypsin ratio of 1:1000 in the presence of K^+. As seen in FIGURE 3, under these conditions a 4-kDa fragment is cleaved off containing the C-terminal phosphorylation motif of PLMS, which is evident from the autoradiogram showing absence of PKA phosphorylation after proteolysis.

The functional consequences of PLMS cleavage on Na,K-ATPase activity is seen in FIGURE 4A in which the activation of ATPase activity by K^+ and by Na^+ is shown. Again, disruption of the PLMS/α interaction leads to activation of the hydrolytic activity at V_{max} conditions. However, the apparent ion affinities for K^+ and Na^+ are un-

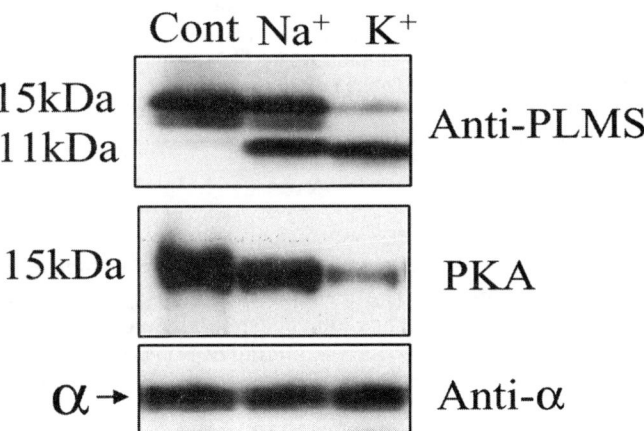

FIGURE 3. PLMS truncation by mild trypsin treatment. *Top panel* shows an immunoblot using anti-PLMS antiserum of control, truncation in Na^+ medium, or truncation in K^+ medium. *Middle panel* shows an autoradiogram after PKA phosphorylation at the same three conditions. *Bottom panel* shows an immunoblot using anti–α antibody with no effects at these mild conditions of proteolysis.

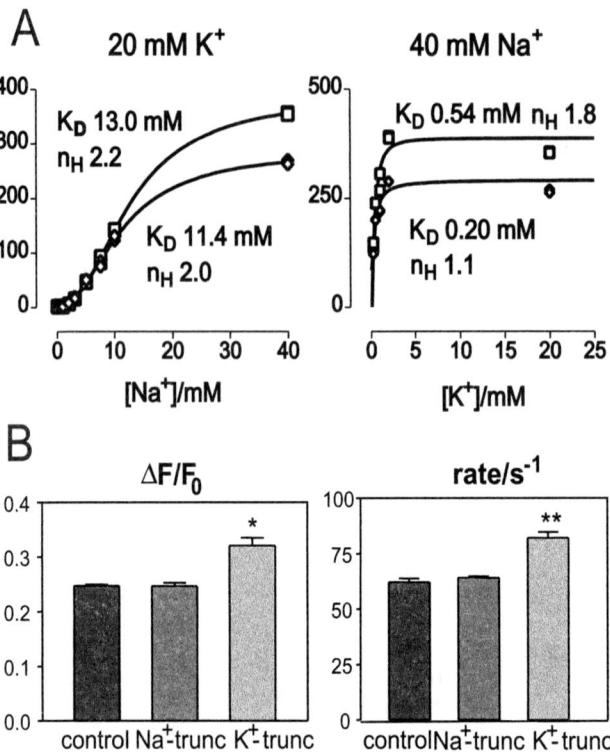

FIGURE 4. Panel A shows Na^+ activation and K^+ activation of control ($\diamond$) and PLMS-truncated enzyme ($\square$) at 3 mM MgATP. Panel B shows results from rapid-mixing stopped-flow experiments of phosphorylation of control and PLMS-truncated enzyme in the presence of 130 mM Na^+ and 3 mM ATP. To the left, the relative fluorescence level is given for controls and enzyme truncated in the presence of Na^+ or K^+. To the right, the rate constant for the phosphorylation reaction is given.

changed. Therefore, other steps along the Na,K-ATPase reaction cycle must be affected. The most likely steps to be affected to increase V_{max} are rate-determining steps. For shark Na,K-ATPase, these are the phosphorylation steps and the E_2 to E_1 conformational step in connection with the binding of Na^+ and ATP.[13,14] Both steps can be measured using the potential sensitive fluorescent probe RH421 in a rapid-mixing stopped-flow experiment. In the phosphorylation reaction, enzyme preincubated with Na^+ is mixed with 3 mM ATP, and the formation of E_2-P is detected by the RH dye. As seen in FIGURE 4B, truncation of PLMS increased both the rate constant for E_2-P formation and the E_2-P level ($\Delta F/F_0$).

In conclusion, it therefore seems that interaction between PLMS and α at the transmembrane level influences the apparent K^+ affinity, whereas cytoplasmic interaction influences the catalytic phosphorylation reaction.

A Hypothetical Model for PLMS/α Interaction

It has been demonstrated that phospholamban (PLN) in SERCA forms pentameric oligomers in the membrane stabilized via leucine/isoleucine zipper (lz) motifs.[15] The Na,K-ATPase γ subunit from pig kidney and PLMS also forms oligomeric structures in the membrane,[7] which may explain their ion-channel properties when incorporated into bilayers. The transmembrane domain of the FXYD proteins contains conserved lz motifs, which may explain their tendency for oligomeric association, although the physiological significance of such channel formation, if any, is unknown.

Actually, several transmembrane segments of the Na,K-ATPase also contain conserved lz motifs. This applies most clearly to M1 and M2, but also to M3 and M7, suggesting that transmembrane association of PLMS and the α subunit could be stabilized via such lz motifs. Interestingly, the closely related gastric H,K-ATPase does not contain similar lz motifs, and the Na,K-ATPase γ subunit is found unable to associate with the H,K-ATPase.[10] Association of PLMS with M1 and/or M2 of the Na,K-ATPase would place its C-terminal part in close contact with the A domain in the model of Toyoshima et al.[6] This domain is believed to move substantially during the phosphorylation reaction, and it is conceivable that any restriction of such movement due to association with PLMS would inhibit the enzyme activity. A working model for the PLMS/α interaction based on these observations therefore is one in which PLMS interaction at the transmembrane level is stabilized via lz interaction with M1 and/or M2 of the Na,K-ATPase, and interaction with the cytoplasmic domain of the Na,K-ATPase, possibly the A domain, restricting its flexibility leading to inhibition. Thus, PLMS is anchored into the membrane part of the Na,K-ATPase, and its association with its cytoplasmic domain is dynamically regulated via protein kinase phosphorylation, which releases the interaction and therefore relieves the inhibition.

ACKNOWLEDGMENTS

This work was supported by the Danish Research Foundation.

REFERENCES

1. THERIEN, A.G. & R. BLOSTEIN. 2000. Mechanisms of sodium pump regulation. Am. J. Physiol. **279**: C541–C566.
2. CORNELIUS, F. AND N. LOGVINENKO. 1996. Functional regulation of reconstituted Na,K-ATPase by protein kinase A phosphorylation. FEBS Lett. **380**: 277–280.
3. EFENDIEV, R., A.M. BERTORELLO, T. PRESSLEY, et al. 2000. Simultaneous phosphorylation of Ser11 and Ser18 in the alpha-subunit promotes the recruitment of Na^+,K^+-ATPase molecules to the plasma membrane. Biochemistry **39**: 9884–9892.
4. MAHMMOUD, Y.M. & F. CORNELIUS. 2002. Protein kinase C phosphorylation of purified Na,K-ATPase: C-terminal phosphorylation sites at the α- and γ-subunits close to the inner face of the plasma membrane. Biophys. J. **82**: 1907–1919.
5. KRÜGER, A., Y.A. MAHMMOUD & F. CORNELIUS. 2003. Protein kinase C phosphorylation directed at novel C-terminal sites in Na,K-ATPase. Ann. N. Y. Acad. Sci. This volume.
6. TOYOSHIMA, C., M. NAKASAKO, H. NOMURA & H. OGAWA. 2000. Crystal structure of the calcium pump of sarcoplasmic reticulum at 2.6 Å resolution. Nature **405**: 647–655.

7. MAHMMOUD, Y.A., H. VORUM & F. CORNELIUS. 2000. Identification of a phospholemman-like protein from shark rectal glands. Evidence for indirect regulation of Na,K-ATPase by protein kinase C via a novel member of the FXYDY family. J. Biol. Chem. **275:** 35969–35977.
8. PALMER, C.J., B.T. SCOTT & L.R. JONES. 1991. Purification and complete sequence determination of the major plasma membrane substrate for cAMP-dependent protein kinase and protein kinase C in myocardium. J. Biol. Chem. **266:** 11126–11130.
9. SWEADNER, K.J. & E. RAEL. 2000. The FXYD gene family of small ion transport regulators or channels: cDNA sequence, protein signature sequence, and expression. Genomics **68:** 41–56.
10. BEGUIN, P., X. WANG, D. FIRSOV, et al. 1997. The γ subunit is a specific component of the Na,K-ATPase and modulates its transport function. EMBO J. **16:** 4250–4260.
11. CORNELIUS, F., Y.A. MAHMMOUD & H.R.Z. CHRISTENSEN. 2001. Modification of Na,K-ATPase by associated small transmembrane regulatory proteins and lipids. J. Bioenerg. Biomembr. **33:** 415–423.
12. STEKHOVEN, F.M., G. FLIK & S.E.W. BONGA. 2001. N-terminal sequences of small ion channels in rectal glands of sharks: a biochemical hallmark for classification and phylogeny? Biochem. Biophys. Res. Commun. **288:** 670–675.
13. CORNELIUS, F. 1999. Rate determination in phosphorylation of shark rectal Na,K-ATPase by ATP: temperature sensitivity and effects of ADP. Biophys. J. **77:** 934–942.
14. LÜPFERT, C., E. GRELL, V. PINTSCHOVIOUS, et al. 2001. Rate limitation of the Na^+,K^+-ATPase pump cycle. Biophys. J. **81:** 2069–2081.
15. SIMMERMAN, H.K.B. & L.R. JONES. 1998. Phospholamban: protein structure, mechanism of action, and role in cardiac function. Physiol. Rev. **78:** 921–947.

Isoform-Specific Regulation of Na$^+$,K$^+$-ATPase Endocytosis and Recruitment to the Plasma Membrane

VERA LUCAS TEIXEIRA,[a] ADRIAN I. KATZ,[b] CARLOS H. PEDEMONTE,[c] AND ALEJANDRO M. BERTORELLO[a]

[a]*Department of Medicine, Atherosclerosis Research Unit, Karolinska Institutet, Karolinska Hospital, 171 76 Stockholm, Sweden*

[b]*Department of Medicine, University of Chicago, Chicago, Illinois 60637*

[c]*College of Pharmacy, University of Houston, Houston, Texas 77204*

ABSTRACT: The Na$^+$,K$^+$-ATPase traffics between the plasma membrane and intracellular compartments in response to acute changes in membrane receptor activation. These effects are accomplished by a time-dependent interaction of the Na$^+$,K$^+$-ATPase α-subunit with specific intracellular signaling molecules either at the plasma membrane (endocytosis) or at the endosome's membranes (recruitment). Most of these studies have been performed in rat renal epithelial cells in which only the α$_1$ isoenzyme is present. Studies in neurons from the neostriatum in which all three α-subunit isoforms are present indicate that neurotransmitter-dependent regulation of Na$^+$,K$^+$-ATPase activity displays isoform specificity and also suggest a more complex organization of the intracellular signaling networks controlling Na$^+$,K$^+$-ATPase traffic in mammalian cells.

KEYWORDS: Na$^+$,K$^+$-ATPase isoforms; endocytosis; catecholamines; neostriatal neurons; protein phosphorylation

INTRODUCTION

Na$^+$,K$^+$-ATPase–mediated ion transport contributes, directly or indirectly, to key physiological processes such as modulation of cardiac and vascular contractility, sodium and potassium transport by the kidney, and neurotransmitter release.[1,2] The electrogenic nature of the transport exercised by the Na$^+$,K$^+$-ATPase (three Na$^+$ are exchanged for two K$^+$) creates an inside negative potential that underlies the electrical properties of the plasma membrane. The electrochemical gradient thus generated by the Na$^+$,K$^+$-ATPase is responsible for maintaining membrane potential and therefore excitability of neural tissues, muscle, and heart. Hormonal regulation of membrane potential is an important mechanism responsible for the activation of voltage-

---

Address for correspondence: Alejandro M. Bertorello, Department of Medicine, King Gustaf V Research Institute, M1, Karolinska Hospital, 171 76 Stockholm, Sweden. Voice: +46-8-5177-3224; fax: +46-8-31-12-98.
  alejandro.bertorello@molmed.ki.se

dependent Ca^{2+} channels and increased intracellular calcium, a process required for synaptic vesicle fusion and neurotransmitter release,[3,4] as well as for exocytosis in several secretory epithelia. In transporting epithelia, the Na^+,K^+-ATPase is confined to the basolateral domain of the cell and thereby contributes to vectorial transport of sodium. Increased Na^+,K^+-ATPase activity in renal proximal tubules has been associated with the development of high blood pressure.

SHORT-TERM REGULATION OF Na^+,K^+-ATPase ACTIVITY IN INTACT CELLS

Regulation of Na^+,K^+-ATPase activity by hormones involves the activation of complex intracellular signaling networks that are not fully understood.[5,6] In epithelial cells, inhibition of Na^+,K^+-ATPase activity by dopamine occurs by removal of the α and β subunits from the plasma membrane and their sequential internalization into early and late endosomes via a clathrin-coated vesicle-dependent mechanism.[7] This process, initiated at the plasma membrane, requires phosphorylation of the Ser-18 residue within the catalytic $α_1$ subunit[8,9] and activation of phosphatidylinositol 3 kinase[10] (FIG. 1). Phosphorylation is mediated by the PKC-ζ isoform,[11] and activation of phosphatidylinositol 3 kinase requires its binding to a proline-rich domain, situated downstream from the PKC phosphorylation site. Phosphatidylinositol 3 kinase is critical for promoting the interaction of the Na^+,K^+-ATPase $α_1$ isoform with AP-1.[12] Binding of AP-2 to the Na^+,K^+-ATPase $α_1$ subunit occurs within a tyrosine-based motif and is necessary for clathrin recruitment and formation of the clathrin-coated pit.[13] AP-2/Na^+,K^+-ATPase $α_1$-isoform interaction appears to be a regulatory point in which agonists (such as oxymetazoline) blocking the action of dopamine prevent Na^+,K^+-ATPase $α_1$-subunit endocytosis and downregulation of Na^+,K^+-ATPase activity.[14]

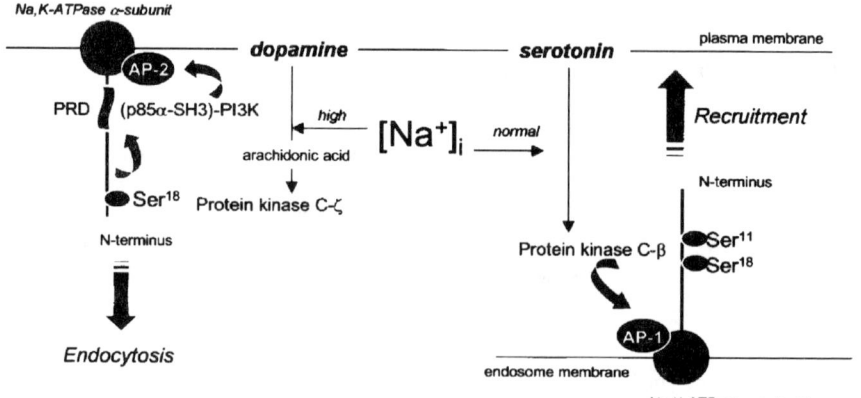

FIGURE 1. Scheme illustrating the mechanisms controlling Na^+,K^+-ATPase activity and removal/incorporation of Na^+,K^+-ATPase $α_1$-subunits from/to the plasma membrane. For explanations see text.

In contrast, agonists, such as isoproterenol,[24] serotonin,[15] and angiotensin II (unpublished observations) that increase Na^+,K^+-ATPase activity do so by promoting the recruitment of new isoforms to the plasma membrane (FIG. 1), rather than by increasing the catalytic activity of those Na^+,K^+-ATPase molecules already present at the plasma membrane. Incorporation of new Na^+,K^+-ATPase molecules requires phosphorylation of both Ser-11 and Ser-18 residues within the α_1-subunit, and this effect is mediated by activating the PKC β isoform.

Interestingly, the balance between stimulation/inhibition of Na^+,K^+-ATPase activity and endocytosis/recruitment of Na^+,K^+-ATPase α_1-subunit in renal epithelial cells is regulated physiologically by changes in the concentration of intracellular sodium[16] (FIG. 1). Basal intracellular sodium favors the action of agonists that stimulate/recruit Na^+,K^+-ATPase molecules, whereas a gradual and modest increase (as low as 5 mM) in intracellular sodium abolishes the stimulation and is sufficient to trigger an inhibition/endocytosis of Na^+,K^+-ATPase molecules. These data indicate the possibility of a sodium-sensing mechanism responsible for controlling the response of renal Na^+,K^+-ATPase to either natriuretic or antinatriuretic hormones during salt load or deprivation.

Whereas these mechanisms of Na^+,K^+-ATPase regulation have been demonstrated in renal epithelial cells, which express only the Na^+,K^+-ATPase α_1-subunit, it remains unclear how such regulation is achieved in tissues that express several Na^+,K^+-ATPase isoforms.

ISOFORM-SPECIFIC REGULATION

The Na^+,K^+-ATPase exists in different isoforms with ubiquitous distribution (reviewed in Blanco and Mercer[17]). Although these isoforms have a specific tissue localization, little is known of their relative contribution to specific physiological processes,[17,18] whether they are selectively regulated, or about the physiological impact of such regulation. Because it appears that changes in Na^+,K^+-ATPase activity in response to several hormones are achieved by removal of active units from the plasma membrane,[7] using isoform-specific antibodies it was possible to study whether Na^+,K^+-ATPase regulation by different neurotransmitters occurs by affecting specific isoforms. The experimental protocols for the studies described below were performed in medium spiny neurons (MSNs) isolated from the rat neostriatum according to the method of Kay and Wong,[19] with few modifications.[20] These neurons were chosen because in this tissue the Na^+,K^+-ATPase is subject to regulation by catecholamines and expresses all major α-subunit isoforms[17,18,20,21] (except for the α_4 isoform that is expressed exclusively in the testis[22]).

In intact cells, the Na^+,K^+-ATPase operates at one third of its maximal capacity, which corresponds to the activity measured at an intracellular sodium concentration of approximately 15 mM. Dopamine inhibited Na^+,K^+-ATPase activity (measured as the rate of ^{32}P-ATP hydrolysis in the presence or absence of ouabain[15]), both at lower (~20 mM) and higher ($\geq$40 mM) sodium concentrations in the assay conditions (FIG. 2). That the inhibitory effect of dopamine is present when the enzyme is operating at V_{max} indicates that its inhibitory action is unlikely to be secondary to changes in sodium permeability. Although basal enzyme activity in the rat striatum was lower than the activity previously reported in isolated neurons from guinea pig,[20] the

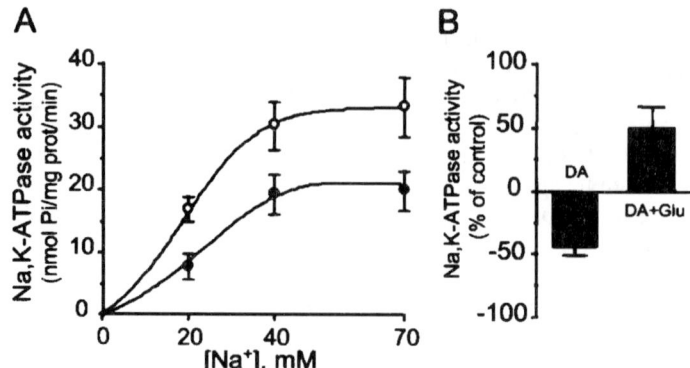

FIGURE 2. Effect of dopamine and glutamate on Na^+,K^+-ATPase activity in MSN isolated from rat striatum. (**A**) Neurons were incubated in the presence (*filled circles*) or absence (*open circles*) of 1 μM dopamine for 10 minutes at 23°C. Enzyme activity was determined as described,[15] except that the Na^+,K^+-ATPase assay solution contained 20, 40, and 70 mM NaCl. Each point is the mean ± SE of eight experiments performed in triplicate. (**B**) MSN were incubated (10 minutes at 23°C) with 1 μM dopamine in the presence or absence of 100 μM glutamate. Each bar represents the mean ± SE of eight experiments performed in triplicate.

magnitude of the response to dopamine under V_{max} conditions was similar. The presence of glutamate (100 μM) not only prevented the inhibitory action of dopamine, but also significantly stimulated Na^+,K^+-ATPase activity. This effect is consistent with other reported antagonistic responses between dopamine and glutamate (see below).

Short-term regulation of Na^+,K^+-ATPase activity in epithelial cells and neurons is accomplished by activation and integration of specific intracellular signaling networks, which ultimately activate protein kinases.[5,6,20] Because dopamine-dependent inhibition of Na^+,K^+-ATPase activity in epithelial cells is associated with endocytosis of the subunits in a protein kinase–dependent manner,[8,9] we examined whether the action of dopamine and glutamate in neostriatal neurons was also the result of a change in α-subunit distribution at the plasma membrane. Isolated MSNs were incubated in the presence or absence of 1 μM dopamine at room temperature after the protocol described for FIGURE 2, and the Na^+,K^+-ATPase isoform abundance was examined by Western blot analysis using specific antibodies (α_1, a gift from Dr. M. J. Caplan and α_2, and α_3, a gift from Dr. K. J. Sweadner). Incubation with dopamine led to a decrease only in α_2 subunits in the plasma membrane (FIG. 3), whereas the presence of 100 μM glutamate partially prevented this effect and significantly increased the abundance of α_1 subunits in the plasma membrane. Neither of these agonists affected the content of Na^+,K^+-ATPase α_3 subunits within the plasma membrane. The short-term incubation with the agonists was not associated with changes in total Na^+,K^+-ATPase content.

Previous studies in rat renal epithelial cells have demonstrated that internalization of the Na^+,K^+-ATPase α_1 subunits in response to dopamine requires phosphorylation of the subunit at the Ser-11/Ser-18 residues.[9] Phosphorylation of the α_1 subunit is necessary for recruitment and activation of phosphoinositide 3-kinase, a process

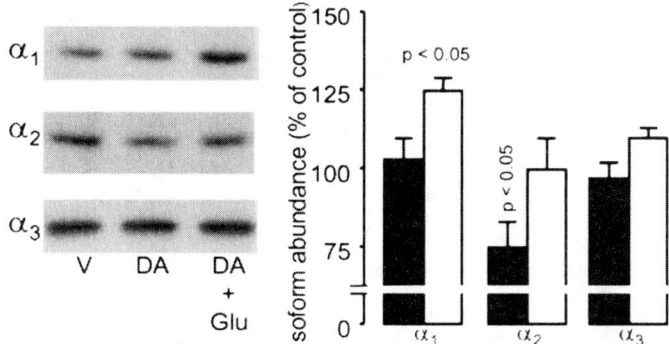

FIGURE 3. Na$^+$,K$^+$-ATPase isoform abundance in plasma membranes prepared from MSN incubated with 1 µM dopamine (10 minutes at 23°C) in the presence or absence of 100 µM glutamate. The incubation was terminated by placing the samples on ice, and membranes were prepared and analyzed by SDS-PAGE and Western blot analysis. A representative Western blot is shown in the *left panel*, and the quantitative determinations of five experiments (mean ± SE) are shown in the *right panel*. (*solid bars*) Cells treated with dopamine; (*open bars*) cells treated with dopamine plus glutamate.

that occurs by its binding to a proline-rich domain located upstream of the PKC phosphorylation site.[12] This motif is conserved among the subunits, and whereas phosphorylation of the α_2 subunit occurs to a lesser magnitude than that of the α_1 subunit,[23] it may be sufficient to trigger the binding and activation of the phosphoinositide 3-kinase, a process that is required for removal of Na$^+$,K$^+$-ATPase subunits from the plasma membrane.

As shown schematically in FIGURE 4, the decreased Na$^+$,K$^+$-ATPase activity in response to dopamine appears to be the result of the specific removal of active molecules from the plasma membrane (this study) and possibly, as described in epithelial cells, their traffic into defined intracellular compartments. However, the mechanisms involved in stimulation of Na$^+$,K$^+$-ATPase activity have not yet been clearly defined. It is possible that the action of glutamate resembles that of catecholamines in alveolar epithelial cells[24] and of phorbol esters in renal epithelia,[25] where stimulation of Na$^+$,K$^+$-ATPase activity is accomplished by increasing the number of α_1 subunits in the plasma membrane that are recruited from intracellular organelles. However, further studies are required to verify this hypothesis in neurons.

The mechanisms by which glutamate blocks the internalization of the α_2 isoform is also not clear. Cellular responses to dopamine and glutamate in the striatum have been extensively documented. The postsynaptic effects on medium spiny neurons are almost exclusively related to the ability of glutamate to increase intracellular calcium and of dopamine to regulate adenylyl cyclase activity.[26] Although some antagonistic crosstalk at the second messengers level has been described, a common site of interaction has been attributed to their actions on protein kinases and protein phosphatases. Thus, opposite effects of glutamate from the corticostriatal pathway and dopamine from the nigrostriatal pathway on the firing rate of striatal neurons have been described.[27] Moreover, dopamine activates adenylyl cyclase and subsequently cAMP-dependent kinase, and the latter phosphorylates DARPP-32, a potent

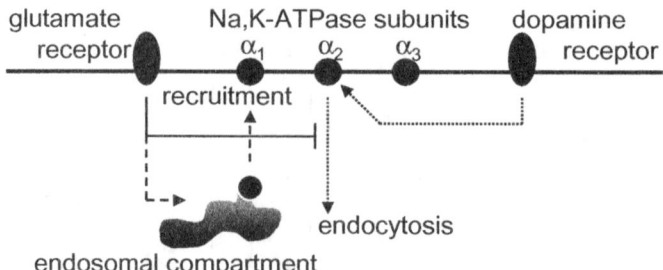

FIGURE 4. Regulation of Na^+,K^+-ATPase isoforms by dopamine and glutamate in the neostriatum. Dopamine inhibits Na^+,K^+-ATPase activity by decreasing the number of α_2 subunits in the plasma membrane (*dotted line*), whereas glutamate counteracts its effect by blocking the endocytosis of α_2 isoforms (*solid line*). In addition, the increased Na^+,K^+-ATPase activity in response to glutamate is the result of an increase in α_1 isoform within the plasma membrane (*dashed line*).

protein phosphatase inhibitor.[28–31] It has been suggested that this effect is responsible for an increase in the state of Na^+,K^+-ATPase subunit phosphorylation and decrease in its catalytic activity,[32] whereas glutamate, by increasing intracellular Ca^{2+}, activates calcineurin, leading to dephosphorylation of DARPP-32, thus reducing the level of Na^+,K^+-ATPase phosphorylation. Although these results could provide an explanation for the antagonistic action of glutamate, it recently has been reported that whereas dopamine phosphorylates DARPP-32, its inhibitory action on Na^+,K^+-ATPase activity may not be directly associated with changes in the state of α-subunit phosphorylation.[21] In another respect, however, the data are in agreement with those reported here, that is, indicating that dopamine regulates the activity of the α_2 but not of the α_1 Na^+,K^+-ATPase isoform.

These studies in neurons provide evidence for the selective regulation of the different Na^+,K^+-ATPase isoforms by different neurotransmitters, as well as additional information on the cellular mechanisms underlying the antagonism between dopamine and glutamate. A different level of interaction, such as regulation of the electrical properties of the plasma membrane through the Na^+,K^+-ATPase, may have an important impact on neuronal physiology, and especially on the role of such regulation in neurotransmitter release.

ACKNOWLEDGMENTS

We thank M. J. Caplan and K. J. Sweadner for providing the Na^+,K^+-ATPase antibodies used in this study. This study was partially supported by funds from the Karolinska Institutet.

REFERENCES

1. SKOU, J.C. & M. ESMANN. 1992. The Na,K-ATPase. J. Bioenerg. Biomembr. **24:** 249–261.

2. GEERING, K. 1997. Na,K-ATPase. Curr. Opin. Nephrol. Hypertens. **6:** 434–439.
3. APPELL, K.C. & D.S. BAREFOOT. 1989. Neurotransmitter release from bradykinin-stimulated PC12 cells. Stimulation of cytosolic calcium and neurotransmitter release. Biochem. J. **263:** 11–18.
4. TAMURA, N., K. YOKOTANI, Y. OKUMA, *et al.* 1995. Properties of the voltage-gated calcium channels mediating dopamine and acetylcholine release from the isolated rat retina. Brain Res. **676:** 363–370.
5. BERTORELLO, A.M. & A.I. KATZ. 1993. Short term regulation of renal Na,K-ATPase activity: physiological relevance and cellular mechanisms. Am. J. Physiol. **265:** F743–F755.
6. EWART, H.S. & A. KLIP. 1995. Hormonal regulation of the Na(+)-K(+)-ATPase: mechanisms underlying rapid and sustained changes in pump activity. Am. J. Physiol. **269:** C295–C311.
7. CHIBALIN, A.V., A.I. KATZ, P.-O. BERGGREN & A.M. BERTORELLO. 1997. Receptor-mediated inhibition of epithelial Na^+,K^+-ATPase is associated with endocytosis of its α- and β-subunits. Am. J. Physiol. **273:** C1458–C1465.
8. CHIBALIN, A.V., C.H. PEDEMONTE, A.I. KATZ, *et al.* 1998. Phosphorylation of the catalytic a subunit constitutes a triggering signal for Na^+,K^+-ATPase endocytosis. J. Biol. Chem. **273:** 8814–8819.
9. CHIBALIN, A.V., G. OGIMOTO, C.H. PEDEMONTE, *et al.* 1999. Dopamine-induced endocytosis of Na^+,K^+-ATPase is initiated by phosphorylation of Ser18 in the rat α-subunit and is responsible for the decreased activity in epithelial cells. J. Biol. Chem. **274:** 1920–1927.
10. CHIBALIN, A.V., J. ZIERATH, A.I. KATZ, *et al.* 1998. Phosphatidylinositol 3-kinase-mediated endocytosis of renal Na^+,K^+-ATPase a subunit in response to dopamine. Mol. Biol. Cell **9:** 1209–1220.
11. EFENDIEV, R., A.M. BERTORELLO & C.H. PEDEMONTE. 1999. PKC-β and PKC-ζ mediate opposing effects on epithelial Na^+,K^+-ATPase activity. FEBS Lett. **456:** 45–48.
12. YUDOWSKI, G.A., R. EFENDIEV, C.H. PEDEMONTE, *et al.* 2000. Phosphoinositide-3 kinase binds to a proline-rich motif in the Na^+,K^+-ATPase α-subunit and regulates its trafficking. Proc. Natl. Acad. Sci. USA **97:** 6556–6561.
13. COTTA DONÉ, S., I.B. LEIBIGER, R. EFENDIEV, *et al.* 2002. Tyrosine-537 within the Na^+,K^+-ATPase α-subunit is essential for AP-2 binding and clathrin-dependent endocytosis. J. Biol. Chem. **277:** 17108–17111.
14. OGIMOTO, G., G.A. YUDOWSKI, C.J. BARKER, *et al.* 2000. G protein-coupled receptors regulate Na^+,K^+-ATPase activity and endocytosis by modulating the recruitment of AP-2 and clathrin. Proc. Natl. Acad. Sci. USA **97:** 3242–3247.
15. BUDU, C., R. EFENDIEV, A.R. CINELLI, *et al.* 2002. Hormonal-dependent recruitment of Na^+,K^+-ATPase to the plasmalemma is mediated by PKC_β and modulated by $[Na+]_i$. Br. J. Pharmacol. **137:** 1380–1386.
16. EFENDIEV, R., A.M. BERTORELLO, R. ZANDOMENI, *et al.* 2002. Agonist-dependent regulation of renal Na,K-ATPase activity is modulated by intracellular sodium. J. Biol. Chem. **277:** 11489–11496.
17. BLANCO, G. & R.W. MERCER. 1998. Isozymes of the Na-K-ATPase: heterogeneity in structure, diversity in function. Am. J. Physiol. **275:** F633–F650.
18. HIEBER, V., G.J. SIEGEL, D.J. FINK, *et al.* 1991. Differential distribution of (Na, K)-ATPase alpha isoforms in the central nervous system. Cell. Mol. Neurobiol. **11:** 253–262.
19. KAY, A.R. & R.K.S. WONG. 1986. Isolation of neurons suitable for patch-clamping from adult mammalian central nervous systems. J. Neurosci. Methods **16:** 227–238.
20. BERTORELLO, A., J.F. HOPFIELD, A. APERIA & P. GREENGARD. 1990. Inhibition by dopamine of (Na^+-K^+)-ATPase activity in neostriatal neurons through D1 and D2 dopamine receptor synergism. Nature **347:** 386–388.
21. NISHI, A., G. FISONE, G.L. SNYDER, *et al.* 1999. Regulation of Na+,K+ATPase isoforms in rat neostriatum by dopamine and protein kinase C. J. Neurochem. **73:** 1492–1501.
22. WOO, A.L., P.F. JAMES & J.B. LINGREL. 1999. Characterization of the fourth isoform of the Na,K-ATPase. J. Membr. Biol. **169:** 39–44.

23. BEGUIN, P., M.C. PEITSCH & K. GEERING. 1996. Alpha 1 but not alpha 2 or alpha 3 isoforms of Na,K-ATPase are efficiently phosphorylated in a novel protein kinase C motif. Biochemistry **35:** 14098–14108.
24. BERTORELLO, A.M., K. RIDGE, A.V. CHIBALIN, *et al.* 1999. Isoproterenol increases Na^+,K^+-ATPase activity by membrane insertion of a1 subunits in lung alveolar cells. Am. J. Physiol. **276:** L20–L27.
25. EFENDIEV, R., A.M. BERTORELLO, T.A. PRESSLEY, *et al.* 2000. Simultaneous phosphorylation of Ser11/Ser18 in the α-subunit promotes the recruitment of Na^+,K^+-ATPase molecules to the plasma membrane. Biochemistry **39:** 9884–9892.
26. PAOLILLO, M., A. MONTECUCCO, P. ZANASSI & S. SCHINELLI. 1998. Potentiation of dopamine-induced cAMP formation by group I metabotropic glutamate receptors via protein kinase C in cultured striatal neurons. Eur. J. Neurosci. **10:** 1937–1945.
27. HALPAIN, S., J.A. GIRAULT & P. GREENGARD. 1990. Activation of NMDA receptors induces dephosphorylation of DARPP-32 in rat striatal slices. Nature **343:** 369–372.
28. NISHI, A., G.L. SNYDER, A.C. NAIRN & P. GREENGARD. 1999. Role of calcineurin and protein phosphatase-2A in the regulation of DARPP-32 dephosphorylation in neostriatal neurons. J. Neurochem. **72:** 2015–2021.
29. HUANG, H.B., A. HORIUCHI, T. WATANABE, *et al.* 1999. Characterization of the inhibition of protein phosphatase-1 by DARPP-32 and inhibitor-2. J. Biol. Chem. **274:** 7870–7878.
30. SCHIFFMANN, S.N., F. DESDOUITS, R. MENU, *et al.* 1998. Modulation of the voltage-gated sodium current in rat striatal neurons by DARPP-32, an inhibitor of protein phosphatase. Eur. J. Neurosci. **10:** 1312–1320.
31. KWON, Y.G., H.B. HUANG, F. DESDOUITS, *et al.* 1997. Characterization of the interaction between DARPP-32 and protein phosphatase 1 (PP-1): DARPP-32 peptides antagonize the interaction of PP-1 with binding proteins. Proc. Natl. Acad. Sci. USA **94:** 3536–3541.
32. APERIA, A., J. FRYCKSTEDT, L. SVENSSON, *et al.* 1991. Phosphorylated Mr 32,000 dopamine- and cAMP-regulated phosphoprotein inhibits Na+,K(+)-ATPase activity in renal tubule cells. Proc. Natl. Acad. Sci. USA **88:** 2798–2801.

The Sodium Pump Keeps Us Going

TORBEN CLAUSEN

Department of Physiology, University of Aarhus, DK-8000 Århus C, Denmark

ABSTRACT: This invited lecture reviews recent evidence that, in skeletal muscle, excitability and contractility depend on the transmembrane distribution of Na^+ and K^+ and the membrane potential, which in turn are determined by the operation of the Na^+-K^+ pump. Action potentials are elicited by passive fluxes of Na^+ and K^+. Because of their size and sudden onset, these transport events constitute the major challenge for the Na^+-K^+ pumps. When the Na^+-K^+ pumps cannot readily restore the Na^+-K^+ gradients, working muscle cells often undergo net loss of K^+ and gain of Na^+. This leads to loss of excitability and force, in particular, in muscles where excitation-induced passive Na^+-K^+ fluxes are large. Thus, excitability depends on the leak/pump ratio for Na^+ and K^+. When this ratio is increased by inhibition or downregulation of the Na^+-K^+ pumps, the force decline seen during continued stimulation is accelerated. This effect is highly significant already within the first seconds of electrical stimulation. Fortunately, electrical stimulation also increases Na^+-K^+ pumping rate within seconds. Thus, maximum increase (20-fold above the resting level) may be reached in 10 seconds, with utilization of all available Na^+-K^+ pumps. In muscles, where excitability was inhibited by exposure to high $[K^+]_o$ (10–12.5 mM), activation of the Na^+-K^+ pumps by hormones or electrical stimulation restored excitability and contractile force. In working muscles, the Na^+-K^+ pumps, because of rapid activation of their large transport capacity, play a dynamic regulatory role in the second-to-second ongoing restoration and maintenance of excitability and force. The Na^+-K^+ pumps become a limiting factor for contractile endurance, in particular, if their capacity is reduced by inactivity or disease.

KEYWORDS: sodium; potassium; skeletal muscle; contractility; excitability; ouabain; Na^+-K^+ATPase

INTRODUCTION

In skeletal muscle, Na^+ and K^+ are exchanged across sarcolemma and t-tubular walls via several pathways. By far the most important are the Na^+ and K^+ channels. The action potentials causing muscle contraction are elicited by a large influx of Na^+, immediately followed by a similar efflux of K^+. This allows prompt activation of muscle contraction and rapid restoration of membrane potential and excitability. The price paid for this rapidity, however, is that before long, the muscle cells are bathed in their own K^+, leaking into the interstitial water space, and, even worse, into the narrow t-tubular lumen.

Address correspondence to: Torben Clausen, Department of Physiology, University of Aarhus, DK-8000 Århus C, Denmark. Voice: +45-89-42-28-22; fax: +45-86-12-90-65.
tc@fi.au.dk

In the rat extensor digitorum longus (EDL) muscle, the K^+ efflux per action potential is approximately 8 nmol/g wet wt. At a stimulation frequency of 60 Hz, therefore, in 1 g of muscle, $60 \times 8 = 480$ nmol K^+ will be released into the extracellular water space every second. With an extracellular space of 20%, or 0.2 ml per gram, the average K^+ concentration will increase by 480 nmol/0.2 mL = 2.4 mM per second. Thus, within 2.5 seconds, extracellular K^+ may increase from the normal level of 4 to 10 mM, sufficient to induce loss of excitability.

During continued stimulation, muscles undergo progressive force decline. Concomitantly, there is a decline in the sum of the action potentials, the compound action potential or M-wave.[1,2] During continued stimulation of isolated rat muscles, we have found a close correlation between force and M-wave area.[3,4] The available evidence indicates that because of rundown of concentration gradients for Na^+ and K^+, accumulation of K^+ in the extracellular water space and the t-tubular lumen and ensuing depolarization, the processes of excitation are self-limiting. The only escape from this progressive loss of excitability is offered by the Na^+-K^+ pumps. This is really the situation, where the expression "all hands to the sodium pump," coined by Ian Glynn, seems most appropriate.[5]

The excitation-induced passive Na^+-K^+ fluxes constitute the largest single challenge to the Na^+-K^+ pumps, sometimes causing them to run full speed. The great advantage of the Na^+-K^+ pumps is that not only do they compensate the loss of K^+ and the intracellular gain of Na^+, but because of their 3:2 exchange ratio, they immediately contribute to repolarization.

EFFECTS OF INHIBITION OR DOWNREGULATION OF THE Na^+-K^+ PUMPS ON CONTRACTILITY

Long ago, it was shown that cardiac glycosides reduce contractile force in isolated skeletal muscles.[6,7] In the isolated rat soleus muscle, ouabain (10^{-3} M) was found to induce almost complete but reversible loss of tetanic force.[8]

Preincubation with ouabain (10^{-5} and 10^{-3} M) induces marked acceleration of the force decline (200–360%) in rat soleus exposed to continuous stimulation at 30 Hz. Moreover, after continuous 90 Hz stimulation to 80% loss of force, the force recovery is abolished at 10^{-3} M ouabain and considerably delayed (by 81% over the first 30 s) at 10^{-5} M ouabain.[9]

In rat soleus, ouabain-induced acceleration of force decline was found to be highly significant ($P < 0.001$) within the first 1–2 s of stimulation.[9,10] Thus, already very early, the transport capacity of the Na^+-K^+ pumps becomes a limiting factor for the maintenance of contractility.

If the inhibitory effect of ouabain on endurance were the result of reduced Na^+-K^+ pump capacity, similar effects should be seen in muscles where the Na^+-K^+ pump content is downregulated. This can be done by maintaining animals on K^+ deficient fodder for a week or two.[11] In rats maintained on K^+-deficient fodder for 10 days, the content of ^{3}H-ouabain binding sites in the soleus muscle had decreased by 69%.[9] In these muscles, both endurance and force recovery after exhaustion were severely impaired. This may explain the fatigue experienced by patients with K^+ deficiency.

It is also possible to induce more graded and defined inhibition of the Na^+-K^+ pumps by longer lasting preincubation at lower concentrations of ouabain (10^{-6}–

2×10^{-6} M for 1–2 h). In a series of observations, the rate of force decline was correlated to the reduction in functional Na^+-K^+ pumps.[8]

In the age interval from 4 to 10 weeks, the content of Na^+-K^+ pumps in rat soleus decreases by 57%. This was associated with a 97% increase in the rate of force decline and a 34% reduction in the rate of force recovery after an exhaustive stimulation.[9]

Taken together, three different types of intervention, all leading to a reduction in the content of functional Na^+-K^+ pumps all caused similar impairment of contractile endurance and force recovery in fatigued muscles. This indicates that the capacity of Na^+-K^+ pumps is an important common determinant for contractile performance.

EFFECTS OF INCREASED LEAKS TO Na^+ AND K^+ ON CONTRACTILITY

Instead of reducing the Na^+-K^+ pump activity or capacity, we may as well examine the completely analogous situation, in which the leak/pump ratio for Na^+-K^+ transport is augmented by increasing the passive Na^+-K^+ fluxes associated with excitation. This can be done by increasing the open time of the Na^+ channels using two plant poisons, aconitine or veratridine, allowing more Na^+ to enter the muscle cells per action potential. In isolated rat soleus muscle, aconitine (10^{-6} M) and veratridine (10^{-6} M) augmented excitation-induced Na^+ influx by 91 and 118%, respectively. When added before a chronic stimulation at 90 Hz, both agents significantly hastened the rate of force decline and severely impaired force recovery upon cessation of the stimulation.[9]

The Na^+ influx per action potential also may be augmented by upregulating the content of Na^+ channels in skeletal muscle. Pretreatment of rats for 8 days with triiodothyronine increased the content of Na^+ channels in soleus muscle by 154%.[12] During chronic stimulation at 90 Hz, these muscles showed a 53% increase in the initial rate of force decline.

Another way of examining the effects of increased passive Na^+-K^+ leaks is to compare two muscles with widely different leak rates. In EDL muscles, excitation-induced Na^+-K^+ leaks are much larger than in soleus muscles, probably because the EDL contains more Na^+ channels.[13]

During continuous or repetitive contractile activity, the force decline is closely correlated to a concomitant decline in M-wave area and amplitude.[2,14] In rat EDL, the amplitude and area of M-waves as well as contractile force decrease four to five times faster than in soleus, indicating that the faster loss of force is caused by a faster loss of excitability.[4]

EFFECTS OF Na^+-K^+ PUMP STIMULATION ON CONTRACTILITY AND EXCITABILITY

Long ago, Tomita showed that, in guinea pig muscle, isometric twitches were considerably reduced by increasing extracellular K^+ to 11 mM.[15] This loss of force could be restored by the β agonist isoproterenol. Because force recovery was suppressed by ouabain, he concluded that it was caused by Na^+-K^+ pump stimulation. In rat soleus, we found that the inhibitory effect of increasing $[K^+]_o$ to 10 mM could

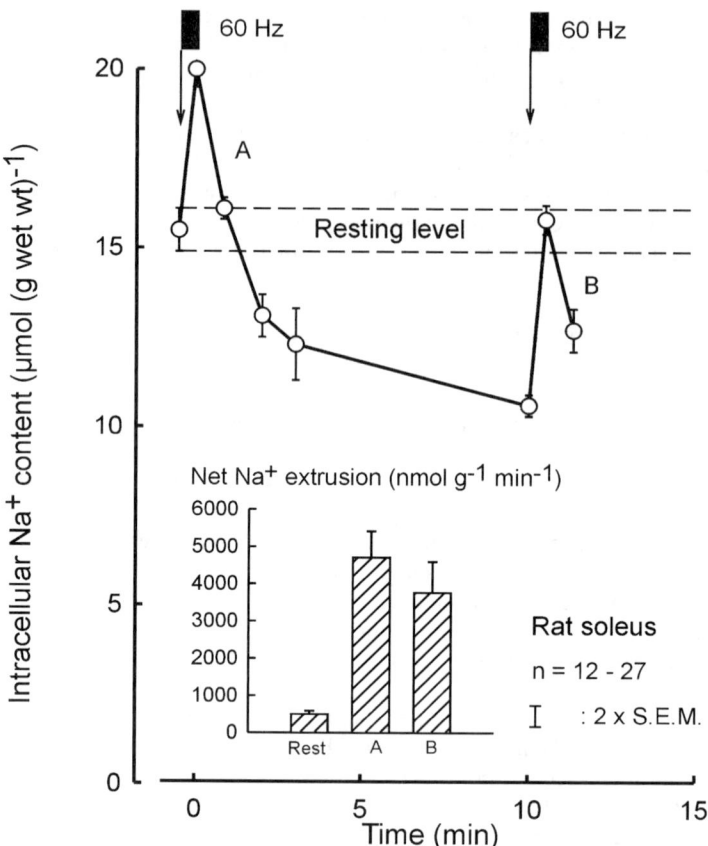

FIGURE 1. Time course of changes in intracellular Na^+ content during and after electrical stimulation of rat soleus contracting isometrically. The muscles were stimulated for 15 s at 60 Hz. The inset shows the mean rate of net Na^+ extrusion during the first 50 s upon cessation of the first (A) and the second (B) stimulation. These values are compared with ouabain-sensitive Na^+ efflux in resting control muscles. All values are means with bars denoting SE. (Reproduced, with permission, from Nielsen and Clausen.[30])

be completely alleviated by epinephrine or the β_2 agonist salbutamol.[17] Other agents known to stimulate the Na^+-K^+ pumps (insulin, calcitonin gene-related peptide, amylin) produced similar effects,[16–18] which were abolished by ouabain. The stimulating effects of such agents on ouabain-suppressible ^{86}Rb uptake were closely correlated to the force recovery elicited at high $[K^+]_o$.[16] There was also a highly significant correlation between the decrease in $[Na^+]_i$ and force recovery. The force recovery seems to be the combined result of repolarization and increased transmembrane Na^+ concentration gradient, allowing reestablishment of excitability and action potential amplitude.

The physiological relevance of these observations is evident from the repeated demonstration that during intense exercise, the K^+ concentrations likely to develop in the extracellular space is in that range. Microdialysis studies indicate that values of approximately 10–11 mM K^+ may be reached,[19] which would be sufficient to cause progressive loss of force. It is important therefore that acute stimulation of the Na^+-K^+ pumps can counterbalance the force decline.

Exercise increases plasma catecholamine levels sufficiently to stimulate the Na^+-K^+ pumps in resting muscles, and this is likely to be an advantage in counterbalancing the inhibitory effect of exercise-induced hyperkalemia. But what about the working muscle with their impending loss of excitability? Fortunately, excitation leads to a rapid and often pronounced stimulation of the Na^+-K^+ pumps.[20,21] After electrical stimulation for 10 s at 120 Hz, net Na^+ extrusion from rat soleus may increase 20-fold within the following 30 s, allowing full utilization of all Na^+-K^+ pumps available.[22,23] This demonstrates that when the muscle cell is working enough to flood its cytoplasm with Na^+ and its surroundings with K^+, all hands are indeed called to the sodium pumps.

The stimulating effect of excitation on the Na^+-K^+ pumps is clearly detectable already after 1 s of electrical stimulation[22,24] and completely blocked by ouabain or cooling. Perhaps the most interesting aspect of this is that even after short-lasting stimulation, $[Na^+]_i$ is only transiently increased and then decreased. As shown in FIGURE 1, this "undershoot" is 30% below the resting control level for $[Na^+]_i$ and may last for up to 25 min.[23] When stimulation is repeated, $[Na^+]_i$ undergoes virtually the same increase and subsequent decrease.[25] Even though $[Na^+]_i$ did not exceed the resting level during the second stimulation, the rate of net Na^+ extrusion during the first 50 s after the cessation of stimulation was eightfold larger than the ouabain-suppressible Na^+ efflux measured in resting muscles. It is well-documented that electrical stimulation both at low and high frequencies leads to an increased rate of active Na^+-K^+ transport despite lowered intracellular Na^+.[22-26] Taken together, these observations indicate that excitation leads to a rapid increase in the affinity of the Na^+-K^+ pumps for intracellular Na^+.

WORKING MUSCLES KEEP GOING BY BOOTSTRAPPING OUT OF IMPENDING LOSS OF EXCITABILITY

We can mimic the working conditions of a skeletal muscle during intense exercise by increasing $[K^+]_o$ to 10 mM. As shown in FIGURE 2, this leads to approximately 80% loss of tetanic force in rat soleus. When 2-s trains of tetanic stimulation were repeated every minute, tetanic force was considerably restored in 5 min[27,28] and there was close correlation between force and M-wave area.

Resting membrane potential and the propagation velocity of the action potentials that had been reduced at 10 mM K^+ also were restored toward the level measured at 4 mM K^+.[28] We named this phenomenon force recovery by bootstrapping. High $[K^+]_o$ paralyses the muscle, but excitation stimulates the Na^+-K^+ pumps, allowing the muscle to lift itself by the bootstraps into restoration of excitability and contractility.

A wide variety of agents or conditions known to elicit acute stimulation of the Na^+-K^+ pumps all produce force recovery in muscles exposed to high K^+. These effects are blocked by ouabain. Thus, Na^+-K^+ pump stimulation seems to be a com-

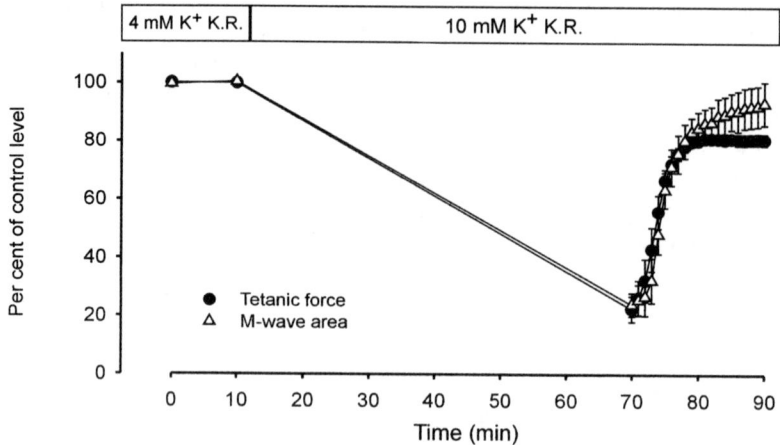

FIGURE 2. Excitation-induced recovery of tetanic force and M-wave area in rat soleus. Tetanic force and M-wave area recorded from the extrajunctional region were measured in standard Krebs-Ringer bicarbonate buffer containing 4 or 10 mM K^+. Values are expressed as percentages of the average level at 4 mM K^+. After 60 min of exposure to 10 mM K^+, tetanic stimulations were applied at 1-min intervals for 20 min. Symbols represent the mean values from six muscles with bars representing SE. (Reproduced, with permission, from Overgaard and Nielsen.[27])

mon mechanism for the restoration of excitability and force. On the basis of these observations as well as several others recently reviewed,[29,30] it is possible to generate a list of indications that in skeletal muscle, contractility depends on the distribution and active transport of Na^+ and K^+. The repeated observation that the total content of Na^+-K^+ pumps in skeletal muscle is upregulated by training and downregulated by inactivity further indicates that the Na^+,K^+ pumps are essential in the maintenance of muscle contractility.

ACKNOWLEDGMENTS

The study was supported by grants from the Danish Center for Biomembrane Research and the Danish Medical Research Council (J. No. 22-01-0189).

REFERENCES

1. BIGLAND-RITCHIE, B., D.A. JONES & J.J. WOODS. 1981. Excitation frequency and muscle fatigue: electrical responses during human voluntary and stimulated contractions. Exp. Neurol. **64:** 414–427.
2. GALEA, V. 2001. Electrical characteristics of human ankle dorsi- and plantal-flexor muscles. Comparative responses during fatiguing stimulation and recovery. Eur. J. Appl. Physiol. **85:** 130–140.
3. OVERGAARD, K., O.B. NIELSEN, J.A. FLATMAN & T. CLAUSEN. 1999. Relations between excitability and contractility in rat soleus muscle: role of the Na^+-K^+ pump and Na^+/K^+ gradients. J. Physiol. **518:** 215–225.

4. CLAUSEN, T., O.B. NIELSEN & K. OVERGAARD. 2003. Na^+-K^+ leaks, Na^+-K^+ pumps, excitability and contactile performance of skeletal muscle. Submitted for publication.
5. GLYNN, I.M. 1993. Annual review prize lecture "All hands to the sodium pump." J. Physiol. **462:** 1–30.
6. AMARANATH, L. & N.B. ANDERSEN. 1976. The effct of general anesthetic agents, ouabain and aldosterone on striated muscle contraction in toad. Anesth. Analg. **55:** 409–414.
7. YAMAMOTO, S., A.A. FOX & K. GREEF. 1981. Inotropic effects and Na^+-K^+-ATPase inhibition of ouabain in isolated guinea-pig atria and diaphragm. Eur. J. Pharmacol. **71:** 437–446.
8. NIELSEN, O.B. & T. CLAUSEN. 1996. The significance of active Na^+,K^+ transport in the maintenance of contractility in rat skeletal muscle. Acta Physiol. Scand. **157:** 199–209.
9. HARRISON, A.P., O.B. NIELSEN & T. CLAUSEN. 1997. Role of Na^+-K^+ pump and Na^+ channel concentrations in the contractility of rat soleus muscle. Am. J. Physiol. **272:** R1402–R1408.
10. CLAUSEN, T., O.B. NIELSEN, A.P. HARRISON, et al. 1998. The Na^+,K^+ pump and muscle excitability. Acta Physiol. Scand. **162:** 183–190.
11. NØRGAARD, A., K. KJELDSEN & T. CLAUSEN. 1981. Potassium depletion decreases the number of ^{3}H-ouabain binding sites and the active Na-K transport in skeletal muscle. Nature **293:** 739–741.
12. HARRISON, A.P. & T. CLAUSEN. 1998. Thyroid hormone-induced upregulation of Na^+ channels and Na^+-K^+ pumps: implications for contractility. Am. J. Physiol. **274:** R864–R867.
13. GISSEL, H. & T. CLAUSEN. 2000. Excitation-induced Ca^{2+} influx in rat soleus and EDL muscle: mechanisms and effects of cellular integrity. Am. J. Physiol. **279:** R917–R924.
14. PAGALA, M.K., K. RAVINDRAN, T. NAMBA & D. GROB. 1998. Skeletal muscle fatigue and physical endurance of young and old mice. Muscle Nerve **21:** 1729–1739.
15. TOMITA, T. 1975. Action of catecholamines on skeletal muscle. *In* Handbook of Physiology. Endocrinology. Adrenal Gland. Sect. 7, Vol. VI, chapt. 34. :537–552. American Physiology Society. Bethesda, MD.
16. CLAUSEN, T., S.V. ANDERSEN & J.A. FLATMAN. 1993. Na^+-K^+ pump stimulation elicits recovery of contratility in K^+ paralysed rat muscle. J. Physiol. **472:** 521–536.
17. ANDERSEN, S.L.V. & T. CLAUSEN. 1993. Calcitonin gene-related peptide stimulates active Na^+-K^+ transport in rat soleus muscle. Am. J. Physiol. **264:** C419–C429.
18. CLAUSEN, T. 2000. Effects of amylin and other peptide hormones on Na^+-K^+ transport and contractility in rat skeletal muscle. J. Physiol. **527:** 121–130.
19. JUEL, C., H. PILEGAARD, J.J. NIELSEN & J. BANGSBO. 2000. Interstitial K^+ in human skeletal muscle during and after dynamic graded exercise determined by microdialysis. Am J. Physiol. **278:** R400–R406.
20. EVERTS, M.E., K. RETTERSTØL & T. CLAUSEN. 1988. Effects of adrenaline on excitation-induced stimulation of the sodium-potassium pump in rat skeletal muscle. Acta Physiol. Scand. **134:** 189–198.
21. HICKS, A. & A.J. MCCOMAS. 1989. Increased sodium pump activity following repetetive stimulation of rat soleus muscles. J. Physiol. **414:** 337–349.
22. EVERTS, M.E. & T. CLAUSEN. 1994. Excitation-induced activation of the Na^+-K^+ pump in rat skeletal muscle. Am. J. Physiol. **266:** C925–C934.
23. NIELSEN, O.B. & T. CLAUSEN. 1997. Regulation of Na^+-K^+ pump in contracting skeletal muscle. J. Physiol. **503:** 571–581.
24. BUCHANAN, R., O.B. NIELSEN & T. CLAUSEN. 2002. Excitation- and β_2-agonist-induced activation of the Na^+-K^+ pump in rat soleus muscle. J. Physiol. **545:** 229–240.
25. EVERTS, M.E. & T. CLAUSEN. 1992. Activation of the Na-K pump by intracellular Na in rat slow- and fast-twitch muscle. Acta Physiol. Scand. **145:** 353–362.
26. NIELSEN, O.B., L. HILSTED & T. CLAUSEN. 1998. Excitation-induced force recovery in potassium-inhibited rat soleus muscle. J. Physiol. **512:** 819–829.
27. OVERGAARD, K. & O.B. NIELSEN. 2001. Activity-induced recovery of excitability in K^+-depressed rat soleus muscle. Am. J. Physiol. **280:** R48–R55.

28. SEJERSTED, O.M. & G. SJØGAAARD. 2000. Dynamics and consequences of potassium shifts in skeletal muscle and heart during exercise. Physiol. Rev. **80:** 1411–1481.
29. CLAUSEN, T. 2003. Na^+-K^+ pump regulation and skeletal muscle contractility. Physiol. Rev. In press.
30. NIELSEN, O.B. & A.P. HARRISON. 1998. The regulation of the Na^+,K^+ pump in contracting skeletal muscle. Acta Physiol. Scand. **162:** 191–200.

Na,K-ATPase and the Significance of Sodium in the Mechanism of Potassium-Induced Relaxation of Rat-Isolated Mesenteric Arteries

DIDIER X. P. BROCHET

Department of Physiology, University of Bristol, Bristol BS8 1TD, United Kingdom

KEYWORDS: Na,K-ATPase; extracellular concentration of potassium; intracellular concentration of sodium; vasorelaxation

In addition to nitric oxide (NO) and prostacyclin, endothelium-derived hyperpolarizing factor (EDHF), the identity and mechanism of action of which remain elusive, is an important mediator of vasorelaxation in various vascular beds, including rat mesenteric arteries. In 1998, Edwards et al.[1] proposed that in rat hepatic and mesenteric arteries potassium ions (K^+) may be EDHF. K^+ released from calcium-sensitive potassium channels on the endothelial cells could accumulate in the myoendothelial space, leading to hyperpolarization of vascular smooth muscle cells, upon activation of the Na,K-ATPase. The authors were able to mimic the effects of EDHF by elevating extracellular concentration of potassium ($[K^+]_o$). However, not all studies agree that K^+ is EDHF in rat mesenteric arteries.[2]

METHODS

Third-order mesenteric arteries of male Wistar rats were dissected free in physiological saline solution containing (mM): NaCl 119; KCl 4.7; $NaHCO_3$ 25; KH_2PO_4 1.18; $CaCl_2$ 1.8; $MgSO_4$ 1.2; glucose 11; EDTA 0.027, pH 7.4, gassed with 95% O_2/ 5% CO_2. All experiments were performed in the presence of L-NAME (100 μM) and indomethacin (2.8 μM), inhibitors of NO-synthase and cyclooxygenase. Segments of mesenteric arteries were mounted in a Mulvany-Halpern wire myograph under normalized tension at 37°C for isometric recording. Forces are expressed as percentage of control before the final K^+ increase. Statistical significance was tested using a Student's *t* test on paired data.

Address for correspondence: Didier X.P. Brochet, Department of Physiology, University of Bristol, University Walk, Bristol BS8 1TD, United Kingdom. Voice: +44-117-928-8371; fax: +44-117-928-8923.
  didier.brochet@bristol.ac.uk

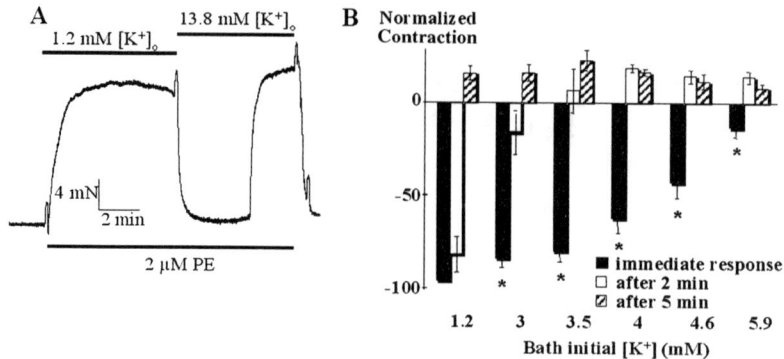

FIGURE 1. (A) An example of an artery that relaxed from a 6-minute period of phenylephrine-induced contraction at 1.2 mM $[K^+]_o$ followed by an increase of $[K^+]_o$ to 13.8 mM. (B) Mean data (±SEM; n = 16) of the amplitude of relaxation to a final $[K^+]_o$ of 13.8 mM from experiments using a range of initial $[K^+]_o$ between 1.2 and 5.9 mM. Asterisks indicate significant difference of a P value less than 0.05.

RESULTS AND DISCUSSION

Increasing $[K^+]_o$ to 13.8 mM caused consistent relaxation of arteries exposed to phenylephrine only if the initial $[K^+]_o$ was less than 5.9 mM (FIG. 1). Decreasing the initial $[K^+]_o$ for 6 minutes between 1.2 and 5.9 mM reduces the activity of the Na,K-ATPase, which has a $K_{0.5}$ for $[K^+]_o$ of approximately 1–1.6 mM at physiological $[Na^+]_i$.[3] Over time, this should increase $[Na^+]_i$ and decrease $[K^+]_i$. The amplitude of relaxation was inversely proportional to the initial $[K^+]_o$ (FIG. 1). This suggests that the relaxation may be proportional to the degree of downturn of Na,K-ATPase activity. Subsequently, when $[K^+]_o$ is restored, the activity of the Na,K-ATPase is increased, leading to hyperpolarization and relaxation of the smooth muscle cells. Direct evidence for involvement of the Na,K-ATPase is demonstrated by the ouabain sensitivity of the relaxation. When $[K^+]_o$ is increased from 1.2 mM for 6 minutes to 13.8 mM, 100 μM ouabain abolished this relaxation.

Substitution of NaCl with either choline or NMDG during the contraction to phenylephrine at 0 mM $[K^+]_o$ for 6 minutes, before increasing $[K^+]_o$ to 13.8 mM, also abolished the relaxation. When the Na,K-ATPase is arrested by the absence of $[K^+]_o$, the decrease of $[Na^+]_o$ from 114 to 26.2 mM should minimize the increase of intracellular concentration of sodium ($[Na^+]_i$). In smooth muscle cells of guinea pig mesenteric artery, the Na,K-ATPase is half maximally stimulated ($K_{0.5}$) at approximately 22 mM $[Na^+]_i$[3]. Therefore, under physiological conditions, with $[Na^+]_i$ between 8 and 12 mM, an increase of $[Na^+]_i$ should strongly activate the Na,K-ATPase. Conversely, increments in $[K^+]_o$ beyond the normal resting concentration of 4 mM should have a small effect on pump rate because of the $K_{0.5}$ for $[K^+]_o$ is approximately 1–1.6 mM at physiological $[Na^+]_i$.[3]

CONCLUSION

The relaxation evoked by an increase of $[K^+]_o$ from a low initial concentration may reflect an accumulation of $[Na^+]_i$ and the subsequent upturn of electrogenic Na,K-ATPase pumping when $[K^+]_o$ subsequently is increased. The outward current generated by increased activity of the pump would hyperpolarize the smooth muscle cells, decrease the open probability of voltage-gated calcium channels, and reduce the intracellular concentration of calcium, thus, leading to relaxation.

REFERENCES

1. Edwards, G., K.A. Dora, M.J. Gardener, *et al.* 1998. K^+ is an endothelium derived hyperpolarizing factor in rat arteries. Nature **396:** 269–272.
2. Doughty, J.M., J.P. Boyle & P.D. Langton. 2000. Potassium does not mimic EDHF in rat mesenteric arteries. Br. J. Pharmacol. **130:** 1174–1182.
3. Nakamura, Y., Y. Ohya, I. Abe & M. Fujishima. 1999. Sodium-potassium pump current in smooth muscle cells from mesenteric resistance arteries of the guinea-pig. J. Physiol. **519:** 203–212.

Molecular Activity of Sodium Pumps in the Kidney of Mammals and Birds

NIGEL TURNER,[a,b] A. J. HULBERT,[a,c] AND PAUL L. ELSE[a,b]

[a]*Metabolic Research Centre, University of Wollongong, New South Wales, Australia 2522*

[b]*Department of Biomedical Science, University of Wollongong, New South Wales, Australia 2522*

[c]*Department of Biological Sciences, University of Wollongong, New South Wales, Australia 2522*

KEYWORDS: metabolism; molar activity; membranes; polyunsaturation; allometry; body mass

Smaller mammals and birds have an increased metabolic intensity compared with their larger counterparts,[1] with the basal metabolism/body mass relationship being proportional to mass$^{-0.25}$. These differences in metabolism appear to result largely from changes at the cellular level, and the goal of this study was to examine whether molecular activity (MA) of sodium pumps from the kidney of these homeotherms reflected their differences in metabolism.

Five mammalian species (30 g to 350 kg; $n = 4$) and eight bird species (13 g to 35 kg; $n = 4$) were examined, with mixed sections of cortex and outer medulla from their kidneys used for measurement. Na$^+$K$^+$ATPase activity (μmol Pi·mg wet wt^{-1}·hr^{-1}) was measured in tissue homogenates (2%, 37°C) and was inversely related to body mass, yielding the allometric equations Na$^+$K$^+$ATPase = 5.7M$^{-0.08}$ and Na$^+$K$^+$ATPase = 6.6M$^{-0.16}$ for mammals and birds, respectively. Kidney sodium pump density (pmol·g wet wt^{-1}), measured via ^{3}H-ouabain binding in tissue biopsies, showed no relationship with increasing body size in mammals; however, in birds there was a significant decrease ($P < 0.05$) with increasing body size. MA (ATP·min^{-1}) of mammalian sodium pumps was significantly higher in smaller mammals, with the relationship described by the equation MA = 34655M$^{-0.14}$ ($P < 0.01$), whereas bird sodium pumps had a fairly constant MA that was on average 6000 ATP·min^{-1}. All allometric equations were determined by linear regression of log-transformed values.

The higher Na$^+$K$^+$ATPase activity seen in smaller mammals and birds is likely a requirement to maintain the appropriate transmembrane concentrations of sodium and potassium, because more metabolically active species tend to have an increased

Address for correspondence: Nigel Turner, University of Wollongong, NSW, Australia 2522. Voice: +61-2-4221-3883; fax: +61-2-4221-4096.
nt05@uow.edu.au

"leakiness" in their membranes.[2] In mammals, the changes seen in $Na^+K^+ATPase$ activity seem to result primarily (62%) from changes in the molecular activity of their sodium pumps, whereas the birds seem to primarily (79%) alter $Na^+K^+ATPase$ activity by changing the number of sodium pumps in their kidneys.

To assess whether sodium pump MA was affected by the lipid composition of the surrounding membrane, we determined fatty acid composition of kidney microsomal membranes. The allometric changes in mammalian MA were positively correlated ($r = 0.95$; $P < 0.01$), with changes in the level of polyunsaturated fatty acids (measured as unsaturation index), whereas the fairly constant MA in birds was associated with nonsignificant changes in the level of membrane polyunsaturation and no correlation was observed. These results are consistent with a recent hypothesis of a "pacemaker" role for membranes in metabolism, via an effect on membrane proteins.[3]

REFERENCES

1. KLEIBER, M. 1961. The Fire of Life. An Introduction to Animal Energetics. Wiley. New York.
2. ELSE, P.L. & A.J. HULBERT. 1987. Evolution of mammalian endothermic metabolism: "leaky" membranes as a source of heat. Am. J. Physiol. **253:** R1–R7.
3. HULBERT, A.J. & P.L. ELSE. 1999. Membranes as possible pacemakers of metabolism. J. Theor. Biol. **199:** 257–274.

Na$^+$,K$^+$-ATPase Subunit Isoforms of the Developing Central Nervous System of the Lizard *Gallotia galloti*

M. F. ARTEAGA,[a] J. AVILA,[a] P. MARTÍN-VASALLO,[a] AND C. M. TRUJILLO[b]

[a]*Laboratorio de Biología del Desarrollo, Departamento de Bioquímica y Biología Molecular, Universidad de La Laguna, 38206 La Laguna, Tenerife, Spain*

[b]*UDI of Biología Celular, Universidad de La Laguna, 38206 La Laguna, Tenerife, Spain*

KEYWORDS: Na,K-ATPase; radial glia; development; central nervous system

In the central nervous system, the gradients generated by the Na$^+$,K$^+$-ATPase are used for electrical membrane potential changes, uptake of neurotransmitters, and, in astrocytes, uptake of K$^+$ from the extracellular space after depolarization of neurons. Radial glia forms the scaffold that guides migrating neurons and the outgrowth of axons during development.[1] In the adult brain, α2 predominates in astrocytes and α3 predominates in neurons, α1 and β1 are ubiquitous, β2 is an adhesion molecule on glia involved in neuron-astrocyte adhesion, and β3 in the central nervous system is expressed only in photoreceptors and oligodendrocytes.[2] The patterns of developmental specification are extremely complex, and both the control of biosynthesis and the intrinsic enzyme properties are affected by the choice of α and β.[3]

We propose to establish the differential expression patterns of the Na$^+$,K$^+$-ATPase isozymes of radial glia along developmental stages of the lizard *Gallotia galloti*. A panel of isoform-specific antibodies (6F, McB2, XVIF9-G10, SpETβ1, SpETβ2, RTN-β3)[2,4] and the glia marker GFAP (glial fibrillary acidic protein)[4] were used for the localization in frozen sections of the developing mesencephalon of the lizard by double-label immunofluorescence.

During ontogeny, immunoreactivity for Na$^+$,K$^+$-ATPase subunits isoforms varied along stages. At embryonic stage 35 (E35), only α2 and β2 isoforms appeared in radial glial processes and glial cell somas (coexpressing with GFAP) in pretectum, optic tectum (in strata), cerebellar ventriculi, and raphe. From E37 to hatching, there is a relative decay of specific immunoreactivity for β2 as the β1 expression signal increases from faint to high. No β3 was found at any stage of development. A faint expression of α1 and α2 appears in isolated cellular groups not yet characterized.

Address for correspondence: M.F. Arteaga, Laboratorio de Biología del Desarrollo, Departamento de Bioquímica y Biología Molecular, Universidad de La Laguna, Avda Astrofísico Sánchez s/n, 38206 La Laguna, Tenerife, Spain. Voice: +34-922-318358; fax: +34-922-318354.
marteaga@ull.es

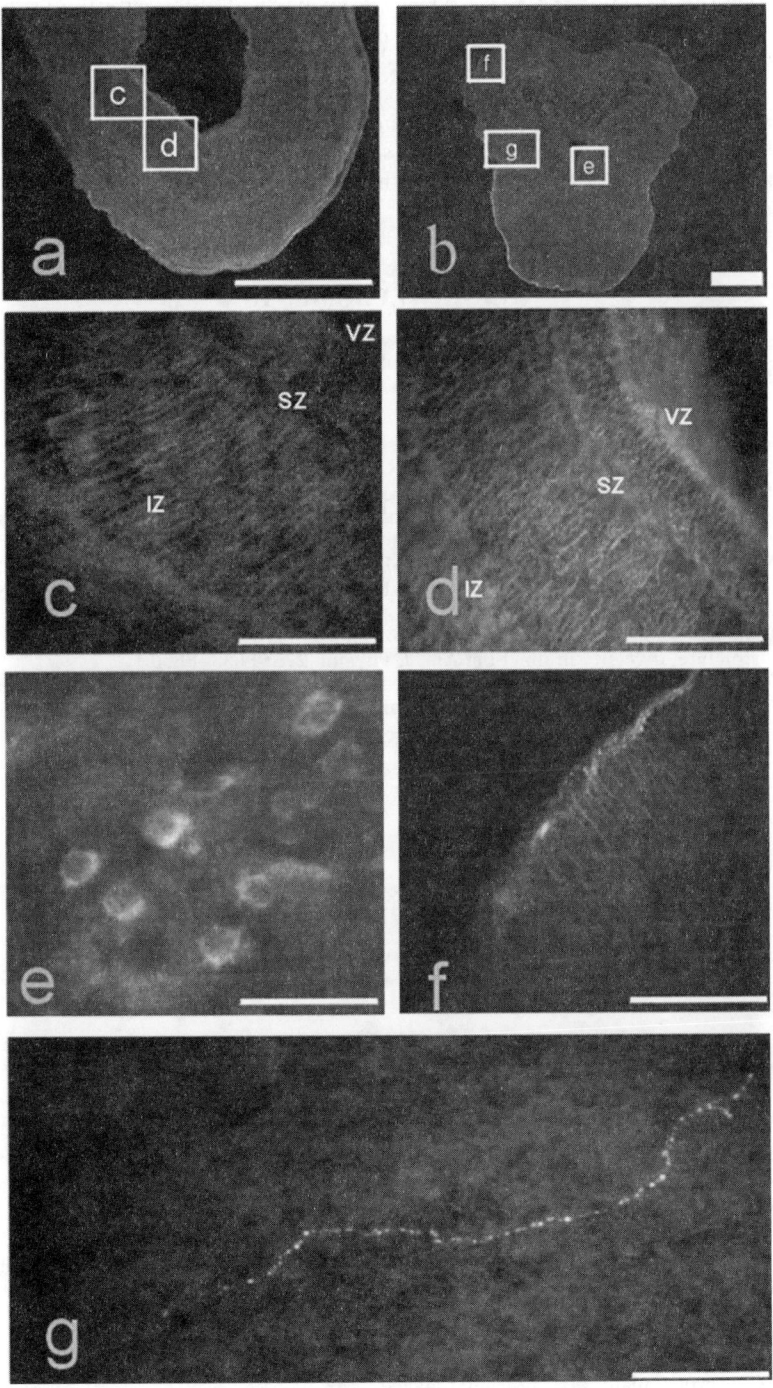

FIGURE 1. *See following page for legend.*

(FIG. 1). The diversity of Na^+,K^+-ATPase isozymes with different affinities for its physiological ligands and their complex spatial and temporal patterns of cellular expression in radial glia suggest that these isozymes are cast along development in such a way that affinity for K^+ increases and affinity for Na^+ decreases and the apparent number of total copies per cell increases.

Noteworthy is the finding of a bright signal for β2, like sparks following a line, from immediately before hatching to further stages in synaptic contacts; this points out a role for β2 in the synaptic setup.

ACKNOWLEDGMENTS

This work was supported by grants from DGICYT (PB97-1488) Spain and COF1999/014 from Gobierno Autónomo de Canarias.

REFERENCES

1. RAKIC, P. 1995. Radial glial cells: scaffolding for brain construction. In Neuroglia. H. Kettenmann & B.R. Ransom, Eds.: 746–762. Oxford University Press. New York.
2. MARTÍN-VASALLO, P., R.K. WETZEL, L.M. GARCÍA-SEGURA, et al. 2000. Oligodendrocytes in brain and optic nerve express the β3 subunit isoform of Na,K-ATPase. Glia **31:** 206–218.
3. BLANCO, G. & R.W. MERCER. 1998. Isozymes of the Na-K-ATPase: heterogeneity in structure, diversity in function. Am. J. Physiol. **275:** F633–F650
4. PENG, L., P. MARTÍN-VASALLO & K.J. SWEADNER. 1997. Isoforms of Na,K-ATPase and subunits in the rat cerebellum and in granule cell cultures. J. Neurosci. **17:** 3488–3502.

FIGURE 1. Immunolocalization of Na^+,K^+-ATPase α and β subunit isoforms in developmental stages of *Gallotia galloti*. Sections of mesencephalon at stages E35 (**a**) and hatching (**b**) marking the squares containing the areas of the corresponding pictures in **c** to **g**. Scale bar = 200 μm. (**c**) α2 immunofluorescent labeling of radial glia processes in E35. Scale bar = 40 μm. (**d**) β2 labeling of radial glia processes and plasma membrane of ventricular cell somas in E35. Scale bar = 50 μm. (**e**) α2 staining of plasma membrane in cells of optic tectum at hatching. Scale bar = 5 μm. (**f**) β1 labeling of radial glia processes in hatching. Scale bar = 25 μm. (**g**) β2 immunofluorescent labeling of synaptic contacts (buttons) in hatching. Scale bar = 10 μm. VZ, ventricular zone; SZ, subventricular zone; IZ, intermedial zone; CP, cortical plate; ML, molecular layer.

Glutamate Receptors Regulate Na/K-ATPase in Cerebellum Neurons

ELENA BULYGINA,[a] OLGA GERASIMOVA,[a] AND ALEXANDER BOLDYREV[a,b]

[a]*Department of Biochemistry, M. V. Lomonosov Moscow State University, 123356 Moscow, Russia*

[b]*Institute of Neurology, Russian Academy of Medical Sciences, 123356 Moscow, Russia*

KEYWORDS: reactive oxygen species; cerebellum granule cells; glutamate; metabotropic receptors; ionotropic receptors; NMDA; ACPD; 3-HPG

There are contradicting data in the literature concerning the action of glutamate on Na/K-ATPase: some authors reported inhibition[1,2] and another reported an activating action of the neuromediator.[3] These data were obtained using synaptosomes or membrane-bound enzyme prepared from cultured neurons. We have studied this problem using viable cerebellum granule cells exposed to several ligands and measured Na/K-ATPase activity simultaneously with intracellular reactive oxygen species (ROS) production, using a flow cytometry technique.[4]

A short preincubation time of cells with glutamate resulted in a 20% activation of Na/K-ATPase, which this changed to inhibition after a longer incubation time (35% after 30-minute incubation). These phenomena are, presumably, attributed to interaction of glutamate with different subtypes of glutamate receptors on the neuronal membrane. To check this assumption, we have measured effects on Na/K-ATPase of NMDA, ACPD and 3-HPG, and activated ionotropic (NMDA) and two groups of metabotropic (ACPD and 3-HPG) receptors for glutamate. Activation of NMDA and ACPD receptors by increasing concentrations of the ligands induced dose-dependent inhibition of Na/K-ATPase, whereas activation of 3-HPG receptors resulted in a dose-dependent activation of Na/K-ATPase (FIG. 1A). The simultaneous presence of NMDA and ACPD during preincubation of the neurons did not increase inhibition of the ATPase, whereas the presence of 3-HPG prevented inhibiting effect of both NMDA and ACPD (FIG. 1B). Inhibition of Na/K-ATPase by NMDA or ACPD was prevented by subsequent addition of 10 µM MK-801 or MCPG. Chelerythrine (10 µM) or indolylmaleimide (50 µM), known inhibitors of protein kinase C, effectively prevented the inhibitory effect of ACPD but were only partially effective in the case of NMDA, the action of which, in turn, was removed by 1 mM cysteine or 0.1 mM DTT.

Address for correspondence: Alexander Boldyrev, Department of Biochemistry, M.V. Lomonosov Moscow State University, 123356 Moscow, Russia. Voice/fax: +7-95-939-1398.
aaboldyrev@mail.ru

TABLE 1. ROS levels in viable cerebellar granule cells exposed to several ligands (0.5 mM each, 30-min exposure)

Conditions	ROS Fluorescence (arb. units)	Percent of Control	P toward Control
Control	42.8 ± 0.7	100	
3-HPG	33.5 ± 0.5	78	<0.01
NMDA	81.1 ± 1.2	190	<0.01
3-HPG+NMDA	45.3 ± 1.1	106	
ACPD	49.3 ± 0.9	115	<0.05
3-HPG+ACPD	44.7 ± 0.8	104	

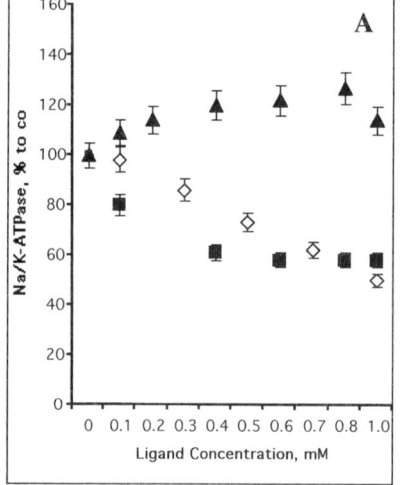 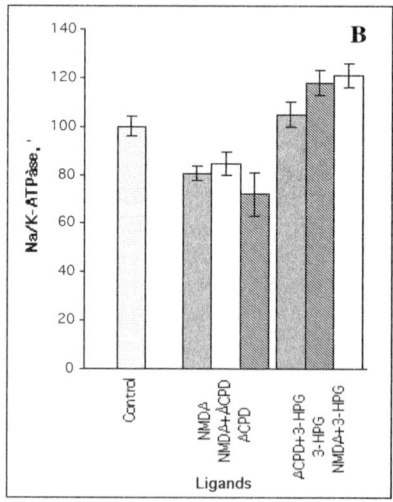

FIGURE 1. Effect of glutamate agonists on Na/K-ATPase of cerebellum granule cells. (**A**) Concentration dependence for 3-HPG (*black triangles*), NMDA (*white lozenges*), and ACPD (*black squares*); in each case, activity was measured after 30-minute preincubation of cells with the ligand. (**B**) Na/K-ATPase activity measured after 30-minute preincubation of the neurons with different ligands (0.5 mM) as indicated.

Flow cytometric analysis demonstrated that both NMDA and ACPD stimulate ROS production, measured as a mean fluorescence of the DCF-preloaded cells, whereas 3-HPG decreased intracellular ROS production and diminished the effect of NMDA or ACPD (TABLE 1).

Thus, activation of both ionotropic NMDA receptors and metabotropic ACPD receptors (associated with cAMP turnover) results in Na/K-ATPase inhibition, which explains, at least partially, their known excitotoxic effect on the neuronal cells. Activation of 3-HPG–dependent receptors (associated with Ca/IP$_3$-dependent protein kinases) decreases the intracellular ROS levels and suppresses their increase induced by NMDA or ACPD. Thus, metabotropic receptors associated with Ca/IP_3 turnover support an active state of Na/K-ATPase and protect the neurons against NMDA and/or ACPD excitotoxicity.

REFERENCES

1. AVROVA, N.F. et al. 1998. Inhibition of glutamate-induced intensification of free radical reactions by gangliosides. Neurochem. Res. **23:** 945–952.
2. STELMASHOOK, E.V. et al. 1999. Short-term block of Na/K-ATPase in neuro-glial cell cultures of cerebellum induces glutamate dependent damage of granule cells. FEBS Lett. **456:** 41–44.
3. MARCAIDA, G. et al. 1996. Glutamate induces a calcineurin-mediated dephosphorylation of Na,K-ATPase that results in its activation in cerebellar neurons in culture. J. Neurochem. **66:** 99–104.
4. BOLDYREV, A., E. BULYGINA, M. YUNEVA & W. SCHONER. 2003. Na/K-ATPase regulates intracellular ROS level in cerebellum neurons. Ann. N. Y. Acad. Sci. This volume.

Na,K-ATPase Isoforms in Pregnant and Nonpregnant Rat Uterus

RACHEL FLOYD,[a] ALI MOBASHERI,[b] PABLO MARTÍN-VASALLO,[c] AND SUSAN WRAY[a]

[a]*Physiological Laboratory, University of Liverpool, United Kingdom*

[b]*Faculty of Veterinary Science, University of Liverpool, United Kingdom*

[c]*Departamento de Bioquimica y Biologia Molecular, Universidad de La Laguna, Spain*

KEYWORDS: uterus; Na,K-ATPase; isoforms; pregnancy

Na,K-ATPase consists of three subunits: α, β, and γ, each of which exist as multiple isoforms, and the expression of these isoforms is tissue and cell type specific.[1] The α isoforms show differential sensitivity to ions and ouabain in a species- and tissue-dependent manner. The control of uterine smooth muscle contraction is critically dependent on the electrogenic action of Na,K-ATPase. Early studies have identified Na,K-ATPase isoforms in the pregnant and nonpregnant rat myometrium, which indicated specific changes in mRNA expression throughout pregnancy.[2] Here, we determined the distribution of $\alpha 1$, $\alpha 2$, $\alpha 3$ and $\beta 1$, $\beta 2$, $\beta 3$ isoforms of Na,K-ATPase and Na^+/Ca^{2+} exchanger (NCX) and Ca^{2+}-ATPase (PMCA1 isoform) in nonpregnant and pregnant rat uterus.

MATERIALS AND METHODS

Immunohistochemical studies were performed on transverse cryosections of whole rat uterus previously fixed in ice-cold methanol. Sections were probed with isoform-specific antibodies using the dilution factors indicated: 6F neat hybridoma supernatant, McB2 1:4, XVIF9G10 1:300, SpEtb1 1:200, SpEtb2 1:200, RNTb3 1:200, PMCA1 1:100, and NCX 1:100 ($n = 4$).

RESULTS AND DISCUSSION

Pregnant Animals

Abundant immunostaining of the $\alpha 1$ subunit was detected in the epithelium and outer longitudinal smooth muscle of the myometrium, with more diffuse staining

Address for correspondence: Ali Mobasheri, Faculty of Veterinary Science, University of Liverpool, Liverpool L69 7ZJ, UK. Voice: +44-151-7944284; fax: +44-151-7944243.
 a.mobasheri@liv.ac.uk

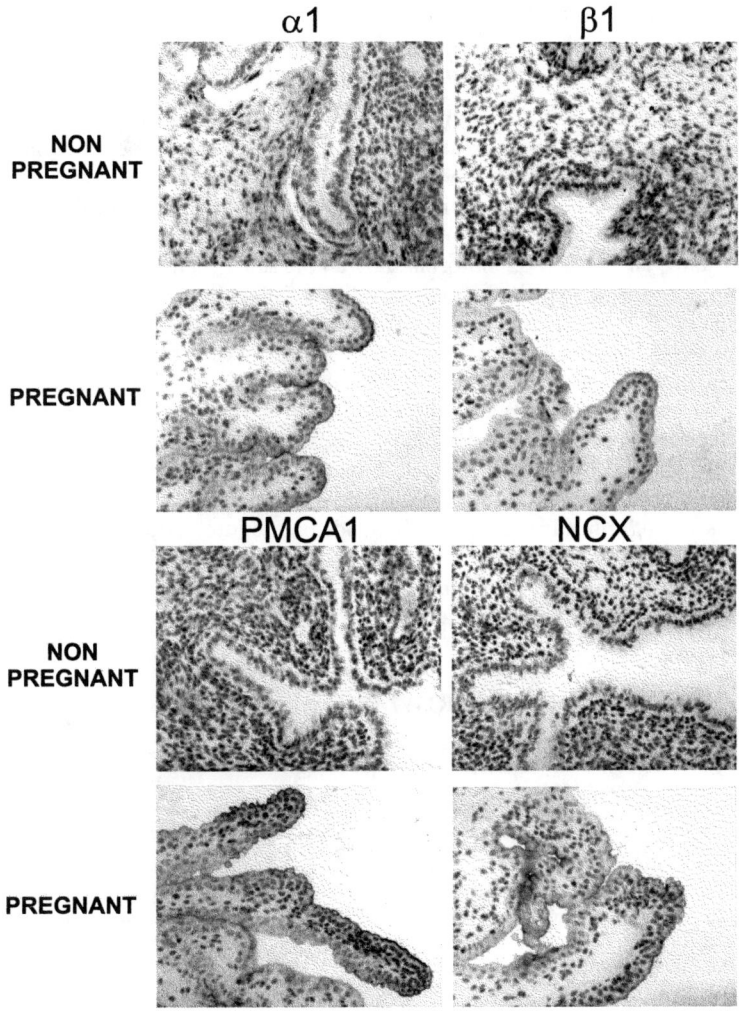

FIGURE 1. Localization of Na,K-ATPase $\alpha 1$ and $\beta 1$ subunits, Ca^{2+}-ATPase (PMCA1 isoform), and Na^+/Ca^{2+} exchanger (NCX) in the epithelia of the rat uterus. For immunostaining, the Fast-Red system was used on cryostat sections with hematoxylin counterstain. The pattern of immunoreactivity for $\alpha 1$ is similar to that of $\beta 1$ (data not shown). Original magnification approximately $\times 100$.

shown in the inner myometrium and endometrium (FIG. 1). This was consistent with the distribution of β1, which was also present in the serosa. Immunoreactivity for α2 was restricted to the same regions that showed positive staining for NCX and PMCA1, namely, the inner circular smooth muscle and the endometrium with distinct staining seen in the epithelium. In contrast, α3 appeared to be localized predominantly in the epithelium, with sparse immunostaining in the endometrium and myometrium. Epithelial localization of β2 and β3 was apparent with further faint staining in the endometrium and myometrium.

Nonpregnant Animals

There was faint immunoreactivity for α1, α2, and α3 and β1 and β2 in the epithelium. Expression of α2 was greater in the nonpregnant serosa, outer myometrium, and lower endometrium than in the pregnant animals. Endometrial expression of α3 was much stronger than all pregnant specimens examined, although the diffuse localization observed in the other regions of the uterus corresponded to that seen in the pregnant state. The β1 and β2 isoforms showed the same regional localization in the outer myometrium, with dense immunoreactivity for α2 in the upper segment of the endometrium. Conversely, expression of β3 subunit and NCX was low throughout the entire tissue. Interestingly, the region of highest expression of PMCA1 in the nonpregnant uterus was the endometrium, which partially corresponded with the localization seen in the pregnant uterus, although during pregnancy there was increased expression of epithelial PMCA1. There was a marked increase in the expression of all isoforms of Na,K-ATPase in pregnant uterine epithelia associated with drastic increase in the expression of NCX and PMCA1.

Uterine smooth muscle contraction depends on the intracellular concentration of calcium. It is important to understand the distribution of Na,K-ATPase isoforms, NCX, and PMCA1 because variable expression of these systems may modulate myometrial contractility during pregnancy.

ACKNOWLEDGMENTS

Supported by grants from the Wellcome Trust and MRC (U.K.).

REFERENCES

1. MOBASHERI, A., J. AVILA, I. CÓZAR-CASTELLANO, et al. 2000. Na^+,K^+-ATPase isozyme diversity; comparative biochemistry and physiological implications of novel functional interactions. Biosci. Rep. **20:** 51–91.
2. TURI, A., Z. MARCSEK, N. MULLNER, et al. 1992. The activity of Na+/K(+)-ATPase and abundance of its mRNA are regulated in rat myometrium during pregnancy. Biochem. Biophys. Res. Commun. **16:** 1191–1197.

Using Na,K-ATPase Itself for Large-Scale Isolation and Purification of Endogenous Digitalis–Like Factors

F. MANDEL,[a,b] A. VASILIEV,[b] AND I. KRIVOI[b]

[a]*Department of Molecular Physiology and Biophysics, Baylor College of Medicine, Houston, Texas 77030, USA*

[b]*St. Petersburg State University, St. Petersburg 199034, Russia*

KEYWORDS: Na,K-ATPase; endogenous digitalis–like factors; OLF; EDLF

Despite numerous studies suggesting the presence of endogenous digitalis–like factors (EDLFs) and intensive efforts to isolate and identify them, low *in vivo* abundance and unknown structure prevented their isolation for almost 40 years. In 1991, our group was the first to isolate an EDLF.[1] However, subnanomolar abundance precluded the isolation of more than 50 μg of ouabain-like factor (OLF) from approximately 1000 liters of human plasma, and, hence, its definitive characterization. A decade later, it is still not absolutely certain whether OLF is ouabain or its isomer. Subsequently, other EDLFs have been isolated, but none have been definitively characterized. Here, we present a protocol capable of isolating sufficient EDLF for definitive characterization via mass spectrometry and/or two-dimensional nuclear magnetic resonance.

Na,K-ATPase AFFINITY EXTRACTION PROTOCOL

Na,K-ATPase affinity extraction (NAE) uses Na,K-ATPase (NKA) itself to extract EDLFs. The essence of this protocol is the nanomolar affinity of NKA for cardiac glycosides in the presence of Mg^{2+} and its micromolar affinity in EDTA. EDLFs may be extracted from almost any media by the addition of NKA and 5 to 10 mM MgP_i. After 3 hours at 37°C, the NKA–EDLF complex formed is centrifuged at 100,000 g and the supernatant is discarded. After one 10 mM MgP_i wash, EDLF is released from NKA/pellet by overnight (22°C) incubation in 5 mM EDTA. Because all nonpelletable material has been eliminated by previous identical centrifugations, the supernatant of the third centrifugation contains, almost exclusively, EDLFs easily separated by HPLC.

Address for correspondence: Frederic Mandel, Department of Molecular Physiology and Biophysics, Baylor College of Medicine, Houston, TX 77030. Voice: 713-798-5711; fax: 713-798-3475.
fmandel@bcm.tmc.edu

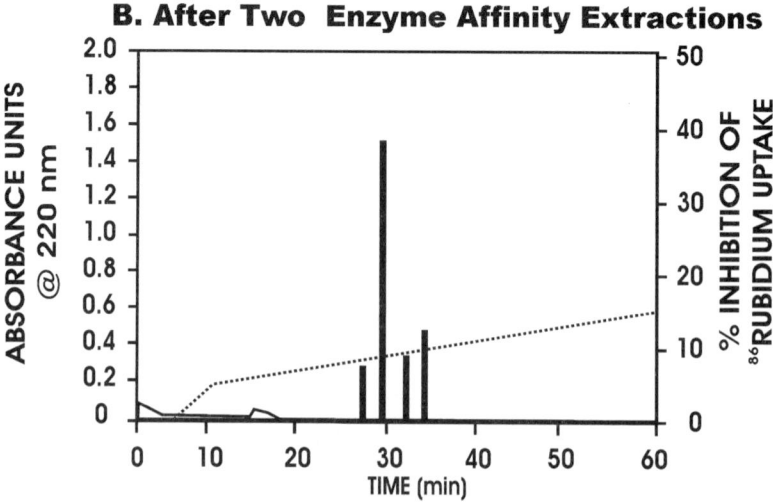

FIGURE 1. Protocol efficacy: change in ^{86}Rb uptake and ultraviolet absorbance after two NAE cycles. (**A**) Kidney extract before NKA addition. (**B**) Same material after two cycles. (*solid lines*) Absorbance (220 nm); (*vertical bars*) percentage of inhibition of ^{86}Rb uptake by RBC; (*dotted lines*) acetonitrile gradient of C-18 column.

EFFICIENCY

Each NAE cycle extracts 65–75% of ^{3}H-ouabain from plasma or tissue homogenates spiked with 10 nM ^{3}H-ouabain and reduces ultraviolet absorbing material by greater than 99%. Thus, two consecutive cycles extract approximately 50% ^{3}H-ouabain or EDLF, while eliminating more than 99.95% of extraneous material. NAE efficacy is demonstrated by comparing kidneys homogenized in ethanol, roto-evaporated, and resuspended in 20 mM Tris-Cl, pH 7.4, before and after NAE (FIG. 1).

ISOLATION

OLF was isolated from human plasma, bovine adrenals, and both porcine and lamb kidneys. Regardless of the source, OLF always appeared in the same HPLC fractions. Depending on the HPLC column used for the final separation, up to three other distinct factors were observed. A second factor exhibits slower off rates than ouabain. The third and fourth factors are as yet uncharacterized.

LARGE-SCALE EXTRACTION

Kidneys were chosen for large-scale isolation because, using NAE, approximately 12-fold as much OLF was isolated from kidneys (1.2 µg/kg) as from plasma (<0.1 µg/L). Approximately 120 kg of porcine kidneys containing approximately 140 µg of OLF was processed. The amounts of the other factors isolated have not been determined yet.

CONCLUSIONS

NAE is 5–10-fold more efficient than HPLC columns alone. It does not isolate only specific subsets of EDLFs as do antibodies and HPLC columns. Keeping volumes constant throughout ensures nothing is concentrated, eliminating false-positives often encountered during HPLC purifications. Finally, 50% efficiency makes the isolation of sufficient quantities of EDLFs for their unequivocal characterization feasible.

ACKNOWLEDGMENTS

This work was supported by the Russian FBR (01-04-49799, 02-04-06957MAC) and Mandel Enterprises.

REFERENCE

1. HAMLYN, J. et al. 1991. Identification and characterization of a ouabain-like compound from human plasma. Proc. Natl. Acad. Sci. USA **81**: 6259–6263.

Inhibition of the Na,K-ATPase by the Antiarrhythmic Drug, Bretylium

CRAIG GATTO,[a] C. THEODORE BARKULIS,[a] WILLIAM R. SCHNEIDER,[a] JEREMY P. HOLDEN,[b] KRISTA L. ARNETT,[c] AND MARK A. MILANICK[c]

[a]*Department of Biological Sciences, Illinois State University, Normal, Illinois 61790-4120, USA*

[b]*Department of Medicine, University of Wisconsin, Madison, Wisconsin 53725, USA*

[c]*Department of Physiology and Dalton Cardiovascular Research Center, University of Missouri, Columbia, Missouri 65211, USA*

KEYWORDS: antiarrhythmic drug; Na pump; quaternary amine; K competition

Bretylium (BrT) is a quaternary amine used extensively as a potassium channel blocker.[1] Here, we report that BrT also inhibits the Na,K-ATPase; the BrT IC_{50} value (at 20 mM K^+) was approximately 5 mM. Interestingly, enzyme activity followed a biphasic pattern in response to BrT; initially, BrT stimulated Na,K-ATPase ([BrT] ≤1 mM), followed by an inhibitory effect. We limit our discussion to BrT inhibition of the Na,K-ATPase.

Increasing concentrations of K (or Rb) were able to overcome the inhibitory effect of BrT, and the maximal velocity for ATPase activity was unchanged. In contrast, increasing Na concentration was unable to compete with BrT and V_{max} was significantly reduced. These observations are in line with reports of BrT inhibition of guinea pig cardiac Na,K-ATPase.[2] We observed that 50 mM BrT (i.e., 10 times the IC_{50} for ATPase) was unable to prevent the formation of E~P from Na and MgATP.[3] These observations, together with K competition, suggest that BrT, a cation, inhibits the Na pump by binding to E_2-P. Based on this and previous work, it appears that BrT inhibits the Na pump by preventing extracelluar K binding.[3]

Clinical Effects of Bretylium

BrT has been used as a class III antiarrhythmic agent. Its primary mode of action has been elucidated as an inhibitor of voltage-dependent potassium channels;[1] thus, BrT slows down the relaxation phase of the cardiac action potential and concomitantly causes an elongated action potential duration (APD) (see FIG.1). More recently, BrT has been shown to chemically convert ventricular fibrillation into ventricular

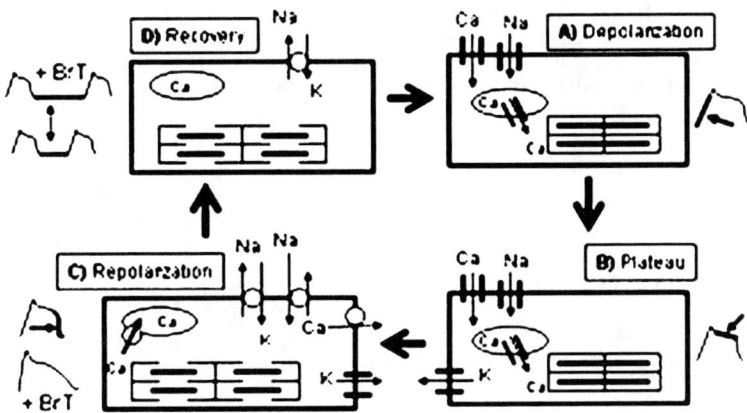

FIGURE 1. Cardiac cycle. The action potential (AP) trace next to each figure depicts the physiological state. (**A**) Depolarization via Na influx facilitates Ca influx via Ca channels and Ca release from sarcoplasmic reticulum (SR), initiating sarcomere shortening. (**B**) K channels open while an influx of Na and Ca persist, thus elongating the AP in the plateau phase. (**C**) During repolarization, Ca is resequestered by the SR and simultaneously removed via 3Na/Ca exchange and PMCA. Also, the K efflux persists and the Na pump actively removes 3Na/2K. At this stage, BrT inhibition would elongate the action potential as Na would build up because of Na/Ca exchange. (**D**) Recovery in which Na and K are reequilibrated. Here, BrT would lengthen this period and decrease the slope of the action potential duration restitution.

tachycardia.[4] Furthermore, this effect of BrT appears to be unrelated to its ability to inhibit potassium channels, because simultaneous application of the K_{ATP}-channel agonist, cromakalim, did not prevent the antifibrillatory effect of BrT.[4] Rather, it was shown that BrT was able to decrease the slope of the action potential duration restitution. We propose that the antiarrhythmic effects of BrT, and possibly cardiac glycosides, are in part caused by Na pump inhibition and that this inhibition effectively lengthens the diastolic interval thereby decreasing the slope of the action potential duration restitution.

ACKNOWLEDGMENTS

This work was supported by American Heart Association Grant 30161N and NIH Grant GM61583 (to C.G.) and NIH Grant DK37512 (to M.A.M.). We thank Phillip Weeks for technical assistance.

REFERENCES

1. GASPAR, R. *et al.* 1994. Effects of bretylium tosylate on voltage-gated potassium channels in human T-lymphocytes. Mol. Pharmacol. **46:** 762–766.
2. DZIMIRI, N. & A.A. ALMOTREFI. 1991. Interaction of bretylium tosylate with guinea-pig myocardial Na,K-ATPase. Gen. Pharmacol. **22:** 935–938.

3. MILANICK, M.A. & L.L. ARNETT. 2002. Extracellular protons regulate the extracellular cation selectivity of the sodium pump J. Gen. Physiol. **120:** 497-508.
4. GARFINKEL, A. *et al.* 2000. Preventing ventricular fibrillation by flattening cardiac restitution. Proc. Natl. Acad. Sci. USA **97:** 6061–6066.

Muscular K-Clearance Capacity *in Vivo* Must Be Evaluated on the Basis of K and Na,K-ATPase Concentrations

HENNING BUNDGAARD AND KELD KJELDSEN

The Heart Centre, National University Hospital, Rigshospitalet, Copenhagen 2100, Denmark

KEYWORDS: skeletal muscles; K homeostasis; K depletion; K supplementation

Skeletal muscle (SM) Na,K-ATPase concentration [Na,K-ATPase] is affected by the level of potassium (K) intake. Thus, prolonged low K intake reduces plasma-K (pK), SM-K, and SM-[Na,K-ATPase], whereas prolonged high K intake has opposite effects.[1] Reduced SM-K uptake has been observed *in vitro* in K-depleted rats. Thus, K depletion in animals and humans reduces K tolerance.[2] On the other hand, long-term high K intake may reduce the risk for K intoxication in response to an acute K load. On this basis, we evaluated K-clearance capacity in SM *in vivo* in rats K depleted for 2 weeks (chow containing 0.01 mmol K/100 g) and rats K supplemented for 2 weeks (200 mmol K/100 g chow) during acute K exposure. Instrumentation of animals has been described previously.[1,3] Briefly, animals were catheterized for infusion of 1.5 ml per 100 g BW/h of 0.5 mol/L KCl and for blood sampling. Before KCl infusion, K-supplemented rats were fasted for 1 day to reduce their high SM-K content level, and because of their very high renal K excretion they were functionally nephrectomized. KCl infusions were continued until animals died. pK was measured by ABL 605 (Radiometer) and SM and chow K contents by FLM3 (Radiometer, Copenhagen, Denmark) and SM ^{3}H-ouabain binding as previously described.

Compared with control rats, pK, soleus, EDL, and gastrocnemius SM ^{3}H-ouabain binding site concentrations and gastrocnemius SM-K content were significantly reduced in K-depleted rats, whereas significant increases were observed in K-supplemented rats. K-depleted rats tolerated a fourfold higher KCl dose than controls, whereas K-supplemented, functionally nephrectomized, fasted rats tolerated a 65% higher KCl dose than controls. At this point gastrocnemius SM-K in K-depleted rats had increased significantly, whereas a minor, although significant, increase was observed in gastrocnemius SM in controls (FIG. 1). In fasted, nephrectomized, K-supplemented rats a significant increase in gastrocnemius SM-K was seen (FIG. 1). However, SM-K uptake per hour was lower in K-depleted rats than in controls and highest in K-supplemented rats. Furthermore, gastrocnemius SM-K-ion transport

Address for correspondence: Henning Bundgaard, The Heart Centre, National University Hospital, Rigshospitalet, Copenhagen 2100, Denmark. Voice: +453-545-2628.
henningbundgaard@dadlnet.dk

Ann. N.Y. Acad. Sci. 986: 623–624 (2003). © 2003 New York Academy of Sciences.

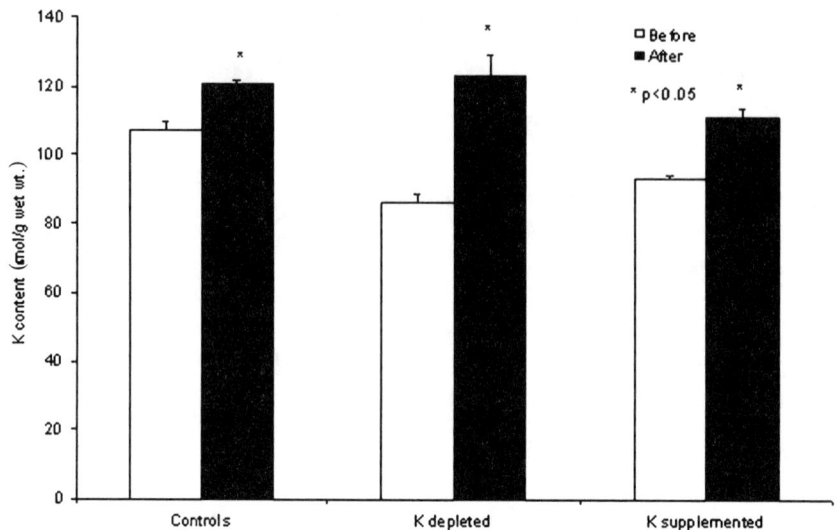

FIGURE 1. Gastrocnemius muscle K content in 2 weeks K-depleted and 2 weeks K-supplemented rats before and after KCl infusion until the animals died. Means and SEM. $n = 6$.

rate per Na,K-ATPase was higher in K-depleted than in controls and lowest in K-supplemented rats. In each group K-transport rate was below 15% of theoretical maximum. Changes were not a result differences in SM protein or water content.

In conclusion, *in vivo* K-depleted rats demonstrated increased K tolerance and K uptake despite reduced SM-[Na,K-ATPase], and K-supplemented rats with increased SM-[Na,K-ATPase] also demonstrated increased K tolerance after fasting. However, a lower K uptake rate was seen in K-depleted rats, confirming the concept of a positive correlation *in vivo* between SM-[Na,K-ATPase] and SM-K-clearance capacity. Thus, *in vivo* the K-clearance capacity cannot be predicted from the size of the SM-Na,K-pump pool only, but is also markedly influenced by the preceding SM-K content. This relationship needs consideration during management of patients with severe disturbances of the K homeostasis.[4]

REFERENCES

1. BUNDGAARD, H., T.A. SCHMIDT, J.S. LARSEN & K. KJELDSEN. 1997. K-supplementation increases muscle [Na,K-ATPase] and improves extrarenal K-homeostasis in rats. J. Appl. Physiol. **82:** 1136–1144.
2. DØRUP, I. 1996. Effects of K, Mg2-deficiency and adrenal steroids on Na, K(+)-pump concentration in skeletal muscle. Acta Physiol. Scand. **156:** 305–311.
3. BUNDGAARD, H. & K. KJELDSEN. 2002. Potassium depletion increases potassium clearance capacity in skeletal muscles *in vivo* during acute repletion. Am. J. Physiol. Cell Physiol. **283:** C1163–C1170.
4. WELFARE, W., P. SASI & M. ENGLISH. 2002. Challenges in managing profound hypokalaemia. Br. Med. J. **324:** 269–270.

Modest K^+ Restriction Provokes Insulin Resistance of Cellular K^+ Uptake without Decrease in Plasma K^+

LI E. YANG, PATRICK K. K. LEONG, JUAN P. GUZMAN, MICHAEL S. RHEE, AND ALICIA A. McDONOUGH

Department of Physiology and Biophysics, University of Southern California, Keck School of Medicine, Los Angeles, California 90089-9142, USA

KEYWORDS: potassium; skeletal muscle; kidney; Na^+,K^+-ATPase; H^+,K^+-ATPase

Extracellular fluid (ECF) K^+ homeostasis is maintained by the coordinate regulation of kidney and muscle. When K^+ output exceeds input, the kidney avidly retains K^+ by shifting from net K^+ excretion to net K^+ absorption, and the skeletal muscle, the biggest intracellular fluid (ICF) K^+ store, redistributes K^+ from ICF to the ECF.

We previously investigated the mechanisms of the K^+ shift from ICF to ECF by feeding rats a K^+-deficient diet for 2–10 days and found that ECF K^+ decreases, cellular K^+ uptake becomes insulin resistant, and muscle Na^+,K^+-ATPase $\alpha 2$ abundance and pump activity decrease.[1] The purpose of this study was to determine whether a decrease in ECF K^+ is a necessary signal to provoke the decrease in cellular K^+ uptake and urinary K^+ excretion when dietary K^+ is reduced to one-third of normal.

Rats were fed with control 1.0% K^+ (CK) or a modest K^+-restricted diet (0.33% K^+, LK). Plasma [K^+] was not reduced after 15 days: 4.10 ± 0.03 mM in CK and 4.02 ± 0.05 in LK, and not reduced after 30 days on the 0.33% K^+ diet (not shown). However, urine K^+ (determined in metabolic cages) was reduced from 2.7 ± 0.1 mEq/18 h to 0.4 ± 0.1 after 15 days of LK diet.

The "K^+ clamp" was applied to measure the insulin-stimulated K^+ uptake *in vivo* (FIG. 1A). The amount of K^+ infused over a defined time period is equivalent to the insulin-activated cellular K^+ uptake plus K^+ excretion; excretion was not significantly changed by insulin infusion in previous studies.[1] Therefore, the K^+ infusion rate (Kinf) required to clamp plasma K^+ provides a good measure of the insulin-stimulated cellular K^+ uptake. Kinf was significantly reduced after 15 days LK diet: 274 ± 41 mEq/h in CK versus 130 ± 63 mEq/h in LK, $P < 0.05$ (glucose uptake unchanged; FIG. 1B).

Address for correspondence: Dr. Alicia A. McDonough, Department of Physiology and Biophysics, University of Southern California, Keck School of Medicine, Los Angeles, CA 90089-9142. Voice: 323-442-1238; fax: 323-442-2283.
mcdonoug@hsc.usc.edu

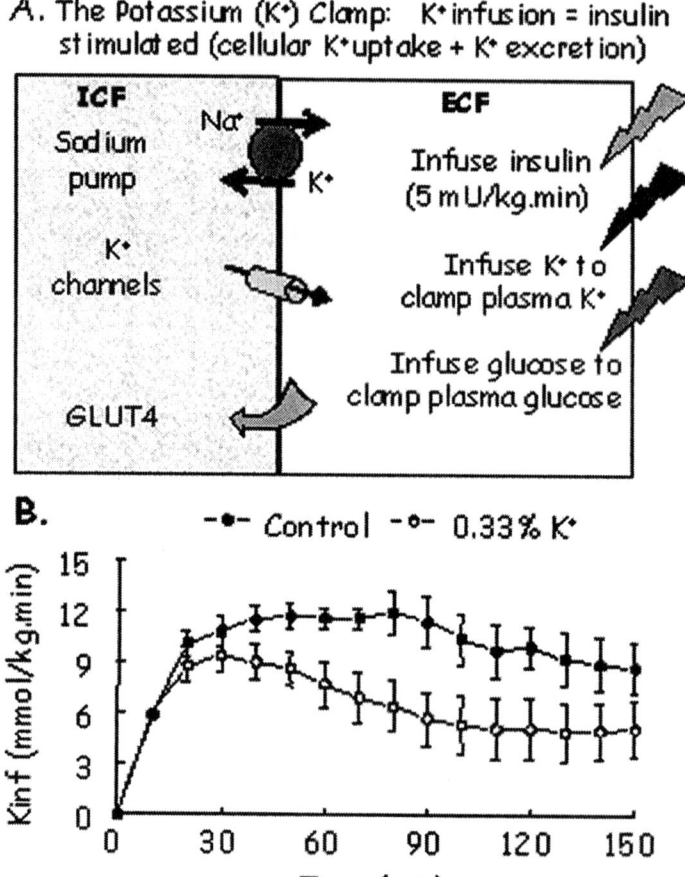

FIGURE 1. (**A**) Hyperinsulnemic euglycemic/K^+ clamp. Conscious rats were infused with insulin (5 mU/kg·min) to stimulate K^+ and glucose uptake. Based on the plasma K^+ and glucose levels measured during infusion, 150 mM KCl and 7.5 mM 20% dextrose were infused to clamp the plasma K^+ and glucose at baseline. (**B**) Fifteen-day 0.33% K^+ diet reduced insulin-stimulated cellular K^+ uptake. Kinf in control (*filled circles*) and 0.33% K^+ diet (*unfilled circles*) fed rats were shown. Values are means ± SE, $n = 6$.

Fifteen-day LK diet did not change in ouabain-sensitive Na^+,K^+-ATPase activity, α2 abundance (Western blotting), or ICF K^+ in whole gastrocnemius muscles. Likewise, abundance of gastric and colonic isoforms of H,K-ATPase in kidney and Na,K-ATPase subunits (Western blotting) were not changed during 15–30 days on LK diet, indicating that active K^+ retention is mediated by either activation of existing H,K-ATPase or inhibition of K^+ secretion.

In conclusion, the molecular mechanisms to preserve plasma K^+ are initiated in kidney and muscle before a decrease in plasma K^+, evidence for outstanding K^+ homeostatic control. The error signal that drives the transport adjustments remains to be determined.

ACKNOWLEDGMENT

This work was supported by NIH Grant DK 57678.

REFERENCE

1. McDonough, A.A. et al. 2002. Skeletal muscle regulates extracellular potassium. Am. J. Physiol. Renal Physiol. **282:** F967–F974.

Na^+,K^+-ATPase in the Marine Alga *Heterosigma akashiwo*

YUKICHI HARA,[a] YUKO MIKAMI,[a] MARIKO SHONO,[b] AND MASATO WADA[c]

[a]*Tokyo Medical and Dental University, Tokyo, Japan*
[b]*Japan International Research Center for Agricultural Sciences, Ishigaki, Japan*
[c]*National Institute of Fruit Tree Sciences, Morioka, Japan*

KEYWORDS: evolution; marine alga; genomic organization; sodium pump; Na,K-ATPase

Plant cells had been believed to extrude intracellular sodium ions primarily with a combination of Na^+/H^+-antiporter and H^+-ATPase, rather than with a sodium pump. However, the cells of the marine alga *Heterosigma akashiwo* were found to have Na^+,K^+-ATPase activity on the plasma membrane.[1] Recently, we cloned a novel Na^+/K^+-ATPase cDNA from *H. akashiwo*.[2] The full-length cDNA of *H. akashiwo* Na^+,K^+-ATPase (HANA) was 4467 bp long and encoded for a 1330–amino acid protein with a molecular weight of 146,306. The deduced product exhibited approximately 40% identity in amino acids with animal Na^+,K^+-ATPase α subunits. The hydropathy profile was quite similar to those of animal Na^+/K^+-ATPase α subunits. The similarities strongly indicate that HANA and animal Na^+/K^+-ATPase α-subunit genes have diverged from a shared ancestor. But there is a possibility that the HANA gene has come from animal Na^+,K^+-ATPase α-subunit cDNA by horizontal gene transfer. To examine this possibility, we analyzed the genomic organization of the HANA gene.

METHODS

Fragments of HANA gene were obtained by means of PCR using specific primers designed based on the HANA cDNA sequence. PCR products were subcloned into pCR4 cloning vector (Invitrogen, Madison, WI), and the sequences were analyzed.

Address for correspondence: Yukichi Hara, Tokyo Medical and Dental University, Tokyo, Japan. Voice: +81-3-5803-5364; fax: +81-3-5803-0161.
y.hara.mbch@tmd.ac.jp

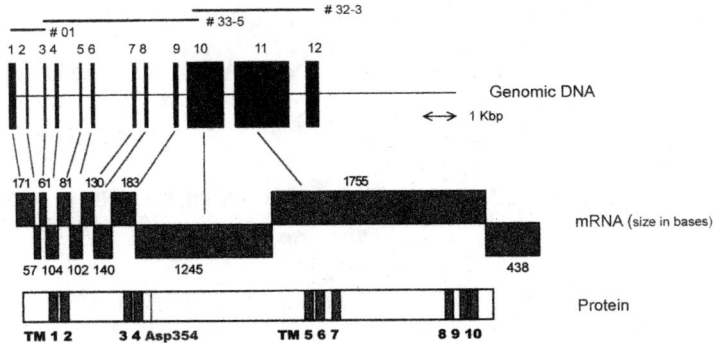

FIGURE 1. Genomic organization of HANA.

RESULTS

Three overlapping PCR products (#01, #33-5, and #32-3) were obtained from *H. akashiwo* genomic DNA as shown in FIGURE 1. Using those clones, we examined the structure of the HANA gene. The gene span was 10.5 kbp, and the gene contained 12 exons and 11 introns. All these introns had canonical GT-AG borders. Because gene transfer is known to result in an intron-less genomic structure, it is unlikely that the HANA gene has been derived from animal Na^+,K^+-ATPase α-subunit gene. Our results strongly indicate that the Na^+,K^+-ATPase gene might be distributed in the plant kingdom as well as the animal kingdom.

REFERENCES

1. WADA, M. *et al.* 1992. A marine algal Na^+-activated ATPase possesses an immunologically identical epitope to Na^+,K^+-ATPase. FEBS Lett. **309:** 272–274.
2. SHONO, M. *et al.* 2001. Molecular cloning of Na^+-ATPase cDNA from a marine alga, *Heterosigma akashiwo*. Biochim. Biophys. Acta **1511:** 193–199.

Positive Inotropic Effect Induced by Sequence-Specific Na$^+$,K$^+$-ATPase Antibody in Intact Cardiac Myocytes

S. Q. WANG,[a,b] H. CHENG,[a,b] A. C. MYERS,[c] B. J. CANNING,[c] AND K. Y. XU[d]

[a]*Laboratory of Cardiovascular Science, National Institute on Aging, Baltimore, Maryland 21224, USA*

[b]*National Laboratory of Biomembrane and Membrane Biotechnology, Peking University, 100871 Beijing, China*

[c]*Department of Medicine, Division of Clinical Immunology, The Johns Hopkins University School of Medicine, Baltimore, Maryland 21224, USA*

[d]*Division of Cardiology, The Johns Hopkins University School of Medicine, Baltimore, Maryland 21224, USA*

KEYWORDS: Na$^+$,K$^+$-ATPase; site-specific antibody; cardiac contraction

The purpose of our study is to identify the essential regions of the Na$^+$,K$^+$-ATPase that participate in the regulation of cardiac contraction. We have generated a site-specific antipeptide antibody (SSAPA) against the KRQPRNPKTDKLVNE sequence of rat α subunit of the Na$^+$,K$^+$-ATPase. Administration of SSAPA markedly increased intracellular Ca^{2+} concentration ($[Ca^{2+}]_i$) and cell shortening in isolated rat cardiac myocytes.[1] Our finding suggests that this structural sequence of the Na$^+$,K$^+$-ATPase may play a critical role in the Na$^+$,K$^+$-ATPase–mediated regulation of cardiac function.

METHODS

Confocal Ca^{2+} imaging and immunofluorescent staining methods were described previously.[1,2]

RESULTS AND DISCUSSION

Membrane location of the KRQPRNPKTDKLVNE region of rat Na$^+$,K$^+$-ATPase was first visualized in frozen sections of CV-1 cells by immunofluorescent staining

---

Address for correspondence: K.Y. Xu, Division of Cardiology, The Johns Hopkins University School of Medicine, Baltimore, MD. Voice: 410-550-2021; fax: 410-550-7480.
kxu@jhmi.edu

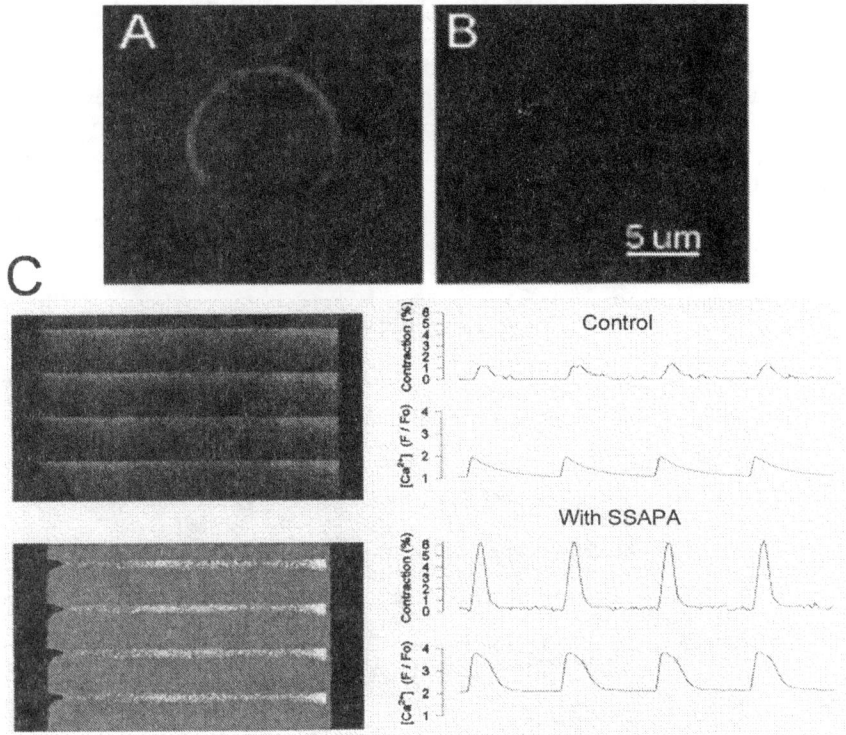

FIGURE 1. Cell staining, contraction, and intracellular Ca^{2+} transient in the presence and absence of SSAPA. Single CV-1 cell staining in the presence of SSAPA (**A**) and in the presence of both PB and SSAPA (**B**). (**C**, *left*) Confocal imaging of $[Ca^{2+}]_i$ transients. (**C**, *right*) Top traces represent time courses of cell shortening; bottom traces are averaged Ca^{2+} transients. The results show that SSAPA bound to its antigenic site of the Na^+,K^+-ATPase on the surface of the CV-1 cell membrane, elevated the diastolic and systolic $[Ca^{2+}]_i$ transient, and enhanced cell shortening in rat myocyte.

with SSAPA. FIGURE 1A shows that SSAPA interacts with its antigen on the cell surface membrane. The specific peptide blocker (PB) eliminated CV-1 cell immunofluorescent staining indicating the specificity of SSAPA binding to the cell membrane (FIG. 1B). We then tested the effect of SSAPA on cardiac contraction and Ca^{2+} transients in isolated rat heart cells. Confocal imaging was used in conjunction with the Ca^{2+}-sensitive indicator, fluo-4, to monitor simultaneously intracellular Ca^{2+} transients and contraction in single rat ventricular myocytes. Cells were continuously excited by electrical stimuli and contracted at 0.5 Hz (FIG. 1C). Administration of SSAPA (20 nM, 10–15 minutes) markedly increased systolic $[Ca^{2+}]_i$ and cell shortenings to $191 \pm 24\%$, ($n = 5$; $P < 0.01$) and $347 \pm 51\%$ ($n = 5$ $P < 0.01$) of controls, respectively (FIG. 1C). The diastolic $[Ca^{2+}]_i$ level also was elevated from 100 nM to 161 ± 13 nM (n = 5; $P < 0.01$) while accompanying a reduction of the resting cell length. Hence, SSAPA exerts a positive inotropic effect in cardiac myocytes via enhancement of diastolic and systolic $[Ca^{2+}]_i$. Our observations suggest that this particular molecular sequence of the Na^+,K^+-ATPase may participate in the regulation of cardiac contraction. The synthetic peptide acted as a SSAPA blocker and largely abolished the increases both in Ca^{2+} transients and contraction. These results confirmed that the binding of SSAPA to its specific antigenic site is necessary and sufficient for Na^+,K^+-ATPase to alter intracellular Ca^{2+} handling and to augment contractility in isolated rat heart cells.

Our study provides a novel approach to establish the important links between biological action and precise molecular structure of the Na^+,K^+-ATPase.

ACKNOWLEDGMENTS

This work was supported by grants HL52175 to K. Y. Xu and HL48198 to A. C. Myers from NIH/NHLBI, USA and NIH Intramural Research grant to H. Cheng.

REFERENCES

1. XU, K.Y., S.-Q. WANG & H. CHENG. 2001. Site-specific antibody of (Na^+-K^+)-ATPase augments cardiac myocytes contraction without inactivating enzyme activity. Biochem. Biophys. Res. Comm. **289:** 167–172.
2. MYERS, A.C. *et al.* 2002. Cell surface expression of a specific antigenic site on the catalytic subunit of Na^+,K^+-ATPase. Biochem. Biophys. Res. Comm. **291:** 111–115.

Inhibition of Purified Human Kidney Na$^+$,K$^+$-ATPase by Cyclosporine A

A Possible Mechanism for Drug Human Nephrotoxicity

M. YOUNES-IBRAHIM,[a] M. BARNESE,[a] P. BURTH,[b] AND M. V. CASTRO-FARIA[c]

[a]*Laboratorio de Fisiopatologia, Renal–Disciplina de Nefrologia, Universidade do Estado do Rio de Janeiro, Rio de Janeiro, Brasil*

[b]*Departamento de Biologia Celular e Molecular, Universidade Federal Fluminense, Niterói, Brasil*

[c]*Departamento de Biologia Celular e Genética—Instituto de Biologia Roberto Alcantara Gomes, Universidade do Estado do Rio de Janeiro, Rio de Janeiro, Brasil*

KEYWORDS: cyclosporine A; nephrotoxicity; Na,K-ATPase

INTRODUCTION

Cyclosporine A (CsA) is a lipophilic polypeptide highly efficient in preventing graft rejection and is also used in the therapy for autoimmune and chronic inflammatory diseases. Notwithstanding, CsA utilization is limited because of its high nephrotoxicity.[1]

Previous reports proposed that inhibition of Na$^+$,K$^+$-ATPase by CsA would be involved in the onset of the acute and chronic nephrotoxicity, but the mechanism involved remains controversial.[2,3]

In this study, we compared tissue and species characteristics of the response of purified Na$^+$,K$^+$-ATPase preparations to CsA.

MATERIALS AND METHODS

Brains and kidney cortical portions from Wistar rats and cortical and medullar portions from one human kidney (which could not be used in transplantation because of technical problems) were used for Na$^+$,K$^+$-ATPase preparation according to Jørgensen.[4] Final enzyme fractions were lyophilized and kept under nitrogen atmosphere at –18°C. The enzyme activity was measured as described elsewhere.[5] The

---

Address for correspondence: M. Younes-Ibrahim, Laboratorio de Fisiopatol, Renal–Disciplina de Nefrologia, UERJ, Rio de Janeiro, Brasil. Voice/fax: +55-21-25876250.
myounes@uerj.br

amount of added enzyme always corresponded to an activity of 19–21 µmol of Pi formed per hour per assay. The mixture without ATP was preincubated 10 min at 37°C in the presence or absence of CsA. Incubation was started by the addition of ATP. Mg^{2+}-ATPase activity was determined in the presence of 3.6 mM ouabain and was less then 5% of the total ATPase activity. Specific activities were 166, 143, 185, and 193 µmol Pi/h/mg of protein for human kidney medulla, human kidney cortex, rat kidney cortex, and rat brain, respectively.

Commercial CsA (50 mg/mL; Sandimmun Sandoz/Novartis) and the CsA vehicle Cremophor (Sigma-Aldrich) were used in all experiments.

RESULTS

Human kidney cortical and medullar Na^+,K^+-ATPases were similarly inhibited by ouabain (IC_{50} of 0.98 and 1.04 mM, respectively), suggesting that α_1 isoform predominates in both fractions. At CsA concentrations of 2.5, 12.5, and 25 µg/mL in the incubation mixture, the human cortical Na^+,K^+-ATPase showed a progressive reduction of the enzymatic activity (FIG. 1). At 25 µg/mL, CsA reduced the enzymatic activity of the human medullar fraction only 19%. At the same concentrations, the CsA effect on rat enzymes were negligible (<5%). Cremophor (3.0–33.5 µg/mL) had no effect on the Na,K-ATPase activity of all tissue preparations studied (results not shown).

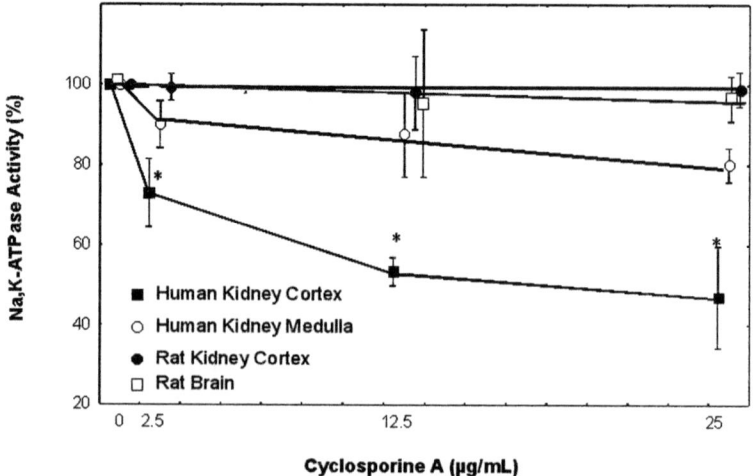

FIGURE 1. Inhibition of purified Na^+, K^+ ATPase preparations from human kidney, rat kidney, and rat brain by cyclosporine A. One hundred percent activity corresponded to 19–21 µM of Pi formed per hour in the assay conditions. Points are mean ± SE of three to four experiments. Comparisons between groups were conducted by the multiple regression technique; P values <0.05 were considered significant (*asterisk*).

DISCUSSION

Tubular alterations in CsA nephrotoxicity may be associated to an inhibition of Na^+,K^+-ATPase activity. Our *in vitro* experiments show a statistically significant dose-dependent CsA inhibition of the human kidney cortex Na^+,K^+-ATPase. However, the enzyme from human kidney medulla was much less responsive, and, curiously, the rat cortical enzyme and the rat brain enzyme were not inhibited, even by high CsA levels, suggesting a species-specific response of Na^+,K^+-ATPase to this drug.

Although *in vitro* CsA inhibitory concentrations were well above that found in blood at therapeutic levels, this finding can have clinical relevance, because renal CsA levels easily reach 10 times that of blood. Because we used purified Na^+,K^+-ATPase enriched-membrane fractions, our data strongly suggest a direct interaction between CsA and human cortical Na^+,K^+-ATPase or its lipid environment.

ACKNOWLEDGMENTS

Grants from CNPq, FAPERJ, and CAPES supported this work.

REFERENCES

1. KAHAN, B.D. 1989. Drug therapy. Cyclosporine. N. Engl. J. Med. **321:** 1725–1738.
2. TUMLIN, J.A. & J.M. SANDS. 1993. Nephron segment-specific inhibition of Na-K-ATPase activity by cyclosporin A. Kidney Int. **43:** 246–251.
3. IHARA, H., S. HOSOKAWA, T. OGINO, *et al.* 1990. Activation of K+ channel and inhibition of Na-K-ATPase of human erythrocytes by cyclosporine; possible role in hiperpotassemia in kidney transplant recipients. Transplant. Proc. **22** (suppl. 4): 1736–1739.
4. JORGENSEN, P.L. 1977. Purification and characterization of (Na^+,K^+)-ATPase. III. Purification from the outer medulla of mammalian kidney after selective removal of membrane components by sodium dodecylsulphate. Biochim. Biophys. Acta **356:** 36–52.
5. BURTH, P., M. YOUNES-IBRAHIM, F.H.F.S. GONÇALVES, *et al.* 1997. Purification and characterization of a Na^+,K^+-ATPase inhibitor found in an endotoxin of *leptospira*. Infect. Immun. **65:** 2557–2560.

Regulation of Na,K-ATPase by cAMP-Dependent Protein Kinase Anchored on Membrane via A-Kinase Anchoring Protein Subtype, AKAP-150, in Rat Parotid Gland

K. KURIHARA[a] AND N. NAKANISHI[b]

[a]*Department of Oral Physiology, Meikai University, School of Dentistry, Saitama 350-0283, Japan*

[b]*Department of Biochemistry, Meikai University, School of Dentistry, Saitama 350-0283, Japan*

KEYWORDS: Na,K-ATPase; PKA; A-kinase anchoring protein

INTRODUCTION

Na^+,K^+-ATPase was shown to be phosphorylated *in vitro* by protein kinase A (PKA), resulting in a decrease in its enzyme activity.[1] However, because the PKA has rather broad substrate specificity, various proteins can be phosphorylated *in vitro* by the kinase regardless of the physiological significance. Therefore, for transducing physiological signals, the cAMP/PKA-dependent phosphorylation system must have some mechanism for the preferential phosphorylation of its specific target substrate *in vivo*. Scott and Carr[2] demonstrated the role of A-kinase anchoring protein (AKAP), a specific protein that anchors the PKA RII regulatory subunit (and thereby the catalytic subunit bound to RII) on the membrane near its specific target proteins. In a previous study, we demonstrated that Na^+,K^+-ATPase in basolateral membranes prepared from acinar cells of rat parotid gland (PG) was much more efficiently phosphorylated by endogenous membrane-bound PKA.[3] In this study, to elucidate the functional relevance of the interaction between saliva secretion and AKAP/PKA-dependent regulation of Na^+,K^+-ATPase, we examined the expressions of AKAP subtypes and membrane-bound RII regulatory subunit, in the three major salivary glands, that is, submandibular gland (SMG), sublingual gland (SLG), and PG, of rats.

Address for correspondence: K. Kurihara, Department of Oral Physiology, Meikai University, School of Dentistry, Saitama 350-0283, Japan. Voice: +81-49-279-2771; fax: +81-49-287-4712.
kkinji@dent.meikai.ac.jp

Ann. N.Y. Acad. Sci. 986: 636–638 (2003). © 2003 New York Academy of Sciences.

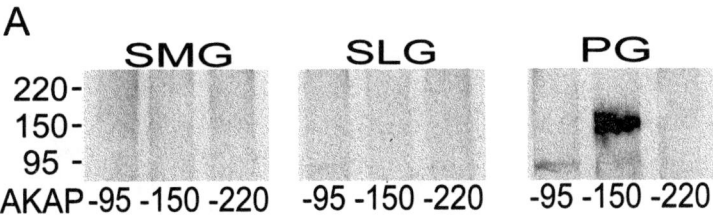

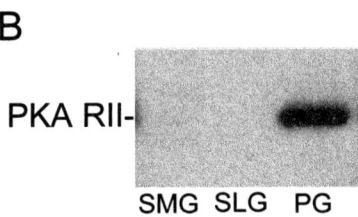

FIGURE 1. (**A**) Western blot analysis of AKAPs and (**B**) membrane-bound RII regulatory subunit of PKA in SMG, SLG, and PG of rat. (A) Proteins in 10 µL of 10% (w/v) homogenates of the salivary glands were separated by SDS-PAGE, transferred to a filter, and immunostained with anti-AKAP-95, -150, or -220 antibodies. (B) Membrane preparations from SMG, SLG, and PG (25, 25, and 5 µg, respectively) were subjected to immunoblot analysis with antibodies against the RII regulatory subunit (55 kDa).

RESULTS AND DISCUSSION

Expressions of AKAPs and Membrane-Bound RII Regulatory Subunit of PKA

We examined mRNA levels for three AKAP subtypes, AKAP-95, AKAP-150, and AKAP-220, in the PG, SMG, and SLG by the RT-PCR method and Western blot analysis. In the PG, both AKAP-150 protein (FIG. 1A) and its mRNA (not shown) were clearly detected, but the other two subtypes were less abundant. On the other hand, these three AKAP subtypes were not detected in SMG and SLG (FIG. 1A). The membrane-bound form of the RII regulatory subunit, an index for the amount of total AKAP protein of any subtypes and for the amount of membrane-anchored PKA holoenzyme, also was undetectable in membranes from SMG and SLG but was in PG (FIG. 1B). Thus, AKAP-150 was expressed specifically in PG. Furthermore, AKAP-150 in PG acinar cells was coimmunoprecipitated with RII regulatory subunit by an anti-RII antiserum (not shown), suggesting that AKAP-150 is, indeed, functional in anchoring RII molecules and therefore the PKA holoenzyme *in vivo*.

Phosphorylation of Na^+,K^+-ATPase in Membranes from PG, SMG, and SLG

Na^+,K^+-ATPase in the PG was phosphorylated by membrane-anchored endogenous PKA triggered by the addition of cAMP, but those in the SMG and SLG mem-

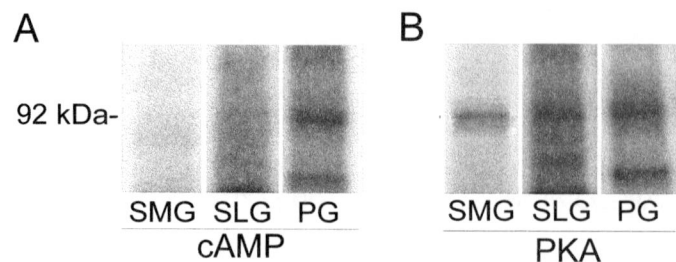

FIGURE 2. Phosphorylation of Na$^+$,K$^+$-ATPase α subunit[3] (92 kDa) in the membranes from SMG, SLG, and PG. (**A**) Phosphorylation by membrane-anchored endogenous PKA. Membranes prepared from SMG, SLG, and PG (40, 40, and 8 µg, respectively) were incubated for 30 s with [γ-^{32}P]ATP (40 µM) in the presence of cAMP (10 µM). (**B**) Phosphorylation by PKA catalytic subunit exogenously added. The membrane preparations were incubated with [γ-^{32}P]ATP for 30 s in the presence of 16 U of PKA catalytic subunit.

branes were not (FIG. 2A). This was caused by neither the lack of substrate protein nor lack of susceptibility of the ATPase in SMG and SLG to PKA. Membrane preparations from any of these glands contained a substantial amount of Na$^+$,K$^+$-ATPase that can be phosphorylated by the exogenously added PKA catalytic subunit (FIG. 2B). Treatment of the PG membranes with HT-31 peptide, an inhibitor of AKAP binding to RII regulatory subunit, decreased the level of cAMP-dependent Na$^+$,K$^+$-ATPase phosphorylation (not shown), further supporting the involvement of AKAP in the regulation of Na$^+$,K$^+$-ATPase by the endogenous PKA in PG.

That the regulation of Na$^+$,K$^+$-ATPase by AKAP-anchored PKA is a characteristic rather specific for PG among the three major salivary glands might be because of the difference in mechanisms regulating saliva production in those glands. There are two types of acinar cells, serous cells and mucous cells, which produce watery and viscous saliva, respectively. AKAP-150/PKA–dependent regulation of Na$^+$,K$^+$-ATPase seemed to be specific for PG acinar cells (serous cells) producing watery saliva. Another property of the PG for saliva secretion is the regulation by neuronal stimuli; that is, the neuronal stimulation caused by food intake in the mouth causes profuse secretion of watery saliva rather than viscous saliva, as observed in Pavlov's experiments. The regulation of Na$^+$,K$^+$-ATPase by AKAP-anchored PKA might reflect a regulatory mechanism for watery saliva secretion in the PG in response to a quick and short-term elevation of intracellular cAMP.

REFERENCES

1. BERTORELLO, A.M. *et al.* 1991. Phosphorylation of the catalytic subunit of Na$^+$,K$^+$-ATPase inhibits the activity of the enzyme. Proc. Natl. Acad. Sci. USA **88:** 11359–11362.
2. SCOTT, J.D. & D.W. CARR. 1992. Subcellular localization of the type II cAMP-dependent protein kinase. News Physiol. Sci. **7:** 143–148.
3. KURIHARA, K. *et al.* 2000. Regulation of Na$^+$,K$^+$-ATPase by cAMP-dependent protein kinase anchored on membrane *via* its anchoring protein. Am. J. Physiol. **279:** C1516–C1527.

Porcine Kidney Extract Contains Factor(s) That Inhibit the Ouabain-Sensitive Isoform of Na,K-ATPase (α2) in Rat Skeletal Muscle

A Convenient Electrophysiological Assay

I. KRIVOI,[a] A. VASILIEV,[a] V. KRAVTSOVA,[a] M. DOBRETSOV,[b] AND F. MANDEL[a,c]

[a]*St. Petersburg State University, St. Petersburg, Russia 199034*

[b]*University of Arkansas for Medical Sciences, Little Rock, Arkansas 72205, USA*

[c]*Baylor College of Medicine, Houston, Texas 77030, USA*

KEYWORDS: Na,K-ATPase isoforms; endogenous digitalis-like factors

A search for a convenient physiological assay for specific inhibitors of the Na,K-ATPase, including endogenous digitalis-like factors (EDLFs), led us to study the effects of such factors on the resting membrane potential (V_m) of rat diaphragm muscle fibers. The value of V_m was recorded using microelectrode techniques in an *in vitro* preparation perfused with a standard oxygenated physiological solution at 28°C. Acetylcholine (ACh) did not change Rb uptake in rat erythrocytes that only express the α_1 isoform of Na,K-ATPase, but hyperpolarized muscle fibers by 4.5 ± 0.8 mV at 100 nM concentration with a $K_{0.5}$ of 36 ± 6 nM. Ouabain preincubation for 1 hour depolarized muscle fibers in a dose-dependent manner that is best fitted with a two-site model of ouabain binding (FIG. 1). This ouabain effect was perceived both in control (no ACh) and after the addition of 100 nM ACh. The range of apparent affinities predicted by the model for the two sites ($K_{0.5}$ = 8–63 nM and $K_{0.5}$ = 13–18 μM) correspond very well to the ouabain affinities of the α_2 and α_1 isoforms of rat Na,K-ATPase, respectively.[1] The predicted contribution to V_m by the α_1 isoform was approximately 15 mV independent of the presence of ACh. In contrast, the α_2 isoform contribution to V_m increased from 4.5 ± 1.2 mV in control to 9.0 ± 1.2 mV in the presence of ACh. Note a slight hyperpolarization at 10–20 nM ouabain (FIG. 1A), presumably because of activation of the α_2 isoform.[2] These data points, marked by arrows, were excluded during the fitting procedure.

Address for correspondence: Igor I. Krivoi, Ph.D., Department of General Physiology, St. Petersburg State University, 7/9 University emb., 199034, St. Petersburg, Russia. Voice: +812-3233842; fax: +812-3232454.

IgorKrivoi@IK4251.spb.edu

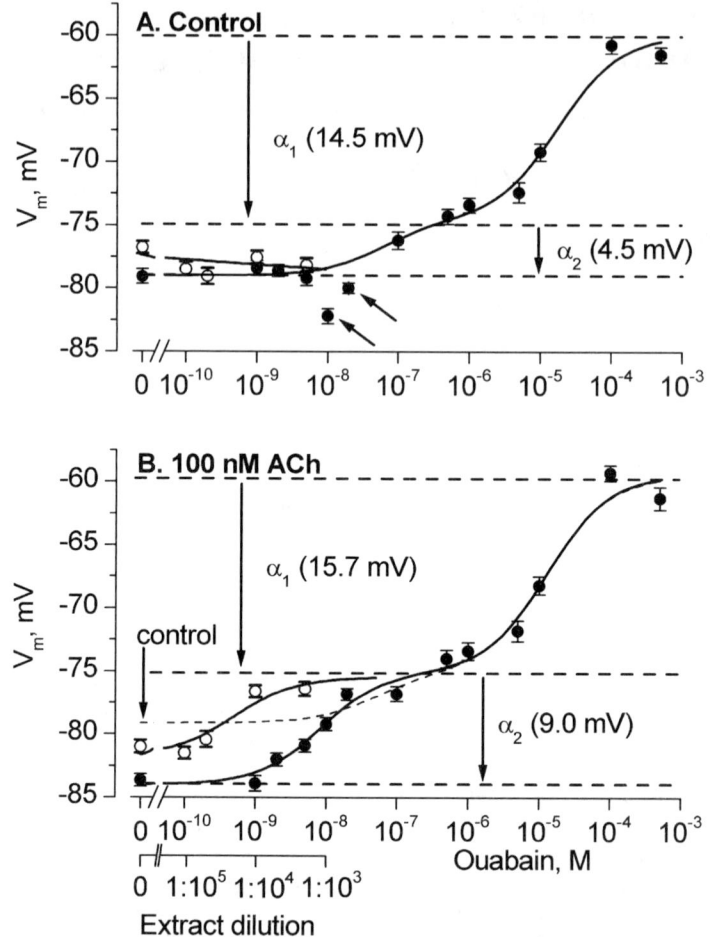

FIGURE 1. V_m as a function of ouabain concentration (*closed circles*) or extract dilution (*open circles*) in the (**A**) absence and (**B**) presence of 100 nM ACh. Each point represents $V_m \pm$ SE measured in more than 100 fibers from four to five muscles.

Preincubation (3 h) in physiological solution with porcine kidney extract containing EDLFs[3] (10 kDa cutoff filtration) slightly hyperpolarized the muscle fibers. Similar to ouabain, in the presence of ACh the extract inhibited the α_2 isoform contribution to V_m in a dose-dependent manner with a $K_{0.5}$ between 10,000- and 100,000-fold dilutions (FIG. 1). The inhibitory activity of extract slowly appeared over a period of 3 h and was still present after 60 min of washing, indicating both a slow on-rate and off-rate.

We suggest (1) that in rat skeletal muscle the α_1 isoform of Na,K-ATPase serves to maintain basal ionic homeostasis, whereas the α_2 isoform provides a "regulatable" component of ion transport and may be under cholinergic and EDLFs control;

and (2) porcine kidney extract contains factor(s) (<10 kDa) that inhibit the rat α_2 isoform. Our experimental approach can be used as a convenient, highly sensitive physiological assay for EDLFs, in particular those EDLFs that selectively inhibit the α_2 isoform of Na,K-ATPase.

ACKNOWLEDGMENTS

This work was supported by RFBR (01-04-49799, 02-04-06957MAC), and by NSF (9904815).

REFERENCES

1. SWEADNER, K.J. 1989. Isozymes of the Na^+/K^+-ATPase. Biochim. Biophys. Acta **988**: 185–220.
2. GAO, J. et al. 2002. Isoform-specific stimulation of cardiac Na/K pumps by nanomolar concentrations of glycosides. J. Gen. Physiol. **119**: 297–312.
3. MANDEL, F. et al. 2002. Using the Na,K-ATPase itself for the large scale isolation and purification of endogenous digitalis-like factors. 10th Int. Conf. on Na,K-ATPase and Related Cation Pumps. Elsinore. Denmark.

Rat Resistance Vessels Preferentially Contain the Ouabain-Insensitive α1 Isoform of Na,K-ATPase

OTTO HANSEN

Department of Physiology, Aarhus University, DK-8000 Århus C, Denmark

KEYWORDS: α isoforms; ouabain-insensitive; resistance vessels

The α isoform of Na,K-ATPase is crucial for the affinity of Na^+. By setting the intracellular Na^+ level, and thus indirectly Ca^{2+}, the distribution of α isoforms in resistance vessels can play an important role in controlling peripheral resistance. Because the rat is the preferred experimental model in vascular research, and because the rat $α_1$ isoform is almost insensitive to cardiac glycosides, for example, also to putative endogenous ouabain, more knowledge of the concentration of individual α isoforms present in rat resistance vessels seems pertinent. A novel immunochemical method[1,2] was used for quantitative isoform determination in Western blots of rat vessel homogenates.

MATERIALS AND METHODS

Segments of mesenteric resistance vessels (second to fourth generation) were taken from 12–14-week male Wistar rats, and pooled biopsies from 20–30 rats were homogenized. Homogenates of vessels were characterized for protein and dry matter. Resistance vessel (RV) homogenate and purified Na,K-ATPase from *rat* kidney ($α_1$ isoform only) or from *rat* brain (all isoforms) were applied to parallel lanes on SDS-gels and blotted to PVDF membranes. Western blots were incubated with isoform-specific antibodies MA3B and MA3915 for $α_1$ and $α_3$, respectively, or with polyclonal antibodies raised to isoform-specific segments of individual rat α isoforms.[1,2] Blots then were incubated in ^{125}I-labeled secondary antibodies, and the radioactive signal from adjacent α spots was compared in an InstantImager (Packard Instrument Company, Meriden, CT, USA). Because enzyme units per milligram protein of purified Na,K-ATPase references are known from ^{32}P-phosphorylation (all isoforms)

Address for correspondence: Otto Hansen, Department of Physiology, Aarhus University, Ole Worms Allé 160, DK-8000 Århus C, Denmark. Voice: +45-8942-2806; fax: +45-8612-9065.
oh@fi.au.dk

and from (^{3}H)ouabain binding ($\alpha_2 + \alpha_3$), enzyme units in RV homogenates can be calculated.

RESULTS AND CONCLUSION

The dominating α isoform of Na,K-ATPase in rat RV appeared to be the ouabain-insensitive α_1 isoform with medium Na$^+$ affinity. The density of α_1 in RV expressed per gram wet weight was 80–125 pmol · g^{-1}. If present at all, very few α_2 and α_3 isoforms were found in RV homogenate, each representing a maximum of 10%, and more likely approximately 5%, of the α_1 concentration. In contrast, rat skeletal muscle preferentially contains α_2 and some α_1. Compared with vessel homogenate, the concentration of α_1 is lower, but the total concentration of pumps is two to four times higher.[2] Because myocytes represent approximately 50% of small vessel segments, the density of α_1-containing Na,K-pumps expressed per volume myocytes is doubled, provided pumps are practically absent from nonmyocyte structures of adventitia and media. The abundance of α_1 and the very modest quantity of α_2 and α_3 in rat RV seem less compatible with a role of putative endogenous ouabain[3] for tension development, for example, in hypertension, and with a role in the postulated signal transduction after occupation of a limited fraction of the Na,K-ATPase in this species.

ACKNOWLEDGMENT

This study was supported by the Danish Biomembrane Research Centre and Aarhus University Research Foundation (Grant F-1999-SUN-1-69).

REFERENCES

1. HANSEN, O. 1998. Isoform of Na$^+$,K$^+$-ATPase from rumen epithelium identified and quantified by immunochemical methods. Acta Physiol. Scand. **163:** 201–208.
2. HANSEN, O. 2001. The α_1 isoform of Na$^+$,K$^+$-ATPase in rat soleus and extensor digitorum longus. Acta Physiol. Scand. **173:** 335–341.
3. HANSEN, O. 1994. Do putative endogenous digitalis-like factors have a physiological role? Hypertension **24:** 640.

Na-Pump Kinetic Properties Are Differently Altered in the Brain Regions of the Cholecystokinin$_2$ Receptor–Deficient Mice

KRISTIINA ROOTS,[a] SULEV KÕKS,[b] CZESLAVA KAIRANE,[a] TIIT SALUM,[a] ELLO KARELSON,[a] EERO VASAR,[b] AND MIHKEL ZILMER[a]

[a]*Department of Biochemistry, University of Tartu, 50411 Tartu, Estonia*

[b]*Department of Physiology, University of Tartu, 50411 Tartu, Estonia*

KEYWORDS: Na pump; CCK$_2$ receptor–deficient mice; sodium cooperativity

The Na,K-ATPase (Na pump) is a membrane protein complex that maintains the internal high K^+ and low Na^+ by using the energy of ATP.[1] Because the Na pump is important in many basic and specialized cellular functions, its regulation has to be highly precise and flexible. The degree of cooperation for Na^+ and K^+ is known to regulate functional properties of the pump.[2] In the nervous system, the Na pump is a target for neuropeptides and classic neurotransmitters, particularly catecholamines.[3] However, the modulation of the pump by cholecystokinin (CCK), the most abundant neuropeptide in the brain, is not completely understood. Besides, possible interactions between the Na pump and CCK$_2$ receptors (CCK2R) in the brain from CCK2R-deficient animals have never been demonstrated. On the basis of these facts, we determined the activity of the Na,K-ATPase and its Na^+ and K^+ cooperativity in the four brain regions from the wild-type, heterozygous (+/–), and homozygous (–/–) CCK2R-deficient mice.

CCK2R-deficient mice were kindly provided by T. Matsui from the Kobe University School of Medicine (Japan). Breeding and genotype analysis were performed in the Department of Physiology, University of Tartu. The Na,K-ATPase membrane preparations were isolated from the cerebral cortex, subcortex, brain stem, and cerebellum. The enzyme-specific activity was measured through the release of inorganic phosphate. The Hill coefficients (n_H) were determined as described by us earlier.[4] The protein content in the enzyme preparations was estimated by Lowry *et al.*[5]

CCK2R gene mutation induced substantial stimulation of the Na,K-ATPase-specific activity in the cerebral cortex (25%), brain stem (12%), and cerebellum (12%) of heterozygous mice compared with wild-type animals. In homozygous mice, only the brain stem exhibited significant increase (22%) in the enzyme activ-

Address for correspondence: Kristiina Roots, Department of Biochemistry, University of Tartu, Ravila 19, 50411 Tartu, Estonia. Voice: +372-7-374313; fax: +372-7-374312.
 kristiinar77@yahoo.com

TABLE 1. Hill coefficients (n_H) for Na^+ and K^+ of the Na,K-ATPase in the membranes from the different brain regions of wild-type and CCK_2 receptor-deficient mice

Brian Region	n_H for Na^+			n_H for K^+		
	Wild-type	Heterozygous	Homozygous	Wild-type	Heterozygous	Homozygous
Cerebral cortex	1.20±0.13	1.30±0.05	1.00±0.09			
Subcortex	1.60±0.06	0.90±0.07*	1.10±0.06	1.60±0.05	1.50±0.05	1.70±0.05
Cerebellum	1.20±0.06	1.20±0.05	1.10±0.09*	1.30±0.05	1.30±0.05	1.20±0.15
Brain stem	1.50±0.15	1.00±0.10*	1.10±0.03*	1.20±0.06	1.30±0.12	1.20±0.15

*$P < 0.05$ versus wild-type. Values are means ± SEM of 2–4 experiments.

ity. These data are suggestive of Na-pump adaptive response to the deficiency of CCK2R and of the related CCK action within the mice brain membranes. Interestingly, the heterozygous mice brain revealed a wider spectrum of the regions with the pump response than the brain in homozygous animals. The latter displayed the pump upregulation only in the brain stem, an area with the vitally important neurophysiological functions.

Next, we found that CCK2R deficiency caused considerable alterations in the allosteric regulation of brain Na,K-ATPase. The cooperativity of the enzyme for Na^+ was reduced, whereas the cooperativity for K^+ remained remarkably unchanged (see TABLE 1). In all of the brain regions of homozygous mice studied, the n_H for Na^+ showed a value of approximately 1.0, suggesting a loss in the enzyme cooperativity for Na^+. In the heterozygous mice brain, the considerable regional differences were detected in the values of n_H for Na^+; the cerebral cortex showed the highest and the subcortex showed the lowest n_H value. Comparison of heterozygous and wild-type mice revealed a significant decrease in the n_H for Na^+ in the subcortex and brain stem of heterozygous animals.

Altogether, these data suggest that lack of CCK2R, and related action of CCK might induce remarkable alterations in the regulation of the neuronal membrane transport systems. Abnormality in the Na,K-ATPase allosteric regulation by sodium might be compensated by substantial elevation in the specific activity of the enzyme.

REFERENCES

1. SKOU, J.C. & M. ESMANN. 1992. The Na,K-ATPase. J. Bioenerg. Biomembr. **24:** 249–261.
2. BOLDYREV, A.A. 1988. Physiological significance of oligomeric associations of ionic pumps in biomembranes. Acta Physiol. Phramacol. Bulg. **14:** 3–9.
3. THERIEN, A.G. & R. BLOSTEIN. 2000. Mechanism of sodium pump regulation. Am. J. Physiol. Cell Physiol. **279:** C541–C566.
4. SALUM, T. et al. 1988. Cooperative interactions of Na,K-ATPase with Na and K and its oligomeric structure. Ukr. Biokhim. Zh. (Russ.) **60:** 47–52.
5. LOWRY, O.H. et al. 1951. Protein measurement with the Folin phenol reagent. J. Biol. Chem. **193:** 265–275.

Electrogenic Na^+/HCO_3^- Cotransporter rkNBC1 Immunolocalized in Rat Eye

HENRIK VORUM,[a] CHRISTIAN AALKJÆR,[b,c] HENRIK HAGER,[b,d]
SØREN NIELSEN,[b,d] AND ARVID B. MAUNSBACH[b,d]

[a]*Department of Ophtalmology, Aarhus DK-8000, Denmark*
[b]*Water and Salt Research Center, Aarhus DK-8000, Denmark*
[c]*Institutes of Physiology, Aarhus DK-8000, Denmark*
[d]*Institute of Anatomy, University of Aarhus, Aarhus DK-8000, Denmark*

KEYWORDS: sodium/bicarbonate cotransporter; rkNBC1; immunolocalization; ciliary body

Electrogenic sodium bicarbonate cotransport plays a central role in the control of intracellular pH and transepithelial transport in ocular tissues. The proteins responsible for mediating electrogenic sodium bicarbonate cotransport in various cell types of the eye has been unknown until recently. The *NBC1* gene[1] encodes a protein that seems to be a candidate for mediating electrogenic sodium bicarbonate cotransport in the eye. To gain further insight into the molecular mechanisms responsible for electrogenic sodium and bicarbonate transport in the eye, we determined the cellular localization of the electrogenic sodium bicarbonate cotransporter rkNBC1 in the rat eye at the microscopic level. For this purpose, a previously made antibody[2] raised against the C-terminus of rkNBC1 was used.

MATERIALS AND METHODS

For immunofluorescence experiments, rat eyes were removed and immediately frozen in dry ice. The rkNBC1 antibody (1:100 dilution) was applied to 5-µm cryostat sections for 45 minutes after several washes in PBS. FITC-conjugated goat anti-rabbit antibody (1:1000 dilution; Alexa 488, Molecular Probes Europe, Leiden, the Netherlands), then was applied for 40 minutes. In control experiments, the primary antibody was preabsorbed with the specific immunizing peptide (10 µg/mL). Microscopy was conducted using a Leica DMRE microscope (Leica, Germany) as previously described.[3]

Address for correspondence: Henrik Vorum, Department of Ophtalmology, Aarhus University Hospital, Nørrebroggade 44, DK-8000 Aarhus, Denmark. Voice: +45-8942-2859; fax: +45-8613-1160.
hv@biokemi.au.dk

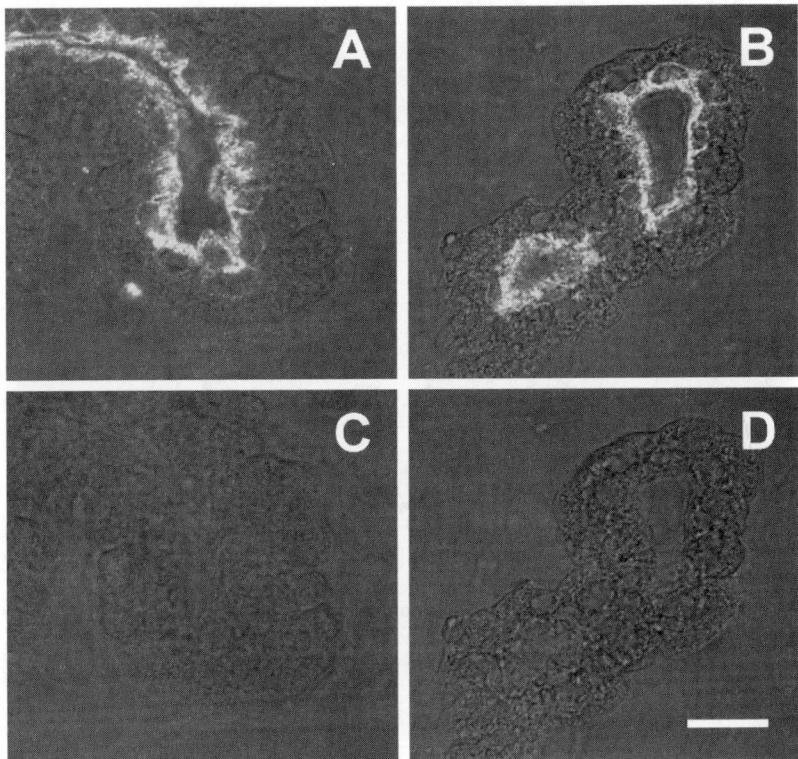

FIGURE 1. Immunolocalization of rkNBC1 in the rat ciliary body. (**A**) Longitudinal section of a ciliary body process. (**B**) Cross section of a ciliary body process. (**C**) Specific peptide blocking of rkNBC1 staining in a longitudinal section of a ciliary process. (**D**) Specific peptide blocking of rkNBC1 staining in a cross section of a ciliary process. *Bar* = 35 μm.

RESULTS

The immunohistochemical analyses of the cellular localization of rkNBC1 in rat eye showed distinct anti-rkNBC1 immunofluorescence labeling in the ciliary body only, whereas no immunoreaction was detectable in the cornea, lens, retina, optic nerve, or sclera (not shown). FIGURE 1A and 1B show the ciliary epithelial cell layer bordering the stroma, the pigmented epithelium, intensely stained for rkNBC1. In contrast with the strong labeling of the pigmented epithelial cells, the anti-rkNBC1 antibody did not label the nonpigmented epithelial cell layer. Furthermore, capillaries and postcapillary venules and the ciliary stroma were unlabeled. Control experiments (FIG. 1C and 1D) demonstrated specific peptide blocking of rkNBC1 staining.

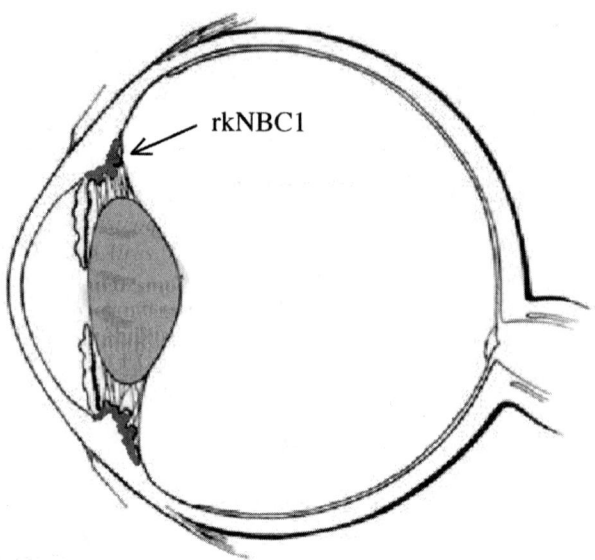

FIGURE 2. Localization of rkNBC1 in the eye.

DISCUSSION

Recent evidence that patients with homozygous mutations in the *NBC1* gene[1] have an ocular phenotype characterized by blindness, glaucoma, cataracts, and band keratopathy[4] underline the importance of electrogenic sodium bicarbonate cotransport for normal ocular function. Characterization of the expression pattern of the rkNBC1 protein in the eye, as described in this study (FIG. 2), may contribute to the understanding of the molecular pathogenesis that causes blindness in this disorder.

REFERENCES

1. ABULADZE, N., M. SONG, A. PUSHKIN, *et al.* 2000. Structural organization of the human NBC1 gene: kNBC1 is transcribed from an alternative promoter in intron 3. Gene **251:** 109–122.
2. MAUNSBACH, A.B., H. VORUM, T.H. KWON, *et al.* 2000. Immunoelectron microscopic localization of the electrogenic Na/HCO$_3$ cotransporter in rat and ambystoma kidney. J. Am. Soc. Nephrol. **11:** 2179–2189.
3. VORUM, H., H. HAGER, B.M. CHRISTENSEN, *et al.* 1999. Human calumenin localizes to the secretory pathway and is secreted to the medium. Exp. Cell Res. **248:** 473–481.
4. IGARASHI, T., J. INATOMI, T. SEKINe, *et al.* 1999. Mutations in SLC4A4 cause permanent isolated proximal renal tubular acidosis with ocular abnormalities. Nat. Genet. **23:** 264–266.

Na,K-ATPase in the Regulation of Epithelial Cell Structure

AYYAPPAN K. RAJASEKARAN, JEGAN GOPAL, AND SIGRID A. RAJASEKARAN

Department of Pathology and Laboratory Medicine and Jonsson Comprehensive Cancer Center, David Geffen School of Medicine, University of California, Los Angeles, California 90095, USA

KEYWORDS: Na,K-ATPase; intracellular sodium homeostasis; tight junctions; epithelial polarity; epithelial-mesenchymal transition; RhoA GTPase

The function of epithelial cells is intimately associated with their cellular structure. Their plasma membrane is divided into functionally and biochemically distinct apical and basolateral plasma membranes separated by tight junctions (polarized phenotype). Na,K-ATPase consisting of α and β subunits is a crucial enzyme that regulates vectorial transport in epithelial cells and that depends largely on the trans-epithelial flow of sodium ions (Na^+) from the apical to the basolateral membrane. Vectorial transport is closely associated with Na,K-ATPase's key function in maintaining low intracellular Na^+ levels.

Earlier studies indicated that the cell-cell adhesion function of E-cadherin is involved in regulating tight junctions and epithelial cell polarity. We now demonstrate that the intracellular sodium homeostasis maintained by Na,K-ATPase plays an important role in the development and maintenance of tight junctions and polarity in epithelial cells. Our results suggest that a functional synergism between E-cadherin and Na,K-ATPase regulates the structure of epithelial cells, and aberration of either E-cadherin or Na,K-ATPase function might be involved in the loss of polarity and epithelial-mesenchymal transition observed in carcinoma.

RESULTS AND CONCLUSIONS

In tumor tissue lysates of patients diagnosed with invasive clear-cell renal cell carcinoma[1] and in Moloney sarcoma virus–transformed Madin-Darby canine kidney (MSV-MDCK) cells,[2] Na,K-ATPase β_1-subunit protein levels were highly reduced, indicating that Na,K-ATPase β-subunit expression levels might be associated with the invasive phenotype of carcinoma cells. MSV-MDCK cells also expressed highly

---

Address for correspondence: Ayyappan K. Rajasekaran, Department of Pathology and Laboratory Medicine, Room 13-344 CHS, University of California, Los Angeles, CA 90095. Voice: 310-825-1199; fax: 310-267-2410.
arajasekaran@mednet.ucla.edu

reduced levels of E-cadherin, and their invasive phenotype had been correlated to reduced E-cadherin expression. Ectopic expression of either E-cadherin or β_1-subunit did not alter the phenotype of MSV-MDCK cells. However, expression of both E-cadherin and β-subunit cells induced a polarized phenotype that was accompanied by significantly reduced intracellular sodium levels.[2] Inhibition of Na,K-ATPase function in epithelial MDCK cells prevented the formation of tight junctions and epithelial polarity in MDCK cells.[3] Reduced activity of RhoA GTPase in Na,K-ATPase-inhibited cells suggested that RhoA is a potential downstream effector of Na,K-ATPase function and is involved in the establishment of tight junctions and induction of polarity in MDCK cells.

Our studies suggest that the low intracellular sodium level maintained by Na,K-ATPase plays an important role in maintaining the polarized phenotype of epithelial cells. Aberrant expression of Na,K-ATPase subunit levels and/or altered Na,K-ATPase activity might be associated with events leading to epithelial-mesenchymal transition observed in carcinoma (FIG. 1).

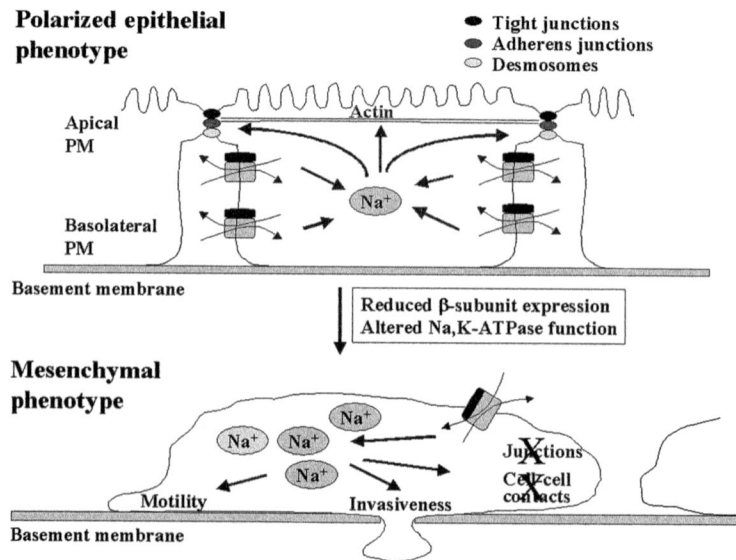

FIGURE 1. A schematic model representing the role of Na,K-ATPase in epithelial to mesenchymal transition. According to this model, normal Na,K-ATPase function maintains low intracellular sodium levels that are necessary to maintain junctional complexes and the polarized phenotype of epithelial cells. Reduced β-subunit expression and/or altered Na,K-ATPase function might lead to increased intracellular sodium levels, resulting in the alteration of actin dynamics and disruption of junctional complexes and leading to a mesenchymal phenotype.

ACKNOWLEDGMENTS

This work was supported by National Institutes of Health Grant No. DK56216, Grant-in-Aid Award No. 1162-G11 from the American Heart Association (Western States Affiliate), and (in part) Department of Defense Grant Nos. PC991140 and PC970546. S. A. Rajasekaran was supported by USHHS Institutional National Research Service Award No. T32CA09056.

REFERENCES

1. RAJASEKARAN, S.A., W.J. BALL, JR., N.H. BANDER, et al. 1999. Reduced expression of β-subunit of Na,K-ATPase in human clear-cell renal cell carcinoma. J. Urol. **162:** 574–580.
2. RAJASEKARAN, S.A., L.G. PALMER, K. QUAN, et al. 2001. Na,K-ATPase β-subunit is required for epithelial polarization, suppression of invasion, and cell motility. Mol. Biol. Cell **12:** 279–295.
3. RAJASEKARAN, S.A., L.G. PALMER, S.Y. MOON, et al. 2001. Na,K-ATPase activity is required for formation of tight junctions, desmosomes, and induction of polarity in epithelial cells. Mol. Biol. Cell **12:** 3717–3732.

Expression of Na,K-ATPase β-Subunit in Transformed MDCK Cells Increases the Translation of the Na,K-ATPase α-Subunit

SIGRID A. RAJASEKARAN, JEGAN GOPAL, AND AYYAPPAN K. RAJASEKARAN

Department of Pathology and Laboratory Medicine and Jonsson Comprehensive Cancer Center, David Geffen School of Medicine, University of California, Los Angeles, California 90095, USA

KEYWORDS: Na,K-ATPase α-subunit; Na,K-ATPase β-subunit; translation on the endoplasmic reticulum

Na,K-ATPase is an oligomeric transmembrane protein consisting of a noncovalently linked α- and β-subunits. The α-subunit (~112 kDa) contains the catalytic and ligand-binding sites. The precise function of the β-subunit is not known, although it is required for the normal activity of the enzyme. Several lines of evidence indicate that the α-subunit and the β-subunit of Na,K-ATPase cotranslationally associate in the endoplasmic reticulum (ER) and are transported to the cell surface as a heterodimer. Geering's group has shown that the β-subunit is necessary for the stability of the newly synthesized α-subunit and may shield a degradation signal in the M7/M8 loop of the α-subunit to protect the α-subunit from ER degradation.[1] These studies collectively suggest that the β-subunit plays a role in the synthesis, stability, and transport of the α-subunit of Na,K-ATPase. In the present study, we provide evidence that the β-subunit of Na,K-ATPase can increase the translation of the α-subunit in the ER.

RESULTS AND CONCLUSIONS

Moloney sarcoma virus–transformed Madin-Darby canine kidney (MSV-MDCK) cells expressed reduced protein levels of β_1-subunit of Na,K-ATPase.[2] Ectopic expression of β_1-subunit of Na,K-ATPase in MSV-MDCK cells increased the levels of the α_1-subunit. We found that this increase in α-subunit level was not caused by an increase in the transcription of the α-subunit gene or by an increased stability of the α-subunit protein. Short radioactive pulse experiments demonstrated a three- to

---

Address for correspondence: Ayyappan K. Rajasekaran, Department of Pathology and Laboratory Medicine, Room 13-344 CHS, University of California, Los Angeles, CA 90095. Voice: 310-825-1199; fax: 310-267-2410.
arajasekaran@mednet.ucla.edu

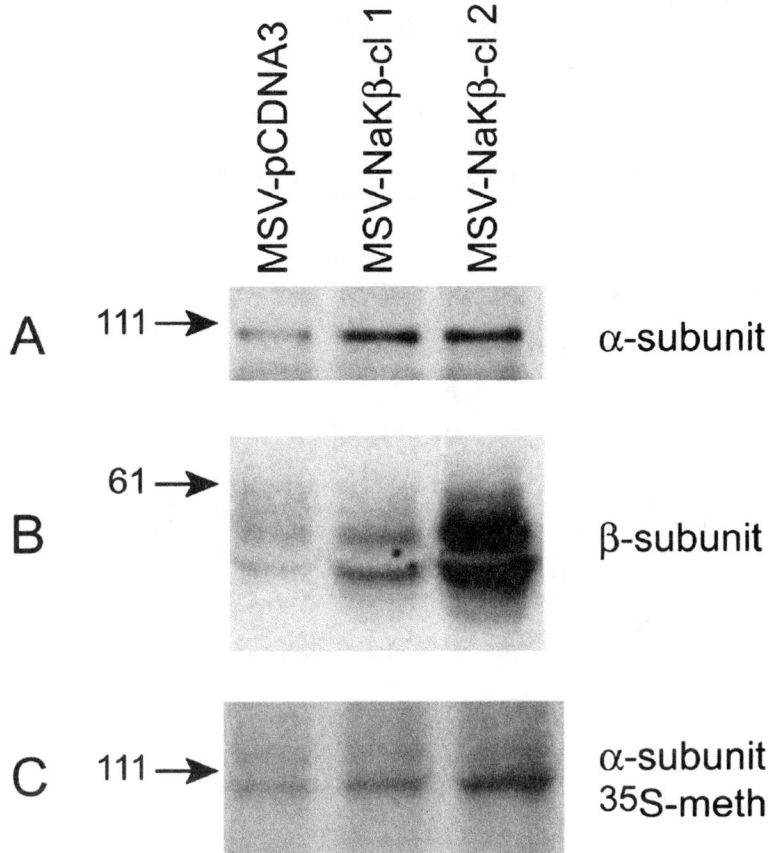

FIGURE 1. Pulse labeling of Na,K-ATPase β_1-subunit expressing MSV-MDCK cells. Canine Na,K-ATPase β_1-subunit cDNA was transfected into MSV-MDCK cells, and stable clones expressing β-subunit were selected as described.[1] (**A**) Immunoblot analysis of α_1-subunit. Note the increased α-subunit protein levels in MSV-NaK β-cells. (**B**) Immunoblot analysis of β_1-subunit. (**C**) ^{35}S-Methionine incorporation into α_1-subunit. Cells were pulsed for 20 minutes with 0.5 mCi/mL of ^{35}S-methionine. Na,K-ATPase α-subunit was immunoprecipitated, the precipitates were separated by SDS-PAGE, and α-subunit was detected by autoradiography. Note the increased ^{35}S-methionine incorporation into Na,K-ATPase α-subunit in MSV-NaK β-cells.

sixfold increase in the incorporation of ^{35}S-methionine into α-subunit in β-subunit-overexpressing cells compared with vector-transfected MSV-MDCK cells (FIG. 1). This increase in translation was specific to the α-subunit because the translation of two unrelated membrane and cytosolic proteins was not affected by β-subunit expression, consistent with the hypothesis that the β-subunit of Na,K-ATPase has a role in increasing the translation of the α-subunit in the ER. Currently, experiments are in progress to understand the mechanism by which the β-subunit increases the translation of the α-subunit in mammalian cells.

ACKNOWLEDGMENTS

This work was supported by National Institutes of Health Grant No. DK56216 and in part by Department of Defense Grant Nos. PC991140 and PC970546. S. A. Rajasekaran was supported by USHHS Institutional National Research Service Award No. T32CA09056. We thank William James Ball, Jr. (University of Cincinnati Medical Center, Cincinnati, OH) for Na,K-ATPase antibodies and Robert Farley (University of Southern California School of Medicine, Los Angeles, CA) for canine Na,K-ATPase β-subunit cDNA.

REFERENCES

1. BEGUIN, P., U. HASLER, O. STAUB & K. GEERING. 2000. Endoplasmic reticulum quality control of oligomeric membrane proteins: topogenic determinants involved in the degradation of the unassembled Na,K-ATPase alpha subunit and in its stabilization by beta subunit assembly. Mol. Biol. Cell **11:** 1657–1672.
2. RAJASEKARAN, S.A., L.G. PALMER, K. QUAN, et al. 2001. Na,K-ATPase β-subunit is required for epithelial polarization, suppression of invasion, and cell motility. Mol. Biol. Cell **12:** 279–295.

Quality Control of Gastric Proton Pump in the Endoplasmic Reticulum by Ubiquitin/Proteasome System

SHINJI ASANO,[a] TOHRU KIMURA,[b] HOKARA ISHIZUKA,[b] MAGOTOSHI MORII,[b] AND NORIAKI TAKEGUCHI[b]

[a]*Life Science Research Center,* [b]*Faculty of Pharmaceutical Sciences, Toyama Medical and Pharmaceutical University, Toyama 930-0194, Japan*

KEYWORDS: gastric proton pump; quality control; proteasome; ubiquitin

The gastric proton pump actively transports proton and K^+ in opposite directions, coupled to hydrolysis of ATP. This pump consists of the catalytic α-subunit and the noncatalytic β-subunit. These two subunits are cotranslationally inserted into the endoplasmic reticulum (ER) membrane and assembled into the functional holoenzyme with a stoichiometry of 1 to 1. However, the genes encoding the α- and β-subunits are located at different chromosomal loci (chromosomes 19q13.11 and 13q34 for human α- and β-subunits, respectively). Therefore, strict control mechanisms seem to regulate the quantity (number) of these subunits in the ER. Here, we studied the quantity and quality control mechanism of the gastric proton pump in the ER (FIG. 1) by using stable expression in HEK-293 cells.[1,2]

Stable cell lines expressing the α-subunit alone and the β-subunit alone and coexpressing the α- and β-subunits were constructed.[1,2] The intracellular localization of the α- and β-subunits in each cell line was studied by immunofluorescence. In the α-expressing cells, the α-subunit was retained in the intracellular compartments, and its expression was low. In the β-expressing cells, the β-subunit was located at the cell surface and in the intracellular compartments. A major portion of the α- and β-subunits was expressed at the cell surface in the α+β-expressing cells. The expression level of the α-subunit was significantly higher than that in the α-expressing cells.

To study the stability of the α- and β-subunits, we performed pulse-chase labeling experiments. Each cell was labeled with [^{35}S]Met/Cys for 1 hour, followed by a cold chase for various periods, and then cell lysates were immunoprecipitated with the anti-α- or anti-β-subunit antibody, respectively.

Address for correspondence: Shinji Asano, Life Science Research Center, Toyama Medical and Pharmaceutical University, 2630 Sugitani, Toyama 930-0194, Japan. Voice: +81-76-434-7187; fax: +81-76-434-5176.
shinji@ms.toyama-mpu.ac.jp

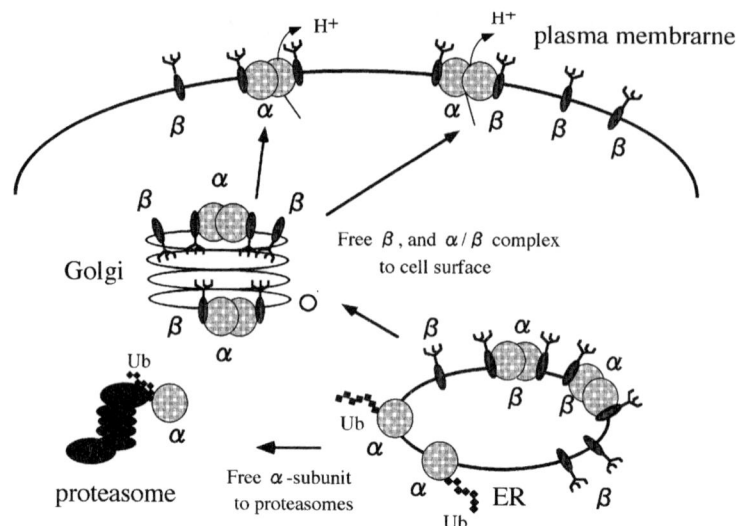

FIGURE 1. Schematic presentation of the quality control mechanism of the gastric proton pump. Unassembled α-subunits are retained in the ER, modified with polyubiquitin chains, and degraded by proteasomes located in the cytoplasm. Both assembled and unassembled β-subunits can leave the ER to travel to the cell surface.

The α-subunit was unstable in the absence of the β-subunit in the α-expressing cells. Degradation of the α-subunit was apparent within 1 hour and was almost complete within 6 hours. The α-subunit was stabilized by the β-subunit in the α+β-expressing cells. The α-subunit in the α-expressing cells was stabilized by a proteasome inhibitor, lactacystin. Lactacystin prevented the degradation of the α-subunit, suggesting that unassembled α-subunits retained in the ER were specifically degraded by proteasomes.

The β-subunit is core-glycosylated in the ER. The carbohydrate chains are further modified to a complex type in the process of intracellular transport from the ER to trans-Golgi and cell surface. This modification process of the carbohydrate chains of the β-subunit was unchanged in the presence or absence of lactacystin. Lactacystin did not affect the stability of the β-subunit in either the β-expressing or α+β-expressing cells. These results suggest that β-subunits were not specifically degraded by proteasomes.

Proteasomes recognize polyubiquitin chains covalently attached to target proteins, acting as a signal for their degradation. The α-subunit was polyubiquitinated in the α-expressing cells as well as in the α+β-expressing cells. The extent of the polyubiquitination was higher in the α-expressing cells and significantly enhanced in the presence of lactacystin. On the other hand, polyubiquitination of the β-subunit was not observed when the cells were cultured even in the presence of lactacystin.

These results indicate that unassembled α-subunits were specifically retained in the ER, modified with polyubiquitin chains, and degraded by proteasomes. The ubiquitin/proteasome system seems to regulate the number of α- and β-subunits in the ER to control the cell surface expression of functional α/β holoenzymes. Similar phenomena were reported for oligomeric membrane proteins such as Na^+,K^+-ATPase and epithelial Na^+ channel, ENaC.[3,4]

REFERENCES

1. KIMURA, T., Y. TABUCHI, N. TAKEGUCHI & S. ASANO. 2002. Mutational study on the roles of disulfide bonds in the β-subunit of gastric H^+,K^+-ATPase. J. Biol. Chem. **277**: 20671–20677.
2. KIMURA, T., A. YOSHIDA, Y. TABUCHI, et al. 2002. Stable expression of gastric proton pump activity at the cell surface. J. Biochem. (Tokyo) **131**: 923–932.
3. COPPI, M.V. & G. GUIDOTTI. 1997. Ubiquitination of Na,K-ATPase α1 and α2 subunits. FEBS Lett. **405**: 281–284.
4. STAUB, O., I. GAUTSCHI, T. ISHIKAWA, et al. 1997. Regulation of stability and function of the epithelial Na^+ channel (ENaC) by ubiquitination. EMBO J. **16**: 6325–6336.

Mg-ATPase from Microsomal Fraction of Rabbit Gastric Mucosa Is Ecto-ATP-Diphosphohydrolase

MARINA SMAGINA, NATALIYA DOLGOVA, ALEXANDER RUBTSOV, AND OLGA LOPINA

Department of Biochemistry, School of Biology, Moscow State University, 119899 Moscow, Russia

KEYWORDS: ecto-ATP-diphosphohydrolase; gastric mucosa; Mg-ATPase

Microsomal fraction from mammalian gastric mucosa possesses the H,K-ATPase that is inhibited by SCH-28080. A significant part of Mg-ATPase activity of this fraction is not sensitive to this inhibitor. To identify the origin of "basal" Mg-ATPase, we studied its properties in microsomal fraction from rabbit gastric mucosa (MFRGM).

METHODS

MFRGM was obtained as described.[1] ATPase activity was measured using an enzyme-coupled ATP release assay.[2] ADP and GDP hydrolysis was determined by measuring the release of P_i.

RESULTS

ATPase activity of MFRGM measured in the presence of Mg^{2+} and K^+ consists of 40.2 ± 17.7 μmol/mg of protein in 1 h. Effects of various inhibitors on the Mg-ATPase activity of this fraction are presented in FIGURE 1. Besides the H,K-ATPase, MFRGM possesses mitochondrial F_1F_0-ATPase activity: SCH-28080 and oligomycin together suppress approximately 45% of the total Mg-ATPase activity of MFRGM. There is no Ca- or Na,K-ATPase in this fraction. Concanavalin A, the specific activator of ecto-ATPase from transverse tubule membrane, does not affect the basal Mg-ATPase.

Address for correspondence: Olga Lopina, Department of Biochemistry, School of Biology, Moscow State University, 119899 Moscow, Russia. Voice: +7-095-9394434; fax: +7-095-9393955.
 od_lopina@yahoo.com

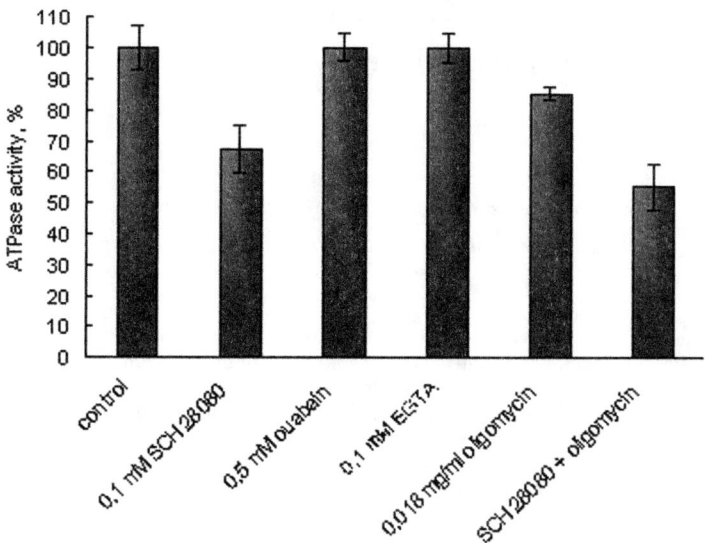

FIGURE 1. Mg-ATPase of MFRGM in the presence of inhibitors of different membrane ATPases. Conditions: 1 mM ATP, 3 mM $MgCl_2$, 10 mM KCl, 25 mM imidazole, pH 7.4.

The release of P_i was observed in the presence of MgADP or MgGDP and MFRGM. ADP breakdown was significantly decreased in the presence of inhibitors of adenylate kinase (AMP or Ap5A). This suggests that adenylate kinase together with ATPase provide the following way of ADP metabolism: $2ADP \rightarrow ATP + AMP$ and $ATP \rightarrow ADP + P_i$. The K_m value for ADP in the presence of Ap5A is comparable with that for GDP. The pH dependence of ATP and ADP hydrolysis and their sensitivity to different inhibitors demonstrate that basal Mg-ATPase of MFRGM is similar to ATP-diphosphohydrolase (ecto-apirase) from chicken oviduct.[3]

ACKNOWLEDGMENTS

This study was supported by Grant Nos. 01-04-48237 from RFBR and 01-0224 from INTAS.

REFERENCES

1. FARLEY, R.A. & L.D. FALLER. 1985. The amino acid sequence of an active site peptide from the H,K-ATPase of gastric mucosa. J. Biol. Chem. **260:** 3839–3901.
2. NORBY, J.G. 1988. Coupled assay of Na,K-ATPase activity. Methods Enzymol. **156:** 116–119.
3. STROBEL, A. & A. NAGY. 1996. Chicken oviductal ecto-ATP-diphosphohydrolase. J. Biol. Chem. **271:** 16323–16331.

Genetics of Hypertension: The Adducin Paradigm

GIUSEPPE BIANCHI[a] AND GRAZIA TRIPODI[b]

[a]*Chair and School of Nephrology, Division of Nephrology, Dialysis, and Hypertension, Università Vita e Salute, San Raffaele Hospital, Milan, Italy*

[b]*Prassis-Sigma Tau Research Institute, Settimo Milanese, Milan, Italy*

> ABSTRACT: The following were investigated: (1) how we became interested in studying adducin genes and what we know about adducin; (2) studies in animals and humans supporting the role of adducin polymorphisms in hypertension, including some methodological problems related to the dissection of the role of a given genetic molecular mechanism in a complex multifactorial polygenic disease like hypertension; (3) biochemical mechanisms underlying the effect of adducin and its interaction with the Na-K pump; and (4) future directions.
>
> KEYWORDS: adducin; genetics; hypertension; Na-K pump; endocytosis; tubular renal reabsorption

The main reasons for discussing the adducin paradigm are as follows: (1) The increase in the constitutive capacity of the renal tubular cells to reabsorb sodium associated with the mutated "hypertensive" variants of adducin is very likely achieved throughout modulation of the Na-K pump activity on the renal basolateral membranes. Thus, the understanding of the various factors modulating the activity of the Na-K pump is instrumental to a better understanding of the molecular mechanism underlying the biological activity of adducin. (2) Here we are interested in understanding how the Na-K pump works and the relevance of its activity modification for the function or dysfunction at cellular, organ, or whole-body levels, which may favor the development of a disorder.

When we try to define the impact of a given molecular mechanism in a complex system, such as cell, organ, or body functions, a variety of strategies may be applied, and the choice of the most appropriate one becomes crucial.

Therefore, this paper is subdivided into the following parts: (1) how we became interested in studying adducin genes and what we know about adducin; (2) studies in animals and humans supporting the role of adducin polymorphisms in hyperten-

---

Address for correspondence: Giuseppe Bianchi, Divisione di Nefrologia, Dialisi, e Ipertensione, Università Vita e Salute, Ospedale San Raffaele, Via Olgettina, 20132 Milano, Italia. Voice: +39-2-26433006; fax: +39-2-26432384.

bianchi.giuseppe@hsr.it

TABLE 1. Comparison of hypertensive humans with rats of the Milan strain

	Human Essential Hypertensives	MHS Hypertensive Rats
Pressor effect of kidney transplantation	↑	↑
Renal blood flow	↑	=
GFR	↑	↑
Na excretion after load	↑	↑
24-h urinary output	↑	↑
Plasma renin	↓	↓
Urine kallikrein	↓	↓
Erythrocyte Na content	↓	↓
Net erythrocyte membrane Na transport	↑	↑
Plasma OLF	↑	↑
Hypertension-associated α-adducin polymorphism	Yes	Yes

NOTE: Higher (↑), lower (↓), or equal (=) in the hypertensive humans and rats compared to the appropriate controls.

sion, including some methodological problems related to the dissection of the role of a given genetic molecular mechanism in a complex multifactorial polygenic disease like hypertension; (3) biochemical mechanisms underlying the effect of adducin and its interaction with the Na-K pump; and (4) future directions.

POINT 1: ADDUCIN

We started our studies in arterial hypertension producing a renal artery constriction to the sole remaining kidney in a conscious dog. After 2–3 weeks from this manipulation, all the initial alterations in plasma renin, body fluids, and cardiac function, which were responsible for the initial increase in blood pressure, disappeared, and the hormonal and hemodynamic patterns were superimposed on those found in primary hypertension.[1-3]

Primary hypertension affects approximately 25% of the adult population, and it is generally believed to arise from the interaction of a predisposing multigenic background with appropriate environmental factors. In other words, hypertension is a typical complex multifactorial polygenic disease. How then did we approach this complexity?

First, we tried to establish whether the parameters listed in TABLE 1, which have been found to be different between a hypertensive (MHS) strain of rats and its normotensive (MNS) control, were similarly modulated in patients who develop primary hypertension when compared with their appropriate controls.[4] As shown in TABLE 1, the similarities between the two species, affected by the same disease, prompted us to consider the rat model as appropriate to give information that could be relevant for the understanding of the human disease. Of course, because of the heterogeneity of the human conditions, such similarities can be applied only to a subset of patients.

Based on the findings on renovascular hypertension mentioned above, the hypothesis that the kidney, despite its normal morphological appearance, could be the carrier of the "genetic message" for hypertension has been tested by kidney cross-transplantation experiments conducted between the two rat strains.[5,6] This study showed that hypertension could be "transplanted" with the kidney. Similar findings also were obtained in humans evaluating the familiarities for hypertension of donors and the antihypertensive drug requirements of the recipient.[7,8]

These results prompted us to set up studies on the whole-kidney function and isolated renal tubules, and also on the ion transport across the renal membrane and erythrocyte. (The erythrocyte was taken as a cellular model mirroring a widespread alteration in cell membrane that also could be easily tested in humans.[9–16]) These studies yielded data supporting the hypothesis that an alteration in cell membrane transport could be responsible for a constitutive increase in tubular sodium reabsorption that, in the long run, could facilitate the development of renal Na retention and hypertension. We demonstrated that such a cell membrane alteration was genetically determined in the stem cells and genetically associated with hypertension both by bone marrow transplantation and by genetic crosses between MHS and MNS.[4] Then we showed that the difference in ion transport between MHS and MNS disappeared after the removal of cytoskeleton, and cross-immunization experiments with cytoskeletal proteins of one strain into the other allowed us to detect an antibody to a protein subsequently identified as adducin.[17]

Adducin functions within the cell as α/β or α/γ heterodimer whose subunits are coded by three genes mapping on different chromosomes. The protein, initially discovered by Van Bennett, favors the binding of actin to spectrin, and it is involved in actin polymerization and bundling.[18–21] Moreover, adducin alone or throughout the actin cytoskeleton may potentially modulate a variety of cell functions such as ion transport, focal contact sites, and cell adhesion molecules.[22,23]

POINT 2: ADDUCIN AND GENETICS OF HYPERTENSION

We and other researchers detected the SNPs on adducin genes in both rats and humans, as illustrated in FIGURE 1. In rats, we identified one SNP for each adducin subunit: α F316Y, β Q529R, γ Q572K.[24,25] We showed that α-adducin MHS allele cosegregates with blood pressure in an F2 (MHS × MNS) segregating population, whereas β- and γ-adducin polymorphisms that "*per se*" do not affect blood pressure were found epistatically interacting with that of α-subunit in modulating blood pressure levels.[24–26] Also, in humans, we demonstrated a significant interaction between α G460W and β C1967T adducin polymorphisms on blood pressure control.[27,28] These subunit interactions, because of the biochemical bases of protein function illustrated above, strongly support the genetic role of adducin in blood pressure regulation. In fact, it is very unlikely that other genes mapping close to adducin loci could encode for proteins with similar strong biochemical and physiological bases of interaction. We also found that the rat α- and β-adducin mutations affect a tyrosine and a serine phosphorylation site, respectively.[24]

In humans, the "hypertensive" α-adducin variant (460W) is associated with hypertension in some (but not all) of the population[30–33] as it occurs for all the other candidate genes so far proposed for hypertension. However, in low renin patients, the

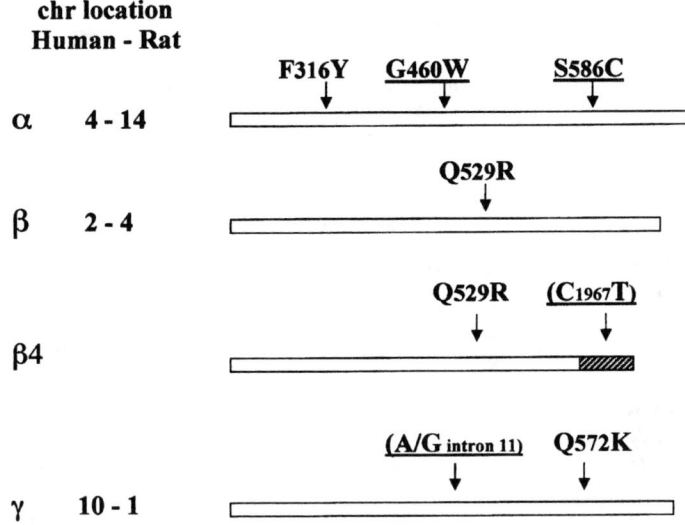

FIGURE 1. Schematic representation of adducin genes with SNPs identified in human and rat hypertension. The chromosomal location in humans and rats (*left*) and the polymorphic sites in humans (*underscore*) and rats are reported for each subunit. The number reported for each polymorphism corresponds to the amino acid (*one-letter code*) or nucleotide (*in parentheses*) change of the coding region (*boxes*). The polymorphism identified in the human β-adducin gene (C1967T) is in the alternative spliced exon 15 that corresponds to the β4 isoform, with a different COOH-terminal region (*hatched area*) compared with β-adducin.[29]

association of α 460W variant to hypertension is consistent among populations.[34–36] We approached this problem by studying two populations from the Milano and Sassari area that diverged approximately 10,000 years ago. The association of the α 460W variant to hypertension was present in the former, but not in the latter population.[37] However, despite this discrepancy, when the hypertensive carriers of the α 460W allele were compared with the corresponding hypertensive carriers of the α 460G allele in each population, the former display a greater decrease in blood pressure with diuretics, a lower plasma renin activity, and a faster Na-K pump activity with lower intracellular Na in the erythrocytes than the latter.[37,38] This demonstrates that the mutated α 460W allele of both populations was associated with similar alteration in cellular and renal sodium handling. This hypothesis is consistent with a biologic stimulating effect of the hypertensive variant of the protein on the Na-K pump in both populations despite the lack of association with hypertension in the Sassari population. Moreover, these results clearly indicate that differences in evolutionary history or in population stratification may affect the statistical genetic criteria based on allele frequency differences, but do not affect the criteria based on physiological genetics. Of course, these last criteria are much more solid than the former ones. This may explain the inconsistency of results among the different populations when the impact of a given individual gene is assessed by statistical genetic analyses.

Besides the differences illustrated above, when the hypertensive carriers of the α 460W allele (taken alone or in combination with the β 1967T allele or the D-ACE allele) were compared with the carriers of α 460G, the former showed a larger blood pressure increase after sodium infusion with a less steep pressure-natriuresis curve;[39] an increased proximal tubular reabsorption of sodium (estimated as a lithium clearance)[40] associated with a larger blood pressure decrease under diuretic therapy;[41] a faster decline of renal function with age (with a greater excretion of albumin in the urine);[42,43] a thicker wall of the femoral artery;[44,45] a larger left ventricular mass;[46] a higher risk of myocardial infarction, stroke,[47] and coronary heart disease;[48] and higher blood pressure.[39,49] Because most of these alterations also have been detected between MHS and MNS rats or in their genetic crosses and, at least in part, are associated with adducin mutations, this spectrum of changes clearly supports the role of adducin polymorphisms as a genetic mechanism favoring cardiovascular and renal diseases.

Therefore, the clarification of the molecular mechanism underlying the clinical effects has an enormous impact on population health also, considering the relatively high frequency of these mutated alleles. This conclusion is also supported by recent findings on a population of 1038 hypertensive subjects followed up for 10 years. In the carriers of α 460W allele, the diuretic therapy reduces by greater than 50% the incidence of strokes and myocardial infarction in comparison with a treatment with ACE inhibitor and β-blocker.[47] This selective effect of diuretics, probably caused by the matching of the pressure mechanism of adducin and the action mechanism of this drug on Na reabsorption, is not present in carriers of α 460G allele.

POINT 3: BIOCHEMICAL MECHANISMS OF ADDUCIN

Our experimental results obtained either at whole animal or at *in vitro* levels allowed us to clarify the contribution of the polymorphisms on adducin subunits on MHS genetic hypertension.

We investigated the effect of adducin polymorphisms on whole animals by the selection of a set of congenic strains in which the α-, β-, and γ-adducin loci have been introgressed from MHS into the MNS genetic background. In the three single congenic strains for adducin loci, the effect of α-adducin gene transfer results in increased blood pressure and renal Na-K pump activity, whereas that of the β- and γ-subunits seems to favor cardiac hypertrophy and glomerulosclerosis. In the double α/β and α/γ congenic strains, we observed a significant interaction of adducin subunits in modulating blood pressure levels and a protective effect of α-subunit in both cardiac hypertrophy and glomerulosclerosis progression[50,51] (also unpublished data).

In vitro studies have been conducted either in "cell-free" systems or in cell lines. In a cell-free system, the MHS variant of the protein (either extracted from tissues or prepared by DNA-recombinant technology) binds to the Na-K pump with greater affinity than the normotensive one.[52] Therefore, despite the difference in mutation sites between rats and humans, the human hypertensive isoforms similarly modify the interaction of adducin with the Na-K pump, which is the key enzyme for tubular Na transport.[52] We also have seen that the MHS variant of the protein stimulates actin polymerization and bundling much more than the normotensive one.[53]

TABLE 2. Major findings

(1) Statistical genetics (clinical and epidemiological) including interactions with either genetics or environmental factors relevant to the gene.

(2) Pathophysiological genetics on the association between the gene variant and the relevant intermediate phenotypes.

(3) Direct link between the gene variant and a plausible cellular mechanism underlying the intermediate phenotypes.

(4) Definition of a clinical entity (or subset of patients) in whom a consistent sequence of events going from the gene variant (with its interacting partners) to the clinical symptoms may be proposed.

(5) Demonstration that a drug, interfering with the cellular mechanism affected by the gene, provides a selective therapeutical advantage over the other competing drugs.

Stable transfection of epithelial tubular cells demonstrated that the overexpression of MHS α-adducin increases the Na-K pump activity and determines a great rearrangement of cytoskeleton compared with the analogue expression of MNS protein.[53] In both isolated proximal tubular cells of MHS and transfected epithelial tubular cells overexpressing MHS α-adducin, the residential time of the Na-K pump onto the plasma membrane is higher than in the correspondent MNS cells. It has been demonstrated that this effect is the result of an impaired removal of the molecules through the control of endocytosis in response to regulatory hormones[54] (also Bertorello et al., manuscript in preparation).

TABLE 2 summarizes the five major groups of findings briefly mentioned so far supporting the direct link between the adducin variants and the clinical syndromes characterized by increased blood pressure and the cardiovascular and renal alterations associated with hypertension. The findings of the last point provide the strongest argument in favor of the adducin clinical impact. In fact, the consistency among the scientific results gathered from various experimental and clinical approaches may provide the necessary approximation to reality in a complex multifactorial disease-like hypertension. This strategy is absolutely necessary because of the weakness of statistical genetic tools to assess the role of a given gene in these pathological conditions.

POINT 4: FUTURE DIRECTIONS

There are many aspects that need further clarification:

(1) The precise molecular mechanism underlying the modulatory influence of adducin on the residential time of some protein in the cell membrane and in particular of the Na-K pump.
(2) The type of interaction between adducin and the Na-K pump.
(3) Whether the adducin effects on renal and cardiovascular phenotype are all secondary to its ability to modulate either actin cytoskeleton or the constitutive capacity of the tubuli to retain Na. (Theoretically, some of these effects may have separate molecular pathways.)

(4) The portion of patients in whom the effect of adducin underlies their clinical symptoms or phenotypes. Many other candidate genes are under study and, on some occasions, their effects seem to be similar to those of adducin at phenotypic levels. The appropriate hierarchical dependency of these different genes must be established to assess which of them are really able to take over the genetic network, at least during some stage of disease or under the appropriate environmental conditions.

REFERENCES

1. BIANCHI, G., L.T. TENCONI & R. LUCCA. 1970. Effect in the conscious dog of constriction of the renal artery to a sole remaining kidney on haemodynamics, sodium balance, body fluid volume, plasma renin concentration, and pressor responsiveness to angiotensin. Clin. Sci. **38**: 741–766.
2. BIANCHI, G., E. BALDOLI, E. LUCCA, et al. 1972. Pathogenesis of arterial hypertension after the constriction of the renal artery leaving the opposite kidney intact both in the anesthetized and in the conscious dog. Clin. Sci. **42**: 651–664.
3. CARAVAGGI, A.M., G. BIANCHI, J.J. BROWN, et al. 1976. Blood pressure and plasma angiotensin II concentration after renal artery constriction and angiotensin in the dog: (5-isoleucine) angiotensin II and its breakdown fragments in dog blood. Circ. Res. **38**: 315–321.
4. BIANCHI, G., et al. 1994. The Milan hypertensive strain of rats. In Textbook of Hypertension, pp. 457–460. Blackwell Scientific Pub. Oxford.
5. BIANCHI, G., U. FOX, G.F. DI FRANCESCO, et al. 1973. The hypertensive role of the kidney in spontaneously hypertensive rats. Clin. Sci. Mol. Med. **45**: 135s–139s.
6. BIANCHI, G., U. FOX, G.F. DI FRANCESCO, et al. 1974. Blood pressure changes produced by kidney cross-transplantation between spontaneously hypertensive rats and normotensive rats. Clin. Sci. Mol. Med. **47**: 435–448.
7. GUIDI, E., G. BIANCHI, E. RIVOLTA, et al. 1985. Hypertension in man with a kidney transplant: role of familial versus other factors. Nephron **41**: 14–21.
8. GUIDI, E., D. MENGHETTI, S. MILANI PHARM, et al. 1996. Hypertension may be transplanted with the kidney in humans. J. Am. Soc. Nephrol. **7**: 1131–1138.
9. BIANCHI, G., D. CUSI, M. GATTI, et al. 1979. A renal abnormality as a possible cause of "essential" hypertension. Lancet **i**: 173–177.
10. BIANCHI, G., D. CUSI, C. BARLASSINA, et al. 1983. Renal dysfunction as a possible cause of essential hypertension in predisposed subjects. Kidney Int. **23**: 870–875.
11. BIANCHI, G., P. BAER, U. FOX, et al. 1975. Changes in renin, water balance, and sodium balance during development of high blood pressure in genetically hypertensive rats. Circ. Res. **36/37**(suppl. I): 153–161.
12. BAER, P. & G. BIANCHI. 1978. Renal micropuncture study of normotensive and Milan hypertensive rats before and after development of hypertension. Kidney Int. **13**: 452–466.
13. PERSSON, A.E., G. BIANCHI & U. BOBERG. 1985. Tubuloglomerular feedback in hypertensive rats of the Milan strain. Acta Physiol. Scand. **123**: 139–146.
14. PARENTI, P., G. HANOZET & G. BIANCHI. 1986. Sodium and glucose transport across renal brush-border membranes of Milan hypertensive rats. Hypertension **8**: 932–939.
15. BIANCHI, G. & P. FERRARI. 1992. Renal factors involved in the pathogenesis of genetic forms of hypertension. In Genetic Hypertension. Vol. 218, pp. 447–458. Colloque INSERM/John Libbey Eurotext, Ltd. Montrouge, France.
16. BIANCHI, G., P. FERRARI, D. TRIZIO, et al. 1985. Red blood cell abnormalities and spontaneous hypertension in the rat: a genetically determined link. Hypertension **7**: 319–325.
17. SALARDI, S., B. SACCARDO, G. BORSANI, et al. 1989. Erythrocyte adducin differential properties in normotensive and hypertensive rats of the Milan strain (characterization of spleen adducin m-RNA). Am. J. Hypertens. **2**: 229–237.

18. BENNETT, V. 1990. Spectrin-based membrane cytoskeleton: a multipotential adaptor between plasma membrane and cytoplasm. Physiol. Rev. **70:** 1029–1065.
19. HUGHES, C.A. & V. BENNETT. 1995. Adducin: a physical model with implications for function in assembly of spectrin-actin complexes. J. Biol. Chem. **270:** 18990–18996.
20. KUHLMAN, P.A., C.A. HUGHES, V. BENNETT, et al. 1996. A new function for adducin. J. Biol. Chem. **271:** 7986–7991.
21. GARDNER, K. & V. BENNETT. 1986. A new erythrocyte membrane-associated protein with calmodulin binding activity: identification and purification. J. Biol. Chem. **261:** 1339–1348.
22. FUKATA, Y., N. OSHIRO, N. KINOSHITA, et al. 1999. Phosphorylation of adducin by Rho-kinase plays a crucial role in cell motility. J. Cell Biol. **145:** 347–361.
23. KURANA, S. 2000. Role of actin cytoskeleton in regulation of ion transport: examples from epithelial cells. J. Membr. Biol. **178:** 73–87.
24. BIANCHI, G., G. TRIPODI, G. CASARI, et al. 1994. Two point mutations in the adducin genes are involved in blood pressure variation. Proc. Natl. Acad. Sci. USA **91:** 3999–4003.
25. TRIPODI, G., C. SZPIRER, C. REINA, et al. 1997. Polymorphism of gamma-adducin gene in genetic hypertension and mapping of the gene to rat chromosome 1q55. Biochem. Biophys. Res. Commun. **237:** 685–689.
26. ZAGATO, L., R. MODICA, M. FLORIO, et al. 2000. Genetic mapping of blood pressure quantitative trait loci in Milan hypertensive rats. Hypertension **36:** 734–737.
27. LANZANI, C., L. TIZZONI, M. JANKARICOVA, et al. 2002. Role of adducin family genes on human essential hypertension. Submitted.
28. WANG, J-G., J.A. STAESSEN, C. BARLASSINA, et al. 2002. Associations of polymorphisms in the alpha and beta-adducin genes with hypertension and blood pressure in white population. Kidney Int. **62:** 2152–2159.
29. SINARD, J.H., G.W. STEWART, P.R. STABACH, et al. 1998. Utilization of an 86 bp exon generates a novel adducin isoform (beta4) lacking the MARCKS homology domain. Biochim. Biophys. Acta **1396:** 57–66.
30. CASARI, G., C. BARLASSINA, D. CUSI, et al. 1995. Association of the alpha-adducin locus with essential hypertension. Hypertension **25:** 320–326.
31. CUSI, D., C. BARLASSINA, T. AZZANI, et al. 1997. Polymorphism of alpha-adducin and salt sensitivity in patients with essential hypertension. Lancet **349:** 1353–1357.
32. BUSJAJN, A., A. AYDIN, N. VON TREUENFELS, et al. 1999. Linkage but lack of association for blood pressure and the alpha-adducin locus in normotensive twins. J. Hypertens. **17:** 1437–1441.
33. BIANCHI, G. & D. CUSI. 2000. Association and linkage analysis of alpha-adducin polymorphism: is the glass half full or half empty? Am. J. Hypertens. **13:** 739–743.
34. MULATERO, P., T.A. WILLIAMS, A. MILAN, et al. 2002. Blood pressure in patients with primary aldosteronism is influenced by bradykinin B2 receptor and alpha-adducin gene polymorphisms. J. Clin. Endocrinol. Metab. **87:** 3337–3343.
35. FISHER, N., S. HURWITZ, X. JEUNEMAITRE, et al. 2002. Familial aggregation of low-renin hypertension. Hypertension **39:** 914–918.
36. SUGIMOTO, K., A. HOZAWA, T. KATSUYA, et al. 2002. Alpha-adducin Gly460Trp polymorphism is associated with low renin hypertension in younger subjects in the Oashama study. J. Hypertens. **20:** 1779–1784.
37. GLORIOSO, N., P. MANUNTA, F. FILIGHEDDU, et al. 1999. The role of alpha-adducin polymorphism in blood pressure and sodium handling regulation may not be excluded by a negative association study. Hypertension **34:** 649–654.
38. GLORIOSO, N., F. FILIGHEDDU, D. CUSI, et al. 2002. Alpha-adducin 460Trp allele is associated with erythrocyte Na transport rate in primary hypertension. Hypertension **39:** 357–362.
39. BARLASSINA, C., N.J. SCHORK, P. MANUNTA, et al. 2000. Synergetic effect of alpha-adducin and ACE genes in causing blood pressure changes with body sodium and volume expansion. Kidney Int. **57:** 1083–1090.
40. MANUNTA, P., M. BURNIER, M. D'AMICO, et al. 1999. Adducin polymorphism affects renal proximal tubule reabsorption in hypertension. Hypertension **33:** 694–697.

41. SCIARRONE, M.T., P. STELLA, C. BARLASSINA, *et al.* 2003. ACE I/D and alpha-adducin Gly460Trp polymorphism as marker of individual response to diuretic treatment. Hypertension. In press.
42. WANG, J.G., J.A. STAESSEN, L. TIZZONI, *et al.* 2001. Renal function in relation to three candidate genes. Am. J. Kidney Dis. **38:** 1158–1168.
43. NICOD, J., B.M. FREY, F.J. FREY, *et al.* 2002. Role of the alpha-adducin genotype on renal disease progression. Kidney Int. **61:** 1270–1275.
44. BALKESTEIN, E.J., J.A. STAESSEN, J. WANG, *et al.* 2002. Carotid and femoral artery stiffness in relation to three candidate genes in a white population. Hypertension **38:** 1190–1197.
45. BALKESTEIN, E.J., J.G. WANG, H.A. STRUIJKER-BOUDIER, *et al.* 2002. Carotid and femoral intima-media thickness in relation to three candidate genes in a Caucasian population. J. Hypertens. **20:** 1551–1561.
46. WINNICKI, M., V.K. SOMERS, V. ACCURSO, *et al.* 2002. Alpha-adducin Gly460Trp polymorphism, left ventricular mass, and plasma renin activity. J. Hypertens. **20:** 1771–1777.
47. PSATY, B.M., N.L. SMITH, S.R. HECKBERT, *et al.* 2002. Diuretic therapy, the alpha-adducin gene variant, and the risk of myocardial infarction or stroke in person with treated hypertension. JAMA **287:** 1680–1689.
48. MORRISON, A.C., M.S. BRAY, A.R. FOLSOM, *et al.* 2002. ADD1 460W allele associated with cardiovascular disease in hypertensive individuals. Hypertension **39:** 1053–1057.
49. STAESSEN, J., J.G. WANG, E. BRAND, *et al.* 2001. Effects of three candidate genes on prevalence and incidence of hypertension in a Caucasian population. J. Hypertens. **19:** 1349–1358.
50. TRIPODI, G., M. FERRANDI, R. MODICA, *et al.* 2001. Effect of adducin gene transfer on blood pressure and Na-K pump activity in the Milan hypertensive rat (MHS). J. Mol. Med. **79:** B30.
51. FERRANDI, M., G. TRIPODI, I. MOLINARI, *et al.* 2001. Protective effect of alpha adducin gene from Milan hypertensive rats on age-dependent progression of proteinuria in Milan normotensive rats. J. Mol. Med. **79:** B4.
52. FERRANDI, M., S. SALARDI, G. TRIPODI, *et al.* 1999. Evidence for an interaction between adducin and Na-K-ATPase: relation to genetic hypertension. Am. J. Physiol. **277:** H1338–H1349.
53. TRIPODI, G., F. VALTORTA, L. TORIELLI, *et al.* 1996. Hypertension-associated point mutations in the adducin alpha and beta subunits affect actin cytoskeleton and ion transport. J. Clin. Invest. **97:** 2815–2822.
54. TORIELLI, L., G. PADOANI, P. TARSINI, *et al.* 2000. Adducin may affect the residential time of the Na/K pump on the plasma membrane. EJCB **79**(suppl. 52): 164.

Mechanisms of Pressure Natriuresis

How Blood Pressure Regulates Renal Sodium Transport

ALICIA A. McDONOUGH, PATRICK K. K. LEONG, AND LI E. YANG

Department of Physiology and Biophysics, University of Southern California, Keck School of Medicine, Los Angeles, California 90089-9142, USA

ABSTRACT: An acute increase in blood pressure provokes a rapid decrease in proximal tubule salt and water reabsorption that is central to tubuloglomerular feedback regulation of renal blood flow and glomerular filtration rate and contributes to pressure natriuresis. The molecular mechanisms responsible for this critical homeostatic adjustment were studied. When blood pressure is acutely elevated, apical proximal tubule NHE3 are rapidly redistributed out of the microvilli to intermicrovillar clefts and then endosomal pools, and Na,K-ATPase activity is suppressed. Depressing apical Na^+ entry without hypertension is not sufficient to decrease Na,K-ATPase activity, and depressing Na,K-ATPase activity alone is not sufficient to decrease proximal tubule Na^+ and water reabsorption; thus, it appears that coordinated decreases in both NHE3 surface distribution and Na,K-ATPase activity may be important for the response to acute hypertension. Clamping plasma angiotensin II levels blunts the retraction of NHE3 from the cell surface to endosomal pools. The increased volume flow of salt and water to the loop of Henle stimulates Na,K-ATPase activity in this region and provides evidence for a downstream shift in sodium transport during acute hypertension. These same responses in the proximal tubule and loop develop and persist in the spontaneously hypertensive rat. These studies demonstrate that sodium transporters along the nephron are very dynamic, responding quickly to normal fluctuations of blood pressure, and are key to generating the macula densa tubuloglomerular feedback signal and for accommodating increased volume flow through the loop of Henle.

KEYWORDS: NHE3; Na,K-ATPase; kidney; trafficking

INTRODUCTION

Although blood pressure fluctuates continuously as a function of daily activity, renal blood flow and thus glomerular filtration rate (GFR) are constant between 100 and 200 mmHg arterial pressure, a response known as autoregulation. Autoregula-

Address for correspondence: Dr. Alicia A. McDonough, Department of Physiology and Biophysics, University of Southern California, Keck School of Medicine, Los Angeles, CA 90089-9142. Voice: 323-442-1238; fax: 323-442-2283.
mcdonoug@hsc.usc.edu

tion of GFR maintains constant delivery of NaCl and volume to the distal nephron where fine control of their transport by hormones such as aldosterone and ADH governs fluid and electrolyte balance. The mechanism for autoregulation during an acute increase in blood pressure is a decrease in afferent arteriolar radius that normalizes flow to the glomerulus. This arteriolar constriction is driven both by a myogenic mechanism (stretch → constriction) as well as by tubuloglomerular feedback, a response that senses NaCl delivery and transport by the apical NaK2Cl cotransporter in the macula densa, a region between the loop of Henle and distal tubule in proximity to the afferent arteriole. The connection between the transport by the NaK2Cl and the constriction of the arteriole is under investigation by several groups.[1,2]

What is the connection between increasing blood pressure and increasing NaCl delivery to the macula densa? An increase in NaCl at the macula densa can result from either increasing the filtered load (increasing GFR) or decreasing tubular reabsorption of NaCl proximal to the macula densa. Because there is no change in GFR during fluctuations in blood pressure (i.e., autoregulation), there must be a decrease in tubular reabsorption. Chou and Marsh provided evidence in rats, using micropuncture, that increasing renal perfusion pressure from 100 mmHg to 150 mmHg (by constricting a set of arterial beds) did not change renal blood flow or GFR, but immediately increased volume flow from the proximal tubule (PT), reaching 140% over baseline at 30 minutes.[3] Because the PT normally reabsorbs 67% of the filtered load of sodium, this change represents a very significant increase in sodium load leaving the PT and, despite load-sensitive adjustments in the loop of Henle, results in an increase in NaCl at the macula densa. The sensitivity of this response to acute hypertension defines the blood pressure set point and is, by definition, altered during chronic hypertension.

One overarching hypothesis connecting hypertension to inhibition of NaCl transport is as follows: acute hypertension is detected by vascular elements in close proximity to the PT and these elements transmit signals to the PT to inhibit NaCl reabsorption by rapidly changing the properties or distribution of sodium transporters in the apical and basolateral membranes. Studies in this laboratory over the past 6 years have focused on defining the cellular mechanisms responsible for the decrease in PT NaCl reabsorption during acute hypertension and defining the extracellular and intracellular signals responsible for the transport inhibition. This short review focuses on the results of these studies.

Sodium is reabsorbed from the PT across the apical membrane via sodium-hydrogen exchangers (NHE3 is the PT isoform responsible for most sodium reabsorption), and sodium is actively pumped out of the cell by the basolateral sodium pump, Na,K-ATPase ($\alpha 1\beta 1$), which generates the gradient for Na^+ entry across the apical membrane. The percentage of the filtered sodium that is reabsorbed can be affected by regulation of sodium entry or exit, or both. Evidence in the literature indicates that internalization as well as inhibition of transporters could effect a rapid decrease in sodium transport in the PT.[4–6] We tested the hypothesis that rapid inhibition or internalization (or both) of sodium transporters mediates the decrease in Na^+ reabsorption during acute hypertension in a rat model.

The methods used were similar in the set of studies discussed. Blood pressure was raised acutely (and in some cases reversibly) to 150 mmHg by constricting arterial beds and the aorta caudal to the renal arteries. At the whole-animal level, volume flow out of the PT was assessed by the clearance of endogenous lithium (C_{Li+}), and urine output was measured gravimetrically over a fixed time (V). NHE3, NaPi2, and Na,K-ATPase (α1 and β1 subunits) subcellular distributions were studied by immunoblots after subcellular fractionation on sorbitol density gradients resolved apical microvilli, intermicrovillar clefts, and basolateral membranes distinct from intracellular endosomes and lysosomes. NHE3 distribution also was studied by confocal microscopy. In total membranes and the subcellular fractions, NHE3 activity was assessed by a quantitative acridine orange quench assay, and Na,K-ATPase activity was measured by ouabain-sensitive ATPase or K^+-dependent pNPPase assays. The effects of acute hypertension were compared with the effects of inhibiting PT sodium transport using the carbonic anhydrase inhibitor, benzolamide (2 mg/kg in 300 mM $NaHCO_3$ at 50 µL/min for 5–7 min), which blocks NHE3 activity by luminal acidification. These measurements allowed us to connect *in vivo* changes in PT Na^+ transport in the whole animal to changes in subcellular distribution of key Na^+ transporters and changes in their activity.

ACUTE CHANGES IN PT APICAL MEMBRANE SODIUM TRANSPORTERS DURING ACUTE HYPERTENSION

The findings at the apical membrane provide evidence for a two-step internalization of NHE3 stimulated by acute hypertension. After 5-minute acute hypertension, there were greater than threefold increases (depending on the study) in both V and C_{Li+}. Two-thirds of the apical NHE3 (20% of total cell NHE3) redistributed to the intermicrovilli cleft region within 5 minutes, and after 30 minutes the NHE3 pool size of the endosomes/lysosomes was doubled (to 20% of the total NHE3; FIG. 1). Confocal microscopy verified the removal of NHE3 from the apical microvilli and the appearance in subapical endosomal stores. There was a parallel redistribution of NHE3 activity indicating that internalized NHE3 are fully active.[7] NaPi2 (Na^+-phosphate cotransporter) showed a similar internalization. If blood pressure was restored to baseline after 5-minute hypertension, NHE3 returned to the apical microvilli, but NaPi2 remained in the heavy density membranes.[8]

Further evidence for a two-step internalization process came from testing the hypothesis that a decrease in plasma angiotensin II (Ang II), provoked indirectly by the increase in NaCl at the macula densa, played a role in decreasing PT NaCl reabsorption. When Ang II levels were clamped during acute hypertension (by coinfusion of an ACE inhibitor, captopril, and a nonpressor dose of Ang II), there was a 50% blunting of the increase in C_{Li+} and urine output, and NHE3 redistributed out of the microvilli to the intermicrovillar cleft region (step 1), but did not internalize to the endosomes/lysosomes (step 2; FIG. 1).[9,10] These findings support the hypothesis that inhibition of renin release by the afferent arteriole during acute hypertension decreases plasma Ang II, which (at least in part) decreases PT Na^+ reabsorption by preventing the second step in NHE3 internalization. This study makes a connection between a vascular element sensing of pressure and a change in PT transport during acute hypertension.

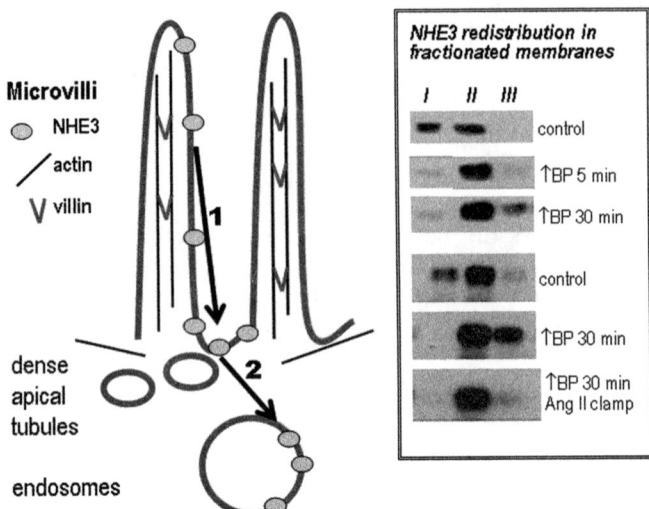

FIGURE 1. Two-step model of internalization of NHE3 in response to acute hypertension. The model is based on results collected from subcellular fractionation and confocal microscopy. At baseline (control), NHE3 is located in microvilli. In response to 5-minute acute hypertension, NHE3 in the microvilli are redistributed to the intermicrovillar cleft region (*step 1 arrow*). After 30-minute hypertension, NHE3 can be found in membrane pools under the apical microvilli (*step 2 arrow*).[7] This second step is significantly blunted if plasma angiotensin II (Ang II) levels are clamped during acute hypertension to prevent the predicted decrease in Ang II driven by increased NaCl delivery to the macula densa.[10] The panel on the right includes typical immunoblots of renal cortical membrane pools fractionated on sorbitol gradients from control, acute hypertension, and acute hypertension during Ang II clamp. Pool I is enriched in apical membranes, pool II is enriched in intermicrovillar cleft and apical membranes, and pool III is enriched in endosomes and lysosomal membranes. The blots show the time-dependent shift of NHE3 out of the apical pool to the intracellular pool and demonstrate the blunting of the internalization during Ang II clamp.[7,10] By dual-labeling confocal microscopy (not shown), there is overlapping staining of the actin bundling protein villin and NHE3 at baseline and a clear separation of the two after 30-minute hypertension.[7]

ACUTE CHANGES IN PT BASOLATERAL MEMBRANE Na,K-ATPase

If there is a change in the first step in Na^+ reabsorption across the apical membrane during acute hypertension, is it necessary to invoke a change in basolateral Na,K-ATPase to effect a decrease in transepithelial Na^+ reabsorption? Evidence demonstrates that, after 5-minute acute hypertension, basolateral membrane Na,K-ATPase activity (V_{max} measured in membrane fractions) decreased 50% in cortex (FIG. 2) accompanied by a smaller (<25%) shift of $\alpha1$ and $\beta1$ subunits to higher density membranes; these changes were reversed with normalization of blood pressure.[8,11] These results provide evidence for parallel regulation of Na,K-ATPase activity or subcellular distribution that coincides with the internalization of apical NHE3. The change in Na,K-ATPase activity per transporter may be linked to the

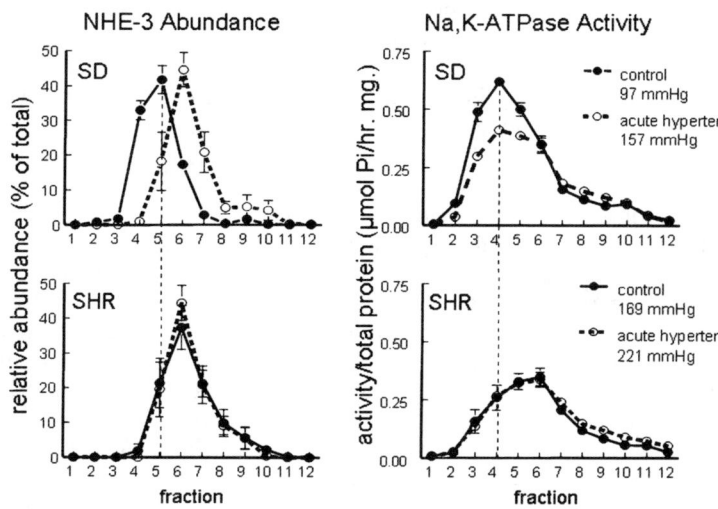

FIGURE 2. Comparison of responses to acute hypertension in the adult Sprague-Dawley rat (SD) and spontaneously hypertensive rat (SHR). In SD (*top panels*), increasing mean arterial pressure from 97 to 157 mmHg for 5 minutes shifts NHE3 to higher density membranes and decreases renal cortical membrane Na,K-ATPase activity. In SHR (*bottom panels*), where mean arterial pressure is elevated to 169 mmHg, the NHE3 is chronically shifted to higher density fractions and the sodium pump activity is decreased (compared with young SHR, not shown).[13] Raising mean arterial pressure further to 221 mmHg for 5 minutes did not cause a further shift of NHE3 or further inhibition of Na,K-ATPase. These results indicate that the proximal tubule responses to chronic hypertension are the same as to acute hypertension.[13]

density change if the redistribution is a marker for a change in membrane environment, associated proteins, or cytoskeletal attachment.

To determine whether the Na,K-ATPase inhibition could be driven by a change in apical sodium entry, independent of a change in blood pressure, we inhibited apical sodium transport with infusion of the carbonic anhydrase inhibitor, benzolamide (described above). Benzolamide increased V and C_{Li+} threefold (same magnitude as acute hypertension), but did not alter Na,K-ATPase activity or distribution, supporting the conclusion that decreasing apical Na^+ uptake is not a sufficient signal to initiate the redistribution or inhibition of basolateral Na,K-ATPase seen in response to acute hypertension.[11] These findings suggest that there is dynamic regulation of PT sodium transport by acute changes in blood pressure mediated by rapid reversible regulation of sodium pump activity as well as relocation of apical sodium transporters.

DOWNSTREAM SHIFT IN Na,K-ATPase DURING ACUTE HYPERTENSION

Acute hypertension increases volume flow out of the PT into the loop of Henle by 30–40%. Because the thick ascending limb of the loop of Henle (TALH) has a load-dependent response to reabsorb more salt when more is delivered (important to normalize delivery to the distal nephron), we tested the hypothesis that this sizable increase in salt and water delivery during acute hypertension would acutely increase TALH Na,K-ATPase activity in the outer medulla (highly enriched in TALH) and that the increase would be independent of the hypertension *per se*.[12]

Volume flow to the TALH was increased by either acutely elevating blood pressure for 5 minutes or inhibiting PT sodium and volume reabsorption with the carbonic anhydrase inhibitor, benzolamide, as described above. Both stimuli increased urine output and lithium clearance three- to fourfold, and both stimuli increased basolateral Na,K-ATPase activity in the renal outer medulla by approximately 40%, supporting the hypothesis. This response to hypertension is reciprocal to that in the cortex; that is, there is a downstream shift in Na,K-ATPase activity that likely reflects a downstream shift in NaCl reabsorption, a decrease in driving force of sodium reabsorption in the proximal nephron, and an increase in the TALH. These studies demonstrate that the ratio of Na,K-ATPase activity in medulla to that in cortex is not a constant, but a fluctuating value that depends on blood pressure: 1.81 ± 0.33 at baseline blood pressure and 3.88 ± 1.15 after 5-minute acute hypertension.[12] This rapid, downstream shift in Na,K-ATPase activity during acute hypertension contributes the driving force for activating tubuloglomerular feedback (by inhibition in the PT) and minimizes changes in distal sodium delivery (by activation in the TALH).

CHANGES IN CHRONIC HYPERTENSION ARE THE SAME AS DURING ACUTE HYPERTENSION

Hypertension develops over 8 weeks in the spontaneously hypertensive rat (SHR), during which there is sodium, water, and potassium retention compared with age-matched controls. With hypertension, there is a shift in the renal function curve, the relationship between blood pressure and sodium excretion. When studying differences between hypertensive and control models, it is not always obvious which differences contribute to hypertension and which are adaptive responses to the hypertension. We tested the hypothesis that development of chronic hypertension in the SHR provokes adaptive persistent changes in NHE3 distribution and Na,K-ATPase activity that mimic those seen in normotensive rats during acute hypertension.[13] The studies conducted supported this hypothesis. At 3–4 weeks of age, NHE3 distribution was the same in SHR as normotensive controls; however, by 12 weeks of age, NHE3 (and NaPi2) moved out of the microvilli to the intermicrovillar cleft region, seen by confocal microscopy and subcellular fractionation (FIG. 2).[13,14] In addition, there was evidence of a downstream shift in Na,K-ATPase activity between 3–4 and 12 weeks of age: Na,K-ATPase activity decreased in cortex and increased in medulla in SHR, whereas there was no detectable change in the normotensive controls in cortex or medulla during this developmental interval. Interestingly, when blood pressure was increased acutely in the SHR (from 140 mmHg to 180 mmHg), there was still an increase in urine output and C_{Li^+}, albeit blunted (shifted pressure natriuresis relationship), yet there was no further detectable change in NHE3 distribution or Na,K-ATPase activity, indicating that additional molecular features of the pressure diuretic response remain to be explored and defined (FIG. 2).

INTRACELLULAR SIGNALING MECHANISMS

We have conducted two studies to address the intracellular signaling mechanisms that govern the redistribution of NHE3 and inhibition of Na,K-ATPase activity. Because cytochrome P450 (cytP450) –dependent arachidonate metabolites participate in the regulation of renal sodium transport and blood pressure, we tested the hypothesis that the renal responses to acute hypertension would be prevented if cytP450 metabolism were inhibited. We used cobalt chloride to increase degradation of cytP450 by increasing heme oxygenase.[15] $CoCl_2$ treatment alone for 2 days did not significantly affect V, C_{Li+}, activity of Na,K-ATPase, or subcellular distribution of NHE3. However, the response to acute hypertension was blunted in $CoCl_2$-treated animals: V and C_{Li+} increased only twofold, and there was no inhibition of Na,K-ATPase activity and no redistribution of NHE3.[15] These findings demonstrate that $CoCl_2$ treatment both attenuates the inhibition of PT sodium reabsorption and diuresis and abolishes Na,K-ATPase inhibition and NHE3 redistribution during acute hypertension, evidence that these responses may be mediated by cytP450 arachidonate metabolites.

In a related line of investigation, we explored the *in vivo* effects of infusing PTH, which also acts on PT Na^+ transporters and causes a natriuresis/diuresis (without changing blood pressure). We compared two analogues with distinct signaling patterns: PTH(1–34) couples to adenylate cyclase (AC), phospholipase C (PLC), and phospholipase A2 (PLA2); and [$Nle^{8,18}$,Tyr^{34}]PTH(3–34) couples to PLC and PLA2, but not to AC. PTH(1–34) infusion increased urinary cAMP excretion, V, and C_{Li+} two- to threefold; caused 20–25% of NHE3 and NaPi2 to redistribute out of the apical microvilli; and inhibited Na,K-ATPase activity by 25%. In contrast, [$Nle^{8,18}$,Tyr^{34}]PTH(3–34) failed to increase V or C_{Li+} and did not change NHE3 or NaPi2 distribution, but did inhibit Na,K-ATPase activity by 25%.[16] From these results, we conclude that cAMP-PKA stimulation is necessary for the natriuresis/diuresis and for NHE3 and NaPi2 internalization, that Na,K-ATPase inhibition is not secondary to depressed apical Na^+ transport, and that PTH(3–34)-driven Na,K-ATPase inhibition is not sufficient to depress transepithelial Na^+ reabsorption.

CONCLUSIONS AND OPEN QUESTIONS

This series of experiments aimed to dissect the molecular mechanisms responsible for a complex homeostatic response in which the detection of a change in blood pressure is translated into a change in sodium transport and urine output. This homeostatic response cannot be studied in cells in culture or even in isolated tubules because it requires the interface between the cardiovascular and renal systems. The use of whole-animal models limits the methods that can be applied to the aim, but has the advantage that the expression of sodium transporters, density of the microvilli, and basal infoldings are much higher than in cultured cell lines and that the discoveries made are, by definition, physiologically relevant.

One of the challenges to surmount with *in vivo* studies is to preserve the enzyme activity, subcellular distribution, and associated proteins present *in vivo* (e.g., during high blood pressure) through biochemical or cellular analyses and to interpret and connect the data collected to events *in vivo*. Relevant to the apical NHE3, we do not know whether the transporter in the intermicrovillar cleft region is transport-competent; results in the Ang II clamp experiments suggest it is significantly blunted, but results from the adult SHR suggest that it can transport Na^+.[10,13] Relevant to this volume on ATPases, we have measured a decrease in Na,K-ATPase activity in a membrane preparation of cortex and an increase in medulla during acute hypertension, but have not connected this to the *in vivo* event. We have to interpret these results very cautiously because many types of changes could alter Na,K-ATPase activity. Acute hypertension (in cortex) and an increase in volume flow (in medulla) may provoke covalent modification of the pump itself or pump-associated proteins or may change the pump environment by redistribution to an endosome or change lipid metabolism in the basolateral membrane. Such changes could affect how readily membranes containing the pump are released and vesiculated during homogenization (thus impact the specific activity of Na,K-ATPase/total protein) or could change enzymatic activity without modifying the pump itself. This array of possibilities provide hypotheses for future investigations and remain to be investigated.

ACKNOWLEDGMENTS

This work was supported by NIH Grant No. DK34316 and by fellowship support to L. E. Yang and P. K. K. Leong from the American Heart Association.

REFERENCES

1. NISHIYAMA, A. & L.G. NAVAR. 2002. ATP mediates tubuloglomerular feedback. Am. J. Physiol. Regul. Integr. Comp. Physiol. **283:** R273–R275; (discussion) R278–R279.
2. SCHNERMANN, J. 2002. Adenosine mediates tubuloglomerular feedback. Am. J. Physiol. Regul. Integr. Comp. Physiol. **283:** R276–R277; (discussion) R278–R279.
3. CHOU, C.L. & D.J. MARSH. 1988. Time course of proximal tubule response to acute arterial hypertension in the rat. Am. J. Physiol. **254:** F601–F607.
4. MURER, H., N. HERNANDO, I. FORSTER & J. BIBER. 2001. Molecular aspects in the regulation of renal inorganic phosphate reabsorption: the type IIa sodium/inorganic phosphate co-transporter as the key player. Curr. Opin. Nephrol. Hypertens. **10:** 555–561.
5. CHIBALIN, A.V., G. OGIMOTO, C.H. PEDEMONTE, *et al.* 1999. Dopamine-induced endocytosis of Na+,K+-ATPase is initiated by phosphorylation of Ser-18 in the rat alpha subunit and is responsible for the decreased activity in epithelial cells. J. Biol. Chem. **274:** 1920–1927.
6. RIBEIRO, C.M., G.R. DUBAY, J.R. FALCK & L.J. MANDEL. 1994. Parathyroid hormone inhibits Na(+)-K(+)-ATPase through a cytochrome P-450 pathway. Am. J. Physiol. **266:** F497–F505.
7. YANG, L., P.K. LEONG, J.O. CHEN, *et al.* 2002. Acute hypertension provokes internalization of proximal tubule NHE3 without inhibition of transport activity. Am. J. Physiol. Renal Physiol. **282:** F730–F740.
8. ZHANG, Y., C.E. MAGYAR, J.M. NORIAN, *et al.* 1998. Reversible effects of acute hypertension on proximal tubule sodium transporters. Am. J. Physiol. **274:** C1090–C1100.
9. LEONG, P.K., Y.B. ZHANG, L.E. YANG, *et al.* 2002. Diuretic response to acute hypertension is blunted during angiotensin II clamp. Am. J. Physiol. Regul. Integr. Comp. Physiol. **283:** R837–R842.
10. LEONG, P.K., L.E. YANG, N.H. HOLSTEIN-RATHLOU & A.A. MCDONOUGH. 2002. Angiotensin II clamp prevents the second step in renal apical NHE3 internalization during acute hypertension. Am. J. Physiol. Renal Physiol. **283:** F1142–F1150.

11. ZHANG, Y., A.K. MIRCHEFF, C.B. HENSLEY, et al. 1996. Rapid redistribution and inhibition of renal sodium transporters during acute pressure natriuresis. Am. J. Physiol. **270:** F1004–F1014.
12. MAGYAR, C.E., Y. ZHANG, N.H. HOLSTEIN-RATHLOU & A.A. MCDONOUGH. 2001. Downstream shift in sodium pump activity along the nephron during acute hypertension. J. Am. Soc. Nephrol. **12:** 2231–2240.
13. MAGYAR, C.E., Y. ZHANG, N.H. HOLSTEIN-RATHLOU & A.A. MCDONOUGH. 2000. Proximal tubule Na transporter responses are the same during acute and chronic hypertension. Am. J. Physiol. Renal Physiol. **279:** F358–F369.
14. YIP, K.P., C.M. TSE, A.A. MCDONOUGH & D.J. MARSH. 1998. Redistribution of Na+/H+ exchanger isoform NHE3 in proximal tubules induced by acute and chronic hypertension. Am. J. Physiol. **275:** F565–F575.
15. ZHANG, Y.B., C.E. MAGYAR, N.H. HOLSTEIN-RATHLOU & A.A. MCDONOUGH. 1998. The cytochrome P-450 inhibitor cobalt chloride prevents inhibition of renal Na,K-ATPase and redistribution of apical NHE-3 during acute hypertension. J. Am. Soc. Nephrol. **9:** 531–537.
16. ZHANG, Y., J.M. NORIAN, C.E. MAGYAR, et al. 1999. In vivo PTH provokes apical NHE3 and NaPi2 redistribution and Na-K-ATPase inhibition. Am. J. Physiol. **276:** F711–F719.

Ouabain as a Mammalian Hormone

WILHELM SCHONER,[a] NATALI BAUER,[b] JOCHEN MÜLLER-EHMSEN,[c]
ULRIKE KRÄMER,[d] NJDE HAMBARCHIAN,[c] ROBERT SCHWINGER,[c]
HANS MOELLER,[d] HOLGER KOST,[a] CHRISTINE WEITKAMP,[a]
THOMAS SCHWEITZER,[a] ULRIKE KIRCH,[a] HORST NEU,[b]
AND ERNST-GÜNTHER GRÜNBAUM[b]

[a]*Institute of Biochemistry and Endocrinology, Justus-Liebig-University Giessen, D-35392 Giessen, Germany*

[b]*Clinic for Small Animal Internal Medicine and Forensic Affairs, Justus-Liebig-University Giessen, D-35392 Giessen, Germany*

[c]*Third Clinic of Internal Medicine, University of Cologne, Cologne, Germany*

[d]*Clinic of Internal Medicine, University of Tübingen, Tübingen, Germany*

ABSTRACT: Endogenous ouabain changes rapidly in humans and dogs upon physical exercise and is under the control of epinephrine and angiotensin II. Hence, the steroid acts as a rapidly acting hormone. A search for a specific binding globulin for cardiac glycosides in bovine plasma resulted in the identification of the d allotype of the μ chain of IgM whose hydrophobic surfaces interact with cardiotonic steroids and cholesterol. Such IgM complexes might be involved in the hepatic elimination of cardiotonic steroids. Thus, differences in the signaling cascade starting at Na^+,K^+-ATPase must explain any differences in the action of ouabain and digoxin in the genesis of arterial hypertension.

KEYWORDS: control of endogenous ouabain; epinephrine; angiotensin II; arterial hypertension; cardiac glycoside binding protein; IgM; μ chain

INTRODUCTION

Cardiac glycosides have been used for at least 200 years to treat heart failure and some kinds of tachycardia. Ouabain, a water-soluble cardiotonic steroid from plants and a specific inhibitor of the plasmalemmal sodium pump,[1] has been identified recently in mammalian tissues.[2] Although cardiotonic steroids can be synthesized in vertebrates and mammals,[3–7] ouabain also can be taken up from the intestine at a low yield and stored in adrenal glands.[8] Thus, the question arises whether, in mammals, epinephrine and the angiotensin II may also lead to ouabain's rapid liberation from such stores as it has been shown *in vitro*.[9,10] We know this is of relevance to understand the role of ouabain in hypertension and heart failure, especially because increased concentrations of this substance have been observed in these diseases.[11,12]

Address for correspondence: Dr. Wilhelm Schoner, Institut für Biochemie und Endokrinologie, Justus-Liebig-Universität Giessen, Frankfurter Strasse 100, D-35392 Giessen, Germany. Voice: +49-641-99-38170; fax: +49-641-99-38179.
wilhelm.schoner@vetmed.uni-giessen.de

Ann. N.Y. Acad. Sci. 986: 678–684 (2003). © 2003 New York Academy of Sciences.

All cardiac glycosides are presumed to act via interaction with the sodium pump.[2] It is astonishing therefore that, in contrast with ouabain, digoxin does not produce arterial hypertension.[13] This points to different mechanisms of action of the two cardiac glycosides. Because steroid hormones seem to be directed to their target cells via binding globulins circulating in blood, it was of interest to learn whether such proteins may also exist for cardiotonic steroids or whether eventually the signaling cascade starting at the sodium pump[14] may differ for both substances. We found that the concentration of endogenous ouabain changes rapidly upon physical exercise and is under the control of epinephrine and angiotensin II. A search for a specific binding globulin for cardiac glycosides resulted in the identification of the μ chain of IgM whose hydrophobic surfaces interact with cardiotonic steroids and cholesterol. Hence, differences in the signaling cascade starting at Na^+,K^+-ATPase[14] must explain any differences in the action of ouabain and digoxin on the arousal of arterial blood pressure.

MATERIALS AND METHODS

Ergometry was performed with 26 healthy human probands (mean age, 33 years) and 21 patients with treated arterial hypertension (mean age, 64 years) for diagnostic or therapeutic reasons. Before medication was given, hypertensive patients ($n = 27$) had a systolic blood pressure exceeding 140 mmHg and/or a diastolic blood pressure exceeding 90 mmHg. Antihypertensive medication included diuretics, β-blockers, ACE inhibitors, AT1 receptor antagonists, and Ca^{2+} antagonists. Written informed consent was given by all participating subjects. The investigation was certified by the Ethics Commission of the Medical Faculties of the Universities of Tübingen and Cologne.

Ergometry in 6 healthy catheterized beagle dogs (vena cephalica antebrachii) was performed for 13 minutes on a treadmill running at 7 km/h and increasing its slope each second minute by 4° up to 20°. The regime for pretreatment of 5.5-year-old dogs (mean body weight, 17.6 kg) with the β1 adrenergic blocker, atenolol, or the ACE inhibitor, benazepril, for 3 weeks has been described.[15] Protocols were approved by the Regierungspräsidium Gießen (Gi/20/15-12/99) in agreement with the federal law of Germany. Endogenous ouabain in venous blood was determined by an ELISA or by inhibition of $^{86}Rb^+$ uptake into human red blood cells.[16] The data were analyzed statistically by use of the PRISM software of GraphPad Software (San Diego, CA) applying one-way ANOVA, Kruskal-Wallis test followed by Dunn's or Bonferroni's multiple comparison test, paired t test, and linear regression analysis.

Purification of a cardiac glycoside binding protein from bovine serum was performed according to a previous report by affinity chromatography using the affinity labeling technique with the protein-reactive HDMA (*N*-hydroxysuccimidyldigoxigenin-3-*O*-methylcarbonyl-ε-aminocaproate) as the detection system,[17] resulting in the isolation of a pure protein of 90 kDa. Affinity-purified polyclonal rabbit antibodies against this 90-kDa protein were used to screen a bovine liver lambda cDNA library (Stratagene 937712). The isolated nucleotide sequence of the d allotype of the bovine μ chain of IgM (GenBank accession number bankit489751 AY145128, gb-admin@ncbi.nim.nih.gov) was expressed in the SOLR strain of *E. coli*, and purified inclusion bodies were solubilized in 8 M urea and dialyzed against 100 mM glycine buffer, pH 8.0, containing 500 mM NaCl. Binding of cardiac glycosides and cholesterol was assayed by affinity labeling with HDMA,[17] by quench of the tryptophan fluorescence (λ_{exc} 280 nm, λ_{em} 340 nm), by increase of fluorescence of 9-anthroylouabain (A-1322) (λ_{exc} 362 nm, λ_{em} 471 nm), and by increase of fluorescence of NBD-cholesterol, 22-[N-(7-nitrobenz-2-oxa-1,3-diazol-4-yl)amino]-23,24-bisnor-5-cholen-3-ol (N-1148, www.probes.com) (λ_{exc} 468 nm, λ_{em} 538 nm), in a Hitachi L-3000 Fluorometer at 5-nm band path and 37°C in 100 mM glycine buffer, pH 8.0, containing 500 mM NaCl, 0.05% NaN_3, and 400 µg/mL protein.

RESULTS AND DISCUSSION

Ouabain Is a Rapidly Changing Hormone of Circulation

When 51 athletes were subjected for 15 minutes to excessive exercise on a bicycle or treadmill ergometer during their training program, a significant 18-fold increase of an inhibitor of the sodium pump (129.8 ± 51 nmol/L vs. 7.3 ± 1.7, $P < 0.05$) was observed in the $^{86}Rb^+$ uptake assay[16] in a serum fraction eluting from a C18 column–like ouabain. In addition, together with the increase of heart rate and lactate concentration, a 36-fold increase of endogenous ouabain from 2.4 ± 0.5 nmol/L to 86.0 ± 27.2 nmol/L ($P < 0.001$) was observed by ELISA. Half-maximal increase of heart rate was seen at a concentration of 4.3 ± 0.5 nmol/L ouabain and that of arterial systolic blood pressure at 6.9 ± 2 nmol/L. Similar observations were made in 6 beagle dogs (FIG. 1). Shortly after stopping physical exercise, endogenous ouabain decreased in both species with a half-life of 2–6 minutes to the values before exercise. Beagle dogs that had been pretreated for 3 weeks with the β-blocker, atenolol, or the angiotensin-converting enzyme inhibitor, benazepril, were not able to increase their endogenous ouabain any more, although they were running as fast on the tread-

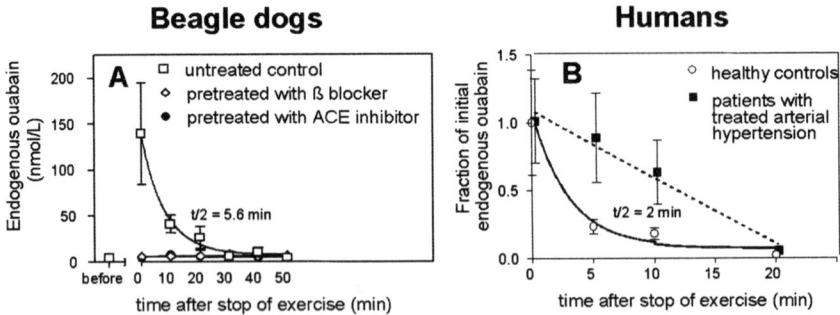

FIGURE 1. Effect of physical exercise for 15 minutes and the subsequent period of rest on the venous concentrations of endogenous ouabain in humans and dogs. (**A**) Effect of pretreatment with atenolol or benazepril on the ergometric increase of endogenous ouabain in 6 beagle dogs. (**B**) Comparison of the decay rate of endogenous ouabain upon ergometric increase in 26 healthy control persons and 21 treated hypertensive patients.

mill as untreated controls (FIG. 1A). Because 50% of Caucasians with essential hypertension have been reported to have elevated levels of endogenous ouabain, patients suffering from this disease were subjected to ergometry, and ouabain's decay was measured. A delay in the rate of decrease of endogenous ouabain from venous blood became evident in patients with treated arterial hypertension (FIG. 1B). We conclude that, in mammals, epinephrine and angiotensin, which increase together with physical exercise, release endogenous ouabain from intracellular stores. Hence, endogenous ouabain behaves like a rapidly acting hormone of circulation whose decay is deranged in patients with essential hypertension.

The μ Chain of IgM Mimics a Cardiac Glycoside Binding Protein

Chronic hypertension is induced in rats by ouabain, but not digoxin. On the contrary, digoxin and digitoxin act antihypertensive.[13] Hence, differences in the long-term action must exist between both groups of cardiotonic steroids. Steroid hormones are transported in blood as complexes with specific binding proteins. Thus, a specific binding protein for cardiac glycosides might exist in blood plasma. Serum albumin is considered so far to fulfill the transport function for cardiotonic steroids in blood,[18] but the dissociation constants of the cardiac glycoside albumin complexes are well above therapeutic digoxin concentrations. Recent search for a specific cardiac glycoside binding protein in serum resulted in the isolation of several proteins with high affinity for ouabain, that is, a protein of 50 kDa with unknown amino acid sequence and function;[17] a 14.4-kDa protein representing the plasmin-cleaved carboxy-terminal end of the heavy chain of IgG2 and IgG1;[19] and an immunoglobulin-like protein with bands of 80, 50, and 25 kDa in SDS-PAGE representing ≤0.5% of the IgG fraction.[20] In our trials to characterize the previously described

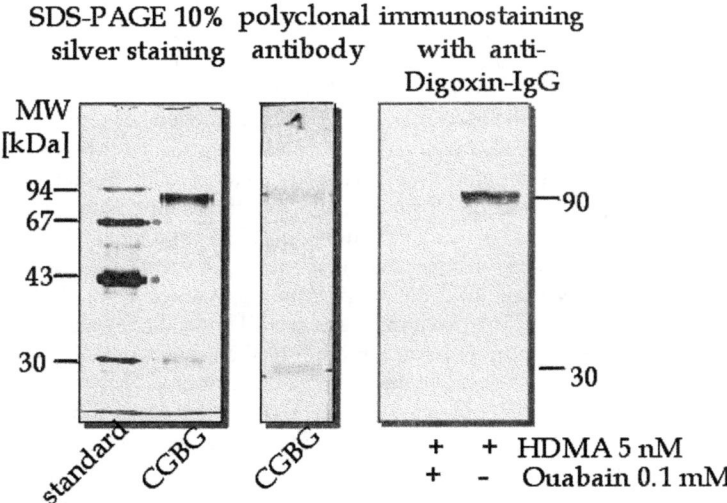

FIGURE 2. Purified cardiac glycoside binding protein (CGBG) from bovine serum can be specifically affinity-labeled by the protein-reactive HDMA.

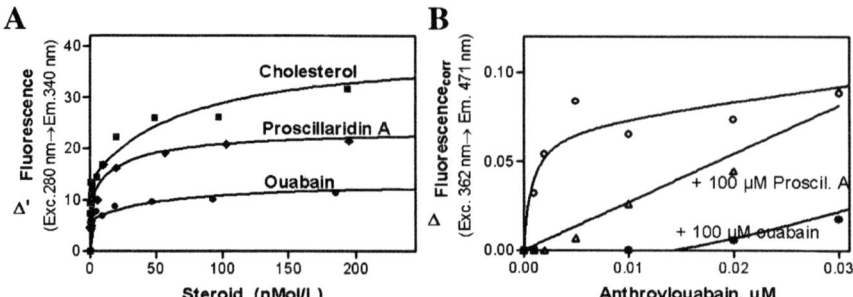

FIGURE 3. Binding of cardiac glycosides and cholesterol to purified IgM (CGBG). **(A)** Analysis of steroid binding by the quench of tryptophan fluorescence. **(B)** Analysis of cardiac glycoside binding by binding of anthroylouabain.

cardiac glycoside binding protein from bovine serum[17] any further, we were confronted with the observation that, in addition to the formerly described 50-kDa band, a 90-kDa protein also became visible in affinity labeling with HDMA. Realizing that the 50-kDa protein is a degradation product of the 90-kDa protein, we raised polyclonal antibodies in rabbits against the latter and isolated this protein by affinity chromatography (FIG. 2). The native glycoprotein reacting with several lectins was a pentadimer of 840 ± 60 kDa. The protein bound ouabain and proscillaridin A with high affinity, as is seen from the quench of its intrinsic tryptophan fluorescence (FIG. 3A). Ouabain competed better with the fluorescent anthroylouabain for its binding site than proscillaridin A (FIG. 3B). However, the protein bound cholesterol as well, as is evident from the quench of fluorescence of intrinsic tryptophan (FIG. 3A) and the increase of fluorescence of NBD-cholesterol (not shown). Unfortunately, the amino acid sequence analysis of several HPLC-purified peptides gave no clear answer on the protein's nature: from the three Lys-C peptides (KDLESHY-LFERH, DLESHYFERH, and GFAPADVFQW), the first one was not listed in SwissProt or TrEMBL data banks, the second one showed sequence similarities with the interleukin receptor and IgMμ, and the third one showed sequence similarities with IgMμc. The sequence obtained from a CNBr peptide GKLFNINKXQQAL showed similarities with an interleukin 1 receptor–like protein. To characterize the cardiac glycoside binding protein further, we measured its tissue concentration applying a competitive ELISA using an affinity-purified polyclonal rabbit antibody. We had to learn that bovine plasma contained such high protein concentrations as 1.5 ± 0.3 mg/mL (SEM) and 42 times more than the average of 11 bovine tissues (0.035 ± 0.02 mg/g wet weight). To understand whether the above-mentioned amino acid sequences represent impurities of a protein or the protein itself, we performed an antibody screening of a bovine liver lambda cDNA library. Nucleotide sequence analysis of the isolated cDNA clone and a BLAST database search resulted in the conclusion that the isolated bovine plasma protein represents the so far unknown d allotype of the bovine μ chain of IgM. Expression of this protein in *E. coli* and its solubilization from purified inclusion bodies allowed us to test whether the d allotype of the μ chain of IgM may bind HDMA, cardiac glycosides, and cholesterol. In fact, HDMA labeling of the expressed μ chain was suppressed by 100 μM ouabain, and cholesterol binding was seen by tryptophan fluorescence quench (not shown).

Apart from the new observation that cardiac glycosides bind to the µ chain of IgM, binding of cholesterol to IgM has been observed earlier: IgM is known to promote the nucleation of cholesterol (better than IgG), a finding relevant for the formation of cholesterol gallstones.[21] IgM is catabolized in liver endothelial cells[22] with a half-life comparable to the elimination rate of cardiac glycosides. Hence, IgM may serve in addition to serum albumin as a transporter for cardiac glycosides to the liver to eliminate the steroids there, especially in cases of analbuminemia.[23] Apparently, differences in long-term action between ouabain and digoxin in increasing arterial hypertension[13] do not seem to be caused by a pathway guiding cardiotonic steroids like other steroid hormones to the nucleus. Any differences in the action of ouabain and digoxin must lie in differences in the intracellular signaling cascade[14] or in the sensitivity of terminal target cells or of cells involved in ouabain's hormonal control.[2]

ACKNOWLEDGMENTS

This work was supported by the Akademie für Tiergesundheit, Bonn; Deutsche Forschungsgemeinschaft, Bonn-Bad Godesberg; Fonds der Chemischen Industrie, Frankfurt/M (to W. Schoner); and Köln Fortune (to J. Müller-Ehmsen).

REFERENCES

1. SCHATZMANN, H.J. 1953. Herzglykoside als Hemmstoffe für den aktiven Kalium- und Natriumtransport durch die Erythrocytenmembran. Helv. Physiol. Pharmacol. Acta **11:** 346–354.
2. SCHONER, W. 2002. Endogenous cardiac glycosides, a new class of steroid hormones. Eur. J. Biochem. **269:** 2440–2448.
3. HAMLYN, J., Z. LU, P. MANUNTA, et al. 1998. Observations on the nature, biosynthesis, secretion, and significance of endogenous ouabain. Clin. Exp. Hypertens. **20:** 523–533.
4. DORIS, P.A., A. HAYWARD-LESTER, D. BOURNE, et al. 1996. Ouabain production by cultured adrenal cells. Endocrinology **137:** 533–539.
5. PERRIN, A., B. BRASMES, E.M. CHAMBAZ, et al. 1997. Bovine adrenocortical cells in culture synthesize an ouabain-like compound. Mol. Cell. Endocrinol. **126:** 7–15.
6. LAREDO, J., B.P. HAMILTON & J.M. HAMLYN. 1994. Ouabain is secreted by bovine adrenocortical cells. Endocrinology **135:** 794–797.
7. KOMIYAMA, Y., N. NISHIMURA, M. MUNAKATA, et al. 2001. Identification of endogenous ouabain in culture supernatant of PC12 cells. J. Hypertens. **19:** 229–236.
8. KITANO, S., S. MORIMOTO, A. NISHIBE, et al. 1998. Exogenous ouabain is accumulated in the adrenals and mimics the kinetics of endogenous digitalis-like factor in rats. Hypertens. Res. **21:** 47–56.
9. HAMLYN, J.M., M.P. BLAUSTEIN, S. BOVA, et al. 1991. Identification and characterization of a ouabain-like compound from human plasma. Proc. Natl. Acad. Sci. USA **88:** 6259–6263.
10. LAREDO, J., J.R. SHAH, Z. LU, et al. 1997. Angiotensin II stimulates secretion of endogenous ouabain from bovine adrenal cortical cells via angiotensin II receptors. Hypertension **29:** 401–407.
11. MANUNTA, P., P. STELLA, R. RIVERA, et al. 1999. Left ventricular mass, stroke volume, and ouabain-like factor in essential hypertension. Hypertension **34:** 450–456.
12. GOTTLIEB, S.S., A.C. ROGOWSKI, M. WEINBERG, et al. 1992. Elevated concentrations of endogenous ouabain in patients with congestive heart failure. Circulation **86:** 420–425.

13. MANUNTA, P., J. HAMILTON, A.C. ROGOWSKI, et al. 2000. Chronic hypertension induced by ouabain, but not digoxin in the rat: antihypertensive effect of digoxin and digitoxin. Hypertens. Res. **23:** S77–S85.
14. XIE, Z. & A. ASKARI. 2002. Na$^+$/K$^+$-ATPase as a signal inducer. Eur. J. Biochem. **269:** 2434–2439.
15. KITTLESON, M.D. & R.D. KIENLE. 1998. Small Animal Cardiovascular Medicine: pp. 95–117. Mosby. St. Louis.
16. SCHNEIDER, R., V. WRAY, M. NIMTZ, et al. 1998. Bovine adrenals contain, in addition to ouabain, a second inhibitor of the sodium pump. J. Biol. Chem. **273:** 784–792.
17. ANTOLOVIC, R., H. KOST, M. MOHADJERANI, et al. 1998. A specific binding protein for cardiac glycosides exists in bovine serum. J. Biol. Chem. **273:** 16259–16264.
18. KRIEGLSTEIN, J. 1981. Plasma protein binding of cardiac glycosides. Handb. Exp. Pharmacol. **56:** 95–104.
19. KOMIYAMA, Y., N. NISHIMURA, N. NISHINO, et al. 1998. Purification and characterization of ouabain-binding protein in human plasma. Clin. Exp. Hypertens. **20:** 683–690.
20. PARHAMI-SEREN, B., R. HABERLY, M.N. MARGOLIES, et al. 2002. Ouabain-binding protein(s) from human plasma. Hypertension **40:** 220–228.
21. UPADHYA, G.A., P.R. HARVEY & S.M. STRASBERG. 1993. Effect of human biliary immunoglobulins on the nucleation of cholesterol. J. Biol. Chem. **268:** 5193–5200.
22. CHRONEOS, Z.C., J.W. BAYNES & S.R. THORPE. 1995. Identification of liver endothelial cells as the primary site of IgM catabolism in the rat. Arch. Biochem. Biophys. **319:** 63–73.
23. WATKINS, S., J. MADISON, M. GALLIANO, et al. 1994. Analbuminemia: three cases resulting from different point mutations in the albumin gene. Proc. Natl. Acad. Sci. USA **91:** 9417–9421.

11-Hydroxylation in the Biosynthesis of Endogenous Ouabain: Multiple Implications

JOHN M. HAMLYN,[a] JAMES LAREDO,[a] JUI R. SHAH,[a] ZHUO REN LU,[a] AND BRUCE P. HAMILTON[b,c]

[a]*Department of Physiology,* [b]*Department of Medicine, University of Maryland, Baltimore, Maryland 21201, USA*

[c]*Veterans Administration Medical Center, Baltimore, Maryland 21201, USA*

ABSTRACT: Accumulating evidence indicates that mammals use steroidal glycosides with "digitalis-like" activity. An endogenous ouabain (EO) has been described and is linked with long-term changes in sodium balance and cardiovascular structure and function. In the adrenal gland, the biosynthesis of EO and similar compounds appears to involve cholesterol side-chain cleavage with sequential metabolism of pregnenolone and progesterone. The more distal events in the biosynthesis have not been elucidated. Preliminary work using primary cell cultures from the bovine adrenal cortex suggests that the biosynthesis of EO is affected by inhibitors of 11β-hydroxylase. Direct participation of 11-hydoxylase in EO synthesis would lead to an 11β isomer of ouabain in mammals and, *in vivo*, an 11β-oriented hydroxyl group would spontaneously form a mixture of two 11–19 hemiketal isomers. The latter isomers would likely be converted back to a single 11β isomer of ouabain during isolation. The existence of an additional ring in the hemiketals, along with reduced flexion of the steroidal A, B, and C rings, raises the possibility that their *in vivo* physiological targets and actions differ from the isolated form of EO.

KEYWORDS: metabolism; metyrapone; adrenal

INTRODUCTION

It has long been known that certain animals and selected plants synthesize cardiotonic steroids with five- and six-member lactone rings.[1,2] For many decades, the existence and chemical nature of putative mammalian analogues of the cardiac glycosides remained unclear until a bioactive compound was isolated from human plasma[3] and was identified as ouabain or a closely related isomer.[4,5] Subsequently, compounds resembling endogenous ouabain (EO) have been described repeatedly in fluids and tissues from numerous mammalian species[6–10] by a variety of means including sophisticated analytical methods.[11,12]

Address for correspondence: John M. Hamlyn, Ph.D., FAHA, Department of Physiology, University of Maryland, 655 West Baltimore Street, Baltimore, MD 21201. Voice: 410-706-3479; fax: 410-706-8341.

jhamlyn@umaryland.edu

We use the term, endogenous ouabain (EO), for the human compound because the circulating form does not have an exogenous origin in humans or rats,[13,14] and there is evidence for biosynthesis mediated by the adrenal glands.[4,15–18] In humans, circulating levels of EO are modulated by salt intake so that EO acts in concert with aldosterone to enhance the sodium-retaining ability of the kidney.[19] In the pathophysiological setting, elevated plasma levels of EO occur with high frequency among patients with congestive heart failure[20] and patients with common hypertension, especially those with coexistent cardiac hypertrophy.[21,22] Significantly, many of the physiological and deleterious effects of high EO and ouabain on the cardiovascular system can be blocked by active or passive immunization or blockade of the renin-angiotensin system.[23–25]

In addition to circulating EO, the EO content of tissues is regulated and is independent of the circulating levels in some (e.g., adrenal cortex, skeletal muscle), but not other (kidney, hypothalamus) tissues.[18] The existence of a complex and as yet not understood system that regulates tissue ouabain not only implies a role for intracellular EO, but may be relevant to the possibility of limited extra-adrenal biosynthesis of EO. It is of interest that, at least in the rat, there is no simple or significant relationship between the tissue content and the nature or number of the expressed Na pump isoform.[26] This suggests that the internalization of the ouabain complexed sodium pump[27] may be less important in setting the ambient tissue level of EO than previously thought and that another mechanism is involved.

EO BIOSYNTHESIS

Many laboratories have presented evidence consistent with EO biosynthesis. First, the constancy of plasma EO in individuals given total parenteral nutrition excludes the diet as a key source.[13] Second, in the rat, the content of EO is highest in the adrenal gland, and blood levels of EO decrease after adrenalectomy.[4,15,18] Third, an inverse arteriovenous gradient for EO was found across the adrenal glands in conscious dogs,[16] and secretions by the intact perfused rat adrenal gland[28] and brain tissue[29] also point to biosynthesis. Fourth, secretion of EO into the media by cultured adrenocortical cells is activated by the classic hormonal stimulators of adrenal function.[17,30,31] Hormone stimulation has no effect on cellular EO levels, while the amount of EO secreted far exceeds its cell content,[17,32] suggesting metabolic transformation of a precursor. In addition, the biosynthesis of EO is normally rate-limited and modulated.

Among the classically recognized adrenocortical steroids, all arise from the rate-limiting side-chain cleavage of cholesterol. Consistent with this view, hydroxycholesterol, pregnenolone, and progesterone stimulate secretion of EO and related compounds.[9,10,33–36] The mechanism of the stimulatory effect, assumed to reflect utilization of these compounds as substrates, has not been proved. Nevertheless, when taken together, the evidence is that EO (and related five-member lactone ring analogues) is a product of an adrenal steroidogenic pathway arising from cholesterol side-chain cleavage. The biosynthesis secretion of bufadienolides (with a six-member lactone ring) may be independent of cholesterol side-chain cleavage.[37]

The synthetic pathway for cardiotonic steroids in plants can use progesterone and deoxycorticosterone,[38,39] although the more distal events, like those in the mammalian pathway, remain unknown. Moreover, the rate of EO biosynthesis in mammals is significantly slower than that for the corticosteroids. Thus, studies based on substrate utilization, inhibitors, and tracer methods in combination with chromatographic and mass spectral analysis will be increasingly important.

METYRAPONE INHIBITS EO SECRETION

The observations that angiotensin II and ACTH stimulate EO secretion[17,30,31] together with the effects of pregnenolone and progesterone[33,35] led us to suspect that EO might be synthesized via an aldosterone-like pathway. The enzyme, 11β-hydroxylase, hydroxylates 11-deoxycorticosterone to form corticosterone and is obligatory in aldosterone biosynthesis. Metyrapone, a selective inhibitor of 11β-hydroxylase, inhibited the angiotensin II–stimulated secretions of EO and aldosterone (FIG. 1). The simplest conclusion is that 11β-hydroxylase activity is involved in EO biosynthesis and that 11-deoxycorticosterone and/or corticosterone are intermediates in EO biosynthesis. However, the inhibitory effect of metyrapone on EO synthesis could be mediated elsewhere in the EO pathway, and additional studies will be required.

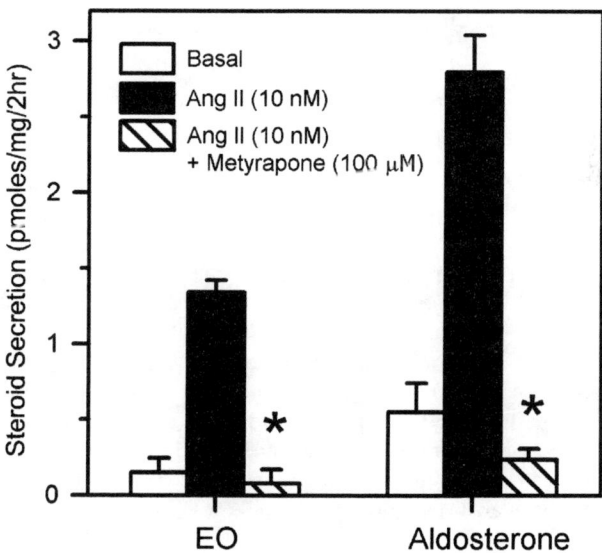

FIGURE 1. Effect of metyrapone on EO and aldosterone secretion by bovine adrenocortical cells. Primary cultures of mixed adrenocortical cells were bathed for 2 hours with serum-free media. The secreted EO and aldosterone were measured by their respective assays. Shown are the steroid secretions under basal (control) conditions and in response to angiotensin II (10 nM) with and without 100 μM metyrapone. No basal secretion of EO or aldosterone was detected in the presence of metyrapone (not shown). Means ± SEM for four replicates; *$P < 0.01$ vs. angiotensin II (Ang II).

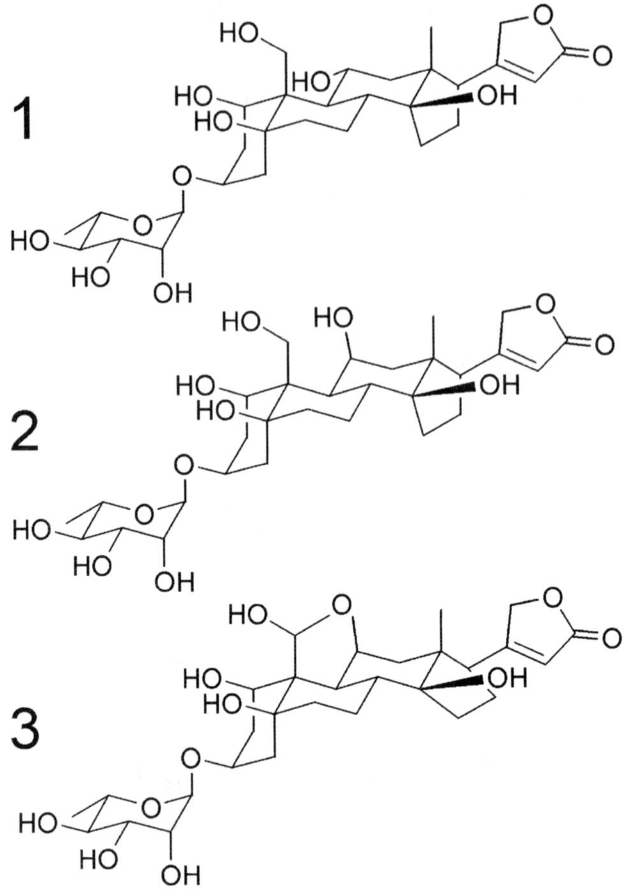

FIGURE 2. Structures of ouabain (1), its 11β isomer (2), and the 11–19 hemiketal (3).

In the interim, there are significant implications concerning the putative involvement of 11-hydroxylase in EO biosynthesis. The most striking of these concerns is the orientation of the hydroxyl group on carbon 11. In plant-derived ouabain, the hydroxyl is in the α orientation (FIG. 2). Two enzymes in the adrenal cortex catalyze 11-hydroxylation in the β orientation, so the secreted EO in FIGURE 1 may be an isomer of ouabain with an 11β-hydroxyl group.

WHAT IS THE ISOLATED FORM OF EO?

From several studies (see TABLE 1) that have focused primarily on the isolated form of ouabain-like compounds from humans and cattle,[4,5,7,10–12,40–44] we can highlight five important points. First, multiple laboratories using different methods and sources confirm the existence of a molecule whose signatures strikingly resem-

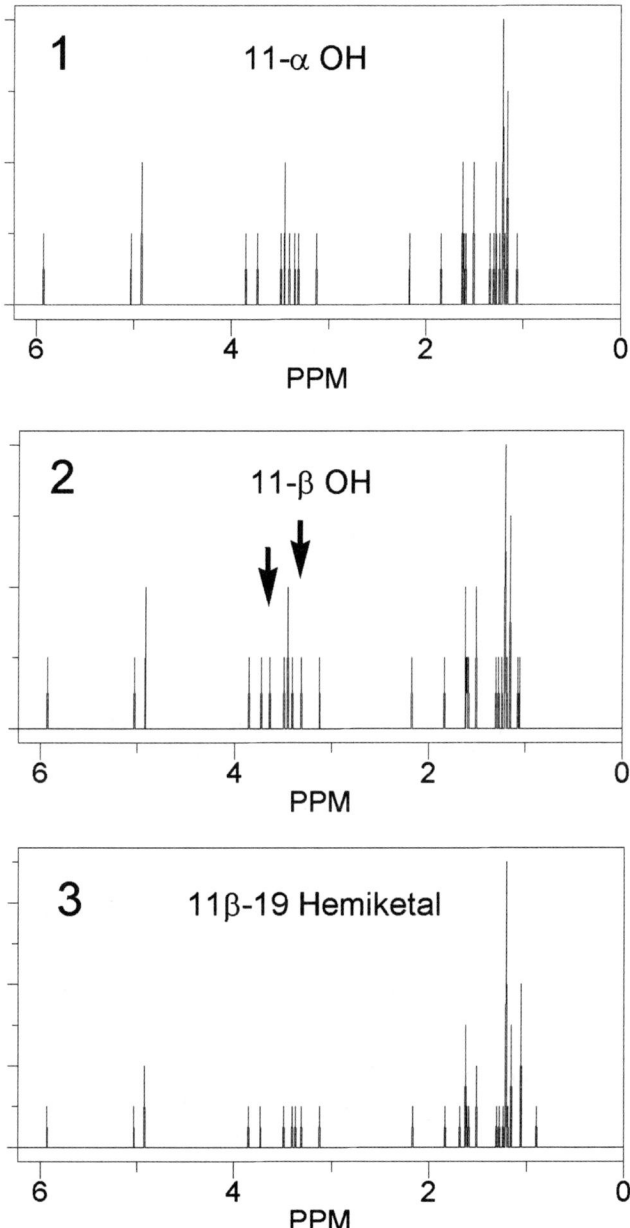

FIGURE 3. Computed proton NMR spectra for ouabain and two analogues. Spectra were calculated from the structures in FIGURE 2 by using NMR Proversion 5 (Chem Draw Ultra, Cambridge Soft, Cambridge, MA).

TABLE 1. Summary of structural studies for endogenous ouabain

Study Number	Source, Amount for Analysis	Analytical Method	Identity of Isolate	Reference
1	Human plasma, 31 µg	LC, FAB-MS native and acetylated derivatives FAB MS-MS	Ouabain or a closely related isomer	4, 5, 40
2	Human plasma	Reanalysis of study 1. See Ref. 42	Ouabain beyond a reasonable doubt	41, 42
3	Bovine hypothalamus, 1 µg	LC-MS circular dicroism	Ouabain isomer	43
4	Bovine hypothalamus, 300 ng	Circular dicroism	Ouabain isomer	44
5	Human plasma	Reanalysis of study 1. See Ref. 5	Ouabain isomer	7
6	Bovine adrenal, 20 µg	LC-ESI-MS-MS ^{1}H-NMR	Ouabain	11
7	Bovine hypothalamus, 3 µg	^{1}H-NMR	Ouabain	12
8	PC-12 cell media, 100 ng	LC-ESI-MS-MS	Ouabain	10

ABBREVIATIONS: FAB, fast atom bombardment; MS, mass spectrometry; MS-MS, tandem mass spectrometry; LC, liquid chromatography; NMR, nuclear magnetic resonance spectrometry.

ble ouabain and not other cardiac glycosides. Second, most of the work has been performed starting from bovine tissue.[11,12,43,44] Third, ^{1}H-NMR spectra are available only for the bovine compound. Fourth, conclusions regarding isomers based on MS or MS-MS analyses may be flawed because the technique is poorly suited for this purpose. NMR spectrometry is a direct means to determine whether or not EO is a positional and/or orientation isomer of ouabain. A drawback of the method is its relative insensitivity requiring the large-scale isolation for low-abundance compounds. Fifth, in studies where NMR was used, the isolate spectra were interpreted as ouabain.[11,12]

Because the NMR conclusions appeared to contradict the inference from the metyrapone studies mentioned above, we computed the ^{1}H-NMR spectra for ouabain, the 11β isomer, and an 11–19 hemiketal analogue to assess possible differences (1, 2, and 3, respectively, in FIG. 3). The spectra for ouabain and the isomer are virtually identical; only with careful inspection does the small change in the shift of the ^{1}H resonance at carbon 11 (ouabain, ~3.3 ppm; isomer, ~3.6 ppm) become apparent (arrowheads in FIG. 3). Inspection of the published NMR spectra for these potentially overlooked changes proved to be interesting. We focused first on the NMR spectra for the bovine adrenal EO from the Giessen group[11] because of their excellent quality. Unique proton resonances with shifts of approximately 3.67, 3.46, and 3.22 ppm were found in the adrenal EO spectra. In addition, the ^{1}H resonance at approximately 4.6 ppm was reduced in intensity and broadened, and that at approximately 1.9 ppm showed heightened intensity relative to ouabain. For the bovine hypothalamic EO, the NMR spectra were noisy and complex because of limited sample and formation of trigonal borate complexes.[12] However, those spectra contained an unusual series of ^{1}H resonances with shifts ranging from 3.5 ppm to 3.65 ppm not observed for ouabain. Thus, selected features of the ^{1}H-NMR spectra observed for the bovine adrenal and hypothalamic EO are consistent with those computed for the 11β isomer of ouabain, and they suggest also that the bovine samples may have been a mixture of the 11β isomer and ouabain, with the latter potentially originating as a contaminant from column calibration.[40–44]

IMPLICATIONS OF 11β-HYDROXYLATION

There are several connotations concerning an 11β-hydroxyl in EO. First, the *in vivo* circulating form of EO would likely be a mixture of the 11–19 hemiketals. Second, formation of the hemiketals may protect EO from 11β-hydroxysteroid dehydrogenase, which converts the more abundant 11-hydroxylated steroids *in vivo* (e.g., cortisol) to their respective ketones, thereby deactivating their mineralocorticoid activity. Third, the hemiketals are likely to have increased rigidity in the steroidal A, B, and C rings. Reduced flexion in the steroidal A ring significantly reduces the biological potency of ouabain on the sodium pump.[45] Thus, the physiological and biochemical mechanism of action of the hemiketals *in vivo* may differ from their isolated form in which all the hydroxyls are free. Finally, the observation that EO may be synthesized in part via the mineralocorticoid pathway points to physiological and pathological roles for EO in many of the same conditions in which aldosterone is normally or inappropriately altered.

REFERENCES

1. REICHSTEIN, T. 1967. Cardenolid- und pregnanglykoside. Die Naturwissenschaften **3**: 53–76.
2. HORIGER, N. *et al.* 1970. Cardenolide hydrogen suberates and other bufadienolide hydrogen suberates in Ch'an Su. Helv. Chim. Acta **53**: 1993–2002.
3. HAMLYN, J.M. *et al.* 1989. Digitalis-like activity in human plasma: purification, affinity, and mechanism. J. Biol. Chem. **264**: 7395–7404.
4. HAMLYN, J.M. *et al.* 1991. Identification and characterization of a ouabain-like compound from human plasma. Proc. Natl. Acad. Sci. USA **88**: 6259–6263.
5. MATHEWS, W.R. *et al.* 1991. Mass spectral characterization of an endogenous digitalislike factor from human plasma. Hypertension **17**: 930–935.
6. GOTO, A. *et al.* 1992. Physiology and pharmacology of endogenous digitalis-like factors. Pharmacol. Rev. **44**: 377–399.
7. HAMLYN, J.M. *et al.* 1996. Endogenous ouabain, sodium balance, and blood pressure: a review and a hypothesis. J. Hypertens. **14**: 151–167.
8. EL-MASRI, M.A. *et al.* 2002. Human adrenal cells in culture produce both ouabain-like (OLF) and dihydroouabain-like (Dh-OLF) factors. Clin. Chem. **48**: 1720–1730.
9. SCHONER, W. 2002. Endogenous cardiac glycosides, a new class of steroid hormones. Eur. J. Biochem. **269**: 2440–2448.
10. KOMIYAMA, Y. *et al.* 2001. Identification of endogenous ouabain in culture supernatant of PC12 cells. J. Hypertens. **19**: 229–236.
11. SCHNEIDER, R. *et al.* 1998. Bovine adrenals contain, in addition to ouabain, a second inhibitor of the sodium pump. J. Biol. Chem. **273**: 784–792.
12. KAWAMURA, A. *et al.* 1999. On the structure of endogenous ouabain. Proc. Natl. Acad. Sci. USA **96**: 6654–6659.

13. HARRIS, D.W. *et al.* 1991. Development of an immunoassay for endogenous digitalis-like factor. Hypertension **17:** 936–943.
14. FERRANDI, M. *et al.* 1995. Age-dependency and dietary influence on the hypothalamic ouabain-like factor in Milan hypertensive rats. J. Hypertens. **13:** 1571–1574.
15. LUDENS, J.H. *et al.* 1992. Rat adrenal cortex is a source of a circulating ouabainlike compound. Hypertension **19:** 721–724.
16. BOULANGER, B.R. *et al.* 1993. Ouabain is secreted by the adrenal gland in awake dogs. Am. J. Physiol. **264:** E413–E419.
17. LAREDO, J. *et al.* 1994. Ouabain is secreted by bovine adrenocortical cells. Endocrinology **135:** 794–797.
18. MANUNTA, P. *et al.* 1994. Ouabain-induced hypertension in the rat: relationships among plasma and tissue ouabain and blood pressure. J. Hypertens. **12:** 549–560.
19. MANUNTA, P. *et al.* 2001. Plasma ouabain-like factor during acute and chronic changes in sodium balance in essential hypertension. Hypertension **38:** 198–203.
20. GOTTLIEB, S.S. *et al.* 1992. Elevated concentrations of endogenous ouabain in patients with congestive heart failure. Circulation **86:** 420–425.
21. PIERDOMENICO, S.D. *et al.* 2001. Endogenous ouabain and hemodynamic and left ventricular geometric patterns in essential hypertension. Am. J. Hypertens. **14:** 44–50.
22. MANUNTA, P. *et al.* 1999. Left ventricular mass, stroke volume, and ouabain-like factor in essential hypertension. Hypertension **34:** 450–456.
23. AILERU, A.A. *et al.* 2001. Synaptic plasticity in sympathetic ganglia from acquired and inherited forms of ouabain-dependent hypertension. Am. J. Physiol. Regul. Integr. Comp. Physiol. **281:** R635–R644.
24. ZHANG, J. & F.H. LEENEN. 2001. AT1 receptor blockers prevent sympathetic hyperactivity and hypertension by chronic ouabain and hypertonic saline. Am. J. Physiol. Heart Circ. Physiol. **280:** H1318–H1323.
25. HUANG, B.S. *et al.* 2000. Chronic blockade of brain "ouabain" prevents sympathetic hyper-reactivity and impairment of acute baroreflex resetting in rats with congestive heart failure. Can. J. Physiol. Pharmacol. **78:** 45–53.
26. LI, S. *et al.* 1998. Bovine adrenals and hypothalamus are a major source of proscillaridin A– and ouabain-immunoreactivities. Life Sci. **62:** 1023–1033.
27. COOK, J.S. *et al.* 1982. Uptake of [^{3}H]ouabain from the cell surface into the lysosomal compartment of HeLa cells. J. Cell. Physiol. **110:** 84–92.
28. HINSON, J.P. *et al.* 1998. Release of ouabain-like compound (OLC) from the intact perfused rat adrenal gland. Endocrinol. Res. **24:** 721–724.
29. DE ANGELIS, C. & G.T. HAUPERT, JR. 1998. Hypoxia triggers release of an endogenous inhibitor of Na(+)-K(+)-ATPase from midbrain and adrenal. Am. J. Physiol. **274:** F182–F188.
30. BECK, M. *et al.* 1996. Production of ouabain by rat adrenocortical cells. Endocrinol. Res. **22:** 845–849.
31. SHAH, J.R. *et al.* 1998. Different signaling pathways mediate stimulated secretion of endogenous ouabain and aldosterone from bovine adrenocortical cells. Hypertension **31:** 463–468.
32. LAREDO, J. *et al.* 2000. Alpha-1 adrenergic receptors stimulate secretion of endogenous ouabain from human and bovine adrenocortical cells. *In* Na/K-ATPase and Related ATPases: 671–679. Elsevier Science. New York.
33. PERRIN, A. *et al.* 1997. Bovine adrenocortical cells in culture synthesize an ouabain-like compound. Mol. Cell. Endocrinol. **126:** 7–15.
34. HAMLYN, J.M. *et al.* 1998. Observations on the nature, biosynthesis, secretion, and significance of endogenous ouabain. Clin. Exp. Hypertens. **20:** 523–533.
35. LU, Z-R. *et al.* 1998. Biosynthesis of endogenous ouabain from pregnenolone and progesterone in bovine adrenal cortical cells. Hypertension **32:** P131.
36. LICHTSTEIN, D. *et al.* 1998. Biosynthesis of digitalis-like compounds in rat adrenal cells: hydroxycholesterol as possible precursor. Life Sci. **62:** 2109–2116.
37. DMITRIEVA, R.I. *et al.* 2000. Mammalian bufadienolide is synthesized from cholesterol in the adrenal cortex by a pathway that is independent of cholesterol side-chain cleavage. Hypertension **36:** 442–448.

38. CASPI, E. & D.O. LEWIS. 1967. Progesterone: its possible role in the biosynthesis of cardenolides in *Digitalis lanata*. Science **156:** 519–520.
39. CASPI, E. *et al.* 1968. The role of deoxycorticosterone in the biosynthesis of cardenolides in *Digitalis lanata*. Biochem. J. **108:** 499–503.
40. LUDENS, J.H. *et al.* 1991. Purification of an endogenous digitalislike factor from human plasma for structural analysis. Hypertension **17:** 923–929.
41. HAMLYN, J.M. & P. MANUNTA. 1992. Ouabain, digitalis-like factors, and hypertension. J. Hypertens. Suppl. **10:** S99–S111.
42. SANDOR, T. & D.R. IDLER. 1972. Steroid methodology. *In* Steroids in Nonmammalian Vertebrates: 6–36. Academic Press. New York.
43. TYMIAK, A.A. *et al.* 1993. Physicochemical characterization of a ouabain isomer isolated from bovine hypothalamus. Proc. Natl. Acad. Sci. USA **90:** 8189–8193.
44. ZHAO, N. *et al.* 1995. Na,K-ATPase inhibitors from bovine hypothalamus and human plasma are different from ouabain: nanogram scale CD structural analysis. Biochemistry **34:** 9893–9896.
45. KAWAMURA, A. *et al.* 2001. Biological implication of conformational flexibility in ouabain: observations with two ouabain phosphate isomers. Biochemistry **40:** 5835–5844.

Antihypertensive Compounds That Modulate the Na-K Pump

P. FERRARI,[a] M. FERRANDI,[a] L. TORIELLI,[a] P. BARASSI,[a] G. TRIPODI,[a] E. MINOTTI,[a] I. MOLINARI,[a] P. MELLONI,[a] AND G. BIANCHI[b]

[a]*Prassis Research Institute, Sigma Tau, Milan, Italy*

[b]*Department of Nephrology, Dialysis, and Hypertension, Vita e Salute University, S. Raffaele Hospital, Milan, Italy*

ABSTRACT: A primary impairment of the kidney sodium excretion has been documented both in hypertensive patients (EH) and genetic animal models (Milan hypertensive rat [MHS]) carrying mutations of the cytoskeletal protein adducin and/or increased plasma levels of endogenous ouabain (EO). Ouabain (OU) itself induces hypertension in rats and both OU and mutated adducin activate the renal Na/K-ATPase function both *in vivo* and in cultured renal cells (NRK). A new antihypertensive agent, PST 2238, able to selectively interact with these alterations has been developed. PST lowers blood pressure (BP) by normalizing the expression and activity of the renal Na-K pump selectively in those rat models carrying the adducin mutation (MHS) and/or increased EO levels (OS) at oral doses of 0.1–10 µg/kg. In NRK cells either transfected with mutated adducin or incubated with 10^{-9} M OU, PST normalizes the Na-K pump activity. Recently, an association between EO and cardiac complications has been observed in both EH and rat models consistent with a prohypertrophic activity of OU. OS rats showed a 10% increase of left ventricle and kidney weights as compared with controls, and PST 2238 (1 µg/kg OS) prevented both ventricle and renal hypertrophy. This effect was associated with the ability of PST to antagonize the OU-dependent activation of growth-related genes, in the membrane subdomains of caveolae. In conclusion, PST is a new antihypertensive agent that may prevent cardiovascular complications associated with hypertension through the selective modulation of the Na-K pump function.

KEYWORDS: Na-K pump; adducin; endogenous ouabain; hypertension; MHS rats; PST 2238

INTRODUCTION

Essential hypertension is a complex disease in which genetic and environmental factors interact to result in a final blood pressure increase and the risk to develop specific cardiac and renal complications.[1,2] Although the efficacy of available antihypertensive drugs is widely recognized, there is much individual variability in the response to a given therapy.[3] A population survey has demonstrated that approximately less than 30% of patients are adequately treated. These problems arise mainly from the lack of a complete understanding of the mechanisms leading to hypertension and its organ complications. Therefore, the success of a new therapy for hypertension will depend on the understanding of these mechanisms operating in a subset of patients and the ability of the new drug to correct them.

For many years, our group has been studying the genetic and hormonal mechanisms that cause hypertension in at least a subgroup of patients characterized by a primary defect of the kidney to excrete sodium. Using the Milan hypertensive rat (MHS) as a model for human disease,[4] we have demonstrated mutations of the genes coding for the cytoskeletal protein adducin[5,6] and increased circulating levels of endogenous ouabain (EO)[7–9] in both species and found them to be associated with increased tubular sodium reabsorption and hypertension. Furthermore, increased expression and function of the renal Na-K pump has been demonstrated both in MHS[10] and in rats made experimentally hypertensive by infusion of low doses of ouabain (OS rats),[11] suggesting that the pump upregulation is a common feature of both adducin- and EO-dependent hypertension. In addition, adducin polymorphism and circulating EO have been demonstrated recently to play a critical role also in favoring cardiac and renal complications associated with hypertension in both rat models[12] and hypertensive subjects, studied at different stages of hypertension.[13–16] All these findings led us to start with a pharmacological project aimed at developing a new class of antihypertensive compounds able to correct the specific genetic-molecular mechanisms, namely, the altered function and expression of the Na-K pump, demonstrated to be involved in both rat and human hypertension. Among more than 800 original molecules, compound PST 2238 has been selected for its ability to lower blood pressure and normalize the Na-K pump function in MHS[17] and OS[11] rats.

METHODS AND RESULTS

PST 2238, the Novel Antihypertensive Compound

PST 2238 [17β-(3-furyl)-5β-androstan-3β,14β,17α-triol] is a digitoxigenin derivative able to displace *in vitro* ^{3}H-ouabain specific binding from Na-K ATPase with an IC_{50} of 2 μM.[11] The specificity of its interaction with the Na-K ATPase was confirmed by the absence of any significant *in vitro* binding to a panel of general and hormonal receptors, known to be involved in blood pressure regulation or hormonal steroid control.[11]

Interaction of PST 2238 with the Na-K Pump in Cell Cultures

The ability of PST 2238 to modulate the Na-K pump was tested in cultured normal rat kidney cells (NRK-52E) either transfected with the mutated variant of rat α-adducin or incubated with nanomolar concentrations of ouabain to mimic the *in vivo* situation of increased levels of EO. In fact, in both conditions (adducin polymorphism[17–19] and high EO[11]), the cell Na-K pump is upregulated as compared with the respective control cells.

NRK cells incubated with 10^{-9} M ouabain for 5 days show an increased V_{max} of the Na-K pump (FIG. 1a).[11] PST 2238 (from 10^{-16} to 10^{-9} M) normalizes the pump

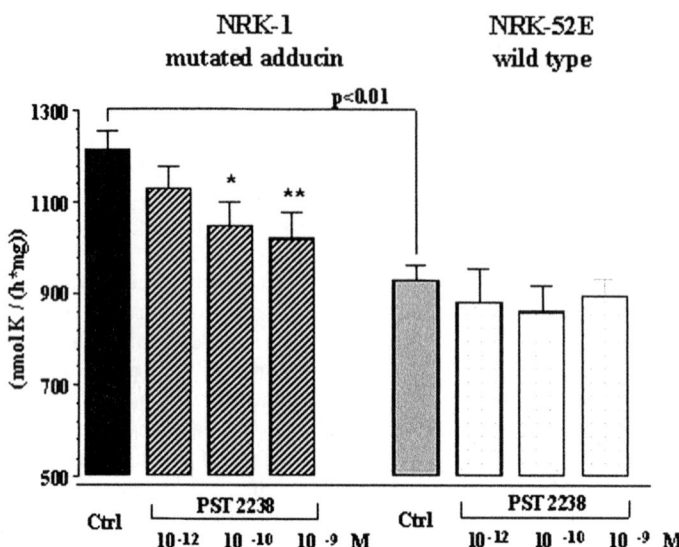

FIGURE 1. Effect of 5-day incubation of both NRK-1 (stably transfected with mutated α-adducin cDNA) and wild-type NRK-52E cells with PST 2238 from 10^{-12} to 10^{-9} M on the Na-K pump activity at V_{max} (Na-loaded cells, ^{86}Rb uptake). Data are mean ± SEM; *$P < 0.05$; **$P < 0.01$ vs. NRK-1 control cells.

activity to the level of the control NRK cells (FIG. 1b),[17] which are not affected by the compound *per se*.[11]

NRK cells stably transfected with the rat "hypertensive" α-adducin variant (F316Y) show an increased surface expression and activity of Na-K pump at V_{max} compared with NRK cells expressing the "normotensive" α-adducin variant (FIG. 2).[19] A 5-day incubation with PST 2238 (10^{-12} to 10^{-9} M) significantly reduces the V_{max} Na-K pump activity in hypertensive NRK-1 cells, but not in the wild-type NRK cells (FIG. 2).[17]

Therefore, these experiments suggest that PST 2238, in a nanomolar range, selectively normalizes the hyperactivation of the Na-K pump either induced by exposure to nanomolar ouabain or consequent to a primary molecular alteration such as that induced by adducin polymorphism.

In Vivo *Activity of PST 2238*

Ouabain-Dependent Hypertension, OS Rats

Chronic infusion of low doses of ouabain in normotensive Sprague-Dawley rats increases blood pressure[11,20] and upregulates the renal Na-K pump.[11] The pressor effect of ouabain is associated with a mild, but significant cardiac hypertrophy and is poorly dose-related because both 15 and 50 μg/kg/day ouabain infusions produce

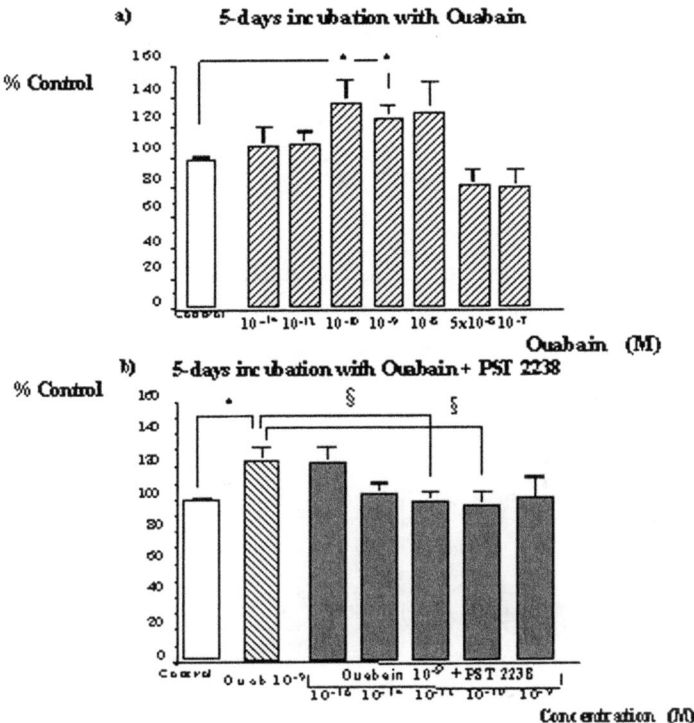

FIGURE 2. Effect of 5-day incubation of NRK cells with (**a**) ouabain or (**b**) ouabain at 10^{-9} M plus PST 2238 from 10^{-16} to 10^{-9} M on the Na-K pump activity at V_{max} (Na-loaded cells, ^{86}Rb uptake). Data (mean ± SEM) are reported as the percentage of control cells run in parallel in the absence of both compounds; *$P < 0.05$ vs. control cells; §$P < 0.05$ vs. ouabain (10^{-9} M) cells.

similar effects on BP (15 µg: +17 mmHg; 50 µg: +20 mmHg vs. controls, $P < 0.001$) and cardiac mass (15 µg: +5.3%; 50 µg: +7% vs. controls, $P < 0.05$). PST 2238, at oral doses of 0.1, 1, 10, and 100 µg/kg, completely antagonizes the effects of ouabain on BP (FIG. 3a),[11] renal Na-K pump (FIG. 3b),[11] and cardiac hypertrophy (not shown) without affecting these parameters in normotensive controls infused with saline (CS rats).[11]

Genetic Hypertension, MHS Rats

In MHS rats, which show both adducin polymorphism[5] and increased levels of EO,[7] PST 2238 at oral doses from 0.1µg/kg to 100 µg/kg for 6 weeks, starting from weaning, reduces BP, with an estimated ED_{50} of 4 µg/kg.[17] Moreover, the renal Na-K pump, which is upregulated in MHS as compared with normotensive MNS controls[10,17] already before the development of hypertension, is normalized by PST

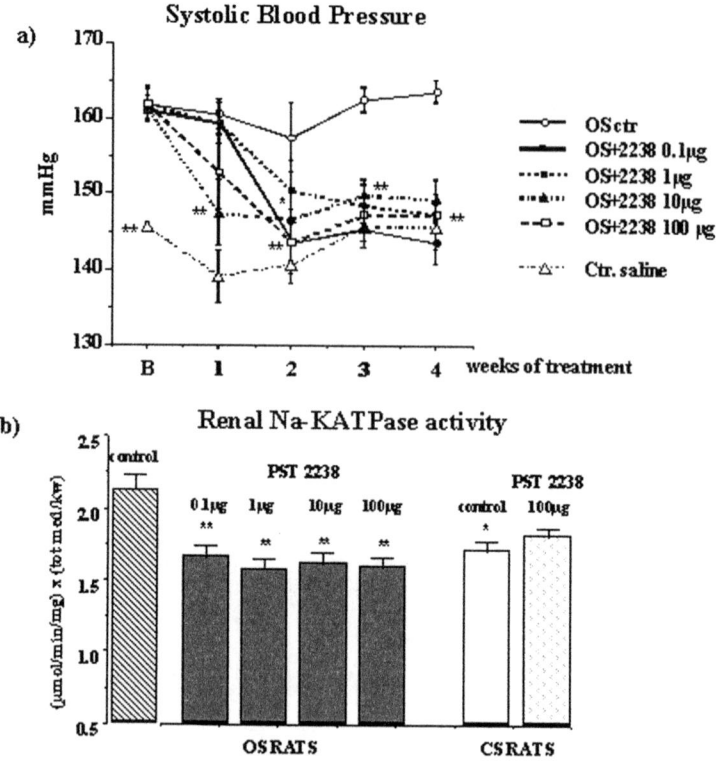

FIGURE 3. (a) Systolic blood pressure in OS and CS (saline-infused) rats during a 4-week oral treatment with PST 2238 (0.1, 1, 10, 100 µg/kg/day). (b) Renal Na-K ATPase activity measured in kidney outer medulla microsomes (^{32}P-ATP hydrolysis) from OS and CS rats after 4 weeks of oral treatment with PST 2238 (0.1, 1, 10, 100 µg/kg/day). Data are mean ± SEM of 8 OS and 7 CS rats for each group; $*P < 0.05$, $**P < 0.01$ vs. OS control rats.

2238 treatment at the same doses effective on BP.[17] Finally, also in MHS, PST 2238 is able to control cardiac hypertrophy, but in this model the compound is effective at doses that are 10- to 50-fold higher than in OS rats,[11,17] suggesting that it may be more selective for the mechanisms supported by high levels of EO than by adducin polymorphism.

Mechanism of Action

To clarify the action mechanism of PST 2238, we have conducted studies on the process of cellular endocytosis of the Na-K pumps. In fact, the internalization of the Na-K pump has been demonstrated to be involved in the regulation of tubular sodium reabsorption, such as during pressure natriuresis[21] and dopamine-dependent natriuresis processes.[22] Both in cells expressing the mutated variant of adducin[17,19] and in normal cells chronically exposed to ouabain,[11] we have demonstrated an increased activity and surface expression of the Na-K pump. Moreover, data have been collecting demonstrating that in both conditions ("mutated" α-adducin, Bertorello et al., unpublished data; and "ouabain infusion"[23]) the increased surface expression of the pump is associated with a reduced rate of internalization through the clathrin-coated vesicles system. Indeed, ouabain-treated cells incubated with nanomolar concentration of PST 2238 show a normalization of the Na-K pump endocytosis.[23]

Another aspect of the PST 2238 action that is currently under investigation concerns its ability to revert both cardiac and renal hypertrophy induced by ouabain in OS rats. Recent data suggest that ouabain can play a role as an intracellular signal transducer[24] that, by activating a complex intracellular signaling pathway triggered by a tyrosine protein kinase–dependent cascade, via EGFr transactivation, leads to transcription of growth-related genes and hypertrophic stimuli.[24–26] In the attempt to verify whether organ hypertrophy in OS rats may be mediated by this ouabain-dependent intracellular signaling and whether PST 2238 specifically interacts with this mechanism, we have investigated in OS rats the role of caveolae, a subpopulation of lipid rafts[27] that are involved in several membrane-linked signal transduction pathways, including those mediated by tyrosine kinase receptor (EGFr), G protein–coupled receptors, and calcium-sensing receptors.[27] Caveolae have been isolated by a detergent-free method followed by a 5–45% sucrose gradient[28] from the kidneys of saline-treated normotensive rats and OS rats treated with either PST 2238 (10 μg/kg OS) or vehicle and were analyzed by Western blot for the presence of the α, β, and γa Na-KATPase subunits, Src, EGFr, and caveolin 1 using specific antibodies. It has been observed that the low buoyant density fractions obtained from control saline-treated rats were highly enriched in caveolin 1, contained Src and EGFr (the specific markers of caveolae), and selectively excluded clathrin. Moreover, the three Na-K ATPase subunits also were detectable in consistent amounts (10% of the protein), supporting the hypothesis that the Na-K pump may be present in caveolae. The same caveolar fractions obtained from OS rats infused with 15 μg/kg/day ouabain showed a significant increase of the α (+34%), β (+44%), and γa (+55%) Na-K ATPases and EGFr (+54%), but not caveolin, as compared with the control fractions, indicating that ouabain may increase the recruitment of these proteins into caveolae. All these effects, however, appeared to be reverted by the parallel treatment of OS rats with PST 2238.

DISCUSSION

Our group has identified both in a subgroup of essential hypertensives and in experimental and genetic rat models common functional, cellular, biochemical, and genetic alterations, all consistent with the hypothesis that hypertension is triggered by a primary increase of tubular reabsorption sustained by an upregulation of renal Na-K pump function.[4,28] This Na-K pump alteration therefore has been considered a functional target for addressing a new causal antihypertensive therapy. PST 2238 represents the prototype of this new class of drugs because it has been selected for its ability to lower blood pressure and antagonize the upregulation of renal Na-K pump, at very low oral doses (μg/kg), in two animal models carrying this alteration: the MHS rats,[17] characterized by α-adducin mutations[5] and increased EO,[7] and the ouabain-hypertensive OS rats.[11] PST 2238 selectively modulates blood pressure and the Na-K pump function only in pathological models, but not in normotensive rats.[11,17] This selectivity of PST 2238 on the Na-K pump is further supported by the data on cultured renal cells showing that the compound normalizes the upregulation of the pump both in NRK cells transfected with the "mutated" adducin variant[17] and in NRK cells incubated with nanomolar ouabain.[11] Moreover, the lack of interference both *in vitro* and *in vivo* with many receptors involved in blood pressure regulation, including mineralocorticoids,[11] further substantiates the PST 2238 selectivity. Even though the molecular mechanism of action of this compound has not been completely elucidated, recent data indicate that PST 2238 may interfere with the processes of internalization and compartmentalization of the Na-K pump on the cell membrane that appear to be altered in the presence of mutated adducin and/or elevated circulating EO. Indeed, PST 2238 appears to be able to accelerate the rate of internalization of the pump, as observed in cultured cells, which is reduced in both conditions.[23] Furthermore, new findings suggest that the prohypertrophic role of ouabain may be mediated by the activation of a mitogenic signal that is triggered by the functional translocation of a pool of Na-K ATPase into caveolae, a membrane subdomain where several plasma membrane–linked signal transduction proteins are organized in an orderly way and activate a specific downstream signaling cascade, leading to growth-related gene transcription. This hypothesis is further sustained by the demonstration that PST 2238, at doses of μg/kg, appears to prevent the Na-K ATPase and EGFr targeting to caveolae, and in parallel normalizes cardiac and kidney hypertrophy, suggesting that this compound can be useful not only to control blood pressure, but also to prevent cardiac and renal complications associated with increased EO and mutated adducin.

The relevance of this compound for the therapy of hypertensive patients is supported by consistent data demonstrating that adducin polymorphisms and increased levels of EO are associated with an increased risk to develop hypertension and organ complications.[13–16] Phase 2 clinical studies on the effects of PST 2238 are currently in progress and indicate that the antihypertensive efficacy of this compound seems to correlate with the levels of circulating EO.

REFERENCES

1. SOUBRIER, F. & G.M. LATHROP. 1995. The genetic basis of hypertension. Curr. Opin. Nephrol. Hypertens. **4:** 177–181.
2. LIFTON, R.P. 1996. Molecular genetics of human blood pressure variation. Science **272:** 676–680.
3. LARAGH, J.H., B. LAMPORT, J. SEALY & M.H. ALDERMAN. 1988. Diagnosis ex juvantibus: individual response patterns to drugs reveal hypertension mechanism and simplify treatment. Hypertension **12:** 223–226.
4. FERRARI, P. & G. BIANCHI. 1995. Lessons from experimental genetic hypertension. *In* Hypertension: Pathophysiology Diagnosis and Management: 1261–1280. Raven Press. New York.
5. BIANCHI, G., G. TRIPODI, G. CASARI, *et al.* 1994. Two point mutations within the adducin genes are involved in blood pressure variation. Proc. Natl. Acad. Sci. USA **91:** 3999–4003.
6. CUSI, D., C. BARLASSINA, T. AZZANI, *et al.* 1997. Polymorphism of α-adducin and salt sensitivity in patients with essential hypertension. Lancet **349:** 1353–1357.
7. FERRANDI, M., E. MINOTTI, S. SALARDI, *et al.* 1992. Ouabainlike factor in Milan hypertensive rats. Am. J. Physiol. **263:** F739–F748.

8. FERRANDI, M., P. MANUNTA, S. BALZAN, et al. 1997. Ouabain-like factor quantification in mammalian tissues and plasma: comparison of two independent assays. Hypertension **30:** 886–896.
9. MANUNTA, P., E. MESSAGGIO, C. BALLABENI, et al. 2001. Plasma ouabain-like factor during acute and chronic changes in sodium balance in essential hypertension. Hypertension **38:** 198–203.
10. FERRANDI, M., G. TRIPODI, S. SALARDI, et al. 1996. Renal Na-KATPase in genetic hypertension. Hypertension **28:** 1018–1025.
11. FERRARI, P., L. TORIELLI, M. FERRANDI, et al. 1998. PST 2238: a new antihypertensive compound that antagonizes the long-term pressor effect of ouabain. J. Pharm. Exp. Ther. **285:** 83–94.
12. FERRANDI, M., G. TRIPODI, I. MOLINARI, et al. 2001. Role of beta and gamma adducin polymorphisms on left ventricle mass and onset of proteinuria in congenic rats from the Milan hypertensive strain. Sixth European Council on Blood Pressure and Cardiovascular Research, American Heart Association, The Netherlands. Abstract 3.03.
13. PIERDOMENICO, S.D., A. BUCCI, P. MANUNTA, et al. 2001. Endogenous ouabain and hemodynamic and left ventricular geometric patterns in essential hypertension. Am. J. Hypertens. **14:** 44–50.
14. MANUNTA, P., P. STELLA, R. RIVERA, et al. 1999. Left ventricular mass, stroke volume, and ouabain-like factor in essential hypertension. Hypertension **34:** 450–456.
15. MORRISON, A.C., M.S. BRAY, A.R. FOLSOM & E. BOERWINKLE. 2002. ADD1 460W allele associated with cardiovascular disease in hypertensive individuals. Hypertension **39:** 1053–1057.
16. WANG, J., J.A. STAESSEN, L. TIZZONI, et al. 2001. Renal function in relation to three candidate genes. Am. J. Kidney Dis. **38:** 1158–1168.
17. FERRARI, P., M. FERRANDI, G. TRIPODI, et al. 1999. PST 2238: a new antihypertensive compound that modulates Na-K ATPase in genetic hypertension. J. Pharm. Exp. Ther. **288:** 1074–1083.
18. TRIPODI, M.G., F. VALTORTA, L. TORIELLI, et al. 1996. Hypertension-associated point mutation in the adducing α and β subunits affect actin cytoskeleton and ion transport. J. Clin. Invest. **97:** 2815–2822.
19. MANUNTA, P., A.C. ROGOWSKI, B.P. HAMILTON & J.M. HAMLYN. 1994. Ouabain-induced hypertension in the rat: relationships among plasma and tissue ouabain and blood pressure. J. Hypertens. **12:** 549–560.
20. ZHANG, Y., A.K. MIRCHEFF, C.B. HENSLEY, et al. 1996. Rapid redistribution and inhibition of renal sodium transporters during acute pressure natriuresis. Am. J. Physiol. **270:** F1004–F1014.
21. CHIBALIN, A.V., G. OGIMOTO, C.H. PEDEMONTE, et al. 1999. Dopamine-induced endocytosis of Na^+,K^+-ATPase is initiated by phosphorylation of Ser-18 in the rat alpha subunit and is responsible for the decreased activity in epithelial cells. J. Biol. Chem. **274:** 1920–1927.
22. TORIELLI, L., G. PADOANI, P. TARSINI, et al. 2000. Adducin may affect the residential time of the Na/K pump on the plasmamembrane. Eur. J. Cell Biol. **79**(Suppl.): 52.
23. LIU, J., J. TIAN, M. HAAS, et al. 2000. Ouabain interaction with cardiac Na^+/K^+-ATPase initiates signal cascades independent of changes in intracellular Na^+ and Ca^{2+} concentrations. J. Biol. Chem. **275:** 27838–27844.
24. HAAS, M., A. ASKARI & Z. XIE. 2000. Involvement of Src and epidermal growth factor receptor in the signal-transducing function of Na^+/K^+-ATPase. J. Biol. Chem. **275:** 27832–27837.
25. SHAUL, P. & R.G. ANDERSON. 1998. Role of plasmalemmal caveolae in signal transduction. Am. J. Physiol. **275:** L843–L851.
26. ANDERSON, R.G. & K. JACOBSON. 2002. A role for lipid shells in targeting proteins to caveolae, rafts, and other lipid domains. Science **296:** 1821–1825.
27. SONG, K.S., L. SHENGWEN, et al. 1996. Copurification and direct interaction of Ras with caveolin, an integral membrane protein of caveolae microdomains: detergent free purification of caveolae microdomains. J. Biol. Chem. **271:** 9690–9697.
28. MANUNTA, P., M. BURNIER, M. D'AMICO, et al. 1999. Adducin polymorphism affects renal proximal tubule reabsorption in hypertension. Hypertension **33:** 694–697.

Myocardial Na,K-ATPase and Digoxin Therapy in Human Heart Failure

KELD KJELDSEN AND HENNING BUNDGAARD

Medical Department B, The Heart Center, Rigshospitalet, University of Copenhagen, DK-2100 Copenhagen, Denmark

ABSTRACT: The specific binding of digitalis glycosides to the Na,K-ATPase is used as a tool for Na,K-ATPase quantification with high accuracy and precision. In myocardial biopsies from patients with heart failure, total Na,K-ATPase concentration is decreased, and the decrease in Na,K-ATPase concentration correlates with a decrease in heart function. During digitalization, a fraction of remaining pumps are occupied by digoxin. No evidence for an endogenous digitalis-like factor of any clinical importance was obtained. It is recommended that digoxin be administered to heart failure patients who still have dyspnea after institution of mortality-reducing therapy.

KEYWORDS: Na,K-ATPase; heart failure; digoxin

QUANTITATIVE ASPECTS OF HUMAN MYOCARDIAL Na,K-ATPase

Digitalis glycosides have been in use for treatment of heart failure for approximately 225 years and are still the safest inotropic drug taken orally that improves hemodynamics. The active Na and K transport is specifically inhibited by cardiac glycosides,[1] and the Na,K pump is the cellular receptor for the inotropic action of digoxin. On this basis, digitalis glycoside binding was developed as a tool for Na,K-ATPase quantification.[2] This method allows quantification of muscular Na,K-ATPase with high accuracy and precision.[3]

The Na,K-ATPase was demonstrated in the human myocardium several years ago[4] and has since then been quantified in both normal and diseased myocardium. Thus, in normal human left ventricular myocardium, an Na,K-ATPase concentration of approximately 700 pmol/g wet weight was found.[5] Hitherto, the absolute amounts of the various isoforms of the human myocardial Na,K-ATPase have not been determined. In human dilated cardiomyopathy, endomyocardial biopsies showed a significant decrease of approximately 40% in total Na,K-ATPase concentration.[6] Later, analyzing data from available studies,[6–9] researchers reported a consistent and significant decrease of 26–32% in Na,K-ATPase in the failing human heart.[10] Furthermore, the decrease in Na,K-ATPase concentration correlates with decrease in heart function,[6,11] indicating that contractile performance decreases in proportion to

---

Address for correspondence: Keld Kjeldsen, Medical Department B 2142, The Heart Center, Rigshospitalet, Blegdamsvej 9, DK-2100 Copenhagen, Denmark. Voice: +45-35452628; fax: +45-35452648.

kjeldsen@rh.dk

the loss of Na,K-ATPase. In the first report of Na,K-ATPase isoform expression in normal and failing human left ventricle, Allen et al. found no significant alteration at the mRNA level.[12] In that study, there was, however, a tendency to a reduction in Na,K-ATPase concentration of only approximately 10%. It was noted that minor changes in protein expression might be missed by studies of mRNA abundancies and that posttranscriptional factors also may be in play. Later, Shamraj et al.[10] reported reductions at mRNA as well as protein levels in human heart failure. Thus, in ventricular tissue, $\alpha1$, $\alpha3$, $\beta1$, maximal ouabain binding, and Na,K-ATPase activity were reduced by approximately one-third; $\alpha2$ showed only a small insignificant tendency to reduction. However, in atrial tissue, $\alpha1$ and $\alpha2$ were expressed at reduced levels.[13]

MYOCARDIAL Na,K-ATPase IN DIGITALIZATION

Occupancy of Na,K-ATPase, that is, percentage of receptors occupied by digitalis glycoside, during digitalization was first evaluated by measurements of membrane potentials in human atrial biopsies, revealing a 38% reduction in electrogenic effect.[14] Later, occupancy was determined in myocardial biopsies from digoxin-treated patients using binding assays. This revealed occupancies of 24–35% in the human heart.[5,15] Because, in the human heart, $\alpha1$ and $\alpha2$ isoforms have almost identical affinity for digitalis glycosides, their occupancy by digoxin during digitalization is probably also of the same order of magnitude. Using genetically manipulated mice, it recently has been found that the $\alpha2$ isoform has a special role in Ca signaling and thus may be of special importance for the inotropic action of digitalis glycosides, whereas the $\alpha1$ isoform may have a special role in maintaining Na and K concentrations.[16] Thus, it is of interest that isoform-specific regulation of the Na,K-ATPase in the heart also has been reported recently.[17]

Some decades ago, development of tolerance to chronic digoxin therapy was suspected from Na,K-ATPase upregulation in various cells, for example, erythrocytes. However, when Na,K-ATPase was later studied in the target organ for the inotropic action of digoxin, the human myocardium of patients with heart failure, no evidence for development of tolerance to long-term digitalization was found at digoxin receptor level.[5,15]

Recent studies using experimental animals indicate that myocardial Na,K-ATPase in addition to digoxin is also influenced by other drugs used for treatment of heart failure. Thus, potassium loss during diuretic therapy has been found to reduce myocardial Na,K-ATPase, whereas angiotensin-converting enzyme inhibitors may stimulate Na,K pump activity.[18,19] Furthermore, hyperaldosteronism induced by heart failure has been found to decrease Na,K-ATPase activity.[20] Accordingly, treatment with aldosterone antagonist, spironolactone, also may influence Na,K-ATPase activity.

ENDOGENOUS DIGITALIS-LIKE FACTOR

A major problem is that the precise chemical structure and properties of the putative factor have not been determined.[21] It has been suggested that the factor could be ouabain; however, no pathway for its synthesis is available in any animal because it is a plant substance.[22] On the other hand, the factor also has been suggested to be a food contaminant. It has been shown that ouabain could be detected in human plasma; however, not all investigators were able to confirm this.[23] Hence, it has not been excluded that unspecific cross-reactions of antibody used for the determination of the digitalis-like factor may be in play. It has been suggested that ouabain could be secreted from the adrenal gland; however, quantitative evaluations later revealed that the production rate was so low that it would take about 1 month to produce enough ouabain to obtain only a 1% occupancy of the Na,K-ATPase in the entire muscular pool.[24] The factor has also been suggested to be digoxin-like. However, although the factor has been closely related to hypertension, digoxin treatment does not increase blood pressure in human subjects. When human myocardial samples were exposed to prolonged wash in buffer containing digoxin antibody and subsequently taken for ouabain binding, no significant occupancy of the Na,K-ATPase was revealed.[5,15] Furthermore, no evidence of an endogenous digitalis-like factor was observed in spontaneously hypertensive rats, neither when ouabain binding to muscular tissue *in vivo* and *in vitro* was compared nor when samples were exposed to prolonged wash before binding.[25,26] Moreover, no significant difference in ouabain binding was observed in resistance vessels between spontaneously hypertensive rats and controls.[25] In comparison, potassium depletion has been found to be associated with significant reductions in ouabain binding site concentrations in resistance vessels of 33%[25] and skeletal muscles of 78%.[27] Taken together, the major conceptual problem for the importance of a digitalis-like factor is that its quantity seems minute as compared with the enormous quantity of Na,K-ATPase in the human subject.

Gao *et al.*, using the whole-cell patch-clamp technique, have reported recently an increase of 35% in Na,K pump–mediated current in myocytes obtained from human subjects within the first minutes of exposure to ouabain in the nanomolar range, and a specific role of this stimulation for the digitalis-like factor was suggested.[28] However, as pointed out by Clausen,[29] this probably does not indicate that the Na,K-ATPase in the human heart is stimulated instead of inhibited during digitalization. Among the many arguments strongly favoring inhibition of Na,K-ATPase[30] are that ouabain infusion induces a significant potassium loss from the human heart,[31] that digitalization is associated with a significant increased exercise-induced potassium increase,[32] and that digitalis intoxication causes hyperkalemia. Thus, studies are in demand to further elucidate the observations of Gao *et al.*, in addition to methodological evaluations ensuring optimum diffusion conditions during experiments; the effect of ouabain exposure beyond a few minutes needs to be assessed.

CLINICAL ASPECTS AND PERSPECTIVES

Heart failure has been deemed the epidemic of cardiology in the twenty-first century in the industrialized world.[33,34] Despite recent improvements in therapy mainly arising from clinical trials, the overall prognosis for heart failure still remains very poor. Hence, clinical trials generally have been able to add only a few months of life to a life expectancy of only a few years in severe heart failure. In this setting, there is a pressing demand for elevation of results from basic research into clinical application in the search for a breakthrough in heart failure research.

One major aspect of reduced myocardial Na,K pump concentration in heart failure and its inhibition during digitalization is its influence on intracellular Na (Na_i) homeostasis. Recently, significant increases of a few millimolars per liter in Na_i have been described in rabbit[35] as well as in human failing heart.[36] Whereas these observations seem in good accord with a decrease in Na,K pump concentration, an increased Na influx also has been found to be of importance.[35] If increased Na influx develops together with decreased Na,K pump concentration, this probably means that remaining Na,K pumps are set at a higher level of activity. This is feasible because at resting conditions only a few percent of the Na,K pumps usually are active, whereas when stimulated all pumps can be recruited.[37] It implies, however, that only a reduced maximal capacity for Na handling is available. Thus, an Na_i increase during, for example, ischemia may secondarily limit Ca and H extrusion, inducing arrhythmias and further progression of heart failure due to cell necrosis.

Another important aspect is impaired handling among heart cells of K coming from myocytes as well as from the bloodstream. Both of these aspects of K homeostasis are affected by regulation of the Na,K pump by physiological and pathophysiological factors as well as medical treatments.[18,38] Thus, digitalization has been shown to affect both parameters.[31,32] Further, in experimental animals, K loading as well as K depletion also is found to significantly affect K handling.[39–41] The effects of K depletion are of special interest because it often occurs in heart failure patients because of diuretic treatment.[42] Thus, because disturbed K homeostasis may induce arrhythmias, there is a need for increased attention to the dynamic aspects of K handling.

The hemodynamic aspects of digitalization in relation to the Na,K pump recently have been reviewed.[34] In heart failure, the reduction in Na,K pumps initially may be considered adaptive, maintaining contractile capacity. In accordance with the Digitalis Investigation Group (DIG) trial showing beneficial effects of digoxin on morbidity without affecting mortality in heart failure patients,[43] it is recommended that digoxin is added to heart failure patients that still have dyspnea after institution of mortality-reducing therapy. A recent reevaluation of the DIG trial indicates decreased mortality in males, but increased mortality in females on digoxin, in part probably because of increased sensitivity to digoxin in females caused by hormonal interactions[44] as well as the more extensive use of diuretics among females than males and the associated potential for K depletion. Thus, digoxin is the safest inotropic drug for oral use that improves hemodynamics, but it should be used in the lowest dose that relieves dyspnea.

REFERENCES

1. SCHATZMANN, H.J. 1953. Herzglycoside als hemmstoffe für den aktiviteten kalium- und natriumtransport durch die erythrocytenmembrane. Helv. Physiol. Pharmacol. Acta **11:** 346–354.
2. HANSEN, O. & T. CLAUSEN. 1988. Quantitative determination of Na,K-ATPase and other sarcolemmal components in muscle cells. Am. J. Physiol. **254:** 1–7.
3. HANSEN, O. & T. CLAUSEN. 1996. Studies on sarcolemma components may be misleading due to inadequate recovery. FEBS Lett. **348:** 203.
4. GIBSON, K. & P. HARRIS. 1970. Na,K-ATPase activity in a preparation from human post-mortem myocardium. Cardiovasc. Res. **4:** 201–206.

5. SCHMIDT, T.A., P.D. ALLEN, W.S. COLUCCI, et al. 1993. No adaptation to digitalization as evaluated by digitalis receptor (Na,K-ATPase) quantification in explanted hearts from donors without heart disease and from digitalized recipients with end-stage heart failure. Am. J. Cardiol. **71:** 110–114.
6. NØRGAARD, A., J.P. BAGGER, P. BJERREGAAD, et al. 1988. Relation of left ventricular function and Na,K-pump concentration in suspected idiopathic dilated cardiomyopathy. Am. J. Cardiol. **61:** 1312–1315.
7. SCHWINGER, R.H.G., M. BÖHM & E. ERDMANN. 1990. Effectiveness of cardiac glycosides in human myocardium with and without downregulated beta-adrenoceptors. J. Cardiovasc. Pharmacol. **15:** 692–697.
8. SCHWINGER, R.H.G., M. BÖHM, K. LA ROSEE, et al. 1992. Na-channel activators increase cardiac glycoside sensitivity in failing human myocardium. J. Cardiovasc. Pharmacol. **19:** 554–561.
9. KJELDSEN, K., P. BJERREGAARD, E.A. RICHTER, et al. 1988. Na,K-ATPase concentration in rodent and human heart skeletal muscle: apparent relation to muscle performance. Cardiovasc. Res. **22:** 95–100.
10. SHAMRAJ, O.I., I.L. GRUPP, G. GRUPP, et al. 1993. Characterization of Na,K-ATPase, its isoforms, and the inotropic response to ouabain in isolated failing human hearts. Cardiovasc. Res. **27:** 2229–2237.
11. ISHINO, K., H.E. BØTKER, T. CLAUSEN, et al. 1999. Myocardial adenine nucleotides, glycogen, and Na,K-ATPase in patients with idiopathic dilated cardiomyopathy requiring mechanical circulatory support. Am. J. Cardiol. **83:** 396–399.
12. ALLEN, P.D., T.A. SCHMIDT, J.D. MARSH, et al. 1992. Na,K-ATPase expression in normal and failing human heart. Basic Res. Cardiol. **87:** 87–94.
13. SCHWINGER, R.H.G., J. WANG, K. FRANK, et al. 1999. Reduced sodium pump alpha 1, alpha 3, and beta 1 isoform protein levels and Na,K-ATPase activity, but unchanged Na-Ca exchanger protein levels in human heart failure. Circulation **99:** 2105–2112.
14. RASMUSSEN, H.H., T.E.N. EICK, G.T. OKITA, et al. 1985. Inhibition of electrogenic Na-pumping attributable to binding of cardiac steroids to high-affinity pump sites in human atrium. J. Pharmacol. Exp. Ther. **235:** 629–635.
15. SCHMIDT, T.A., P. HOLM-NIELSEN & K. KJELDSEN. 1991. No upregulation of digitalis glycoside receptor (Na,K-ATPase) concentration in human heart left ventricle samples obtained at necropsy after long term digitalization. Cardiovasc. Res. **25:** 684–691.
16. JAMES, P.F., I.L. GRUPP, G. GRUPP, et al. 1999. Identification of a specific role for the Na,K-ATPase alpha 2 isoform as a regulator of calcium in the heart. Mol. Cell **3:** 555–563.
17. MATHIAS, R.T., I.S. COHEN, J. GAO, et al. 2000. Isoform-specific regulation of the Na,K-pump in heart. News Physiol. Sci. **15:** 176–180.
18. CLAUSEN, T. 1998. Clinical and therapeutic significance of the Na,K-pump. Clin. Sci. **95:** 3–17.
19. HOOL, L.C., D.W. WHALLEY, M.M. DOOHAN, et al. 1995. Angiotensin-converting enzyme inhibition, intracellular Na, and Na-K pumping in cardiac myocytes. Am. J. Physiol. **268:** C366–C375.
20. MIHAILIDOU, A.S., H. BUNDGAARD, M. MARDINI, et al. 2000. Hyperaldosteronemia in rabbits inhibits the cardiac sarcolemmal Na,K-pump. Circ. Res. **86:** 37–42.
21. THERIEN, A.G. & R. BLOSTEIN. 2000. Mechanisms of sodium pump regulation. Am. J. Physiol. **279:** C541–C566.
22. GOTO, A., K. YAMADA, N. YAGI, et al. 1992. Physiology and pharmacology of endogenous digitalis-like factors. Pharmacol. Rev. **44:** 377–399.
23. LEWIS, L.K., T.G. YANDLE, J.G. LEWIS, et al. 1994. Ouabain is not detectable in human plasma. Hypertension **24:** 549–555.
24. HANSEN, O. 1994. Do putative endogenous digitalis-like factors have a physiological role? Hypertension **24:** 640.
25. AALKJÆR, C., K. KJELDSEN, A. NØRGAARD, et al. 1985. Ouabain binding sites and Na content in resistance vessels and skeletal muscles of SHR and K-depleted rats. Hypertension **7:** 277–286.
26. CLAUSEN, T., O. HANSEN, K. KJELDSEN, et al. 1982. Effect of age, potassium depletion, and denervation on specific displaceable ^{3}H-ouabain binding in rat skeletal muscle in vivo. J. Physiol. **333:** 367–381.

27. NØRGAARD, A., K. KJELDSEN & T. CLAUSEN. 1981. Potassium depletion decreases the number of ouabain binding sites and the active sodium-potassium-transport in skeletal muscle. Nature **293:** 739–741.
28. GAO, J.R.S., Y. WYMORE, G.R. WANG, *et al.* 2002. Isoform specific stimulation of cardiac Na,K-pumps by nM concentrations of glycosides. J. Gen. Physiol. **119:** 297–312.
29. CLAUSEN, T. 2002. Acute stimulation of Na,K-pumps by cardiac glycosides in the nanomolar range. J. Gen. Physiol. **119:** 295–296.
30. LEVI, A.J., M.R. BOYETT & C.O. LEE. 1994. The cellular actions of digitalis glycosides on the heart. Prog. Biophys. Mol. Biol. **62:** 1–54.
31. BRENNAN, F.J., J.L. MCCANS, M.A. CHIONG, *et al.* 1972. Effects of ouabain on myocardial potassium and sodium balance in man. Circulation **45:** 107–113.
32. SCHMIDT, T.A., H. BUNDGAARD, H.L. OLESEN, *et al.* 1995. Digoxin affects potassium homeostasis during exercise in patients with heart failure. Cardiovasc. Res. **29:** 506–511.
33. GHEORGHIADE, M. & R.O. BONOW. 1998. Chronic heart failure in the United States. Circulation **97:** 282–289.
34. KJELDSEN, K., A. NØRGAARD & M. GHEORGHIADE. 2002. Myocardial Na,K-ATPase: the molecular basis for the hemodynamic effect of digoxin therapy in congestive heart failure. Cardiovasc. Res. **55:** 710–713.
35. DESPA, S., M.A. ISLAM, S.M. POGWIZD, *et al.* 2002. Intracellular Na concentration is elevated in heart failure, but Na,K-pump function is unchanged. Circulation **105:** 2543–2548.
36. PIESKE, B., L.S. MAIER, V. PIACENTINO, *et al.* 2002. Rate dependence of $[Na^+]_i$ and contractility in nonfailing and failing human myocardium. Circulation **106:** 447–453.
37. CLAUSEN, T., M. EVERTS & K. KJELDSEN. 1987. Quantification of maximum capacity for active sodium-potassium transport in rat skeletal muscle. J. Physiol. **388:** 163–181.
38. DOOHAN, M.M. & H.H. RASMUSSEN. 1997. Myocardial cation transport. J. Hypertens. **11:** 683–691.
39. BUNDGAARD, H., T.A. SCHMIDT, J.S. LARSEN, *et al.* 1997. K supplementation increases muscle Na,K-ATPase and improves extrarenal K homeostasis in rats. J. Appl. Physiol. **82:** 1136–1144.
40. BUNDGAARD, H., M.T. ENEVOLDSEN & K. KJELDSEN. 1998. Chronic K-supplementation decreases myocardial Na,K-ATPase and net K-uptake capacity in rodents. J. Mol. Cell. Cardiol. **30:** 2037–2046.
41. BUNDGAARD, H. & K. KJELDSEN. 2002. Potassium depletion increases potassium clearance capacity in skeletal muscles *in vivo* during acute repletion. Am. J. Physiol. Cell. Physiol. **283:** C1163–C1170.
42. DØRUP, I., K. SKAJAA, T. CLAUSEN, *et al.* 1988. Reduced concentration of K, Mg, and Na,K-pumps in human skeletal muscle during diuretic treatment. Br. Med. J. **296:** 455–458.
43. THE DIGITALIS INVESTIGATION GROUP. 1997. The effect of digoxin on mortality and morbidity in patients with heart failure. N. Engl. J. Med. **336:** 525–533.
44. RATHORE, S.S., Y. WANG & H.M. KRUMHOLZ. 2002. Sex-based differences in the effect of digoxin for the treatment of heart failure. N. Engl. J. Med. **347:** 1403–1411.

Expression and Cellular Localization of Na,K-ATPase Isoforms in Dog Prostate in Health and Disease

A. MOBASHERI,[a] I. EVANS,[a] P. MARTÍN-VASALLO,[b] AND C. S. FOSTER[c]

[a]*Department of Veterinary Preclinical Sciences, Faculty of Veterinary Science, University of Liverpool, Liverpool L69 7ZJ, United Kingdom*

[b]*Departamento de Bioquímica y Biología Molecular, Universidad de La Laguna, 38201 La Laguna, Tenerife, Spain*

[c]*Department of Cellular and Molecular Pathology, Faculty of Medicine, University of Liverpool, Liverpool L69 3GA, United Kingdom*

KEYWORDS: Na,K-ATPase; isoform; prostate; benign prostatic hyperplasia; prostate cancer; citrate metabolism

The principal physiological function of the prostate gland is the synthesis, accumulation, and secretion of citrate.[1] Ouabain-sensitive Na,K-ATPase-mediated transport is critical for citrate biosynthesis because the Na^+ gradient established by Na,K-ATPase is essential for the Na^+-dependent uptake of aspartate (a precursor for citrate) to create a high cytosolic aspartate concentration.[2] Furthermore, androgen activation of Na,K-ATPase serves as a metabolic pacemaker in the prostate,[3] exerting transcriptional control over the expression of Na,K-ATPase β subunits, which play a critical role in the biogenesis of Na,K-ATPase.[4] In prostate cancer, citrate production is significantly reduced as a result of altered cellular metabolism and bioenergetics.[1] Despite the importance of Na,K-ATPase function for citrate production, there is no information about its expression patterns in hyperplastic or neoplastic prostate. The objectives of this study were to establish the localization of Na,K-ATPase; to compare expression of its isoforms in normal canine prostate, benign prostatic hyperplasia (BPH), and prostatic adenocarcinoma (PCa); and to determine whether reduced citrate levels in PCa also are accompanied by changes in Na,K-ATPase.

Address for correspondence: A. Mobasheri, Department of Veterinary Preclinical Sciences, Faculty of Veterinary Science, University of Liverpool, Liverpool L69 7ZJ, UK. Voice: +44-0-151-794-4284; fax: +44-0-151-794-4243.
 a.mobasheri@liverpool.ac.uk

MATERIALS AND METHODS

Normal BPH and PCa prostates were dissected from canine cadavers after euthanasia and fixed in formaldehyde. Immunohistochemical studies were conducted using a panel of well-characterized antibodies to the α and β subunits of Na,K-ATPase. Immunohistochemical micrographs were analyzed using Scion Image (version 4.0.2).

RESULTS

Na,K-ATPase α1 and β1 isoforms were localized to the basolateral margins of epithelial cells in normal and BPH prostates. The α1 isoform was expressed in abundance, but there was no evidence of α2 or α3 expression (data not shown). The abundant immunostaining observed in normal and BPH tissue was significantly reduced in advanced PCa as determined by immunohistochemistry and image analysis (FIG. 1).

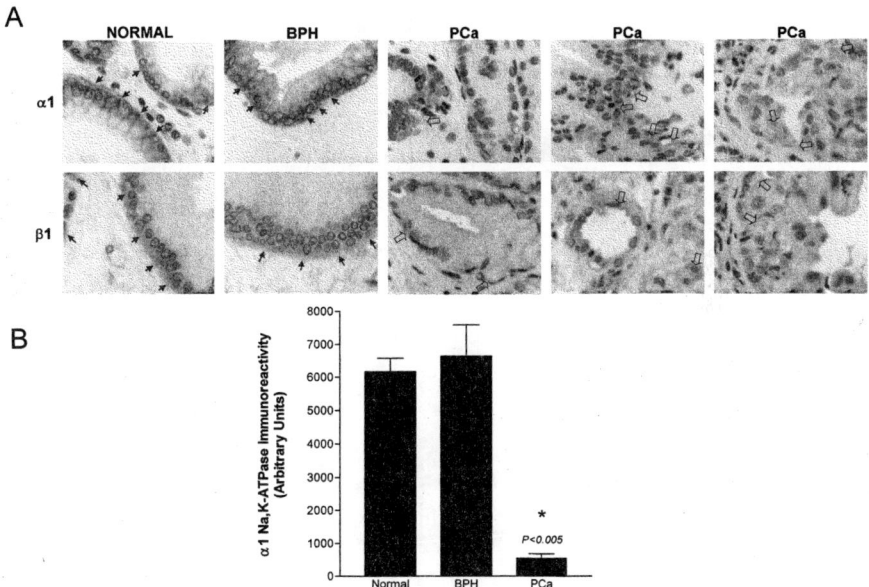

FIGURE 1. (A) Immunohistochemical localization of Na,K-ATPase α1 and β1 subunits in canine normal, BPH, and PCa tissues. Fast-Red TR/Naphthol AS-MX was used as precipitating substrate for the secondary alkaline phosphatase conjugated secondary antibody to reveal sites of Na,K-ATPase immunoreactivity. In normal and BPH sections, Na,K-ATPase was detected in the basolateral membranes of epithelial cells (*solid arrows*). In three different PCa tumor explants, Na,K-ATPase expression was significantly downregulated compared with normal and BPH prostates and was diffusely spread in selected areas of sheets of neoplastic cells (*open arrows*). Nuclei were counterstained with hematoxylin. Original magnifications: ×400. (B) Quantitative analysis of Na,K-ATPase immunoreactivity in normal, BPH, and PCa canine prostate.

DISCUSSION AND CONCLUSIONS

The altered intermediary metabolism of prostate cells has been implicated in the pathogenesis of prostate adenocarcinoma and the progression of malignancy.[1] Previous studies have indicated that the β subunit of Na,K-ATPase is downregulated *in vitro* in the prolonged presence of a synthetic androgen at a transcriptional level, resulting in a reduction of functional Na,K-ATPase in androgen-dependent prostate cell lines.[4] This study confirms that Na,K-ATPase expression is also downregulated in prostate tumors *in vivo*. This epigenetic alteration may be intricately associated with the impaired citrate production and secretion that accompany neoplastic development in the prostate.

ACKNOWLEDGMENTS

This work was supported by research grants from the Pet Plan Charitable Trust and the Wellcome Trust (United Kingdom).

REFERENCES

1. COSTELLO, L.C. & R.B. FRANKLIN. 2000. The intermediary metabolism of the prostate: a key to understanding the pathogenesis and progression of prostate malignancy. Oncology **59:** 269–282.
2. FRANKLIN, R.B. *et al.* 1990. Evidence for two aspartate transport systems in prostate epithelial cells. Prostate **16:** 137–145.
3. FARNSWORTH, W.E. 1970. Androgen regulation of prostatic membrane ATPase. Biol. Reprod. **3:** 218–222.
4. BLOK, L.J. *et al.* 1999. Regulation of expression of Na^+,K^+-ATPase in androgen-dependent and androgen-independent prostate cancer. Br. J. Cancer **81:** 28–36.

Molecular Characterization of a Putative Sodium/Iodide Symporter in the South African Clawed Frog, *Xenopus laevis*

D. L. CARR, F. LAHARRAGUE, B. KAHN, T. A. PRESSLEY, AND J. A. CARR

Department of Physiology, Texas Tech University Health Sciences Center, and Department of Biological Sciences, Texas Tech University, Lubbock, Texas, USA

KEYWORDS: sodium/iodide symporter (NIS); iodide; *Xenopus laevis*

INTRODUCTION

Iodide is a critical nutrient in the synthesis of thyroid hormone (TH), yet is found only in trace amounts in the environment. Obtaining iodide is particularly important for larval amphibians, as they live in a dilute aqueous environment with an unpredictable iodide supply. The Na^+/K^+-ATPase pump provides the driving force for iodide uptake in the thyroid gland and other absorptive tissues. This uptake is mediated by the sodium/iodide symporter (NIS). Ionic perchlorate is the most potent known competitive inhibitor of thyroidal iodide uptake and, as such, disrupts the normal accumulation of iodide in the thyroid.[1-3] Environmental contaminants that competitively inhibit the NIS, such as ammonium perchlorate (AP) and sodium chlorate, may pose a significant threat to developing amphibians, which require TH for metamorphosis and reproductive development. Unfortunately, nothing is known about the structure, tissue distribution, or regulation of the amphibian NIS. In this study, we examined tissue-specific distribution of a putative NIS in the South African clawed frog, *Xenopus laevis*.

METHODS

Juvenile *X. laevis* (8–15 g) were maintained in 40-L glass aquaria containing 18 L of dechlorinated tap water. Experiment 1: Frogs were injected with 0.006 IU of bovine TSH in 100 mL of 0.6% NaCl via the dorsal lymph sac for 4–6 days prior to euthanasia and tissue collection. An equal number of frogs were injected with 0.6% NaCl and served as vehicle controls. Experiment 2: Frogs were maintained for 7 days (RNA analyses) or 14 days (radiolabeled iodide uptake) in the presence (1 mM) or absence of AP dissolved in their tank water.

Address for correspondence: Deborah L. Carr, Department of Physiology, Texas Tech University HSC, 3601 4th Street, STOP 6551, Lubbock, TX 79430-6551. Voice: 806-743-4056; fax: 806-743-1512.
Deborah.Carr@ttuhsc.edu

Total RNA was isolated from whole Nieuwkoop-Faber stage 60 tadpoles.[4] Using primers based on the structure of the putative *Xenopus* NIS (xNIS), RT-PCR was used to amplify a cDNA fragment with considerable sequence homology to the mammalian transporter. The sequence of the cDNA product was confirmed and the fragment cloned into bacterial vector for storage and stability. This cDNA product was radiolabeled and used as a probe for subsequent Northern blots.

Tissue Collection

Frogs were rapidly decapitated and various tissues harvested for analysis. In order to assess 24-h iodide uptake, frogs were injected ip with 3.6 mCi of $Na^{125}I$ at 24 h prior to euthanasia. Tissues collected included stomach, kidney, gonads, skin (RNA only), heart, liver, small intestine, large intestine, fat bodies, and brain (RNA only).

SUMMARY AND CONCLUSIONS

We have identified a putative NIS that is expressed in the adult and developing *X. laevis* and is homologous to the mammalian NIS. The sequence is specifically expressed in stomach, kidney, ovaries, and fat bodies. We did not see expression in brain, heart, liver, small intestine, skeletal muscle, testes, or skin. Expression of the putative xNIS in the stomach is particularly interesting, given the fact that the stomach possessed the greatest radiolabeled iodide uptake of any nonthyroidal tissue examined. Furthermore, radiolabeled iodide uptake in the stomach was significantly blocked in frogs exposed for 2 weeks to AP. Further characterization of the xNIS will yield a promising tool for investigating iodide transport in amphibians, including how frogs sequester iodide from their environment and the mechanisms underlying perchlorate disruption of thyroid function and reproduction.

REFERENCES

1. MIRANDA, L.A., A. PISANO & V. CASCO. 1996. Ultrastructural study of thyroid glands of *Bufo renarum* larvae kept in potassium perchlorate solution. Biocell **20:** 147–153.
2. MANZON, R.G. & J.H. YOUSON. 1997. The effects of exogenous thyroxine (T4) or triiodothyronine (T3), in the presence and absence of potassium perchlorate, on the incidence of metamorphosis and on serum T4 and T3 concentrations in larval sea lampreys (*Petromyzon marinus* L.). Gen. Comp. Endocrinol. **106:** 211–220.
3. WOLFF, J. 1998. Perchlorate and the thyroid gland. Pharmacol. Rev. **50(1):** 89–105.
4. NIEUWKOOP, P.D. & J. FABER, Eds. 1967. Normal Table of *Xenopus laevis* (Daudin): A Systematical and Chronological Survey of the Development from the Fertilized Egg till the End of Metamorphosis. 2nd ed. North-Holland. Amsterdam.

Index of Contributors

Aalkjær, C., 646–648
Abe, K., 278–280, 281–282
Abramowitz, J., 504–508
Adamo, H.P., 552–553
Adams, G., 183–187
Aizman, O., 489–496
Aizman, R., 395–400, 530–531
Akimova, O., 527–529
Albers, R.W., 338–340
Al-Khalili, L., 449–452
Allen, J.C., 504–508
Allen, K.E., 168–174
Alonso, G.L., 320–322
Altendorf, K., 351–353
Amler, E., 242–244
Andersen, J.P., 72–81, 101–105, 310–311, 335–337
Apell, H.-J., 133–140, 159–162, 229–231, 252–254, 325–326, 327–329
Aperia, A., 489–496
Argüello, J.M., 212–218, 224–225
Arnett, K.L., 620–622
Arruda, A.P., 481–488
Arteaga, M.F., 608–610
Artigas, P., 116–126
Arystarkhova, E., 382–387, 416–419
Asahi, M., 472–480
Asano, S., 655–657
Asher, C., 395–400
Avila, J., 608–610
Ayuyan, A.G., 252–254

Ball, W.J., Jr., 296–297
Bamberg, E., 31–38, 150–154
Bar Shimon, M., 287–289
Barassi, P., 694–701
Barkulis, C.T., 620–622
Barnese, M., 633–635
Barrientos, G., 315–317
Bauer, N., 678–684
Beaugé, L., 287–289
Becker, A., 31–38
Béguin, P., 388–394, 444–448
Berl, T., 410–415
Berlin, J.R., 141–149

Berman, M.C., 323–324
Bertorello, A.M., 587–594
Bianchi, G., 660–668, 694–701
Bibi, E., 247–248
Birkelund, S., 546–547
Blanco, G., 536–538
Blaustein, M., 509–513
Blostein, R., 58–62
Boldyrev, A., 519–521, 611–613
Boutry, M., 198–203
Bramkamp, M., 351–353
Bredeston, L.M., 552–553
Brochet, D.X.P., 603–605
Buch-Pedersen, M.J., 188–197, 349–350
Bulygina, E., 519–521, 611–613
Bundgaard, H., 623–624, 702–707
Burnay, M., 127–132
Burth, P., 633–635

Canning, B.J., 630–632
Capasso, J.M., 410–415
Caplan, M.J., 360–368
Carr, D.L., 258–259, 260–262, 711–712
Carr, J.A., 711–712
Carvalho, D.P., 481–488
Castello, P.R., 283–286
Castro-Faria, M.V., 633–635
Cavieres, J.D., 265–266, 270–272, 315–317
Champeil, P., 17–19, 82–89, 263–264
Chauvet, V., 360–368
Cheng, H., 630–632
Chibalin, A.V., 449–452
Clarke, R.J., 159–162
Clausen, J.D., 72–81, 101–105, 335–337
Clausen, T., 595–602
Cluzeaud, F., 428–436
Cornelius, F., 159–162, 541–542, 548–549, 579–586
Corradi, G.R., 552–553
Corre, F., 90–95, 312–314
Cougnon, M., 354–359
Crambert, G., 183–187, 226–228, 304–305, 388–394, 444–448

Dambly, S., 198–203
da-Silva, W.S., 481–488
De Jong, J.C., 437–443
de Meis, L., 481–488
De Pont, J.J.H.H.M., 150–154, 175–182, 255–257, 308–309, 437–443
de Tezanos Pinto, F., 552–553
Dediu, O., 293–295
Delavoie, F., 17–19
DeLisle, R.K., 296–297
DeMarco, S.J., 461–471
Denawa, M., 219–223
Denevich, S., 111–115
Deschênes, G., 570–578
Dobretsov, M., 639–641
Dolgova, N., 527–529, 658–659
Donnet, C., 249–251, 382–387
Dostanic, I., 354–359
Doucet, A., 570–578
Drabkin, A., 198–203
Duffield, A., 360–368
Duran, M.-J., 258–259, 260–262

Einholm, A.P., 72–81, 310–311
Ellis, D.Z., 534–535
Else, P.L., 525–526, 606–607
Enomoto, L.M., 410–415
Esmann, M., 238–239, 263–264, 290–292
Ettrich, R., 242–244
Evans, I., 708–710

Faller, L.D., 96–100, 275–277
Falson, P., 90–95, 312–314, 333–334
Farley, R.A., 96–100, 275–277
Farman, N., 428–436
Farr, C.D., 296–297
Fay, M., 428–436
Fedosova, N.U., 238–239, 263–264
Féraille, E., 554–561, 570–578
Ferrandi, M., 694–701
Ferrari, P., 694–701
Flecha, F.L.G., 283–286
Floyd, R., 614–616
Foster, C.S., 708–710
Friedrich, T., 31–38, 150–154
Froehlich, J.P., 338–340
Fuentes, J., 90–95
Fukushima, Y., 330–332

Füzesi, M., 388–394, 395–400, 530–531, 532–533

Gadsby, D.C., 116–126
Garrahan, P.J., 155–158, 287–289, 298–300, 301–303
Garty, H., 388–394, 395–400, 401–409, 444–448, 530–531, 532–533
Gatto, C., 620–622
Geering, K., 127–132, 183–187, 226–228, 304–305, 388–394, 444–448
Geibel, S., 31–38, 150–154
Gerasimova, O., 611–613
Giraldo, A.M.V., 283–286
Goellner, G.M., 461–471
Goldshleger, R., 247–248, 395–400, 532–533
Golovina, V., 509–513
Gonin, S., 570–578
González, D.A., 320–322
González-Lebrero, R.M., 155–158, 287–289, 298–300
Goormaghtigh, E., 347–348
Gopal, J., 649–651, 652–654
Grabsch, E., 150–154
Grell, E., 245–246
Grünbaum, E.-G., 678–684
Guennoun, S., 127–132
Guzman, J.P., 625–627

Haddow, J., 265–266
Hager, H., 646–648
Hagiwara, E., 232–234, 235–237
Håkansson, K.O., 163–167
Hambarchian, N., 678–684
Hamilton, B.P., 685–693
Hamlyn, J.M., 685–693
Hansen, O., 642–643
Hara, Y., 628–629
Hasler, U., 226–228
Hayashi, Y., 232–234, 235–237, 278–280
He, S., 354–359
Hebert, H., 9–16
Heitzmann, D., 554–561
Henao, F., 17–19
Hidalgo, C., 315–317
Hinsen, K., 344–346
Holden, J.P., 620–622

INDEX OF CONTRIBUTORS

Horisberger, J.-D., 127–132, 388–394, 444–448
Houghton-Larsen, J., 369–377
Hu, Y.-K., 31–38
Hua, S., 63–71
Hulbert, A.J., 525–526, 606–607
Humphrey, P.A., 159–162

Imagawa, T., 240–241, 281–282
Inesi, G., 63–71
Ishizuka, H., 655–657

Jacobsen, M.D., 369–377
Jakobsen, L.O., 290–292
James, P., 354–359, 509–513
Jaxel, C., 90–95
Jørgensen, J.R., 369–377
Jorgensen, P.L., xix–xxi, 22–30, 163–167, 546–547
Juul, B.S., 82–89, 318–319

Kahn, B., 711–712
Kairane, C., 644–645
Kanczewska, J., 198–203
Kane, D.J., 96–100, 275–277
Kaplan, J.H., 31–38, 273–274
Karelson, E, 644–645
Karlish, S.J.D., xix–xxi, 39–49, 106–110, 247–248, 287–289, 323–324, 388–394, 395–400, 401–409, 530–531, 532–533
Katoh, T., 278–280
Katz, A.I., 587–594
Kaufman, S.B., 155–158, 301–303
Kawase, T., 522–524
Kaya, S., 278–280, 281–282
Keenan, S.M., 296–297
Kerek, F., 327–329
Kikumoto, M., 278–280
Kim, G.-H., 562–569
Kimura, T., 360–368, 655–657
Kirch, U., 678–684
Kjeldsen, K., 623–624, 702–707
Klodos, I., 306–307
Knepper, M.A., 562–569
Knoers, N.V.A.M., 437–443
Kobayashi, T., 235–237
Koenderink, J.B., 31–38, 150–154, 175–182, 255–257, 308–309, 437–443

Kõks, S., 644–645
Koksoy, A., 504–508
Kondrashev-Lugovskii, A.S., 550–551
Korneenko, T.V., 183–187, 304–305
Kost, H., 678–684
Kozel, P., 453–460
Krämer, U., 678–684
Kravtsova, V., 639–641
Kristensen, B., 546–547
Krivoi, I., 617–619, 639–641
Krook, A., 449–452
Krüger, A., 541–542
Krumscheid, R., 242–244
Kurihara, K., 636–638
Kutchai, H., 338–340

Lacapère, J.-J. 17–19, 320–322, 344–346
Laharrague, F., 711–712
Lane, L.K., 58–62
Laredo, J., 685–693
Laughery, M.D., 273–274
Le Maire, M., 82–89, 90–95, 312–314, 333–334
Lenoir, G., 82–89, 312–314, 333–334
Lenz, A.A., 229–231
Leong, P.K.K., 625–627, 669–677
Levi, V., 283–286
Levine, B.A., 90–95
Lewitzki, E., 245–246
Li, C., 226–228
Lifshitz, Y., 530–531
Lindzen, M., 395–400, 530–531
Lingrel, J., 354–359, 509–513
Liu, L.H., 453–460
Løcke, E.-M., 401–409
Lopina, O.D., 527–529, 550–551, 658–659
Lorenz, J.N., 453–460
Lu, Z.R., 685–693
Lüpfert, C., 159–162
Lutsenko, S., 204–211

Ma, H., 63–71
MacLennan, D.H., 101–105, 472–480
Mahaney, J.E., 338–340
Mahmmoud, Y.A., 541–542, 548–549, 579–586
Malysheva, A.N., 550–551
Mana-Capelli, S., 212–218

Mandal, A.K., 212–218, 224–225
Mandel, F., 617–619, 639–641
Martin, P.-Y., 570–578
Martín-Vasallo, P., 608–610, 614–616, 708–710
Masilamani, S., 562–569
Mast, N., 527–529
Maudoux, O., 198–203
Maunsbach, A.B., xix–xxi, 9–16, 401–409, 646–648
McDonough, A.A., 625–627, 669–677
McIntosh, D.B., 17–19, 101–105
McLoud, S., 273–274
Meij, I.C., 437–443
Melloni, P., 694–701
Menguy, T., 90–95, 312–314
Mikami, Y., 628–629
Mikhailova, L., 224–225
Milanick, M.A., 620–622
Mimura, K., 235–237
Minotti, E., 694–701
Miranda, M., 168–174
Mobasheri, A., 614–616, 708–710
Modyanov, N.N., 183–187, 304–305
Moeller, H., 678–684
Molinari, I., 694–701
Møller, A.L., 188–197, 349–350
Møller, J.V., 82–89, 90–95, 318–319
Monnens, L.A.H., 437–443
Monnet-Tschudi, F., 444–448
Montes, M.R., 298–300
Montigny, C., 312–314, 333–334
Mordasini, D., 554–561, 570–578
Morii, M., 655–657
Moseley, A., 354–359
Müller-Ehmsen, J., 678–684
Munson, K., 106–110, 111–115
Mutz, M., 245–246
Myers, A.C., 630–632

Nagy, A.K., 275–277
Nakamura, J., 341–343
Nakanishi, N., 636–638
Neu, H., 678–684
Neumann, J., 354–359
Nielsen, N. Chr., 290–292
Nielsen, S., 646–648
Nienhaus, G.U., 293–295
Nomura, H., 1–8

O'Connor, K., 354–359
Ohniwa, R., 219–223
Oiwa, K., 278–280
Okamura, H., 219–223
Okunade, G., 453–460
Olden-Stahl, N., 293–295
Ostuni, M.A., 320–322
Owen, D., 270–272

Pagel, P., 360–368
Palmgren, M.G., 188–197, 349–350
Pardo, J.P., 168–174
Paula, S., 296–297
Pedemonte, C.H., 587–594
Pedersen, P.A., 312–314, 369–377, 539–540
Peinelt, C., 325–326
Peluffo, R.D., 141–149
Peranzi, G., 17–19
Perismy, M., 453–460
Pestov, N.B., 183–187, 304–305
Pierre, S.V., 258–259, 260–262
Pihakaski-Maunsbach, K., 401–409
Post, R.L., 20–21
Prasad, V., 453–460
Pratap, P.R., 293–295
Pressley, T.A., 258–259, 260–262, 711–712
Pu, H.X., 420–427
Purhonen, P., 9–16

Qiu, L.Y., 255–257

Radresa, O., 347–348
Rajasekaran, A.K., 649–651, 652–654
Rajasekaran, S.A., 649–651, 652–654
Rajendran, V., 360–368
Raussens, V., 347–348
Reis, M., 481–488
Reuter, N., 344–346
Rhee, M.S., 625–627
Rivard, C.J., 410–415
Roots, K., 644–645
Rossi, J.P.F.C., 283–286
Rossi, R.C., 155–158, 287–289, 298–300, 301–303
Rubtsov, A.M., 527–529, 550–551, 658–659
Ruysschaert, J-M., 347–348

INDEX OF CONTRIBUTORS

Sachs, G., 106–110, 111–115
Salum, T., 644–645
Sato, C., 341–343
Scanzano, R., 420–427
Schick, E., 245–246
Schneider, W.R., 620–622
Schoner, W., 242–244, 519–521, 678–684
Schwarz, W., 543–545
Schweitzer, T., 678–684
Schwinger, R., 678–684
Segall, L., 58–62
Shah, J.R., 685–693
Shakhparonov, M.I., 183–187, 304–305
Shinji, N., 232–234, 235–237
Shono, M., 628–629
Shull, G.E., 453–460
Skou, J.C., xxiii–xxv
Slayman, C.W., 168–174
Smagina, M., 658–659
Sokolov, V.S., 229–231, 252–254
Song, H., 509–513
Spielmann, A., 543–545
Steffensen, L., 539–540
Stimac, R., 327–329
Stokes, D., 17–19
Stolz, M., 245–246
Storey, K.B., 550–551
Storlien, L.H., 525–526
Strehler, E.E., 461–471
Strugatsky, D., 247–248
Suánková, K., 242–244
Sugita, Y., 1–8
Summa, V., 554–561, 570–578
Swarts, H.G.P., 175–182, 255–257, 308–309
Sweadner, K.J., 249–251, 382–387, 534–535

Tahara, Y., 232–234, 235–237
Tajima, G., 341–343
Takeguchi, N., 655–657
Takenaka, H., 232–234, 235–237
Takeyasu, K., 219–223, 378–381, 522–524
Taniguchi, K., 240–241, 278–280, 281–282
Tariq, A., 270–272
Taylor, M., 270–272, 315–317
Teisinger, J., 242–244
Teixeira, V.L., 587–594

Teramachi, S., 240–241
Thinès, D., 312–314
Thomsen, K., 9–16
Tillekeratne, M., 183–187
Tonosaki, K., 636–638
Torielli, L., 694–701
Toustrup-Jensen, M., 50–57, 267–269
Toyoshima, C., 1–8, 63–71
Tripodi, G., 660–668, 694–701
Trujillo, C.M., 608–610
Tsivkovskii, R., 204–211
Tupling, A.R., 472–480
Turner, N., 606–607

Uhlén, P., 514–518
Uldry, M., 444–448
Ushimaru, M., 330–332

Vagin, O., 106–110, 111–115
Van Den Heuvel, L.P.W.J., 437–443
Vandewalle, A., 554–561, 570–578
Vasar, E., 644–645
Vasilets, L.A., 543–545
Vasiliev, A., 617–619, 639–641
Verrey, F., 554–561, 570–578
Vilsen, B., 50–57, 72–81, 101–105, 267–269, 310–311
Vinciguerra, M., 570–578
Vorum, H., 9–16, 401–409, 646–648

Wada, M., 628–629
Walker, J.M., 204–211
Wang, S.Q., 630–632
Weitkamp, C., 678–684
Welsh, W.J., 296–297
Wetzel, R.K., 382–387, 416–419
Willems, P.H.G.M., 175–182, 255–257, 308–309
Woloszynska, M., 198–203
Woo, A., 354–359
Woolley, D.G., 101–105
Wray, S., 614–616
Wu, B.J., 525–526

Xie, Z., 497–503
Xu, G., 96–100
Xu, K.Y., 630–632

Yang, L.E., 625–627, 669–677
Yazawa, M., 278–280
Yoshimura, S.H., 378–381, 522–524
Younes-Ibrahim, M., 633–635
Yudowski, G.A., 287–289
Yuneva, M., 519–521

Zatti, A., 360–368
Zecevic, M., 554–561
Zhao, H., 183–187, 304–305
Zifarelli, G., 31–38
Zilmer, M., 644–645
Zimmermann, D., 31–38
Zouzoulas, A., 420–427

OHIO UNIVERSITY LIBRARY
Please return this book as soon as you have finished with it. In order to avoid a fine it must be returned by the latest date stamped below. All books are subject to recall after two weeks or immediately if needed for reserve.

JUN 1 4 2005

NOV 2 8 2005
TODAY

CF